de Gruyter Studies in Mathematics 18

de Gruyter Studies in Mathematics

1 Riemannian Geometry, *Wilhelm Klingenberg*
2 Semimartingales, *Michel Métivier*
3 Holomorphic Functions of Several Variables, *Ludger Kaup and Burchard Kaup*
4 Spaces of Measures, *Corneliu Constantinescu*
5 Knots, *Gerhard Burde and Heiner Zieschang*
6 Ergodic Theorems, *Ulrich Krengel*
7 Mathematical Theory of Statistics, *Helmut Strasser*
8 Transformation Groups, *Tammo tom Dieck*
9 Gibbs Measures and Phase Transitions, *Hans-Otto Georgii*
10 Analyticity in Infinite Dimensional Spaces, *Michel Hervé*
11 Elementary Geometry in Hyperbolic Space, *Werner Fenchel*
12 Transcendental Numbers, *Andrei B. Shidlovskii*
13 Ordinary Differential Equations, *Herbert Amann*
14 Dirichlet Forms and Analysis on Wiener Space, *Nicolas Bouleau and Francis Hirsch*
15 Nevanlinna Theory and Complex Differential Equations, *Ilpo Laine*
16 Rational Iteration, *Norbert Steinmetz*
17 Korovkin-type Approximation Theory and its Applications, *Francesco Altomare and Michele Campiti*

V. G. Turaev

Quantum Invariants of Knots and 3-Manifolds

With 206 illustrations

Walter de Gruyter
Berlin · New York 1994

Author
Vladimir G. Turaev
Institut de Recherche Mathématique Avancée
Université Louis Pasteur – C.N.R.S.
7, rue René Descartes
67084 Strasbourg Cedex, France
e-mail: turaev@math.u-strasbg.fr

1991 Mathematics Subject Classification: 57-02, 18-02; 17B37, 81Txx, 82B23

Key words: Topological field theory, knots, 3-manifold invariants, state sums, monoidal categories

∞ Printed on acid-free paper which falls within the guidelines of the ANSI to ensure permanence and durability.

Library of Congress Cataloging-in-Publication Data

Turaev, V. G. (Vladimir G.), 1954–
Quantum invariants of knots and 3-manifolds / V. G. Turaev.
p. cm. – (De Gruyter studies in mathematics : 18)
Includes bibliographical references and index.
ISBN 3-11-013704-6 (acid-free paper)
1. Quantum field theory. 2. Knot theory. 3. Three-manifolds (Topology) 4. Invariants. 5. Mathematical physics. I. Title. II. Series.
QC174.52.C66T87 1994
512'.55–dc20 94-21249

Die Deutsche Bibliothek – Cataloging-in-Publication Data

Turaev, Vladimir G.:
Quantum invariants of knots and 3-manifolds / V. G. Turaev. – Berlin ; New York : de Gruyter, 1994
(De Gruyter studies in mathematics ; 18)
ISBN 3-11-013704-6
NE: GT

Disk conversion: Lewis & Leins, Berlin. Printing: Arthur Collignon GmbH, Berlin. Binding: Dieter Mikolai, Berlin. Cover design: Rudolf Hübler, Berlin.
Illustrations: DTP L. C. & R. Hartmann, Haierweg 47, D-79114 Freiburg

Dedicated to my parents

Contents

Introduction

In the last decade we have witnessed the birth of a fascinating new mathematical theory. It is often called by algebraists the theory of quantum groups and by topologists quantum topology. These terms, however, seem to be too restrictive and do not convey the breadth of this new domain which is closely related to the theory of von Neumann algebras, the theory of Hopf algebras, the theory of representations of semisimple Lie algebras, the topology of knots, etc. The most spectacular achievements in this theory are centered around quantum groups and invariants of knots and 3-dimensional manifolds.

The whole theory has been, to a great extent, inspired by ideas that arose in theoretical physics. Among the relevant areas of physics are the theory of exactly solvable models of statistical mechanics, the quantum inverse scattering method, the quantum theory of angular momentum, 2-dimensional conformal field theory, etc. The development of this subject shows once more that physics and mathematics intercommunicate and influence each other to the profit of both disciplines.

Three major events have marked the history of this theory. A powerful original impetus was the introduction of a new polynomial invariant of classical knots and links by V. Jones (1984). This discovery drastically changed the scenery of knot theory. The Jones polynomial paved the way for an intervention of von Neumann algebras, Lie algebras, and physics into the world of knots and 3-manifolds.

The second event was the introduction by V. Drinfel'd and M. Jimbo (1985) of quantum groups which may roughly be described as 1-parameter deformations of semisimple complex Lie algebras. Quantum groups and their representation theory form the algebraic basis and environment for this subject. Note that quantum groups emerged as an algebraic formalism for physicists' ideas, specifically, from the work of the Leningrad school of mathematical physics directed by L. Faddeev.

In 1988 E. Witten invented the notion of a topological quantum field theory and outlined a fascinating picture of such a theory in three dimensions. This picture includes an interpretation of the Jones polynomial as a path integral and relates the Jones polynomial to a 2-dimensional modular functor arising in conformal field theory. It seems that at the moment of writing (beginning of 1994), Witten's approach based on path integrals has not yet been justified mathematically. Witten's conjecture on the existence of non-trivial 3-dimensional TQFT's has served as a major source of inspiration for the research in this area.

The development of the subject (in its topological part) has been strongly influenced by the works of M. Atiyah, A. Joyal and R. Street, L. Kauffman,

A. Kirillov and N. Reshetikhin, G. Moore and N. Seiberg, N. Reshetikhin and V. Turaev, G. Segal, V. Turaev and O. Viro, and others (see References). Although this theory is very young, the number of relevant papers is overwhelming. We do not attempt to give a comprehensive history of the subject and confine ourselves to sketchy historical remarks in the chapter notes.

In this monograph we focus our attention on the topological aspects of the theory. Our goal is the construction and study of invariants of knots and 3-manifolds. There are several possible approaches to these invariants, based on Chern-Simons field theory, 2-dimensional conformal field theory, and quantum groups. We shall follow the last approach. The fundamental idea is to derive invariants of knots and 3-manifolds from algebraic objects which formalize the properties of modules over quantum groups at roots of unity. This approach allows a rigorous mathematical treatment of a number of ideas considered in theoretical physics.

This monograph is addressed to mathematicians and physicists with a knowledge of basic algebra and topology. We do not assume that the reader is acquainted with the theory of quantum groups or with the relevant chapters of mathematical physics.

Besides an exposition of the material available in published papers, this monograph presents new results of the author, which appear here for the first time. Indications to this effect and priority references are given in the chapter notes.

The fundamental notions discussed in the monograph are those of modular category, modular functor, and topological quantum field theory (TQFT). The mathematical content of these notions may be outlined as follows.

Modular categories are tensor categories with certain additional algebraic structures (braiding, twist) and properties of semisimplicity and finiteness. The notions of braiding and twist arise naturally from the study of the commutativity of the tensor product. Semisimplicity means that all objects of the category may be decomposed into "simple" objects which play the role of irreducible modules in representation theory. Finiteness means that such a decomposition can be performed using only a finite stock of simple objects.

The use of categories should not frighten the reader unaccustomed to the abstract theory of categories. Modular categories are defined in algebraic terms and have a purely algebraic nature. Still, if the reader wants to avoid the language of categories, he may think of the objects of a modular category as finite dimensional modules over a Hopf algebra.

Modular functors relate topology to algebra and are reminiscent of homology. A modular functor associates projective modules over a fixed commutative ring K to certain "nice" topological spaces. When we speak of an n-dimensional modular functor, the role of "nice" spaces is played by closed n-dimensional manifolds (possibly with additional structures like orientation, smooth structure, etc.). An

n-dimensional modular functor $\mathcal{T}$ assigns to a closed n-manifold (with a certain additional structure) Σ, a projective K-module $\mathcal{T}(\Sigma)$, and assigns to a homeomorphism of n-manifolds (preserving the additional structure), an isomorphism of the corresponding modules. The module $\mathcal{T}(\Sigma)$ is called the module of states of Σ. These modules should satisfy a few axioms including multiplicativity with respect to disjoint union: $\mathcal{T}(\Sigma \amalg \Sigma') = \mathcal{T}(\Sigma) \otimes_K \mathcal{T}(\Sigma')$. It is convenient to regard the empty space as an n-manifold and to require that $\mathcal{T}(\emptyset) = K$.

A modular functor may sometimes be extended to a topological quantum field theory (TQFT), which associates homomorphisms of modules of states to cobordisms ("spacetimes"). More precisely, an $(n+1)$-dimensional TQFT is formed by an n-dimensional modular functor $\mathcal{T}$ and an operator invariant of $(n+1)$-cobordisms τ. By an $(n+1)$-cobordism, we mean a compact $(n+1)$-manifold M whose boundary is a disjoint union of two closed n-manifolds $\partial_- M$, $\partial_+ M$ called the bottom base and the top base of M. The operator invariant τ assigns to such a cobordism M a homomorphism

$$\tau(M) : \mathcal{T}(\partial_- M) \to \mathcal{T}(\partial_+ M).$$

This homomorphism should be invariant under homeomorphisms of cobordisms and multiplicative with respect to disjoint union of cobordisms. Moreover, τ should be compatible with gluings of cobordisms along their bases: if a cobordism M is obtained by gluing two cobordisms M_1 and M_2 along their common base $\partial_+(M_1) = \partial_-(M_2)$ then

$$\tau(M) = k\, \tau(M_2) \circ \tau(M_1) : \mathcal{T}(\partial_-(M_1)) \to \mathcal{T}(\partial_+(M_2))$$

where $k \in K$ is a scalar factor depending on M, M_1, M_2. The factor k is called the anomaly of the gluing. The most interesting TQFT's are those which have no gluing anomalies in the sense that for any gluing, $k = 1$. Such TQFT's are said to be anomaly-free.

In particular, a closed $(n+1)$-manifold M may be regarded as a cobordism with empty bases. The operator $\tau(M)$ acts in $\mathcal{T}(\emptyset) = K$ as multiplication by an element of K. This element is the "quantum" invariant of M provided by the TQFT $(\mathcal{T}, \tau)$. It is denoted also by $\tau(M)$.

We note that to speak of a TQFT $(\mathcal{T}, \tau)$, it is necessary to specify the class of spaces and cobordisms subject to the application of $\mathcal{T}$ and τ.

In this monograph we shall consider 2-dimensional modular functors and 3-dimensional topological quantum field theories. Our main result asserts that every modular category gives rise to an anomaly-free 3-dimensional TQFT:

$$\text{modular category} \quad \mapsto \quad \text{3-dimensional TQFT}.$$

In particular, every modular category gives rise to a 2-dimensional modular functor:

$$\text{modular category} \quad \mapsto \quad \text{2-dimensional modular functor.}$$

The 2-dimensional modular functor $\mathcal{T}_\mathcal{V}$, derived from a modular category $\mathcal{V}$, applies to closed oriented surfaces with a distinguished Lagrangian subspace in 1-homologies and a finite (possibly empty) set of marked points. A point of a surface is marked if it is endowed with a non-zero tangent vector, a sign ± 1, and an object of $\mathcal{V}$; this object of $\mathcal{V}$ is regarded as the "color" of the point. The modular functor $\mathcal{T}_\mathcal{V}$ has a number of interesting properties including nice behavior with respect to cutting surfaces out along simple closed curves. Borrowing terminology from conformal field theory, we say that $\mathcal{T}_\mathcal{V}$ is a rational 2-dimensional modular functor.

We shall show that the modular category $\mathcal{V}$ can be reconstructed from the corresponding modular functor $\mathcal{T}_\mathcal{V}$. This deep fact shows that the notions of modular category and rational 2-dimensional modular functor are essentially equivalent; they are two sides of the same coin formulated in algebraic and geometric terms:

$$\text{modular category} \quad \Longleftrightarrow \quad \text{rational 2-dimensional modular functor.}$$

The operator invariant τ, derived from a modular category $\mathcal{V}$, applies to compact oriented 3-cobordisms whose bases are closed oriented surfaces with the additional structure as above. The cobordisms may contain colored framed oriented knots, links, or graphs which meet the bases of the cobordism along the marked points. (A link is colored if each of its components is endowed with an object of $\mathcal{V}$. A link is framed if it is endowed with a non-singular normal vector field in the ambient 3-manifold.) For closed oriented 3-manifolds and for colored framed oriented links in such 3-manifolds, this yields numerical invariants. These are the "quantum" invariants of links and 3-manifolds derived from $\mathcal{V}$. Under a special choice of $\mathcal{V}$ and a special choice of colors, we recover the Jones polynomial of links in the 3-sphere S^3 or, more precisely, the value of this polynomial at a complex root of unity.

An especially important class of 3-dimensional TQFT's is formed by so-called unitary TQFT's with ground ring $K = \mathbb{C}$. In these TQFT's, the modules of states of surfaces are endowed with positive definite Hermitian forms. The corresponding algebraic notion is the one of a unitary modular category. We show that such categories give rise to unitary TQFT's:

$$\text{unitary modular category} \quad \mapsto \quad \text{unitary 3-dimensional TQFT.}$$

Unitary 3-dimensional TQFT's are considerably more sensitive to the topology of 3-manifolds than general TQFT's. They can be used to estimate certain classical numerical invariants of knots and 3-manifolds.

To sum up, we start with a purely algebraic object (a modular category) and build a topological theory of modules of states of surfaces and operator invariants of 3-cobordisms. This construction reveals an algebraic background to 2-dimensional modular functors and 3-dimensional TQFT's. It is precisely because there are non-trivial modular categories, that there exist non-trivial 3-dimensional TQFT's.

The construction of a 3-dimensional TQFT from a modular category $\mathcal{V}$ is the central result of Part I of the book. We give here a brief overview of this construction.

The construction proceeds in several steps. First, we define an isotopy invariant F of colored framed oriented links in Euclidean space $\mathbb{R}^3$. The invariant F takes values in the commutative ring $K = \mathrm{Hom}_{\mathcal{V}}(\mathbb{1}, \mathbb{1})$, where $\mathbb{1}$ is the unit object of $\mathcal{V}$. The main idea in the definition of F is to dissect every link $L \subset \mathbb{R}^3$ into elementary "atoms". We first deform L in $\mathbb{R}^3$ so that its normal vector field is given everywhere by the vector (0,0,1). Then we draw the orthogonal projection of L in the plane $\mathbb{R}^2 = \mathbb{R}^2 \times 0$ taking into account overcrossings and undercrossings. The resulting plane picture is called the diagram of L. It is convenient to think that the diagram is drawn on graph paper. Stretching the diagram in the vertical direction, if necessary, we may arrange that each small square of the paper contains either one vertical line of the diagram, an X-like crossing of two lines, a cap-like arc $\cap$, or a cup-like arc $\cup$. These are the atoms of the diagram. We use the algebraic structures in $\mathcal{V}$ and the colors of link components to assign to each atom a morphism in $\mathcal{V}$. Using the composition and tensor product in $\mathcal{V}$, we combine the morphisms corresponding to the atoms of the diagram into a single morphism $F(L) : \mathbb{1} \to \mathbb{1}$. To verify independence of $F(L) \in K$ on the choice of the diagram, we appeal to the fact that any two diagrams of the same link may be related by Reidemeister moves and local moves changing the position of the diagram with respect to the squares of graph paper.

The invariant F may be generalized to an isotopy invariant of colored graphs in $\mathbb{R}^3$. By a coloring of a graph, we mean a function which assigns to every edge an object of $\mathcal{V}$ and to every vertex a morphism in $\mathcal{V}$. The morphism assigned to a vertex should be an intertwiner between the objects of $\mathcal{V}$ sitting on the edges incident to this vertex. As in the case of links we need a kind of framing for graphs, specifically, we consider ribbon graphs whose edges and vertices are narrow ribbons and small rectangles.

Note that this part of the theory does not use semisimplicity and finiteness of $\mathcal{V}$. The invariant F can be defined for links and ribbon graphs in $\mathbb{R}^3$ colored over arbitrary tensor categories with braiding and twist. Such categories are called ribbon categories.

Next we define a topological invariant $\tau(M) = \tau_{\mathcal{V}}(M) \in K$ for every closed oriented 3-manifold M. Present M as the result of surgery on the 3-sphere $S^3 = \mathbb{R}^3 \cup \{\infty\}$ along a framed link $L \subset \mathbb{R}^3$. Orient L in an arbitrary way and vary the colors of the components of L in the finite family of simple objects of $\mathcal{V}$ appearing in the definition of a modular category. This gives a finite family of colored (framed oriented) links in $\mathbb{R}^3$ with the same underlying link L. We define $\tau(M)$ to be a certain weighted sum of the corresponding invariants $F \in K$. To verify independence on the choice of L, we use the Kirby calculus of links allowing us to relate any two choices of L by a sequence of local geometric transformations.

The invariant $\tau(M) \in K$ generalizes to an invariant $\tau(M, \Omega) \in K$ where M is a closed oriented 3-manifold and Ω is a colored ribbon graph in M.

At the third step we define an auxiliary 3-dimensional TQFT that applies to parametrized surfaces and 3-cobordisms with parametrized bases. A surface is parametrized if it is provided with a homeomorphism onto the standard closed surface of the same genus bounding a standard unknotted handlebody in $\mathbb{R}^3$. Let M be an oriented 3-cobordism with parametrized boundary (this means that all components of ∂M are parametrized). Consider first the case where $\partial_+ M = \emptyset$ and $\Sigma = \partial_- M$ is connected. Gluing the standard handlebody to M along the parametrization of Σ yields a closed 3-manifold $\tilde{M}$. We consider a certain canonical ribbon graph R in the standard handlebody in $\mathbb{R}^3$ lying there as a kind of core and having only one vertex. Under the gluing used above, R embeds in $\tilde{M}$. We color the edges of R with arbitrary objects from the finite family of simple objects appearing in the definition of $\mathcal{V}$. Coloring the vertex of R with an intertwiner we obtain a colored ribbon graph $\tilde{R} \subset \tilde{M}$. Denote by $\mathcal{T}(\Sigma)$ the K-module formally generated by such colorings of R. We can regard $\tau(\tilde{M}, \tilde{R}) \in K$ as a linear functional $\mathcal{T}(\Sigma) \to K$. This is the operator $\tau(M)$. The case of a 3-cobordism with non-connected boundary is treated similarly: we glue standard handlebodies (with the standard ribbon graphs inside) to all the components of ∂M and apply τ as above. This yields a linear functional on the tensor product $\otimes_i \mathcal{T}(\Sigma_i)$ where Σ_i runs over the components of ∂M. Such a functional may be rewritten as a linear operator $\mathcal{T}(\partial_- M) \to \mathcal{T}(\partial_+ M)$.

The next step is to define the action of surface homeomorphisms in the modules of states and to replace parametrizations of surfaces with a less rigid structure. The study of homeomorphisms may be reduced to a study of 3-cobordisms with parametrized bases. Namely, if Σ is a standard surface then any homeomorphism $f : \Sigma \to \Sigma$ gives rise to the 3-cobordism $(\Sigma \times [0, 1], \Sigma \times 0, \Sigma \times 1)$ whose bottom base is parametrized via f and whose top base is parametrized via id_Σ. The operator invariant τ of this cobordism yields an action of f in $\mathcal{T}(\Sigma)$. This gives a projective linear action of the group $\mathrm{Homeo}(\Sigma)$ on $\mathcal{T}(\Sigma)$. The corresponding 2-cocycle is computed in terms of Maslov indices of Lagrangian spaces in $H_1(\Sigma; \mathbb{R})$. This computation implies that the module $\mathcal{T}(\Sigma)$ does not depend on the choice of parametrization, but rather depends on the Lagrangian space in $H_1(\Sigma; \mathbb{R})$ determined by this parametrization. This fact allows us to define a TQFT based on closed oriented surfaces endowed with a distinguished Lagrangian space in 1-homologies and on compact oriented 3-cobordisms between such surfaces. Finally, we show how to modify this TQFT in order to kill its gluing anomalies.

The definition of the quantum invariant $\tau(M) = \tau_{\mathcal{V}}(M)$ of a closed oriented 3-manifold M is based on an elaborate reduction to link diagrams. It would be most important to compute $\tau(M)$ in intrinsic terms, i.e., directly from M rather than from a link diagram. In Part II of the book we evaluate in intrinsic terms the product $\tau(M)\,\tau(-M)$ where $-M$ denotes the same manifold M with the opposite orientation. More precisely, we compute $\tau(M)\,\tau(-M)$ as a state sum on a triangulation of M. In the case of a unitary modular category,

$$\tau(M)\,\tau(-M) = |\tau(M)|^2 \in \mathbb{R}$$

so that we obtain the absolute value of $\tau(M)$ as the square root of a state sum on a triangulation of M.

The algebraic ingredients of the state sum in question are so-called $6j$-symbols associated to $\mathcal{V}$. The $6j$-symbols associated to the Lie algebra $sl_2(\mathbb{C})$ are well known in the quantum theory of angular momentum. These symbols are complex numbers depending on 6 integer indices. We define more general $6j$-symbols associated to a modular category $\mathcal{V}$ satisfying a minor technical condition of unimodality. In the context of modular categories, each $6j$-symbol is a tensor in 4 variables running over so-called multiplicity modules. The $6j$-symbols are numerated by tuples of 6 indices running over the set of distinguished simple objects of $\mathcal{V}$. The system of $6j$-symbols describes the associativity of the tensor product in $\mathcal{V}$ in terms of multiplicity modules. A study of $6j$-symbols inevitably appeals to geometric images. In particular, the appearance of the numbers 4 and 6 has a simple geometric interpretation: we should think of the 6 indices mentioned above as sitting on the edges of a tetrahedron while the 4 multiplicity modules sit on its 2-faces. This interpretation is a key to applications of $6j$-symbols in 3-dimensional topology.

We define a state sum on a triangulated closed 3-manifold M as follows. Color the edges of the triangulation with distinguished simple objects of $\mathcal{V}$. Associate to each tetrahedron of the triangulation the $6j$-symbol determined by the colors of its 6 edges. This $6j$-symbol lies in the tensor product of 4 multiplicity modules associated to the faces of the tetrahedron. Every 2-face of the triangulation is incident to two tetrahedra and contributes dual multiplicity modules to the corresponding tensor products. We consider the tensor product of $6j$-symbols associated to all tetrahedra of the triangulation and contract it along the dualities determined by 2-faces. This gives an element of the ground ring K corresponding to the chosen coloring. We sum up these elements (with certain coefficients) over all colorings. The sum does not depend on the choice of triangulation and yields a homeomorphism invariant $|M| \in K$ of M. It turns out that for oriented M, we have

$$|M| = \tau(M)\,\tau(-M).$$

Similar state sums on 3-manifolds with boundary give rise to a so-called simplicial TQFT based on closed surfaces and compact 3-manifolds (without additional structures). The equality $|M| = \tau(M)\,\tau(-M)$ for closed oriented 3-manifolds generalizes to a splitting theorem for this simplicial TQFT.

The proof of the formula $|M| = \tau(M)\,\tau(-M)$ is based on a computation of $\tau(M)$ inside an arbitrary compact oriented piecewise-linear 4-manifold bounded by M. This result, interesting in itself, gives a 4-dimensional perspective to quantum invariants of 3-manifolds. The computation in question involves the fundamental notion of shadows of 4-manifolds. Shadows are purely topological objects intimately related to $6j$-symbols. The theory of shadows was, to a great extent, stimulated by a study of 3-dimensional TQFT's.

The idea underlying the definition of shadows is to consider 2-dimensional polyhedra whose 2-strata are provided with numbers. We shall consider only so-called simple 2-polyhedra. Every simple 2-polyhedron naturally decomposes into a disjoint union of vertices, 1-strata (edges and circles), and 2-strata. We say that a simple 2-polyhedron is shadowed if each of its 2-strata is endowed with an integer or half-integer, called the gleam of this 2-stratum. We define three local transformations of shadowed 2-polyhedra (shadow moves). A shadow is a shadowed 2-polyhedron regarded up to these moves.

Being 2-dimensional, shadows share many properties with surfaces. For instance, there is a natural notion of summation of shadows similar to the connected summation of surfaces. It is more surprising that shadows share a number of important properties of 3-manifolds and 4-manifolds. In analogy with 3-manifolds they may serve as ambient spaces of knots and links. In analogy with 4-manifolds they possess a symmetric bilinear form in 2-homologies. Imitating surgery and cobordism for 4-manifolds, we define surgery and cobordism for shadows.

Shadows arise naturally in 4-dimensional topology. Every compact oriented piecewise-linear 4-manifold W (possibly with boundary) gives rise to a shadow $\mathrm{sh}(W)$. To define $\mathrm{sh}(W)$, we consider a simple 2-skeleton of W and provide every 2-stratum with its self-intersection number in W. The resulting shadowed polyhedron considered up to shadow moves and so-called stabilization does not depend on the choice of the 2-skeleton. In technical terms, $\mathrm{sh}(W)$ is a stable integer shadow. Thus, we have an arrow

$$\text{compact oriented PL 4-manifolds} \;\mapsto\; \text{stable integer shadows.}$$

It should be emphasized that this part of the theory is purely topological and does not involve tensor categories.

Every modular category $\mathcal{V}$ gives rise to an invariant of stable shadows. It is obtained via a state sum on shadowed 2-polyhedra. The algebraic ingredients of this state sum are the $6j$-symbols associated to $\mathcal{V}$. This yields a mapping

$$\text{stable integer shadows} \xrightarrow{\text{state sum}} K = \mathrm{Hom}_{\mathcal{V}}(\mathbb{1}, \mathbb{1}).$$

Composing these arrows we obtain a K-valued invariant of compact oriented PL 4-manifolds. By a miracle, this invariant of a 4-manifold W depends only on ∂W and coincides with $\tau(\partial W)$. This gives a computation of $\tau(\partial W)$ inside W.

The discussion above naturally raises the problem of existence of modular categories. These categories are quite delicate algebraic objects. Although there are elementary examples of modular categories, it is by no means obvious that there exist modular categories leading to deep topological theories. The source of interesting modular categories is the theory of representations of quantum groups at roots of unity. The quantum group $U_q(\mathfrak{g})$ is a Hopf algebra over $\mathbb{C}$ obtained by a 1-parameter deformation of the universal enveloping algebra of a simple Lie algebra $\mathfrak{g}$. The finite dimensional modules over $U_q(\mathfrak{g})$ form a semisimple tensor category with braiding and twist. To achieve finiteness, we take the deformation parameter q to be a complex root of unity. This leads to a loss of semisimplicity which is regained under the passage to a quotient category. If $\mathfrak{g}$ belongs to the series A, B, C, D and the order of the root of unity q is even and sufficiently big then we obtain a modular category with ground ring $\mathbb{C}$:

$$\text{quantum group at a root of 1} \quad \mapsto \quad \text{modular category.}$$

Similar constructions may be applied to exceptional simple Lie algebras, although some details are yet to be worked out. It is remarkable that for $q = 1$ we have the classical theory of representations of a simple Lie algebra while for non-trivial complex roots of unity we obtain modular categories.

Summing up, we may say that the simple Lie algebras of the series A, B, C, D give rise to 3-dimensional TQFT's via the q-deformation, the theory of representations, and the theory of modular categories. The resulting 3-dimensional TQFT's are highly non-trivial from the topological point of view. They yield new invariants of 3-manifolds and knots including the Jones polynomial (which is obtained from $\mathfrak{g} = sl_2(\mathbb{C})$) and its generalizations.

At earlier stages in the theory of quantum 3-manifold invariants, Hopf algebras and quantum groups played the role of basic algebraic objects, i.e., the role of modular categories in our present approach. It is in this book that we switch to categories. Although the language of categories is more general and more simple, it is instructive to keep in mind its algebraic origins.

There is a dual approach to the modular categories derived from the quantum groups $U_q(sl_n(\mathbb{C}))$ at roots of unity. The Weyl duality between representations of $U_q(sl_n(\mathbb{C}))$ and representations of Hecke algebras suggests that one should study the categories whose objects are idempotents of Hecke algebras. We shall treat the simplest but most important case, $n = 2$. In this case instead of Hecke algebras we may consider their quotients, the Temperley-Lieb algebras. A study of idempotents in the Temperley-Lieb algebras together with the skein theory of tangles gives a construction of modular categories. This construction is elementary and self-contained. It completely avoids the theory of quantum groups but yields the same modular categories as the representation theory of $U_q(sl_2(\mathbb{C}))$ at roots of unity.

Until now main efforts have been spent to construct 2-dimensional modular functors and 3-dimensional TQFT's. Little is known about their role in low-dimensional topology. Relationships between the TQFT's and the classical invariants of 3-manifolds, for instance, the fundamental group are poorly understood. Topological properties and applications of the modular functors derived from quantum groups are yet to be studied. The most important problem is to relate the 3-dimensional TQFT's to the Donaldson invariants of 4-manifolds and the Floer homologies of 3-manifolds. (Very interesting results in this direction have been recently obtained by H. Murakami.)

The book consists of three parts. Part I (Chapters I - V) is concerned with the construction of a 2-dimensional modular functor and 3-dimensional TQFT from a modular category. Part II (Chapters VI - X) deals with $6j$-symbols, shadows, and state sums on shadows and 3-manifolds. Part III (Chapters XI, XII) is concerned with constructions of modular categories.

It is possible but not at all necessary to read the chapters in their linear order. The reader may start with Chapter III or with Chapters VIII, IX which are independent of the previous material. It is also possible to start with Part III which is almost independent of Parts I and II, one needs only to be acquainted with the definitions of ribbon, modular, semisimple, Hermitian, and unitary categories given in Section I.1 (i.e., Section 1 of Chapter I) and Sections II.1, II.4, II.5.

The interdependence of the chapters is presented in the following diagram. An arrow from A to B indicates that the definitions and results of Chapter A are essential for Chapter B. Weak dependence of chapters is indicated by dotted arrows.

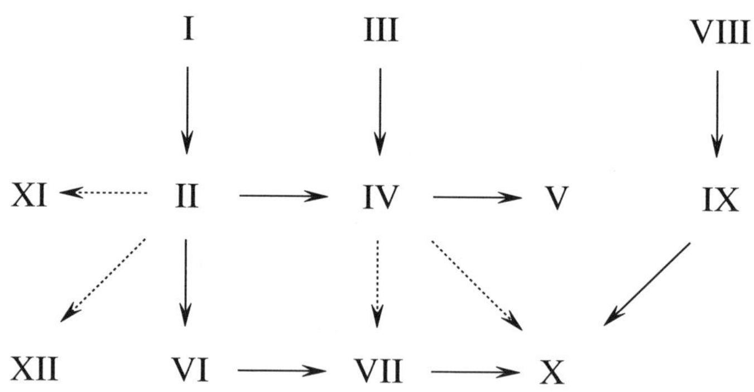

The content of the chapters should be clear from the headings. The following remarks give more directions to the reader.

Chapter I starts off with ribbon categories and invariants of colored framed graphs and links in Euclidean 3-space. The relevant definitions and results, given in the first two sections of Chapter I, will be used throughout the book. They contain the seeds of main ideas of the theory. Sections I.3 and I.4 are concerned with the proof of Theorem I.2.5 and may be skipped without much loss.

Chapter II starts with two fundamental sections. In Section II.1 we introduce modular categories which are the main algebraic objects of the monograph. In Section II.2 we introduce the invariant τ of closed oriented 3-manifolds. In Section II.3 we prove that τ is well defined. The ideas of the proof are used in the same section to construct a projective linear action of the group $SL(2, \mathbb{Z})$. This action does not play an important role in the book, rather it serves as a precursor for the actions of modular groups of surfaces on the modules of states introduced in Chapter IV. In Section II.4 we define semisimple ribbon categories and establish an analogue of the Verlinde-Moore-Seiberg formula known in conformal field theory. Section II.5 is concerned with Hermitian and unitary modular categories.

Chapter III deals with axiomatic foundations of topological quantum field theory. It is remarkable that even in a completely abstract set up, we can establish meaningful theorems which prove to be useful in the context of 3-dimensional TQFT's. The most important part of Chapter III is the first section where we give an axiomatic definition of modular functors and TQFT's. The language introduced in Section III.1 will be used systematically in Chapter IV. In Section III.2 we establish a few fundamental properties of TQFT's. In Section III.3 we introduce the important notion of a non-degenerate TQFT and establish sufficient conditions for isomorphism of non-degenerate anomaly-free TQFT's. Section III.5 deals with Hermitian and unitary TQFT's, this study will be continued in the 3-dimensional setting at the end of Chapter IV. Sections III.4 and III.6 are more or less isolated from the rest of the book; they deal with the abstract notion of a quantum invariant of topological spaces and a general method of killing the gluing anomalies of a TQFT.

In Chapter IV we construct the 3-dimensional TQFT associated to a modular category. It is crucial for the reader to get through Section IV.1, where we define the 3-dimensional TQFT for 3-cobordisms with parametrized boundary. Section IV.2 provides the proofs for Section IV.1; the geometric technique of Section IV.2 is probably one of the most difficult in the book. However, this technique is used only a few times in the remaining part of Chapter IV and in Chapter V. Section IV.3 is purely algebraic and independent of all previous sections. It provides generalities on Lagrangian relations and Maslov indices. In Sections IV.4 - IV.6 we show how to renormalize the TQFT introduced in Section IV.1 in order to replace parametrizations of surfaces with Lagrangian spaces in 1-homologies. The 3-dimensional TQFT $(\mathcal{T}^e, \tau^e)$, constructed in Section IV.6 and further studied in Section IV.7, is quite suitable for computations and applications. This TQFT has anomalies which are killed in Sections IV.8 and IV.9 in two different ways. The anomaly-free TQFT constructed in Section IV.9 is the final product of Chapter IV. In Sections IV.10 and IV.11 we show that the TQFT's derived from Hermitian (resp. unitary) modular categories are themselves Hermitian (resp. unitary). In the purely algebraic Section IV.12 we introduce the Verlinde algebra of a modular category and use it to compute the dimension of the module of states of a surface.

The results of Chapter IV shall be used in Sections V.4, V.5, VII.4, and X.8.

Chapter V is devoted to a detailed analysis of the 2-dimensional modular functors (2-DMF's) arising from modular categories. In Section V.1 we give an axiomatic definition of 2-DMF's and rational 2-DMF's independent of all previous material. In Section V.2 we show that each (rational) 2-DMF gives rise to a (modular) ribbon category. In Section V.3 we introduce the more subtle notion of a weak rational 2-DMF. In Sections V.4 and V.5 we show that the constructions of Sections IV.1 - IV.6, being properly reformulated, yield a weak rational 2-DMF.

Chapter VI deals with $6j$-symbols associated to a modular category. The most important part of this chapter is Section VI.5, where we use the invariants of ribbon graphs introduced in Chapter I to define so-called normalized $6j$-symbols. They should be contrasted with the more simple-minded $6j$-symbols defined in Section VI.1 in a direct algebraic way. The approach of Section VI.1 generalizes the standard definition of $6j$-symbols but does not go far enough. The fundamental advantage of normalized $6j$-symbols is their tetrahedral symmetry. Three intermediate sections (Sections VI.2 - VI.4) prepare different kinds of preliminary material necessary to define the normalized $6j$-symbols.

In the first section of Chapter VII we use $6j$-symbols to define state sums on triangulated 3-manifolds. Independence on the choice of triangulation is shown in Section VII.2. Simplicial 3-dimensional TQFT is introduced in Section VII.3. Finally, in Section VII.4 we state the main theorems of Part II; they relate the theory developed in Part I to the state sum invariants of closed 3-manifolds and simplicial TQFT's.

Chapters VIII and IX are purely topological. In Chapter VIII we discuss the general theory of shadows. In Chapter IX we consider shadows of 4-manifolds, 3-manifolds, and links in 3-manifolds. The most important sections of these two chapters are Sections VIII.1 and IX.1 where we define (abstract) shadows and shadows of 4-manifolds. The reader willing to simplify his way towards Chapter X may read Sections VIII.1, VIII.2.1, VIII.2.2, VIII.6, IX.1 and then proceed to Chapter X coming back to Chapters VIII and IX when necessary.

In Chapter X we combine all the ideas of the previous chapters. We start with state sums on shadowed 2-polyhedra based on normalized $6j$-symbols (Section X.1) and show their invariance under shadow moves (Section X.2). In Section X.3 we interpret the invariants of closed 3-manifolds $\tau(M)$ and $|M|$ introduced in Chapters II and VII in terms of state sums on shadows. These results allow us to show that $|M| = \tau(M)\,\tau(-M)$. Sections X.4 - X.6 are devoted to the proof of a theorem used in Section X.3. Note the key role of Section X.5 where we compute the invariant F of links in $\mathbb{R}^3$ in terms of $6j$-symbols. In Sections X.7 and X.8 we relate the TQFT's constructed in Chapters IV and VII. Finally, in Section X.9 we use the technique of shadows to compute the invariant τ for graph 3-manifolds.

In Chapter XI we explain how quantum groups give rise to modular categories. We begin with a general discussion of quasitriangular Hopf algebras, ribbon Hopf algebras, and modular Hopf algebras (Sections XI.1 - XI.3 and XI.5). In order to derive modular categories from quantum groups we use more general quasimodular categories (Section XI.4). In Section XI.6 we outline relevant results from the theory of quantum groups at roots of unity and explain how to obtain modular categories. For completeness, we also discuss quantum groups with generic parameter; they give rise to semisimple ribbon categories (Section XI.7).

In Chapter XII we give a geometric construction of the modular categories determined by the quantum group $U_q(sl_2(\mathbb{C}))$ at roots of unity. The corner-stone of this approach is the skein theory of tangle diagrams (Sections XII.1 and XII.2) and a study of idempotents in the Temperley-Lieb algebras (Sections XII.3 and XII.4). After some preliminaries in Sections XII.5 and XII.6 we construct modular skein categories in Section XII.7. These categories are studied in the next two sections where we compute multiplicity modules and discuss when these categories are unitary.

A part of this monograph grew out of the joint papers of the author with N. Reshetikhin and O. Viro written in 1987 - 1990. Their collaboration is gratefully appreciated.

The author would like to thank H. Andersen, J. Birman, L. Crane, I. Frenkel, V. Jones, C. Kassel, L. Kauffman, D. Kazhdan, A. Kirillov, W.B.R. Lickorish, G. Masbaum, H. Morton, M. Rosso, K. Walker, and H. Wenzl for useful discussions and comments. Particular thanks are due to P. Deligne for stimulating correspondence.

The author is sincerely grateful to Matt Greenwood who read a preliminary version of the manuscript and made important suggestions and corrections. It is a pleasure to acknowledge the valuable assistance of M. Karbe and I. Zimmermann with the editing of the book. The meticulous work of drawing the pictures for the book was done by R. Hartmann, to whom the author wishes to express gratitude for patience and cooperation.

Parts of this book were written while the author was visiting the University of Geneva, the Aarhus University, the University of Stanford, the Technion (Haifa), the University of Liverpool, the Newton Mathematical Institute (Cambridge), and the University of Göttingen. The author is indebted to these institutions for their invitations and hospitality.

Part I

Towards Topological Field Theory

Chapter I
Invariants of graphs in Euclidean 3-space

1. Ribbon categories

1.0. Outline. We introduce ribbon categories forming the algebraic base of the theory presented in this book. These are monoidal categories (i.e., categories with tensor product) endowed with braiding, twist, and duality. All these notions are discussed here in detail; they will be used throughout the book. We also introduce an elementary graphical calculus allowing us to use drawings in order to present morphisms in ribbon categories.

As we shall see in Section 2, each ribbon category gives rise to a kind of "topological field theory" for links in Euclidean 3-space. In order to extend this theory to links in other 3-manifolds and to construct 3-dimensional TQFT's we shall eventually restrict ourselves to more subtle modular categories.

The definition of ribbon category has been, to a great extent, inspired by the theory of quantum groups. The reader acquainted with this theory may notice that braiding plays the role of the universal R-matrix of a quantum group (cf. Chapter XI).

1.1. Monoidal categories. The definition of a monoidal category axiomatizes the properties of the tensor product of modules over a commutative ring. Here we recall briefly the concepts of category and monoidal category, referring for details to [Ma2].

A category $\mathcal{V}$ consists of a class of objects, a class of morphisms, and a composition law for the morphisms which satisfy the following axioms. To each morphism f there are associated two objects of $\mathcal{V}$ denoted by source(f) and target(f). (One uses the notation f : source(f) $\to$ target(f).) For any objects V, W of $\mathcal{V}$, the morphisms $V \to W$ form a set denoted by Hom(V, W). The composition $f \circ g$ of two morphisms is defined whenever target(g) = source(f). This composition is a morphism source(g) $\to$ target(f). Composition is associative:

$$(f \circ g) \circ h = f \circ (g \circ h) \tag{1.1.a}$$

whenever both sides of this formula are defined. Finally, for each object V, there is a morphism $\mathrm{id}_V : V \to V$ such that

$$f \circ \mathrm{id}_V = f, \quad \mathrm{id}_V \circ g = g \tag{1.1.b}$$

for any morphisms $f : V \to W$, $g : W \to V$.

A tensor product in a category $\mathcal{V}$ is a covariant functor $\otimes : \mathcal{V} \times \mathcal{V} \to \mathcal{V}$ which associates to each pair of objects V, W of $\mathcal{V}$ an object $V \otimes W$ of $\mathcal{V}$ and to each pair of morphisms $f : V \to V'$, $g : W \to W'$ a morphism $f \otimes g : V \otimes W \to V' \otimes W'$. To say that $\otimes$ is a covariant functor means that we have the following identities

(1.1.c) $$(f \circ f') \otimes (g \circ g') = (f \otimes g) \circ (f' \otimes g'),$$

(1.1.d) $$\mathrm{id}_V \otimes \mathrm{id}_W = \mathrm{id}_{V \otimes W}.$$

A strict monoidal category is a category $\mathcal{V}$ equipped with a tensor product and an object $\mathbb{1} = \mathbb{1}_{\mathcal{V}}$, called the unit object, such that the following conditions hold. For any object V of $\mathcal{V}$, we have

(1.1.e) $$V \otimes \mathbb{1} = V, \quad \mathbb{1} \otimes V = V$$

and for any triple U, V, W of objects of $\mathcal{V}$, we have

(1.1.f) $$(U \otimes V) \otimes W = U \otimes (V \otimes W).$$

For any morphism f in $\mathcal{V}$,

(1.1.g) $$f \otimes \mathrm{id}_{\mathbb{1}} = \mathrm{id}_{\mathbb{1}} \otimes f = f$$

and for any triple f, g, h of morphisms in $\mathcal{V}$,

(1.1.h) $$(f \otimes g) \otimes h = f \otimes (g \otimes h).$$

More general (not necessarily strict) monoidal categories are defined similarly to strict monoidal categories though instead of (1.1.e), (1.1.f) one assumes that the right-hand sides and left-hand sides of these equalities are related by fixed isomorphisms. (A morphism $f : V \to W$ of a category is said to be an isomorphism if there exists a morphism $g : W \to V$ such that $fg = \mathrm{id}_W$ and $gf = \mathrm{id}_V$). These fixed isomorphisms should satisfy two compatibility conditions called the pentagon and triangle identities, see [Ma2]. These isomorphisms should also appear in (1.1.g) and (1.1.h) in the obvious way. For instance, the category of modules over a commutative ring with the standard tensor product of modules is monoidal. The ground ring regarded as a module over itself plays the role of the unit object. Note that this monoidal category is not strict. Indeed, if U, V, and W are modules over a commutative ring then the modules $(U \otimes V) \otimes W$ and $U \otimes (V \otimes W)$ are canonically isomorphic but not identical.

We shall be concerned mainly with strict monoidal categories. This does not lead to a loss of generality because of MacLane's coherence theorem which establishes equivalence of any monoidal category to a certain strict monoidal category. In particular, the category of modules over a commutative ring is equivalent to a strict monoidal category. Non-strict monoidal categories will essentially appear only in this section, in Section II.1, and in Chapter XI. Working with non-strict monoidal categories, we shall supress the fixed isomorphisms relating the right-hand sides and left-hand sides of equalities (1.1.e), (1.1.f). (Such abuse of notation is traditional in linear algebra.)

1.2. Braiding and twist in monoidal categories. The tensor multiplication of modules over a commutative ring is commutative in the sense that for any modules V, W, there is a canonical isomorphism $V \otimes W \to W \otimes V$. This isomorphism transforms any vector $v \otimes w$ into $w \otimes v$ and extends to $V \otimes W$ by linearity. It is called the flip and denoted by $P_{V,W}$. The system of flips is compatible with the tensor product in the obvious way: for any three modules U, V, W, we have

$$P_{U,V\otimes W} = (\mathrm{id}_V \otimes P_{U,W})(P_{U,V} \otimes \mathrm{id}_W), \quad P_{U\otimes V,W} = (P_{U,W} \otimes \mathrm{id}_V)(\mathrm{id}_U \otimes P_{V,W}).$$

The system of flips is involutive in the sense that $P_{W,V}P_{V,W} = \mathrm{id}_{V\otimes W}$. Axiomatization of these properties of flips leads to the notions of a braiding and a twist in monoidal categories. From the topological point of view, braiding and twist (together with the duality discussed below) form a minimal set of elementary blocks necessary and sufficient to build up a topological field theory for links in $\mathbb{R}^3$.

A braiding in a monoidal category $\mathcal{V}$ consists of a natural family of isomorphisms

$$c = \{c_{V,W} : V \otimes W \to W \otimes V\}, \tag{1.2.a}$$

where V, W run over all objects of $\mathcal{V}$, such that for any three objects U, V, W, we have

$$c_{U,V\otimes W} = (\mathrm{id}_V \otimes c_{U,W})(c_{U,V} \otimes \mathrm{id}_W), \tag{1.2.b}$$

$$c_{U\otimes V,W} = (c_{U,W} \otimes \mathrm{id}_V)(\mathrm{id}_U \otimes c_{V,W}). \tag{1.2.c}$$

(The reader is recommended to draw the corresponding commutative diagrams.) The naturality of the isomorphisms (1.2.a) means that for any morphisms $f : V \to V'$, $g : W \to W'$, we have

$$(g \otimes f)\, c_{V,W} = c_{V',W'}(f \otimes g). \tag{1.2.d}$$

Applying (1.2.b), (1.2.c) to $V = W = \mathbb{1}$ and $U = V = \mathbb{1}$ and using the invertibility of $c_{V,\mathbb{1}}, c_{\mathbb{1},V}$, we get

$$c_{V,\mathbb{1}} = c_{\mathbb{1},V} = \mathrm{id}_V \tag{1.2.e}$$

for any object V of $\mathcal{V}$. In Section 1.6 we shall show that any braiding satisfies the following Yang-Baxter identity:

$$(\mathrm{id}_W \otimes c_{U,V})\,(c_{U,W} \otimes \mathrm{id}_V)\,(\mathrm{id}_U \otimes c_{V,W}) = \tag{1.2.f}$$

$$= (c_{V,W} \otimes \mathrm{id}_U)\,(\mathrm{id}_V \otimes c_{U,W})\,(c_{U,V} \otimes \mathrm{id}_W).$$

Axiomatization of the involutivity of flips is slightly more involved. It would be too restrictive to require the composition $c_{W,V}c_{V,W}$ to be equal to $\mathrm{id}_{V\otimes W}$. What suits our aims better is to require this composition to be a kind of coboundary. This suggests the notion of a twist as follows. A twist in a monoidal category $\mathcal{V}$ with a braiding c consists of a natural family of isomorphisms

$$\theta = \{\theta_V : V \to V\}, \tag{1.2.g}$$

where V runs over all objects of $\mathcal{V}$, such that for any two objects V, W of $\mathcal{V}$, we have

$$\theta_{V\otimes W} = c_{W,V}\, c_{V,W}\, (\theta_V \otimes \theta_W). \tag{1.2.h}$$

The naturality of θ means that for any morphism $f : U \to V$ in $\mathcal{V}$, we have $\theta_V f = f\theta_U$. Using the naturality of the braiding, we may rewrite (1.2.h) as follows:

$$\theta_{V\otimes W} = c_{W,V}\, (\theta_W \otimes \theta_V)\, c_{V,W} = (\theta_V \otimes \theta_W)\, c_{W,V}\, c_{V,W}.$$

Note that $\theta_{\mathbb{1}} = \mathrm{id}_{\mathbb{1}}$. This follows from invertibility of $\theta_{\mathbb{1}}$ and the formula

$$(\theta_{\mathbb{1}})^2 = (\theta_{\mathbb{1}} \otimes \mathrm{id}_{\mathbb{1}})(\mathrm{id}_{\mathbb{1}} \otimes \theta_{\mathbb{1}}) = \theta_{\mathbb{1}} \otimes \theta_{\mathbb{1}} = \theta_{\mathbb{1}}.$$

These equalities follow respectively from (1.1.g), (1.1.c) and (1.1.b), (1.2.h) and (1.2.e) where we substitute $V = W = \mathbb{1}$.

1.3. Duality in monoidal categories. Duality in monoidal categories is meant to axiomatize duality for modules usually formulated in terms of non-degenerate bilinear forms. Of course, the general definition of duality should avoid the term "linear". It rather axiomatizes the properties of the evaluation pairing and copairing (cf. Lemma III.2.2).

Let $\mathcal{V}$ be a monoidal category. Assume that to each object V of $\mathcal{V}$ there are associated an object V^* of $\mathcal{V}$ and two morphisms

$$b_V : \mathbb{1} \to V \otimes V^*, \quad d_V : V^* \otimes V \to \mathbb{1}. \tag{1.3.a}$$

The rule $V \mapsto (V^*, b_V, d_V)$ is called a duality in $\mathcal{V}$ if the following identities are satisfied:

$$(\mathrm{id}_V \otimes d_V)(b_V \otimes \mathrm{id}_V) = \mathrm{id}_V, \tag{1.3.b}$$

$$(d_V \otimes \mathrm{id}_{V^*})(\mathrm{id}_{V^*} \otimes b_V) = \mathrm{id}_{V^*}. \tag{1.3.c}$$

Note that we do not require that $(V^*)^* = V$.

We need only one axiom relating the duality morphisms b_V, d_V with braiding and twist. We say that the duality in $\mathcal{V}$ is compatible with the braiding c and the twist θ in $\mathcal{V}$ if for any object V of $\mathcal{V}$, we have

$$(\theta_V \otimes \mathrm{id}_{V^*})\, b_V = (\mathrm{id}_V \otimes \theta_{V^*})\, b_V. \tag{1.3.d}$$

The compatibility leads to a number of implications pertaining to duality. In particular, we shall show in Section 2 that any duality in $\mathcal{V}$ compatible with braiding and twist is involutive in the sense that $V^{**} = (V^*)^*$ is canonically isomorphic to V.

1.4. Ribbon categories. By a ribbon category, we mean a monoidal category $\mathcal{V}$ equipped with a braiding c, a twist θ, and a compatible duality $(*, b, d)$. A ribbon category is called strict if its underlying monoidal category is strict.

Fundamental examples of ribbon categories are provided by the theory of quantum groups: Finite-dimensional representations of a quantum group form a ribbon category. For details, see Chapter XI.

To each ribbon category $\mathcal{V}$ we associate a mirror ribbon category $\overline{\mathcal{V}}$. It has the same underlying monoidal category and the same duality $(*, b, d)$. The braiding $\overline{c}$ and the twist $\overline{\theta}$ in $\overline{\mathcal{V}}$ are defined by the formulas

$$\overline{c}_{V,W} = (c_{W,V})^{-1} \text{ and } \overline{\theta}_V = (\theta_V)^{-1} \tag{1.4.a}$$

where c and θ are the braiding and the twist in $\mathcal{V}$. The axioms of ribbon category for $\overline{\mathcal{V}}$ follow directly from the corresponding axioms for $\mathcal{V}$.

MacLane's coherence theorem that establishes equivalence of any monoidal category to a strict monoidal category works in the setting of ribbon categories as well (cf. Remark XI.1.4). This enables us to focus attention on strict ribbon categories: all results obtained below for these categories directly extend to arbitrary ribbon categories.

1.5. Traces and dimensions. Ribbon categories admit a consistent theory of traces of morphisms and dimensions of objects. This is one of the most important features of ribbon categories sharply distinguishing them from arbitrary monoidal categories. We shall systematically use these traces and dimensions.

Let $\mathcal{V}$ be a ribbon category. Denote by $K = K_{\mathcal{V}}$ the semigroup $\mathrm{End}(\mathbb{1})$ with multiplication induced by the composition of morphisms and the unit element $\mathrm{id}_{\mathbb{1}}$. The semigroup K is commutative because for any morphisms $k, k' : \mathbb{1} \to \mathbb{1}$, we have

$$kk' = (k \otimes \mathrm{id}_{\mathbb{1}})(\mathrm{id}_{\mathbb{1}} \otimes k') = k \otimes k' = (\mathrm{id}_{\mathbb{1}} \otimes k')(k \otimes \mathrm{id}_{\mathbb{1}}) = k'k.$$

The traces of morphisms and the dimensions of objects which we define below take their values in K.

For an endomorphism $f : V \to V$ of an object V, we define its trace $\mathrm{tr}(f) \in K$ to be the following composition:

$$\mathrm{tr}(f) = d_V\, c_{V,V^*}((\theta_V f) \otimes \mathrm{id}_{V^*})\, b_V : \mathbb{1} \to \mathbb{1}. \tag{1.5.a}$$

For an object V of $\mathcal{V}$, we define its dimension $\dim(V)$ by the formula

$$\dim(V) = \mathrm{tr}(\mathrm{id}_V) = d_V\, c_{V,V^*}(\theta_V \otimes \mathrm{id}_{V^*})\, b_V \in K.$$

The main properties of the trace are collected in the following lemma which is proven in Section 2.

1.5.1. Lemma. (i) *For any morphisms* $f : V \to W$, $g : W \to V$, *we have* $\mathrm{tr}(fg) = \mathrm{tr}(gf)$.

(ii) *For any endomorphisms* f, g *of objects of* $\mathcal{V}$, *we have* $\mathrm{tr}(f \otimes g) = \mathrm{tr}(f)\mathrm{tr}(g)$.

(iii) *For any morphism* $k : \mathbb{1} \to \mathbb{1}$, *we have* $\mathrm{tr}(k) = k$.

The first claim of this lemma implies the naturality of the trace: for any isomorphism $g : W \to V$ and any $f \in \mathrm{End}(V)$,

$$\mathrm{tr}(g^{-1} f g) = \mathrm{tr}(f). \tag{1.5.b}$$

Lemma 1.5.1 implies fundamental properties of dim:

(i)′ isomorphic objects have equal dimensions,

(ii)′ for any objects V, W, we have $\dim(V \otimes W) = \dim(V)\dim(W)$, and

(iii)′ $\dim(\mathbb{1}) = 1$.

We shall show in Section 2 that $\dim(V^*) = \dim(V)$.

1.6. Graphical calculus for morphisms. Let $\mathcal{V}$ be a strict ribbon category. We describe a pictorial technique used to present morphisms in $\mathcal{V}$ by plane diagrams. This pictorial calculus will allow us to replace algebraic arguments involving commutative diagrams by simple geometric reasoning. This subsection serves as an elementary introduction to operator invariants of ribbon graphs introduced in Section 2.

A morphism $f : V \to W$ in the category $\mathcal{V}$ may be represented by a box with two vertical arrows oriented downwards, see Figure 1.1.

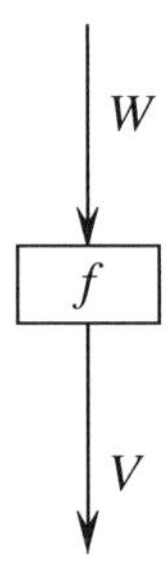

Figure 1.1

Here V, W should be regarded as "colors" of the arrows and f should be regarded as a color of the box. (Such boxes are called coupons.) More generally, a morphism $f : V_1 \otimes \ldots \otimes V_m \to W_1 \otimes \ldots \otimes W_n$ may be represented by a picture as in Figure 1.2. We do not exclude the case $m = 0$, or $n = 0$, or $m = n = 0$; by definition, for $m = 0$, the tensor product of m objects of $\mathcal{V}$ is equal to $\mathbb{1} = \mathbb{1}_{\mathcal{V}}$.

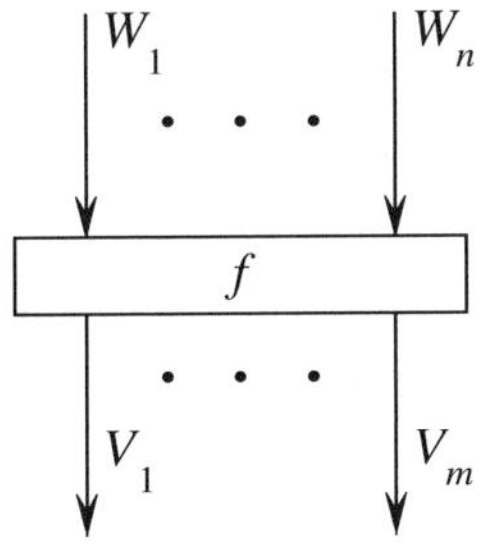

Figure 1.2

We shall use also vertical arrows oriented upwards under the convention that the morphism sitting in a box attached to such an arrow involves not the color of the arrow but rather the dual object. For example, a morphism $f : V^* \to W^*$ may be represented in four different ways, see Figure 1.3. From now on the symbol $\doteq$ displayed in the figures denotes equality of the corresponding morphisms in $\mathcal{V}$.

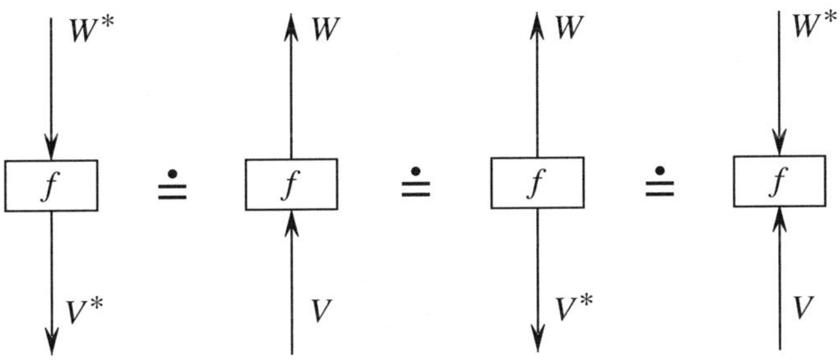

Figure 1.3

The identity endomorphism of any object V will be represented by a vertical arrow directed downwards and colored with V. A vertical arrow directed upwards and colored with V represents the identity endomorphism of V^*, see Figure 1.4.

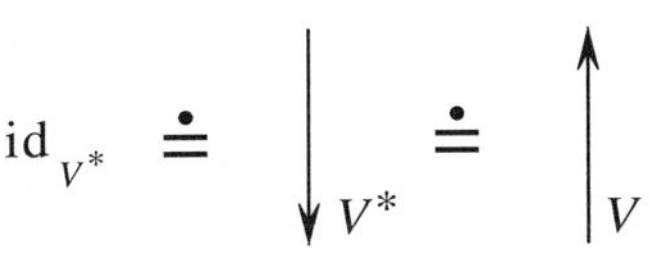

Figure 1.4

Note that a vertical arrow colored with $\mathbb{1}$ may be effaced from any picture without changing the morphism represented by this picture. We agree that the empty picture represents the identity endomorphism of $\mathbb{1}$.

The tensor product of two morphisms is presented as follows: just place a picture of the first morphism to the left of a picture of the second morphism. A picture for the composition of two morphisms f and g is obtained by putting a picture of f on the top of a picture of g and gluing the corresponding free ends of arrows. (Of course, this procedure may be applied only when the numbers of arrows, as well as their directions and colors are compatible.) In order to make this gluing smooth we should draw the arrows so that their ends are strictly vertical. For example, for any morphisms $f : V \to V'$ and $g : W \to W'$, the identities

$$(f \otimes \mathrm{id}_{W'})(\mathrm{id}_V \otimes g) = f \otimes g = (\mathrm{id}_{V'} \otimes g)(f \otimes \mathrm{id}_W)$$

have a graphical incarnation shown in Figure 1.5.

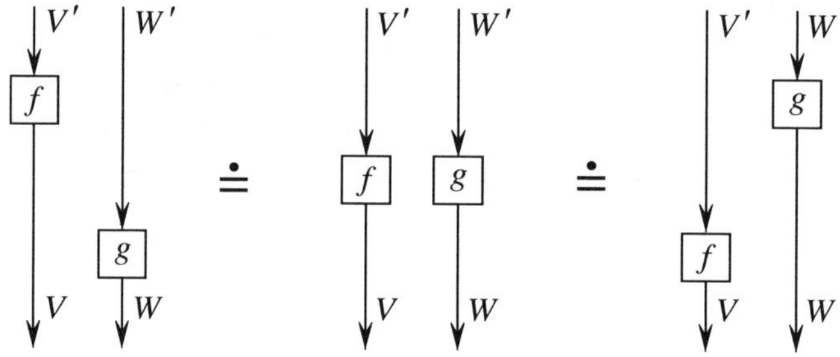

Figure 1.5

The braiding morphism $c_{V,W} : V \otimes W \to W \otimes V$ and the inverse morphism $c_{V,W}^{-1} : W \otimes V \to V \otimes W$ are represented by the pictures in Figure 1.6. Note that the colors of arrows do not change when arrows pass a crossing. The colors may change only when arrows hit coupons.

A graphical form of equalities (1.2.b), (1.2.c), (1.2.d) is given in Figure 1.7.

Using this notation, it is easy to verify the Yang-Baxter identity (1.2.f), see Figure 1.8 where we apply twice (1.2.b) and (1.2.d). Here is an algebraic form of the same argument:

$$(\mathrm{id}_W \otimes c_{U,V})(c_{U,W} \otimes \mathrm{id}_V)(\mathrm{id}_U \otimes c_{V,W}) = c_{U,W\otimes V}(\mathrm{id}_U \otimes c_{V,W}) =$$

$$= (c_{V,W} \otimes \mathrm{id}_U)\, c_{U,V\otimes W} = (c_{V,W} \otimes \mathrm{id}_U)(\mathrm{id}_V \otimes c_{U,W})(c_{U,V} \otimes \mathrm{id}_W).$$

Using coupons colored with identity endomorphisms of objects, we may give different graphical forms to the same equality of morphisms in $\mathcal{V}$. In Figure 1.9 we give two graphical forms of (1.2.b). Here $\mathrm{id} = \mathrm{id}_{V\otimes W}$. For instance, the upper picture in Figure 1.9 presents the equality

$$c_{U,V\otimes W}(\mathrm{id}_U \otimes \mathrm{id}_{V\otimes W}) = (\mathrm{id}_{V\otimes W} \otimes \mathrm{id}_U)(\mathrm{id}_V \otimes c_{U,W})(c_{U,V} \otimes \mathrm{id}_W)$$

which is equivalent to (1.2.b). It is left to the reader to give similar reformulations of (1.2.c) and to draw the corresponding figures.

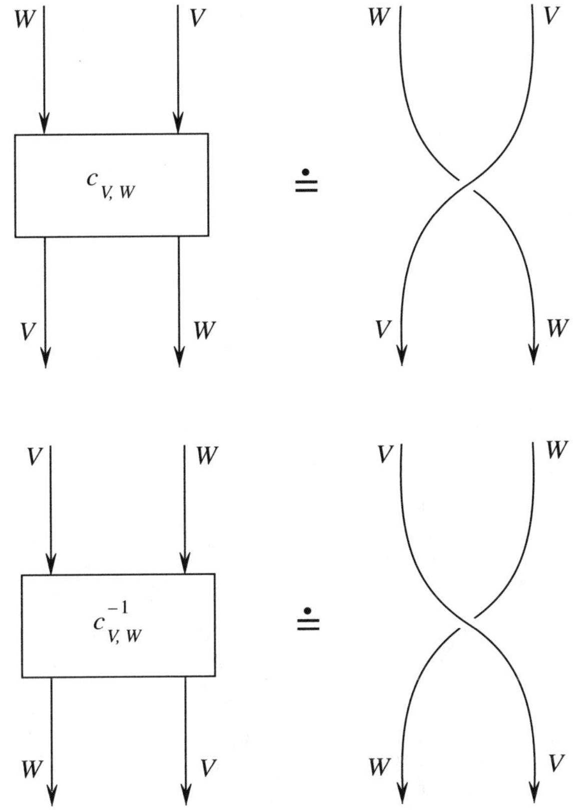

Figure 1.6

Duality morphisms $b_V : \mathbb{1} \to V \otimes V^*$ and $d_V : V^* \otimes V \to \mathbb{1}$ will be represented by the right-oriented cup and cap shown in Figure 1.10. For a graphical form of the identities (1.3.b), (1.3.c), see Figure 1.11.

The graphical technique outlined above applies to diagrams with only right-oriented cups and caps. In Section 2 we shall eliminate this constraint, describe a standard picture for the twist, and further generalize the technique. More importantly, we shall transform this pictorial calculus from a sort of skillful art into a concrete mathematical theorem.

1.7. Elementary examples of ribbon categories. We shall illustrate the concept of ribbon category with two simple examples. For more elaborate examples, see Chapters XI and XII.

1. Let K be a commutative ring with unit. By a projective K-module, we mean a finitely generated projective K-module, i.e., a direct summand of K^n with finite $n = 0, 1, 2, \ldots$ For example, free K-modules of finite rank are projective. It is obvious that the tensor product of a finite number of projective modules is

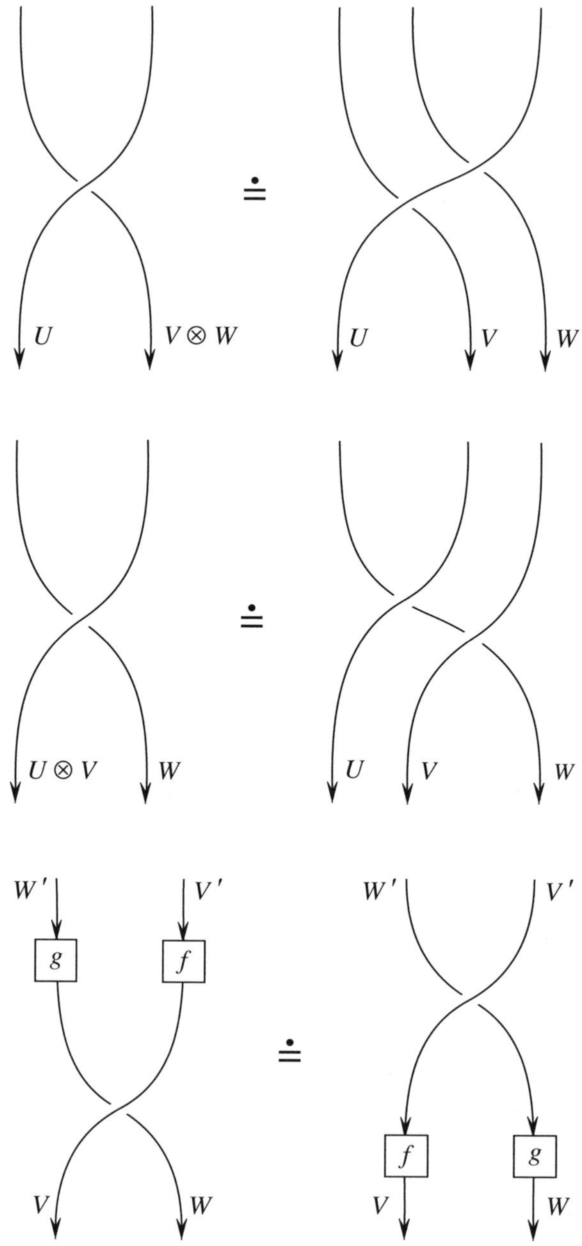

Figure 1.7

projective. For any projective K-module V, the dual K-module $V^\star = \mathrm{Hom}_K(V, K)$ is also projective and the canonical homomorphism $V \to V^{\star\star}$ is an isomorphism.

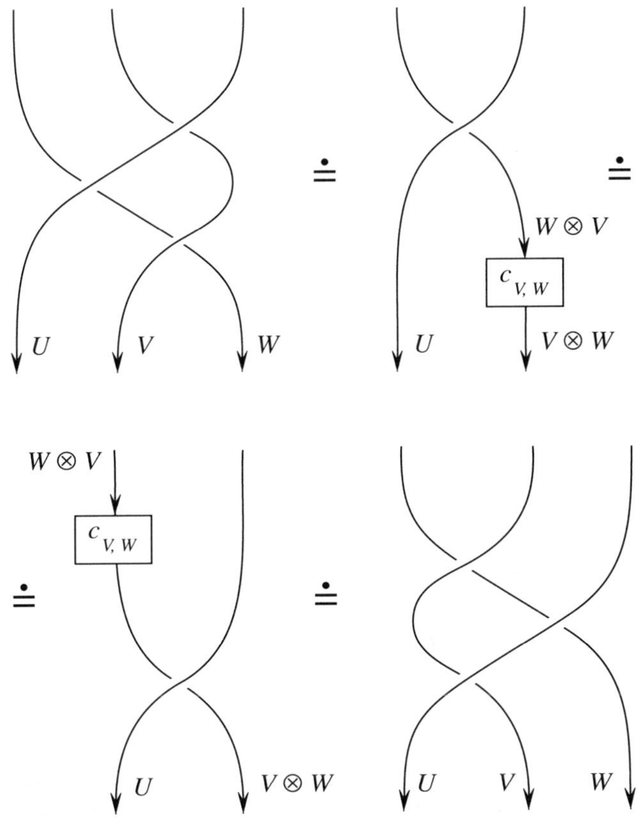

Figure 1.8

Let Proj(K) be the category of projective K-modules and K-linear homomorphisms. Provide Proj(K) with the usual tensor product over K. Set $\mathbb{1} = K$. It is obvious that Proj(K) is a monoidal category. We provide this category with braiding, twist, and duality. The braiding in Proj(K) is given by flips described in Section 1.2. The twist is given by the identity endomorphisms of objects. For any projective K-module V, set $V^* = V^\star = \mathrm{Hom}_K(V, K)$ and define d_V to be the evaluation pairing $v \otimes w \mapsto v(w) : V^\star \otimes V \to K$. Finally, define b_V to be the homomorphism $K \to V \otimes V^\star$ dual to $d_V : V^\star \otimes V \to K$ where we use the standard identifications $K^\star = K$ and $(V^\star \otimes V)^\star = V^{\star\star} \otimes V^\star = V \otimes V^\star$. The last two equalities follow from projectivity of V. (If V is a free module with a basis $\{e_i\}_i$ and $\{e^i\}_i$ is the dual basis of V^* then $b_V(1) = \sum_i e_i \otimes e^i$.) All axioms of ribbon categories are easily seen to be satisfied. Verification of (1.3.b) and (1.3.c) is an exercise in linear algebra, it is left to the reader.

The ribbon category Proj(K) is not interesting from the viewpoint of applications to knots. Indeed, we have $c_{V,W} = (c_{W,V})^{-1}$ so that the morphisms associ-

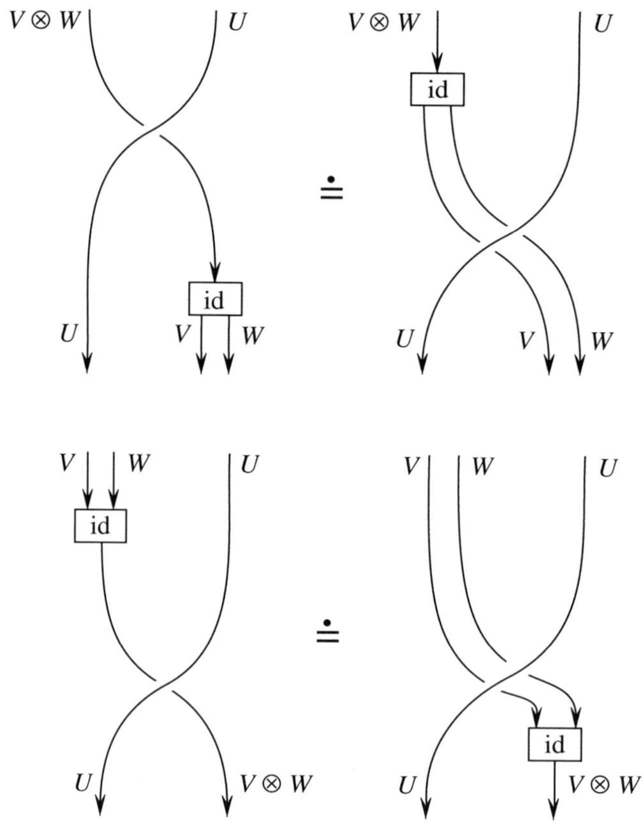

Figure 1.9

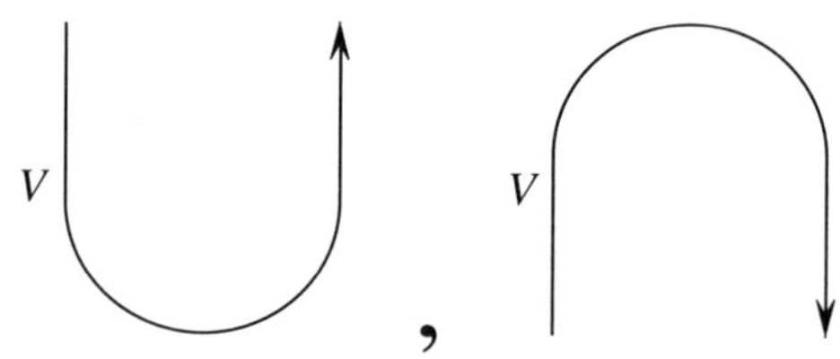

Figure 1.10

ated with diagrams are preserved under trading overcrossings for undercrossings, which kills the 3-dimensional topology of diagrams (cf. Figure 1.6).

Applying the definitions of Section 1.5 to the morphisms and objects of $\mathrm{Proj}(K)$ we get the notions of a dimension for projective K-modules and a trace for K-endomorphisms of projective K-modules. We shall denote these dimension and trace by Dim and Tr respectively. They generalize the usual dimension and trace for free modules and their endomorphisms (cf. Lemma II.4.3.1).

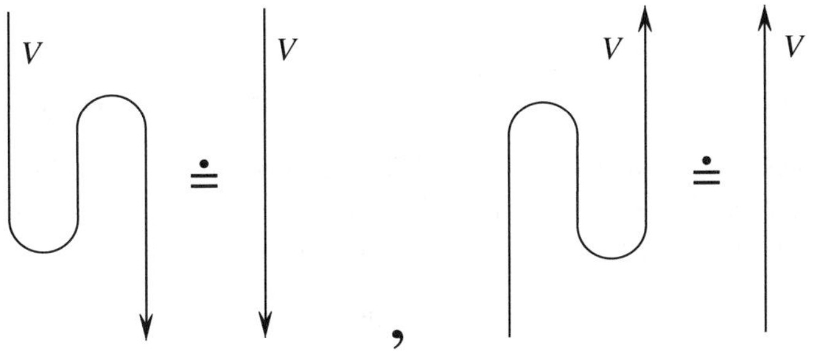

Figure 1.11

2. Let G be a multiplicative abelian group, K a commutative ring with unit, c a bilinear pairing $G \times G \to K^*$ where K^* is the multiplicative group of invertible elements of K. Thus $c(gg', h) = c(g, h)\, c(g', h)$ and $c(g, hh') = c(g, h)\, c(g, h')$ for any $g, g', h, h' \in G$. Using these data, we construct a ribbon category $\mathcal{V}$. The objects of $\mathcal{V}$ are elements of G. For any $g \in G$, the set of morphisms $g \to g$ is a copy of K. For distinct $g, h \in G$ the set of morphisms $g \to h$ consists of one element called zero. The composition of two morphisms $g \to h \to f$ is the product of the corresponding elements of K if $g = h = f$ and zero otherwise. The unit of K plays the role of the identity endomorphism of any object. The tensor product of $g, h \in G$ is defined to be their product $gh \in G$. The tensor product $gg' \to hh'$ of two morphisms $g \to h$ and $g' \to h'$ is the product of the corresponding elements of K if $g = h$ and $g' = h'$ and zero otherwise. It is easy to check that $\mathcal{V}$ is a strict monoidal category with the unit object being the unit of G. For $g, h \in G$, we define the braiding $gh \to hg = gh$ to be $c(g, h) \in K$ and the twist $g \to g$ to be $c(g, g) \in K$. Equalities (1.2.b), (1.2.c), and (1.2.h) follow from bilinearity of c. The naturality of the braiding and twist is straightforward. For $g \in G$, the dual object g^* is defined to be the inverse $g^{-1} \in G$ of g. Morphisms (1.3.a) are endomorphisms of the unit of G represented by $1 \in K$. Equalities (1.3.b) and (1.3.c) are straightforward. Formula (1.3.d) follows from the identity $c(g^{-1}, g^{-1}) = c(g, g)$. Thus, $\mathcal{V}$ is a ribbon category.

We may slightly generalize the construction of $\mathcal{V}$. Besides G, K, c, fix a group homomorphism $\varphi : G \to K^*$ such that $\varphi(g^2) = 1$ for all $g \in G$. We define the braiding and duality as above but define the twist $g \to g$ to be $\varphi(g)\, c(g, g) \in K$. It is easy to check that this yields a ribbon category. (The assumption $\varphi(g^2) = 1$ ensures (1.3.d).) This ribbon category is denoted by $\mathcal{V}(G, K, c, \varphi)$. The case considered above corresponds to $\varphi = 1$.

1.8. Exercises. 1. Use the graphical calculus to show that for any three objects U, V, W of a ribbon category, the homomorphisms

$$f \mapsto (d_V \otimes \mathrm{id}_W)(\mathrm{id}_{V^*} \otimes f) : \mathrm{Hom}(U, V \otimes W) \to \mathrm{Hom}(V^* \otimes U, W)$$

and

$$g \mapsto (\mathrm{id}_V \otimes g)(b_V \otimes \mathrm{id}_U) : \mathrm{Hom}(V^* \otimes U, W) \to \mathrm{Hom}(U, V \otimes W)$$

are mutually inverse. This establishes a bijective correspondence between the sets $\mathrm{Hom}(U, V \otimes W)$ and $\mathrm{Hom}(V^* \otimes U, W)$. Write down similar formulas for a bijective correspondence between $\mathrm{Hom}(U \otimes V, W)$ and $\mathrm{Hom}(U, W \otimes V^*)$.

2. Define the dual $f^* : V^* \to U^*$ of a morphism $f : U \to V$ by the formula

$$f^* = (d_V \otimes \mathrm{id}_{U^*})(\mathrm{id}_{V^*} \otimes f \otimes \mathrm{id}_{U^*})(\mathrm{id}_{V^*} \otimes b_U).$$

Give a pictorial interpretation of this formula. Use it to show that $(\mathrm{id}_V)^* = \mathrm{id}_{V^*}$ and $(fg)^* = g^* f^*$ for composable morphisms f, g. Show that (1.3.d) is equivalent to the formula

$$\theta_{V^*} = (\theta_V)^*. \tag{1.8.a}$$

3. Show that every duality in a monoidal category $\mathcal{V}$ is compatible with the tensor product in the sense that for any objects V, W of $\mathcal{V}$, the object $(V \otimes W)^*$ is isomorphic to $W^* \otimes V^*$. Set $U = V \otimes W$. Use the graphical calculus to show that the following morphisms are mutually inverse isomorphisms:

$$(d_U \otimes \mathrm{id}_{W^*} \otimes \mathrm{id}_{V^*})(\mathrm{id}_{U^*} \otimes \mathrm{id}_V \otimes b_W \otimes \mathrm{id}_{V^*})(\mathrm{id}_{U^*} \otimes b_V) : U^* \to W^* \otimes V^*,$$

$$(d_W \otimes \mathrm{id}_{U^*})(\mathrm{id}_{W^*} \otimes d_V \otimes \mathrm{id}_W \otimes \mathrm{id}_{U^*})(\mathrm{id}_{W^*} \otimes \mathrm{id}_{V^*} \otimes b_U) : W^* \otimes V^* \to U^*.$$

(Hint: use coupons colored with id_U.) Show that modulo these isomorphisms we have $(f \otimes g)^* = g^* \otimes f^*$ for any morphisms f, g in $\mathcal{V}$.

4. Use the graphical calculus to show that if $f : U \to V$ and $g : V \to U$ are mutually inverse morphisms in a ribbon category then $(f \otimes g^*)\, b_U = b_V$ and $d_V(g^* \otimes f) = d_U$.

2. Operator invariants of ribbon graphs

2.0. Outline. The objective of this section is to relate the theory of ribbon categories to the theory of links in Euclidean space $\mathbb{R}^3$. For technical reasons, it is convenient to deal with the strip $\mathbb{R}^2 \times [0, 1]$ rather than with $\mathbb{R}^3$. This does not lead to a loss of generality because any link in $\mathbb{R}^3$ may be deformed into $\mathbb{R}^2 \times [0, 1]$.

In generalization of links and braids we shall consider graphs embedded in $\mathbb{R}^2 \times [0, 1]$. In fact, instead of usual graphs formed by vertices and edges we shall consider ribbon graphs formed by small rectangles (coupons) and long bands. It is understood that the bands are attached to the bases of coupons and, possibly, to certain intervals in the planes $\mathbb{R}^2 \times 0$ and $\mathbb{R}^2 \times 1$. The bands attached to the last intervals are called free ends of the graph.

The next step is to marry the topology of ribbon graphs with the algebra of ribbon categories. To this end we introduce colorings of ribbon graphs by objects and morphisms of a given ribbon category $\mathcal{V}$. The bands are colored with objects whilst the coupons are colored with morphisms. The ribbon graphs with such colorings (or rather their isotopy classes) form a monoidal category $\mathrm{Rib}_{\mathcal{V}}$. The definition and study of the category of ribbon graphs marks one of the major steps towards the 3-dimensional topological field theory.

The main result of this section (Theorem 2.5) establishes the existence of a certain covariant functor $F : \mathrm{Rib}_{\mathcal{V}} \to \mathcal{V}$. The functor $F = F_{\mathcal{V}}$ should be viewed as a "topological field theory" in Euclidean 3-space. This functor will play a fundamental role in the book. It is instrumental in the construction of 3-manifold invariants in Chapter II, in the construction of 3-dimensional TQFT's in Chapter IV, as well as in the definition of normalized $6j$-symbols in Chapter VI. The functor F also provides a solid grounding for the pictorial calculus of Section 1.6.

Note that ribbon graphs (and not just links) are really important for the constructions in Chapters IV - VI. We demonstrate utility of coupons at the end of this section where we discuss a few simple properties and applications of F.

The definition of ribbon graphs and related notions are somewhat technical. They involve a number of small arrangements concerned with orientations, bases of coupons, free ends of graphs, etc. The reader should not focus all his attention on these details, it is more important to catch the general idea rather than technicalities.

2.1. Ribbon graphs and their diagrams. Roughly speaking, ribbon graphs are oriented compact surfaces in $\mathbb{R}^3$ decomposed into elementary pieces: bands, annuli, and coupons. We start with the formal definition of these pieces.

A band is the square $[0, 1] \times [0, 1]$ or a homeomorphic image of this square. The images of the intervals $[0, 1] \times 0$ and $[0, 1] \times 1$ are called bases of the band. The image of the interval $(1/2) \times [0, 1]$ is called the core of the band. An annulus is the cylinder $S^1 \times [0, 1]$ or a homeomorphic image of this cylinder. The image of the circle $S^1 \times (1/2)$ is called the core of the annulus. A band or an annulus is said to be directed if its core is oriented. The orientation of the core is called the direction of the band (resp. annuli). A coupon is a band with a distinguished base. This distinguished base is called the bottom base of the coupon, the opposite base is said to be the top one.

Let k, l be non-negative integers. We define ribbon graphs with k inputs and l outputs or, briefly, ribbon (k, l)-graphs. A ribbon (k, l)-graph in $\mathbb{R}^3$ is an oriented surface Ω embedded in the strip $\mathbb{R}^2 \times [0, 1]$ and decomposed into a union of a finite number of annuli, bands, and coupons such that

(i) Ω meets the planes $\mathbb{R}^2 \times 0$, $\mathbb{R}^2 \times 1$ orthogonally along the following segments which are bases of certain bands of Ω:

(2.1.a) $$\{ [i-(1/10), i+(1/10)] \times 0 \times 0 \mid i = 1, ..., k \},$$

(2.1.b) $$\{ [j-(1/10), j+(1/10)] \times 0 \times 1 \mid j = 1, ..., l \}.$$

In the points of these segments the orientation of Ω is determined by the pair of vectors $(1,0,0)$, $(0,0,1)$ tangent to Ω;

(ii) other bases of bands lie on the bases of coupons; otherwise the bands, coupons, and annuli are disjoint;

(iii) the bands and annuli of Ω are directed.

The surface Ω with the splitting into annuli, bands, and coupons forgotten is called the surface of the ribbon (k, l)-graph Ω. The intervals (2.1.a) (resp. (2.1.b)) are called bottom (resp. top) boundary intervals of the graph.

Each band should be thought of as a narrow strip or ribbon with short bases. The coupons lie in $\mathbb{R}^2 \times (0, 1)$, each coupon should be thought of as a small rectangle with a distinguished base. Note that we impose no conditions on the geometric position of coupons in $\mathbb{R}^2 \times (0, 1)$. In particular, the distinguished (bottom) bases of coupons may actually lie higher than the opposite bases. (Since we shall consider ribbon graphs up to isotopy we shall be able to avoid this in our pictures.)

The choice of orientation for the surface Ω of a ribbon graph is equivalent to a choice of a preferred side of Ω. (We fix the right-handed orientation in $\mathbb{R}^3$.) The orientation condition in (i) means that near the boundary intervals the preferred side of Ω is the one turned up, i.e., towards the reader.

By a ribbon graph, we mean a ribbon (k, l)-graph with $k, l \geq 0$. Examples of ribbon graphs are given in Figure 2.1 where the bottom bases of coupons are their lower horizontal bases and the preferred side of Ω is the one turned up.

By isotopy of ribbon graphs, we mean isotopy in the strip $\mathbb{R}^2 \times [0, 1]$ constant on the boundary intervals and preserving the splitting into annuli, bands, and coupons, as well as preserving the directions of bands and annuli, and the orientation of the graph surface. Note that in the course of isotopy the bases of bands lying on the bases of coupons may move along these bases (not touching each other) but can not slide to the sides of coupons. Note also that when we rotate an annulus in $\mathbb{R}^3$ around its core by the angle of π we get the same annulus with the opposite orientation. Therefore, orientations of annuli are immaterial when we consider ribbon graphs up to isotopy.

There is a convenient technique enabling us to present ribbon graphs by plane pictures generalizing the standard knot diagrams. The idea is to deform the graph in $\mathbb{R}^2 \times [0, 1]$ into a "standard position" so that it lies almost parallel and very close to the plane $\mathbb{R} \times 0 \times \mathbb{R}$ as in Figures 2.1.a and 2.1.b. (The plane $\mathbb{R} \times 0 \times \mathbb{R}$ is identified with the plane of the pictures.) In particular, the coupons should be

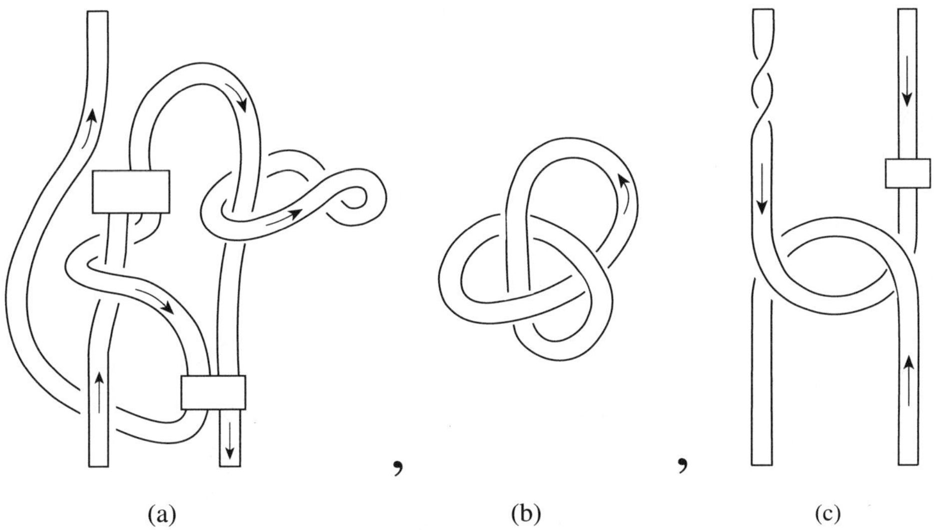

Figure 2.1

plane rectangles parallel to $\mathbb{R} \times 0 \times \mathbb{R}$. The bases of coupons should be parallel to the horizontal line $\mathbb{R} \times 0 \times 0$ and the top base of each coupon should lie higher than the bottom one. The orientation of coupons induced by the orientation of Ω should correspond to the counterclockwise orientation in $\mathbb{R} \times 0 \times \mathbb{R}$ (so that the preferred side of each coupon is turned towards the reader). The bands and annuli of the graph should go close and "parallel" to this plane. The projections of the cores of bands and annuli in the plane $\mathbb{R} \times 0 \times \mathbb{R}$ should have only double transversal crossings and should not overlap with the projections of coupons. After having deformed the graph in such a position we draw the projections of the coupons and the cores of the bands and annuli in $\mathbb{R} \times 0 \times \mathbb{R}$ taking into account the overcrossings and undercrossings of the cores. The projections of the cores of bands and annuli are oriented in accordance with their directions. The resulting picture is called a diagram of the ribbon graph.

Looking at such a diagram we may reconstruct the original ribbon graph (up to isotopy) just by letting the bands and annuli go "parallel" to the plane of the picture along their cores. One may think that arcs in our diagrams have some small width so that actually we draw very thin bands and annuli. For example, the graph diagrams in Figure 2.2 present the same ribbon graphs as in Figure 2.1, (a) and (b).

The technique of graph diagrams is sufficiently general: any ribbon graph is isotopic to a ribbon graph lying in a standard position (as described above) and therefore presented by a graph diagram. To see this, we first deform the graph so that its coupons lie in a standard position and then we deform the bands so that they go "parallel" to the plane of the picture. The only problem which we

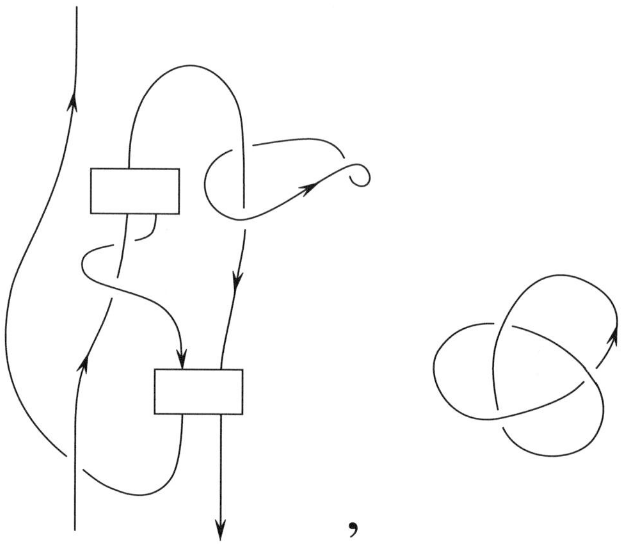

Figure 2.2

may encounter is that the bands may be twisted several times around their cores. However, both positive and negative twists in a band are isotopic to cirls which go "parallel" to the plane. See Figure 2.3 which presents positive and negative twists in a band. (The symbol $\approx$ denotes isotopy.) Note that positivity of the twist does not depend on the direction of the band and depends solely on the orientation of the ambient 3-manifold; we use everywhere the right-handed orientation in $\mathbb{R}^3$. Annuli are treated in a similar way.

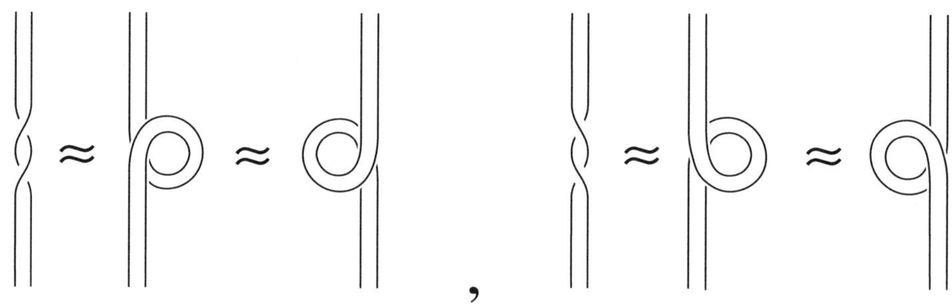

Figure 2.3

The theory of ribbon graphs generalizes the more familiar theory of framed oriented links in $\mathbb{R}^3$. A link L in $\mathbb{R}^3$ is a finite collection of smooth disjoint embedded circles $L_1, ..., L_m \subset \mathbb{R}^3$. The link L is oriented if its components $L_1, ..., L_m$ are oriented. The link L is framed if it is endowed with a homotopy class of non-singular normal vector fields on $L_1, \ldots, L_m$ in $\mathbb{R}^3$. Note that the homotopy class of a non-singular normal vector field on a component L_i is completely determined by the rotation number of the field around L_i. (This is an integer defined as the linking number of L_i with the longitude $L_i' \subset \mathbb{R}^3 \setminus L_i$ obtained by pushing L_i along the normal vector field on L_i; to compute this linking number we need to orient L_i and provide L_i' with the induced orientation; the resulting linking number does not depend on the choice of orientation in L_i.) Therefore, in order to specify a framing on a link it suffices to assign an integer to each component. These integers are called framing numbers or framings.

To each ribbon (0,0)-graph Ω consisting of annuli we may associate the link of circles in $\mathbb{R}^3$ formed by the oriented cores of the annuli. These circles are provided with a normal vector field transversal to Ω. The resulting framing is correctly defined since different choices of the normal vector field lead to homotopic vector fields on the link. In this way we get a bijective correspondence between isotopy classes of ribbon (0,0)-graphs consisting of annuli and isotopy classes of framed oriented links in $\mathbb{R}^2 \times (0, 1)$. For instance, the ribbon graph drawn in Figure 2.1.b corresponds to the trefoil knot with the framing number -3.

2.2. Ribbon graphs over $\mathcal{V}$. Fix a strict monoidal category with duality $\mathcal{V}$. A ribbon graph is said to be colored (over $\mathcal{V}$) if each band and each annulus of the graph is equipped with an object of $\mathcal{V}$. This object is called the color of the band (the annulus).

The coupons of a ribbon graph may be colored by morphisms in $\mathcal{V}$. Let Q be a coupon of a colored ribbon graph Ω. Let $V_1, \ldots, V_m$ be the colors of the bands of Ω incident to the bottom base of Q and encountered in the order induced by the orientation of Ω restricted to Q (see Figure 2.4 where Q is oriented counterclockwise). Let $W_1, \ldots, W_n$ be the colors of the bands of Ω incident to the top base of Q and encountered in the order induced by the opposite orientation of Q. Let $\varepsilon_1, \ldots, \varepsilon_m \in \{+1, -1\}$ (resp. $\nu_1, \ldots, \nu_n \in \{1, -1\}$) be the numbers determined by the directions of these bands: $\varepsilon_i = +1$ (resp. $\nu_j = -1$) if the band is directed "out" of the coupon and $\varepsilon_i = -1$ (resp. $\nu_j = +1$) in the opposite case. A color of the coupon Q is an arbitrary morphism

$$f : V_1^{\varepsilon_1} \otimes \ldots \otimes V_m^{\varepsilon_m} \to W_1^{\nu_1} \otimes \ldots \otimes W_n^{\nu_n}$$

where for an object V of $\mathcal{V}$, we set $V^{+1} = V$ and $V^{-1} = V^*$. A ribbon graph is v-colored (over $\mathcal{V}$) if it is colored and all its coupons are provided with colors as above. It is in the definition of colorings of coupons that we need to distinguish bottom and top bases of coupons.

For example, Figure 2.4 presents a v-colored ribbon (m, n)-graph containing one coupon, $m + n$ vertical untwisted unlinked bands incident to this coupon, and no annuli. As above the signs $\varepsilon_1, \ldots, \varepsilon_m, \nu_1, \ldots, \nu_n \in \{+1, -1\}$ determine the directions of the bands (the band is directed downwards if the corresponding sign is $+1$ and upwards if the sign is -1). We shall call this ribbon graph an elementary

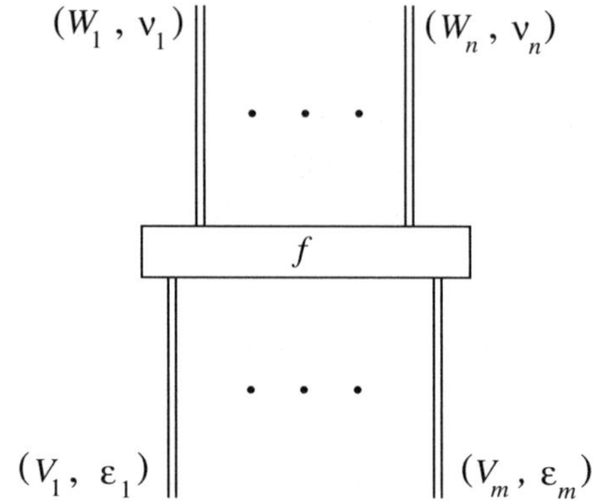

Figure 2.4

v-colored ribbon graph. Figure 2.1 gives a few examples of non-elementary ribbon graphs.

By isotopy of colored (resp. v-colored) ribbon graphs, we mean color-preserving isotopy.

The technique of diagrams readily extends to colored and v-colored ribbon graphs. To present a colored ribbon graph by a diagram, we attach an object of $\mathcal{V}$ to the cores of bands and annuli. To present a v-colored ribbon graph, we additionally assign colors to all coupons.

The notions of colored and v-colored ribbon graphs at first glance seem to be artificial and eclectic. These notions mix topological and algebraic concepts in a seemingly arbitrary way. In particular, links may be colored in many different ways, leading to numerous link invariants (constructed below). However, it is precisely in this mix of topology and algebra that lies the novelty and strength of the theory. As a specific justification of this approach, note that the invariants of a framed link $L \subset \mathbb{R}^3$ corresponding to essentially all colorings of L may be combined to produce a single invariant of the 3-manifold obtained by surgery along L (see Chapter II).

2.3. Category of ribbon graphs over $\mathcal{V}$. Let $\mathcal{V}$ be a strict monoidal category with duality. The v-colored ribbon graphs over $\mathcal{V}$ may be organized into a strict monoidal category denoted by $\text{Rib}_{\mathcal{V}}$. The objects of $\text{Rib}_{\mathcal{V}}$ are finite sequences $((V_1, \varepsilon_1), ..., (V_m, \varepsilon_m))$ where $V_1, ..., V_m$ are objects of $\mathcal{V}$ and $\varepsilon_1, ..., \varepsilon_m \in \{+1, -1\}$. The empty sequence is also considered as an object of $\text{Rib}_{\mathcal{V}}$. A morphism $\eta \to \eta'$ in $\text{Rib}_{\mathcal{V}}$ is an isotopy type of a v-colored ribbon graph (over $\mathcal{V}$) such that η (resp. η') is the sequence of colors and directions of those bands which hit the bottom (resp. top) boundary intervals. As usual, $\varepsilon = 1$ corresponds to the downward direction near the corresponding boundary interval and $\varepsilon = -1$ corresponds to the band directed up. For example, the ribbon graph drawn in Figure 2.4 represents a morphism $((V_1, \varepsilon_1), ..., (V_m, \varepsilon_m)) \to ((W_1, \nu_1), ..., (W_n, \nu_n))$. It should be emphasized that isotopic v-colored ribbon graphs present the same morphism in $\mathrm{Rib}_{\mathcal{V}}$.

The composition of two morphisms $f : \eta \to \eta'$ and $g : \eta' \to \eta''$ is obtained by putting a v-colored ribbon graph representing g on the top of a ribbon graph representing f, gluing the corresponding ends, and compressing the result into $\mathbb{R}^2 \times [0, 1]$. The identity morphisms are represented by ribbon graphs which have no annuli and no coupons, and consist of untwisted unlinked vertical bands. The identity endomorphism of the empty sequence is represented by the empty ribbon graph.

We provide $\mathrm{Rib}_{\mathcal{V}}$ with a tensor multiplication. The tensor product of objects η and η' is their juxtaposition η, η'. The tensor product of morphisms f, g is ob-

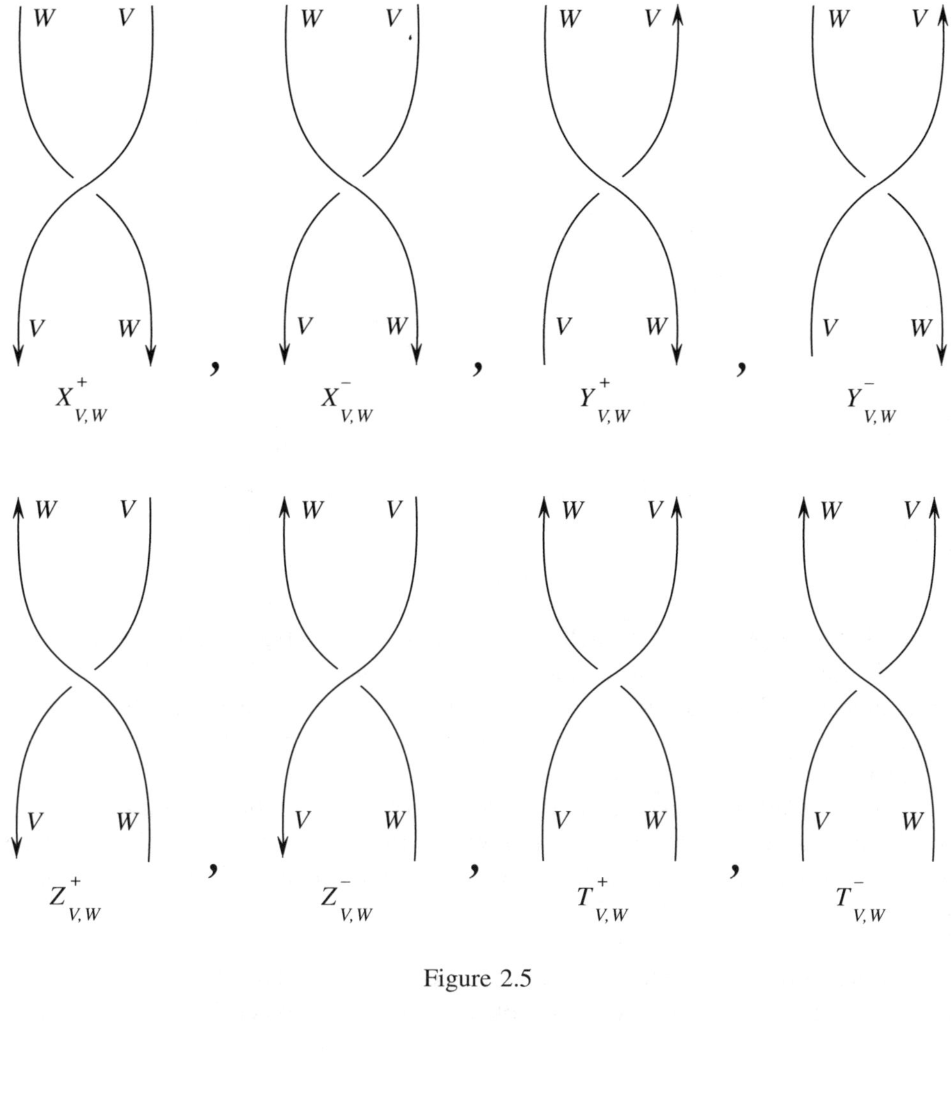

Figure 2.5

tained by placing a v-colored ribbon graph representing f to the left of a v-colored ribbon graph representing g so that there is no mutual linking or intersection. It is obvious that this tensor multiplication makes $\mathrm{Rib}_{\mathcal{V}}$ a strict monoidal category.

We shall need certain specific morphisms in $\mathrm{Rib}_{\mathcal{V}}$ presented by graph diagrams in Figure 2.5 where we also specify notation for these morphisms. Here the colors of the strings V, W run over objects of $\mathcal{V}$. The morphisms in $\mathrm{Rib}_{\mathcal{V}}$ presented by the diagrams in Figure 2.6 will be denoted by $\downarrow_V, \uparrow_V, \varphi_V, \varphi'_V, \cap_V, \cap_V^-, \cup_V, \cup_V^-$, respectively.

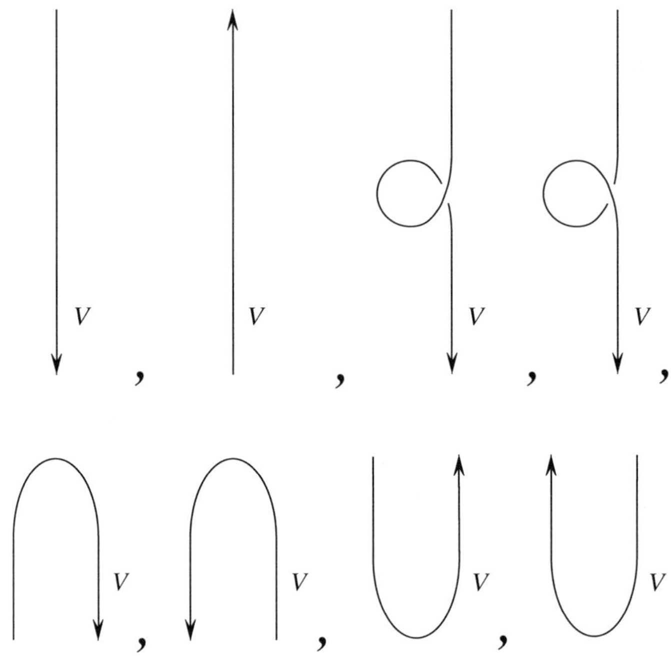

Figure 2.6

A ribbon graph over $\mathcal{V}$ which has no coupons is called a ribbon tangle over $\mathcal{V}$. It is obvious that ribbon tangles form a subcategory of $\mathrm{Rib}_{\mathcal{V}}$ which has the same objects as $\mathrm{Rib}_{\mathcal{V}}$ but less morphisms. This subcategory is called the category of colored ribbon tangles. It is a strict monoidal category under the same tensor product.

As the reader may have guessed, the category $\mathrm{Rib}_{\mathcal{V}}$ admits a natural braiding, twist, and duality and becomes in this way a ribbon category. We shall not use these structures in $\mathrm{Rib}_{\mathcal{V}}$ and do not discuss them. (For similar structures in a related setting, see Chapter XII.)

2.4. Digression on covariant functors. A covariant functor F of a category $\mathcal{X}$ into a category $\mathcal{Y}$ assigns to each object V of $\mathcal{X}$ an object $F(V)$ of $\mathcal{Y}$ and to each morphism $f: V \to W$ in $\mathscr{X}$ a morphism $F(f): F(V) \to F(W)$ in $\mathscr{Y}$ so that $F(\mathrm{id}_V) = \mathrm{id}_{F(V)}$ for any object V of $\mathscr{X}$ and $F(fg) = F(f)F(g)$ for any two composable morphisms f, g in $\mathscr{X}$. If $\mathscr{X}$ and $\mathscr{Y}$ are monoidal categories then the covariant functor $F: \mathscr{X} \to Y$ is said to preserve the tensor product if $F(\mathbb{1}_{\mathscr{X}}) = \mathbb{1}_{\mathscr{Y}}$ and for any two objects or morphisms f, g of $\mathscr{X}$ we have $F(f \otimes g) = F(f) \otimes F(g)$.

2.5. Theorem. *Let $\mathscr{V}$ be a strict ribbon category with braiding c, twist θ, and compatible duality $(*, b, d)$. There exists a unique covariant functor $F = F_{\mathscr{V}}$: $\mathrm{Rib}_{\mathscr{V}} \to \mathscr{V}$ preserving the tensor product and satisfying the following conditions:*

(1) *F transforms any object $(V, +1)$ into V and any object $(V, -1)$ into V^*;*

(2) *for any objects V, W of $\mathscr{V}$, we have*

$$F(X^+_{V,W}) = c_{V,W}, \ \ F(\varphi_V) = \theta_V, \ \ F(\cup_V) = b_V, \ \ F(\cap_V) = d_V;$$

(3) *for any elementary v-colored ribbon graph Γ, we have $F(\Gamma) = f$ where f is the color of the only coupon of Γ.*

The functor F has the following properties:

(2.5.a) $$F(X^-_{V,W}) = (c_{W,V})^{-1}, \ \ F(Y^+_{V,W}) = (c_{W,V^*})^{-1}, \ \ F(Y^-_{V,W}) = c_{V^*,W},$$

$$F(Z^+_{V,W}) = (c_{W^*,V})^{-1}, \ \ F(Z^-_{V,W}) = c_{V,W^*},$$

$$F(T^+_{V,W}) = c_{V^*,W^*}, \ \ F(T^-_{V,W}) = (c_{W^*,V^*})^{-1}, \ \ F(\varphi'_V) = (\theta_V)^{-1}.$$

Theorem 2.5 plays a fundamental role in this monograph. It may be regarded from several complementary viewpoints. First of all, it yields isotopy invariants of v-colored ribbon graphs and, in particular, invariants of colored framed links in $\mathbb{R}^3$. Indeed, by definition of $\mathrm{Rib}_{\mathscr{V}}$, isotopic v-colored ribbon graphs Ω and Ω' represent the same morphism in $\mathrm{Rib}_{\mathscr{V}}$ and therefore $F(\Omega) = F(\Omega')$. As we shall see in Chapter XII these invariants form a far-reaching generalization of the Jones polynomial of links. Secondly, Theorem 2.5 elucidates the role of braiding, twist, and duality exhibiting them as elementary blocks sufficient to build up a consistent theory of isotopy invariants of links. Theorem 2.5 renders rigorous and amplifies the graphical calculus described in Section 1.6. The main new feature is the isotopy invariance of the morphisms in $\mathscr{V}$ associated to ribbon graphs. This makes Theorem 2.5 a useful tool in the study of ribbon categories. Theorem 2.5 may also be viewed as a machine extracting morphisms in $\mathscr{V}$ from ribbon graphs in $\mathbb{R}^3$.

The morphism $F(\Omega)$ associated to a v-colored ribbon graph Ω is called the operator invariant of Ω. The term "operator invariant" does not mean that $F(\Omega)$ is linear in any sense. This term is intended to remind of the following multiplicativity properties of F. Since F is a covariant functor we have

(2.5.b) $$F(\downarrow_V) = \mathrm{id}_V, \ \ F(\uparrow_V) = \mathrm{id}_{V^*}, \ \text{ and } F(\Omega\,\Omega') = F(\Omega)F(\Omega')$$

for any two composable v-colored ribbon graphs Ω and Ω'. Since F preserves the tensor product we have

(2.5.c) $$F(\Omega \otimes \Omega') = F(\Omega) \otimes F(\Omega')$$

for any two v-colored ribbon graphs Ω and Ω'. Note also that for any v-colored ribbon (0,0)-graph Ω, we have $F(\Omega) \in K = \mathrm{End}(\mathbb{1}_{\mathcal{V}})$.

The values of F on $\cup_V^-$ and $\cap_V^-$ may be computed from the formulas

(2.5.d) $$\cup_V^- = (\uparrow_V \otimes \varphi'_V) \circ Z_{V,V}^+ \circ \cup_V;$$

(2.5.e) $$\cap_V^- = \cap_V \circ Z_{V,V}^- \circ (\varphi_V \otimes \uparrow_V).$$

The proof of Theorem 2.5 occupies Sections 3 and 4. The idea of the proof may be roughly described as follows. We shall use the tensor product and the composition in $\mathrm{Rib}_{\mathcal{V}}$ in order to express any ribbon graph via the ribbon graphs mentioned in the items (2) and (3) of the theorem. Such an expression allows us to define the value of F for any ribbon graph. Although every ribbon graph admits different expressions of this kind, they may be obtained from each other by elementary local transformations. To show that F is correctly defined, we verify the invariance of F under these transformations.

To demonstrate the power of Theorem 2.5 we devote the rest of Section 2 to applications of this theorem to duality, traces, and dimensions in ribbon categories. We also study the behavior of $F(\Omega)$ under simple transformations of ribbon graphs.

Up to the end of Section 2, the symbol $\mathcal{V}$ denotes a strict ribbon category. By coloring and v-coloring of ribbon graphs, we mean coloring and v-coloring over $\mathcal{V}$. As in Section 1, we shall write $\Omega \doteq \Omega'$ for v-colored ribbon graphs Ω, Ω' such that $F(\Omega) = F(\Omega')$. For instance, if $\Omega \approx \Omega'$, i.e., Ω and Ω' are isotopic, then $\Omega \doteq \Omega'$. Similarly, for a v-colored ribbon graph Ω and a morphism f in $\mathcal{V}$, we write $\Omega \doteq f$ and $f \doteq \Omega$ whenever $f = F(\Omega)$. For example, $X_{V,W}^+ \doteq c_{V,W}$, $\varphi_V \doteq \theta_V$, etc.

2.6. Applications to duality

2.6.1. Corollary. *For any object V of $\mathcal{V}$, the object V^{**} is canonically isomorphic to V.*

Proof. Consider the morphisms $\alpha_V : V \to V^{**}$, $\beta_V : V^{**} \to V$ corresponding under the functor F to the v-colored ribbon graphs in Figure 2.7 where $\mathrm{id} = \mathrm{id}_{V^*}$. Thus,

$$\alpha_V = (F(\cap_V^-) \otimes \mathrm{id}_{V^{**}})(\mathrm{id}_V \otimes b_{V^*}), \quad \beta_V = (d_{V^*} \otimes \mathrm{id}_V)(\mathrm{id}_{V^{**}} \otimes F(\cup_V^-)).$$

The argument in Figure 2.8 shows that $\beta_V \alpha_V = \mathrm{id}_V$. (We use the isotopy invariance and other properties of F established in Theorem 2.5; as an exercise the

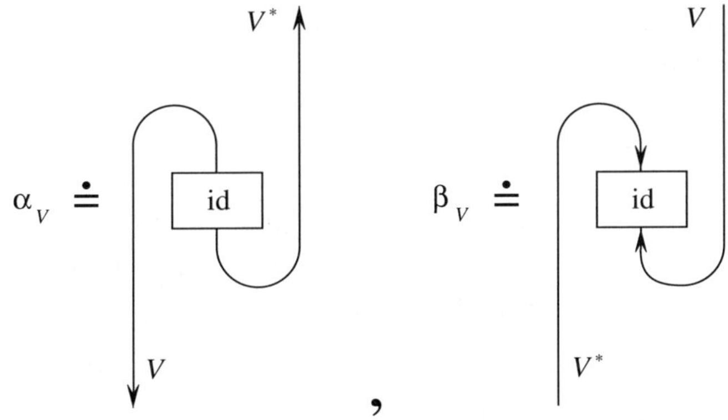

Figure 2.7

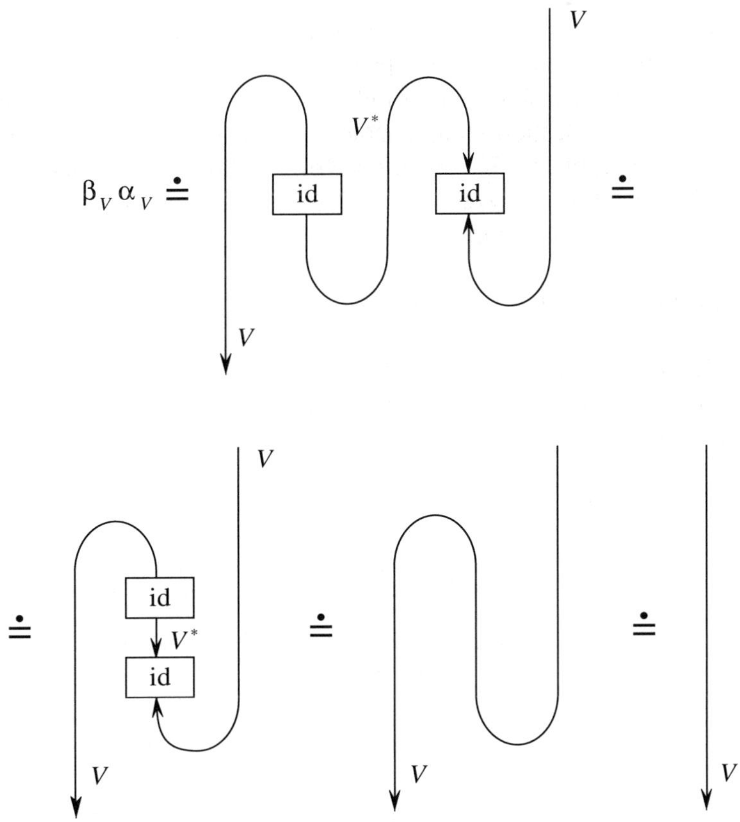

Figure 2.8

reader may rewrite the equalities in Figure 2.8 in the algebraic form.) A similar argument shows that $\alpha_V \beta_V = \mathrm{id}_{V^{**}}$.

2.6.2. Corollary. *The morphisms* $b_{\mathbb{1}} : \mathbb{1} \to \mathbb{1}^*$ *and* $d_{\mathbb{1}} : \mathbb{1}^* \to \mathbb{1}$ *are mutually inverse isomorphisms.*

Proof. The existence of isomorphism $\mathbb{1}^{**} \approx \mathbb{1}$ implies that $\mathbb{1}^*$ is isomorphic to $\mathbb{1}$. Indeed,

$$\mathbb{1}^* = \mathbb{1}^* \otimes \mathbb{1} \approx \mathbb{1}^* \otimes \mathbb{1}^{**} \approx (\mathbb{1}^* \otimes \mathbb{1})^* = (\mathbb{1}^*)^* \approx \mathbb{1}.$$

Formula (1.3.b) applied to $V = \mathbb{1}$ yields $d_{\mathbb{1}} b_{\mathbb{1}} = \mathrm{id}_{\mathbb{1}}$. This equality and the commutativity of $\mathrm{End}(\mathbb{1})$ imply that for any isomorphism $g : \mathbb{1} \to \mathbb{1}^*$, we have

$$(g^{-1} b_{\mathbb{1}})(d_{\mathbb{1}} g) = (d_{\mathbb{1}} g)(g^{-1} b_{\mathbb{1}}) = d_{\mathbb{1}} b_{\mathbb{1}} = \mathrm{id}_{\mathbb{1}}.$$

Therefore $b_{\mathbb{1}} d_{\mathbb{1}} = \mathrm{id}_{\mathbb{1}^*}$.

2.6.3. Remark. The proof of Corollary 2.6.1 may seem to certain readers a bit light-minded and not quite convincing. In fact, the proof is complete albeit based on somewhat unusual ideas. We introduce the morphisms α_V, β_V using the functor F. Then we compute their composition using the fact that F is a covariant functor invariant under a few simple modifications of v-colored ribbon graphs. The modifications in question include isotopy and cancelling of coupons colored with the identity morphisms. The invariance of F follows from Theorem 2.5. The reader would do well to analyze the proof of Corollary 2.6.1 in detail; we shall systematically use similar arguments.

2.7. Applications to trace and dimension

2.7.1. Corollary. *Let* f *be an endomorphism of an object* V *of* $\mathcal{V}$. *Let* Ω_f *be the ribbon* $(0, 0)$*-graph consisting of one* f*-colored coupon and one* V*-colored band and presented by the diagram in Figure 2.9. Then* $F(\Omega_f) = \mathrm{tr}(f)$.

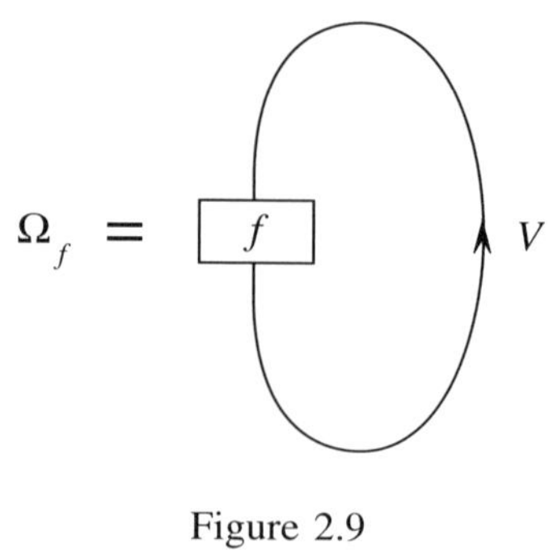

Figure 2.9

This Corollary gives a geometric interpretation of the trace of morphisms introduced in Section 1.5. Applying Corollary 2.7.1 to $f = \mathrm{id}_V$ we get $\dim(V) = F(\Omega_V)$ where Ω_V is an unknotted untwisted annulus of color V with an arbitrary direction of the core. (The annulus Ω_V is obtained from Ω_{id_V} by elimination of the coupon. This does not change the operator invariant because this coupon is colored with the identity morphism.) Note that there is an isotopy of Ω_V onto itself reversing direction of the core. This fact and the isotopy invariance of $F(\Omega_V)$ explain why $F(\Omega_V)$ does not depend on this direction.

Proof of Corollary. Let Γ_f be the v-colored ribbon (1,1)-graph presented by the diagram in Figure 1.1 with $W = V$. It is obvious that $\Omega_f \approx \cap_V^- \circ (\Gamma_f \otimes \uparrow_V) \circ \cup_V$ where the symbol $\approx$ denotes isotopy. It follows from (2.5.e) that

$$\Omega_f \approx \cap_V \circ Z_{V,V}^- \circ (\varphi_V \otimes \uparrow_V) \circ (\Gamma_f \otimes \uparrow_V) \circ \cup_V .$$

Theorem 2.5 implies that $F(\Omega_f)$ is equal to the expression used to define $\mathrm{tr}(f)$.

2.7.2. Corollary (a generalization of Corollary 2.7.1). *Let Ω be a v-colored ribbon graph determining an endomorphism of a certain object of* $\mathrm{Rib}_{\mathcal{V}}$. *Let $\overline{\Omega}$ be the v-colored ribbon* $(0,0)$*-graph obtained by closing the free ends of Ω (see Figure* 2.10 *where the box bounded by broken line substitutes a diagram of Ω). Then* $\mathrm{tr}(F(\Omega)) = F(\overline{\Omega})$.

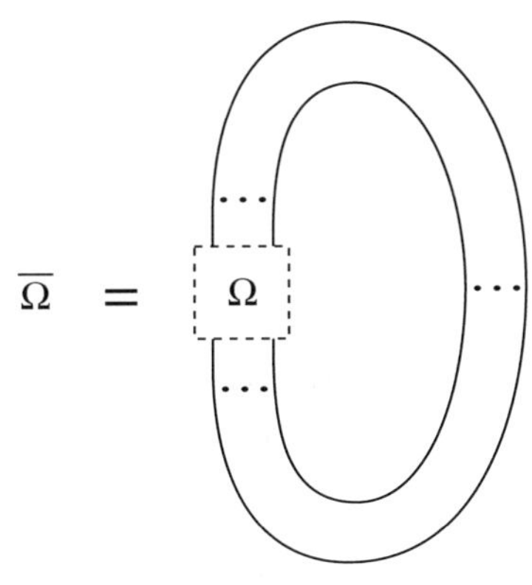

Figure 2.10

Proof. Note that the ribbon $(0,0)$-graph $\overline{\Omega}$ is obtained by connecting the top free ends of Ω to the bottom free ends of Ω in the way indicated in Figure 2.10. The v-coloring of Ω determines a v-coloring of $\overline{\Omega}$ in the obvious way. Set $V = F(\eta)$ where η is the object of $\mathrm{Rib}_{\mathcal{V}}$ which is both the source and the target of Ω. The box bounded by broken line with a diagram of Ω inside may be replaced with a coupon colored by $F(\Omega) : V \to V$ without changing the operator invariant. This yields the first equality in Figure 2.11. The second equality follows from the properties of F specified in Theorem 2.5. The isotopy in Figure 2.11 is obtained by pulling the id_V-colored coupon along the strands so that it comes close to

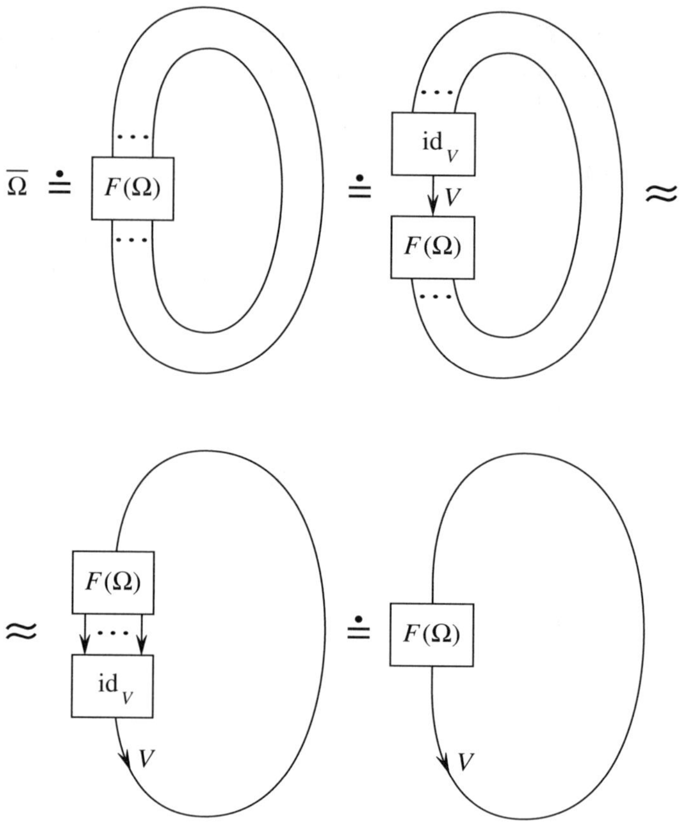

Figure 2.11

the $F(\Omega)$-colored coupon from below. The last equality in Figure 2.11 follows from the properties of F. It remains to apply Corollary 2.7.1 to deduce that $\overline{\Omega} \doteq \mathrm{tr}(F(\Omega))$.

2.7.3. Corollary. *For any objects V, W of $\mathcal{V}$, we have the equality in Figure* 2.12.

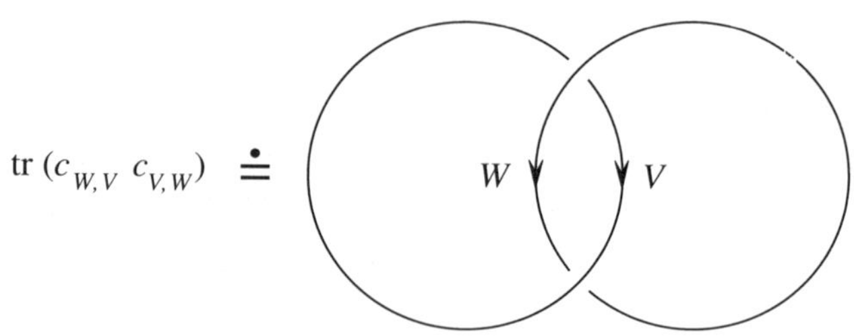

Figure 2.12

This assertion follows from Corollary 2.7.2 applied to $\Omega = X^+_{W,V} X^+_{V,W}$ and the isotopy invariance of F. The framed link presented by the diagram in Figure 2.12 is called the Hopf link with zero framing.

2.7.4. Proof of Lemma 1.5.1. The proof of the equality $\mathrm{tr}(fg) = \mathrm{tr}(gf)$ is given in Figure 2.13. The proof of the equality $\mathrm{tr}(f \otimes g) = \mathrm{tr}(f)\,\mathrm{tr}(g)$ is given in Figure 2.14. Here we use Corollary 2.7.2 and the isotopy invariance of F.

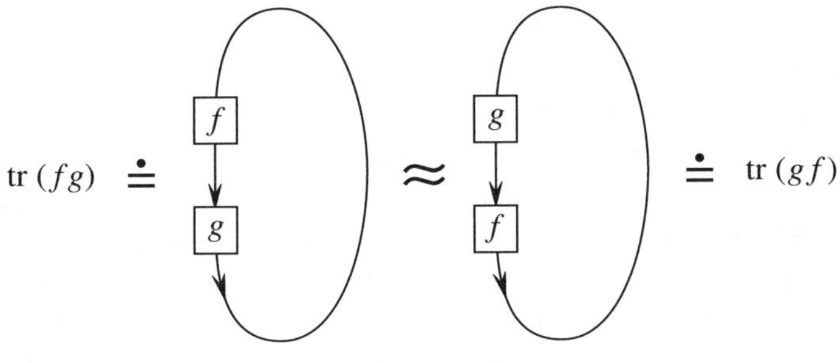

Figure 2.13

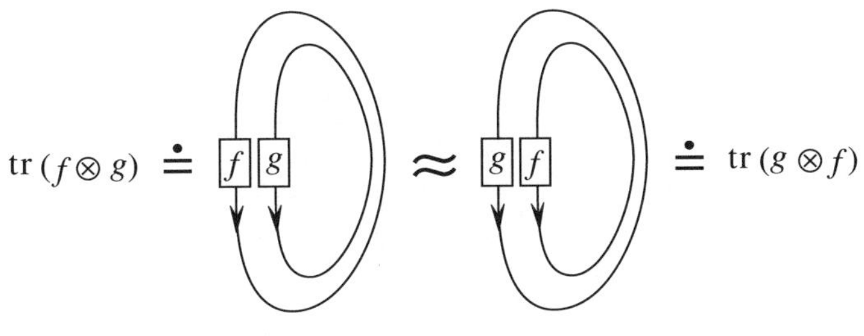

Figure 2.14

Let us show that for any $k \in K$, we have $\mathrm{tr}(k) = k$. It follows from (1.2.e), the equality $\theta_{\mathbb{1}} = \mathrm{id}_{\mathbb{1}}$, and Corollary 2.6.2 that

$$\mathrm{tr}(k) = d_{\mathbb{1}}(k \otimes \mathrm{id}_{\mathbb{1}^*})b_{\mathbb{1}} = (\mathrm{id}_{\mathbb{1}} \otimes d_{\mathbb{1}})(k \otimes \mathrm{id}_{\mathbb{1}} \otimes \mathrm{id}_{\mathbb{1}^*})(\mathrm{id}_{\mathbb{1}} \otimes b_{\mathbb{1}}) =$$

$$= k \otimes (d_{\mathbb{1}}(\mathrm{id}_{\mathbb{1}} \otimes \mathrm{id}_{\mathbb{1}^*})b_{\mathbb{1}}) = k \otimes d_{\mathbb{1}} b_{\mathbb{1}} = k \otimes \mathrm{id}_{\mathbb{1}} = k.$$

2.8. Transformations of ribbon graphs. We describe three simple geometric transformations of ribbon graphs and discuss the behavior of the operator invariant under these transformations. The first transformation is applied to an annulus component of a ribbon graph. It reverses the direction of (the core of) the annulus and replaces its color by the dual object. The second transformation is applied to an annulus component colored with the tensor product of two objects of $\mathcal{V}$; the annulus is split into two parallel annuli colored with these two objects. Finally, the third transformation is the mirror reflection of the ribbon graph with respect to the plane of our pictures $\mathbb{R} \otimes 0 \otimes \mathbb{R}$. (This reflection keeps the boundary ends of the graph.) We shall see that the first two transformations preserve the operator invariant whereas the third one involves the passage to the mirror ribbon category.

2.8.1. Corollary. *Let Ω be a v-colored ribbon graph containing an annulus component ℓ. Let Ω' be the v-colored ribbon graph obtained from Ω by reversing the direction of ℓ and replacing the color of ℓ with its dual object. Then* $F(\Omega') = F(\Omega)$.

Proof. Denote the color of ℓ by V. Choose a small vertical segment of ℓ directed upwards and replace it by the composition of two coupons Q_1 and Q_2 both colored with id_{V^*} (see Figure 2.15 where the distinguished (bottom) bases of Q_1 and Q_2 are the lower horizontal bases). This transformation does not change the operator invariant of Ω. Now pulling the coupon Q_1 along ℓ we deform the graph in $\mathbb{R}^3$ so that at the end Q_1 comes close to Q_2 from below. Since the colors of Q_1 and Q_2 are the identity endomorphisms of V^* we may eliminate Q_1 and Q_2 in this final position without changing the operator invariant. This yields Ω'. Hence $F(\Omega') = F(\Omega)$.

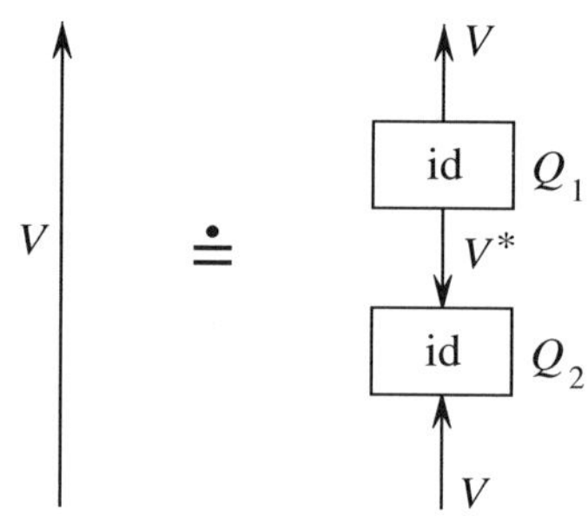

Figure 2.15

2.8.2. Corollary. *For any object V of $\mathcal{V}$, we have* $\dim(V^*) = \dim(V)$.

Proof. The oriented trivial knot is isotopic to the same knot with the opposite orientation. Therefore Corollary 2.8.1 and the remarks following the statement of Corollary 2.7.1 imply Corollary 2.8.2.

2.8.3. Corollary. *Let Ω be a v-colored ribbon graph containing an annulus component ℓ of color $U \otimes V$ where U and V are two objects of $\mathcal{V}$. Let Ω' be the v-colored ribbon graph obtained from Ω by cutting ℓ off along its core and coloring two newly emerging annuli with U and V. Then $F(\Omega) = F(\Omega')$.*

Proof. The idea of the proof is the same as in the proof of the previous corollary. Take a small vertical segment of ℓ directed downwards and replace it by two coupons as in Figure 2.16 where $\mathrm{id} = \mathrm{id}_{U\otimes V}$. It is obvious that this modification does not change the operator invariant. Now, pushing the upper coupon along ℓ until it approaches the lower coupon from below we deform our ribbon graph in the position where we may cancel these two coupons. This does not change the operator invariant and results in Ω'. Hence $F(\Omega) = F(\Omega')$.

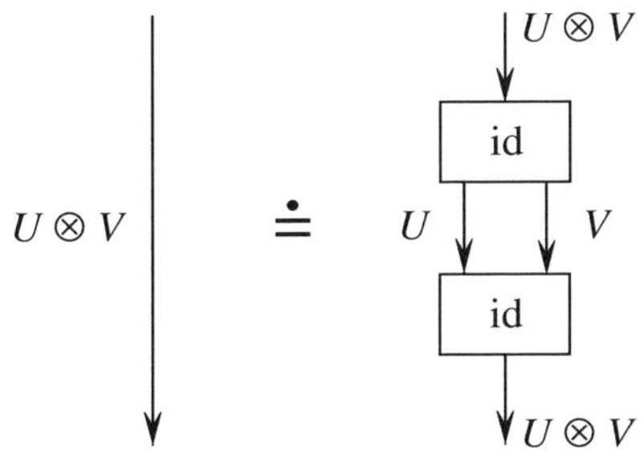

Figure 2.16

2.8.4. Corollary. *Let Ω be a v-colored ribbon graph over $\mathcal{V}$. Let $\overline{\Omega}$ be its mirror image with respect to the plane $\mathbb{R} \times 0 \times \mathbb{R}$. Then*

$$F_{\overline{\mathcal{V}}}(\overline{\Omega}) = F_{\mathcal{V}}(\Omega).$$

Note that to get a diagram of $\overline{\Omega}$ from a diagram of Ω we should simply trade all overcrossings for undercrossings. For instance, the mirror images of $X^+_{V,W}, Y^+_{V,W}, Z^+_{V,W}, T^+_{V,W}$ are $X^-_{V,W}, Y^-_{V,W}, Z^-_{V,W}, T^-_{V,W}$ respectively.

Proof of Corollary. Consider the covariant functor $G : \mathrm{Rib}_{\mathcal{V}} \to \mathcal{V}$ which coincides on the objects with $F_{\mathcal{V}}$ and transforms the morphism represented by a v-colored ribbon graph Ω into $F_{\overline{\mathcal{V}}}(\overline{\Omega})$. (We are allowed to regard $F_{\overline{\mathcal{V}}}(\overline{\Omega})$ as a morphism in the category $\mathcal{V}$ because $\overline{\mathcal{V}}$ and $\mathcal{V}$ have the same underlying monoidal category.) It is straightforward to see that G satisfies conditions (1) - (3) of Theorem 2.5. In particular,

$$G(X^+_{V,W}) = F_{\overline{\mathcal{V}}}(X^-_{V,W}) = (\overline{c}_{W,V})^{-1} = c_{V,W} = F_{\mathcal{V}}(X^+_{V,W})$$

and

$$G(\varphi_V) = F_{\overline{\mathcal{V}}}(\varphi'_V) = (\overline{\theta}_V)^{-1} = \theta_V = F_{\mathcal{V}}(\varphi_V).$$

The uniqueness in Theorem 2.5 implies that $G = F_{\mathcal{V}}$. This yields our claim.

2.8.5. Corollary. *The dimensions of any object of $\mathcal{V}$ with respect to $\mathcal{V}$ and $\overline{\mathcal{V}}$ are equal.*

This follows from Corollary 2.8.4 since the mirror image of a plane annulus is the same annulus.

2.9. Exercises. 1. Let Ω be a v-colored ribbon graph containing an annulus component. Let Ω' be the v-colored ribbon graph obtained from Ω by replacing the color of this component with an isomorphic object. Show that $\Omega \doteq \Omega'$. What is the natural analogue of this assertion for bands? What are the analogues of Corollaries 2.8.1 and 2.8.3 for bands?

2. Let Ω be a v-colored ribbon graph containing an annulus component or a band of color $\mathbb{1}$. Let Ω' be the v-colored ribbon graph obtained from Ω by eliminating this annulus (resp. band). Show that $\Omega \doteq \Omega'$.

3. Show that for any three objects U, V, W of a ribbon category, the formulas $f \mapsto (\mathrm{id}_V \otimes F(\cap_W^-))(f \otimes \mathrm{id}_{W^*})$ and $g \mapsto (g \otimes \mathrm{id}_W)(\mathrm{id}_U \otimes F(\cup_W^-))$ establish mutually inverse bijective correspondences between the sets $\mathrm{Hom}(U, V \otimes W)$ and $\mathrm{Hom}(U \otimes W^*, V)$. Write down similar formulas for a bijective correspondence between $\mathrm{Hom}(U \otimes V, W)$ and $\mathrm{Hom}(V, U^* \otimes W)$.

4. Show that if $\theta_V = \mathrm{id}_V$ for all objects V of $\mathcal{V}$ then for any colored ribbon graph Ω consisting of m annuli, we have $F(\Omega) = \prod_{i=1}^m \dim(V_i)$ where $V_1, ..., V_m$ are the colors of these annuli. This applies, for example, to the ribbon category constructed in Section 1.7.1.

5. Let $\mathcal{V}$ be the ribbon category $\mathcal{V}(G, K, c, \varphi)$ constructed in Section 1.7.2. Use formulas (2.5.d), (2.5.e) to show that for any $g \in G$, we have $F(\cup_g^-) = \varphi(g)$ and $F(\cap_g^-) = \varphi(g)$. Deduce from these equalities that $\dim(g) = \varphi(g)$ for any $g \in G$. Show that for a framed m-component link $L = L_1 \cup ... \cup L_m$ whose components are colored with $g_1, ..., g_m \in G$ respectively, we have

$$F(L) = \prod_{1 \leq j < k \leq m} (c(g_j, g_k)\, c(g_k, g_j))^{l_{jk}} \times \prod_{j=1}^{m} c(g_j, g_j)^{l_j} \varphi(g_j)^{l_j+1}$$

where $l_{jk} \in \mathbb{Z}$ is the linking number of L_j and L_k, and $l_j \in \mathbb{Z}$ is the framing number of L_j.

6. Let V be an object of a ribbon category $\mathcal{V}$ such that any endomorphism of V has the form $k \otimes \mathrm{id}_V$ for certain $k \in K = \mathrm{End}(\mathbb{1}_{\mathcal{V}})$. Let Ω be a v-colored ribbon $(0, 0)$-graph containing an annulus of color V. Show that $F(\Omega)$ is divisible by $\dim(V)$ in the semigroup K. (Hint: present Ω as the closure of a v-colored ribbon (1,1)-graph which is an endomorphism of $(V, 1)$.)

7. Show that if duality in a strict monoidal category $\mathcal{V}$ is compatible with a braiding and a twist then the square of the duality functor $(V \mapsto V^*, f \mapsto f^*) : \mathcal{V} \to \mathcal{V}$ is canonically equivalent to the identity functor $\mathcal{V} \to \mathcal{V}$ (cf.

Exercise 1.8.2). Show that for any endomorphism f of any object of $\mathcal{V}$, we have $\mathrm{tr}(f^*) = \mathrm{tr}(f)$.

3. Reduction of Theorem 2.5 to lemmas

3.0. Outline. The functor F may be regarded as a "linear representation" of the category $\mathrm{Rib}_{\mathcal{V}}$. This point of view allows us to appeal to the standard technique of group theory: in order to define a linear representation of a group one assigns matrices to generators and checks definining relations. Following this line we shall introduce generators and relations for $\mathrm{Rib}_{\mathcal{V}}$ and use them to construct F.

The material of Sections 3 and 4 will not be used in the remaining part of the book and may be skipped without harm for what follows. Still, the author finds the arguments given in these two sections beautiful and instructive in themselves.

3.1. Generators for $\mathbf{Rib}_{\mathcal{V}}$. Our immediate aim is to describe the category of v-colored ribbon graphs $\mathrm{Rib}_{\mathcal{V}}$ and its subcategory of colored ribbon tangles in terms of generators and relations.

We say that a family of morphisms in a strict monoidal category $\mathcal{X}$ generates $\mathcal{X}$ if any morphism in $\mathcal{X}$ may be obtained from these generators and the identity endomorphisms of objects of $\mathcal{X}$ using composition and tensor product. A system of relations between the generating morphisms is said to be complete if any relation between these morphisms may be deduced from the given ones using the axioms of strict monoidal category. For a more detailed discussion of generators and relations in monoidal categories, see Section 4.2.

Recall the morphisms in $\mathrm{Rib}_{\mathcal{V}}$ introduced at the end of Section 2.3.

3.1.1. Lemma. *The colored ribbon tangles*

$$X^{\nu}_{V,W},\ Z^{\nu}_{V,W},\ \varphi_V,\ \varphi'_V,\ \cup_V,\ \cap_V \tag{3.1.a}$$

where V, W run over objects of $\mathcal{V}$ and ν runs over $+1, -1$ generate the category of ribbon tangles. The same ribbon tangles together with all elementary v-colored ribbon graphs generate $\mathrm{Rib}_{\mathcal{V}}$.

Proof. To prove the lemma we need the notion of a generic tangle diagram. Let $D \subset \mathbb{R} \times [0, 1]$ be a diagram of a ribbon tangle. By the height function on D, we mean the projection $\mathbb{R} \times [0, 1] \mapsto [0, 1]$ restricted to D. By an extremal point of D, we mean a point of D lying in $\mathbb{R} \times (0, 1)$ (i.e., distinct from the end points of D) such that the height function on D attains its local maximum or local minimum in this point. By singular points on D, we mean extremal points and crossing points of D. We say that D is generic if its extremal points are distinct from its crossing points, the singular points of D are finite in number and lie on different levels of the height function, and the height function is non-degenerate in all extremal points. The last condition means that in a neighborhood of any extremal point the diagram D looks like a cup or a cap, i.e., like the graph of the function x^2 or $-x^2$ near $x = 0$.

It is obvious that a small deformation transforms any tangle diagram into a generic tangle diagram. Therefore every ribbon tangle may be presented by a generic tangle diagram.

Take an arbitrary ribbon tangle and present it by a generic diagram $D \subset \mathbb{R} \times [0, 1]$. Consider the boundary lines of the strip $\mathbb{R} \times [0, 1]$ and draw several parallel horizontal lines in this strip so that between any two adjacent lines lies no more than one singular point of D. It is clear that the part of D lying between such adjacent lines represents the tensor product of several identity morphisms and one morphism from the family of morphisms drawn in Figures 2.5 and 2.6 (except φ, φ'). The ribbon tangle presented by D is decomposed in this way in a composition of such tensor products. To prove the first assertion of the lemma it remains to express the tangles drawn in Figures 2.5 and 2.6 via the tangles (3.1.a) and the "identity" tangles $\uparrow_V$, $\downarrow_V$. Such expressions are provided by (2.5.d), (2.5.e), and

(3.1.b) $$Y^{\nu}_{V,W} = (\cap_V \otimes \downarrow_W \otimes \uparrow_V)(\uparrow_V \otimes X^{\nu}_{W,V} \otimes \uparrow_V)(\uparrow_V \otimes \downarrow_W \otimes \cup_V),$$

(3.1.c) $$T^{\nu}_{V,W} = (\cap_V \otimes \uparrow_W \otimes \uparrow_V)(\uparrow_V \otimes Y^{\nu}_{W,V} \otimes \uparrow_V)(\uparrow_V \otimes \uparrow_W \otimes \cup_V)$$

where $\nu = \pm 1$. In the last formula we substitute (3.1.b) to get an expression for $T^{\nu}_{V,W}$ via the generators. (The reader is urged to draw the corresponding pictures.)

The second assertion of Lemma is proven similarly: in addition to crossing points and local extrema on a diagram we should single out the coupons and apply the same argument.

3.2. Relations between generating tangles. Here is a list of fundamental relations between the tangles (3.1.a):

(3.2.a) $$(\downarrow_W \otimes X^{+}_{U,V})\,(X^{+}_{U,W} \otimes \downarrow_V)\,(\downarrow_U \otimes X^{+}_{V,W}) =$$
$$= (X^{+}_{V,W} \otimes \downarrow_U)\,(\downarrow_V \otimes X^{+}_{U,W})\,(X^{+}_{U,V} \otimes \downarrow_W),$$

(3.2.b) $$\downarrow_V = (\downarrow_V \otimes \cap_V)\,(\cup_V \otimes \downarrow_V),$$

(3.2.c) $$\uparrow_V = (\cap_V \otimes \uparrow_V)\,(\uparrow \otimes \cup_V),$$

(3.2.d) $$X^{-}_{V,W} = (X^{+}_{W,V})^{-1},$$

(3.2.e) $$\varphi'_V = (\varphi_V)^{-1},$$

(3.2.f) $$X^{\nu}_{V,W}\,(\downarrow_V \otimes\varphi_W) = (\varphi_W\otimes \downarrow_V)\, X^{\nu}_{V,W},$$

(3.2.g) $$Z^{\nu}_{V,W} = [(\cap_W\otimes \downarrow_V \otimes \uparrow_W)\,(\uparrow_W \otimes X^{-\nu}_{V,W}\otimes \uparrow_W)\,(\uparrow_W \otimes \downarrow_V \otimes\cup_W)]^{-1},$$

(3.2.h) $$(\varphi_V)^2 = (\cap_V\otimes \downarrow_V)\,(\uparrow_V \otimes X^{+}_{V,V})\,(Z^{+}_{V,V}\otimes \downarrow_V)\,(\cup_V\otimes \downarrow_V).$$

Here U, V, W run over arbitrary objects of $\mathcal{V}$ and ν runs over $\{+1, -1\}$. Each relation of the form $f = g^{-1}$ (specifically (3.2.d), (3.2.e), (3.2.g)) replaces the two relations $fg = \mathrm{id}$ and $gf = \mathrm{id}$. It is a pleasant geometric exercise for the reader to draw the pictures corresponding to (3.2.a) - (3.2.h) and to verify that these relations do hold in the category of ribbon tangles. Note that relations (3.2.a) - (3.2.c) already appeared in Section 1 at least in the graphical form (see Figures 1.8 and 1.11). Relations (3.2.d) and (3.2.g) represent the second Reidemeister move (cf. Section 4). Relation (3.2.e) reflects the fact that the left-hand twist and the right-hand twist in a band cancel each other. Relation (3.2.f) is obvious: a small curl on a branch of tangle diagram may be pulled over (or under) a crossing point. Relation (3.2.h) is less obvious but also true, one may prove it by inspecting a pictorial form of (3.2.h) or playing with a piece of elastic band.

As an exercise the reader may show that relations (3.2.d) - (3.2.f) imply the relations

(3.2.i) $$X^{\nu}_{V,W}(\varphi_V\otimes \downarrow_W) = (\downarrow_W \otimes\varphi_V)X^{\nu}_{V,W},$$

(3.2.j) $$X^{\nu}_{V,W}(\downarrow_V \otimes\varphi'_W) = (\varphi'_W\otimes \downarrow_V)X^{\nu}_{V,W},$$

(3.2.k) $$X^{\nu}_{V,W}(\varphi'_V\otimes \downarrow_W) = (\downarrow_W \otimes\varphi'_V)X^{\nu}_{V,W}.$$

3.3. Lemma. *Relations* (3.2.a) - (3.2.h) *form a complete set of relations between the generators* (3.1.a) *of the category of colored ribbon tangles.*

This lemma is proven in Section 4. The reader may notice that our choice of generators and relations for the category of tangles is somewhat arbitrary. There may exist presentations of this category with fewer generators and relations.

To extend Lemma 3.3 to the category of ribbon graphs $\mathrm{Rib}_{\mathcal{V}}$ we have to specify relations involving elementary ribbon graphs. These relations form an infinite sequence already on the geometric level (not speaking about colors) since the elementary ribbon graphs may have arbitrary numbers of bands. We describe these relations in a graphical form, a translation into the algebraic language is left to the reader. There are three families of relations shown in Figures 3.1 and 3.2. Here each arc represents (a piece of) the core of a band which go parallel to the plane of the picture. The numbers of arcs incident to the bottom and top bases of the coupon vary independently. The directions of these arcs are arbitrary except the downward direction of the long arc in Figure 3.1. The colors of the arcs and the coupon are also arbitrary. Of course, the numbers of arcs, their directions and colors, as well as the color of the coupon are the same on the right-hand side and the left-hand side of the equalities. By the r-twist (resp. l-twist) in Figure 3.2, we mean a full right-hand (resp. left-hand) twist of the bunch of the vertical bands. The equality in Figure 3.2 indicates that the full right-hand twist of the bunch of the top bands of an elementary ribbon graph cancels (up to isotopy) with the full left-hand twist of the bunch of the bottom bands of this ribbon graph. Note that the full right-hand twist on the bunch of n colored arcs may be presented as a composition of $n(n-1)$ ribbon tangles of type $\mathrm{id}_U \otimes X^+_{V,W} \otimes \mathrm{id}_{U'}$ and of n ribbon tangles of type $\mathrm{id}_U \otimes \varphi_V \otimes \mathrm{id}_{U'}$. Similarly, the full left-hand twist on the bunch of n colored arcs may be presented as a composition of $n(n-1)$ ribbon tangles of type $\mathrm{id}_U \otimes X^-_{V,W} \otimes \mathrm{id}_{U'}$ and of n ribbon tangles of type $\mathrm{id}_U \otimes \varphi'_V \otimes \mathrm{id}_{U'}$. It is in this way that we express these twists via the generators.

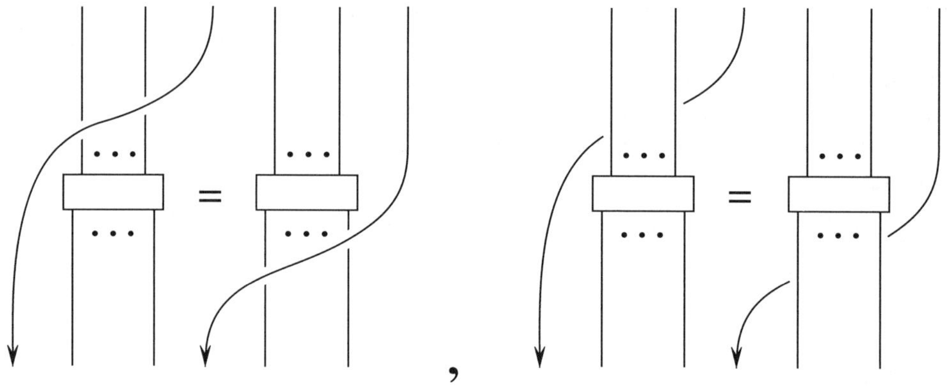

Figure 3.1

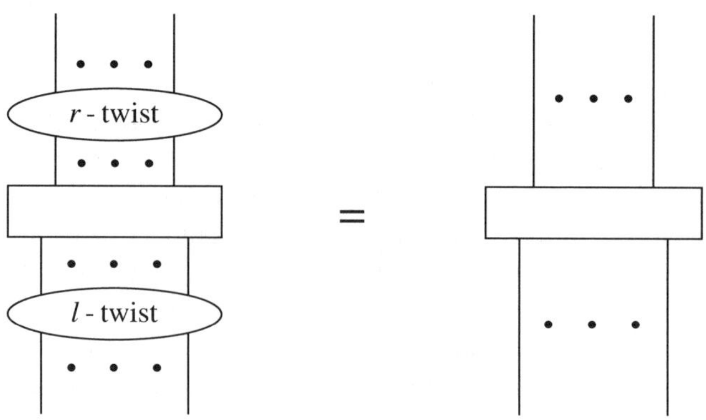

Figure 3.2

3.4. Lemma. *Relations* (3.2.a) - (3.2.h) *together with the relations presented in Figures* 3.1, 3.2 *form a complete set of relations between the generators of the category* $\mathrm{Rib}_{\mathcal{V}}$ *listed in Lemma* 3.1.1.

This lemma is proven in Section 4.

3.5. Deduction of Theorem 2.5 from Lemmas 3.3 and 3.4. In view of the relations (3.2.d), (3.2.e), (3.2.g) the conditions (2), (3) of Theorem 2.5 enable us to reconstruct the values of F on all generators of $\mathrm{Rib}_{\mathcal{V}}$ listed in Lemma 3.1.1. This implies uniqueness of F.

Now we construct F. The value of F on any object $((V_1, \varepsilon_1), \ldots, (V_m, \varepsilon_m))$ of $\mathrm{Rib}_{\mathcal{V}}$ is the object $V_1^{\varepsilon_1} \otimes \ldots \otimes V_m^{\varepsilon_m}$ of $\mathcal{V}$ (where $V^{+1} = V$ and $V^{-1} = V^*$). The conditions (2), (3) of Theorem 2.5 together with the following formulas define F on all generators:

$$F(X^-_{V,W}) = (c_{W,V})^{-1}, F(Z^+_{V,W}) = (c_{W^*,V})^{-1}, F(Z^-_{V,W}) = c_{V,W^*}, F(\varphi'_V) = \theta_V^{-1}.$$

For these values of F, we shall check the identities between morphisms in $\mathcal{V}$ which correspond to the generating relations of $\mathrm{Rib}_{\mathcal{V}}$ listed in Lemma 3.4. With these identities verified, we may uniquely extend F to a covariant functor $\mathrm{Rib}_{\mathcal{V}} \to \mathcal{V}$ preserving the tensor product and satisfying conditions (1) - (3) of the theorem.

The identity in $\mathcal{V}$ corresponding to (3.2.a) is the Yang-Baxter equation (1.2.f) verified in Section 1.6. The equalities (3.2.b), (3.2.c) correspond to (1.3.b), (1.3.c) respectively. The identities in $\mathcal{V}$ corresponding to (3.2.d) and (3.2.e) follow from definitions. The identity corresponding to (3.2.f) follows from the naturality of braiding.

Let us check the identity in $\mathcal{V}$ corresponding to (3.2.g) with $\nu = -1$. Consider the following commutative diagram:

(3.5.a)
$$\begin{array}{ccccc}
V = V \otimes \mathbb{1} & \xrightarrow{\mathrm{id}_V \otimes b_W} & V \otimes W \otimes W^* & \xrightarrow{c_{V,W} \otimes \mathrm{id}_{W^*}} & W \otimes V \otimes W^* \\
\downarrow{\scriptstyle \mathrm{id}_V} & & \downarrow{\scriptstyle c_{V,W \otimes W^*}} & & \downarrow{\scriptstyle \mathrm{id}_W \otimes c_{V,W^*}} \\
V = \mathbb{1} \otimes V & \xrightarrow{b_W \otimes \mathrm{id}_V} & W \otimes W^* \otimes V & = = = & W \otimes W^* \otimes V.
\end{array}$$

Commutativity of the left square follows from the naturality of the braiding and the equality $\mathrm{id}_V = c_{V,\mathbb{1}}$. Commutativity of the right square follows from the definition of braiding. Thus,

$$b_W \otimes \mathrm{id}_V = (\mathrm{id}_W \otimes c_{V,W^*})(c_{V,W} \otimes \mathrm{id}_{W^*})(\mathrm{id}_V \otimes b_W).$$

This equality is rewritten in graphical notation in Figure 3.3.

Multiplying the equality in Figure 3.3 from the left by id_{W^*} and from above by $d_W \otimes \mathrm{id}_{W^*} \otimes \mathrm{id}_V$ we get the equality in Figure 3.4. Therefore $\mathrm{id}_{W^* \otimes V}$ equals the composition of c_{V,W^*} with

(3.5.b)
$$(d_W \otimes \mathrm{id}_V \otimes \mathrm{id}_{W^*})(\mathrm{id}_{W^*} \otimes c_{V,W} \otimes \mathrm{id}_{W^*})(\mathrm{id}_{W^*} \otimes \mathrm{id}_V \otimes b_W).$$

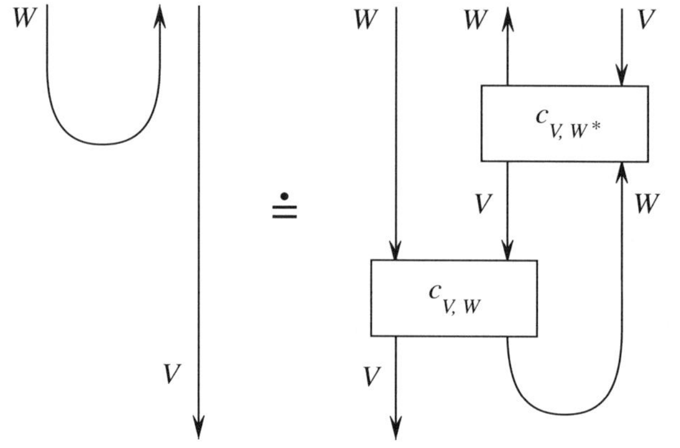

Figure 3.3

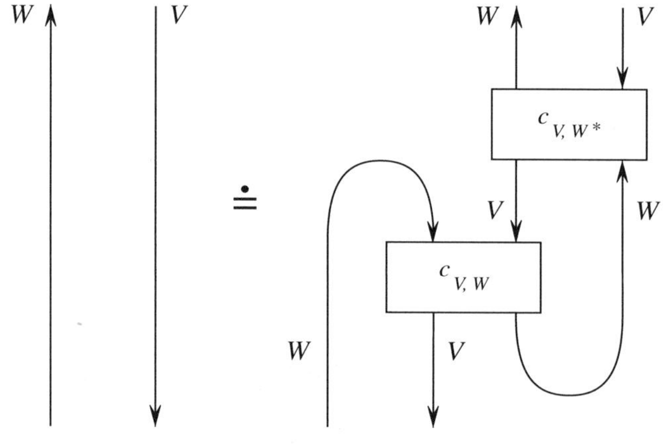

Figure 3.4

A similar argument based on the commutativity of the diagram

$$
\begin{array}{ccccc}
W^* \otimes V \otimes W & \xleftarrow{c_{V,W^*} \otimes \mathrm{id}_W} & V \otimes W^* \otimes W & \xrightarrow{\mathrm{id}_V \otimes d_W} & V \otimes \mathbb{1} = V \\
{\scriptstyle \mathrm{id}_{W^*} \otimes c_{V,W}} \downarrow & & {\scriptstyle c_{V,W^* \otimes W}} \downarrow & & {\scriptstyle \mathrm{id}_V} \downarrow \\
W^* \otimes W \otimes V & = & W^* \otimes W \otimes V & \xrightarrow{d_W \otimes \mathrm{id}_V} & \mathbb{1} \otimes V = V
\end{array} \tag{3.5.c}
$$

shows that the composition of the same two morphisms taken in the opposite order is the identity endomorphism of $V \otimes W^*$. The morphism (3.5.b) corresponds to the expression in the square brackets on the right-hand side of (3.2.g) where $\nu = -1$. This yields the equality in $\mathcal{V}$ corresponding to (3.2.g) with $\nu = -1$. This equality applied to the braiding (1.4.a) gives the equality in $\mathcal{V}$ corresponding to (3.2.g) with $\nu = 1$.

Let us check the equality in $\mathscr{V}$ corresponding to (3.2.h). Let V be an object of $\mathscr{V}$. Since $\theta_{\mathbb{1}} = \mathrm{id}_{\mathbb{1}}$, the naturality of the twist and (1.2.h) imply that

$$b_V = b_V \theta_{\mathbb{1}} = \theta_{V \otimes V^*}\, b_V = c_{V^*,V} c_{V,V^*} (\theta_V \otimes \theta_{V^*})\, b_V.$$

Since $(\theta_V \otimes \theta_{V^*})\, b_V = ((\theta_V)^2 \otimes \mathrm{id}_{V^*})\, b_V$ we obtain

$$((\theta_V)^2 \otimes \mathrm{id}_{V^*})\, b_V = (c_{V,V^*})^{-1} (c_{V^*,V})^{-1} b_V.$$

Present this equality in the graphical way and multiply it from the right by id_V and from above by $\mathrm{id}_V \otimes d_V$. This gives the formula in Figure 3.5. The results of the preceding paragraph show that the morphism $(c_{V,V^*})^{-1}$ is equal to expression (3.5.b) where $W = V$. Therefore we obtain the formula in Figure 3.6. This is exactly the formula (in the graphical notation) corresponding to (3.2.h).

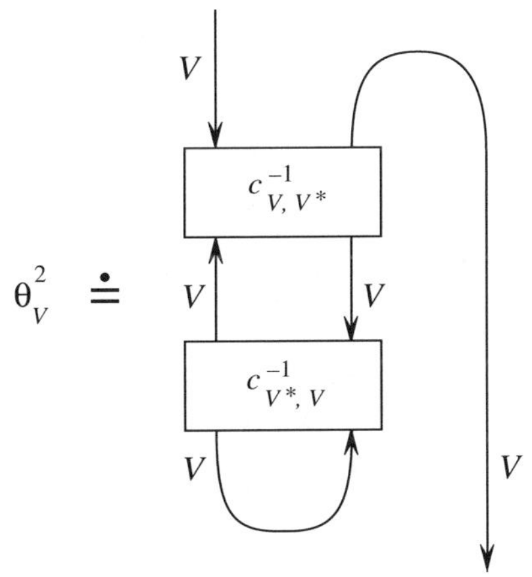

Figure 3.5

Lemma 3.3 implies that we have a correctly defined functor F from the category of ribbon tangles into $\mathscr{V}$. We may already verify the formulas for the values of F given in the statement of Theorem 2.5. The formulas for $F(X^\nu_{V,W}), F(Z^\nu_{V,W}), F(Y^\nu_{V,W})$ with $\nu = \pm 1$ follow from the definition of F and the equality $Y^\nu_{V,W} = (Z^{-\nu}_{W,V})^{-1}$. Since the tangle $T^+_{V,W}$ is the inverse of $T^-_{W,V}$ it suffices to check that $F(T^-_{V,W}) = (c_{W^*,V^*})^{-1}$. Formula (3.1.c) and the formula $F(Y^-_{W,V}) = c_{W^*,V}$ imply the equality in Figure 3.7. Thus we have to prove that the morphism in $\mathscr{V}$ represented by the diagram in Figure 3.7 is inverse to c_{W^*,V^*}. This is proven similarly to the equalities in $\mathscr{V}$ corresponding to (3.2.g). One may use diagrams (3.5.a), (3.5.c) with V and W replaced by W^* and V respectively.

Let us verify the identity in $\mathscr{V}$ corresponding to the first relation in Figure 3.1. This relation involves one elementary ribbon graph and the tangle generators

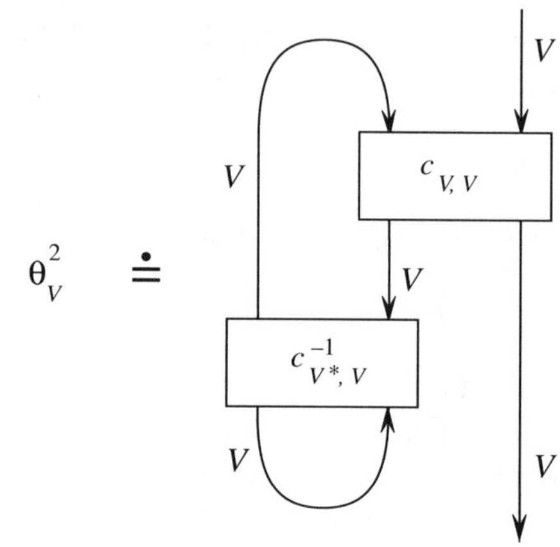

Figure 3.6

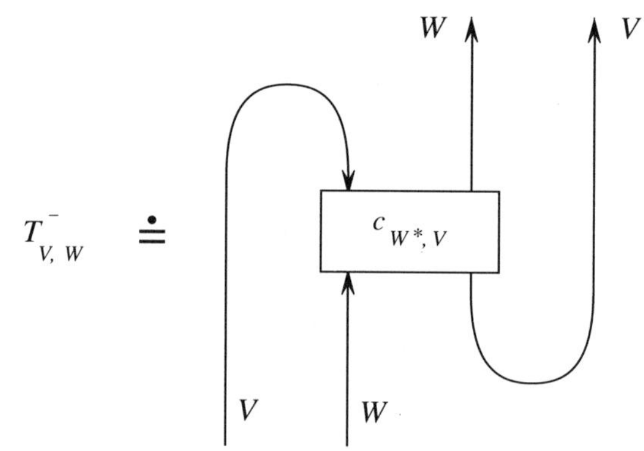

Figure 3.7

$\{X^+_{V,W}, Z^-_{V,W}\}$ corresponding to the crossing points of the diagram. The generators $\{Z^-_{V,W}\}$ appear in the crossing points of the long band with the short bands directed upwards. Redirect all such short bands downwards and replace their colors with the dual objects. (This transformation does not affect the color of the coupon.) The resulting geometric relation between the generators of $\mathrm{Rib}_{\mathcal{V}}$ differs from the original one but the corresponding identity in $\mathcal{V}$ is the same. This follows from the equalities $F(Z^-_{V,W}) = c_{V,W^*} = F(X^+_{V,W^*})$. Therefore we may restrict ourselves to the case where all bands are directed downwards. In this case the identity in question follows directly from properties of braiding (cf. the proof of the Yang-Baxter equality in Section 1.6).

The identity in $\mathcal{V}$ which corresponds to the second relation in Figure 3.1 is proven by the same argument applied to the braiding (1.4.a).

It remains to verify the identity in $\mathcal{V}$ which corresponds to the relation in Figure 3.2. We first compute the value of F on the colored ribbon tangle obtained as the full right-hand twist of a bunch of m vertical untwisted unlinked bands directed downwards and colored with objects $V_1, \ldots, V_m$ of $\mathcal{V}$. (This ribbon tangle appears in the left part of Figure 3.2 just above the coupon.) Denote this colored ribbon tangle by $r(V_1, \ldots, V_m)$. We claim that

$$F(r(V_1, \ldots, V_m)) = \theta_{V_1 \otimes V_2 \otimes \ldots \otimes V_m}.$$

Indeed, if $m = 1$ then $r(V) = \varphi_V$ and $F(r(V)) = F(\varphi_V) = \theta_V$ where $V = V_1$. Assume that $m \geq 2$. We may obtain the ribbon tangle $r(V_1, \ldots, V_m)$ from $r(V_1, \ldots, V_{m-1}) \otimes r(V_m)$ by winding the left $m-1$ bands around the m-th band. Set $V = V_1 \otimes \ldots \otimes V_{m-1}$ and $W = V_m$. By induction

$$F(r(V_1, \ldots, V_m)) = c_{W,V}\, c_{V,W}\, (F(r(V_1, \ldots, V_{m-1})) \otimes F(r(W))) =$$

$$= c_{W,V}\, c_{V,W}\, (\theta_V \otimes \theta_W) = \theta_{V \otimes W} = \theta_{V_1 \otimes V_2 \otimes \ldots \otimes V_m}.$$

Similarly, denote by $l(V_1, \ldots, V_m)$ the colored ribbon tangle obtained as the full left-hand twist of a bunch of m vertical untwisted unlinked bands directed downwards and colored with objects $V_1, \ldots, V_m$ of $\mathcal{V}$. It is obvious that $l(V_1, \ldots, V_m)$ and $r(V_1, \ldots, V_m)$ represent mutually inverse morphisms in the category of colored ribbon tangles. Therefore

$$F(l(V_1, \ldots, V_m)) = \theta^{-1}_{V_1 \otimes \ldots \otimes V_m}.$$

Now we may prove the identity in $\mathcal{V}$ which corresponds to the relation in Figure 3.2. The same trick as above enables us to confine ourselves to the case where all bands are directed downwards. Let $V_1, \ldots, V_m$ (resp. $W_1, \ldots, W_n$) be the objects of $\mathcal{V}$ coloring the inputs (resp. outputs) of the coupon. Let $f : V_1 \otimes \ldots \otimes V_m \to W_1 \otimes \ldots \otimes W_n$ be the color of the coupon. The identity in $\mathcal{V}$ corresponding to the relation in Figure 3.2 has the form

$$\theta_{W_1 \otimes \ldots \otimes W_n}\, f\, \theta^{-1}_{V_1 \otimes \ldots \otimes V_m} = f.$$

This identity follows directly from the naturality of twist in $\mathcal{V}$.

4. Proof of lemmas

In this section we prove Lemmas 3.3 and 3.4.

4.1. Outline of the proof. The arguments used to prove Lemma 3.1.1 show that any generic diagram of a ribbon tangle gives rise to an expression of this tangle via the generators (3.1.a). This expression involves two basic operations on tangles: tensor product and composition. Thus, any generic diagram of a tangle yields a presentation of this tangle as a word in our generators (and operations). We shall consider the changes in this word under local transformations of the diagram. The transformations in question are the Reidemeister moves described below and the local isotopies modifying the position of the diagram with respect to the height function. We show that under these transformations the word associated with the diagram is replaced by an equivalent one where the equivalence relation is generated by (3.2.a) - (3.2.h). This assertion together with general properties of generators and relations in strict monoidal categories would imply Lemma 3.3. The proof of Lemma 3.4 is analogous.

Here is the plan of the proof. In Section 4.2 we treat in detail generators and relations for strict monoidal categories. In Section 4.3 we associate to each generic tangle diagram a word in generators (3.1.a). In Sections 4.4 and 4.5 we trace the changes in this word under the second and third Reidemeister moves. In Section 4.6 we consider ambient isotopies of diagrams and corresponding changes in the associated words. In Section 4.7 we treat the remaining Reidemeister move. The proof of Lemma 3.3 is accomplished in Section 4.8. In Section 4.9 we prove Lemma 3.4.

4.2. Generators and relations for categories. Let $\mathcal{X}$ be a strict monoidal category, and let $\mathcal{W}$ be a collection of morphisms in $\mathcal{X}$. We define "words in the alphabet $\mathcal{W}$". A word of rank ≤ 1 is a symbol f where $f \in \mathcal{W}$ or a symbol id_V where V is an object of $\mathcal{X}$. For a word a of rank ≤ 1, we denote by $\langle a \rangle$ the morphism in $\mathcal{X}$ represented by a (that is the morphism f or id_V respectively). The only subword of a word of rank ≤ 1 is this word itself. Inductively assume that the words of rank $\leq n$ have been already defined, and that for every word a of rank $\leq n$, we have a morphism $\langle a \rangle$ in $\mathcal{X}$ and a collection of subwords of a. Then the words of rank $\leq n+1$ are defined to be the symbols of the form $a \otimes b$ where a and b are words of rank $\leq n$, and the symbols of the form $a \circ b$ where a and b are words of rank $\leq n$ such that $\mathrm{source}(\langle a \rangle) = \mathrm{target}(\langle b \rangle)$. Set $\langle a \otimes b \rangle = \langle a \rangle \otimes \langle b \rangle$ and $\langle a \circ b \rangle = \langle a \rangle \circ \langle b \rangle$. By a subword of $a \otimes b$ or $a \circ b$, we mean this word itself together with all subwords of a and b. By a word in the alphabet $\mathcal{W}$, we mean any word of rank $\leq n$ for $n = 1, 2, 3, \ldots$.

Let $\{c_j, d_j\}_{j \in J}$ be a family of words in the alphabet $\mathcal{W}$ such that $\langle c_j \rangle = \langle d_j \rangle$ for all $j \in J$. We say that two words a and b in the alphabet $\mathcal{W}$ are equivalent modulo the relations $\{c_j = d_j\}_{j \in J}$ if there exists a finite sequence of words $a_1 = a, a_2, \ldots, a_k = b$ such that for each $i = 1, 2, \ldots, k-1$, the word a_{i+1} is obtained from a_i by replacing a subword with another subword such that either for some $j \in J$ these subwords are c_j and d_j (up to permutation), or they are the sides of one of equalities (1.1.a) - (1.1.d), (1.1.g), (1.1.h). Here it is understood that the entries f, g, h, f', g' of these equalities are words in the alphabet $\mathcal{W}$ and V, W are objects of $\mathcal{X}$. Equivalence of words is denoted by the symbol $\sim$.

We say that $\langle \mathcal{W} : c_j = d_j, j \in J \rangle$ is a presentation of $\mathcal{X}$ by generators and relations if

(i) $\mathcal{W}$ generates $\mathcal{X}$, i.e., each morphism of $\mathcal{X}$ is equal to $\langle a \rangle$ for some word a in the alphabet $\mathcal{W}$, and

(ii) for any words a and b in the alphabet $\mathcal{W}$, the equality $\langle a \rangle = \langle b \rangle$ holds if and only if a is equivalent to b modulo the relations $\{c_j = d_j\}_{j \in J}$.

We shall need the following three lemmas. In these lemmas, $\mathcal{W}$ is a collection of morphisms in a strict monoidal category $\mathcal{X}$. By a "word", we mean a word in the alphabet $\mathcal{W}$. By equivalence of words, we mean equivalence modulo the empty set of relations.

4.2.1. Lemma. *If $f_1, f_2, ..., f_n \in \mathcal{W}$ and the composition $f_1 \circ f_2 \circ ... \circ f_n$ is defined then for any objects V, W of $\mathcal{X}$, we have*

$$(\mathrm{id}_V \otimes f_1 \otimes \mathrm{id}_W) \circ ... \circ (\mathrm{id}_V \otimes f_n \otimes \mathrm{id}_W) \sim \mathrm{id}_V \otimes (f_1 \circ ... \circ f_n) \otimes \mathrm{id}_W.$$

Proof. The general case follows from the case $n = 2$. We have

$$(\mathrm{id}_V \otimes f_1 \otimes \mathrm{id}_W) \circ (\mathrm{id}_V \otimes f_2 \otimes \mathrm{id}_W) \sim [(\mathrm{id}_V \otimes f_1) \circ (\mathrm{id}_V \otimes f_2)] \otimes (\mathrm{id}_W \circ \mathrm{id}_W) \sim$$

$$\sim (\mathrm{id}_V \circ \mathrm{id}_V) \otimes (f_1 \circ f_2) \otimes (\mathrm{id}_W \circ \mathrm{id}_W) \sim \mathrm{id}_V \otimes (f_1 \circ f_2) \otimes \mathrm{id}_W.$$

4.2.2. Lemma. *If $f : X \to U$ and $g : V \to W$ are morphisms from the collection $\mathcal{W}$ then the words $(f \otimes \mathrm{id}_W) \circ (\mathrm{id}_X \otimes g)$ and $(\mathrm{id}_U \otimes g) \circ (f \otimes \mathrm{id}_V)$ are equivalent.*

Proof.

$$(f \otimes \mathrm{id}_W) \circ (\mathrm{id}_X \otimes g) \sim (f \circ \mathrm{id}_X) \otimes (\mathrm{id}_W \circ g) \sim f \otimes g \sim$$

$$\sim (\mathrm{id}_U \circ f) \otimes (g \circ \mathrm{id}_V) \sim (\mathrm{id}_U \otimes g) \circ (f \otimes \mathrm{id}_V).$$

4.2.3. Lemma. *Every word in the alphabet $\mathcal{W}$ is equivalent either to a word* id_V *where V is an object of $\mathcal{X}$, or to a word of the form*

$$(4.2.a) \qquad (\mathrm{id}_{V_1} \otimes f_1 \otimes \mathrm{id}_{W_1}) \circ (\mathrm{id}_{V_2} \otimes f_2 \otimes \mathrm{id}_{W_2}) \circ ... \circ (\mathrm{id}_{V_k} \otimes f_k \otimes \mathrm{id}_{W_k})$$

where $k \geq 1$, $f_i \in \mathcal{W}$, and $V_1, ..., V_k, W_1, ..., W_k$ are objects of $\mathcal{X}$.

Proof. For a word of rank ≤ 1, the claim is obvious: if $f \in \mathcal{W}$ then $f \sim \mathrm{id}_{\mathbb{1}} \otimes f \otimes \mathrm{id}_{\mathbb{1}}$. Assume that the claim holds for words of rank $\leq n$ and prove it for the words of rank $\leq n+1$. Let a be a word of rank $\leq n+1$. Then either $a = b \otimes c$ or $a = b \circ c$ where b and c are certain words of rank $\leq n$. By the induction hypothesis, we may assume that b and c are words of the form (4.2.a) or of the form id_V where V is an object of $\mathcal{X}$.

The case $a = b \circ c$ is obvious: if b and c both have the form (4.2.a) then so does a, if $b = \mathrm{id}_V$ or $c = \mathrm{id}_V$ then $a \sim c$ and $a \sim b$ respectively. Suppose that $a = b \otimes c$. Let $b = b_1 \circ ... \circ b_k$ and $c = c_1 \circ ... \circ c_l$ where all b_i and c_j are words of the form $\mathrm{id}_V \otimes f \otimes \mathrm{id}_W$. Set $S = \mathrm{source}(b_k)$ and $T = \mathrm{target}(c_1)$. Then

$$a = b \otimes c \sim b(\mathrm{id}_S)^l \otimes (\mathrm{id}_T)^k c \sim (b_1 \otimes \mathrm{id}_T) \circ ... \circ (b_k \otimes \mathrm{id}_T) \circ (\mathrm{id}_S \otimes c_1) \circ ... \circ (\mathrm{id}_S \otimes c_l).$$

Thus, a is equivalent to a word of the form (4.2.a). The case where $b = \mathrm{id}_V$ and/or $c = \mathrm{id}_W$ is treated similarly.

4.3. Words assigned to generic diagrams. Recall the notions of singular points of diagrams and generic diagrams introduced in the proof of Lemma 3.1.1. Let $D \subset \mathbb{R} \times [0, 1]$ be a generic diagram of a colored ribbon tangle. We assign to D a word $a(D)$ in the generators (3.1.a). If D does not have singular points then $a(D) = \mathrm{id}_V$ where V is the object of $\mathrm{Rib}_{\mathcal{V}}$ which is both the source and the target of the colored ribbon tangle represented by D. Suppose that D has $n \geq 1$ singular points. As in the proof of Lemma 3.1.1 we consider the lines $\mathbb{R}\times 0$, $\mathbb{R}\times 1$ and draw $n-1$ parallel horizontal lines in $\mathbb{R} \times (0, 1)$ so that between any two adjacent lines lies exactly one singular point of D. The part of D lying between such adjacent lines corresponds in a natural way to the tensor product $\mathrm{id}_V \otimes f \otimes \mathrm{id}_W$ where V, W are objects of $\mathrm{Rib}_{\mathcal{V}}$ and f is a morphism of this category presented by one of the diagrams of Figure 2.5 or one of the four last diagrams of Figure 2.6. If f does not belong to the set of generators (3.1.a) then we use (2.5.d), (2.5.e), (3.1.b), (3.1.c) to replace f with a word in the generators (3.1.a). Thus, to the part of D lying between adjacent lines we have associated a word in the generators (3.1.a). We assign to the diagram D the composition $a(D)$ of these n words written down from the left to the right in the order of decreasing height. It is clear that the morphism $\langle a(D)\rangle$ in $\mathrm{Rib}_{\mathcal{V}}$ is exactly the isotopy class of the colored ribbon tangle represented by D.

4.4. The Reidemeister move Ω_2. We consider the local transformations of generic diagrams shown in Figure 4.1. These transformations are oriented versions of the second Reidemeister move. We shall show that under these transformations the word associated to a diagram is replaced with an equivalent word. More exactly, for any generic diagram D' obtained from a generic diagram D by an application of $\Omega_{2,i}$ with $i = 1, ..., 8$, the word $a(D')$ is equivalent to $a(D)$ modulo relations (3.2.a) - (3.2.h). It is important to keep in mind that in our pictures the height function on diagrams is the orthogonal projection on the vertical axis.

To prove the equivalence $a(D') \sim a(D)$ it suffices to consider the model case when D and D' are the diagrams in the picture of $\Omega_{2,i}$. (The reduction of the general case to the model one is straightforward, although a detailed argument should use Lemma 4.2.1.) Denote by V and W the objects of $\mathcal{V}$ coloring the two strings of D. (These colors are omitted in Figure 4.1.) If $i = 1$ then $a(D) = \downarrow_V \otimes \downarrow_W$ is equivalent to $a(D') = X^-_{W,V} \circ X^+_{V,W}$ modulo (3.2.d). The case $i = 2$ is similar. If $i = 3$ then $a(D') = Z^+_{W,V} \circ y^-_{V,W}$ where $y^-_{V,W}$ denotes the right-hand side of (3.1.b) with $\nu = -1$. The word $a(D')$ is equivalent to $a(D) = \uparrow_V \otimes \downarrow_W$ modulo (3.2.g) with $\nu = 1$. The cases $i = 4, 5, 6$ are similar: we apply (3.2.g) with $\nu = -1, 1, -1$ respectively. The pictorial argument given in Figure 4.2 works for $i = 7$. (The colors V, W are omitted in Figure 4.2.) Here the first two equalities follow from the definitions given in Section 4.3. The four equivalences

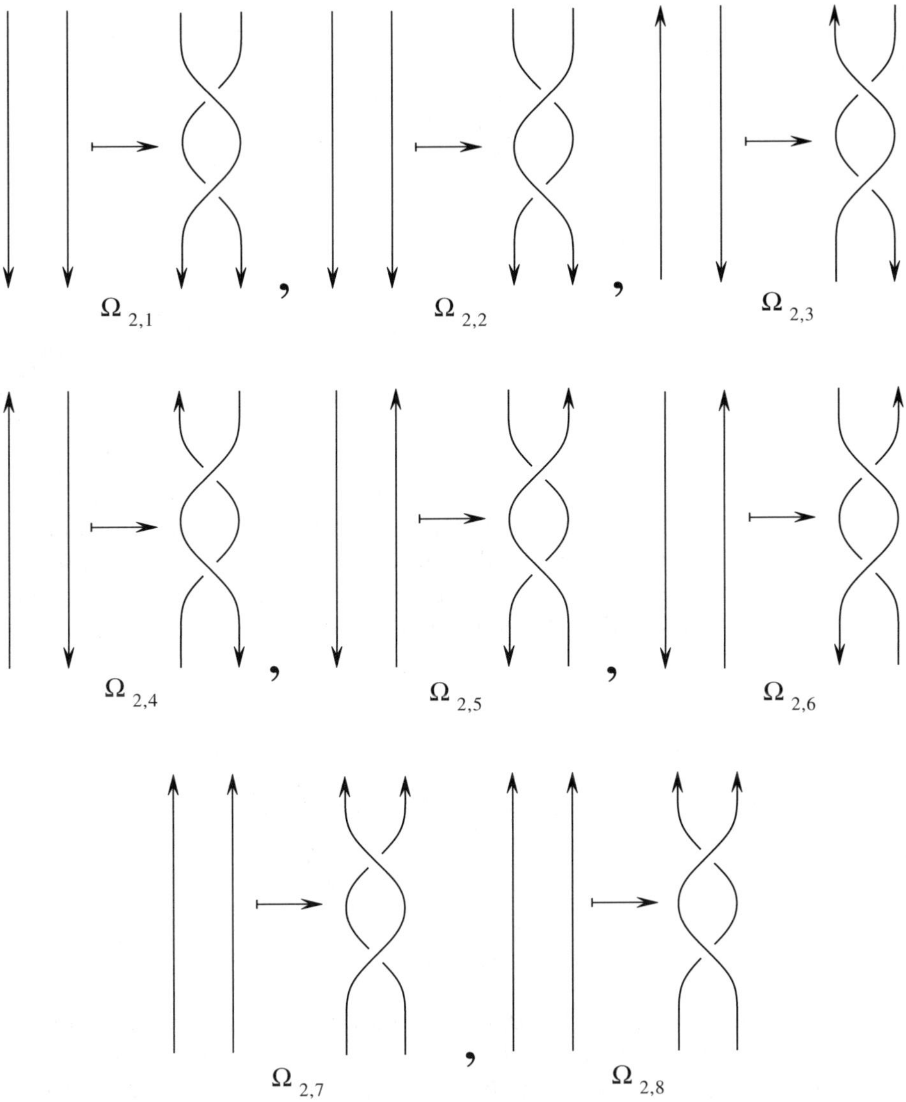

Figure 4.1

in Figure 4.2 are ensured by (1.1.c), (3.2.b), (3.2.d), (3.2.c), respectively. The case $i = 8$ is treated similarly using the mirror image of Figure 4.2.

4.5. The Reidemeister move Ω_3. The Reidemeister move Ω_3 is schematically shown in Figure 4.3. Here each symbol A, B, C stands for one of the crossings $X^{\nu}_{V,W}, Y^{\nu}_{V,W}, Z^{\nu}_{V,W}, T^{\nu}_{V,W}$ where $\nu \in \{\pm 1\}$ and V, W are objects of $\mathcal{V}$, see Figure 2.5. (The colors and directions of the strands of A, B, C are omitted in Figure

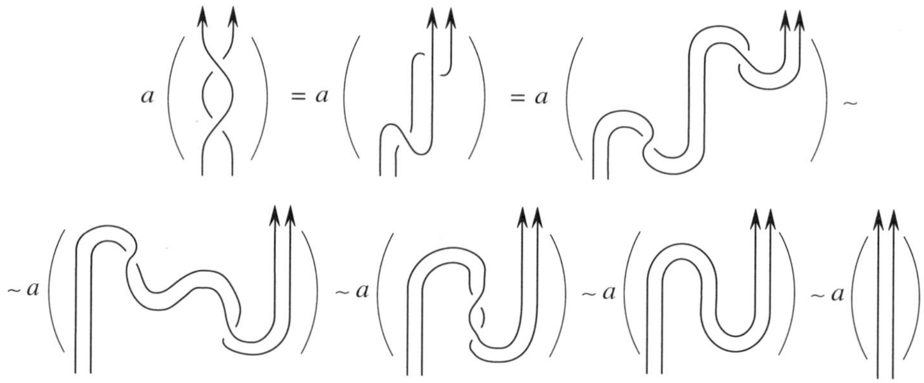

Figure 4.2

4.3.) We say that the triple (A, B, C) is compatible if one of the three straight lines drawn on the left-hand side of Figure 4.3 goes over the other two and the directions and colors of the strands of A, B, C are induced by certain directions and colors of these three straight lines. The three lines in question will be denoted by AB, AC, BC; the line AB is the one traversing A and B, etc.

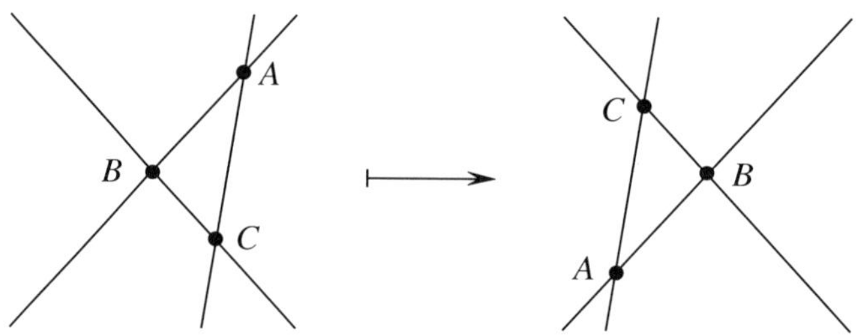

Figure 4.3

For each compatible triple (A, B, C), we may replace the crossing points in Figure 4.3 with the pictures of A, B, C and obtain diagrams of colored tangles. Denote by $D(A, B, C)$ and $D'(A, B, C)$ the diagrams arising in this way in the left and right parts of Figure 4.3. The transformation $D(A, B, C) \mapsto D'(A, B, C)$ is denoted by $\Omega_3(A, B, C)$ and called the third Reidemeister move.

Let us say that a compatible triple (A, B, C) is good if the words $a(D(A, B, C))$ and $a(D'(A, B, C))$ are equivalent modulo relations (3.2.a) - (3.2.h).

4.5.1. Claim. *All compatible triples are good.*

This claim shows that the third Reidemeister move preserves the equivalence class of the associated word of a diagram. To prove Claim 4.5.1 we define certain transformations on compatible triples which preserve goodness. Then we show that these transformations relate any compatible triple to a good one.

Define an involution $A \mapsto \overline{A}$ in the set of crossings $\{X^\nu_{V,W}, Y^\nu_{V,W}, Z^\nu_{V,W}, T^\nu_{V,W}\}$ as follows. The picture of $\overline{A}$ is the mirror image of the picture of A with respect to the vertical line. More precisely, if $A = x^\nu_{V,W}$ with $\nu \in \{\pm 1\}$ and $x = X, Y, Z, T$, then $\overline{A} = y^{-\nu}_{\overline{W},V}$ where $y = X, Z, Y, T$ respectively. The diagrams $A \circ \overline{A}$ and $\overline{A} \circ A$ obtained by putting A on the top of $\overline{A}$ or vice versa are the images under the second Reidemeister move of identity diagrams.

Let (A, B, C) be a compatible triple of crossings. Multiplying both diagrams $D(A, B, C)$ and $D'(A, B, C)$ from above by $\overline{C} \otimes |$ and from below by $| \otimes \overline{C}$ and applying the corresponding second Reidemeister moves we get $D'(B, A, \overline{C})$ and $D(B, A, \overline{C})$ respectively. Therefore, the triple $(B, A, \overline{C})$ is also compatible. Moreover, this argument together with the results of Section 4.4 shows that goodness of (A, B, C) implies goodness of $(B, A, \overline{C})$. Applying the same argument to the triple $(B, A, \overline{C})$ we observe that goodness of (A, B, C) is equivalent to goodness of $(B, A, \overline{C})$. Similar arguments show that the triple $(\overline{A}, C, B)$ is compatible and that (A, B, C) is good if and only if $(\overline{A}, C, B)$ is good.

Consider an arbitrary compatible triple (A, B, C) such that A, B, C are crossings of type X^ν with $\nu = \pm 1$. Let us show that this triple is good. Replacing if necessary (A, B, C) by $(B, A, \overline{C})$ or $(\overline{A}, C, B)$ we may assume that the signs ν of the (first and third) crossings A and C are equal. Compatibility of the triple (A, B, C) implies that B has the same sign. If this common sign of A, B, C is $+1$ then $a(D(A, B, C)) \sim a(D'(A, B, C))$ modulo (3.2.a). If the sign of A, B, C is -1 then $a(D(A, B, C)) \sim a(D'(A, B, C))$ modulo (3.2.a) and (3.2.d). Therefore (A, B, C) is good.

To accomplish the proof of Claim 4.5.1 we need more transformations on compatible triples. For any crossing A of type T or Y, we define a crossing A^* as follows. If $A = T^\nu_{V,W}$ then $A^* = Y^\nu_{W,V}$. If $A = Y^\nu_{V,W}$ then $A^* = X^\nu_{W,V}$. It follows from the definition of the word associated to a diagram that for any crossing A of type T or Y, we have the equality in Figure 4.4.

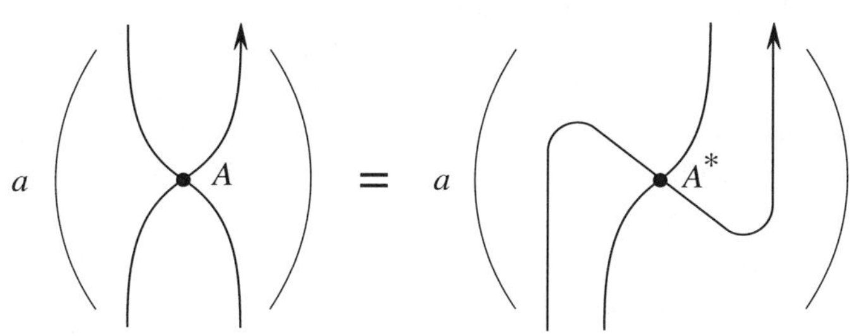

Figure 4.4

Consider any compatible triple (A, B, C) such that the line AB traversing A and B in Figure 4.3 is oriented upwards. In this case A and B are crossings of type T^ν or Y^ν with $\nu = \pm 1$. The equality in Figure 4.4 implies the equivalence modulo (3.2.b) shown in Figure 4.5. Similarly, we have the equivalence in Figure

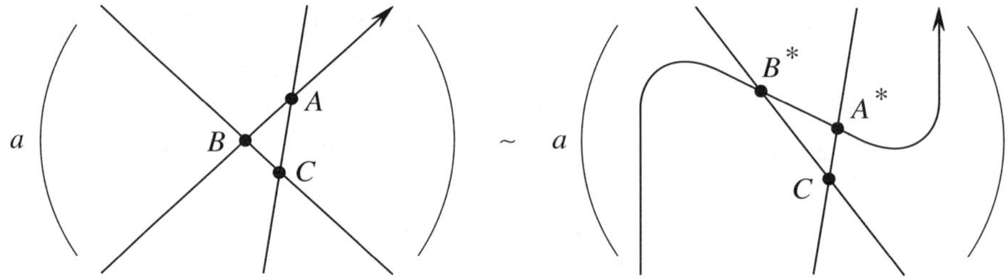

Figure 4.5

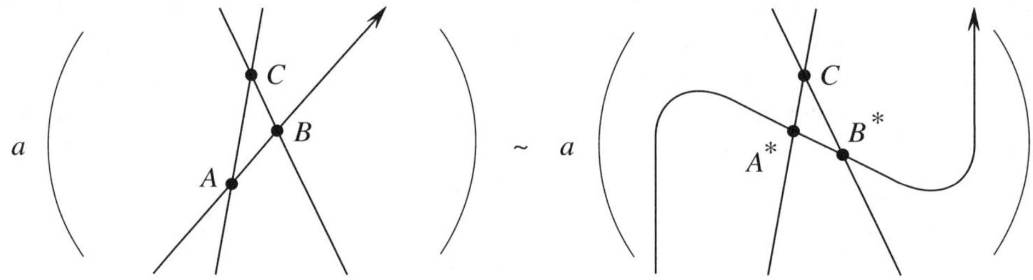

Figure 4.6

4.6. Therefore the triple (C, A^*, B^*) is compatible and goodness of (C, A^*, B^*) would imply goodness of (A, B, C).

Now we can complete the proof of Claim 4.5.1. Consider an arbitrary compatible triple (A, B, C). Let us show first that applying the transformations $(A, B, C) \mapsto (B, A, \overline{C})$ and $(A, B, C) \mapsto (\overline{A}, C, B)$ we may ensure that neither A nor C is a crossing of type Z^ν. Applying the second transformation we may ensure that A does not have type Z^ν. If C has not type Z^ν then we are done. If C has type Z^ν then the first transformation would do the job except in the case where both B and C have type Z^ν. In this case A has type X^ν and applying the first transformation and then the second one we get a triple with the desired property. Compatibility of (A, B, C) implies that if A and C do not have type Z^ν then B also does not have this type. Therefore it remains to consider the case where A, B, C are of types X^ν, Y^ν, T^ν (perhaps, with different ν). If the line AB traversing A and B in Figure 4.3 is oriented downwards then two other lines are also oriented downwards. (Otherwise either A or B would be of type Z^ν.) In this case A, B, C have type X^ν, and the assertion follows from the results above. Assume that the line AB is oriented upwards. Set $A' = C, B' = A^*, C' = B^*$. In view of the results obtained above it suffices to show that the triple $(A', B', C') = (C, A^*, B^*)$ is good. In this way we have reduced the problem to the case of a compatible triple (A', B', C') where the line $B'C'$ is oriented downwards. Note that we stay in the class of crossings of types X^ν, Y^ν, T^ν. If the line $A'B'$ is oriented downwards then we are done. If this line is oriented upwards then we again apply the same transformation bending $A'B'$ as above. In the resulting diagram all lines will be oriented downwards which is the case considered above. This completes the proof of the equivalence $a(D(A, B, C)) \sim a(D'(A, B, C))$.

4.6. Deformations of diagrams. By isotopy of diagrams, we mean their ambient isotopy in the strip $\mathbb{R} \times [0, 1]$ constant on the boundary of the strip. It is easy to see that isotopic generic diagrams may be obtained from each other by a finite sequence of the following transformations.

(I) An isotopy in the class of generic diagrams.

(II) An isotopy interchanging the order of two singular points with respect to the height function. (Such isotopies have already appeared in Figure 4.2, see, for example, the first equivalence in this figure.)

(III) Birth or annihilation of a pair of local extrema (cf. Figure 1.11).

(IV) Isotopies shown in Figure 4.7.

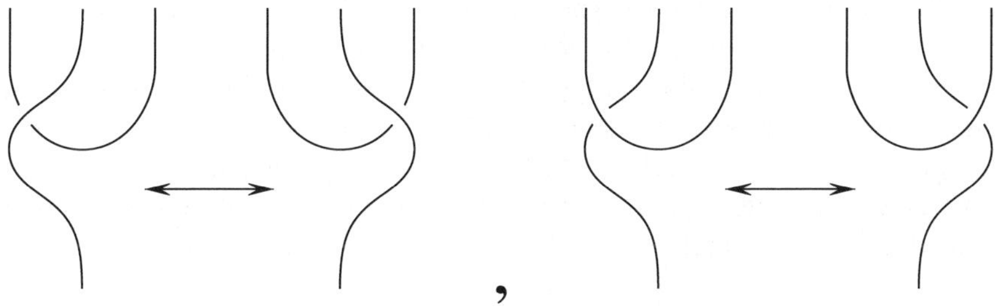

Figure 4.7

Analogous isotopies near points of maximum of the height function are not included in this list since they can be presented as compositions of the deformations (I) - (IV), see Figure 4.8.

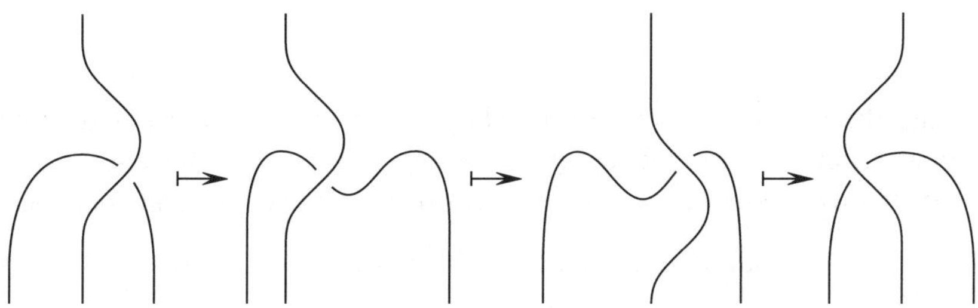

Figure 4.8

For the same reason, the list does not include deformations near crossing points similar to the one in Figure 4.9.

We shall show that the transformations of types (I) - (IV) do not change the equivalence class of the word associated to a generic diagram. The transformations

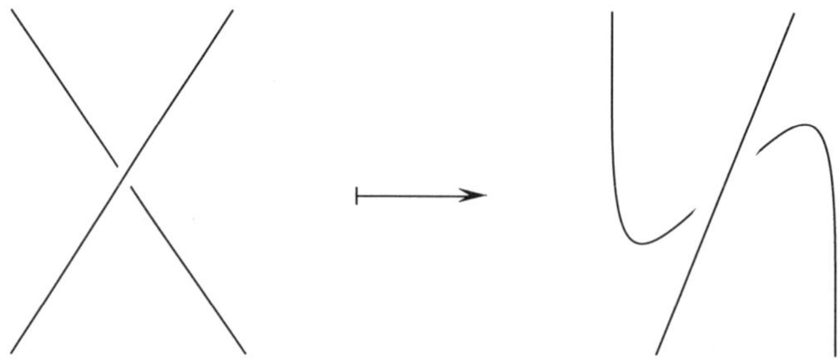

Figure 4.9

of type (I) do not change the word at all. For transformations of type (II) the claim follows from Lemma 4.2.2. The transformations of type (III) with the strands oriented as in Figure 1.11 do not change the word up to relations (3.2.b), (3.2.c). The transformations of type (III) with the orientations opposite to those in Figure 1.11 will be considered at the end of this subsection.

There are 8 transformations of type (IV) which are determined by the choice of orientations in the strands. In the case where the bent strand (the cup) is oriented to the right the desired equivalence follows from definitions and (3.2.b). In Figure 4.10 we give a sample argument pertaining to the case where the vertical strand is oriented downwards and goes under the cup.

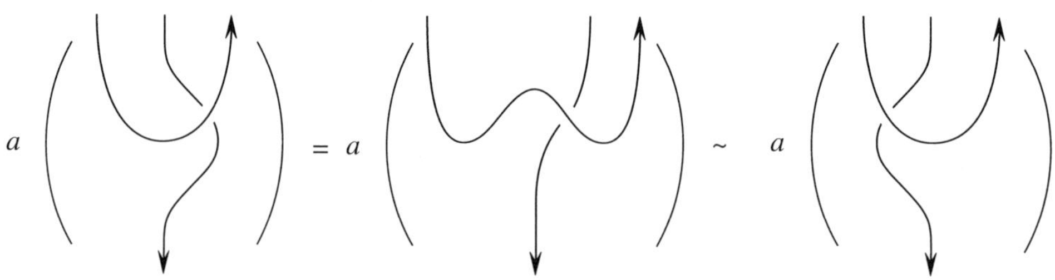

Figure 4.10

Consider the transformation of type (IV) where the cup is oriented to the left, the vertical strand is oriented downwards and goes under the cup (see Figure 4.11). Let us prove equivalence of the corresponding words. Consider the words

$$\Phi = \mathrm{id}_V \otimes \mathrm{id}_{W^*} \otimes \varphi'_W, \qquad \Psi = \mathrm{id}_{W^*} \otimes \varphi'_W \otimes \mathrm{id}_V.$$

We have equalities and equivalences of words in Figure 4.12. Here the equalities at the beginning and at the end follow from the definition of the word a. The first three equivalences follow from the already established equivalences under the Reidemeister moves and transformations of type (IV) with the cup oriented to the right. The fourth equivalence uses (3.2.k) and the obvious fact that if two generic diagrams D and D' are composable then $a(D \circ D') = a(D) \circ a(D')$. Multiplying the first and last diagrams in Figure 4.12 from above by $Z^+_{V,W} \otimes \downarrow_W$ and applying

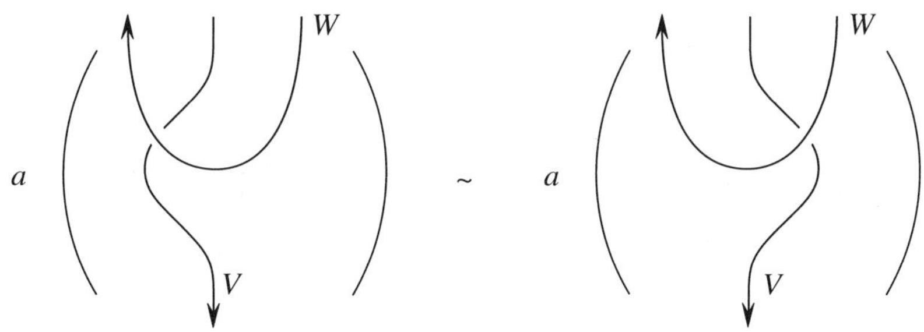

Figure 4.11

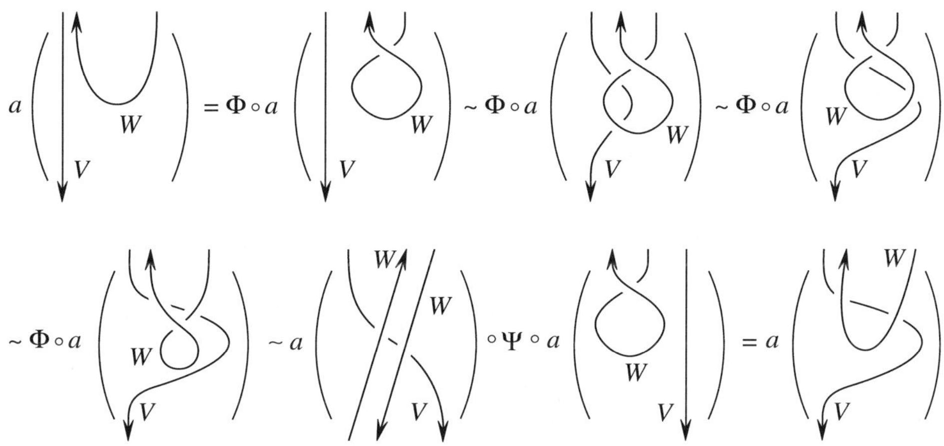

Figure 4.12

$\Omega_{2,3}^{-1}$ we get the equivalence in Figure 4.11. Other transformations corresponding to the left-oriented cup are handled by similar arguments.

It remains to treat the transformations of type (III) with orientations opposite to those in Figure 1.11. We start with the equivalence in Figure 4.13 which follows from the definition of the word $a(D)$. Pulling the bottom end of the second diagram to the right we may transform it into the last diagram in Figure 4.15. This may be done using the second and third Reidemeister moves and the transformations of types (I), (IV). Pulling the top end of the last diagram in Figure 4.15 to the left we may transform it into $\downarrow_V$. This may be done using the second Reidemeister moves, the transformations of types (I), (II), (IV), and transformations of type (III) corresponding to (3.2.b), (3.2.c). Therefore the word associated to the original diagram is equivalent to $\varphi'_V \circ \downarrow_V \circ \varphi_V \sim \varphi'_V \circ \varphi_V \sim \downarrow_V$. The equivalence $a((\uparrow_V \otimes \cap_V^-)(\cup_V^- \otimes \uparrow_V)) \sim \uparrow_V$ is proven similarly.

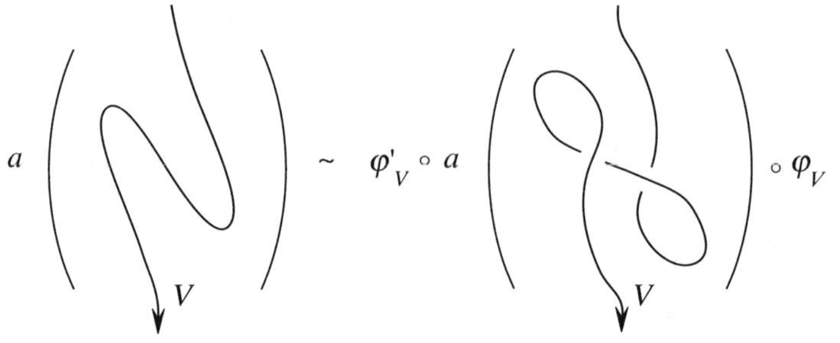

Figure 4.13

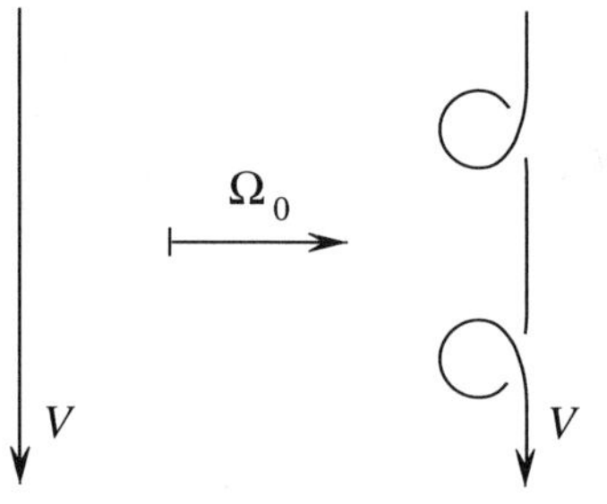

Figure 4.14

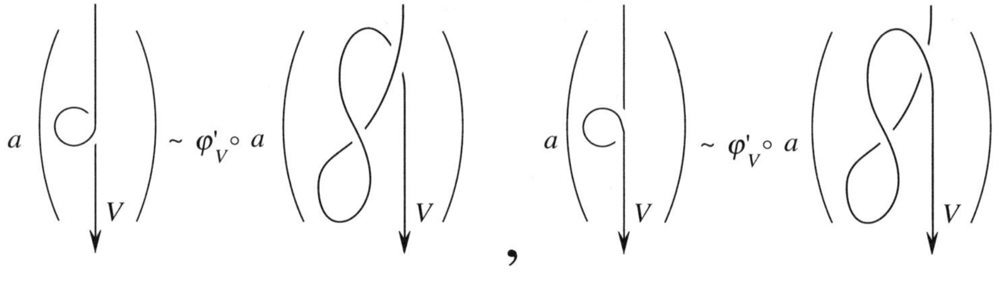

Figure 4.15

4.7. The move Ω_0. We need one more move on diagrams Ω_0 which preserves the isotopy type of the ribbon tangle represented by a diagram. This move inserts one positive and one negative cirls in a strand, see Figure 4.14.

We claim that Ω_0 does not change the equivalence class of the word associated to a generic diagram. We have equivalences of words in Figure 4.15, they follow from the definition of the word a and (3.2.k). The word on the right-hand side of the first equivalence may be easily computed to be $\varphi'_V \circ \Xi$ where Ξ is the right-hand side of (3.2.h). Therefore this word is equivalent to φ_V modulo (3.2.e) and (3.2.h). The right-hand side of the second formula in Figure 4.15 is equivalent to φ'_V, this follows from the invariance of the associated word under the second

Reidemeister move and ambient isotopies. Combining these computations with the equivalence $\varphi_V \circ \varphi'_V \sim \downarrow_V$ we prove the claim.

4.8. Proof of Lemma 3.3. We shall show that any two words a, a' in the generators (3.1.a) representing isotopic ribbon tangles are equivalent modulo relations (3.2.a) - (3.2.h). In view of Lemma 4.2.3 we may restrict ourselves to the case where a and a' are either identity morphisms or words of the form (4.2.a). In this case both a and a' are the words associated to certain generic tangle diagrams D and D'. Our assumptions imply that these diagrams represent isotopic tangles. A well known theorem, due essentially to Reidemeister (see also [Tr]) affirms that such diagrams considered up to ambient isotopy may be related by a sequence of Reidemeister moves $\Omega_0, \Omega_2, \Omega_3$ and their inverses. The results of the previous subsections imply that the words $a = a(D)$ and $a' = a(D')$ are equivalent modulo relations (3.2.a) - (3.2.h).

4.9. Proof of Lemma 3.4. To prove Lemma 3.4 we describe a family of elementary moves on diagrams of ribbon graphs such that diagrams of isotopic ribbon graphs may be related by these moves. Assume that we have two diagrams representing isotopic ribbon graphs. Note that the coupons of these ribbon graphs are parallel to the plane of the page and their bases are horizontal. Let us show that these graphs may be related by an isotopy which maintains these conditions on coupons. Take an arbitrary isotopy between these graphs. We may assume that in the course of this isotopy each coupon moves in $\mathbb{R}^3$ as a solid rectangle. Thus, the isotopy of each coupon gives rise to a loop in the group $SO(3)$ beginning and ending in the unit element. All such loops may be deformed into $SO(2)$. This implies that we may deform the original isotopy to another isotopy relating the same graphs and keeping all coupons parallel to the plane. Moreover, since the fundamental group of $SO(2)$ is generated by a 2π-rotation we may relate our graphs by a composition of isotopies of the following two kinds: (i) isotopies keeping the bases of all coupons horizontal, and (ii) isotopies rotating a coupon in the plane by the angle of $\pm 2\pi$. Figure 4.16 presents an isotopy of type (ii) in the case of a coupon with one top and one bottom bands.

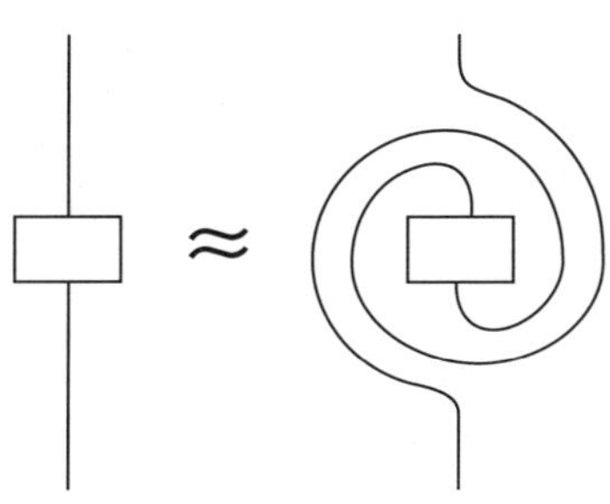

Figure 4.16

Each isotopy of type (ii) may be presented as a composition of isotopies of type (i) and isotopies relating the left-hand and right-hand sides of the equality in Figure 3.2. See, for example, Figure 4.17.

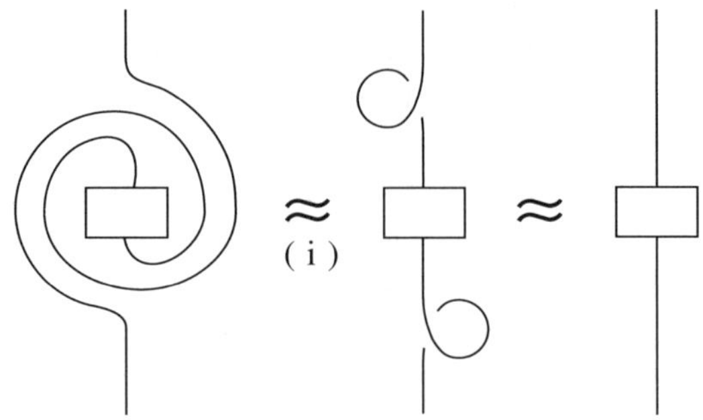

Figure 4.17

Isotopies of graph diagrams of type (i) may be presented as compositions of the following transformations: (iii) ambient isotopies in $\mathbb{R} \times [0, 1]$ keeping the bases of coupons horizontal, (iv) the Reidemeister moves described in Sections 4.4, 4.5, 4.7, proceeding far away from coupons, and (v) isotopies which push a strand of the diagram over or under a coupon. If this strand is oriented downwards then the isotopy of type (v) is the one shown in Figure 3.1. If the strand in question is oriented upwards then the isotopy of type (v) may be presented as a composition of isotopies of types (iii), (iv), and isotopies shown in Figure 3.1. An example of such a decomposition is given in Figure 4.18. We conclude that diagrams of isotopic ribbon graphs may be related by a sequence of transformations of types (iii), (iv), and the transformations shown in Figures 3.1 and 3.2. Using this fact, we prove Lemma 3.4 by repeating the proof of Lemma 3.3 with obvious changes.

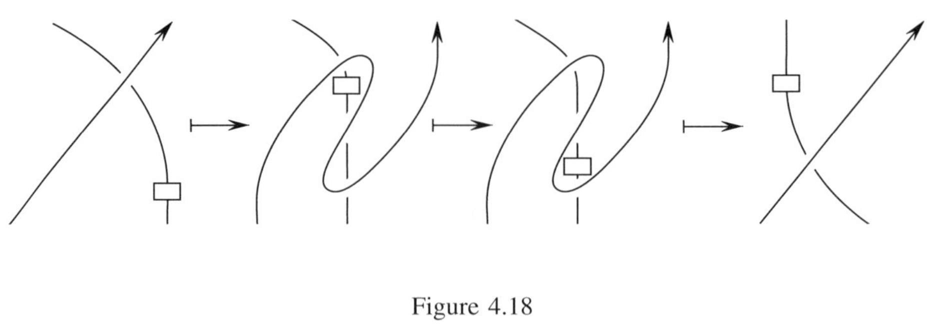

Figure 4.18

Notes

Section 1. The notion of a braiding in a monoidal category was introduced by Joyal and Street [JS1] - [JS3]. The definition of a twist in a braided category was given by Shum [Sh]. These authors use the term balanced tensor category for a monoidal category with braiding and twist and the term tortile tensor category for a monoidal category with braiding, twist, and compatible duality.

In the theory of Hopf algebras the structures parallel to braiding and twist were introduced by Drinfel'd [Dr1], [Dr2], Jimbo [Ji1], and Reshetikhin and Turaev [RT1]. Hopf algebras with these structures are called quasitriangular and ribbon Hopf algebras (for precise definitions, see Chapter XI). To conform with this terminology we prefer the term ribbon category to other existing terms.

Duality in monoidal categories has been discussed by several authors (see [KeL], [FY1], [JS1] - [JS3]). The treatment given here follows [JS2]. The definition of the trace generalizes the one given in [KeL], [FY1].

The graphical notation for tensors over the Lie algebra $sl_2(\mathbb{C})$ was first used by Penrose [Pe]. For a concise but fascinating discussion of the history of this calculus, see the introduction to [JS3]. It should be remarked that the modern art of playing with plane diagrams has been strongly influenced by the work of Louis Kauffman, cf. Chapter XII.

Section 2. The categories of graphs and tangles were invented by Yetter [Ye1] (see also Freyd and Yetter [FY1], [FY2]) and independently by the author, see [Tu6].

Theorem 2.5 is a generalized version of the main result of [RT1].

Sections 3 and 4. The idea to describe the categories of graphs and tangles via generators and relations is due to Yetter [Ye1] and the author [Tu6]. Here we follow [Tu6].

Chapter II
Invariants of closed 3-manifolds

1. Modular tensor categories

1.0. Outline. As we know, ribbon categories give rise to invariants of links in Euclidean 3-space. Unfortunately, they are far too general to yield similar invariants of links in arbitrary 3-manifolds. This leads us to a concept of modular category which is the key algebraic concept of this monograph. In Sections 2 and 3 we shall see that modular categories give rise to invariants of closed oriented 3-manifolds and links in such manifolds. Moreover, as we show in Chapter IV, modular categories give rise to 3-dimensional TQFT's.

Here we define modular categories and discuss their elementary properties. Modular categories are ribbon categories satisfying several extra conditions. The most important conditions are (i) a rudimentary additivity condition (we want to add morphisms with the same source and target but need neither images or kernels of morphisms nor direct sums of objects) and (ii) existence of a finite number of simple objects which in a sense dominate the category. The simple objects play here the role of irreducible modules in the representation theory, the domination replaces usual decompositions of modules into direct sums of irreducible modules. The reader should pay attention to the non-degeneracy axiom which will be crucial in Section 3.

1.1. Ab-categories. A category $\mathcal{V}$ is said to be an Ab-category if for any pair of its objects V, W, the set $\mathrm{Hom}(V, W)$ of $\mathcal{V}$-morphisms $V \to W$ is an additive abelian group and the composition of morphisms is bilinear (cf. [Ma2]; Ab-categories are also called pre-abelian categories). For example, the category of modules over a commutative ring with the usual addition of homomorphisms is an Ab-category. Speaking about monoidal Ab-categories we shall always assume that the tensor product of morphisms is bilinear.

Let $\mathcal{V}$ be a monoidal Ab-category. The composition of morphisms, considered as multiplication in $\mathrm{End}(\mathbb{1}) = \mathrm{Hom}(\mathbb{1}, \mathbb{1})$, renders this abelian group a ring with unit $\mathrm{id}_{\mathbb{1}}$. According to the results of Section I.1.5 this ring is commutative. It is called the ground ring of $\mathcal{V}$ and denoted by $K_{\mathcal{V}}$ or by K.

For any objects V, W of $\mathcal{V}$, the abelian group $\mathrm{Hom}(V, W)$ acquires the structure of a left K-module by the formula $kf = k \otimes f$ where $k \in K = \mathrm{End}(\mathbb{1})$ and $f \in \mathrm{Hom}(V, W)$. This K-linear structure is compatible with composition of morphisms in $\mathcal{V}$ in the sense that this composition is K-bilinear. Indeed,

$$(kf) \circ g = (k \otimes f)(\mathrm{id}_{\mathbb{1}} \otimes g) = k \otimes fg = k(f \circ g)$$

and similarly $f \circ kg = k(f \circ g)$.

It follows from the associativity of the tensor product that the tensor product of two morphisms in $\mathcal{V}$ is K-linear with respect to the first factor. The K-linearity with respect to the second factor is equivalent to the equality $k \otimes \mathrm{id}_V = \mathrm{id}_V \otimes k$ for every $k \in K$ and every object V of $\mathcal{V}$. It seems that there is no reason for this equality to hold for general monoidal Ab-categories. However, in the case which will be of interest to us, i.e., the case where $\mathcal{V}$ is a ribbon category, the equality $k \otimes \mathrm{id}_V = \mathrm{id}_V \otimes k$ is satisfied. This follows from the naturality of the braiding c in $\mathcal{V}$ and the formula $c_{\mathbb{1},V} = \mathrm{id}_V$. Thus, in the case of ribbon Ab-categories both composition and tensor product of morphisms are bilinear with respect to the module structure on the sets of morphisms.

Combining the definition of Ab-category with the definitions of Section I.1 we come to the notion of a ribbon Ab-category. This is a monoidal Ab-category equipped with braiding, twist, and compatible duality. For example, both categories constructed in Section I.1.7 (with the obvious additive structure on the sets of morphisms) are ribbon Ab-categories.

1.2. Simple objects. Let $\mathcal{V}$ be a ribbon Ab-category. For any $k \in K$ and any object V of $\mathcal{V}$, the morphism $k \otimes \mathrm{id}_V : V \to V$ is called multiplication by k in V. An object V of $\mathcal{V}$ is said to be simple if the formula $k \mapsto k \otimes \mathrm{id}_V$ defines a bijection $K \to \mathrm{End}(V)$. For example, the unit object $\mathbb{1}$ is simple.

Here is a convenient characterization of simple objects: an object V of $\mathcal{V}$ is simple if and only if $\mathrm{End}(V)$ is a free K-module of rank 1. Indeed, if V is simple then $\mathrm{End}(V) \simeq K$ with the generator id_V. Conversely, if $\mathrm{End}(V) \simeq K$ with a free generator x then $\mathrm{id}_V = kx$ and $x^2 = k'x$ with $k, k' \in K$. Hence $x = \mathrm{id}_V x = kx^2 = kk'x$. Therefore k is invertible in K and id_V is a free generator of $\mathrm{End}(V)$.

It is easy to see that an object isomorphic or dual to a simple object is also simple. The tensor product of simple objects may be non-simple, for instance, this is usually the case in the categories of representations of quantum groups (see Chapter XI).

1.3. Domination. In the representation theory of algebras one is often concerned with decompositions of modules into direct sums of irreducible modules. In the setting of ribbon categories it is tempting to decompose objects into direct sums of simple objects. However, in general, ribbon categories do not admit direct sums. It turns out that instead of decompositions of objects in direct sums we may decompose their identity endomorphisms. This leads to the following notion of domination.

Let $\{V_i\}_{i\in I}$ be a family of objects of a ribbon Ab-category $\mathcal{V}$. An object V of $\mathcal{V}$ is dominated by the family $\{V_i\}_{i\in I}$ if there exist a finite set $\{V_{i(r)}\}_r$ of objects of this family (possibly with repetitions which means that the same object may appear several times) and a family of morphisms $\{f_r : V_{i(r)} \to V, g_r : V \to V_{i(r)}\}_r$ such that

$$\mathrm{id}_V = \sum_r f_r g_r. \tag{1.3.a}$$

(Here $i(r) \in I$ for all r.) The definition of domination may be reformulated as follows: V is dominated by $\{V_i\}_{i\in I}$ if the images of the pairings

$$\{(g, f) \mapsto fg : \mathrm{Hom}(V, V_i) \otimes_K \mathrm{Hom}(V_i, V) \to \mathrm{End}(V)\}_{i\in I}$$

additively generate $\mathrm{End}(V)$.

In the case where the category $\mathcal{V}$ admits direct sums it is easy to show that V is dominated by the family $\{V_i\}_{i\in I}$ if and only if for some object W of $\mathcal{V}$, the direct sum $V \oplus W$ splits as a direct sum of a finite number of objects from this family (possibly with multiplicities). We shall not use this fact and shall never require ribbon categories at hand to admit direct sums.

1.4. Modular categories. A modular category is a pair consisting of a ribbon Ab-category $\mathcal{V}$ and a finite family $\{V_i\}_{i\in I}$ of simple objects of $\mathcal{V}$ satisfying the following four axioms.

(1.4.1) (Normalization axiom). There exists $0 \in I$ such that $V_0 = \mathbb{1}$.

(1.4.2) (Duality axiom). For any $i \in I$, there exists $i^* \in I$ such that the object V_{i^*} is isomorphic to $(V_i)^*$.

(1.4.3) (Axiom of domination). All objects of $\mathcal{V}$ are dominated by the family $\{V_i\}_{i\in I}$.

To formulate the next and last axiom we need some notation. For $i, j \in I$, set

$$\dim(i) = \dim(V_i) \in K \quad \text{and} \quad S_{i,j} = \mathrm{tr}(c_{V_j,V_i} \circ c_{V_i,V_j}) \in K$$

where K is the ground ring of $\mathcal{V}$. In view of Lemma I.1.5.1 we have $S_{i,j} = S_{j,i}$. (This is also clear from the geometric interpretation of $S_{i,j}$ as the invariant of the Hopf link shown in Figure I.2.12 where $V = V_i, W = V_j$.) Thus, $S = [S_{i,j}]_{i,j\in I}$ is a symmetric square matrix over K. Note that

$$S_{0,i} = S_{i,0} = \mathrm{tr}(\mathrm{id}_{V_i}) = \dim(i).$$

(1.4.4) (Non-degeneracy axiom). The square matrix $S = [S_{i,j}]_{i,j\in I}$ is invertible over K.

This completes the list of axioms for modular categories. For the sake of brevity, we shall sometimes denote the modular category $(\mathcal{V}, \{V_i\}_{i\in I})$ simply by $\mathcal{V}$. The (finite) set I will be denoted by $I_{\mathcal{V}}$.

Since the ring K is commutative, the non-degeneracy axiom amounts to saying that $\det(S)$ is invertible in K. Note an important implication of this axiom often used below. It follows from the result of Exercise I.2.9.6 that all terms of the i-th row of the matrix S are divisible by $\dim(i)$. Therefore, $\det(S)$ is divisible by $\dim(i)$. The invertibility of S implies that $\dim(i)$ is invertible in K for all $i \in I$. Of course, if K is a field then the invertibility of an element $a \in K$ means just that $a \neq 0$.

Note that if the objects V_i, V_j were isomorphic then the i-th and j-th rows of S would be equal. By the non-degeneracy axiom, the objects V_i, V_j with distinct i, j are not isomorphic. This implies that there is exactly one $0 \in I$ such that $V_0 = \mathbb{1}$ and that for any $i \in I$, there exists exactly one $i^* \in I$ such that V_{i^*} is isomorphic to $(V_i)^*$. The formula $i \mapsto i^*$ defines an involution in I. Corollary I.2.6.2 implies that $0^* = 0$.

A modular category is called strict if its underlying monoidal category is strict. The coherence theorem of MacLane establishing equivalence of any monoidal category to a certain strict monoidal category works in the setting of modular categories as well. Therefore, without loss of generality we may restrict ourselves to strict modular categories. In the sequel, by modular category we mean a strict modular category unless explicitly stated to the contrary.

Though we are not yet prepared to study deep properties of modular categories we may apply the technique of Chapter I to establish the following analogue of the Schur lemma.

1.5. Lemma. *Let $(\mathcal{V}, \{V_i\}_{i\in I})$ be a modular category. For any distinct $j, r \in I$, we have* $\mathrm{Hom}(V_j, V_r) = 0$.

Proof. For $i \in I$, consider the v-colored ribbon graph H^i_j presented by the diagram in Figure 1.1. (Here and in the figures to follow the symbols $i, j, \ldots$ stand for $V_i, V_j, \ldots$.) Since the object V_j is simple, the operator invariant $F(H^i_j) : V_j \to V_j$ is multiplication by a certain $x(i, j) \in K$. Hence $\mathrm{tr}(F(H^i_j)) = x(i, j) \dim(j)$. Corollary I.2.7.1 and the geometric interpretation of $S_{i,j}$ given above imply that $\mathrm{tr}(F(H^i_j)) = S_{i,j}$. Therefore

$$x(i, j) = S_{i,j}(\dim(j))^{-1}. \tag{1.5.a}$$

Let $f \in \mathrm{Hom}(V_j, V_r)$. Denote by Γ_f the v-colored ribbon graph presented in Figure I.1.1 where $V = V_j$ and $W = V_r$. It is obvious that the v-colored ribbon (1,1)-graphs $\Gamma_f \circ H^i_j$ and $H^i_r \circ \Gamma_f$ are isotopic. Therefore their operator invariants are equal, i.e.,

$$S_{i,j}(\dim(j))^{-1} f = S_{i,r}(\dim(r))^{-1} f. \tag{1.5.b}$$

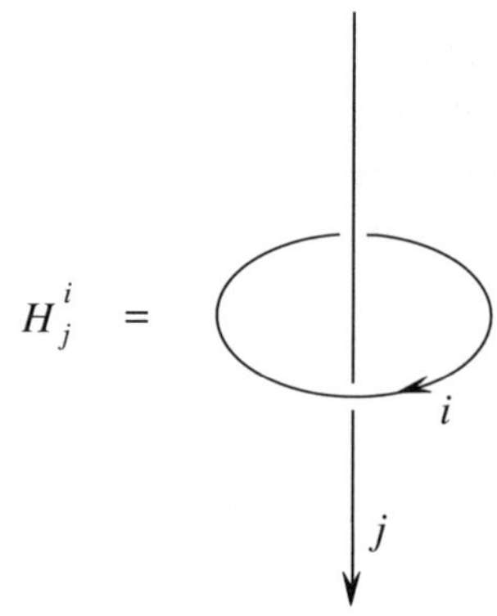

Figure 1.1

Since S is invertible and $j \neq r$ there exist certain $\{S'_i\}_{i\in I} \in K$ such that $\sum_i S_{i,j}S'_i = 1$ and $\sum_i S_{i,r}S'_i = 0$. Multiplying formula (1.5.b) by S'_i and summing up over all $i \in I$ we get $(\dim(j))^{-1}f = 0$. Therefore $f = 0$. This shows that $\mathrm{Hom}(V_j, V_r) = 0$.

1.6. Elements $\mathcal{D}$ and Δ of K. To define invariants of 3-manifolds associated with a modular category $(\mathcal{V}, \{V_i\}_{i\in I})$ we shall need an element $\mathcal{D}$ of the ground ring $K = K_{\mathcal{V}}$ such that

$$\mathcal{D}^2 = \sum_{i\in I} (\dim(i))^2. \tag{1.6.a}$$

Each such element of K is called a rank of $\mathcal{V}$. In general, $\mathcal{D}$ may not exist and if it exists it may be non-unique. (For instance, if $\mathcal{D}$ is a rank of $\mathcal{V}$ then $-\mathcal{D}$ is also a rank of $\mathcal{V}$.) The 3-manifold invariant defined below depends on the choice of $\mathcal{D}$.

The existence of a rank is a technical condition which essentially does not reduce the range of our constructions. Indeed, we may always formally add to K an element satisfying (1.6.a). More exactly, set $\Sigma = \sum_{i\in I}(\dim(i))^2$, $\tilde{K} = K[x]/(x^2 - \Sigma)$, and pass to the category $\tilde{\mathcal{V}}$ which has the same objects as $\mathcal{V}$ but more morphisms: for any objects V, W of $\mathcal{V}$, set

$$\mathrm{Hom}_{\tilde{\mathcal{V}}}(V, W) = \tilde{K} \otimes_K \mathrm{Hom}_{\mathcal{V}}(V, W)$$

(with the obvious $\tilde{K}$-bilinear extension of composition and tensor product). It is easy to verify that $\tilde{\mathcal{V}}$ is a modular category simultaneously with $\mathcal{V}$ and that $x \in \tilde{K}$ is its rank.

Besides the rank we shall need another element $\Delta = \Delta_{\mathcal{V}}$ of K. In contrast to $\mathcal{D}$ it is uniquely determined by the modular category as follows. Since V_i is a simple object, the twist acts in V_i as multiplication by a certain $v_i \in K$. Since the twist acts via isomorphisms, v_i is invertible in K. The elements $\{v_i\}_{i\in I}$ of K

will be often used below. In geometric language, their role may be described as follows: when we insert one full right-hand twist in a band of a ribbon graph in $\mathbb{R}^3$ colored with V_i the operator invariant of this graph is multiplied by v_i. Set

$$\Delta_{\mathcal{V}} = \sum_{i \in I} v_i^{-1} (\dim(i))^2 \in K.$$

We shall show in Section 3 that $\Delta_{\mathcal{V}}$ and any rank of $\mathcal{V}$ are invertible in K.

1.7. Examples. We give here two "toy" examples of modular categories. For deeper examples, see Chapters XI and XII.

1. If K is a field then the ribbon category $\mathrm{Proj}(K)$ constructed in Section I.1.7.1 is modular. The set $\{V_i\}_{i \in I}$ consists of one element K. The matrix S is the unit (1×1)-matrix.

2. Consider the ribbon category $\mathcal{V} = \mathcal{V}(G, K, c, \varphi)$ constructed in Section I.1.7.2. Endow $\mathcal{V}$ with the family $\{V_i\}_{i \in I}$ consisting of all objects of this category (they are all simple). Here $I = G$. This category is modular if and only if G is a finite group and the matrix $[c(g, h)\, c(h, g)]_{g,h \in G}$ is invertible over K. Indeed, the axioms (1.4.1) - (1.4.3) are straightforward. For any $g, h \in G$,

$$S_{g,h} = c(g, h)\, c(h, g)\, \varphi(g)\, \varphi(h) \in K.$$

Therefore axiom (1.4.4) holds if and only if the matrix $[c(g, h)\, c(h, g)]_{g,h \in G}$ is invertible over K. For example, if G is a finite non-trivial abelian group and $c(G, G) = 1$ then this matrix is not invertible. This shows that axiom (1.4.4) is independent from the other axioms. If G is finite and $K = \mathbb{C}$ then invertibility of S is equivalent to any of the following two conditions: (i) when g runs over G the homomorphism $h \mapsto c(g, h)\, c(h, g) : G \to \mathbb{C}^* \setminus \{0\}$ runs over all group homomorphisms $G \to \mathbb{C}^*$; (ii) for any $g \in G$, there exists $h \in G$ such that $c(g, h)\, c(h, g) \neq 1$. Thus, invertibility of S is equivalent to non-degeneracy of the symmetrized form c.

1.8. Remark. In the case where the ground ring is a field, the definition of modular category may be reformulated in more invariant terms. It is easy to show that in this case any simple object V of $\mathcal{V}$ dominated by a finite family of simple objects is actually isomorphic to one of them. Therefore in this case the family $\{V_i\}_{i \in I}$ may be described as a set of representatives of the isomorphism classes of simple objects of $\mathcal{V}$. This shows that a modular category over a field K (considered up to choice of objects $\{V_i\}_{i \in I}$ in their isomorphism classes) may be invariantly characterized as a ribbon Ab-category over K such that all objects are dominated by simple ones, the set I of isomorphism classes of simple objects is finite, and axiom (1.4.4) is fulfilled. The same remarks apply to a larger class of commutative rings K, namely, when the non-invertible elements of K form an additive subgroup of K.

1.9. Exercises. 1. Show that in any ribbon Ab-category the dualization of morphisms $f \mapsto f^*$ is linear over the ground ring K. Show that the trace of endomorphisms introduced in Section I.1.5 is linear over K.

2. Let $(\mathcal{V}, \{V_i\}_{i\in I})$ be a modular category. Show that the mirror category $\overline{\mathcal{V}}$ endowed with the same family of objects $\{V_i\}_{i\in I}$ is a modular category. Any rank of $\mathcal{V}$ is a rank of $\overline{\mathcal{V}}$ and vice versa. The matrix S determined by $(\overline{\mathcal{V}}, \{V_i\}_{i\in I})$ is equal to $[S_{i^*,j}]_{i,j\in I}$.

2. Invariants of 3-manifolds

Fix a strict modular category $(\mathcal{V}, \{V_i\}_{i\in I})$ with ground ring K and rank $\mathcal{D} \in K$.

2.0. Outline. For every closed oriented 3-manifold M, we define a K-valued invariant $\tau(M) = \tau_{(\mathcal{V},\mathcal{D})}(M) \in K$. This achieves one of the main goals of this book: a derivation of a homeomorphism invariant of 3-manifolds from any modular category. The remaining part of the book is, to a great extent, devoted to a study of this invariant from various viewpoints.

The construction of $\tau(M)$ is based on a reduction to the theory of framed links in Euclidean 3-space. To this end we use the technique of surgery, well known in topology. We present M as the result of surgery on the 3-sphere S^3 along a framed link in $\mathbb{R}^3 = S^3 \setminus \{\infty\}$. This allows us to build up an invariant of M from invariants of this link corresponding to different colorings of its components by the simple objects $\{V_i\}_{i\in I}$. Similar constructions apply to v-colored ribbon graphs in M and yield their K-valued invariants.

2.1. Surgery on links in the 3-sphere. Let L be a framed link in S^3 with m components $L_1, ..., L_m$ (cf. Section I.2.1), the link L is not assumed to be oriented or colored. Let U be a closed regular neighborhood of L in S^3. It consists of m disjoint solid tori $U_1, ..., U_m$ whose cores are the corresponding components of L. (A solid torus is a topological space homeomorphic to $S^1 \times B^2$ where B^2 is the closed 2-disk.) For each $n = 1, ..., m$, we identify U_n with $S^1 \times B^2$ so that L_n is identified with $S^1 \times 0$ where 0 is the center of B^2 and the given framing of L_n is identified with a constant normal vector field on $S^1 \times 0 \subset S^1 \times B^2$.

Let B^4 be a closed 4-ball bounded by S^3 so that $U \subset S^3 = \partial B^4$. Let us glue m copies of the 2-handle $B^2 \times B^2$ to B^4 along the identifications $\{U_n = S^1 \times B^2 = \partial B^2 \times B^2\}_{n=1}^m$. This gluing results in a compact connected 4-manifold denoted by W_L. The (closed connected) 3-manifold ∂W_L is formed by $S^3 \setminus \mathrm{Int}(U)$ and m copies of $B^2 \times \partial B^2$ glued to $S^3 \setminus \mathrm{Int}(U)$ along the boundary. We provide ∂W_L with the orientation extending the right-handed orientation in U. We say that the oriented 3-manifold ∂W_L is obtained by surgery on S^3 along L.

The manifold W_L is easily seen to be orientable. We provide it with the orientation induced by the orientation of ∂W_L defined above. (We use everywhere the "outward vector first" convention. Thus, in any point of ∂W_L the orientation of W_L is determined by the tuple (a tangent vector directed outwards, a positive basis in the tangent space of ∂W_L)). The signature of W_L, i.e., the signature of the intersection form on $H_2(W_L; \mathbb{R})$, will be denoted by $\sigma(L)$. If $L = \emptyset$ then, by definition, $W_L = B^4$ and $\sigma(L) = 0$.

A classical theorem due to Lickorish [Li1] and Wallace [Wal] asserts that any closed connected oriented 3-manifold (considered up to degree 1 homeomorphisms) can be obtained by surgery on S^3 along a certain framed link. Essentially this follows from the elementary theory of handles and the theorem of V.A. Rokhlin asserting that every closed oriented 3-manifold bounds a compact oriented piecewise-linear 4-manifold.

2.2. Invariants of closed 3-manifolds. To combine surgery with the technique of Chapter I we have to pass from links in $S^3 = \mathbb{R}^3 \cup \{\infty\}$ to links in $\mathbb{R}^3$. Fortunately, any link in S^3 may be deformed into $\mathbb{R}^3$ and isotopic links in S^3 give rise to isotopic links in $\mathbb{R}^3$. Therefore we may apply the invariant F to (colored framed oriented) links in S^3. Similar remarks hold for ribbon (0,0)-graphs.

Now we are in a position to define the invariant of 3-manifolds associated to $(\mathcal{V}, \mathcal{D})$. Let M be a closed connected oriented 3-manifold obtained by surgery on S^3 along a framed link L. Let $L_1, \ldots, L_m$ be the components of L. Fix an arbitrary orientation of L. Denote by $\operatorname{col}(L)$ the set of all mappings from the set of components of L into I. Clearly, $\operatorname{col}(L)$ is a finite set consisting of $(\operatorname{card}(I))^m$ elements. For each $\lambda \in \operatorname{col}(L)$, the pair (L, λ) determines a colored ribbon (0,0)-graph $\Gamma(L, \lambda)$ formed by m annuli. The cores of these annuli are the oriented circles $L_1, \ldots, L_m$, the normal vector field on the cores transversal to the annuli represents the given framing. The color of the i-th annuli is $V_{\lambda(L_i)}$. According to Theorem I.2.5 and remarks above we may consider the invariant $F(\Gamma(L, \lambda)) \in K$. Set

$$\{L\}_\lambda = \dim(\lambda)\, F(\Gamma(L, \lambda)) \in K \quad \text{where } \dim(\lambda) = \prod_{n=1}^{m} \dim(\lambda(L_n)).$$

(Recall that $\dim(i) = \dim(V_i)$ for all $i \in I$.) Set

$$\{L\} = \sum_{\lambda \in \operatorname{col}(L)} \{L\}_\lambda \in K.$$

It is clear that $\{L\}$ does not depend on the numeration of components of L.

2.2.1. Lemma. *$\{L\}$ does not depend on the choice of orientation in L.*

Proof. It suffices to show that $\{L\}$ is preserved when we reverse the orientation of a component of L. Let L' be obtained from L by reversing the orientation of a component L_n. For a coloring $\lambda \in \operatorname{col}(L)$, denote by λ' the coloring of L' which coincides with λ on all components of $L \setminus L_n$ and equals $(\lambda(L_n))^* \in I$ on L_n. Corollary I.2.8.1 and the equality $\dim(i) = \dim(i^*)$ imply that $\{L'\}_{\lambda'} = \{L\}_\lambda$. When λ runs over $\operatorname{col}(L)$ the coloring λ' runs over $\operatorname{col}(L')$. Therefore summing up the last equality over all $\lambda \in \operatorname{col}(L)$ we get $\{L'\} = \{L\}$. This completes the proof of the lemma.

Recall the element $\Delta = \Delta_{\mathcal{V}}$ defined in Section 1.6. Set

$$\tau(M) = \tau_{(\mathcal{V},\mathcal{D})}(M) = \Delta^{\sigma(L)}\mathcal{D}^{-\sigma(L)-m-1}\{L\} \in K. \tag{2.2.a}$$

2.2.2. Theorem. *$\tau(M)$ is a topological invariant of M.*

Theorem 2.2.2 will be generalized in Section 2.3 to ribbon graphs in 3-manifolds and proven (in the generalized form) in Section 3. The idea of the proof is as follows. We ought to show that the right-hand side of (2.2.a) does not depend on the choice of L. It is known that one and the same 3-manifold may be obtained from the 3-sphere by surgery along different framed links. Such links are related by so-called Kirby moves. It suffices to show the invariance of the right-hand side of (2.2.a) under these moves. The author does not know a direct proof of this fact. Instead, we introduce a slightly different expression $\tau'(M)$ and show its invariance under the Kirby moves. Then we use this invariance to relate $\tau'(M)$ to $\tau(M)$ and to deduce the invariance of $\tau(M)$.

We caution that we use in (2.2.a) invertibility of Δ and $\mathcal{D}$ proven in Section 3.

Theorem 2.2.2 is the central result of Chapter II. It establishes a deep connection between modular categories and the topology of 3-manifolds. The main feature of this connection is that a purely algebraic object (modular category) gives rise to a topological invariant of 3-manifolds. At the moment of writing there is no known way to obtain this invariant via homological or homotopical constructions.

The definition of $\tau(M)$ given above raises several questions. It is natural to ask if this definition is flexible, i.e., if we can vary the formula for $\tau(M)$ using other colorings of L or using some other expressions in place of $\dim(\lambda)$. To the author's knowledge the definition of $\tau(M)$ is not flexible. Apart from combining $\tau(M)$ with some other invariants of M there is no known way to produce analogous topological invariants of M by varying the expression for $\tau(M)$. In particular, it will be clear from the results of Section 4 that non-simple objects used as colors do not lead to more general invariants. Another natural question is concerned with the role of the factor $\dim(\lambda)$. In the framework of ideas developed in this chapter the only reasonable though tautological answer is that this factor is a component of the construction indispensable to build up a topological invariant of M. In Chapter IV we shall shed more light on the nature of this factor. It is also reasonable to ask to what extent $\tau(M)$ depends on the choice of rank $\mathcal{D}$ of $\mathcal{V}$. In contrast to the first two questions there is a simple and complete answer to this one. If the first Betti number $b_1(M) = \dim H_1(M;\mathbb{R})$ of M is odd then $\tau(M)$ does not depend on the choice of $\mathcal{D}$. If $b_1(M)$ is even then the product $\mathcal{D}\tau(M)$ does not depend on the choice of $\mathcal{D}$. This follows from the fact that $\mathcal{D}^2$ is determined by $\{V_i\}_{i\in I}$ and $\sigma(L) + m = b_1(M) \pmod 2$. The last equality follows from the well known computation of $b_1(M)$ as the nullity of the intersection form in $H_2(W_L;\mathbb{R})$. Note also that the invariant $\tau(M)$ does not depend on the choice of objects $\{V_i\}_{i\in I}$ in their isomorphism classes.

The normalization of $\tau(M)$ is chosen here to be the same as in Witten's theory of 3-manifold invariants given by path integrals (see [Wi2]). In particular,

$$\tau(S^1 \times S^2) = 1 \quad \text{and} \quad \tau(S^3) = \mathcal{D}^{-1}. \tag{2.2.b}$$

These formulas follow directly from definitions, it suffices to note that $S^1 \times S^2$ is obtained by surgery on S^3 along the trivial knot with zero framing and S^3 is obtained by surgery on S^3 along the empty link. In the case of modular categories associated with quantum groups at roots of unity the invariant τ coincides with the invariant constructed in [RT2], [TW] (up to normalization).

The definition of the invariant $\tau = \tau_{(\mathcal{V},\mathcal{D})}$ is ready for explicit computations. A simplest example is provided by the lens spaces of type $(n, 1)$ with integer $n \geq 2$. The lens space $L(n, 1)$ may be obtained from S^3 by surgery along the trivial knot with framing n. It follows from definitions that

$$\tau(L(n, 1)) = \Delta\mathcal{D}^{-3} \sum_{i\in I} v_i^n (\dim(i))^2.$$

The same manifold with the opposite orientation $-L(n, 1)$ may be obtained from S^3 by surgery along the trivial knot with framing $-n$ (recall that n is non-negative). Therefore

$$\tau(-L(n, 1)) = \Delta^{-1}\mathcal{D}^{-1} \sum_{i\in I} v_i^{-n} (\dim(i))^2.$$

It is well known that the lens space $L(n, 1)$ is the total space of a circle bundle over S^2. In Section X.9 we obtain similar formulas for circle bundles over arbitrary closed orientable surfaces.

2.3. Invariants of ribbon graphs in closed 3-manifolds. The definition of ribbon (0,0)-graphs in $\mathbb{R}^3$ directly extends to arbitrary 3-manifolds. By a ribbon graph (without free ends) in a 3-manifold M, we mean an oriented surface embedded in M and decomposed in a union of a finite number of directed annuli, directed bands, and coupons such that the bases of bands lie on the bases of coupons and the bands, coupons, and annuli are disjoint otherwise. The definitions of ribbon graphs in M colored or v-colored over $\mathcal{V}$ are straightforward. By isotopy of ribbon graphs in M, we mean an isotopy in M preserving the splitting into annuli, bands, and coupons, the directions of the annuli and bands, and the orientation of the graph surface. By isotopy of colored or v-colored ribbon graphs, we mean color-preserving isotopy.

The invariant of 3-manifolds defined in Section 2.2 generalizes to closed oriented 3-manifolds with v-colored ribbon graphs sitting inside. Let M be a closed connected oriented 3-manifold. Let Ω be a v-colored ribbon graph over $\mathcal{V}$ in M. Present M as the result of surgery on S^3 along a framed link L with components $L_1, \ldots, L_m$. Fix an orientation in L. Applying isotopy to Ω we can deform it into $S^3 \setminus U \subset M$ where U is a closed regular neighborhood of L in S^3. Thus, we may assume that $\Omega \subset S^3 \setminus U$. For any $\lambda \in \operatorname{col}(L)$, we form a v-colored ribbon (0,0)-graph $\Gamma(L, \lambda) \subset U$ as in Section 2.2. The union $\Gamma(L, \lambda) \cup \Omega$ is a v-colored ribbon (0,0)-graph in S^3. Set

$$\{L, \Omega\} = \sum_{\lambda \in \operatorname{col}(L)} \dim(\lambda)\ F(\Gamma(L, \lambda) \cup \Omega) \in K.$$

Here $\dim(\lambda)$ is defined by the same formula as in Section 2.2.

2.3.1. Lemma. $\{L, \Omega\}$ *does not depend on the choice of orientation in* L.

The proof is analogous to the proof of Lemma 2.2.1.

Set

$$\tau(M, \Omega) = \tau_{(\mathcal{V},\mathcal{D})}(M, \Omega) = \Delta^{\sigma(L)} \mathcal{D}^{-\sigma(L)-m-1} \{L, \Omega\}.$$

2.3.2. Theorem. $\tau(M, \Omega)$ *is a topological invariant of the pair* (M, Ω).

Theorem 2.3.2 is proven in Section 3. It includes Theorem 2.2.2 as a special case because $\tau(M, \emptyset) = \tau(M)$.

Theorem 2.3.2 implies that $\tau(M, \Omega)$ is an isotopy invariant of Ω. Using ribbon graphs consisting of annuli we obtain an isotopy invariant of colored framed oriented links in M.

If $M = S^3$ then $\tau(M, \Omega) = \mathcal{D}^{-1} F(\Omega)$ (we may take $L = \emptyset$ to compute $\tau(S^3, \Omega)$). The invariant $\tau(M, \Omega)$ satisfies the following multiplicativity law:

(2.3.a) $$\tau(M_1 \# M_2, \Omega_1 \amalg \Omega_2) = \mathcal{D}\, \tau(M_1, \Omega_1)\, \tau(M_2, \Omega_2)$$

where Ω_1, Ω_2 are v-colored ribbon graphs in closed connected oriented 3-manifolds M_1, M_2 respectively. The properties of the operator invariants of ribbon graphs established in Section I.2.8 generalize in the obvious way to ribbon graphs in closed 3-manifolds, precise formulations and proofs are left to the reader. Note also that the invariant $M \mapsto \mathcal{D}\, \tau(M)$ is multiplicative with respect to connected summation. Under certain circumstances this modified invariant may be more convenient than τ.

The invariant τ extends to v-colored ribbon graphs in any non-connected closed oriented 3-manifold M by the formula $\tau(M, \Omega) = \prod_r \tau(M_r, \Omega_r)$ where M_r runs over connected components of M and Ω_r denotes the part of Ω lying in M_r.

2.4. Remarks and examples. 1. Presenting S^3 as the result of surgery on itself along the trivial knot with framing 1 or along the empty link we get two expressions for $\tau(S^3)$:

$$\tau(S^3) = \mathcal{D}^{-1} = \Delta \mathcal{D}^{-3} \sum_i v_i (\dim(i))^2.$$

Therefore

$$\Delta_{\mathcal{V}} \Delta_{\overline{\mathcal{V}}} = \Delta \sum_i v_i (\dim(i))^2 = \mathcal{D}^2. \tag{2.4.a}$$

For instance, in the case $\mathcal{V} = \mathcal{V}(G, K, c, \varphi)$ (assuming that this category is modular) formula (2.4.a) yields

$$\left(\sum_{g \in G} c(g, g) \right) \left(\sum_{g \in G} (c(g, g))^{-1} \right) = \mathcal{D}^2 = \operatorname{card}(G).$$

If $K = \mathbb{C}$ then $(c(g, g))^{-1} = \overline{c(g, g)}$ so that we get the well known computation of the absolute value of a Gauss sum. This suggests that a study of modular categories may involve number-theoretic considerations.

2. The invariant τ of a v-colored ribbon graph in a 3-manifold is polylinear with respect to the colors of coupons. This polylinear function of the colors of coupons may be considered as an isotopy invariant of the underlying colored (but not v-colored) ribbon graph. More exactly, let $\Omega \subset M$ be a colored ribbon graph with n coupons $Q_1, \ldots, Q_n$. The colorings of Q_i form a K-module of type $\operatorname{Hom}(V, W)$ where V and W are objects of $\mathcal{V}$ determined by colors and directions of the bands of Ω attached to Q_i as in Section I.2.2. Denote this K-module by G_i. Choosing a color of each coupon $Q_1, \ldots, Q_n$ we transform Ω to a v-colored ribbon graph. Its invariant τ yields a polylinear mapping $G_1 \times G_2 \times \ldots \times G_n \to K$ or, what is equivalent, a K-homomorphism $G_1 \otimes G_2 \otimes \ldots \otimes G_n \to K$. This homomorphism is an isotopy invariant of Ω.

3. Let K be a field and let $\mathcal{V}$ be the category $\operatorname{Proj}(K)$ constructed in Section I.1.7.1. Then $\mathcal{D} = 1$ is a rank of $\mathcal{V}$ and $\tau_{(\mathcal{V},1)}(M) = 1$ for any closed connected oriented 3-manifold M.

It would be interesting to compute explicitly the invariant $\tau(M)$ corresponding to the modular category described in Section I.1.7.2. Conjecturally, it is computable from the linking form in $\operatorname{Tors} H_1(M)$ and the Rokhlin invariants of spin structures on M.

2.5. Exercise. Let Ω be a v-colored ribbon graph in a closed connected oriented 3-manifold M. Show that the product $\mathcal{D}^{b_1(M)+1}\tau_{(\mathcal{V},\mathcal{D})}(M,\Omega)$ does not depend on the choice of $\mathcal{D}$,

$$\tau_{(\mathcal{V},-\mathcal{D})}(M,\Omega)=(-1)^{b_1(M)+1}\tau_{(\mathcal{V},\mathcal{D})}(M,\Omega),$$

and

$$\tau_{(\mathcal{V},\mathcal{D})}(-M,\Omega)=\tau_{(\overline{\mathcal{V}},\mathcal{D})}(M,\Omega). \tag{2.5.a}$$

Here $-M$ denotes the manifold M with opposite orientation. (Hint: if M is obtained by surgery on S^3 along a framed link L then $-M$ is obtained by surgery on S^3 along the mirror image of L.)

3. Proof of Theorem 2.3.2. Action of $SL(2,\mathbb{Z})$

3.0. Outline. The main body of this section is devoted to the proof of Theorem 2.3.2. As a corollary of this theorem and its proof we establish a number of useful identities involving the matrix S. These identities will be crucially used in Section 4 and in Chapters VII and X. At the end of this section we use these identities to derive from $(\mathcal{V},\{V_i\}_{i\in I})$ a projective linear representation of the matrix group $SL(2,\mathbb{Z})$. The nature of this representation will be clarified in Chapter IV where we define more general representations of the modular groups of surfaces.

The proof of Theorem 2.3.2 given below reveals intimate connections between the topology of 3-manifolds and the algebra of modular categories. As a geometric counterpart the proof involves the Kirby calculus of links in S^3. It should be noted that this calculus and other ideas involved in the proof of Theorem 2.3.2 will not essentially be used in the remaining part of the book.

3.1. The Kirby calculus. It has been known for a long time that a closed oriented 3-manifold can be obtained from the 3-sphere by surgery along different framed links. For example, the surgery along a trivial link with any number of components and with framing numbers of components ±1 always produces S^3. This phenomenon gives rise to an equivalence relation in the class of framed links in S^3: two links are equivalent if the surgery on S^3 along these links yield homeomorphic 3-manifolds. (We work here in the category of oriented 3-manifolds; by homeomorphisms we mean orientation-preserving homeomorphisms.) Kirby [Ki] introduced certain transformations (or moves) on framed links generating this equivalence relation. Another system of moves on framed links generating the same equivalence relation was introduced by Fenn and Rourke [FR]. Their moves have an advantage over the original approach of Kirby: they follow certain standard patterns and therefore can be described algebraically in terms of tangle generators. In this respect these moves are similar to Reidemeister moves on link diagrams. It was shown in [RT2] that some of the Kirby-Fenn-Rourke moves (in fact, almost half of them) can be presented as compositions of remaining moves.

We describe the Kirby moves (in the form of Fenn and Rourke) in terms of link diagrams. Recall that framed links in $S^3 = \mathbb{R}^3 \cup \{\infty\}$ are represented by link diagrams in the plane $\mathbb{R}^2 \subset \mathbb{R}^3$ where the framing is determined by the vector field normal to $\mathbb{R}^2$. Two framed links L, L_1 are related by a Kirby 1-move if they can be presented by diagrams in $\mathbb{R}^2$ which are identical except for pieces shown in Figure 3.1. On the right we have a system of strictly vertical untwisted unlinked bands (only two bands are shown in the figure). On the left we have the same system which undergoes a full right-hand twist (r-twist) and acquires an additional unknotted component C. In other words, C is an unknotted component of L which disappears in L_1 at the price of dropping out a full right-hand twist in the bunch of bands linked by C. The small positive curl in the loop presenting C shows that the framing number of C is equal to 1; this is why we use the term 1-move. More technically, the Kirby 1-move $L \mapsto L_1$ shown in Figure 3.1 is called the 1-move along C. The inverse transformation $L_1 \mapsto L$ is called the inverse Kirby move along C.

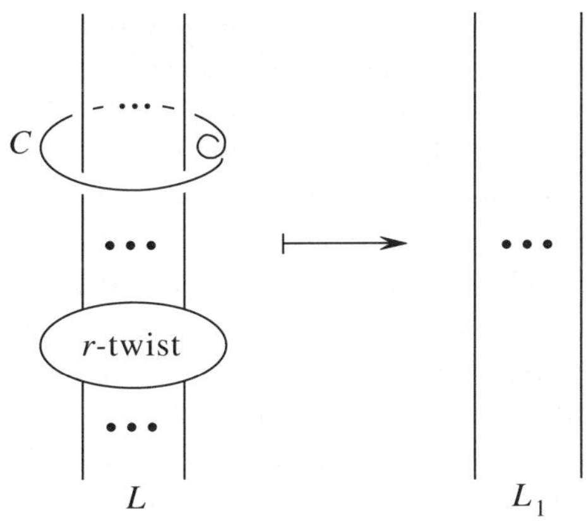

Figure 3.1

A few examples of Kirby 1-moves are given in Figure 3.2 where the number of bands linked by C is equal to 0,1,2, respectively. (In the case of two bands we use the obvious fact that, up to isotopy, exclusion of a full right-hand twist amounts to insertion of a full left-hand twist.) The first move in Figure 3.2, i.e., elimination or insertion of an unknotted component with framing 1 separated from other components is called a special Kirby 1-move. We shall also need special Kirby (-1)-moves which eliminate or insert unknotted components with framing -1 separated from other components.

It is well known that for any framed links L, L' in S^3 related by a Kirby 1-move or a special Kirby (-1)-move, the results of surgery on S^3 along L and L'

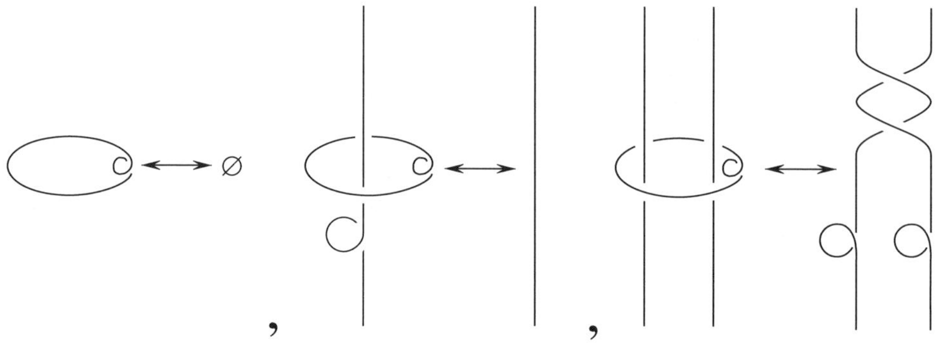

Figure 3.2

are homeomorphic (see [Rolf]). A deep theorem of Kirby [Ki] as refined in [FR] and [RT2] affirms the converse. This gives a description of the homeomorphism classes of closed oriented 3-manifolds in terms of links in S^3 and Kirby moves. Namely, two closed oriented 3-manifolds obtained from S^3 by surgery along framed links L, L' are homeomorphic if and only if L' can be obtained from L by a finite sequence of Kirby 1-moves and special Kirby (-1)-moves.

The Kirby calculus can be generalized to describe ribbon graphs in 3-manifolds. Let L be a framed link in S^3 and let Ω be a ribbon graph in $S^3 \setminus L$. Surgering S^3 along L we get a 3-manifold M with $\Omega \subset M$. We say that the pair (M, Ω) is obtained by surgery on S^3 along (L, Ω). We define generalized Kirby moves on (L, Ω) preserving (up to homeomorphism) the result of surgery. Note first that the union $L \cup \Omega$ is a ribbon graph in S^3 which may be presented by plane diagrams. Applying to any diagram of $L \cup \Omega$ the transformation shown in Figure 3.1 we get a diagram presenting a new pair $L_1 \cup \Omega_1 \subset S^3$. It is understood that this transformation proceeds far away from coupons, which correspond to coupons of Ω. As above, the circle C is an unknotted component of L with framing 1. Each string linked by C corresponds either to a component of L or to a band or an annulus of Ω. We call the transformation $(L, \Omega) \mapsto (L_1, \Omega_1)$ and the inverse transformation $(L_1, \Omega_1) \mapsto (L, \Omega)$ (generalized) Kirby 1-moves. In the case where C is not linked with the remaining part of $L \cup \Omega$ we call this move a special Kirby 1-move. Similarly, eliminating or inserting an unknotted component of L with framing -1 separated from other components of L and from Ω we get a special Kirby (-1)-move. Using the mirror image of Figure 3.1 with respect to the plane of the picture we get a generalized Kirby (-1)-move. The same definitions apply if Ω is colored or v-colored. In this case Kirby moves preserve the colors of bands, annuli, and coupons of Ω.

As it was explained in the second paragraph of Section 2.3, any pair (a closed connected oriented 3-manifold M, a ribbon graph in M) considered up to homeomorphism may be obtained by surgery on S^3 along a pair (L, Ω) as above. The result of surgery on (L, Ω) is preserved under the Kirby moves on (L, Ω). Conversely, if the results of surgery along two such pairs are homeomorphic then these pairs may be related by a finite sequence of Kirby 1-moves and special Kirby (-1)-moves (for a proof, see [RT2, Section 7.3]). In this way we reduce the study of ribbon graphs in closed oriented 3-manifolds to a study of pairs (L, Ω) in S^3.

3.2. Reduction of Theorem 2.3.2 to lemmas. Let $(\mathcal{V}, \{V_i\}_{i\in I})$ be a modular category. Recall the matrix $S = [S_{i,j}]_{i,j\in I}$ and the elements $v_i, \dim(i) \in K = \mathrm{End}(\mathbb{1})$ introduced in Sections 1.4 and 1.6. Since S is invertible over K there exists a unique solution $\{d_i\}_{i\in I} \in K$ of the following system of card(I) linear equations:

$$\sum_{i\in I} d_i v_i v_j S_{i,j} = \dim(j). \tag{3.2.a}$$

(Here j runs over I.) In particular, substituting $j = 0$ in (3.2.a) we get

$$\sum_{i\in I} d_i v_i \dim(i) = 1. \tag{3.2.b}$$

To prove Theorem 2.3.2 we shall replace in the definition of τ the factor $\dim(\lambda)$ by $\prod_n d_{\lambda(L_n)}$. Formula (3.2.a) will enable us to show that this modified invariant is preserved under the Kirby moves. Then we shall use this result to prove that $\{d_i\}_{i\in I}$ are proportional to $\{\dim(i)\}_{i\in I}$ so that the modified invariant is actually equivalent to τ.

Let L be a framed link in S^3 and let $\Omega \subset S^3 \setminus L$ be a v-colored ribbon graph over $\mathcal{V}$. Using $\{d_i\}_{i\in I}$ instead of $\{\dim(i)\}_{i\in I}$ and following the lines of Sections 2.2 and 2.3 we define $\{L, \Omega\}' \in K$ as follows. Let $L_1, \ldots, L_m$ be the components of L. Fix an orientation of L. For any $\lambda \in \mathrm{col}(L)$, consider the v-colored ribbon (0,0)-graph $\Gamma(L, \lambda) \cup \Omega$ defined as in Sections 2.2, 2.3 and set

$$\{L, \Omega\}'_\lambda = \left(\prod_{n=1}^{m} d_{\lambda(L_n)}\right) F(\Gamma(L, \lambda) \cup \Omega), \qquad \{L, \Omega\}' = \sum_{\lambda\in \mathrm{col}(L)} \{L, \Omega\}'_\lambda \in K.$$

In the case $\Omega = \emptyset$ we shall write simply $\{L\}'_\lambda$ and $\{L\}'$.

3.2.1. Lemma. *The element $\{L, \Omega\}' \in K$ does not depend on the choice of orientation in L. This element is invariant under the generalized Kirby* 1*-moves on (L, Ω).*

This lemma is the heart of the proof. With this lemma proven it will remain only to check the invariance under the special (-1)-move and to relate $\{L, \Omega\}'$ to $\{L, \Omega\}$. To this end we have to study the elements $\{d_i\}_{i\in I}$ in more detail.

Set $x = \{\mathbb{O}\}' = \{\mathbb{O}, \emptyset\}'$ where $\mathbb{O}$ is the trivial knot in S^3 with zero framing. It is clear that

$$x = \sum_{i \in I} d_i \dim(i).$$

3.2.2. Lemma. (i) *For any $j \in I$, we have*

$$\sum_{i \in I} d_i S_{i,j} = x \delta_0^j, \tag{3.2.c}$$

where δ_k^j is the Kronecker delta equal to 1 *if $j = k$ and to* 0 *otherwise.*

(ii) *For any $j, r \in I$, we have*

$$\sum_{i \in I} d_i (\dim(i))^{-1} S_{j,i} S_{i,r} = x \delta_{j^*}^r. \tag{3.2.d}$$

(iii) *For any $j \in I$, we have*

$$\sum_{i \in I} d_i v_i^{-1} v_j^{-1} S_{i,j} = x d_j. \tag{3.2.e}$$

Note that (3.2.c) is the case $r = 0$ of (3.2.d). It is convenient for us to separate this case because we shall first prove (3.2.c) and then (3.2.d). The substitution $j = 0$ transforms (3.2.e) into the formula

$$\sum_{i \in I} d_i v_i^{-1} \dim(i) = x d_0. \tag{3.2.f}$$

We may draw a few conclusions from Lemma 3.2.2 concerned with invertibility of $\{d_i\}_{i \in I}$ and x in K. It follows from the very definition of the elements $\{d_i\}_{i \in I}$ that not all of them are equal to zero. Otherwise we would have $\dim(j) = 0$ for all $j \in I$ which contradicts invertibility of $\dim(j)$. A similar argument shows that, moreover, any common divisor of $\{d_i\}_{i \in I}$ in K is invertible in K. Formula (3.2.c) and the invertibility of the matrix S imply that x is a common divisor of $\{d_i\}_{i \in I}$ in K. Hence x is invertible in K. Formula (3.2.d) and the invertibility of x imply that the matrix

$$\{x^{-1} d_i (\dim(i))^{-1} S_{i,j^*}\}_{i,j \in I}$$

is inverse to S. Since the i-th row of this inverse matrix is divisible by d_i in K we may conclude that all $\{d_i\}_{i \in I}$ are invertible in K.

Following along the lines of Section 2 we assign to an arbitrary pair (a closed connected oriented 3-manifold M, a v-colored ribbon graph Ω in M) a certain element $\tau'(M, \Omega) \in K$. Present M as the result of surgery on S^3 along a framed m-component link L. Set $\sigma_-(L) = (m - \dim H_1(M; \mathbb{R}) - \sigma(L))/2$. This is an integer equal to the number of negative eigenvalues of the intersection form of the 4-manifold W_L introduced in Section 2.1. Push Ω into $S^3 \setminus L$. Set $\tau'(M, \Omega) = (x d_0)^{-\sigma_-(L)} \{L, \Omega\}'$. (Here we use invertibility of x and d_0.) Thus,

$$\tau'(M, \Omega) = (xd_0)^{-\sigma_-(L)} \sum_{\lambda \in \operatorname{col}(L)} \left(\prod_{n=1}^{m} d_{\lambda(L_n)} \right) F(\Gamma(L, \lambda) \cup \Omega). \tag{3.2.g}$$

Lemma 3.2.1 affirms that any Kirby 1-move applied to (L, Ω) keeps $\{L, \Omega\}'$ intact. It is easy to observe that such a move preserves the topological type of W_L and keeps $-\sigma_-(L)$. (Indeed, a Kirby 1-move on L either simultaneously diminishes by 1 or simultaneously increases by 1 the number of components of L and the signature $\sigma(L)$.) Thus, $\tau'(M, \Omega)$ is invariant under the Kirby 1-moves on (L, Ω).

If a pair (L_{-1}, Ω) is obtained from (L, Ω) by adding to L a distant unknotted circle with framing -1 then $\sigma_-(L_{-1}) = \sigma_-(L) + 1$. It follows from (3.2.f) that

$$\{L_{-1}, \Omega\}' = \{L, \Omega\}' \sum_{i \in I} d_i v_i^{-1} \dim(i) = \{L, \Omega\}' x d_0.$$

Therefore $\tau'(M, \Omega)$ is invariant under special Kirby (-1)-moves. In view of the results quoted in Section 3.1 we conclude that $\tau'(M, \Omega)$ is a topological invariant of (M, Ω).

We shall use the topological invariance of $\tau'(M, \Omega)$ to compute $\{d_i\}_{i \in I}$.

3.2.3. Lemma. *For any $j \in I$, we have*

$$d_j = d_0 \dim(j). \tag{3.2.h}$$

Lemmas 3.2.1 - 3.2.3 imply invertibility of the elements Δ and $\mathcal{D}$ of K defined in Section 1.6. Indeed, substituting the expression (3.2.h) for d_j into (3.2.f) and factorizing out d_0 we get

$$x = \sum_{i \in I} v_i^{-1} (\dim(i))^2 = \Delta. \tag{3.2.i}$$

Therefore Δ is invertible. Substituting (3.2.h) in the expression for x we get

$$\Delta = x = \sum_{i \in I} d_i \dim(i) = d_0 \sum_{i \in I} (\dim(i))^2 = d_0 \mathcal{D}^2. \tag{3.2.j}$$

Therefore $\mathcal{D}$ is invertible and $d_0 = \Delta \mathcal{D}^{-2}$.

Invertibility of Δ and $\mathcal{D}$ shows that the formulas used in Section 2 to define $\tau(M)$ and $\tau(M, \Omega)$ make sense. Now we may compare $\tau'(M, \Omega)$ and $\tau(M, \Omega)$ and complete the proof of Theorem 2.3.2. Substituting in (3.2.g) the expressions $d_j = d_0 \dim(j)$, $x = \Delta$, and $d_0 = \Delta \mathcal{D}^{-2}$ we get

$$\tau'(M, \Omega) = \Delta^{b_1(M)} \mathcal{D}^{-b_1(M)+1} \tau(M, \Omega). \tag{3.2.k}$$

This formula together with the topological invariance of $\tau'(M, \Omega)$ implies Theorem 2.3.2.

The next five subsections are concerned with the proof of Lemmas 3.2.1, 3.2.2, and 3.2.3.

3.3. Proof of Lemma 3.2.1. Let $H_{i,j}$ be the Hopf link drawn in Figure I.2.12 with V and W being V_i and V_j respectively. According to Corollary I.2.7.3 we have $S_{i,j} = F(H_{i,j})$. Inverting the orientation of both components of $H_{i,j}$ and trading i, j for i^*, j^* we get H_{i^*,j^*}. Corollary I.2.8.1 implies that

$$S_{i,j} = S_{i^*,j^*}. \tag{3.3.a}$$

Formula (I.1.3.d) (or (I.1.8.a)) applied to $V = V_i$ shows that

$$v_{i^*} = v_i. \tag{3.3.b}$$

Formulas (3.2.a), (3.3.a), and (3.3.b) imply that $d_{i^*} = d_i$ for all $i \in I$. This fact together with Corollary I.2.8.1 implies the first claim of Lemma 3.2.1 (cf. the proof of Lemma 2.2.1).

Let us show that for each $j \in I$, we have equality (3.3.c) in Figure 3.3. As above, the symbol $\doteq$ denotes equality of the associated morphisms $V_j \to V_j$. Recall that the operator invariant $F(H^i_j)$ of the colored ribbon tangle H^i_j shown in Figure 1.1 is multiplication by $x(i, j) \in K$ given by (1.5.a). Therefore the morphism $V_j \to V_j$ associated to the tangle on the left part of Figure 3.3 is multiplication by

$$\sum_{i \in I} d_i v_i v_j x(i, j) = \sum_{i \in I} d_i v_i v_j S_{i,j} (\dim(j))^{-1} = 1$$

where the last equality follows from (3.2.a). This proves (3.3.c).

(3.3.c)

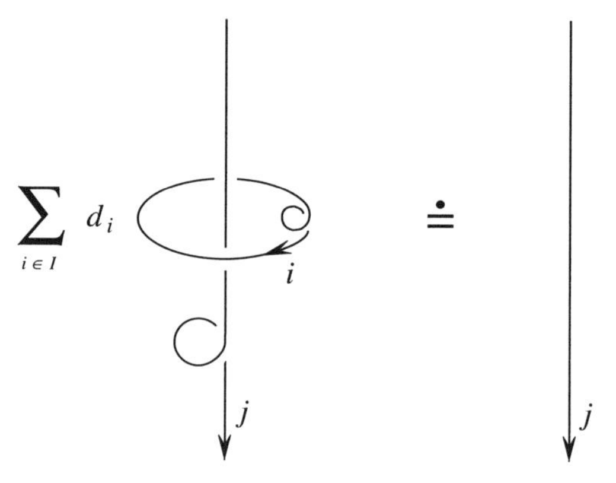

Figure 3.3

Let (L, Ω) be a pair consisting of a framed link $L \subset S^3$ and a v-colored ribbon graph $\Omega \subset S^3 \setminus L$. Let (L_1, Ω_1) be another such pair obtained from (L, Ω) by a

Kirby 1-move along an unknotted component C of L. Let us prove that

$$\{L,\Omega\}' = \{L_1,\Omega_1\}'. \tag{3.3.d}$$

Present $L \cup \Omega$ and $L_1 \cup \Omega_1$ by diagrams D, D_1 identical except for pieces shown in Figure 3.1. Denote by r the number of strands of D linked by C. Some of these strands represent components of L, the others represent bands and annuli of Ω. If $r = 0$ then (3.3.d) follows directly from (3.2.b). Consider the case $r \geq 1$. The diagram D, considered up to ambient isotopy in the plane, may be presented as the closure of the tangle diagram in the left part of Figure 3.4 where w stands for a tangle diagram with r inputs and r outputs. Therefore it suffices to prove the following claim.

Claim. *Let $W_1, \ldots, W_r$ be objects of $\mathcal{V}$, $\varepsilon_1, \ldots, \varepsilon_r \in \{\pm 1\}$, and w an arbitrary endomorphism of*

$$W = W_1^{\varepsilon_1} \otimes \ldots \otimes W_r^{\varepsilon_r}$$

(where $W_k^1 = W_k$ and $W_k^{-1} = W_k^*$ for $k = 1, \ldots, r$). Then we have equality (3.3.e) *in Figure* 3.4 *where the k-th vertical strand attached to the coupon from above or from below is colored with W_k and is directed downwards if $\varepsilon_k = 1$ and upwards otherwise.*

(3.3.e) 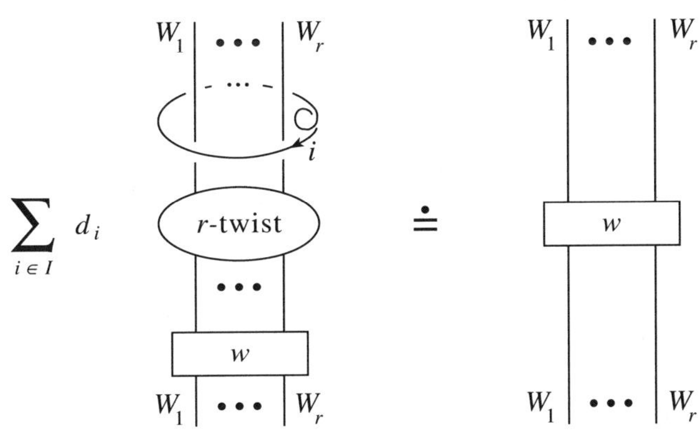

Figure 3.4

The axiom of domination applied to the identity endomorphism of W implies that $w = w \circ \mathrm{id}_W$ is a sum of several morphisms of the form fg where f and g are certain morphisms $g : W \to V_j$ and $f : V_j \to W$ with $j \in I$. It is clear that both sides of (3.3.e) are linear with respect to w. Therefore it suffices to prove (3.3.e) in the case $w = fg$ with f, g as above. We replace the coupon in the left part of Figure 3.4 by two coupons colored with f and g and one short vertical band colored with V_j and directed downward. This does not change the associated morphism in $\mathcal{V}$. Pulling up the f-colored coupon we get the v-colored ribbon graph drawn in the left part of Figure 3.5. (The curl on the j-colored band is equivalent up to isotopy to a full right-hand twist in the bunch of r strings attached to the f-colored coupon from above.) The first equality in Figure 3.5 follows from (3.3.c), the second equality is obvious since $w = fg$. This implies (3.3.e), (3.3.d), and the lemma.

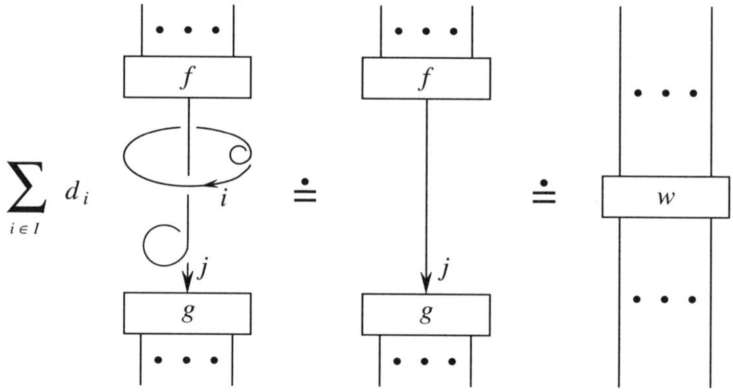

Figure 3.5

3.4. Proof of Lemma 3.2.2, claim (i). Fix $j, k \in I$. Set

$$e_j = \sum_{i \in I} d_i S_{i,j} \in K$$

and

$$\omega = \sum_{i \in I} d_i F(H_j^k \circ H_j^i) : V_j \to V_j$$

where H_j^i is the v-colored ribbon graph in Figure 1.1. It follows from (1.5.a) that

$$\omega = \sum_{i \in I} d_i S_{i,j} S_{k,j} (\dim(j))^{-2} \mathrm{id}_{V_j} = e_j S_{j,k} (\dim(j))^{-2} \mathrm{id}_{V_j}. \tag{3.4.a}$$

We shall compute ω in a different way. Namely, we transform a diagram of ω by a sequence of Kirby 1-moves into another diagram, see Figure 3.6. The first two transformations are the inverse Kirby 1-moves along the components C_1 and C_2 respectively. (To visualize this, regard the inverse moves which are just the Kirby 1-moves eliminating C_1 and C_2.) The next two transformations are the Kirby 1-moves along the components C_3 and C_4 respectively.

To each tangle diagram in Figure 3.6 we associate a morphism $V_j \to V_j$ as follows. We color the oriented circle marked by k with V_k and the string leading from the top to the bottom with V_j. We orient all other circles in an arbitrary way and color them with certain $V_{r_1}, \ldots, V_{r_n}$ where n is the number of unoriented circles in the diagram at hand and $r_1, \ldots, r_n \in I$. This yields a diagram of a colored

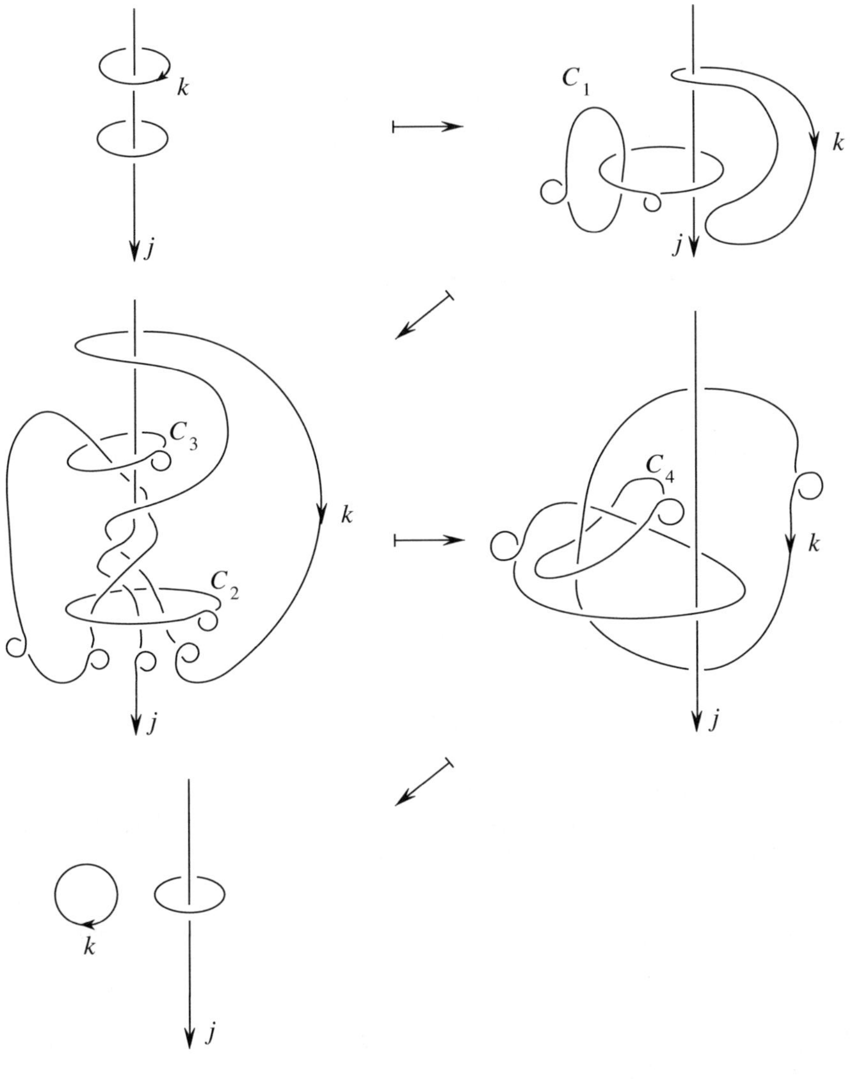

Figure 3.6

ribbon tangle. Consider the associated morphism $V_j \to V_j$ and multiply it by $d_{r_1} d_{r_2} ... d_{r_n}$. Summing up the resulting morphisms over all sequences $r_1, ..., r_n \in I$ we get a morphism $V_j \to V_j$. For instance, the morphisms obtained in this way from the first and last diagrams in Figure 3.6 are equal respectively to ω and

$$\dim(k) \sum_{i \in I} d_i F(H^i_j) = \dim(k) \sum_{i \in I} d_i S_{i,j} (\dim(j))^{-1} \mathrm{id}_{V_j} =$$
$$= e_j \dim(k) (\dim(j))^{-1} \mathrm{id}_{V_j}.$$

The arguments used to prove Lemma 3.2.1 show that the morphism constructed in this way is preserved under the Kirby 1-moves. (Instead of the proof of Lemma 3.2.1 we may refer to the result of this Lemma, it suffices to pass from the tangles in Figure 3.6 to their closures as defined in Section I.2.7.) Therefore all tangle diagrams in Figure 3.6 give rise to the same morphism. Hence,

$$\omega = e_j \dim(k)(\dim(j))^{-1}\mathrm{id}_{V_j}.$$

Comparing this formula with (3.4.a) we get

$$e_j = e_j S_{j,k}(\dim(j)\dim(k))^{-1} = e_j S'_{j,k}$$

where $S'_{j,k} = S_{j,k}(\dim(j)\dim(k))^{-1}$. The non-degeneracy axiom in the definition of modular categories implies that the matrix $S' = [S'_{j,k}]_{j,k\in I}$ is invertible. Denote the inverse matrix by S''. Since the row $\{S'_{0,k}\}_k$ of S' consists of unities, we have

$$\sum_{k\in I} S''_{k,0} = 1.$$

Therefore

$$e_j = \sum_{k\in I} e_j S''_{k,0} = \sum_{k\in I} e_j S'_{j,k} S''_{k,0} = \delta_0^j e_0.$$

The formula $e_0 = x$ follows from definitions.

3.5. Lemma. *If $r \neq j^*$ then* $\mathrm{Hom}(\mathbb{1}, V_j \otimes V_r) = \mathrm{Hom}(V_j \otimes V_r, \mathbb{1}) = 0$. *If $r = j^*$ and $q : V_j^* \to V_r$ is an isomorphism then the module* $\mathrm{Hom}(\mathbb{1}, V_j \otimes V_r)$ *is generated by* $(\mathrm{id}_{V_j} \otimes q)b_{V_j}$ *and the module* $\mathrm{Hom}(V_j \otimes V_r, \mathbb{1})$ *is generated by* $d^-_{V_j}(\mathrm{id}_{V_j} \otimes q^{-1})$. *Here b_{V_j} is the duality morphism $\mathbb{1} \to V_j \otimes V_j^*$ and $d^-_{V_j}$ is the operator invariant $F(\cap^-_{V_j}) : V_j \otimes V_j{}^* \to \mathbb{1}$ of the left-oriented cap colored with V_j.*

Proof. Using the graphical calculus it is easy to see that any $f \in \mathrm{Hom}(\mathbb{1}, V_j \otimes V_r)$ may be presented in the form $f = (\mathrm{id}_{V_j} \otimes h_f)b_{V_j}$ where

$$h_f = (d_{V_j} \otimes \mathrm{id}_{V_r})(\mathrm{id}_{V_j^*} \otimes f) : V_j^* \to V_r.$$

(cf. Exercise I.1.8.1). Therefore the formula $h \mapsto (\mathrm{id}_{V_j} \otimes h)b_{V_j}$ defines a surjection $\mathrm{Hom}(V_j^*, V_r) \to \mathrm{Hom}(\mathbb{1}, V_j \otimes V_r)$. If $r \neq j^*$ then $\mathrm{Hom}(V_j^*, V_r) = 0$ by Lemma 1.5. If $r = j^*$ then $\mathrm{Hom}(V_j^*, V_r) = Kq$ because V_r is simple. This implies the claim of the lemma concerning $\mathrm{Hom}(\mathbb{1}, V_j \otimes V_r)$. The assertion concerning $\mathrm{Hom}(V_j \otimes V_r, \mathbb{1})$ is proven similarly.

3.6. Proof of Lemma 3.2.2, claims (ii) and (iii). The axiom of domination implies that the identity endomorphism of $V_j \otimes V_r$ may be presented as a sum of a finite number of compositions fg where f, g are certain morphisms $V_l \to V_j \otimes V_r$,

$V_j \otimes V_r \to V_l$ respectively. Therefore

$$\sum_{i \in I} d_i (\dim(i))^{-1} S_{j,i} S_{i,r} \doteq$$

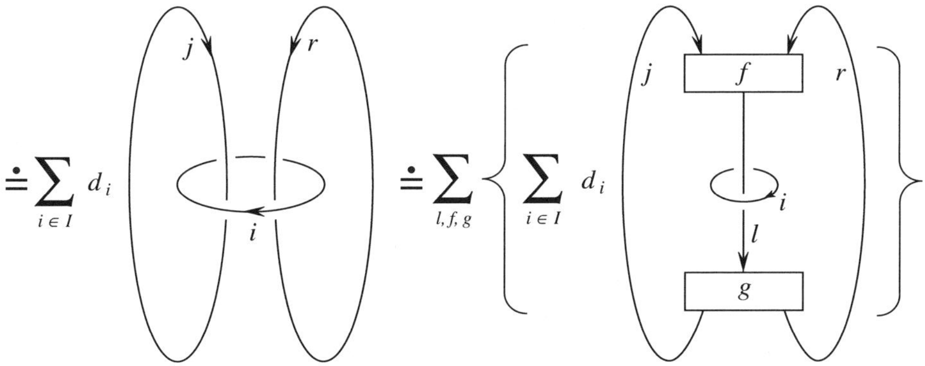

Figure 3.7

where in the last expression the summation goes over all triples $(l \in I, f, g)$ in question. Claim (i) of Lemma 3.2.2 implies that if $l \neq 0$ then the sum in parentheses is equal to zero. It remains to consider the summands corresponding to $l = 0$. Lemma 3.5 implies that in the case $r \neq j^*$ all these summands are equal to zero. This yields (3.2.d) for $r \neq j^*$. Let $r = j^*$. Lemma 3.5 shows that the morphisms f, g corresponding to $l = 0$ are proportional to $(\mathrm{id}_{V_j} \otimes q) b_{V_j}$ and $d_{\overline{V}_j}(\mathrm{id}_{V_j} \otimes q^{-1})$ where q is an isomorphism $V_j^* \to V_r$. Therefore without loss of generality we may assume that the decomposition of the identity endomorphism of $V_j \otimes V_r$ involves only one pair $(f : \mathbb{1} \to V_j \otimes V_r,\, g : V_j \otimes V_r \to \mathbb{1})$ corresponding to $l = 0$. We have

(3.6.a) $$f = k(\mathrm{id}_{V_j} \otimes q) b_{V_j} \text{ and } g = k' d_{\overline{V}_j}(\mathrm{id}_{V_j} \otimes q^{-1})$$

with $k, k' \in K$. Let us show first that $kk' = (\dim(j))^{-1}$. Composing this decomposition of the identity endomorphism of $V_j \otimes V_r$ from the left with $d_{\overline{V}_j}(\mathrm{id}_{V_j} \otimes q^{-1})$ and from the right with $(\mathrm{id}_{V_j} \otimes q) b_{V_j}$ we get a decompositon of $d_{\overline{V}_j} b_{V_j} : \mathbb{1} \to \mathbb{1}$ into a sum of several morphisms $\mathbb{1} \to \mathbb{1}$ passing through certain $\{V_l\}_l$ with $l \neq 0$ and the morphism

$$d_{\overline{V}_j}(\mathrm{id}_{V_j} \otimes q^{-1}) f g (\mathrm{id}_{V_j} \otimes q) b_{V_j} = kk'(d_{\overline{V}_j} b_{V_j})^2 : \mathbb{1} \to \mathbb{1}.$$

The endomorphisms of $\mathbb{1}$ of the first type are equal to zero. Hence $d_{\overline{V}_j} b_{V_j} = kk'(d_{\overline{V}_j} b_{V_j})^2$. It follows from definitions that $d_{\overline{V}_j} b_{V_j} = \dim(j)$. Therefore $kk' = (\dim(j))^{-1}$.

In the case $l = 0$ the l-colored band in Figure 3.7 may be safely erased. Substituting the expressions (3.6.a) for f and g we easily compute that the morphism in $\mathcal{V}$ presented by the right picture in Figure 3.7 equals

$$kk'\dim(j)\sum_{i\in I} d_i\dim(i) = x.$$

This completes the proof of (3.2.d).

The claim (iii) of the lemma follows from the claim (ii): multiplying both sides of (3.2.a) by

$$v_j^{-1}v_r^{-1}d_j(\dim(j))^{-1}S_{j,r},$$

summing up over all $j \in I$, and applying (3.2.d) and (3.3.b) we get a formula equivalent to (3.2.e).

3.7. Proof of Lemma 3.2.3. Let Ω^j be the ribbon graph in S^3 consisting of one unknotted untwisted annulus with a color $j \in I$. We may consider S^3 as the result of surgery on S^3 along the empty link, and use this presentation to compute $\tau'(S^3, \Omega^j)$. It follows from the definition of τ' that

$$\tau'(S^3, \Omega^j) = \{\emptyset, \Omega^j\}' = \dim(j).$$

We may obtain the same pair (S^3, Ω^j) by surgery on S^3 along the pair (L, Ω) shown in Figure 3.8. Here L is an unknotted circle with framing -1, and Ω is an annulus with twisting number -1 and color j. It is clear that $\sigma_-(L) = 1$. Therefore

$$\tau'(S^3, \Omega^j) = (xd_0)^{-1}\{L, \Omega\}' = (xd_0)^{-1}\sum_{i\in I} d_i v_i^{-1}v_j^{-1}S_{i,j}.$$

Comparing these two expressions for $\tau'(S^3, \Omega^j)$ we get

$$\sum_{i\in I} d_i v_i^{-1}v_j^{-1}S_{i,j} = xd_0\dim(j).$$

Combining the last formula with (3.2.e) we get $xd_j = xd_0\dim(j)$. Since x is inverible in K we conclude that $d_j = d_0\dim(j)$.

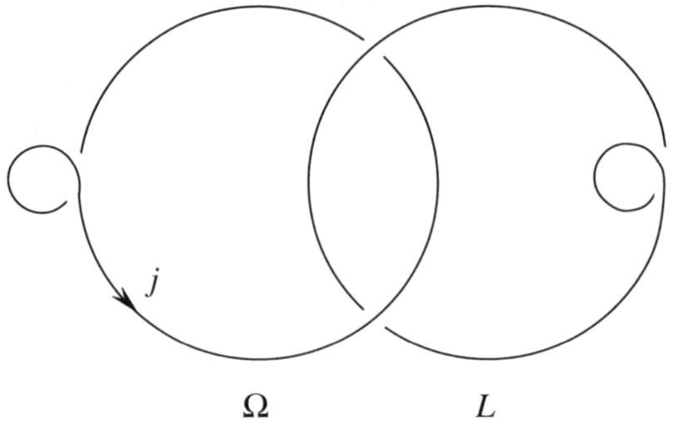

Figure 3.8

3.8. Corollaries. Theorem 2.3.2 and its proof allow us to establish several useful identities involving the matrix S and the elements $\mathcal{D}$, Δ, $\{v_i\}_{i\in I}$ of K. It follows directly from (3.2.d), (3.2.h), and the formulas $d_0 = \Delta\mathcal{D}^{-2} = x\mathcal{D}^{-2}$ that for any $r, j \in I$, we have

$$\sum_{i\in I} S_{j,i}\, S_{i,r} = \mathcal{D}^2 \delta^r_{j^*}. \tag{3.8.a}$$

In other words $S^2 = \mathcal{D}^2[\delta^r_{j^*}]_{j,r}$. Note the following curious corollary of this formula: $(\det(S))^2 = \mathcal{D}^{2\,\mathrm{card}(I)}$. Substituting $r = 0 \in I$ in (3.8.a) we get

$$\sum_{i\in I} \dim(i)\, S_{i,j} = \mathcal{D}^2 \delta^0_j. \tag{3.8.b}$$

Another important identity is given in the following corollary.

3.8.1. Corollary. *For any $r, j \in I$, we have*

$$\Delta\mathcal{D}^{-2} \sum_{i\in I} v_i S_{j,i} S_{i,r} = v_j^{-1} v_r^{-1} S_{j,r^*}. \tag{3.8.c}$$

Substituting here $j = 0 \in I$ we get

$$\Delta\mathcal{D}^{-2} \sum_{i\in I} v_i \dim(i) S_{i,r} = v_r^{-1} \dim(r). \tag{3.8.d}$$

Proof of Corollary. Let H_{j,r^*} be the Hopf link drawn in Figure I.2.12 with V and W being V_j and V_{r^*} respectively. We compute $\tau(S^3, H_{j,r^*})$ in two different ways. We have

$$\tau(S^3, H_{j,r^*}) = \mathcal{D}^{-1} F(H_{j,r^*}) = \mathcal{D}^{-1} S_{j,r^*}. \tag{3.8.e}$$

The pair (S^3, H_{j,r^*}) may be obtained by surgery on S^3 along the pair (L, Ω) drawn in Figure 3.9. Here L is a trivial knot with framing 1 linking two other components which form Ω. It is easy to see that the Kirby 1-move along the only component of L transforms (L, Ω) into $(\emptyset, H_{j,r^*})$. Therefore the surgery

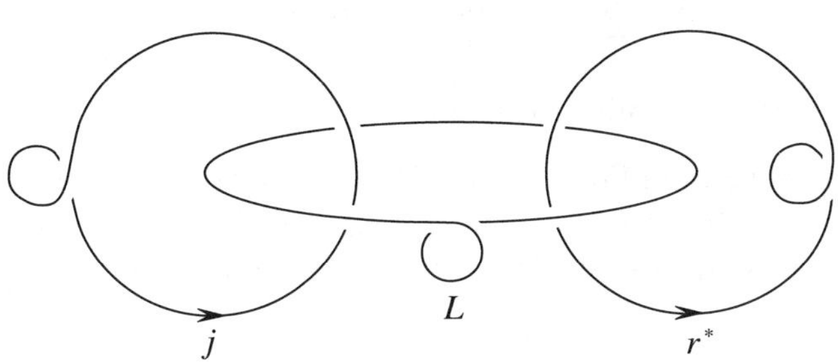

Figure 3.9

on S^3 along (L, Ω) produces (S^3, H_{j,r^*}) and we may employ (L, Ω) to compute $\tau(S^3, H_{j,r^*})$. To this end we use (1.5.a) and Corollary I.2.8.1. This gives

$$\tau(S^3, H_{j,r^*}) = \Delta\mathcal{D}^{-3} \sum_{i\in I} v_i v_j v_r S_{j,i} S_{i,r}.$$

Comparing this formula with (3.8.e) we get (3.8.c).

3.9. Action of $SL(2, \mathbb{Z})$. The matrix $S = [S_{i,j}]_{i,j\in I}$ plays an important role in the definition of modular categories. Here we show that S may be used to construct a projective linear representation of the matrix group $SL(2, \mathbb{Z})$. This enlightens the role of the non-degeneracy axiom (1.4.4).

Consider the diagonal square matrix $T = [\delta_i^j v_i]_{i,j\in I}$ and the involutive square matrix $J = [\delta_{i^*}^j]_{i,j\in I}$ of order card(I). The rows and columns of these matrices are numerated by elements of the set I. It is easy to verify that both $S = [S_{i,j}]_{i,j\in I}$ and T commute with J. The formulas (3.8.a) and (3.8.c) may be rewritten as follows: $S^2 = \mathcal{D}^2 J$ and $\Delta\mathcal{D}^{-2} STS = T^{-1} SJT^{-1}$.

The group $SL(2, \mathbb{Z})$ is generated by two matrices

$$s = \begin{bmatrix} 0 & -1 \\ 1 & 0 \end{bmatrix}, \quad t = \begin{bmatrix} 1 & 1 \\ 0 & 1 \end{bmatrix}.$$

Here is a generating system of relations: $s^4 = 1$, $(ts)^3 = s^2$. The assignements $s \mapsto \mathcal{D}^{-1} S$ and $t \mapsto T^{-1}$ define a projective linear representation of $SL(2, \mathbb{Z})$ in the group of invertible matrices over K of order card(I). Indeed, $(\mathcal{D}^{-1}S)^4 = 1$ and

$$(\mathcal{D}^{-1}T^{-1}S)^3 = \mathcal{D}^{-3}(T^{-1}SJT^{-1})J^{-1}ST^{-1}S = \Delta\mathcal{D}^{-5} STSJ^{-1}ST^{-1}S =$$

$$= \Delta\mathcal{D}^{-3} S^2 = \Delta\mathcal{D}^{-1}(\mathcal{D}^{-1}S)^2.$$

Because of the factor $\Delta\mathcal{D}^{-1}$ we get a projective representation and not just a linear one. The nature of this representation will become clearer in Chapter IV.

3.10. Exercises. 1. Prove that for any $r \in I$,

$$\Delta^{-1} \sum_{i\in I} v_i^{-1} \dim(i)\, S_{i,r} = v_r \dim(r). \tag{3.10.a}$$

Hint: apply (3.8.d) to the mirror category or use (3.2.e).

2. Verify the equality in Figure 3.10 where i is an arbitrary element of I. Show that if the color of the left vertical strand is replaced with $j \neq i$ then the sum on the left-hand side is equal to 0. (Hint: use the same argument as in Section 3.6.)

3. Verify that the projective linear representation of $SL(2, \mathbb{Z})$ determined by the mirror modular category $\overline{\mathcal{V}}$ transforms the generators s, t into $\mathcal{D}^{-1}SJ = [\mathcal{D}^{-1}S(i, j^*)]_{i,j\in I}$ and T respectively.

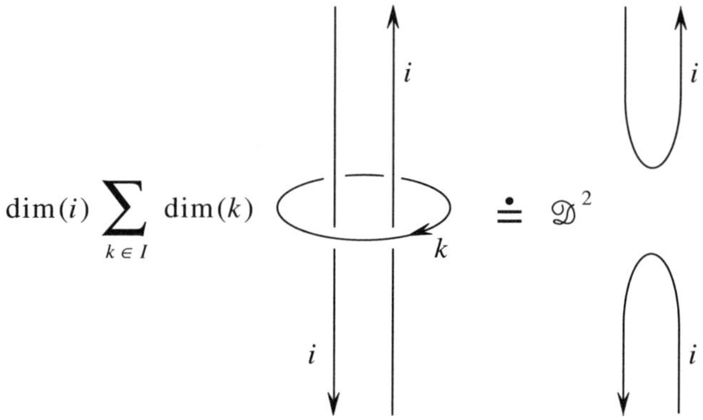

Figure 3.10

4. Computations in semisimple categories

4.0. Outline. We introduce semisimple (ribbon) categories generalizing modular categories. The main difference is that the family of simple objects dominating a semisimple category may be infinite. Semisimple categories share many algebraic features of modular categories and form a natural setting for the study of modules of morphisms, multiplicities, $6j$-symbols, etc.

Semisimple categories, being ribbon categories, give rise to invariants of links and ribbon graphs in $\mathbb{R}^3$. However, they do not lead to invariants of 3-manifolds. This makes them less interesting for a topologist than modular categories. The reader interested mainly in 3-manifold invariants may restrict himself everywhere to modular categories.

We establish here a number of useful results concerned with modules of morphisms in semisimple and modular categories. In particular, we establish the Verlinde-Moore-Seiberg formula computing multiplicities of tensor products in terms of the matrix S. This and other formulas obtained in this section will be used in Chapters VII and X.

4.1. Semisimple categories. Usually, by a semisimple category one means an abelian category whose objects split as direct sums of simple objects. We shall use the term "semisimple category" in a related but different sense. We restrict ourselves to ribbon categories and do not involve direct sums of objects. Here is a precise definition. A semisimple category is a pair consisting of a ribbon Ab-category $\mathcal{V}$ and a family $\{V_i\}_{i\in I}$ of simple objects of $\mathcal{V}$ satisfying axioms (1.4.1), (1.4.2), (1.4.3), and the following axiom:

(4.1.1) (Schur's axiom). For any distinct $i, j \in I$, we have $\mathrm{Hom}(V_i, V_j) = 0$.

It follows from Lemma 1.5 that all modular categories are semisimple. It is obvious that a semisimple category $(\mathcal{V}, \{V_i\}_{i\in I})$ is modular if and only if the set I is finite and axiom (1.4.4) is fulfilled. For instance, the category constructed in Section I.1.7.2 with infinite G is semisimple but not modular.

A part of results pertaining to modular categories directly extends to semisimple categories. It follows from the axioms that for any distinct $i, j \in I$, the objects V_i, V_j are not isomorphic. Therefore for each $i \in I$, there exists exactly one $i^* \in I$ such that V_{i^*} is isomorphic to V_i^*. The formula $i \mapsto i^*$ defines an involution in the set I. Corollary I.2.6.2 implies that $0^* = 0$. We shall use the notation $\dim(i) = \dim(V_i)$ and $S_{i,j} = \mathrm{tr}(c_{V_j,V_i} \circ c_{V_i,V_j})$ for $i, j \in I$.

Semisimple categories over a field (considered up to the choice of objects $\{V_i\}_{i\in I}$ in their isomorphism classes) may be invariantly characterized as ribbon Ab-categories over a field such that all objects are dominated by simple ones, and Schur's axiom (4.1.1) is fulfilled (cf. Remark 1.8).

A semisimple category is called strict if its underlying monoidal category is strict.

4.2. Modules of morphisms. Let $(\mathcal{V}, \{V_i\}_{i\in I})$ be a semisimple category with ground ring K. For objects V, W of $\mathcal{V}$, we may consider the K-module $\mathrm{Hom}(V, W)$. The next four lemmas establish basic properties of these modules.

4.2.1. Lemma. *For any objects V, W of $\mathcal{V}$, the K-module* $\mathrm{Hom}(V, W)$ *is projective. For any object W of $\mathcal{V}$, there is only a finite number of $i \in I$ such that* $\mathrm{Hom}(V_i, W) \neq 0$.

Recall that by projective modules we mean finitely generated projective modules.

Proof of Lemma. By the axiom of domination, the object W is dominated by the family $\{V_i\}_{i\in I}$. Let $\{V_{i(r)}\}_r$ be a finite set consisting of objects of this family (possibly with repetitions) and let $\{f_r : V_{i(r)} \to W, g_r : W \to V_{i(r)}\}_r$ be a family of morphisms such that $\mathrm{id}_W = \sum_r f_r g_r$.

Choose $i \in I$. Set $H_i = \oplus_r \mathrm{Hom}(V_i, V_{i(r)})$ and $J_i = \mathrm{Hom}(V_i, W)$. Consider the K-linear homomorphisms

$$h \mapsto \oplus_r(g_r h) : J_i \to H_i \quad \text{and} \quad \oplus_r h_r \mapsto \sum_r f_r h_r : H_i \to J_i.$$

The composition $J_i \to J_i$ of these two homomorphisms transforms $h \in J_i$ into $\sum_r f_r g_r h = \mathrm{id}_W h = h$. Therefore these homomorphisms are nothing but an embedding of J_i into H_i and a left inverse to this embedding. Thus, J_i is a direct summand of H_i. If $i \neq i(r)$ for all r then $H_i = 0$ and therefore $J_i = 0$. This proves the second claim of the lemma.

Let us prove the first claim. It follows from the axioms that $\mathrm{Hom}(V_0, V_{i(r)}) = 0$ if $i(r) \neq 0$ and $\mathrm{Hom}(V_0, V_{i(r)}) = K$ if $i(r) = 0$. Thus, $H_0 = \oplus_r \mathrm{Hom}(V_0, V_{i(r)})$ is a free K-module of finite rank and its direct summand $J_0 = \mathrm{Hom}(V_0, W)$ is a projective module. This proves the first claim of the lemma in the case $V = V_0 = \mathbb{1}$. The case of arbitrary V follows from the equalities

$$\mathrm{Hom}(V, W) = \mathrm{Hom}(\mathbb{1} \otimes V, W) = \mathrm{Hom}(\mathbb{1}, W \otimes V^*)$$

(cf. Exercise I.1.8.1).

4.2.2. Lemma. *Let V, W be objects of $\mathcal{V}$. For each $i \in I$, the composition of morphisms defines a K-homomorphism*

$$\mathrm{Hom}(V, V_i) \otimes_K \mathrm{Hom}(V_i, W) \to \mathrm{Hom}(V, W). \tag{4.2.a}$$

The direct sum of these homomorphisms is an isomorphism

$$\bigoplus_{i \in I} (\mathrm{Hom}(V, V_i) \otimes_K \mathrm{Hom}(V_i, W)) \to \mathrm{Hom}(V, W). \tag{4.2.b}$$

Proof. Denote the homomorphisms (4.2.a) and (4.2.b) by u_i and u respectively. The semisimplicity of $\mathcal{V}$ implies that the identity endomorphism of W can be decomposed into a sum of morphisms passing through $\{V_i\}_{i \in I}$. Every morphism $f : V \to W$ equals $\mathrm{id}_W f$ and therefore can be decomposed into a sum of morphisms passing through $\{V_i\}_{i \in I}$. It is clear that each composition $V \to V_i \to W$ belongs to the image of u_i. This implies surjectivity of u.

Let us prove that every homomorphism u_j with $j \in I$ is injective. Fix a finite set $\{V_{i(r)}\}_r$ of objects of the family $\{V_i\}_{i \in I}$ (possibly with repetitions) and fix a family of morphisms $\{f_r : V_{i(r)} \to W, g_r : W \to V_{i(r)}\}_r$ such that $\mathrm{id}_W = \sum_r f_r g_r$. Here r runs over a certain finite set R.

Let $\{\alpha_q : V \to V_j, \beta_q : V_j \to W\}_q$ be an arbitrary (finite) family of morphisms such that $\sum_q \alpha_q \otimes \beta_q \in \mathrm{Ker}(u_j)$. In other words $\sum_q \beta_q \alpha_q = 0 \in \mathrm{Hom}(V, W)$. For any q, we have

$$\beta_q = (\sum_{r \in R} f_r g_r) \beta_q = \sum_{r \in R, i(r) = j} f_r g_r \beta_q.$$

Note that when $i(r) = j$ the composition $g_r \beta_q : V_j \to V_j$ is multiplication by a certain $z(r, q) \in K$. Thus,

$$\beta_q = \sum_{r \in R, i(r) = j} z(r, q) f_r.$$

Therefore

$$\sum_q \alpha_q \otimes \beta_q = \sum_q \sum_{r\in R, i(r)=j} z(r,q)\alpha_q \otimes f_r = \sum_q \sum_{r\in R, i(r)=j} g_r\beta_q\alpha_q \otimes f_r.$$

The assumption $\sum_q \beta_q\alpha_q = 0$ implies that the last expression equals 0. Hence $\mathrm{Ker}(u_j) = 0$.

Now we can prove injectivity of u. Let $h \in \mathrm{Ker}(u)$. Present h as a sum $\sum_{j\in I} h_j$ where $h_j \in \mathrm{Hom}(V, V_j) \otimes_K \mathrm{Hom}(V_j, W)$. It suffices to show that $g_r \circ u_j(h_j) = 0$ for all j, r. Indeed, this would imply that $u_j(h_j) = \mathrm{id}_W \circ u_j(h_j) = 0$ and since u_j is injective we would have $h_j = 0$ for all j.

We have $0 = u(h) = \sum_{j\in I} u_j(h_j)$. If $i(r) \neq j$ then $g_r \circ u_j(h_j) = 0$ since any morphism $V_j \to V_{i(r)}$ is equal to zero. If $i(r) = j$ then

$$g_r \circ u_j(h_j) = g_r \circ (u(h) - \sum_{i\in I, i\neq j} u_i(h_i)) = g_r \circ u(h) - \sum_{i\in I, i\neq j} g_r \circ u_i(h_i) = 0.$$

This completes the proof of the lemma.

4.2.3. Lemma. *For any objects V, W of $\mathcal{V}$, the pairing $(x, y) \mapsto \mathrm{tr}(y \circ x)$: $\mathrm{Hom}(V, W) \otimes_K \mathrm{Hom}(W, V) \to K$ is non-degenerate.*

We say that a K-bilinear pairing $H_1 \otimes_K H_2 \to K$ is non-degenerate if the adjoint homomorphisms $H_1 \to \mathrm{Hom}(H_2, K)$ and $H_2 \to \mathrm{Hom}(H_1, K)$ are isomorphisms. Recall the notation $H^\star = \mathrm{Hom}(H, K)$ where H is a K-module and $\mathrm{Hom}(H, K)$ is considered as a K-module in the usual way.

Proof of Lemma. Set $H_1 = \mathrm{Hom}(V, W)$ and $H_2 = \mathrm{Hom}(W, V)$. Denote by w the homomorphism $H_1 \to H_2^\star$ adjoint to the pairing $(x, y) \mapsto \mathrm{tr}(yx) : H_1 \otimes_K H_2 \to K$. We prove below that w is an isomorphism. Exchanging the roles of V and W we get that the homomorphism $H_2 \to H_1^\star$ adjoint to the same pairing is also an isomorphism.

Consider first the case $V = \mathbb{1}$ so that $H_1 = \mathrm{Hom}(\mathbb{1}, W)$ and $H_2 = \mathrm{Hom}(W, \mathbb{1})$. Present id_W in the form $\mathrm{id}_W = \sum_r f_r g_r$ where r runs over a finite set R, and $f_r : V_{i(r)} \to W$, $g_r : W \to V_{i(r)}$ are certain morphisms with $i(r) \in I$. For $x \in H_1$, we have

$$x = \sum_r f_r g_r x = \sum_{r, i(r)=0} f_r g_r x = \sum_{r, i(r)=0} f_r\, \mathrm{tr}(g_r x).$$

Here the second equality follows from the Schur axiom (4.1.1) and the third equality follows from Lemma I.1.5.1. Therefore, if $x \neq 0$ then $\mathrm{tr}(gx) \neq 0$ for a certain $g \in H_2$. This yields injectivity of w. Let us prove surjectivity. Let $z \in H_2^\star$. Set

$$x = \sum_{r, i(r)=0} z(g_r) f_r \in H_1.$$

We shall prove that $w(x) = z$. It suffices to show that $\mathrm{tr}(yx) = z(y)$ for any $y \in H_2$. We have

$$\mathrm{tr}(yx) = \sum_{r,i(r)=0} z(g_r)\,\mathrm{tr}(yf_r) = \sum_{r,i(r)=0} z(\mathrm{tr}(yf_r)g_r) =$$

$$= \sum_{r,i(r)=0} z(yf_rg_r) = z(y \sum_{r,i(r)=0} f_rg_r) = z(y).$$

In the case of arbitrary V we have canonical isomorphisms

$$\alpha : H_1 \to \mathrm{Hom}(\mathbb{1}, W \otimes V^*), \quad \beta : H_2 \to \mathrm{Hom}(W \otimes V^*, \mathbb{1})$$

(see Exercises I.1.8.1 and I.2.9.3). It is easy to deduce from the geometric interpretation of the trace that for any $x \in H_1$, $y \in H_2$, we have $\mathrm{tr}(yx) = \mathrm{tr}(\beta(y)\alpha(x))$. Therefore the general case of Lemma follows from the case $V = \mathbb{1}$.

4.2.4. Lemma. *For any simple object V of $\mathcal{V}$, its dimension* $\dim(V)$ *is invertible in K.*

Proof. With respect to the generator id_V of $\mathrm{Hom}(V, V) = K$ the pairing $(x, y) \mapsto \mathrm{tr}(yx) : \mathrm{Hom}(V, V) \otimes_K \mathrm{Hom}(V, V) \to K$ is presented by the 1×1-matrix $[\mathrm{tr}(\mathrm{id}_V)]$. Non-degeneracy of this pairing implies that $\dim(V) = \mathrm{tr}(\mathrm{id}_V)$ is invertible in K.

4.3. Dimension of modules of morphisms. Recall the dimension Dim of projective modules over a commutative ring introduced in Section I.1.7.1. Lemma I.1.5.1 implies that Dim is multiplicative with respect to tensor multiplication of modules. Corollary I.2.8.2 implies that $\mathrm{Dim}(X^\star) = \mathrm{Dim}(X)$ for any projective module X. It is easy to deduce from definitions that Dim is additive with respect to the direct sum of modules. At the end of this subsection we shall show that the dimension Dim and the trace Tr of endomorphisms of projective modules, introduced in Section I.1.7.1, generalize the usual dimension and trace of endomorphisms for free modules.

Let $(\mathcal{V}, \{V_i\}_{i\in I})$ be a semisimple category with ground ring K. Lemma 4.2.1 allows us to consider for any objects V, W of $\mathcal{V}$ the dimension $\mathrm{Dim}\,(\mathrm{Hom}(V, W)) \in K$. Lemma 4.2.3 implies that $\mathrm{Dim}\,(\mathrm{Hom}(V, W)) = \mathrm{Dim}\,(\mathrm{Hom}(W, V))$.

For any $i \in I$ and any object W of $\mathcal{V}$, set

$$\mathrm{Dim}_i(W) = \mathrm{Dim}\,(\mathrm{Hom}(V_i, W)) = \mathrm{Dim}\,(\mathrm{Hom}(W, V_i)) \in K.$$

By Lemma 4.2.1, for any W, there is only a finite number of $i \in I$ such that $\mathrm{Dim}_i(W) \neq 0$. Lemma 4.2.2 implies that for any objects V, W,

$$\mathrm{Dim}\,(\mathrm{Hom}(V, W)) = \sum_{i\in I} \mathrm{Dim}_i(V)\,\mathrm{Dim}_i(W).$$

4.3.1. Lemma. *For any integer $n \geq 0$, we have* $\mathrm{Dim}(K^n) = n \cdot 1_K \in K$. *For any K-linear homomorphism $f : K^n \to K^n$, the trace* $\mathrm{Tr}(f)$ *is the standard trace of f defined as the sum of the diagonal entries of a matrix of f.*

We note that the distinction between $n \in \mathbb{Z}$ and $n \cdot 1_K \in K$ may be essential when K has non-zero characteristic.

Proof of Lemma. Set $X = K^n$. Since $\mathrm{Dim}(X) = \mathrm{Tr}(\mathrm{id}_X)$ it suffices to prove the part of the lemma concerned with Tr. Let f be an endomorphism of X and let $\{f_{rs}\}$ be the matrix of f with respect to a basis $e_1, \ldots, e_n$ of X so that $f(e_r) = \sum_s f_{rs} e_s$ for all $r = 1, \ldots, n$. Let $e_1^\star, \ldots, e_n^\star$ be the dual basis of $X^\star = \mathrm{Hom}_K(X, K)$. By definition $\mathrm{Tr}(f) = d_X P_{X,X^\star}(f \otimes \mathrm{id}_{X^\star}) b_X$. Here the homomorphism $b_X : K \to X \otimes X^\star$ transforms 1 into $\sum_{r=1}^n e_r \otimes e_r^\star$, the homomorphism $P_{X,X^\star} : X \otimes X^\star \to X^\star \otimes X$ is the flip $x \otimes y \mapsto y \otimes x$, and the homomorphism $d_X : X^\star \otimes X \to K$ transforms $e_r^\star \otimes e_s$ into δ_s^r. A direct computation shows that $\mathrm{Tr}(f) = \sum_{r=1}^n f_{rr}$.

4.4. Lemma. *Let $(\mathcal{V}, \{V_i\}_{i \in I})$ be a strict semisimple category. Let Ω be a v-colored ribbon $(0,0)$-graph in S^3 over $\mathcal{V}$ containing an annulus component ℓ of color V. Let Ω^k be the same v-colored ribbon graph with the color of ℓ replaced by V_k with $k \in I$. Then*

$$F(\Omega) = \sum_{k \in I} \mathrm{Dim}_k(V)\, F(\Omega^k).$$

The sum on the right-hand side contains only a finite number of non-zero summands and is well-defined even when the set I is infinite. Applying Lemma 4.4 to the trivial knot in $\mathbb{R}^3$ colored with an object V of $\mathcal{V}$ we obtain

$$\mathrm{dim}(V) = \sum_{k \in I} \mathrm{Dim}_k(V)\, \mathrm{dim}(k). \tag{4.4.a}$$

Proof of Lemma. Let $\{V_{i(r)}\}_{r \in R}$ be a finite set of objects of the family $\{V_i\}_{i \in I}$ and let $\{f_r : V_{i(r)} \to V, g_r : V \to V_{i(r)}\}_{r \in R}$ be a family of morphisms such that $\mathrm{id}_V = \sum_{r \in R} f_r g_r$. For each $r \in R$, the morphism $g_r f_r : V_{i(r)} \to V_{i(r)}$ is multiplication by a certain element of K. Denote this element by z_r.

Choose $k \in I$. Denote by R_k the subset of the set R consisting of $r \in R$ such that $i(r) = k$. Let us prove that

$$\mathrm{Dim}_k(V) = \sum_{r \in R_k} z_r. \tag{4.4.b}$$

Note first that the K-module $H_k = \oplus_{r \in R} \mathrm{Hom}(V_k, V_{i(r)})$ is free; it has a basis numerated by elements of R_k. As in the proof of Lemma 4.2.1 the formula $h \mapsto \oplus_{r \in R} g_r h$ defines a K-linear embedding $\mathrm{Hom}(V_k, V) \to H_k$, say α_k, and the formula $\oplus_{r \in R} h_r \mapsto \sum_{r \in R} f_r h_r$ defines its left inverse $H_k \to \mathrm{Hom}(V_k, V)$, say

β_k. Therefore

$$\mathrm{Dim}_k(V) = \mathrm{Dim}\,(\mathrm{Hom}(V_k, V)) = \mathrm{Tr}(\mathrm{id}_{\mathrm{Hom}(V_k,V)}) = \mathrm{Tr}(\beta_k\alpha_k) = \mathrm{Tr}(\alpha_k\beta_k).$$

Lemma 4.3.1 allows us to compute the trace Tr of $\alpha_k\beta_k : H_k \to H_k$ in the usual way using the basis of H_k mentioned above. It is easy to see that $\mathrm{Tr}(\alpha_k\beta_k) = \sum_{r\in R_k} z_r$. This implies (4.4.b).

To compute $F(\Omega)$ we use the same trick as in the proof of Corollaries I.2.8.1 and I.2.8.3. Namely, we take a vertical strand of ℓ directed downwards and replace a small segment of this strand by a coupon colored with id_V. This transformation does not change the operator invariant F of the ribbon graph. Using linearity of F with respect to colors of coupons and the equality $\mathrm{id}_V = \sum_{r\in R} f_r g_r$ we get $F(\Omega) = \sum_{r\in R} F(\Omega_r)$ where Ω_r is the v-colored ribbon graph obtained from Ω by replacing the same segment of ℓ with two coupons colored by f_r, g_r and a short vertical strand between them colored by $V_{i(r)}$ and directed downwards. Pushing the upper coupon (colored with f_r) along ℓ we deform Ω_r so that in the final position this coupon lies below the coupon colored with g_r. Denote the resulting v-colored ribbon graph by Ω'_r. Apart from the position of the two coupons on ℓ, the ribbon graph Ω'_r differs from Ω_r only in that the color of ℓ has been changed to $V_{i(r)}$. The isotopy invariance of F implies that $F(\Omega'_r) = F(\Omega_r)$. Finally we replace our two coupons on Ω'_r by one coupon colored with the composition $g_r f_r : V_{i(r)} \to V_{i(r)}$. Denote the resulting v-colored ribbon graph by Ω''_r. Clearly, $F(\Omega''_r) = F(\Omega'_r) = F(\Omega_r)$. On the other hand, $g_r f_r$ is multiplication by $z_r \in K$ and therefore $F(\Omega''_r) = z_r F(\Omega^{i(r)})$. (Recall that Ω^k is obtained from Ω by replacing the color of ℓ with V_k.) Combining these computations we get

$$F(\Omega) = \sum_{r\in R} F(\Omega''_r) = \sum_{r\in R} z_r F(\Omega^{i(r)}) = \sum_{k\in I}\sum_{r\in R_k} z_r F(\Omega^k) = \sum_{k\in I} \mathrm{Dim}_k(V) F(\Omega^k).$$

4.5. Applications to multiplicities. Let $(\mathcal{V}, \{V_i\}_{i\in I})$ be a strict semisimple category with ground ring K. For $i, j, k \in I$, the multiplicity of V_k in $V_i \otimes V_j$ is defined to be

$$h_k^{ij} = \mathrm{Dim}_k(V_i \otimes V_j) = \mathrm{Dim}\,(\mathrm{Hom}(V_k, V_i \otimes V_j)) \in K. \tag{4.5.a}$$

Multiplicities will play an important technical role in this book.

It is obvious that

$$h_k^{ij} = h_k^{ji} = h_{i^*}^{k^*j} = h_{i^*}^{jk^*} = h_{j^*}^{ik^*} = h_{j^*}^{k^*i}.$$

Lemma 4.2.1 shows that for given i, j, there exist only a finite number of $k \in I$ such that $h_k^{ij} \neq 0$. A more symmetric version of multiplicities is defined by the formula

$$h^{ijk} = \mathrm{Dim}(\mathrm{Hom}(\mathbb{1}, V_i \otimes V_j \otimes V_k)) \in K.$$

It is clear that h^{ijk} is an invariant of the unordered triple $\{i, j, k\}$ and that $h^{ijk} = h_{k^*}^{ij}$.

Multiplicities are closely related to the dimensions of objects of $\mathcal{V}$. A simple result in this direction may be obtained by application of formula (4.4.a) to $V = V_i \otimes V_j$. This gives

(4.5.b) $$\dim(i)\dim(j) = \sum_{k\in I} h_k^{ij}\dim(k).$$

Here is another useful formula: for any $i, j \in I$ and any object V of $\mathcal{V}$, we have

(4.5.c) $$\mathrm{Dim}_i(V_j \otimes V) = \sum_{k\in I} h_i^{jk}\mathrm{Dim}_k(V).$$

Indeed, Lemma 4.2.2 implies that

$$\mathrm{Hom}(V_i, V_j\otimes V) = \mathrm{Hom}(V_j^*\otimes V_i, V) = \bigoplus_{k\in I}\big(\mathrm{Hom}(V_j^* \otimes V_i, V_k) \otimes \mathrm{Hom}(V_k, V)\big).$$

Computing the dimension and taking into account that $\mathrm{Hom}(V_j^* \otimes V_i, V_k) = \mathrm{Hom}(V_i, V_j \otimes V_k)$ we get (4.5.c).

4.5.1. Theorem (the Verlinde-Moore-Seiberg formula). *Let $(\mathcal{V}, \{V_i\}_{i\in I})$ be a strict semisimple category. For any $i, j, r \in I$, we have*

(4.5.d) $$\sum_{k\in I} h_k^{ij} S_{k,r} = (\dim(r))^{-1} S_{i,r} S_{j,r}.$$

This formula generalizes formula (4.5.b) which corresponds to $r = 0$. Invertibility of $\dim(r)$ used in (4.5.d) follows from Lemma 4.2.4.

Proof of Theorem. Denote by L the colored 3-component link in S^3 with zero framing presented in Figure 4.1. It is easy to compute that $F(L) = (\dim(r))^{-1} S_{i,r} S_{j,r}$ (cf. the proof of Lemma 1.5). On the other hand, Corollary I.2.8.3 implies that $F(L) = F(L')$ where L' is the Hopf link with zero framing whose components are colored with V_r and $V_i \otimes V_j$. Lemma 4.4 implies that

$$F(L) = F(L') = \sum_{k\in I} h_k^{ij} S_{k,r}.$$

These equalities imply (4.5.d).

4.5.2. Theorem. *Let $(\mathcal{V}, \{V_i\}_{i\in I})$ be a modular category. Then for any $i, j, k \in I$,*

(4.5.e) $$h_k^{ij} = \mathcal{D}^{-2} \sum_{r\in I} (\dim(r))^{-1} S_{i,r} S_{j,r} S_{k^*,r}.$$

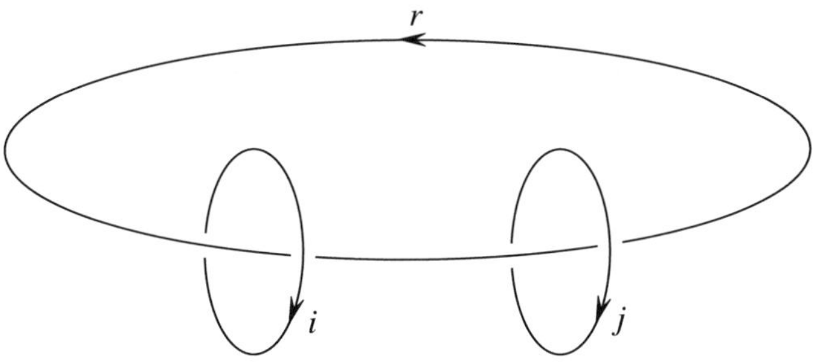

Figure 4.1

This equality implies the following more symmetric formula

$$h^{ijk} = \mathcal{D}^{-2} \sum_{r \in I} (\dim(r))^{-1} S_{i,r} S_{j,r} S_{k,r}.$$

In contrast to formulas (4.5.a) - (4.5.d), formula (4.5.e) can not be directly extended to semisimple categories with infinite I. Indeed, if the set I is infinite then the sum in (4.5.e) can involve an infinite number of non-trivial summands.

Proof of Theorem. Multiplying both sides of (4.5.d) by $\mathcal{D}^{-2} S_{r,k^*} = \mathcal{D}^{-2} S_{k^*,r}$, summing up over all $r \in I$, and taking into account (3.8.a) we get (4.5.e).

4.5.3. Theorem. *Let $(\mathcal{V}, \{V_i\}_{i \in I})$ be a modular category. Then for any $k \in I$,*

$$\mathcal{D}^{-2} \sum_{i,j \in I} h_k^{ij} \dim(i) \dim(j) = \dim(k). \tag{4.5.f}$$

This formula will be used in the proof of the invariance of state sums on triangulated 3-manifolds under subdivisions of triangulations (see Chapter VII).

Proof of Theorem. We have

$$\sum_{i \in I} h_k^{ij} \dim(i) = \sum_{i \in I} h_{i^*}^{jk^*} \dim(i^*) = \dim(j) \dim(k)$$

where the last formula follows from (4.5.b) and the formula $\dim(k^*) = \dim(k)$. Multiplying the resulting equality by $\dim(j)$ and summing up over all $j \in I$ we get a formula equivalent to (4.5.f).

4.5.4. Theorem. *Let $(\mathcal{V}, \{V_i\}_{i \in I})$ be a modular category. Then for any $k \in I$,*

$$\mathcal{D}^{-2} \sum_{i,j \in I} h_k^{ij} v_i v_j^{-1} \dim(i) \dim(j) = \delta_k^0. \tag{4.5.g}$$

This formula will play an important role in the study of state sum invariants of shadows in Chapter X. It will guarantee that these invariants are preserved under cobordisms of shadows.

Proof of Theorem. Multiplying both parts of (4.5.e) by $\mathcal{D}^{-2}v_j^{-1}\dim(j)$, summing up over all $j \in I$, and using consecutively (3.10.a) (with i replaced by j) and (3.8.c) we get

$$\mathcal{D}^{-2}\sum_{j\in I} h_k^{ij} v_j^{-1}\dim(j) = \Delta\mathcal{D}^{-4}\sum_{r\in I} v_r S_{i,r} S_{k^*,r} = \mathcal{D}^{-2} v_i^{-1} v_k^{-1} S_{i,k}.$$

Multiplying by $v_i\dim(i)$, summing up over all $i \in I$, and taking into account (3.8.b) we get (4.5.g).

4.6. Exercises. 1. Generalize Lemma 4.4 to ribbon graphs in 3-manifolds as follows. Let Ω be a v-colored ribbon graph in a closed oriented 3-manifold M over a modular category $(\mathcal{V}, \{V_i\}_{i\in I})$. Let ℓ be an annulus component of Ω of color V. For $k \in I$, denote by Ω^k the same v-colored ribbon graph with the color of ℓ replaced by V_k. Then $\tau(M,\Omega) = \sum_{k\in I} \mathrm{Dim}_k(V)\,\tau(M,\Omega^k)$.

2. Let $(\mathcal{V}, \{V_i\}_{i\in I})$ be a strict semisimple category. Show that for any object V of $\mathcal{V}$ and any $i \in I$, we have $\mathrm{Dim}_i(V^*) = \mathrm{Dim}_{i^*}(V)$. Show that for any $i, j, k \in I$, we have

$$h^{ijk} = h^{i^*j^*k^*}.$$

3. Let $(\mathcal{V}, \{V_i\}_{i\in I})$ be a strict semisimple category with finite set I. Let $\mathcal{D}$ be an invertible element of K such that (3.8.b) holds for each $j \in I$. Show that the matrix $S = [S_{i,j}]_{i,j}$ is invertible and that $(\mathcal{V}, \{V_i\}_{i\in I})$ is a modular category with rank $\mathcal{D}$. (Hint: use Lemma 4.4 to prove that

$$S_{j,i}S_{i,r} = \sum_{k\in I} h_k^{jr}\dim(i) S_{i,k},$$

then verify (3.8.a).)

5. Hermitian and unitary categories

5.0. Outline. In the framework of representation theory, Hermitian forms on modules V, W induce a pairing $(f, g) \mapsto \mathrm{tr}(f\overline{g})$ in $\mathrm{Hom}(V, W)$ where $\overline{g} : W \to V$ is adjoint to $g : V \to W$. It does not make sense to speak of Hermitian forms on objects of ribbon categories because these objects, in general, have no intrinsic linear structure. On the other hand, the pairing $(f, g) \mapsto \mathrm{tr}(f\overline{g})$ may be axiomatized in the setting of ribbon categories. Moreover, since there is a trace in any ribbon category, it suffices to axiomatize the involution $g \mapsto \overline{g}$. These considerations lead us to notions of a Hermitian ribbon category and a Hermitian modular category. In the case when the ground ring is $\mathbb{C}$ we define unitary modular categories.

The geometric part of this section is concerned with the invariant F of links and the invariant τ of 3-manifolds derived from Hermitian and unitary modular categories. For framed links in $\mathbb{R}^3$ colored over a unitary modular category, we establish an estimate of the absolute value of F. The principal result concerned with 3-manifolds says that the invariant τ derived from a Hermitian modular category conjugates under inversion of orientation.

The results of this section will be used in Sections IV.10 and IV.11 where we show that Hermitian (resp. unitary) modular categories give rise to Hermitian (resp. unitary) 3-dimensional TQFT's. The estimate mentioned above will be used in Section IV.12.7.

5.1. Hermitian ribbon categories. Let $\mathcal{V}$ be a (strict) monoidal category. A conjugation in $\mathcal{V}$ assigns to each morphism $f : V \to W$ in $\mathcal{V}$ a morphism $\overline{f} : W \to V$ so that the following identities hold:

$$\overline{\overline{f}} = f, \quad \overline{f \otimes g} = \overline{f} \otimes \overline{g}, \quad \overline{f \circ g} = \overline{g} \circ \overline{f}.$$

In the second formula f and g are arbitrary morphisms in $\mathcal{V}$ and in the third formula f and g are arbitrary composable morphisms in $\mathcal{V}$.

For any object V of $\mathcal{V}$, we have

$$\overline{\mathrm{id}_V} = \overline{\mathrm{id}_V}\mathrm{id}_V = \overline{\mathrm{id}_V}\,\overline{\overline{\mathrm{id}_V}} = \overline{\overline{\overline{\mathrm{id}_V}}\mathrm{id}_V} = \overline{\overline{\mathrm{id}_V}} = \mathrm{id}_V.$$

A Hermitian ribbon category is a ribbon monoidal category $\mathcal{V}$ endowed with a conjugation $f \mapsto \overline{f}$ satisfying the following conditions:

(5.1.1) for any objects V, W of $\mathcal{V}$, we have

$$\overline{c_{V,W}} = (c_{V,W})^{-1},$$

(5.1.2) for any object V of $\mathcal{V}$, we have

$$\overline{\theta_V} = (\theta_V)^{-1}, \quad \overline{b_V} = d_V c_{V,V^*}(\theta_V \otimes \mathrm{id}_{V^*}), \quad \overline{d_V} = (\mathrm{id}_{V^*} \otimes \theta_V^{-1})\, c_{V^*,V}^{-1} b_V.$$

We rewrite these formulas in the following more compact form:

$$\overline{c_{V,W}}c_{V,W} = \mathrm{id}_{V\otimes W}, \quad \overline{\theta_V}\theta_V = \mathrm{id}_V, \quad \overline{b_V} = F(\cap_V^-), \quad \overline{d_V} = F(\cup_V^-).$$

Here $\cap_V^-$ and $\cup_V^-$ are the v-colored ribbon graphs in $\mathbb{R}^3$ defined in Section I.2.3. As usual, F is the operator invariant of v-colored ribbon graphs in $\mathbb{R}^3$ defined in Section I.2.

To give a geometric interpretation of conditions (5.1.1), (5.1.2) we define a negation for v-colored ribbon graphs in $\mathbb{R}^3$. Let Ω be a v-colored ribbon graph in $\mathbb{R}^3$ over a Hermitian ribbon category $\mathcal{V}$. Apply to Ω the composition of the following transformations (T1), (T2), (T3): the transformation (T1) inverts orientation in the surface of Ω and reverses directions of the bands and annuli of Ω, (T2) proclaims the top and bottom bases of each coupon bottom and top bases respectively and replaces the color of each coupon by the conjugate morphism (while keeping the colors of bands and annuli), (T3) replaces the ribbon graph with its mirror image with respect to the plane $\mathbb{R}^2 \times 1/2 \subset \mathbb{R}^3$ so that the bottom and top lines of Ω exchange their places. It is easy to see that the composition of (T1), (T2), (T3) produces a v-colored ribbon graph in $\mathbb{R}^3$. It is denoted by $-\Omega$. Clearly, $-(-\Omega) = \Omega$. Note that under the passage from Ω to $-\Omega$ the source and the target of Ω in the category $\text{Rib}_{\mathcal{V}}$ exchange their roles. For instance, $-X^+_{V,W} = X^-_{W,V}$ and $-Y^+_{V,W} = Z^-_{W,V}$.

If Ω is presented by a diagram $D \subset \mathbb{R} \times [0,1]$ then $-\Omega$ can be presented by a diagram $-D$ obtained as follows: reverse the orientations of all arcs and loops of D, proclaim the top and bottom bases of each coupon bottom and top bases respectively and replace the color of the coupon by the conjugate morphism (while keeping the colors of arcs and loops), finally take the mirror image of the diagram with respect to the line $\mathbb{R} \times 1/2 \subset \mathbb{R} \times [0,1]$ so that the bottom and top lines exchange their places.

5.1.3. Lemma. *Let Ω be a v-colored ribbon graph in $\mathbb{R}^3$ over a Hermitian ribbon category $\mathcal{V}$. Then*

$$F(-\Omega) = \overline{F(\Omega)}.$$

Proof. Both formulas $\Omega \mapsto F(-\Omega)$ and $\Omega \mapsto \overline{F(\Omega)}$ define contravariant functors $\text{Rib}_{\mathcal{V}} \to \mathcal{V}$. Therefore it suffices to verify the coincidence of these two functors on the generators of $\text{Rib}_{\mathcal{V}}$ specified in Lemma I.3.1.1. The generators $Z^v_{V,W}$ and φ'_V may be expressed via other generators using formulas (I.3.2.e) and (I.3.2.g). The coincidence of the two functors in question on the remaining generators follows directly from (5.1.1) and (5.1.2).

5.1.4. Corollary. *For any endomorphism f of an object of a Hermitian ribbon category, we have* $\text{tr}(\overline{f}) = \overline{\text{tr}(f)}$.

This follows from Lemma 5.1.3 and the geometric interpretation of the trace given in Corollary I.2.7.1. Indeed, $\text{tr}(f) = F(\Omega_f)$ where Ω_f is the v-colored ribbon graph drawn in Figure I.2.9. Therefore it suffices to note that $-\Omega_f = \Omega_{\overline{f}}$.

The next corollary follows from Corollary 5.1.4 and the formula $\dim(V) = \text{tr}(\text{id}_V)$.

5.1.5. Corollary. *For any object V of a Hermitian ribbon category, we have* $\overline{\dim(V)} = \dim(V)$.

There is a useful formula computing $\dim(V)$ in terms of conjugation. Namely,

$$\dim(V) = F(\cap_{\overline{V}}^{-} \cup_V) = F(\cap_{\overline{V}}^{-})F(\cup_V) = \overline{b_V} b_V = \operatorname{tr}(\overline{b_V} b_V) = \operatorname{tr}(b_V \overline{b_V}). \tag{5.1.a}$$

5.2. Hermitian ribbon Ab-categories. A conjugation in a monoidal Ab-category is said to be additive if for any morphisms $f, g : V \to W$, we have $\overline{f+g} = \overline{f} + \overline{g}$. A Hermitian ribbon Ab-category is a ribbon Ab-category equipped with an additive conjugation satisfying (5.1.1) and (5.1.2).

We shall always view the ground ring $K = \operatorname{Hom}_{\mathcal{V}}(\mathbb{1}, \mathbb{1})$ of the Hermitian ribbon Ab-category $\mathcal{V}$ as a ring with involution $f \mapsto \overline{f}$. Note that the conjugation $f \mapsto \overline{f} : \operatorname{Hom}(V, W) \to \operatorname{Hom}(W, V)$ is linear over this ring involution: for any $k \in K$ and any $f \in \operatorname{Hom}(V, W)$, we have

$$\overline{kf} = \overline{k \otimes f} = \overline{k} \otimes \overline{f} = \overline{k}\,\overline{f}.$$

The next lemma justifies the term "Hermitian" in this setting.

5.2.1. Lemma. *Let $\mathcal{V}$ be a Hermitian ribbon* Ab-*category. For any objects V, W of $\mathcal{V}$, the pairing $(f, g) \mapsto \operatorname{tr}(f\overline{g}) : \operatorname{Hom}(V, W) \times \operatorname{Hom}(V, W) \to K$ is a non-degenerate Hermitian pairing.*

Proof. This pairing is sesquilinear and non-degenerate because of Lemma 4.2.3. The Hermitian symmetry $\operatorname{tr}(f\overline{g}) = \overline{\operatorname{tr}(g\overline{f})}$ follows from Corollary 5.1.4.

5.3. Hermitian modular categories. A Hermitian modular category is a modular category equipped with an additive conjugation satisfying (5.1.1) and (5.1.2). We shall see in Chapter IV that each Hermitian modular category gives rise to a 3-dimensional Hermitian topological field theory.

Our next aim is to generalize Lemma 5.1.3 to v-colored ribbon graphs in closed oriented 3-manifolds. To this end we define a negation for such ribbon graphs. Let M be a closed oriented 3-manifold and $\Omega \subset M$ be a v-colored ribbon graph over a Hermitian modular category. By $-M$ we denote the manifold M with the opposite orientation, by $-\Omega$ we denote the v-colored ribbon graph in $-M$ obtained from Ω by the transformations (T1) and (T2) described in Section 5.1. Clearly, $-(-\Omega) = \Omega$.

5.4. Theorem. *Let $\mathcal{V}$ be a Hermitian modular category with rank $\mathcal{D} \in K$ such that $\overline{\mathcal{D}} = \mathcal{D}$. For any closed oriented 3-manifold M and any v-colored ribbon graph $\Omega \subset M$ over $\mathcal{V}$, we have*

$$\tau_{(\mathcal{V},\mathcal{D})}(-M, -\Omega) = \overline{\tau_{(\mathcal{V},\mathcal{D})}(M, \Omega)}. \tag{5.4.a}$$

Note the case $\Omega = \emptyset$:

$$\tau_{(\mathcal{V},\mathcal{D})}(-M) = \overline{\tau_{(\mathcal{V},\mathcal{D})}(M)}.$$

Proof of Theorem. To prove (5.4.a) we have to study ribbon categories somewhat more extensively. To each ribbon category $\mathcal{W}$ we associate an opposite ribbon category $\mathcal{W}^o$. The objects of $\mathcal{W}^o$ are just the objects of $\mathcal{W}$. A morphism $V \to W$ in $\mathcal{W}^o$ is, by definition, a morphism $W \to V$ in $\mathcal{W}$. For a morphism $f : W \to V$ in $\mathcal{W}$, the corresponding morphism $V \to W$ in $\mathcal{W}^o$ will be denoted by f^{op}. The composition and tensor product of morphisms in $\mathcal{W}^o$ is defined by the formulas $f^{op} \circ g^{op} = (gf)^{op}$ and $f^{op} \otimes g^{op} = (f \otimes g)^{op}$. The morphism $(\mathrm{id}_V)^{op}$ is the identity endomorphism of V in $\mathcal{W}^o$. The braiding c^o, twist θ^o, and duality b^o, d^o in $\mathcal{W}^o$ are defined by the formulas

$$c^o_{V,W} = (c_{W,V})^{op}, \;\; \theta^o_V = (\theta_V)^{op}, \;\; b^o_V = (F(\cap^-_V))^{op}, \;\; d^o_V = (F(\cup^-_V))^{op}.$$

It is a nice exercise on definitions to check that $\mathcal{W}^o$ is a ribbon category. We verify identity (I.1.2.b) for the braiding in $\mathcal{W}^o$ leaving other identities to the reader:

$$c^o_{U,V\otimes W} = (c_{V\otimes W,U})^{op} = ((c_{V,U} \otimes \mathrm{id}_W)(\mathrm{id}_V \otimes c_{W,U}))^{op} = (\mathrm{id}_V \otimes c^o_{U,W})(c^o_{U,V} \otimes \mathrm{id}_W).$$

Note also that the formula $f \mapsto f^{op}$ defines an isomorphism of commutative semigroups $\mathrm{End}_{\mathcal{W}}(\mathbb{1}) \to \mathrm{End}_{\mathcal{W}^o}(\mathbb{1})$. This is, of course, a ring isomorphism if $\mathcal{W}$ is a ribbon Ab-category.

Each v-colored ribbon graph $\Omega \subset \mathbb{R}^3$ over $\mathcal{W}$ gives rise to an opposite v-colored ribbon graph $\Omega^o \subset \mathbb{R}^3$ over $\mathcal{W}^o$. It is obtained from Ω by the composition of the transformation (T1) inverting the orientation in the surface of Ω and reversing the directions of the bands and annuli of Ω, the transformation (T2)$'$ proclaiming the top and bottom bases of each coupon bottom and top bases respectively and replacing the color f of each coupon by the morphism f^{op} in $\mathcal{W}^o$ (while keeping the colors of bands and annuli), and the transformarion (T3)$'$ rotating Ω in $\mathbb{R}^3$ around the line $\mathbb{R} \times 0 \times (1/2)$ to the angle 180° so that the bottom and top lines of Ω exchange their places. (The rotation (T3)$'$ is redundant when Ω has no free ends.) Analogously to Lemma 5.1.3 we have

$$F_{\mathcal{W}^o}(\Omega^o) = (F_{\mathcal{W}}(\Omega))^{op}. \tag{5.4.b}$$

Applying this formula to the trivial knot with zero framing colored with an object V of $\mathcal{W}$ we obtain $\dim_{\mathcal{W}^o}(V) = (\dim_{\mathcal{W}}(V))^{op} : \mathbb{1} \to \mathbb{1}$.

It is obvious that the construction $\Omega \mapsto \Omega^o$ extends to v-colored ribbon graphs in 3-manifolds: we just perform (T1) and (T2)$'$ skipping (T3)$'$. If $\mathcal{W}$ is a modular category with rank $\mathcal{D} \in \mathrm{End}_{\mathcal{W}}(\mathbb{1})$ then the opposite ribbon category $\mathcal{W}^o$ is also modular with rank $\mathcal{D}^{op} \in \mathrm{End}_{\mathcal{W}^o}(\mathbb{1})$. It follows from (5.4.b) and definitions that

$$\tau_{(\mathcal{W}^o,\mathcal{D}^{op})}(M, \Omega^o) = (\tau_{(\mathcal{W},\mathcal{D})}(M, \Omega))^{op} \in \mathrm{End}_{\mathcal{W}^o}(\mathbb{1}).$$

We shall use this formula in the following equivalent form

$$\tau_{(\mathcal{W},\mathcal{D})}(M, \Omega) = (\tau_{(\mathcal{W}^o,\mathcal{D}^{op})}(M, \Omega^o))^{op} \in \mathrm{End}_{\mathcal{W}}(\mathbb{1}). \tag{5.4.c}$$

Consider the case where $\mathcal{W}$ is a Hermitian ribbon category. Recall the mirror ribbon category $\overline{\mathcal{W}}$ (see Section I.1.4). It is clear that the formula $f \mapsto \overline{f}^{op}$, where

f is a morphism in $\mathcal{W}$, defines a covariant functor $\mathcal{W} \to \overline{\mathcal{W}}^o$ identical on objects. Involutivity of the conjugation implies that this functor is an isomorphism. It is easy to verify that this is an isomorphism of ribbon categories. In particular, it follows from (5.1.1) that the braiding $c_{V,W} : V \otimes W \to W \otimes V$ in $\mathcal{W}$ corresponds under this isomorphism to the braiding $(c_{V,W}^{-1})^{op} : V \otimes W \to W \otimes V$ in $\overline{\mathcal{W}}^o$. Similarly, twist and duality in $\mathcal{W}$ correspond under this isomorphism $\mathcal{W} \to \overline{\mathcal{W}}^o$ to twist and duality in $\overline{\mathcal{W}}^o$.

Now we can prove (5.4.a). The results of Exercise 1.9.2 and formulas (2.5.a) and (5.4.c) (with $\mathcal{W} = \overline{\mathcal{V}}$) imply that

$$\tau_{(\mathcal{V},\mathcal{D})}(-M, -\Omega) = \tau_{(\overline{\mathcal{V}},\mathcal{D})}(M, -\Omega) = (\tau_{(\overline{\mathcal{V}}^o,\mathcal{D}^{op})}(M, (-\Omega)^o))^{op}. \tag{5.4.d}$$

It is easy to see that the v-colored ribbon graph $(-\Omega)^o$ over $\overline{\mathcal{V}}^o$ geometrically coincides with Ω and has the same colors of bands and annuli. These ribbon graphs differ only in the colors of their coupons: if a coupon of Ω is colored with a morphism f in $\mathcal{V}$, then the corresponding coupon of $(-\Omega)^o$ is colored with the morphism $\overline{f}^{op}$ in $\overline{\mathcal{V}}^o$. In other words, Ω is transformed into $(-\Omega)^o$ by the functor $\mathcal{V} \to \overline{\mathcal{V}}^{op}$ determined by the Hermitian structure as in the preceding paragraph. Therefore

$$\tau_{(\overline{\mathcal{V}}^o,\mathcal{D}^{op})}(M, (-\Omega)^o) = (\overline{\tau_{(\mathcal{V},\mathcal{D})}(M, \Omega)})^{op}.$$

Substituting this expression in (5.4.d) we get (5.4.a).

5.5. Unitary modular categories. A unitary modular category is a Hermitian modular category $\mathcal{V} = (\mathcal{V}, \{V_i\}_{i\in I})$ over $K = \mathbb{C}$ (with the usual complex conjugation in $\mathbb{C}$) such that for any morphism f in $\mathcal{V}$, we have $\operatorname{tr}(f\overline{f}) \geq 0$. In other words, the Hermitian form $(f, g) \mapsto \operatorname{tr}(f\overline{g})$ on $\operatorname{Hom}(V, W)$ should be positive definite for any objects V, W of $\mathcal{V}$. Positivity implies a number of important estimates for link invariants and 3-manifold invariants. A simplest estimate of this kind follows from (5.1.a): for any object V of a unitary modular category, we have $\dim(V) \geq 0$. (Without positivity condition we can assure only that $\dim(V)$ is a real number.)

We establish here an inequality pertaining to operator invariants of framed braids. A framed braid is a ribbon tangle in $\mathbb{R}^2 \times [0, 1]$ consisting solely of bands such that the projection $\mathbb{R}^2 \times [0, 1] \to [0, 1]$ transforms the cores of these bands homeomorphically onto $[0, 1]$. A framed braid can be presented by a diagram which has no points of local maxima or local minima of the height function. A framed braid is said to be colored (over $\mathcal{V}$) if each of its bands is provided with an object of $\mathcal{V}$. Every colored framed braid α regarded as a colored framed tangle represents a morphism in the category $\mathrm{Rib}_{\mathcal{V}}$ (see Section I.2.3). We shall consider the case where $\operatorname{source}(\alpha) = \operatorname{target}(\alpha)$ so that the colors and directions of the bottom ends of α form the same sequence as the colors and directions of the top ends of α. For example, this is the case when α is a pure braid, i.e., when the top end of each string lies exactly over its bottom end. Another example: all strings of α are directed downwards and have the same color.

5.5.1. Theorem. *Let $\mathcal{V}$ be a unitary modular category. Let α be a colored framed braid such that* $\text{source}(\alpha) = \text{target}(\alpha) = ((W_1, \nu_1), \ldots, (W_m, \nu_m))$ *where m is the number of bands of α, $W_1, \ldots, W_m$ are objects of $\mathcal{V}$, and $\nu_1, \ldots, \nu_m \in \{+1, -1\}$. Then*

$$|\operatorname{tr}(F(\alpha))| \leq \prod_{r=1}^{m} \dim(W_r).$$

Applying this theorem to the 2-string pure braid $\alpha = X^+_{W,V} \circ X^+_{V,W}$ we obtain the following important corollary.

5.5.2. Corollary. *For any objects V, W of a unitary modular category, we have*

$$|\operatorname{tr}(c_{W,V} \circ c_{V,W})| \leq \dim(V)\dim(W).$$

This implies a useful estimate for the entries of the matrix S introduced in Section 1: for a unitary modular category $(\mathcal{V}, \{V_i\}_{i\in I})$ and any $i, j \in I$, we have

$$|S_{i,j}| \leq \dim(i)\dim(j).$$

Combining Theorem 5.5.1 with the geometric interpretation of the trace provided by Corollary I.2.7.2 we get an estimate for the absolute value of the link invariant $F(L) \in \mathbb{C}$ where L is the colored framed link obtained as the closure of α. This yields the following estimate of the braid index of links. (A braid index of a link in $\mathbb{R}^3$ is the minimal integer m such that this link may be obtained as the closure of a braid of m strings.)

5.5.3. Corollary. *Let $\mathcal{V}$ be a unitary modular category. Let L be a framed link in $\mathbb{R}^3$ with braid index m. Let L_V be the same link L with components colored by an object V of $\mathcal{V}$. Then $|F(L_V)| \leq (\dim(V))^m$.*

To prove Theorem 5.5.1 we need the following lemma concerned with arbitrary modular categories.

5.5.4. Lemma. *Let $(\mathcal{V}, \{V_i\}_{i\in I})$ be a modular category. Let $f: V \to V$ be an endomorphism of an object V of $\mathcal{V}$. For $j \in I$, denote by f^j the homomorphism $x \mapsto fx : \operatorname{Hom}(V_j, V) \to \operatorname{Hom}(V_j, V)$. Then*

$$\operatorname{tr}(f) = \sum_{j\in I} \operatorname{Tr}(f^j)\dim(j). \tag{5.5.a}$$

Proof. If $f = \mathrm{id}_V$ then $\operatorname{tr}(f) = \dim(V)$ and $\operatorname{Tr}(f^j) = \operatorname{Dim}(\operatorname{Hom}(V_j, V)) = \operatorname{Dim}_j(V)$. In this case (5.5.a) follows directly from (4.4.a). Similar arguments apply in the case where f is proportional to id_V, i.e., $f \in K \cdot \mathrm{id}_V$. In the general case we present $f : V \to V$ as a finite linear combination of morphisms of type gh where $h : V \to V_i$ and $g : V_i \to V$ are morphisms with a certain $i \in I$. Since both parts of (5.5.a) are K-linear with respect to f it suffices to consider the case $f = gh$. We have $\mathrm{tr}(f) = \mathrm{tr}(gh) = \mathrm{tr}(hg)$. Since hg is an endomorphism of V_i it is proportional to the identity endomorphism and satisfies (5.5.a) with f replaced by hg and f^j replaced by composition of $h^j : \mathrm{Hom}(V_j, V) \to \mathrm{Hom}(V_j, V_i)$ and $g^j : \mathrm{Hom}(V_j, V_i) \to \mathrm{Hom}(V_j, V)$. Hence,

$$\mathrm{tr}(f) = \mathrm{tr}(hg) = \sum_{j \in I} \mathrm{Tr}(h^j g^j)\dim(j) = \sum_{j \in I} \mathrm{Tr}(g^j h^j)\dim(j) = \sum_{j \in I} \mathrm{Tr}(f^j)\dim(j).$$

5.5.5. Proof of Theorem 5.5.1. Set

$$V = F(\mathrm{source}(\alpha)) = W_1^{\nu_1} \otimes \ldots \otimes W_m^{\nu_m}.$$

Set $f = F(\alpha) : V \to V$. We claim that $\overline{f} = f^{-1}$. To see this, we decompose α into a product of elementary framed braids. A framed braid is elementary if it is either presented by a diagram with exactly one crossing point or obtained from the trivial framed braid by twisting one of the bands around its core. Therefore the morphism $f = F(\alpha)$ may be expressed as a composition of morphisms of type $\mathrm{id} \otimes c_{U,W}^{\pm 1} \otimes \mathrm{id}$ and $\mathrm{id} \otimes \theta_U^{\pm 1} \otimes \mathrm{id}$. Every such morphism g satisfies the equality $\overline{g} = g^{-1}$ because of (5.1.1) and (5.1.2). Therefore $\overline{f} = f^{-1}$.

For any $j \in I$, consider the homomorphism $f^j : \mathrm{Hom}(V_j, V) \to \mathrm{Hom}(V_j, V)$ defined by the formula $x \mapsto fx$. This homomorphism preserves the unitary form $\langle x, y \rangle = \mathrm{tr}(x\overline{y})$ on $\mathrm{Hom}(V_j, V)$. Indeed, for any morphisms $x, y : V_j \to V$, we have

$$\langle f^j(x), f^j(y) \rangle = \langle fx, fy \rangle = \mathrm{tr}(fx\overline{y}\overline{f}) = \mathrm{tr}(fx\overline{y}f^{-1}) = \mathrm{tr}(x\overline{y}).$$

Therefore

$$|\mathrm{Tr}(f^j)| \leq \dim_{\mathbb{C}}(\mathrm{Hom}(V_j, V)) = \mathrm{Dim}_j(V).$$

Together with Lemma 5.5.4 this gives

$$|\mathrm{tr}(f)| = |\sum_{j \in I} \mathrm{Tr}(f^j)\dim(j)| \leq \sum_{j \in I} \mathrm{Dim}_j(V)\dim(j) = \dim(V) = \prod_{r=1}^{m} \dim(W_r).$$

5.5.6. Remarks. 1. Let $(\mathcal{V}, \{V_i\}_{i \in I})$ be a Hermitian modular category over $\mathbb{C}$. Corollary 5.1.5 implies that $\sum_{i \in I}(\dim(i))^2$ is a positive real number. Therefore the positive square root $(\sum_{i \in I}(\dim(i))^2)^{1/2}$ is a rank of $\mathcal{V}$.

2. Combining the definitions of Hermitian ribbon category and semisimple category we get the notions of Hermitian and unitary semisimple categories. Theorem 5.5.1 and Lemma 5.5.4 directly generalize to this setting.

5.6. Exercises. 1. Let $(\mathcal{V}, \{V_i\}_{i\in I})$ be a Hermitian modular category with rank $\mathcal{D}$. Show that for any $i \in I$, we have $\overline{v_i} = v_i^{-1}$. (Hint: apply Lemma 5.1.3 to the tangle φ_{V_i} introduced in Section I.2.3.) Show that

$$\overline{\Delta_{\mathcal{V}}} = \sum_{i\in I} v_i(\dim(i))^2 = \Delta_{\overline{\mathcal{V}}}.$$

This formula and (2.4.a) imply that

$$\mathcal{D}^2 = \Delta_{\mathcal{V}}\overline{\Delta_{\mathcal{V}}}. \tag{5.6.a}$$

2. Let $\mathcal{V}$ be a Hermitian modular category over $K = \mathbb{C}$ (with the usual complex conjugation in $\mathbb{C}$) such that for any object V of $\mathcal{V}$ and any morphism $f : \mathbb{1} \to V$, we have $\operatorname{tr}(f\overline{f}) \geq 0$. Show that $\mathcal{V}$ is a unitary modular category. (Hint: use the same argument as at the end of the proof of Lemma 4.2.3.)

3. Consider the ribbon Ab-category $\mathcal{V} = \mathcal{V}(G, K, c, \varphi)$ constructed in Section I.1.7.2. Any ring involution $k \mapsto \overline{k}$ in K induces an involution in the set of morphisms in $\mathcal{V}$ because this set consists of copies of K. Show that this involution satisfies (5.1.1) and (5.1.2) if and only if $\varphi(G) = 1$ and $\overline{c(g,h)} = (c(g,h))^{-1}$ for any $g, h \in G$.

4. Show that the constructions of opposite and mirror categories commute: for any ribbon category $\mathcal{V}$, we have $\overline{\mathcal{V}^o} = \overline{\mathcal{V}}^o$.

Notes

Section 1. The notion of a modular category was introduced in [Tu11]. Here we present this notion in a slightly modified form.

The parallel notion in the theory of Hopf algebras, namely the one of a modular Hopf algebra was introduced by Reshetikhin and Turaev, see [RT2]. We shall discuss modular Hopf algebras in Chapter XI. After appearance of [RT2] it has become clear that the technique of Hopf algebras may and should be replaced by a more general technique of monoidal categories. Still, it remained to define the appropriate class of categories and to overcome technical difficulties. Personally, the author got a strong impetus in this direction from Deligne [De]. Similar ideas were expressed by D. Kazhdan, see also Lyubashenko [Ly1] and Yetter [Ye4].

Our modular categories should not be confused with modular tensor categories introduced by Moore and Seiberg [MS1], [MS2]. Their modular tensor categories are not categories in the usual sense of the word but rather systems of tensors similar to the $6j$-symbols studied in Chapter VI.

Section 2. We follow the general lines of [RT2] where the surgery presentations of 3-manifolds were first used to derive invariants of 3-manifolds from modular Hopf algebras. The formulas used in [RT2] and in Section 2 to define the invariant τ are inspired by properties of Witten's path integral invariants of 3-manifolds, see [Wi2]. (Reshetikhin and the author started working on quantum invariants of 3-manifolds in the early spring of

1988, before appearance of [Wi2]. This paper became available to us in September, 1988 and sped up our study.) For further work in this direction, based on surgery presentations of 3-manifolds, see Kirby and Melvin [KM], Lickorish [Li2] - [Li5], Walker [Wa], Blanchet, Habegger, Masbaum, Vogel [BHMV1], [BHMV2].

Section 3. The proof of Theorem 2.3.2 consists of two essentially independent parts: (i) the construction of τ' and (ii) equivalence of τ' and τ. Part (i) and in particular Lemma 3.2.1 follow along the lines of [RT2]. (The construction of [RT2] produces τ' rather than τ.) The part (ii) is new. The proof of the crucial claim (ii) of Lemma 3.2.2 was obtained during a conversation of the author with Kevin Walker.

A counterpart of the projective representation of $SL(2, \mathbb{Z})$ is well known in 2-dimensional conformal field theory, see [MS2, page 285].

Section 4. Semisimple categories, as defined here, were introduced in [Tu9]. Similar categories are studied by Lyubashenko [Ly1], [Ly2] and Yetter [Ye4].

Formulas (4.5.d) and (4.5.e) were conjectured in the context of 2-dimensional conformal field theory by Verlinde [Ver] and proven by G. Moore and N. Seiberg (see [MS2, page 287]). Our proof follows an idea of Witten [Wi2]. We note that the formula (4.5.e) differs from the formula in [MS2] by an additional factor of $\mathcal{D}^{-2}$. This is due to the fact that the matrix S used in [MS2] corresponds to our $\mathcal{D}^{-1}S$. Formulas (4.5.f) and (4.5.g) seem to be new.

Section 5. The material of this section is new.

Chapter III
Foundations of topological quantum field theory

1. Axiomatic definition of TQFT's

1.0. Outline. We give an axiomatic definition of modular functors and topological quantum field theories (TQFT's). The reader will notice that our approach to TQFT's has a strong flavour of abstract nonsense. However, the relevant abstract notions form a natural and important part of the theory. They create a general set up for the topological quantum field theory and put the 3-dimensional TQFT's introduced in Chapters IV and VII into a proper perspective.

In order to define modular functors and topological quantum field theories we need to formalize the notion of an additional structure on a topological space. For such formalized additional structures, we use the term space-structures. One may use the language of space-structures on a heuristical level without going into the details of our axiomatic approach. The reader willing to cut short the abstract theory of space-structures may skip Sections 1.1 and 1.3 restricting himself to the case where $\mathfrak{A}$-spaces are closed oriented n-manifolds and $\mathfrak{B}$-spaces are compact oriented $(n+1)$-manifolds possibly with some reasonable additional structures.

1.1. Space-structures. Before giving a definition of space-structures we briefly analyze one of the simplest examples, specifically the orientation in n-dimensional topological manifolds. Fix an integer $n \geq 0$. For every n-dimensional topological manifold X we have a set $\mathrm{Ori}_n(X)$ (possibly void) of orientations in X. For technical reasons, we shall agree that the empty space is a topological manifold of dimension n (for any n) and admits a unique orientation. It is convenient to extend Ori_n to arbitrary topological spaces: set $\mathrm{Ori}_n(X) = \emptyset$ if the space X is not an n-dimensional topological manifold. It is clear that any homeomorphism $f : X \to Y$ induces a bijection $\mathrm{Ori}_n(X) \to \mathrm{Ori}_n(Y)$ denoted by $\mathrm{Ori}_n(f)$. Obviously, $\mathrm{Ori}_n(\mathrm{id}_X) = \mathrm{id}$ and $\mathrm{Ori}_n(fg) = \mathrm{Ori}_n(f)\,\mathrm{Ori}_n(g)$ for any composable homeomorphisms f, g. We may express these properties by saying that the formulas $(X \mapsto \mathrm{Ori}_n(X), f \mapsto \mathrm{Ori}_n(f))$ define a covariant functor from the category of topological spaces and their homeomorphisms into the category of sets and bijections. Axiomatizing these properties we get the notion of a space-structure.

A space-structure is a covariant functor from the category of topological spaces and their homeomorphisms into the category of sets and bijections such that the value of this functor on the empty space is a one-element set. Such a functor $\mathfrak{A}$ assigns to every topological space X a set $\mathfrak{A}(X)$ and to every homeomorphism

$f : X \to Y$ a bijection $\mathfrak{A}(f) : \mathfrak{A}(X) \to \mathfrak{A}(Y)$ such that $\mathfrak{A}(\mathrm{id}_X) = \mathrm{id}_{\mathfrak{A}(X)}$ for any topological space X and $\mathfrak{A}(fg) = \mathfrak{A}(f)\,\mathfrak{A}(g)$ for any composable homeomorphisms f, g. By assumption, $\mathfrak{A}(\emptyset)$ is a one-element set.

Elements of $\mathfrak{A}(X)$ are called $\mathfrak{A}$-structures on X. Any pair (a topological space X, an $\mathfrak{A}$-structure $\alpha \in \mathfrak{A}(X)$) is called an $\mathfrak{A}$-space. An $\mathfrak{A}$-homeomorphism of an $\mathfrak{A}$-space (X, α) onto an $\mathfrak{A}$-space (X', α') is a homeomorphism $f : X \to X'$ such that $\mathfrak{A}(f)(\alpha) = \alpha'$. It is clear that the composition of $\mathfrak{A}$-homeomorphisms is an $\mathfrak{A}$-homeomorphism and that the identity self-homeomorphisms of $\mathfrak{A}$-spaces are $\mathfrak{A}$-homeomorphisms.

We shall also consider more general space-structures $\mathfrak{A}$ which are defined as above with the only difference that the values of $\mathfrak{A}$ on topological spaces are not sets but rather classes. The set-theoretic concept of class is broader and more convenient in this abstract setting. An example of a space-structure involving classes is provided by the structure of a topological space with several distinguished points endowed with objects of a certain category. We shall meet similar examples in Chapter IV.

A space-structure $\mathfrak{A}$ is said to be compatible with disjoint union if the disjoint union of a finite family of $\mathfrak{A}$-spaces acquires the structure of an $\mathfrak{A}$-space in a natural way. Here is a more meticulous formulation. A space-structure $\mathfrak{A}$ is compatible with disjoint union if for any topological spaces X and Y there is a canonical mapping $\mathfrak{A}(X) \times \mathfrak{A}(Y) \to \mathfrak{A}(X \amalg Y)$ so that the following four conditions are satisfied.

(1.1.1) The diagram

$$\begin{array}{ccc} \mathfrak{A}(X) \times \mathfrak{A}(Y) & \longrightarrow & \mathfrak{A}(X \amalg Y) \\ \text{Perm} \downarrow & & \downarrow = \\ \mathfrak{A}(Y) \times \mathfrak{A}(X) & \longrightarrow & \mathfrak{A}(Y \amalg X) \end{array}$$

is commutative. Here Perm is the flip $x \times y \mapsto y \times x$ and the horizontal arrows represent the canonical mappings.

(1.1.2) For arbitrary $\mathfrak{A}$-homeomorphisms $f : X \to X'$ and $g : Y \to Y'$, the diagram

$$\begin{array}{ccc} \mathfrak{A}(X) \times \mathfrak{A}(Y) & \longrightarrow & \mathfrak{A}(X \amalg Y) \\ \mathfrak{A}(f) \times \mathfrak{A}(g) \downarrow & & \downarrow \mathfrak{A}(f \amalg g) \\ \mathfrak{A}(X') \times \mathfrak{A}(Y') & \longrightarrow & \mathfrak{A}(X' \amalg Y') \end{array}$$

is commutative.

(1.1.3) For any three topological spaces X, Y, Z, the diagram of canonical mappings

$$\begin{array}{ccc} \mathfrak{A}(X) \times \mathfrak{A}(Y) \times \mathfrak{A}(Z) & \longrightarrow & \mathfrak{A}(X \amalg Y) \times \mathfrak{A}(Z) \\ \downarrow & & \downarrow \\ \mathfrak{A}(X) \times \mathfrak{A}(Y \amalg Z) & \longrightarrow & \mathfrak{A}(X \amalg Y \amalg Z) \end{array}$$

is commutative.

(1.1.4) The canonical mapping $\mathfrak{A}(\emptyset) \times \mathfrak{A}(Y) \to \mathfrak{A}(\emptyset \amalg Y) = \mathfrak{A}(Y)$ is induced by the identity $\mathrm{id} : \mathfrak{A}(Y) \to \mathfrak{A}(Y)$.

Following the remarks made at the beginning of this section the orientation of n-dimensional topological manifolds may be regarded as a space-structure Ori_n. It is compatible with disjoint union in the obvious way. Inversion of orientation defines an involution in the set $\mathrm{Ori}_n(X)$ for any topological space X. This suggests the following definition.

A space-structure $\mathfrak{A}$ is involutive if for any topological space X, the set (or, more generally, the class) $\mathfrak{A}(X)$ is provided with an involution such that:

(i) for any topological spaces X, Y, the canonical mapping $\mathfrak{A}(X) \times \mathfrak{A}(Y) \to \mathfrak{A}(X \amalg Y)$ is equivariant and

(ii) for any homeomorphism $f : X \to Y$, the induced bijection $\mathfrak{A}(f) : \mathfrak{A}(X) \to \mathfrak{A}(Y)$ is equivariant.

The $\mathfrak{A}$-structure on X which is the image of $\alpha \in \mathfrak{A}(X)$ under the given involution is said to be opposite to α and denoted by $-\alpha$. For an $\mathfrak{A}$-space X, denote by $-X$ the same space with the opposite $\mathfrak{A}$-structure. Clearly $-(-X) = X$. For an $\mathfrak{A}$-homeomorphism $f : X \to Y$, denote by $-f$ the same mapping viewed as an $\mathfrak{A}$-homeomorphism $-X \to -Y$.

An example of a space-structure is provided by the structure of a smooth n-dimensional manifold with fixed $n \geq 0$. The corresponding functor Diff_n assigns to a topological space X the class of C^∞-smooth structures on X if X is an n-dimensional topological manifold, $\mathrm{Diff}_n(X) = \emptyset$ otherwise. We shall agree that $\mathrm{Diff}_n(\emptyset)$ is a one-element set. The space-structure Diff_n is obviously compatible with disjoint union. It is an involutive space-structure in a trivial way: we provide each class $\mathrm{Diff}_n(X)$ with the identity involution. We may similarly formalize the structure of oriented smooth n-dimensional manifolds, the structure of oriented n-dimensional piecewise-linear manifolds, the structure of triangulated spaces, the structure of cell spaces, etc. All these space-structures are compatible with disjoint union.

1.2. Modular functors. A modular functor assigns modules to topological spaces with a certain structure and isomorphisms to the structure-preserving homeomorphisms of these spaces. We give an axiomatic definition of modular functors.

Let $\mathfrak{A}$ be a space-structure compatible with disjoint union. Let K be a commutative ring with unit. A modular functor $\mathcal{T}$ with ground ring K based on $\mathfrak{A}$ assigns to every $\mathfrak{A}$-space X a projective K-module (of finite type) $\mathcal{T}(X)$ and to every $\mathfrak{A}$-homeomorphism $f : X \to Y$ a K-isomorphism $f_\# : \mathcal{T}(X) \to \mathcal{T}(Y)$ satisfying the following axioms (1.2.1) - (1.2.3).

(1.2.1) For any $\mathfrak{A}$-homeomorphisms $g : X \to Y, f : Y \to Z$, we have $(fg)_\# = f_\# g_\#$.

This axiom allows us to view $\mathcal{T}$ as a covariant functor from the category of $\mathfrak{A}$-spaces and $\mathfrak{A}$-homeomorphisms into the category of projective K-modules and K-isomorphisms. For the identity homeomorphism $\mathrm{id} = \mathrm{id}_X : X \to X$ of an $\mathfrak{A}$-space X, we have $\mathrm{id}_\# = \mathrm{id}_{\mathcal{T}(X)}$. Proof: apply (1.2.1) to $f = g = \mathrm{id}_X$. For any $\mathfrak{A}$-homeomorphism f, we have $f_\#^{-1} = (f_\#)^{-1}$. Proof: apply (1.2.1) to $g = f^{-1}$.

(1.2.2) For (disjoint) $\mathfrak{A}$-spaces X, Y, there is an identification isomorphism $\mathcal{T}(X \amalg Y) = \mathcal{T}(X) \otimes_K \mathcal{T}(Y)$ satisfying the following three conditions.

(i) (Commutativity). The diagram

$$\begin{array}{ccc} \mathcal{T}(X \amalg Y) & = & \mathcal{T}(X) \otimes_K \mathcal{T}(Y) \\ {\scriptstyle =}\downarrow & & \downarrow{\scriptstyle \mathrm{Perm}} \\ \mathcal{T}(Y \amalg X) & = & \mathcal{T}(Y) \otimes_K \mathcal{T}(X) \end{array}$$

is commutative. (Here Perm is the flip $x \otimes y \mapsto y \otimes x$.)

(ii) (Associativity). For any $\mathfrak{A}$-spaces X, Y, Z, the composition of identifications

$$(\mathcal{T}(X) \otimes_K \mathcal{T}(Y)) \otimes_K \mathcal{T}(Z) = \mathcal{T}(X \amalg Y) \otimes_K \mathcal{T}(Z) = \mathcal{T}(X \amalg Y \amalg Z) =$$

$$= \mathcal{T}(X) \otimes_K \mathcal{T}(Y \amalg Z) = \mathcal{T}(X) \otimes_K (\mathcal{T}(Y) \otimes_K \mathcal{T}(Z))$$

is the standard identification $(x \otimes y) \otimes z = x \otimes (y \otimes z)$.

(iii) (Naturality). For any $\mathfrak{A}$-homeomorphisms $f : X \to X', g : Y \to Y'$, the diagram

$$\begin{array}{ccc} \mathcal{T}(X \amalg Y) & \xrightarrow{(f \amalg g)_\#} & \mathcal{T}(X' \amalg Y') \\ {\scriptstyle =}\downarrow & & \downarrow{\scriptstyle =} \\ \mathcal{T}(X) \otimes_K \mathcal{T}(Y) & \xrightarrow{f_\# \otimes g_\#} & \mathcal{T}(X') \otimes_K \mathcal{T}(Y') \end{array}$$

is commutative.

(1.2.3) $\mathcal{T}(\emptyset) = K$ and for any $\mathfrak{A}$-space Y, the identification $\mathcal{T}(Y) = \mathcal{T}(\emptyset \amalg Y) = \mathcal{T}(\emptyset) \otimes_K \mathcal{T}(Y)$ is induced by $\mathcal{T}(\emptyset) = K$.

We shall call the module $\mathcal{T}(X)$ the module of states of X. Its elements are called $\mathcal{T}$-states on X.

The simplest example of a modular functor is provided by the trivial modular functor which assigns K to all topological spaces and id_K to all homeomorphisms. The underlying space-structure assigns a one-element set to all topological spaces. For more interesting examples of modular functors see Section 1.5.

A modular functor $\mathcal{T}$ based on an involutive space-structure $\mathfrak{A}$ is said to be self-dual if it satisfies the following condition.

(1.2.4) For any $\mathfrak{A}$-space X, there is a non-degenerate bilinear pairing $d_X : \mathcal{T}(X) \otimes_K \mathcal{T}(-X) \to K$. The system of pairings $\{d_X\}_X$ is natural with respect to $\mathfrak{A}$-homeomorphisms, multiplicative with respect to disjoint union, and symmetric in the sense that $d_{-X} = d_X \circ \mathrm{Perm}_{\mathcal{T}(-X),\mathcal{T}(X)}$ for any $\mathfrak{A}$-space X.

It follows from (1.2.2) that every modular functor $\mathcal{T}$ assigns to the disjoint union of a finite family of $\mathfrak{A}$-spaces $\{X_j\}_j$ the tensor product of the modules $\{\mathcal{T}(X_j)\}_j$. There is a subtle point here: if the family $\{X_j\}_j$ is not ordered then the tensor product in question is the non-ordered tensor product. Recall briefly its definition. Consider all possible total orders in the given (finite) family of modules, form the corresponding tensor products over K and identify them via the canonical isomorphisms induced by permutations of factors. This results in a module canonically isomorphic to any of these ordered tensor products but itself independent of the choice of ordering. For instance, for any two K-modules G, H, their non-ordered tensor product is the K-module F consisting of all pairs $(f \in G \otimes_K H,\ f' \in H \otimes_K G)$ such that $f' = \mathrm{Perm}(f)$. The formulas $(f, f') \mapsto f$ and $(f, f') \mapsto f'$ define isomorphisms $F \to G \otimes_K H$ and $F \to H \otimes_K G$. For any $g \in G, h \in H$, the pair $(g \otimes h, h \otimes g)$ is an element of F. As is customary in algebra, we denote both the ordered and non-ordered tensor products by the same symbol $\otimes$.

1.3. Cobordisms of $\mathfrak{A}$-spaces. The notion of cobordism for $\mathfrak{A}$-spaces is motivated by the needs of topological quantum field theory. In the realm of manifolds an $(n+1)$-dimensional TQFT associates modules to closed n-manifolds and homomorphisms to cobordisms between closed n-manifolds. (Recall that an $(n+1)$-dimensional cobordism is a compact $(n+1)$-manifold whose boundary is split as a disjoint union of two closed n-manifolds, called the bases of the cobordism.) The main idea in the definition of TQFT's is a functorial behavior of these homomorphisms with respect to gluing of cobordisms along their bases. Similar ideas may be applied in a more general context provided there are suitable notions of boundary and gluing. This suggests a definition of a cobordism theory for space-structures. Such a theory will serve as a basis for TQFT's.

Let $\mathfrak{A}$ and $\mathfrak{B}$ be space-structures compatible with disjoint union. We suppose that $\mathfrak{A}$ is involutive (but we do not need to suppose $\mathfrak{B}$ to be involutive). Suppose that any $\mathfrak{B}$-space M contains a distinguished topological subspace equipped with an $\mathfrak{A}$-structure. We call this subspace with its $\mathfrak{A}$-structure the boundary of M and denote it by ∂M. Thus, ∂M is an $\mathfrak{A}$-space. Warning: the underlying topological space of ∂M and the $\mathfrak{A}$-structure on this space may depend on the choice of $\mathfrak{B}$-structure on M. We assume that the boundary is natural with respect to $\mathfrak{B}$-homeomorphisms, i.e., that any $\mathfrak{B}$-homeomorphism $M_1 \to M_2$ restricts to an $\mathfrak{A}$-homeomorphism $\partial M_1 \to \partial M_2$. We also assume that the boundary commutes with disjoint union, i.e., that $\partial(M_1 \amalg M_2) = \partial M_1 \amalg \partial M_2$.

The space-structures $(\mathfrak{B}, \mathfrak{A})$ form a cobordism theory if the following four axioms (1.3.1) - (1.3.4) are satisfied.

(1.3.1) The $\mathfrak{B}$-spaces are subject to gluing as follows. Let M be a $\mathfrak{B}$-space whose boundary is the disjoint union of $\mathfrak{A}$-spaces X, Y, Z such that X is $\mathfrak{A}$-homeomorphic to $-Y$. Let M' be the topological space obtained from M by gluing of X to Y along an $\mathfrak{A}$-homeomorphism $f : X \to -Y$. Then the $\mathfrak{B}$-structure on M gives rise in a canonical way to a $\mathfrak{B}$-structure in M' such that $\partial M' = Z$. The $\mathfrak{B}$-structures on M' corresponding in this way to $f : X \to -Y$ and $(-f)^{-1} : Y \to -X$ coincide.

We say that the topological space M' with the $\mathfrak{B}$-structure provided by this axiom is obtained from M by gluing of X to Y along f.

(1.3.2) The gluings of $\mathfrak{B}$-spaces are natural with respect to $\mathfrak{B}$-homeomorphisms and commute with disjoint union. The gluing along disjoint $\mathfrak{A}$-subspaces of the boundary coincides with the gluing along their union.

This condition means that if, in the previous axiom, X is the disjoint union of a finite number of $\mathfrak{A}$-spaces $\{X_r\}_r$ then the $\mathfrak{B}$-structure on M' obtained by gluing of X to Y along $f : X \to -Y$ coincides with the $\mathfrak{B}$-structure on M' obtained by consecutive gluings of X_r to $f(X_r)$ along f for all r. In particular, the last structure is independent of the order in which the gluings are performed.

(1.3.3) Each $\mathfrak{A}$-structure α on a topological space X gives rise in a canonical way to a $\mathfrak{B}$-structure on $X \times [0, 1]$ denoted by $\alpha \times [0, 1]$ and such that

$$\partial(X \times [0, 1], \alpha \times [0, 1]) = ((X, -\alpha) \times 0) \amalg ((X, \alpha) \times 1).$$

The homeomorphism $(x, t) \mapsto (x, 1 - t) : X \times [0, 1] \to X \times [0, 1]$ transforms $\alpha \times [0, 1]$ into $(-\alpha) \times [0, 1]$. The correspondence $\alpha \mapsto \alpha \times [0, 1]$ is natural with respect to homeomorphisms and commutes with disjoint union.

Thus, for any $\mathfrak{A}$-space X, we may regard the cylinder $X \times [0, 1]$ as a $\mathfrak{B}$-space with the boundary $(-X \times 0) \amalg (X \times 1)$. We say that $X \times 0$ and $X \times 1$ are bases of $X \times [0, 1]$.

(1.3.4) Let X, X' be two copies of the same $\mathfrak{A}$-space. Gluing the $\mathfrak{B}$-spaces $X \times [0, 1]$, $X' \times [0, 1]$ along the identity $x \times 1 \mapsto x \times 0 : X \times 1 \to X' \times 0$ yields a $\mathfrak{B}$-space homeomorphic to the same cylinder $X \times [0, 1]$ via a $\mathfrak{B}$-homeomorphism which is the identity on the bases.

Example: $\mathfrak{B}$ is the structure of an oriented compact smooth $(n+1)$-manifold possibly with boundary, $\mathfrak{A}$ is the structure of an oriented closed smooth n-manifold, and ∂ is the standard boundary. These space-structures form a cobordism theory in the usual way.

1.4. Definition of a TQFT. Let $\mathfrak{A}$ be an involutive space-structure and $\mathfrak{B}$ be a space-structure both compatible with disjoint union. Assume that the space-structures $(\mathfrak{B}, \mathfrak{A})$ form a cobordism theory. A $(\mathfrak{B}, \mathfrak{A})$-cobordism is a triple (M, X, Y) where M is a $\mathfrak{B}$-space, X and Y are $\mathfrak{A}$-spaces, and $\partial M = (-X) \amalg Y$. The $\mathfrak{A}$-spaces X and Y are called the bottom and top bases of this cobordism and denoted by $\partial_- M$ and $\partial_+ M$ respectively. We say that M is a cobordism between X and Y. For instance, for any $\mathfrak{A}$-space X, the cylinder $(X \times [0, 1], X \times 0, X \times 1)$ is a cobordism between two copies of X. It is clear that we may form disjoint unions of cobordisms. We may also glue cobordisms M_1, M_2 along any $\mathfrak{A}$-homeomorphism $\partial_+(M_1) \to \partial_-(M_2)$.

A topological quantum field theory (TQFT) based on $(\mathfrak{B}, \mathfrak{A})$ consists of a modular functor $\mathcal{T}$ with ground ring K based on $\mathfrak{A}$ and a map τ which assigns to every $(\mathfrak{B}, \mathfrak{A})$-cobordism (M, X, Y) a K-homomorphism

$$\tau(M) = \tau(M, X, Y) : \mathcal{T}(X) \to \mathcal{T}(Y)$$

which satisfies the following axioms (1.4.1) - (1.4.4).

(1.4.1) (Naturality). If M_1 and M_2 are $(\mathfrak{B}, \mathfrak{A})$-cobordisms and $f : M_1 \to M_2$ is a $\mathfrak{B}$-homeomorphism preserving the bases then the diagram

$$\begin{array}{ccc} \mathcal{T}(\partial_-(M_1)) & \xrightarrow{\tau(M_1)} & \mathcal{T}(\partial_+(M_1)) \\ {\scriptstyle (f|_{\partial_-(M_1)})_\#}\downarrow & & \downarrow{\scriptstyle (f|_{\partial_+(M_1)})_\#} \\ \mathcal{T}(\partial_-(M_2)) & \xrightarrow{\tau(M_2)} & \mathcal{T}(\partial_+(M_2)) \end{array}$$

is commutative.

(1.4.2) (Multiplicativity). If a cobordism M is the disjoint union of cobordisms M_1, M_2 then under the identifications (1.2.2) we have $\tau(M) = \tau(M_1) \otimes \tau(M_2)$.

(1.4.3) (Functoriality). If a $(\mathfrak{B}, \mathfrak{A})$-cobordism M is obtained from the disjoint union of two $(\mathfrak{B}, \mathfrak{A})$-cobordisms M_1 and M_2 by gluing along an $\mathfrak{A}$-homeomorphism $f : \partial_+(M_1) \to \partial_-(M_2)$ then for some invertible $k \in K$,

$$\tau(M) = k\, \tau(M_2) \circ f_\# \circ \tau(M_1).$$

(1.4.4) (Normalization). For any $\mathfrak{A}$-space X, we have

$$\tau(X \times [0, 1], X \times 0, X \times 1) = \mathrm{id}_{\mathcal{T}(X)}.$$

The homomorphism $\tau(M, X, Y) : \mathcal{T}(X) \to \mathcal{T}(Y)$ is called the operator invariant of the cobordism (M, X, Y). The factor k in (1.4.3) is called the anomaly or the gluing anomaly of the triple (M_1, M_2, f). This factor may depend on the choice of this triple. Note that the gluing anomaly of (M_1, M_2, f) is not determined uniquely. For instance, it may happen that $\tau(M) = 0$ and $\tau(M_1) = 0$ in which case any invertible element of K is the anomaly of (M_1, M_2, f).

We say that the TQFT $(\mathcal{T}, \tau)$ is anomaly-free if the factor k in (1.4.3) may be always taken to be 1.

Every $\mathfrak{B}$-space M gives rise to a cobordism $(M, \emptyset, \partial M)$. The corresponding K-homomorphism $\mathcal{T}(\emptyset) \to \mathcal{T}(\partial M)$ is completely determined by its value on the unit $1_K \in K = \mathcal{T}(\emptyset)$. This value is denoted by $\tau(M)$. It follows from the axioms that $\tau(M) \in \mathcal{T}(\partial M)$ is natural with respect to $\mathfrak{B}$-homeomorphisms and multiplicative with respect to disjoint union.

We say that a $\mathfrak{B}$-space M is closed if its boundary is empty. (In contrast to the usual terminology pertaining to manifolds we do not assume closed $\mathfrak{B}$-spaces to be compact.) For a closed $\mathfrak{B}$-space M, we have $\tau(M) \in \mathcal{T}(\emptyset) = K$ so that $\tau(M)$ is a K-valued $\mathfrak{B}$-homeomorphism invariant of M. For instance, the empty set is a closed $\mathfrak{B}$-space and $\tau(\emptyset) = 1$. This follows from (1.4.4).

For any TQFT's $(\mathcal{T}_1, \tau_1)$ and $(\mathcal{T}_2, \tau_2)$ based on $(\mathfrak{B}, \mathfrak{A})$ we define their tensor product by the formulas $(\mathcal{T}_1 \otimes \mathcal{T}_2)(X) = \mathcal{T}_1(X) \otimes_K \mathcal{T}_2(X)$ and $(\tau_1 \otimes \tau_2)(M) = \tau_1(M) \otimes \tau_2(M)$. The action of homeomorphisms and the identification isomorphisms for disjoint unions are determined by the corresponding data for $(\mathcal{T}_1, \tau_1)$, $(\mathcal{T}_2, \tau_2)$ in the obvious way. Thus, $(\mathcal{T}_1 \otimes \mathcal{T}_2, \tau_1 \otimes \tau_2)$ is a TQFT based on the same cobordism theory $(\mathfrak{B}, \mathfrak{A})$.

1.5. Examples of TQFT's. We give a few elementary examples of TQFT's illustrating the concepts introduced above. In these examples $\mathfrak{A}$ is the structure of finite cell space (with the involution in each set $\mathfrak{A}(X)$ being the identity) and $\mathfrak{B}$ is the structure of finite cell space with a fixed cell subspace which plays the role of the boundary. Thus, $(\mathfrak{B}, \mathfrak{A})$-cobordisms are just finite cell triples (M, X, Y) with $X \cap Y = \emptyset$. The gluing of cell spaces and the cell structures on cylinders are defined in the standard way. The TQFT's in the following examples are based on this cobordism theory $(\mathfrak{B}, \mathfrak{A})$. The ground ring K is an arbitrary commutative ring with unit.

1. Let $\mathcal{T}$ be the trivial modular functor restricted to finite cell spaces. Fix an invertible element $q \in K$. For any finite cell triple (M, X, Y), define the operator $\tau(M, X, Y) : K \to K$ to be the multiplication by $q^{\chi(M,X)}$ where χ is the Euler characteristic. This yields an anomaly-free TQFT.

2. Fix an integer $i \geq 0$ and a finite abelian group G whose order is invertible in K. For any finite cell space X set $\mathcal{T}(X) = K[H_i(X; G)]$. Thus, $\mathcal{T}(X)$ is the module of formal linear combinations of elements of $H_i(X; G)$ with coefficients in K. (We ignore the structure of group ring in this module.) Note that $H_i(\emptyset; G) = 0$ so that

$\mathcal{T}(\emptyset) = K$. The action of cell homeomorphisms is induced by their action in H_i. Additivity of homologies with respect to disjoint union yields identifications satisfying axioms (1.2.2), (1.2.3). The operator invariant $\tau = \tau(M, X, Y)$ of a finite cell triple (M, X, Y) carries any $g \in H_i(X; G)$ into the formal sum of those $h \in H_i(Y; G)$ which are homological to g in M. (If there is no such h then $\tau(g) = 0$.) For instance, $\tau(M, \emptyset, \emptyset) = 1 \in K$. The axioms of topological field theory are straightforward. In axiom (1.4.3), k^{-1} is the order of the group $f_*(F_1) \cap F_2$ where F_1 and F_2 are the kernels of the inclusion homomorphisms $H_i(\partial_+(M_1); G) \to H_i(M_1; G)$ and $H_i(\partial_-(M_2); G) \to H_i(M_2; G)$ respectively. This yields a TQFT with non-trivial anomalies. Instead of homologies with coefficients in G we may use any homology or cohomology theory (or homotopy groups), provided some finiteness assumptions are imposed to assure that the modules $\{\mathcal{T}(X)\}$ are finitely generated.

1.6. Remark. The problem with any axiomatic definition is that it should be sufficiently general but not too abstract. It is especially hard to find a balance in axiomatic systems for TQFT's because the available stock of non-trivial examples is very limited. One may vary the axioms of TQFT's given above in several different directions. First of all the axioms may be extended to $\mathfrak{A}$-spaces with boundary, which should be preserved under $(\mathfrak{B}, \mathfrak{A})$-cobordisms. Such an extension could be useful, although it would definitely make the exposition more heavy. In the theory of 3-manifolds such an extension may be avoided because we can always eliminate the boundary of surfaces and 3-cobordisms by gluing in 2-disks and solid tubes respectively. Another possible generalization of TQFT's abandons the condition that the operator invariant τ is defined for all $(\mathfrak{B}, \mathfrak{A})$-cobordisms. A guiding example in this direction could be the "Reidemeister TQFT" defined only for acyclic cobordisms. This "TQFT" involves the trivial modular functor and assigns to every finite cell triple (M, X, Y) equipped with a flat K-linear bundle ξ with $H_*(M, X; \xi) = 0$, the multiplication by the Reidemeister torsion $\tau(M, X; \xi) \in K$. To eliminate the indeterminacy in the definition of torsion the pair (M, X) should be endowed with the homological orientation and Euler structure as defined in [Tu3], [Tu5]. Another idea would be to replace the module $\mathcal{T}(X)$ in the definition of a modular functor by a vector bundle over a certain topological space which itself may depend on X. There is no doubt that further experiments with axioms for TQFT's will follow.

1.7. Conventions. Up to the end of this chapter the symbol K denotes a commutative ring with unit and the symbol $(\mathfrak{B}, \mathfrak{A})$ denotes a cobordism theory where $\mathfrak{A}$ is an involutive space-structure and $\mathfrak{B}$ is a space-structure both compatible with disjoint union.

2. Fundamental properties

2.0. Outline. We establish here three fundamental properties of TQFT's. The first and the most important property says that the modular functor underlying a TQFT is always self-dual. We construct a canonical self-duality pairing for any such modular functor, this pairing will be systematically used in this chapter.

The second property allows us to compute the invariant of a closed $\mathfrak{B}$-space glued from two $\mathfrak{B}$-spaces with common boundary. This computation involves the canonical self-duality pairing in the module of states of the boundary.

The third property allows us to compute the invariant of a closed $\mathfrak{B}$-space obtained by multiplication of an $\mathfrak{A}$-space and the circle. In contrast to the first property, the second and third ones apply only to anomaly-free TQFT's.

2.1. Statements. Fix a TQFT $(\mathcal{T}, \tau)$ with ground ring K based on the cobordism theory $(\mathfrak{B}, \mathfrak{A})$.

2.1.1. Theorem. *The modular functor $\mathcal{T}$ is self-dual.*

In the proof of Theorem 2.1.1 given in Section 2.3 we shall explicitly construct for any $\mathfrak{A}$-space X a non-degenerate bilinear pairing $d_X = d_X^{\mathcal{T}} : \mathcal{T}(X) \otimes_K \mathcal{T}(-X) \to K$ satisfying (1.2.4). The existence of such a pairing implies that for any $\mathfrak{A}$-space X, we have $\operatorname{Dim}(\mathcal{T}(-X)) = \operatorname{Dim}(\mathcal{T}(X))$.

2.1.2. Theorem. *Let M be a closed $\mathfrak{B}$-space obtained from the disjoint union of two $\mathfrak{B}$-spaces M_1, M_2 by gluing along an $\mathfrak{A}$-homeomorphism $g : \partial M_1 \to -\partial M_2$. Let $-g$ be the same mapping g considered as an $\mathfrak{A}$-homeomorphism $-\partial M_1 \to \partial M_2$. If $(\mathcal{T}, \tau)$ is anomaly-free then*

$$\tau(M) = d_{\partial M_1}(\tau(M_1) \otimes (-g)_{\#}^{-1}(\tau(M_2))) = d_{\partial M_2}(\tau(M_2) \otimes g_{\#}(\tau(M_1))).$$

For an $\mathfrak{A}$-space X, consider the cylinder $X \times [0, 1]$ (with its $\mathfrak{B}$-structure) and glue its bases along the homeomorphism $x \times 0 \mapsto x \times 1 : X \times 0 \to X \times 1$. This yields a $\mathfrak{B}$-structure on the topological space $X \times S^1$. The resulting $\mathfrak{B}$-space is denoted by the same symbol $X \times S^1$. It follows from axioms (1.3.1) and (1.3.3) that $\partial(X \times S^1) = \emptyset$.

2.1.3. Theorem. *If $(\mathcal{T}, \tau)$ is anomaly-free then for any $\mathfrak{A}$-space X, we have $\operatorname{Dim}(\mathcal{T}(X)) = \tau(X \times S^1) \in K$.*

The remaining part of the section is concerned with the proof of Theorems 2.1.1 - 2.1.3.

2.2. Lemma. *Let P, Q be K-modules and let $b : K \to Q \otimes_K P$, $d : P \otimes_K Q \to K$ be K-homomorphisms satisfying the identities*

$$(\mathrm{id}_Q \otimes d)(b \otimes \mathrm{id}_Q) = k\,\mathrm{id}_Q, \quad (d \otimes \mathrm{id}_P)(\mathrm{id}_P \otimes b) = k'\mathrm{id}_P$$

where k and k' are invertible elements of K. Then $k = k'$ and both b and d are non-degenerate in the sense that the homomorphisms

$$Q \to P^\star = \mathrm{Hom}_K(P, K), \quad P \to Q^\star = \mathrm{Hom}_K(Q, K)$$

induced by d and the homomorphisms $P^\star \to Q, Q^\star \to P$ induced by b are isomorphisms.

Proof. Denote the homomorphisms $Q \to P^\star, P \to Q^\star, P^\star \to Q, Q^\star \to P$ mentioned in the statement of the lemma by f_1, f_2, f_3, f_4 respectively. The first identity between b and d implies that $f_3 f_1 = k\,\mathrm{id}_Q$. Indeed, if $b(1) = \sum_i q_i \otimes p_i$ with $q_i \in Q$, $p_i \in P$ then this identity indicates that for any $q \in Q$,

$$\sum_i d(p_i, q) q_i = kq.$$

The homomorphism f_1 carries q into the linear functional $p \mapsto d(p, q)$ and the homomorphism f_3 carries this functional $f_1(q)$ into

$$\sum_i f_1(q)(p_i) q_i = \sum_i d(p_i, q) q_i = kq.$$

A similar argument deduces from the same identity that $f_2 f_4 = k\,\mathrm{id}_{Q^\star}$. The second identity between b and d similarly implies that $f_1 f_3 = k'\mathrm{id}_{P^\star}$ and $f_4 f_2 = k'\mathrm{id}_P$. Therefore the homomorphisms f_1, f_2, f_3, f_4 are isomorphisms. The equalities $k f_3 = (f_3 f_1) f_3 = f_3(f_1 f_3) = k' f_3$ imply that $k = k'$.

2.3. Proof of Theorem 2.1.1. Let $(\mathcal{T}, \tau)$ be a TQFT. Let X be an $\mathfrak{A}$-space. Set $P = \mathcal{T}(X)$ and $Q = \mathcal{T}(-X)$. Denote by J the cylinder $X \times [0, 1]$ with the $\mathfrak{B}$-structure given by axiom (1.3.3) so that $\partial J = (-X) \amalg X$. This $\mathfrak{B}$-space gives rise to two cobordisms: $(J, \emptyset, \partial J)$ and $(J, -\partial J, \emptyset)$. Denote the corresponding operators $K \to \mathcal{T}(\partial J) = Q \otimes_K P$ and $P \otimes_K Q = \mathcal{T}(-\partial J) \to K$ by b_X and d_X respectively. We shall prove that the operators $\{d_X\}_X$ satisfy the self-duality axiom (1.2.4).

It follows from the definition of d_X and the axioms that this pairing is natural with respect to $\mathfrak{A}$-homeomorphisms and multiplicative with respect to disjoint union. Let us verify that $d_{-X} = d_X \mathrm{Perm}_{Q,P}$ where $\mathrm{Perm}_{Q,P}$ is the flip $Q \otimes P \to P \otimes Q$. Consider the homeomorphism $g : X \times [0, 1] \to X \times [0, 1]$ defined by the formula $g(x, t) = (x, 1 - t)$ where $x \in X, t \in [0, 1]$. Axiom (1.3.3) implies that g is a $\mathfrak{B}$-homeomorphism $X \times [0, 1] \to (-X) \times [0, 1]$. It is clear that g yields a homeomorphism of cobordisms

$$(X \times [0, 1], (X \times 0) \cup (-X \times 1), \emptyset) \to ((-X) \times [0, 1], (-X \times 0) \cup (X \times 1), \emptyset).$$

The homomorphism

$$P \otimes Q = \mathcal{T}((X \times 0) \cup (-X \times 1)) \to \mathcal{T}((-X \times 0) \cup (X \times 1)) = Q \otimes P$$

induced by g is just the flip $\mathrm{Perm}_{P,Q}$. The naturality of τ with respect to $\mathfrak{B}$-homeomorphisms of cobordisms implies that $d_{-X}\mathrm{Perm}_{P,Q} = d_X$. Therefore $d_{-X} = d_X \mathrm{Perm}_{Q,P}$. A similar argument shows that $b_{-X} = \mathrm{Perm}_{Q,P} b_X$.

It remains to prove non-degeneracy of d_X. We shall prove that d_X and b_X satisfy the conditions of Lemma 2.2. Let us take four copies J_1, J_2, J_3, J_4 of the cylinder $J = X \times [0, 1]$. Clearly, $\partial J_i = -X_i^- \amalg X_i^+$ where X_i^-, X_i^+ are copies of X with $i = 1, 2, 3, 4$. Consider the cobordisms $(J_1 \amalg J_2, -\partial J_1 \amalg X_2^-, X_2^+)$ and $(J_3 \amalg J_4, X_3^-, X_3^+ \amalg \partial J_4)$. The operators τ corresponding to these two cobordisms may be identified with $d_X \otimes \mathrm{id}_P$ and $\mathrm{id}_P \otimes b_X$ respectively. Gluing these two cobordisms along the identification $X_3^+ \amalg \partial J_4 = X \amalg (-X) \amalg X = -\partial J_1 \amalg X_2^-$ we get a cobordism, say M, between X_3^- and X_2^+. The functoriality axiom (1.4.3) implies that $\tau(M) = k(d_X \otimes \mathrm{id}_P)(\mathrm{id}_P \otimes b_X)$ for some invertible $k \in K$. On the other hand, applying axiom (1.3.4) three times we obtain that M is $\mathfrak{B}$-homeomorphic to the cylinder J via a homeomorphism extending the identifications of their bases $X_3^- = X$, $X_2^+ = X$ (see Figure 2.1). The naturality of τ with respect to $\mathfrak{B}$-homeomorphisms and the normalization axiom (1.4.4) imply that $\tau(M) = \tau(J) = \mathrm{id}_P$. Therefore $(d_X \otimes \mathrm{id}_P)(\mathrm{id}_P \otimes b_X) = k^{-1}\mathrm{id}_P$. Replacing in this formula X, P with $-X, Q$ respectively and using the expressions for d_{-X}, b_{-X} established above, we get an equality equivalent to the formula $(\mathrm{id}_Q \otimes d_X)(b_X \otimes \mathrm{id}_Q) = k'\mathrm{id}_Q$ with invertible $k' \in K$. Now, Lemma 2.2 implies that the pairing $d_X : \mathcal{T}(X) \otimes \mathcal{T}(-X) \to K$ is non-degenerate. This completes the proof of Theorem 2.1.1.

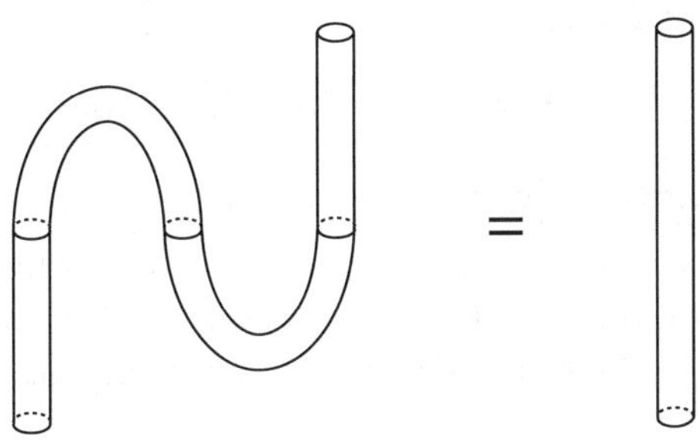

Figure 2.1

2.4. Lemma. *If $(\mathcal{T}, \tau)$ is anomaly-free then for any $\mathfrak{B}$-space M,*

$$\tau(M, -\partial M, \emptyset) = d_{\partial M}(\tau(M) \otimes id_{\mathcal{T}(-\partial M)}) : \mathcal{T}(-\partial M) \to K.$$

Proof. Set $X = \partial M$. Let X' be a copy of X. Let J and $\overline{J}$ be the cylinders $X \times [0, 1]$ and $(-X') \times [0, 1]$ respectively. Consider the cobordisms

$$(M \amalg \overline{J}, -X' \times 0, X \amalg (-X' \times 1)) \quad \text{and} \quad (J, (X \times 0) \amalg (-X \times 1), \emptyset).$$

The operators τ corresponding to these two cobordisms may be identified with $\tau(M) \otimes \mathrm{id}_{\mathcal{T}(-X)}$ and d_X respectively. Gluing these two cobordisms along $X \amalg (-X' \times 1) = (X \times 0) \amalg (-X \times 1)$ we get a cobordism M' between $-X' \times 0 = -X$ and $\emptyset$. It follows from the functoriality axiom (1.4.3) (with $k = 1$) that $\tau(M', -X, \emptyset) = d_X(\tau(M) \otimes \mathrm{id}_{\mathcal{T}(-X)})$.

Consider the cobordism $(M'', -X, \emptyset)$ obtained by gluing of the cylinder $(-X) \times [0, 1]$ to $(M, -X, \emptyset)$ along $-X \times 1 = -X$. It is easy to deduce from the axiom (1.3.4) that the cobordisms $(M'', -X, \emptyset)$ and $(M', -X, \emptyset)$ are $\mathfrak{B}$-homeomorphic via a homeomorphism extending the identity $-X = -X$ (see Figure 2.2). Therefore

$$\tau(M, -X, \emptyset) = \tau(M'', -X, \emptyset) = \tau(M', -X, \emptyset) = d_X(\tau(M) \otimes \mathrm{id}_{\mathcal{T}(-X)}).$$

Here the first equality follows from the functoriality and normalization axioms and the second equality follows from the naturality of τ.

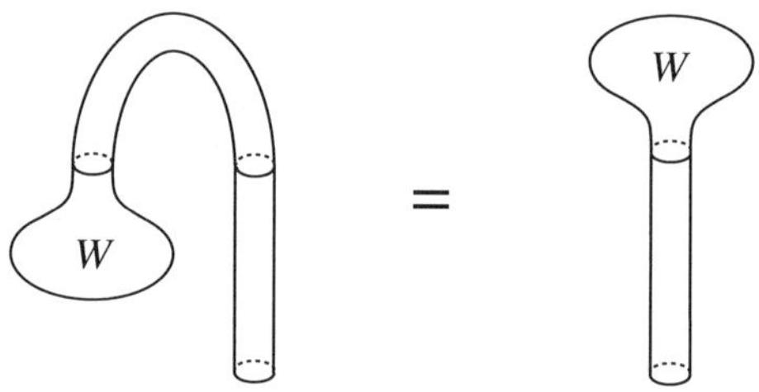

Figure 2.2

2.5. Proof of Theorem 2.1.2. We have

$$\tau(M, \emptyset, \emptyset) = \tau(M_2, -\partial M_2, \emptyset) \circ g_\# \circ \tau(M_1, \emptyset, \partial M_1) = d_{\partial M_2}(\tau(M_2) \otimes g_\#(\tau(M_1))) =$$

$$= d_{-\partial M_1}(((-g)_\#^{-1}(\tau(M_2)) \otimes \tau(M_1)) = d_{\partial M_1}(\tau(M_1) \otimes (-g)_\#^{-1}(\tau(M_2))).$$

Here the first equality follows from axiom (1.4.3) (with $k = 1$), the second equality follows from Lemma 2.4, the third equality follows from the naturality of d_X, and the last equality follows from the symmetry of d_X.

2.6. Lemma. *If under the conditions of Lemma 2.2 the modules P and Q are projective then*

$$\mathrm{Dim}(P) = \mathrm{Dim}(Q) = k^{-1}(d \circ \mathrm{Perm}_{Q,P} \circ b)(1) \in K.$$

Proof. If we identify P with $Q^\star$ via the isomorphism $P \to Q^\star$ induced by d then d is identified with the evaluation pairing $Q^\star \otimes Q \to K$ and b is identified with a homomorphism $K \to Q \otimes Q^\star$ which satisfies the identities $(\mathrm{id}_Q \otimes d)(b \otimes \mathrm{id}_Q) = k\, \mathrm{id}_Q$ and $(d \otimes \mathrm{id}_{Q^\star})(\mathrm{id}_{Q^\star} \otimes b) = k\, \mathrm{id}_{Q^\star}$. Such a homomorphism $b : K \to Q \otimes Q^\star$ is uniquely determined by the evaluation pairing d and equals $k\, b_Q$ where $b_Q =$

$d^\star : K \to Q \otimes Q^\star$ (cf. Section 1.7.1 of Chapter I). It follows directly from the definition of $\mathrm{Dim}(Q)$ that

$$\mathrm{Dim}(Q) = d\,\mathrm{Perm}_{Q,P}\, b_Q(1) = k^{-1}(d \circ \mathrm{Perm}_{Q,P} \circ b)(1).$$

The equality $\mathrm{Dim}(P) = \mathrm{Dim}(Q)$ follows from the duality of the modules P, Q.

2.7. Proof of Theorem 2.1.3. Let d_X and b_X be the linear operators defined in Section 2.3. The proof of Theorem 2.1.1 shows that these operators satisfy the equalities of Lemma 2.2. Since $(\mathcal{T}, \tau)$ is anomaly-free we may assume that $k = k' = 1$. The previous lemma shows that

$$\mathrm{Dim}(\mathcal{T}(X)) = (d_X \mathrm{Perm}_{\mathcal{T}(-X),\mathcal{T}(X)} b_X)(1) = d_{-X} b_X(1).$$

It follows from the definition of d_X, b_X and the functoriality axiom that $d_{-X} b_X = \tau(Z)$ where Z is the $\mathfrak{B}$-space obtained by gluing of $X \times [0, 1]$ to $(-X) \times [0, 1]$ along the identity homeomorphism of the boundaries. Gluing first along $X \times 0$ we obtain a $\mathfrak{B}$-space which is $\mathfrak{B}$-homeomorphic to $X \times [0, 1]$ because of the naturality of gluing with respect to $\mathfrak{B}$-homeomorphisms and axioms (1.3.3), (1.3.4). Therefore Z is $\mathfrak{B}$-homeomorphic to $X \times S^1$. This implies Theorem 2.1.3.

2.8. Exercises. Let $(\mathcal{T}, \tau)$ be an anomaly-free TQFT based on a cobordism theory $(\mathfrak{B}, \mathfrak{A})$.

1. Establish the following generalization of Theorem 2.1.3. Let X be an $\mathfrak{A}$-space and g be an $\mathfrak{A}$-homeomorphism $X \to X$. Gluing the top base of the cylinder $X \times [0, 1]$ to its bottom base along g we get a $\mathfrak{B}$-structure on the mapping torus M_g of g. Clearly, $\partial M_g = \emptyset$. Show that $\tau(M_g) = \mathrm{Tr}(g_\#)$ (for the definition of Tr, see Section I.1.7.1).

2. Let (M, X, Y) be a $(\mathfrak{B}, \mathfrak{A})$-cobordism. Show that under the identifications $\mathcal{T}(-X) = (\mathcal{T}(X))^\star$, $\mathcal{T}(-Y) = (\mathcal{T}(Y))^\star$ induced by d_X, d_Y we have

$$\tau(M, -Y, -X) = (\tau(M, X, Y))^\star.$$

3. Assume that the cobordism theory $(\mathfrak{B}, \mathfrak{A})$ is involutive in the sense that $\mathfrak{B}$ is involutive and the boundary commutes with negation: $\partial(-M) = -\partial M$ for any $\mathfrak{B}$-space M. We define the dual TQFT $(\mathcal{T}^\star, \tau^\star)$ by the formulas $\mathcal{T}^\star(X) = (\mathcal{T}(X))^\star$ for any $\mathfrak{A}$-space X and $\tau^\star(M, X, Y) = (\tau(-M, Y, X))^\star$ for any $(\mathfrak{B}, \mathfrak{A})$-cobordism (M, X, Y). (The action of homeomorphisms and the identification isomorphisms for disjoint unions are determined in the obvious way by the corresponding data for $(\mathcal{T}, \tau)$.) Verify that $(\mathcal{T}^\star, \tau^\star)$ is a TQFT. Show that for any $\mathfrak{B}$-space M, we have $\tau^\star(M) = \tau(-M)$ where $(\mathcal{T}(\partial M))^\star$ is identified with $\mathcal{T}(-\partial M)$ via $d^{\mathcal{T}}_{\partial M}$.

3. Isomorphisms of TQFT's

3.0. Outline. We discuss a natural non-degeneracy condition on TQFT's. It says roughly that the module of states of any $\mathfrak{A}$-space X is generated by the vectors $\tau(M)$ where M runs over $\mathfrak{B}$-spaces bounded by X. The non-degenerate anomaly-free TQFT's are much easier to work with than general TQFT's. In particular, we can solve the isomorphism problem for such TQFT's. We show that the K-valued function $M \mapsto \tau(M)$ on the class of closed $\mathfrak{B}$-spaces is the only isomorphism invariant of a non-degenerate anomaly-free TQFT. This result will play an important role in our study of 3-dimensional TQFT's, specifically in the proof of Theorem VII.4.2.

3.1. Non-degenerate TQFT's. We need a few preliminary definitions. Let X be a $\mathfrak{A}$-space. A $\mathfrak{B}$-space with X-parametrized boundary is a pair (a $\mathfrak{B}$-space M, an $\mathfrak{A}$-homeomorphism $f : X \to \partial M$). We shall call f the parametrizing homeomorphism. Sometimes, by abuse of notation, we shall denote such a pair (M, f) simply by M. If (M, f) is a $\mathfrak{B}$-space with X-parametrized boundary and $(N, g : -X \to \partial N)$ is a $\mathfrak{B}$-space with $(-X)$-parametrized boundary then $M \cup_X N$ denotes the closed $\mathfrak{B}$-space obtained by gluing of M to N along $(-g)f^{-1} : \partial M \to -\partial N$. It is clear that $M \cup_X N = N \cup_{-X} M$.

Let $(\mathcal{T}, \tau)$ be a TQFT based on $(\mathfrak{B}, \mathfrak{A})$. For any $\mathfrak{A}$-space X and any $\mathfrak{B}$-space with X-parametrized boundary $(M, f : X \to \partial M)$, set

$$\tau^X(M, f) = f_{\#}^{-1}(\tau(M)) \in \mathcal{T}(X).$$

To simplify the notation we shall sometimes denote $\tau^X(M, f)$ by $\tau^X(M)$ where presence of f is implicitly understood.

The TQFT $(\mathcal{T}, \tau)$ is said to be non-degenerate if for any $\mathfrak{A}$-space X, the elements $\{\tau^X(M, f) \in \mathcal{T}(X)\}_{(M,f)}$ determined by $\mathfrak{B}$-spaces with X-parametrized boundary generate $\mathcal{T}(X)$ over the ground ring K.

3.2. Homomorphisms and isomorphisms of TQFT's. Let $(\mathcal{T}_1, \tau_1)$ and $(\mathcal{T}_2, \tau_2)$ be two TQFT's with the same ground ring K based on the same cobordism theory $(\mathfrak{B}, \mathfrak{A})$. A homomorphism $(\mathcal{T}_1, \tau_1) \to (\mathcal{T}_2, \tau_2)$ assigns to every $\mathfrak{A}$-space X a K-homomorphism $\mathcal{T}_1(X) \to \mathcal{T}_2(X)$ which commutes with the action of $\mathfrak{A}$-homeomorphisms, with the identification isomorphisms for disjoint unions, and with the operators corresponding to cobordisms. Examples: zero homomorphism $(\mathcal{T}_1, \tau_1) \to (\mathcal{T}_2, \tau_2)$, identity homomorphism $(\mathcal{T}_1, \tau_1) \to (\mathcal{T}_1, \tau_1)$.

A homomorphism of TQFT's $g : (\mathcal{T}_1, \tau_1) \to (\mathcal{T}_2, \tau_2)$ is an isomorphism iff for any $\mathfrak{A}$-space X, the homomorphism $g(X) : \mathcal{T}_1(X) \to \mathcal{T}_2(X)$ is an isomorphism.

3.2.1. Lemma. *Let $g : (\mathcal{T}_1, \tau_1) \to (\mathcal{T}_2, \tau_2)$ be an isomorphism. Then $g(\emptyset) = \mathrm{id}_K$ and for any $\mathfrak{B}$-space M, we have $\tau_2(M) = g(\partial M)\,(\tau_1(M)) \in \mathcal{T}_2(\partial M)$.*

Lemma 3.2.1 implies that if the TQFT's $(\mathcal{T}_1, \tau_1)$ and $(\mathcal{T}_2, \tau_2)$ are isomorphic then for any closed $\mathfrak{B}$-space M, we have $\tau_1(M) = \tau_2(M)$. This lemma also implies that the non-degeneracy of TQFT's is preserved under isomorphism.

Proof of Lemma. The K-linear homomorphism

$$g(\emptyset) : K = \mathcal{T}_1(\emptyset) \to \mathcal{T}_2(\emptyset) = K$$

is the multiplication by a certain $k \in K$. Since $g(\emptyset)$ is an isomorphism, k is invertible in K. The obvious formula $\emptyset = \emptyset \amalg \emptyset$ and the compatibility of g with disjoint union imply that $k^2 = k$. Therefore $k = 1$. Thus, $g(\emptyset) = \mathrm{id}_K$. The equality $\tau_2(M) = g(\partial M)\,(\tau_1(M))$ follows from definitions and commutativity of the diagram

$$\begin{array}{ccc} \mathcal{T}_1(\emptyset) & \xrightarrow{g(\emptyset)} & \mathcal{T}_2(\emptyset) \\ {\scriptstyle \tau_1(M,\emptyset,\partial M)}\downarrow & & \downarrow{\scriptstyle \tau_2(M,\emptyset,\partial M)} \\ \mathcal{T}_1(\partial M) & \xrightarrow{g(\partial M)} & \mathcal{T}_2(\partial M). \end{array}$$

The commutativity of this diagram follows from the assumption that g commutes with the operator invariants of cobordisms.

3.3. Theorem. *Let $(\mathcal{T}_1, \tau_1)$ and $(\mathcal{T}_2, \tau_2)$ be non-degenerate anomaly-free TQFT's with the same ground ring K based on the same cobordism theory $(\mathfrak{B}, \mathfrak{A})$. If for every closed $\mathfrak{B}$-space M, we have $\tau_1(M) = \tau_2(M)$ then the TQFT's $(\mathcal{T}_1, \tau_1)$ and $(\mathcal{T}_2, \tau_2)$ are isomorphic.*

This theorem reduces the isomorphism problem for non-degenerate anomaly-free TQFT's to a simpler problem concerned with K-valued functions.

At the end of this section we give a version of Theorem 3.3 where we require that only one of the two TQFT's at hand is non-degenerate at the price of introducing certain constraints on the ground ring K.

Theorem 3.3 is proven in Section 3.6 using the results of Sections 3.4 and 3.5.

3.4. Modules α and β. For any TQFT $(\mathcal{T}, \tau)$ with ground ring K based on $(\mathfrak{B}, \mathfrak{A})$ and any $\mathfrak{A}$-space X, we denote by $\alpha_\tau(X)$ the submodule of $\mathcal{T}(X)$ generated over K by the set $\{\tau^X(M, f)\}_{(M,f)}$ where (M, f) runs over all $\mathfrak{B}$-spaces with X-parametrized boundary. (In particular, f runs over all X-parametrizations of ∂M.) It is obvious that the construction $X \mapsto \alpha_\tau(X)$ is natural with respect to $\mathfrak{A}$-homeomorphisms.

We define a quotient module $\beta_\tau(X)$ by the formula

$$\beta_\tau(X) = \alpha_\tau(X)/(\alpha_\tau(X) \cap \mathrm{Ann}(\alpha_\tau(-X)))$$

where $\mathrm{Ann}(\alpha_\tau(-X)) \subset \mathcal{T}(X)$ denotes the annihilator of $\alpha_\tau(-X) \subset \mathcal{T}(-X)$ with respect to the pairing $d_X^\tau : \mathcal{T}(X) \otimes_K \mathcal{T}(-X) \to K$ constructed in Section 2.

It is clear that the TQFT $(\mathcal{T}, \tau)$ is non-degenerate if and only if $\alpha_\tau(X) = \mathcal{T}(X)$ for any $\mathfrak{A}$-space X. If $(\mathcal{T}, \tau)$ is non-degenerate then $\beta_\tau(X) = \alpha_\tau(X) = \mathcal{T}(X)$. This follows from the non-degeneracy of the pairing d_X^τ (Theorem 2.1.1).

3.5. Lemma. *Let $(\mathcal{T}_1, \tau_1)$ and $(\mathcal{T}_2, \tau_2)$ be two anomaly-free TQFT's with ground ring K based on $(\mathfrak{B}, \mathfrak{A})$. Assume that for any closed $\mathfrak{B}$-space M, we have $\tau_1(M) = \tau_2(M)$. Then for any $\mathfrak{A}$-space X, there is a canonical isomorphism $\eta_X : \beta_{\tau_1}(X) \to \beta_{\tau_2}(X)$.*

Proof. Let X be an $\mathfrak{A}$-space. For $i = 1, 2$, set $A_i = \operatorname{Ann}(\alpha_{\tau_i}(-X)) \subset \mathcal{T}_i(X)$. For an arbitrary $\mathfrak{B}$-space with X-parametrized boundary M, set

$$\eta_X(\tau_1^X(M) \pmod{A_1}) = \tau_2^X(M) \pmod{A_2}.$$

Extend η_X by K-linearity to the whole module $\beta_{\tau_1}(X)$. To show that this gives a well-defined homomorphism $\beta_{\tau_1}(X) \to \beta_{\tau_2}(X)$ it suffices to show the following assertion. Let $\{M_r\}_r$ be a finite family of $\mathfrak{B}$-spaces with X-parametrized boundary such that for certain $\{k_r \in K\}_r$, we have $\sum_r k_r \tau_1^X(M_r) \in A_1$. Then $\sum_r k_r \tau_2^X(M_r) \in A_2$. This assertion follows from the fact that for any $\mathfrak{B}$-space with $(-X)$-parametrized boundary N, we have

$$d_X^{\tau_2}\Big(\sum_r k_r \tau_2^X(M_r) \otimes \tau_2^{-X}(N)\Big) = \sum_r k_r \tau_2(M_r \cup_X N) =$$

$$= \sum_r k_r \tau_1(M_r \cup_X N) = d_X^{\tau_1}\Big(\sum_r k_r \tau_1^X(M_r) \otimes \tau_1^{-X}(N)\Big) = 0.$$

Here the first and third equalities follow from Theorem 2.1.2, the second equality follows from the assumptions of the lemma, and the last equality follows from the assumption $\sum_r k_r \tau_1^X(M_r) \in A_1$. Therefore η_X is a well-defined homomorphism $\beta_{\tau_1}(X) \to \beta_{\tau_2}(X)$. Exchanging the roles of $(\mathcal{T}_1, \tau_1)$ and $(\mathcal{T}_2, \tau_2)$ we get a homomorphism $\beta_{\tau_2}(X) \to \beta_{\tau_1}(X)$. By the very definition of these homomorphisms they are mutually inverse. Hence, η_X is an isomorphism.

3.6. Proof of Theorem 3.3. Non-degeneracy of $(\mathcal{T}_i, \tau_i)$ implies that for every $\mathfrak{A}$-space X, we have $\beta_{\tau_i}(X) = \mathcal{T}_i(X)$ where $i = 1, 2$. Lemma 3.5 yields an isomorphism $\eta_X : \mathcal{T}_1(X) \to \mathcal{T}_2(X)$. Let us check that the isomorphisms $\{\eta_X\}_X$ constitute an isomorphism $(\mathcal{T}_1, \tau_1) \to (\mathcal{T}_2, \tau_2)$. It follows from definitions that these isomorphisms commute with the action of $\mathfrak{A}$-homeomorphisms. It is clear that $\eta_\emptyset = \mathrm{id}_K : K \to K$. It follows from the multiplicativity axiom (1.4.2) that for any $\mathfrak{A}$-spaces X, Y, we have $\eta_{X \amalg Y} = \eta_X \otimes \eta_Y$. It remains to establish that the isomorphisms $\{\eta_X\}_X$ commute with the operator invariants of cobordisms. Let (W, X, Y) be a $(\mathfrak{B}, \mathfrak{A})$-cobordism. We should establish commutativity of the following diagram

$$\begin{array}{ccc} \mathcal{T}_1(X) & \xrightarrow{\eta_X} & \mathcal{T}_2(X) \\ {\scriptstyle \tau_1(W)}\big\downarrow & & \big\downarrow{\scriptstyle \tau_2(W)} \\ \mathcal{T}_1(Y) & \xrightarrow{\eta_Y} & \mathcal{T}_2(Y). \end{array} \tag{3.6.a}$$

Since the TQFT $(\mathcal{T}_1, \tau_1)$ is non-degenerate it suffices to verify that the homomorphisms $\eta_Y \circ \tau_1(W)$ and $\tau_2(W) \circ \eta_X$ coincide on any $\tau_1^X(M) \in \mathcal{T}_1(X)$ where M is a $\mathfrak{B}$-space with X-parametrized boundary. Denote by W' the $\mathfrak{B}$-space obtained by gluing of M to W along the $\mathfrak{A}$-homeomorphism $\partial M \to X$ inverse to the given parametrization $X \to \partial M$. It is clear that $\partial W' = Y$. In the absense of anomalies the functoriality axiom (1.4.3) implies that for $i = 1, 2$,

$$\tau_i(W') = \tau_i(W)(\tau_i^X(M)) \in \mathcal{T}_i(Y).$$

By the definition of η we have $\eta_X(\tau_1^X(M)) = \tau_2^X(M)$ and $\eta_Y(\tau_1(W')) = \tau_2(W')$. Therefore

$$(\eta_Y \circ \tau_1(W))(\tau_1^X(M)) = \eta_Y(\tau_1(W')) = \tau_2(W') =$$

$$= \tau_2(W)(\tau_2^X(M)) = (\tau_2(W) \circ \eta_X)(\tau_1^X(M)).$$

This proves commutativity of diagram (3.6.a) and completes the proof of the theorem.

3.7. Theorem. *Let $(\mathcal{T}_1, \tau_1)$ and $(\mathcal{T}_2, \tau_2)$ be anomaly-free TQFT's with ground ring K based on $(\mathfrak{B}, \mathfrak{A})$. Assume that:* (i) *at least one of these two TQFT's is non-degenerate;* (ii) *for any closed $\mathfrak{B}$-space M, we have $\tau_1(M) = \tau_2(M)$;* (iii) *the ring K is a field of zero characteristic. Then the TQFT's $(\mathcal{T}_1, \tau_1)$ and $(\mathcal{T}_2, \tau_2)$ are isomorphic.*

Proof. Assume that $(\mathcal{T}_1, \tau_1)$ is non-degenerate. Let us show that $\beta_{\tau_2}(X) = \mathcal{T}_2(X)$. We have

$$\mathrm{Dim}(\beta_{\tau_2}(X)) = \mathrm{Dim}(\beta_{\tau_1}(X)) = \mathrm{Dim}(\mathcal{T}_1(X)) =$$

$$= \tau_1(X \times S^1) = \tau_2(X \times S^1) = \mathrm{Dim}(\mathcal{T}_2(X)).$$

These equalities follow respectively from Lemma 3.5, the equality $\beta_{\tau_1}(X) = \mathcal{T}_1(X)$, Theorem 2.1.3, the assumption (ii), and again Theorem 2.1.3. Since $\beta_{\tau_2}(X)$ is a quotient of a linear subspace of $\mathcal{T}_2(X)$ the equality of their dimensions implies that $\beta_{\tau_2}(X) = \mathcal{T}_2(X)$. (Here we use the assumption (iii).) Now we just repeat the proof of the previous theorem.

3.8. Exercises. 1. Show that the TQFT dual to a non-degenerate TQFT (over an involutive cobordism theory) is also non-degenerate.

2. Show that the TQFT's defined in Sections 1.5.1 and 1.5.2 are non-degenerate and degenerate respectively.

4. Quantum invariants

4.0. Outline. According to Theorem 3.3 the invariants of closed $\mathfrak{B}$-spaces classify non-degenerate anomaly-free TQFT's up to isomorphism. Here we give an internal characterization of these invariants. To this end we introduce an abstract notion of a quantum invariant. It is defined using only the definitions of space-structure and cobordism theory given in Sections 1.1 and 1.3. We show that quantum invariants are exactly the invariants of closed $\mathfrak{B}$-spaces provided by non-degenerate anomaly-free TQFT's.

The material of this section will not be used in the remaining part of the book.

4.1. Quantum invariants of closed $\mathfrak{B}$-spaces. We need a few preliminary definitions. We say that a $\mathfrak{B}$-space has a collar if it is $\mathfrak{B}$-homeomorphic to the result of gluing of a certain $\mathfrak{B}$-space M and the cylinder $\partial M \times [0, 1]$ along $\partial M = \partial M \times 0$. (As usual, the $\mathfrak{B}$-structure on $\partial M \times [0, 1]$ is induced by the $\mathfrak{A}$-structure on ∂M determined by the $\mathfrak{B}$-structure on M.) The gluing of the cylinder $\partial M \times [0, 1]$ to M along $\partial M = \partial M \times 0$ is called the gluing of a collar to M. It is clear that such a gluing yields a $\mathfrak{B}$-space having a collar. Axiom (1.3.4) implies that if M already has a collar then gluing of a collar to M yields a $\mathfrak{B}$-space homeomorphic to M via a $\mathfrak{B}$-homeomorphism which is the identity on the boundary. The disjoint union of $\mathfrak{B}$-spaces with collars gives a $\mathfrak{B}$-space with a collar. Note that in standard cobordism theories of manifolds all $\mathfrak{B}$-spaces have collars. In the general cobordism theory $(\mathfrak{B}, \mathfrak{A})$ this is not necessarily the case.

Let τ_0 be a function assigning to every closed $\mathfrak{B}$-space M a certain $\tau_0(M) \in K$. Assume that τ_0 is a $\mathfrak{B}$-homeomorphism invariant so that for any $\mathfrak{B}$-homeomorphic closed $\mathfrak{B}$-spaces M, M', we have $\tau_0(M) = \tau_0(M')$. Let X be a $\mathfrak{A}$-space. Assume that we have a finite family of triples $\{(k_s, M_s, N_s)\}_s$ where $k_s \in K$, M_s is a $\mathfrak{B}$-space with X-parametrized boundary and a collar, and N_s is a $\mathfrak{B}$-space with $(-X)$-parametrized boundary and a collar. We call this family of triples a splitting system for X (with respect to τ_0) if, for an arbitrary pair (a $\mathfrak{B}$-space with X-parametrized boundary M, a $\mathfrak{B}$-space with $(-X)$-parametrized boundary N),

$$\tau_0(M \cup_X N) = \sum_s k_s\, \tau_0(M \cup_X N_s)\, \tau_0(M_s \cup_X N). \tag{4.1.a}$$

For example, the triple $(1, M_1 = \emptyset, N_1 = \emptyset)$ is a splitting system for $X = \emptyset$ if and only if the invariant τ_0 is multiplicative with respect to disjoint union.

A quantum invariant of closed $\mathfrak{B}$-spaces is a K-valued $\mathfrak{B}$-homeomorphism invariant τ_0 of closed $\mathfrak{B}$-spaces satisfying the following three axioms.

(4.1.1) (Normalization). $\tau_0(\emptyset) = 1$.

(4.1.2) (Multiplicativity). For any two closed $\mathfrak{B}$-spaces M, N, we have $\tau_0(M \amalg N) = \tau_0(M)\, \tau_0(N)$.

(4.1.3) (Splitting). Any $\mathfrak{A}$-space X has a splitting system with respect to τ_0. Any $\mathfrak{A}$-spaces X and Y have splitting systems $\{(k_s, M_s, N_s)\}_s$ and $\{(k'_t, M'_t, N'_t)\}_t$ respectively such that the family $\{(k_s k'_t, M_s \amalg M'_t, N_s \amalg N'_t)\}_{(s,t)}$ is a splitting system for $X \amalg Y$.

We call the last family the product of the families $\{(k_s, M_s, N_s)\}_s$ and $\{(k'_t, M'_t, N'_t)\}_t$. As an exercise the reader may verify that if the product of certain splitting systems of X and Y yields a splitting system for $X \amalg Y$ then the same is true for arbitrary splitting systems of X and Y. (We shall not use this fact.)

Although quantum invariants are defined for $\mathfrak{B}$-spaces without boundary, their definition explicitly involves $\mathfrak{A}$-spaces. Thus, to speak of a quantum invariant of closed $\mathfrak{B}$-spaces we must involve the space-structure $\mathfrak{A}$ as well as $\mathfrak{B}$-spaces with boundary.

The next two lemmas clarify the relationships between quantum invariants and topological quantum field theories.

4.2. Lemma. *Let $(\mathcal{T}, \tau)$ be a non-degenerate anomaly-free TQFT (with ground ring K based on $(\mathfrak{B}, \mathfrak{A})$). Then the function $M \mapsto \tau(M)$ on the class of closed $\mathfrak{B}$-spaces is a quantum invariant.*

We shall say that the quantum invariant $M \mapsto \tau(M)$ underlies the TQFT $(\mathcal{T}, \tau)$ and that $(\mathcal{T}, \tau)$ extends this invariant.

Proof of Lemma. The only non-trivial point is to verify the splitting axiom (4.1.3). Let X be a $\mathfrak{A}$-space. Let $b_X : K \to \mathcal{T}(-X) \otimes_K \mathcal{T}(X)$ be the homomorphism constructed at the beginning of Section 2.3. Since the TQFT $(\mathcal{T}, \tau)$ is non-degenerate we may present $b_X(1)$ in the following form:

$$b_X(1) = \sum_s k_s \tau^{-X}(N_s) \otimes \tau^X(M_s) \tag{4.2.a}$$

where s runs over a certain finite set of indices, $k_s \in K$, M_s is a $\mathfrak{B}$-space with X-parametrized boundary, and N_s is a $\mathfrak{B}$-space with $(-X)$-parametrized boundary. Gluing collars to these spaces we get $\mathfrak{B}$-spaces with X-parametrized (resp. $(-X)$-parametrized) boundary and collars. This new spaces still satisfy (4.2.a) because of (1.4.3) and (1.4.4). We denote these $\mathfrak{B}$-spaces by the same symbols M_s, N_s. We claim that the family $\{(k_s, M_s, N_s)\}_s$ is a splitting system for X.

Let us first show that

$$\mathrm{id}_{\mathcal{T}(X)} = \sum_s k_s\, \tau^X(M_s, \emptyset, X)\, \tau^X(N_s, X, \emptyset) : \mathcal{T}(X) \to \mathcal{T}(X). \tag{4.2.b}$$

Here $\tau^X(N_s, X, \emptyset)$ denotes the linear operator $\mathcal{T}(X) \to K$ obtained by composing $\tau(N_s, -\partial N_s, \emptyset) : \mathcal{T}(-\partial N_s) \to K$ with the parametrization isomorphism $\mathcal{T}(X) \to \mathcal{T}(-\partial N_s)$. The homomorphism $\tau^X(M_s, \emptyset, X)$ is defined in a similar way and coincides with the homomorphism $k \mapsto k\,\tau^X(M_s) : K \to \mathcal{T}(X)$.

We shall deduce (4.2.b) from the formula $\mathrm{id}_{\mathcal{T}(X)} = (d_X \otimes \mathrm{id}_{\mathcal{T}(X)})(\mathrm{id}_{\mathcal{T}(X)} \otimes b_X)$ established in Section 2.3. Note that since the TQFT at hand is anomaly-free we have no additional factors in this formula. Substituting on the right-hand side of this formula the expression (4.2.a) for $b_X(1)$ we get

$$\mathrm{id}_{\mathcal{T}(X)} = \sum_s k_s\, d_X(\mathrm{id}_{\mathcal{T}(X)} \otimes \tau^{-X}(N_s))\, \tau^X(M_s).$$

It follows from the symmetry of the self-duality pairings and Lemma 2.4 that

$$d_X(\mathrm{id}_{\mathcal{T}(X)} \otimes \tau^{-X}(N_s)) = d_{-X}(\tau^{-X}(N_s) \otimes \mathrm{id}_{\mathcal{T}(X)}) = \tau^X(N_s, X, \emptyset).$$

Combining these formulas together we get (4.2.b). This proof of (4.2.b) is reproduced in a schematic form in Figure 4.1 where the symbol $\doteq$ indicates the equality of the corresponding operator invariants.

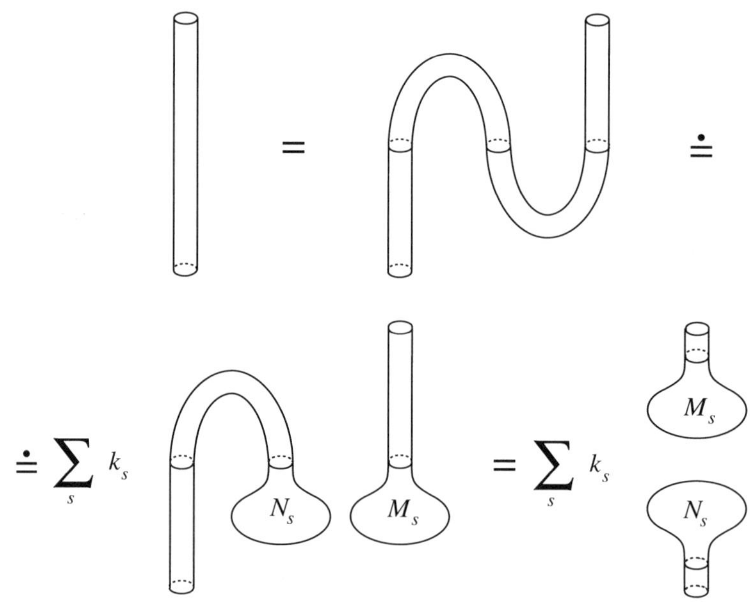

Figure 4.1

Let us show that the family $\{(k_s, M_s, N_s)\}_s$ is a splitting system for X. Let M be a $\mathfrak{B}$-space with X-parametrized boundary and N be a $\mathfrak{B}$-space with

$(-X)$-parametrized boundary. Using the functoriality axiom (1.4.3) and formula (4.2.b) we get

$$\tau(M \cup_X N) = \tau(N, X, \emptyset)\, \tau(M, \emptyset, X) = \tau(N, X, \emptyset)\, \mathrm{id}_{\mathcal{T}(X)}\, \tau(M, \emptyset, X) =$$

$$= \sum_s k_s\, \tau(N, X, \emptyset)\, \tau^X(M_s, \emptyset, X)\, \tau^X(N_s, X, \emptyset)\, \tau(M, \emptyset, X) =$$

$$= \sum_s k_s\, \tau(M_s \cup_X N)\, \tau(M \cup_X N_s).$$

Let us verify the second part of the splitting axiom. Let X and Y be $\mathfrak{A}$-spaces. Let $\{(k_s, M_s, N_s)\}_s$ be a family of triples associated to X as above and satisfying (4.2.b). Let $\{(k'_t, M'_t, N'_t)\}_t$ be a family of triples associated to Y in a similar way and providing a decomposition of $\mathrm{id}_{\mathcal{T}(Y)}$ analogous to (4.2.b). Since $\mathcal{T}(X \amalg Y) = \mathcal{T}(X) \otimes_K \mathcal{T}(Y)$ we have $\mathrm{id}_{\mathcal{T}(X \amalg Y)} = \mathrm{id}_{\mathcal{T}(X)} \otimes \mathrm{id}_{\mathcal{T}(Y)}$. This implies that the family $\{(k_s k'_t, M_s \amalg M'_t, N_s \amalg N'_t)\}_{(s,t)}$ yields a decomposition of $\mathrm{id}_{\mathcal{T}(X \amalg Y)}$ analogous to (4.2.b). Therefore this family is a splitting system for $X \amalg Y$.

4.3. Lemma. *Every quantum invariant τ_0 of closed $\mathfrak{B}$-spaces extends to a non-degenerate anomaly-free TQFT based on* $(\mathfrak{B}, \mathfrak{A})$.

Proof. For every $\mathfrak{A}$-space X, we define $T(X)$ to be the K-module freely generated by $\mathfrak{B}$-spaces with X-parametrized boundary. Each $\mathfrak{B}$-space with X-parametrized boundary M represents a generator of $T(X)$. Consider the bilinear pairing $\omega_X : T(X) \times T(-X) \to K$ defined on the generators by the formula $\omega_X(M, N) = \tau_0(M \cup_X N)$. Let

$$\mathrm{Ann}_X = \{t \in T(X) \mid \omega_X(t, T(-X)) = 0\}$$

be the left annihilator of this pairing. Set $\mathcal{T}(X) = T(X)/\mathrm{Ann}_X$. For any $\mathfrak{B}$-space with X-parametrized boundary M, set $[M] = M \pmod{\mathrm{Ann}_X} \in \mathcal{T}(X)$.

Below we shall use one simple observation concerned with $\mathcal{T}(X)$. Consider $\mathfrak{B}$-spaces with X-parametrized boundary M, M' which are homeomorphic via a $\mathfrak{B}$-homeomorphism $M \to M'$ commuting with the parametrizations of $\partial M, \partial M'$. Then $[M] = [M']$. Indeed, for any $\mathfrak{B}$-space with $(-X)$-parametrized boundary N, the closed $\mathfrak{B}$-spaces $M \cup_X N$ and $M' \cup_X N$ are $\mathfrak{B}$-homeomorphic. Therefore $\omega_X(M, N) = \omega_X(M', N)$ and $M - M' \in \mathrm{Ann}_X$.

Let us show that the module $\mathcal{T}(X)$ is projective. Let $\{(k_s, M_s, N_s)\}_{s=1}^r$ be a splitting system for X where r is a positive integer. We define a homomorphism $i = i_X : K^r \to \mathcal{T}(X)$ by the formula $i((x_1, \ldots, x_r)) = \sum_{s=1}^r k_s x_s [M_s]$ where $x_1, \ldots, x_r \in K$. We define a homomorphism $j = j_X : \mathcal{T}(X) \to K^r$ by its values on the generators:

$$j([M]) = (\tau_0(M \cup_X N_1), \tau_0(M \cup_X N_2), \ldots, \tau_0(M \cup_X N_r)).$$

It follows from the definition of $\mathcal{T}(X)$ that j is a well-defined homomorphism. Let us show that $i \circ j = \mathrm{id}_{\mathcal{T}(X)}$. This equality is equivalent to the fact that for any $\mathfrak{B}$-space with X-parametrized boundary M, the formal linear combination $M - \sum_{s=1}^{r} k_s \tau_0(M \cup_X N_s)[M_s]$ lies in Ann_X. This fact follows directly from (4.1.a). Thus $i \circ j = \mathrm{id}_{\mathcal{T}(X)}$ so that $K^r = \mathcal{T}(X) \oplus \mathrm{Ker}(i)$. This implies projectivity of $\mathcal{T}(X)$.

It is obvious that the construction $X \mapsto \mathcal{T}(X)$ is functorial with respect to $\mathfrak{A}$-homeomorphisms. We claim that $\mathcal{T}$ is a modular functor. Let us check axioms (1.2.1) - (1.2.3) of modular functor. Axiom (1.2.1) is straightforward. To define the isomorphism $\mathcal{T}(X \amalg Y) = \mathcal{T}(X) \otimes_K \mathcal{T}(Y)$ note that for any $\mathfrak{B}$-spaces M, N whose boundaries are parametrized by X and Y respectively, the disjoint union $M \amalg N$ is a $\mathfrak{B}$-space with boundary parametrized by $X \amalg Y$. Moreover, if W is a $\mathfrak{B}$-space with boundary parametrized by $-(X \amalg Y) = (-X) \amalg (-Y)$ then by the very definition of the form ω we have

$$\omega_{X \amalg Y}(M \amalg N, W) = \omega_X(M, N \cup_Y W) = \omega_Y(N, M \cup_X W).$$

These formulas imply that the disjoint union induces a homomorphism $\mathcal{T}(X) \otimes_K \mathcal{T}(Y) \to \mathcal{T}(X \amalg Y)$. Denote this homomorphism by u. Let us show that u is an isomorphism. We define a homomorphism v going in the opposite direction. Let $\{(k_s, M_s, N_s)\}_s$ be a splitting system for X and let $\{(k'_t, M'_t, N'_t)\}_t$ be a splitting system for Y such that their product is a splitting system for $X \amalg Y$. For any $\mathfrak{B}$-space V with boundary parametrized by $X \amalg Y$, set

$$v([V]) = \sum_{s,t} k_s\, k'_t\, \omega_{X \amalg Y}(V, N_s \amalg N'_t)\, [M_s] \otimes [M'_t].$$

The formula $[V] \mapsto v([V])$ defines a homomorphism

$$v : \mathcal{T}(X \amalg Y) \to \mathcal{T}(X) \otimes_K \mathcal{T}(Y).$$

The multiplicativity of τ_0 (which we use here for the first time) implies that for any generators $[M] \in \mathcal{T}(X)$, $[M'] \in \mathcal{T}(Y)$, we have

$$vu\,([M] \otimes [M']) = v\,([M \amalg M']) =$$

$$= \sum_{s,t} k_s\, k'_t\, \tau_0(M \cup_X N_s)\, \tau_0(M' \cup_Y N'_t)\, [M_s] \otimes [M'_t].$$

The last expression is equal to $(i_X \circ j_X)([M]) \otimes (i_Y \circ j_Y)([M'])$ where i, j are the homomorphisms defined in the previous paragraph. Since $i \circ j = \mathrm{id}$ we get $vu = \mathrm{id}$. Let us show that $uv = \mathrm{id}$ as well. It suffices to check that for any generator $[V]$ of $\mathcal{T}(X \amalg Y)$, we have $uv([V]) = [V]$. For any $\mathfrak{B}$-space W with the boundary parametrized by $(-X) \amalg (-Y)$, we have

$$\omega_{X \amalg Y}(uv([V]), W) = \sum_{s,t} k_s\, k'_t\, \tau_0(V \cup_{X \amalg Y} (N_s \amalg N'_t))\, \tau_0((M_s \amalg M'_t) \cup_{X \amalg Y} W).$$

Since the family $\{(k_s k'_t, M_s \amalg M'_t, N_s \amalg N'_t)\}_{s,t}$ is a splitting system for $X \amalg Y$, the right-hand side of the last formula is equal to $\tau_0(V \cup_{X\amalg Y} W) = \omega_{X\amalg Y}(V, W)$. This shows that $uv([V]) = [V]$. Therefore u and v are mutually inverse isomorphisms. It is straightforward to check that u is natural, commutative, and associative in the sense specified in the definition of a modular functor. This completes the verification of (1.2.2).

The formula $M \mapsto \tau_0(M)$ defines a homomorphism $\mathcal{T}(\emptyset) \to K$. Injectivity and surjectivity of this homomorphism follow from (4.1.2) and (4.1.1) respectively. Therefore $\mathcal{T}(\emptyset) = K$. Note that under this identification the class of the empty space $[\emptyset] \in \mathcal{T}(\emptyset)$ corresponds to the unit $1_K \in K$. This implies that for any $\mathfrak{A}$-space Y, the identification $\mathcal{T}(Y) = \mathcal{T}(\emptyset \amalg Y) = \mathcal{T}(\emptyset) \otimes_K \mathcal{T}(Y)$ is induced by $\mathcal{T}(\emptyset) = K$. Therefore $\mathcal{T}$ is a modular functor.

Let us extend $\mathcal{T}$ to a TQFT. For any $(\mathfrak{B}, \mathfrak{A})$-cobordism (M, X, Y), we define the homomorphism $\tau(M) : \mathcal{T}(X) \to \mathcal{T}(Y)$ as follows. For a generator $[V] \in \mathcal{T}(X)$ represented by a $\mathfrak{B}$-space with X-parametrized boundary V, set

$$\tau(M, X, Y)\,([V]) = [V \cup_X M] \in \mathcal{T}(Y)$$

where the boundary of $V \cup_X M$ is provided with the identity parametrization $Y = \partial(V \cup_X M)$. It follows from definitions that this yields a well-defined homomorphism $\mathcal{T}(X) \to \mathcal{T}(Y)$. It is easy to verify axioms (1.4.1) - (1.4.3) with the trivial anomaly $k = 1$. (To verify (1.4.1) we need the observation made after the definition of $\mathcal{T}(X)$.) Let us verify (1.4.4) for an $\mathfrak{A}$-space X. Take a splitting system $\{(k_s, M_s, N_s)\}_s$ for X. The arguments given above show that the module $\mathcal{T}(X)$ is generated by the elements $\{[M_s]\}_s$. By the definition of splitting system each M_s has a collar. Gluing $X \times [0, 1]$ to M_s along the parametrization $X \times 0 = X \to \partial M_s$ we get a $\mathfrak{B}$-space with X-parametrized boundary M'_s homeomorphic to M_s via a $\mathfrak{B}$-homeomorphism commuting with the parametrizations of boundaries. The observation made at the beginning of the proof shows that $\tau(X \times [0, 1])([M_s]) = [M'_s] = [M_s]$. Therefore $\tau(X \times [0, 1]) = \mathrm{id}_{\mathcal{T}(X)}$.

4.4. Theorem. *Assigning to each non-degenerate anomaly-free TQFT its underlying quantum invariant we get a bijective correspondence between isomorphism classes of non-degenerate anomaly-free TQFT's based on* $(\mathfrak{B}, \mathfrak{A})$ *and quantum invariants of closed* $\mathfrak{B}$*-spaces.*

This theorem follows from two previous lemmas and Theorem 3.3. It solves the isomorphism problem for non-degenerate anomaly-free TQFT's in terms of quantum invariants.

4.5. Remarks. 1. A direct verification of the splitting axiom (4.1.3) seems to be quite hard even when we have a good candidate for a quantum invariant. In order to verify this axiom we may extend the given invariant to a TQFT and apply Lemma 4.2.

2. The invariant τ of closed 3-manifolds with v-colored ribbon graphs introduced in Chapter II is a quantum invariant. Here the space-structure $\mathfrak{B}$ is the structure of a closed oriented topological 3-manifold with a numerical weight and a v-colored ribbon graph inside, $\mathfrak{A}$ is the structure of a closed oriented surface with a distinguished Lagrangian subspace in real 1-homologies. For more details, see Chapter IV. It would be very interesting to verify the splitting axiom for τ directly without appealing to TQFT's.

3. The proof of Lemma 4.3 shows that under favorable circumstances we may dispense with the verification of the normalization axiom (1.4.4). Namely, if we have a candidate $(\mathcal{T}, \tau)$ for TQFT satisfying all axioms except possibly (1.4.4) and satisfying the non-degeneracy condition in the class of $\mathfrak{B}$-spaces with collars then $(\mathcal{T}, \tau)$ satisfies (1.4.4).

5. Hermitian and unitary TQFT's

5.0. Outline. We define and study Hermitian structures on TQFT's. The main result (Theorem 5.3) provides a necessary and sufficient condition for the existence of a Hermitian structure on a TQFT. This result will be used in Section IV.10 where we study Hermitian 3-dimensional TQFT's derived from Hermitian modular categories.

5.1. Algebraic preliminaries. Let K be a commutative ring with unit equipped with a ring involution $k \mapsto \overline{k} : K \to K$. (We assume that $\overline{1} = 1$.) By antilinear K-homomorphism we mean an additive homomorphism of K-modules $f : P \to Q$ such that $f(kx) = \overline{k} f(x)$ for any $k \in K, x \in P$. The formula $kf(x) = f(kx)$ defines a K-module structure in the set of antilinear K-homomorphisms $P \to Q$. The resulting K-module is denoted by $\mathrm{Hom}_{\overline{K}}(P, Q)$.

A Hermitian form on a K-module P is a pairing $(x, y) \mapsto \langle x, y\rangle : P \otimes_K P \to K$ linear with respect to x, antilinear with respect to y, satisfying the identity $\langle y, x\rangle = \overline{\langle x, y\rangle}$, and non-degenerate in the sense that the adjoint homomorphisms $P \to \mathrm{Hom}_K(P, K)$ and $P \to \mathrm{Hom}_{\overline{K}}(P, K)$ are bijective.

5.2. Hermitian structures. Let $(\mathcal{T}, \tau)$ be a TQFT with ground ring K based on an involutive cobordism theory $(\mathfrak{B}, \mathfrak{A})$ (see Exercise 2.8.3). Assume that K is provided with a ring involution $k \mapsto \overline{k} : K \to K$. A Hermitian structure on $(\mathcal{T}, \tau)$ assigns to each $\mathfrak{A}$-space X a non-degenerate Hermitian form $\langle ., .\rangle_X : \mathcal{T}(X) \otimes_K \mathcal{T}(X) \to K$ satisfying the following conditions:

(5.2.1) The pairing $\langle ., .\rangle_X$ is natural with respect to $\mathfrak{A}$-homeomorphisms and multiplicative with respect to disjoint union; for $X = \emptyset$ the pairing $\langle ., .\rangle_X$ is determined by the unit 1×1-matrix over K.

(5.2.2) For any $\mathfrak{B}$-cobordism (M, X, Y) and any $x \in \mathcal{T}(X)$, $y \in \mathcal{T}(Y)$, we have

$$\langle \tau(M, X, Y)(x), y \rangle_Y = \langle x, \tau(-M, Y, X)(y) \rangle_X. \tag{5.2.a}$$

A TQFT with a Hermitian structure is called a Hermitian TQFT. In the case where $K = \mathbb{C}$ with the usual complex conjugation and the Hermitian form $\langle ., . \rangle_X$ is positive definite for every X, we say that the Hermitian TQFT is unitary.

The following two features of Hermitian TQFT's are especially important. First of all, if M is a closed $\mathfrak{B}$-space then $\tau(-M) = \overline{\tau(M)}$. This follows from (5.2.2). Secondly, the group of $\mathfrak{A}$-self-homeomorphisms of any $\mathfrak{A}$-space X acts in $\mathcal{T}(X)$ preserving the Hermitian form $\langle ., . \rangle_X$. For instance, in the case of unitary TQFT we get a unitary action. This observation allows us to estimate the absolute value of the invariant τ of the mapping torus M_g of an $\mathfrak{A}$-homeomorphism $g : X \to X$. The result of Exercise 2.8.1 implies that for a unitary TQFT $(\mathcal{T}, \tau)$, we have $|\tau(M_g)| \leq \dim_{\mathbb{C}} \mathcal{T}(X)$.

There is a simple description of Hermitian non-degenerate anomaly-free TQFT's as follows.

5.3. Theorem. *Let K be a commutative ring with unit equipped with a ring involution $k \mapsto \overline{k} : K \to K$. Let $(\mathcal{T}, \tau)$ be a non-degenerate anomaly-free TQFT with ground ring K based on an involutive cobordism theory $(\mathfrak{B}, \mathfrak{A})$. The TQFT $(\mathcal{T}, \tau)$ admits a Hermitian structure if and only if $\tau(-M) = \overline{\tau(M)}$ for any closed $\mathfrak{B}$-space M. If a Hermitian structure exists on $(\mathcal{T}, \tau)$ then it is unique.*

Proof. Let us first prove uniqueness. To this end we compute the Hermitian form $\langle ., . \rangle_X$ on the generators specified in the definition of non-degenerate TQFT's. Consider an $\mathfrak{A}$-space X and a $\mathfrak{B}$-space with X-parametrized boundary $(M, f : X \to \partial M)$. Set $x = \tau^X(M, f) \in \mathcal{T}(X)$. Take an arbitrary $y \in \mathcal{T}(X)$. The naturality of the pairing $\langle ., . \rangle_X$ and axiom (5.2.2) imply that

$$\langle x, y \rangle_X = \langle f_\#^{-1}(\tau(M)), y \rangle_X = \langle \tau(M), f_\#(y) \rangle_{\partial M} =$$

$$= \langle \tau(M, \emptyset, \partial M)(1_K), f_\#(y) \rangle_{\partial M} = \langle 1_K, \tau(-M, \partial M, \emptyset)(f_\#(y)) \rangle_\emptyset.$$

The pairing $\langle ., . \rangle_\emptyset$ is specified by axiom (5.2.1) so that $\langle x, y \rangle_X$ is determined by the operator $\tau(-M, \partial M, \emptyset)$. Since the TQFT $(\mathcal{T}, \tau)$ is non-degenerate, the elements x as above generate $\mathcal{T}(X)$ so that the pairing $\langle ., . \rangle_X$ is determined by $\tau(-M, \partial M, \emptyset)$. This establishes uniqueness of the Hermitian structure.

It remains to show that if $\tau(-M) = \overline{\tau(M)}$ for any closed $\mathfrak{B}$-space M then $(\mathcal{T}, \tau)$ has a Hermitian structure. We first define for an arbitrary $\mathfrak{A}$-space X an antilinear homomorphism $J_X : \mathcal{T}(X) \to \mathcal{T}(-X)$. Let $x = \tau^X(M, f) \in \mathcal{T}(X)$ be the element of $\mathcal{T}(X)$ determined by a $\mathfrak{B}$-space with X-parametrized boundary $(M, f : X \to \partial M)$. Set $J_X(x) = \tau^{-X}(-M, -f) \in \mathcal{T}(-X)$. Extend the mapping $x \mapsto J_X(x)$ to an antilinear homomorphism $J_X : \mathcal{T}(X) \to \mathcal{T}(-X)$. To show that J_X is well-defined it suffices to show the following assertion. Let $\{M_r\}_r$ be a finite family of $\mathfrak{B}$-spaces with X-parametrized boundary such that for certain $\{k_r \in K\}_r$, we have $\sum_r k_r \tau^X(M_r) = 0$. Then $\sum_r \overline{k_r} \tau^{-X}(-M_r) = 0$. To prove this assertion consider an arbitrary $\mathfrak{B}$-space with X-parametrized boundary N. We have

$$\sum_r \overline{k_r}\, d_X(\tau^X(N) \otimes \tau^{-X}(-M_r)) = \sum_r \overline{k_r}\, \tau(N \cup_X -M_r) =$$

$$= \overline{\sum_r k_r\, \tau(M_r \cup_X -N)} = \overline{\sum_r k_r\, d_X(\tau^X(M_r) \otimes \tau^{-X}(-N))} = 0.$$

Here the first and third equalities follow from Theorem 2.1.2, the second equality follows from the conditions of Theorem 5.3 because $-(N \cup_X -M_r) = M_r \cup_X -N$, and the last equality follows from the assumption $\sum_r k_r \tau^X(M_r) = 0$. Therefore the linear combination $\sum_r \overline{k_r} \tau^{-X}(-M_r) \in \mathcal{T}(-X)$ annihilates under d_X all elements of $\mathcal{T}(X)$ of the form $\tau^X(N)$. Since such elements generate $\mathcal{T}(X)$ and the pairing d_X is non-degenerate, the linear combination in question is equal to zero. Thus, J_X is a well-defined antilinear homomorphism $\mathcal{T}(X) \to \mathcal{T}(-X)$. It is obvious that the homomorphisms J_X and J_{-X} are mutually inverse. Therefore they are isomorphisms.

For $x, y \in \mathcal{T}(X)$, set

$$\langle x, y\rangle_X = d_X(x \otimes J_X(y)).$$

It is obvious that the pairing $\langle .,.\rangle_X : \mathcal{T}(X) \otimes_K \mathcal{T}(X) \to K$ is linear with respect to x and antilinear with respect to y. Since d_X is non-degenerate and J_X is an isomorphism, the pairing $\langle .,.\rangle_X$ is also non-degenerate. Let us show that for any $x, y \in \mathcal{T}(X)$, we have $\langle y, x\rangle = \overline{\langle x, y\rangle}$. It suffices to prove this for generators $x = \tau^X(M)$, $y = \tau^X(N)$ where M, N are $\mathfrak{B}$-spaces with X-parametrized boundary. We have

$$\text{(5.3.a)} \qquad \langle x, y\rangle = d_X(x \otimes J_X(y)) = d_X(\tau^X(M) \otimes \tau^{-X}(-N)) = \tau(M \cup_X -N).$$

Similarly, $\langle y, x\rangle = \tau(N \cup_X -M)$. The equality $M \cup_X -N = -(N \cup_X -M)$ implies that $\langle y, x\rangle = \overline{\langle x, y\rangle}$. Hence the pairing $\langle .,.\rangle_X$ is Hermitian.

It is straightforward to verify that the pairing $\langle .,.\rangle_X$ satisfies (5.2.1). Let us verify (5.2.2). Let (M, X, Y) be a $\mathfrak{B}$-cobordism. To prove (5.2.a) it suffices to consider the case of generators $x = \tau^X(A, f)$, $y = \tau^X(B, g)$ where $(A, f : X \to \partial A)$ and $(B, g : Y \to \partial B)$ are $\mathfrak{B}$-spaces with parametrized boundary. Denote by W the $\mathfrak{B}$-space obtained by gluing M, A, and $-B$ along f and g. We claim that

$$\text{(5.3.b)} \qquad \langle \tau(M, X, Y)(x), y\rangle_Y = \tau(W) = \langle x, \tau(-M, Y, X)(y)\rangle_X.$$

To prove the first equality denote by A' the result of gluing A and M along $f^{-1} : \partial A \to X$. Clearly, $\partial A' = Y$. The functoriality axiom (1.4.3) ensures that $\tau(A') = \tau(M, X, Y)(x)$. Since W is the result of gluing of A' and $-B$ along

$g : Y \to \partial B$ we have

$$\tau(W) = d_Y(\tau(A') \otimes (-g)_{\#}^{-1}(\tau(-B))) = d_Y(\tau(A') \otimes \tau^{-Y}(-B)) =$$

$$= d_Y(\tau(M, X, Y)(x) \otimes J_Y(y)) = \langle \tau(M, X, Y)(x), y \rangle_Y .$$

Since $-W$ is obtained by gluing $-M, B, -A$ along $g : Y \to \partial B$ and $f : X \to \partial A$, a similar argument applied to the cobordism $(-M, Y, X)$ shows that $\tau(-W) = \langle \tau(-M, Y, X)(y), x \rangle_X$. Therefore

$$\tau(W) = \overline{\tau(-W)} = \overline{\langle \tau(-M, Y, X)(y), x \rangle_X} = \langle x, \tau(-M, Y, X)(y) \rangle_X .$$

This completes the proof of (5.3.b), (5.2.a), and the theorem.

5.4. Exercises. Let $(\mathcal{T}, \tau, \langle ., . \rangle)$ be a Hermitian anomaly-free TQFT based on an involutive cobordism theory $(\mathfrak{B}, \mathfrak{A})$.

1. Let M and N be $\mathfrak{B}$-spaces and $f : \partial M \to \partial N$ be an $\mathfrak{A}$-homeomorphism. Show that $\tau(M \cup_f -N) = \langle f_{\#}(\tau(M)), \tau(N) \rangle_{\partial N}$.

2. Denote by $\overline{K}$ the ground ring K of $(\mathcal{T}, \tau, \langle ., . \rangle)$ viewed as a K-bimodule where the left (resp. right) K-action is determined by the ring multiplication (resp. by the formula $k'k = k'\overline{k}$). For an $\mathfrak{A}$-space X and $y \in \mathcal{T}(X)$, denote the K-homomorphism $x \mapsto \langle x, y \rangle_X : \mathcal{T}(X) \to K$ by $\tilde{y}$. The formula $y \mapsto 1 \otimes \tilde{y}$ determines an isomorphism of left K-modules $\mathcal{T}(X) \to \overline{K} \otimes_K (\mathcal{T}(X))^\star$. Show that these isomorphisms form an isomorphism of TQFT's $(\mathcal{T}, \tau) \to \overline{K} \otimes_K (\mathcal{T}^\star, \tau^\star)$. This implies that $(\mathcal{T}^\star, \tau^\star) = \overline{K} \otimes_K (\mathcal{T}, \tau)$.

6. Elimination of anomalies

6.0. Outline. We introduce a renormalization technique for TQFT's. Under favorable conditions this technique transforms a TQFT $(\mathcal{T}, \tau)$ with anomalies into an anomaly-free TQFT. The condition in question amounts to the existence of a so-called anomaly 2-cocycle of $(\mathcal{T}, \tau)$. To renormalize $(\mathcal{T}, \tau)$ we replace the underlying cobordism theory with another one determined by this cocycle.

The renormalization technique will be applied to 3-dimensional TQFT's in Chapter IV. We reproduce a simplified version of this technique in Chapter IV so that the reader interested mainly in 3-manifolds may skip this section.

6.1. Gluing patterns. The notion of a gluing pattern plays a technical role in the definiton of anomaly cocycles. A gluing pattern is a triple (M, N, f) where M, N are $(\mathfrak{B}, \mathfrak{A})$-cobordisms and $f : \partial_+ M \to \partial_- N$ is an $\mathfrak{A}$-homeomorphism. Any gluing pattern $P = (M, N, f)$ gives rise to a cobordism $(M \cup_f N, \partial_- M, \partial_+ N)$ obtained by gluing of M to N along f. We shall say that this cobordism is obtained by gluing of M to N from below.

By a homeomorphism of gluing patterns $(M, N, f) \to (M', N', f')$ we mean a pair of $\mathfrak{B}$-homeomorphisms $g : M \to M'$ and $h : N \to N'$ transforming bases onto the corresponding bases and such that $f' g|_{\partial_+ M} = hf : \partial_+ M \to \partial_-(N')$.

6.2. Two-cocycles. We define 2-cocycles of the cobordism theory $(\mathfrak{B}, \mathfrak{A})$. This definition is inspired by the properties of anomalies of TQFT's. The relationship with anomalies will be discussed in Section 6.4.

A 2-cocycle of $(\mathfrak{B}, \mathfrak{A})$ with values in an abelian group G is a function assigning to each gluing pattern P a certain $g(P) \in G$ satisfying the following four axioms.

(6.2.1) (Naturality). If the gluing patterns P and P' are homeomorphic then $g(P) = g(P')$.

(6.2.2) (Multiplicativity). If the gluing pattern P is the disjoint union of gluing patterns P_1 and P_2 then $g(P) = g(P_1)\, g(P_2)$.

Here we use multiplicative notation for the group operation in G.

(6.2.3) (Normalization). For any $(\mathfrak{B}, \mathfrak{A})$-cobordism M, we have

$$g(M, \partial_+ M \times [0, 1], x \to x \times 0 : \partial_+ M \to \partial_+ M \times 0) = 1.$$

(6.2.4) (Compatibility with disjoint union). Let M, N, W be $(\mathfrak{B}, \mathfrak{A})$-cobordisms such that $\partial_- W$ is the disjoint union of $\mathfrak{A}$-spaces X and Y. Let $e : \partial_+ M \to X$ and $f : \partial_+ N \to Y$ be $\mathfrak{A}$-homeomorphisms. Then

$$g(M \amalg N, W, e \amalg f) =$$

$$= g(M, (W, X, \partial_+ W \amalg -Y), e)\; g(N, (M \cup_e W, Y, -\partial_- M \amalg \partial_+ W), f).$$

Axiom (6.2.3) means that if the upper cobordism of a gluing pattern is a cylinder then the value of g on this pattern is equal to 1. Axiom (6.2.4) stipulates that when the disjoint union of two cobordisms $M \amalg N$ is glued from below to a cobordism W the associated value of g is equal to the product of the values of g associated to the gluing of M and W and the gluing of the resulting cobordism to N.

Note that exchanging the roles of M and N, we get a different decomposition of $g(M \amalg N, W, e \amalg f)$ as the product of two values of g. Therefore the last axiom implies the identity

$$g(M, (W, X, \partial_+ W \amalg -Y), e)\; g(N, (M \cup_e W, Y, -\partial_- M \amalg \partial_+ W), f) =$$

$$= g(N, (W, Y, \partial_+ W \amalg -X), f)\; g(M, (N \cup_f W, X, -\partial_- N \amalg \partial_+ W), e).$$

We shall say that the 2-cocycle g is symmetric if the value of g on any gluing pattern (M, N, f) coincides with the value of g on the "opposite" gluing pattern

$$((N, -\partial_+ N, -\partial_- N), (M, -\partial_+ M, -\partial_- M), -f^{-1} : -\partial_- N \to -\partial_+ M).$$

Any symmetric cocycle satisfies the following version of (6.2.4) in which the disjoint union $M \amalg N$ is glued to W from above.

(6.2.4)′ Let M, N, W be $(\mathfrak{B}, \mathfrak{A})$-cobordisms such that $\partial_+ W$ is the disjoint union of $\mathfrak{A}$-spaces X and Y. Let $e : X \to \partial_- M$ and $f : Y \to \partial_- N$ be $\mathfrak{A}$-homeomorphisms. Then

$$g(W, M \amalg N, e \amalg f) =$$

$$= g((W, \partial_- W \amalg -Y, X), M, e)\ g((W \cup_e M, \partial_- W \amalg -\partial_+ M, Y), N, f).$$

An example of a symmetric 2-cocycle is given in Section 6.6.

6.3. Cobordism theory $(\mathfrak{B}_G, \mathfrak{A})^g$. The idea behind elimination of anomalies is as follows: we provide $\mathfrak{B}$-spaces with an additional structure which allows us to renormalize the operator invariants of cobordisms by scalar factors. The additional structure in question amounts to attaching to $\mathfrak{B}$-spaces certain elements of an abelian group G. This leads to a space-structure $\mathfrak{B}_G$ defined below. We use a G-valued 2-cocycle g of $(\mathfrak{B}, \mathfrak{A})$ in order to define the gluing of $\mathfrak{B}_G$-spaces and a cobordism theory $(\mathfrak{B}_G, \mathfrak{A})^g$. It is this tricky gluing which will eventually allow us to kill anomalies.

Fix an abelian group G and a G-valued 2-cocycle g of $(\mathfrak{B}, \mathfrak{A})$. For any non-empty topological space X, set $\mathfrak{B}_G(X) = \mathfrak{B}(X) \times G$. Thus, a non-empty $\mathfrak{B}_G$-space is a pair (a non-empty $\mathfrak{B}$-space M, an element $k \in G$). The last element is called the (multiplicative) weight of the $\mathfrak{B}_G$-space (M, k). The set of $\mathfrak{B}_G$-structures on the empty space by definition consists of one element; we assume that the weight of the empty space is equal to $1 \in G$.

For a homeomorphism of topological spaces $f : X \to Y$, the induced bijection $\mathfrak{B}(f) : \mathfrak{B}(X) \to \mathfrak{B}(Y)$ uniquely lifts to a weight-preserving bijection $\mathfrak{B}_G(f) : \mathfrak{B}_G(X) \to \mathfrak{B}_G(Y)$. For any topological spaces X, Y, we define a mapping $\mathfrak{B}_G(X) \times \mathfrak{B}_G(Y) \to \mathfrak{B}_G(X \amalg Y)$ to be the product of the canonical mapping $\mathfrak{B}(X) \times \mathfrak{B}(Y) \to \mathfrak{B}(X \amalg Y)$ and the group multiplication $G \times G \to G$. This makes $\mathfrak{B}_G$ a space-structure compatible with disjoint union.

It remains to define the boundary of $\mathfrak{B}_G$-spaces, the cylinder structures, and the gluing of $\mathfrak{B}_G$-spaces (it is here that we use g). The boundary $\partial(M, k)$ of a $\mathfrak{B}_G$-space (M, k) is the $\mathfrak{A}$-space ∂M. (The boundary does not depend on the weight.) The naturality of the boundary with respect to $\mathfrak{B}_G$-homeomorphisms and the appropriate behavior with respect to disjoint union follow from the corresponding properties of $(\mathfrak{B}, \mathfrak{A})$. For any $\mathfrak{A}$-space X, we define the $\mathfrak{B}_G$-space $X \times [0, 1]$ to be the pair $(X \times [0, 1], 1)$ where $X \times [0, 1]$ is the cylinder over X provided by (1.3.3) and 1 is the unit of G. The gluing of $\mathfrak{B}_G$-spaces is defined as follows. Let (M, k) be a $\mathfrak{B}_G$-space whose boundary is the disjoint union of $\mathfrak{A}$-spaces X, Y, Z and let $f : X \to -Y$ be an $\mathfrak{A}$-homeomorphism. Let M' be the $\mathfrak{B}$-space obtained from M by gluing of X to Y along f. Denote by $P(M, X, Y, Z, f)$

the gluing pattern

$$((X \times [0,1], \emptyset, (-X \times 0) \amalg (X \times 1)), \ (M, -X \amalg -Y, Z),$$

$$\mathrm{id}_{-X} \amalg f : (-X \times 0) \amalg (X \times 1) \to -X \amalg -Y).$$

We define the $\mathfrak{B}_G$-space obtained from (M, k) by gluing of X to Y along f to be the pair $(M', k \cdot g(P(M, X, Y, Z, f)) \in G)$.

6.3.1. Lemma. *For any 2-cocycle g, the space-structures $\mathfrak{B}_G$ and $\mathfrak{A}$ with boundary, gluing, and cylinder structures defined above form a cobordism theory (in the sense of Section 1.3) denoted by $(\mathfrak{B}_G, \mathfrak{A})^g$.*

Proof. We have to check axioms (1.3.1) - (1.3.4). Let us first verify that the gluing is independent of the order X, Y, i.e., that (in the notation used above) the $\mathfrak{B}_G$-spaces obtained by gluing along $f : X \to -Y$ and $-f^{-1} : Y \to -X$ coincide. It suffices to show that the values of g on the gluing patterns $P(M, X, Y, Z, f)$ and $P(M, Y, X, Z, -f^{-1})$ are equal to each other. In fact, these two patterns are homeomorphic. Their homeomorphism is induced by the identity $\mathrm{id}_M : M \to M$ and the composition of $\mathfrak{B}$-homeomorphisms $f \times \mathrm{id}_{[0,1]} : X \times [0,1] \to (-Y) \times [0,1]$ and $(y, t) \mapsto (y, 1-t) : (-Y) \times [0,1] \to Y \times [0,1]$. Therefore the desired equality follows from (6.2.1).

The naturality and multiplicativity of g imply that the gluing of $\mathfrak{B}_G$-spaces is natural with respect to $\mathfrak{B}_G$-homeomorphisms and commutes with disjoint union.

Let us show that successive gluings along disjoint $\mathfrak{A}$-subspaces of ∂M give the same result as the gluing along their union. Assume that ∂M is the disjoint union of $\mathfrak{A}$-spaces X, Y, Z and that $f : X \to -Y$ is a $\mathfrak{A}$-homeomorphism. Assume that X is the disjoint union of $\mathfrak{A}$-spaces X_1 and X_2. Set $Y_r = f(X_r)$ for $r = 1, 2$. Denote by N the $\mathfrak{B}$-space obtained by gluing of $X_1 \times [0,1]$ to M along

$$\mathrm{id} \amalg f : (-X_1 \times 0) \amalg (X_1 \times 1) \to -X_1 \amalg -Y_1.$$

We have to show that

$$g(P(M, X, Y, Z, f)) = g(P(M, X_1, Y_1, Z \amalg X_2 \amalg Y_2, f|_{X_1}))\ g(P(N, X_2, Y_2, Z, f|_{X_2})).$$

This equality follows directly from (6.2.4) where the role of W is played by M and the role of two cobordisms glued to W from below is played by

$$(X_1 \times [0,1], \emptyset, (-X_1 \times 0) \amalg (X_1 \times 1)) \quad \text{and} \quad (X_2 \times [0,1], \emptyset, (-X_2 \times 0) \amalg (X_2 \times 1)).$$

Axioms (1.3.3) and (1.3.4) follow from the corresponding axioms for $(\mathfrak{B}, \mathfrak{A})$ and (6.2.3).

6.4. Anomaly cocycles for TQFT's. Let $(\mathcal{T}, \tau)$ be a TQFT with ground ring K based on the cobordism theory $(\mathfrak{B}, \mathfrak{A})$. According to the functoriality axiom (1.4.3) any $(\mathfrak{B}, \mathfrak{A})$-cobordisms M_1, M_2 and any $\mathfrak{A}$-homeomorphism

$f : \partial_+(M_1) \to \partial_-(M_2)$ give rise to certain gluing anomalies. We shall be interested in the case where each such triple has a specified anomaly which yields a cocycle in the sense of Section 6.2.

Denote by K^* the (abelian) group of invertible elements of the ring K. An anomaly cocycle for the TQFT $(\mathcal{T}, \tau)$ is a K^*-valued symmetric 2-cocycle g of $(\mathfrak{B}, \mathfrak{A})$ such that for any $(\mathfrak{B}, \mathfrak{A})$-cobordisms M_1, M_2 and any $\mathfrak{A}$-homeomorphism $f : \partial_+(M_1) \to \partial_-(M_2)$, the element $k = g(M_1, M_2, f) \in K$ satisfies (1.4.3) and the following axiom holds:

(6.4.1) For any two $(\mathfrak{B}, \mathfrak{A})$-cobordisms M, N and any $\mathfrak{A}$-homeomorphism $f : \partial_+ M \to \partial_- N$, the value of g on the gluing pattern

$$(M \cup_{\mathrm{id}:\partial_+ M \to \partial_+ M \times 0} (\partial_+ M \times [0,1]), N, f : \partial_+ M \times 1 \to \partial_- N) \tag{6.4.a}$$

is equal to $g(M, N, f)$.

Note that if M has a collar (in the sense of Section 4.1) then this equality holds for the obvious reason: the gluing pattern (6.4.a) is $\mathfrak{B}$-homeomorphic to the gluing pattern (M, N, f). Therefore in the case when all $\mathfrak{B}$-spaces have collars, axiom (6.4.1) is superfluous.

The existence of an anomaly cocycle seems to be a natural and not very restrictive assumption on TQFT's. In interesting topological quantum field theories the gluing anomalies are usually computed by an explicit formula. It is reasonable to expect that such a formula yields an anomaly cocycle.

Note finally that the trivial cocycle assigning $1 \in K^*$ to all cobordisms is an anomaly cocycle for any anomaly-free TQFT. Therefore the class of TQFT's admitting anomaly cocycles includes anomaly-free TQFT's.

6.5. Anomaly-free renormalization. Let $(\mathcal{T}, \tau)$ be a TQFT based on the cobordism theory $(\mathfrak{B}, \mathfrak{A})$. Let g be an anomaly cocycle for $(\mathcal{T}, \tau)$. We shall define an anomaly-free TQFT $(\mathcal{T}, \tau^g)$ based on the cobordism theory $(\mathfrak{B}_{K^*}, \mathfrak{A})^g$. The modular functor of $(\mathcal{T}, \tau^g)$ is the same modular functor $\mathcal{T}$ based on $\mathfrak{A}$. For a $\mathfrak{B}_{K^*}$-cobordism $(M, k \in K^*)$ (where M is a $\mathfrak{B}$-cobordism), set

$$\tau^g(M, k) = k^{-1}\tau(M) : \mathcal{T}(\partial_- M) \to \mathcal{T}(\partial_+ M).$$

6.5.1. Theorem. *For any anomaly cocycle g of $(\mathcal{T}, \tau)$, the modular functor $\mathcal{T}$ together with τ^g forms an anomaly-free TQFT based on $(\mathfrak{B}_{K^*}, \mathfrak{A})^g$.*

Proof. Axioms (1.4.1), (1.4.2), and (1.4.4) follow directly from the corresponding properties of $(\mathcal{T}, \tau)$ and definitions. Let us verify (1.4.3) with $k = 1$. Let $(M, r \in K^*)$ be a $(\mathfrak{B}_{K^*}, \mathfrak{A})^g$-cobordism obtained from the disjoint union of two $(\mathfrak{B}_{K^*}, \mathfrak{A})^g$-cobordisms $(P, p \in K^*)$ and $(Q, q \in K^*)$ by gluing along an $\mathfrak{A}$-homeomorphism

$f : \partial_+ P \to \partial_- Q$. By the definition of an anomaly cocycle, we have

$$\tau(M) = g(P, Q, f)\, \tau(Q) \circ f_\# \circ \tau(P).$$

By the definition of gluing for $\mathfrak{V}_{K^*}$-spaces, we have $r = pqA$ with $A = g(N_1, N_2, h)$ where N_1 and N_2 are the gluing patterns $(\partial_+ P \times [0,1], \emptyset, (-\partial_+ P \times 0) \amalg (\partial_+ P \times 1))$ and $(P \amalg Q, -\partial_+ P \amalg \partial_- Q, -\partial_- P \amalg \partial_+ Q)$ respectively, and

$$h = -\mathrm{id} \amalg f : (-\partial_+ P \times 0) \amalg (\partial_+ P \times 1) \to -\partial_+ P \amalg \partial_- Q.$$

The property (6.2.4)$'$ of symmetric 2-cocycles implies that $A = g_1 g_2$ where g_1 and g_2 are the values of the cocycle g on the gluing patterns

$$((\partial_+ P \times [0,1], -\partial_+ P \times 1, -\partial_+ P \times 0), (P, -\partial_+ P, \partial_- P), \mathrm{id} : -\partial_+ P \times 0 \to -\partial_+ P)$$

and

$$(P \cup_{\mathrm{id}: \partial_+ P \to \partial_+ P \times 0} (\partial_+ P \times [0,1]), Q, f : \partial_+ P \times 1 \to \partial_- Q).$$

The symmetry of the anomaly cocycle and axiom (6.2.3) imply that $g_1 = 1$. The axiom (6.4.1) implies that $A = g_2 = g(P, Q, f)$. Combining these formulas we get

$$\tau^g(M) = r^{-1} \tau(M) = p^{-1} q^{-1} A^{-1} \tau(M) =$$
$$= p^{-1} q^{-1} \tau(Q) \circ f_\# \circ \tau(P) = \tau^g(Q) \circ f_\# \circ \tau^g(P).$$

6.6. Example. Consider the TQFT constructed in Section 1.5.2. It is easy to see that the formula for its anomalies given in Section 1.5.2 yields an anomaly 2-cocycle. Therefore this TQFT admits an anomaly-free renormalization. The underlying quantum invariant of the renormalized TQFT assigns to every pair (a finite cell space, $k \in K^*$) the element k^{-1}. This invariant does not look very exciting. Still, this example indicates that the concepts introduced in this section do work.

6.7. Exercises. 1. If a TQFT $(\mathcal{T}, \tau)$ is non-degenerate then the TQFT $(\mathcal{T}, \tau^g)$ is also non-degenerate.

2. Any anomaly cocycle for a TQFT, based on an involutive cobordism theory (cf. Exercise 2.8.3), is also an anomaly cocycle for the dual TQFT.

Notes

Section 1. The notion of a modular functor arose in the framework of 2-dimensional rational conformal field theory (see [MS1], [MS2]). An axiomatic treatment of 2-dimensional modular functors was first given by Segal [Se] and Moore and Seiberg [MS1]. Here we give a generalized version of this notion abandoning all connections with the topology of surfaces and conformal structures.

The concept of topological quantum field theory is inspired by the work of E. Witten, see [Wi1], [Wi2]. Basic axioms of unitary TQFT's were first formulated by Atiyah [At1] extending G. Segal's axioms for the modular functor. Quinn [Qu1], [Qu2] has carried out a systematic study of axiomatic foundations of TQFT's in an abstract set up. Our approach differs considerably from that of Quinn. We use space-structures rather than domain categories as in [Qu1], [Qu2]. See also Walker [Wa] for an axiomatic definition of unitary 3-dimensional TQFT's.

The notion of a space-structure seems to be new, albeit not very interesting in itself.

Example 1.5.1 is borrowed from [Qu1].

Section 2. Theorem 2.1.1 has been independently obtained by Quinn [Qu1], [Qu2] and the author [Tu14]. Theorem 2.1.3 is well known in physical literature. Theorem 2.1.2 is also known in a disguised form: physicists consider only unitary TQFT's and use the corresponding Hermitian pairings instead of the self-duality pairing d (cf. Exercise 5.4). Our ability to state and to prove such theorems in an abstract setting shows that our mathematics adequately formalizes physical ideas.

Section 3. The results of this section are new.

Section 4. The definitions and results of this section are new. The construction of the module $\mathcal{T}(X)$ in the proof of Lemma 4.3 is inspired by the construction of a 3-dimensional TQFT due to P. Vogel and his collaborators, see [BHMP1], [BHMP2].

Section 5. Hermitian structures appeared in the very first works on topological quantum field theory. Theorem 5.3 is new.

Section 6. The definitions of this section formalize ideas well known to physicists.

Chapter IV
Three-dimensional topological quantum field theory

Fix up to the end of this chapter a strict modular category $(\mathcal{V}, \{V_i\}_{i\in I})$ with ground ring K and rank $\mathcal{D} \in K$.

1. Three-dimensional TQFT: preliminary version

1.0. Outline. In this section we define a preliminary 3-dimensional TQFT $(\mathcal{T}, \tau) = (\mathcal{T}_{\mathcal{V}}, \tau_{\mathcal{V}})$. It is non-degenerate but has anomalies which will be studied in further sections. The heart of the construction is the definition of an operator invariant of 3-cobordisms with embedded v-colored ribbon graphs. (The ribbon graphs may meet the bases of cobordisms along bases of some bands.) The main idea is to reduce the general case to the case of closed 3-manifolds and apply the invariant τ introduced in Chapter II. With this in mind we shall consider 3-cobordisms with a very strong structure on the boundary. The boundary is assumed to be parametrized in the sense that each of its components is provided with a homeomorphism onto a standard surface. The standard surfaces are the boundaries of standard unknotted handlebodies in $\mathbb{R}^3$. Each standard handlebody contains a canonical ribbon graph with one uncolored coupon, a few uncolored bands, and a few colored bands. This ribbon graph may be viewed as the core of the handlebody.

The operator invariant of 3-cobordisms is constructed roughly as follows. Consider a 3-cobordism with parametrized boundary M and a v-colored ribbon graph $\Omega \subset M$. Glue to M (copies of) standard handlebodies along the parametrization homeomorphisms of the components of ∂M. This gives a closed 3-manifold $\tilde{M}$. Under appropriate assumptions on the parametrizations, the canonical ribbon graphs contained in the standard handlebodies match well with Ω. Their union is a ribbon graph $\tilde{\Omega} \subset \tilde{M}$. The uncolored bands and coupons of the canonical ribbon graphs give rise to uncolored bands and coupons of $\tilde{\Omega}$. Choosing their colors we get a v-colored ribbon graph in $\tilde{M}$. The invariant τ yields a K-valued function of the varying colors of uncolored bands and coupons of $\tilde{\Omega}$. This function is polylinear with respect to the colors of coupons. Assembling together the coupons associated to the bottom (resp. top) base of M we rewrite this polylinear function as an operator invariant of (M, Ω). This train of ideas leads to a TQFT based on compact oriented 3-cobordisms with parametrized decorated boundary and with embedded v-colored ribbon graphs. Here we need the notion of a decorated surface which formalizes the properties of the system of arcs $\Omega \cap \partial M$.

A considerable part of this section is devoted to the language of decorated and parametrized surfaces, standard handlebodies, ribbon graphs in 3-manifolds, etc. One may first read this section ignoring distinguished arcs on surfaces and ribbon graphs in 3-manifolds. This would considerably simplify the reading and lead to a TQFT based on compact oriented 3-cobordisms with parametrized boundary (without decorations and ribbon graphs). Note however that even in this special case our constructions use the full power of the theory of ribbon graphs developed in Chapter I.

1.1. Decorated surfaces and decorated types. By an arc on a surface Σ we mean a simple oriented arc lying in $\Sigma \setminus \partial\Sigma$. (The arcs are not assumed to be parametrized.) An arc on Σ endowed with an object W of $\mathcal{V}$ and a sign $\nu = \pm 1$ is said to be marked. The pair (W, ν) is called the mark of this arc. The object W and the number ν are called the label and the sign of this arc respectively.

A closed connected orientable surface is said to be decorated if it is oriented and endowed with a finite totally ordered set of disjoint marked arcs. (To fix a total order in this set it suffices to number the arcs by 1, 2, ...) A non-connected surface is said to be decorated if its connected components are decorated. For the sake of brevity, decorated surfaces will be also called d-surfaces. By their very definition, d-surfaces are oriented and closed. Note that the distinguished family of arcs on a decorated surface may be empty.

A d-homeomorphism of d-surfaces is a degree 1 homeomorphism of the underlying surfaces preserving the distinguished arcs together with their orientations, marks, and order (on each component).

There is a natural negation of the structure on a d-surface. For a d-surface Σ, the opposite d-surface $-\Sigma$ is obtained from Σ by reversing the orientation of Σ, reversing the orientation of its distinguished arcs, and multiplying the signs of all distinguished arcs by -1 while keeping the labels and the order of these arcs. Clearly, $-(-\Sigma) = \Sigma$. The transformation $\Sigma \mapsto -\Sigma$ is natural in the sense that any d-homeomorphism $f : \Sigma \to \Sigma'$ induces a d-homeomorphism $-f : -\Sigma \to -\Sigma'$ which coincides with f as a mapping.

A decorated type or, briefly, a type is a tuple $(g; (W_1, \nu_1), ..., (W_m, \nu_m))$ where g is a non-negative integer, $W_1, ..., W_m$ are objects of the category $\mathcal{V}$, and $\nu_1, ..., \nu_m \in \{1, -1\}$. For a type $t = (g; (W_1, \nu_1), ..., (W_m, \nu_m))$, we define the opposite type $-t = (g; (W_1, -\nu_1), ..., (W_m, -\nu_m))$. It is clear that $-t = t$ if and only if $m = 0$.

The type $t(\Sigma)$ of a connected non-empty d-surface Σ is the tuple formed by the genus g of Σ and the sequence of marks of distinguished arcs on Σ written down in the given order. It is clear that $t(-\Sigma) = -t(\Sigma)$. Note that the d-homeomorphism class of a connected non-empty d-surface is completely determined by its type.

This follows from the fact that for any closed connected oriented surfaces of the same genus Σ, Σ', there is an orientation-preserving homeomorphism $\Sigma \to \Sigma'$.

1.2. Standard *d*-surfaces. For every decorated type $t = (g; (W_1, \nu_1), \ldots, (W_m, \nu_m))$, we construct a canonical d-surface Σ_t of type t. It will play the role of a surface of reference, a standard d-surface of type t.

Denote by R_t the ribbon $(0, m)$-graph in $\mathbb{R}^3$ presented by the diagram in Figure 1.1. This ribbon graph lies in the plane $\mathbb{R} \times 0 \times \mathbb{R}$ which is identified with the plane of the picture. The graph is formed by one coupon and $m + g$ bands (we only draw their cores). The preferred side of this graph is the one turned towards the reader (in other words, the surface of R_t is oriented counterclockwise). The m vertical bands are untwisted and unlinked. The r-th vertical band is colored with W_r and directed down if $\nu_r = 1$ and up otherwise. The g cap-like bands are not colored, their cores are oriented from right to left. The bottom base of the coupon is its lower horizontal base. The coupon is not colored. Note that the broken horizontal lines in Figure 1.1 are the lines $\mathbb{R} \times 0 \times 0$ and $\mathbb{R} \times 0 \times 1$.

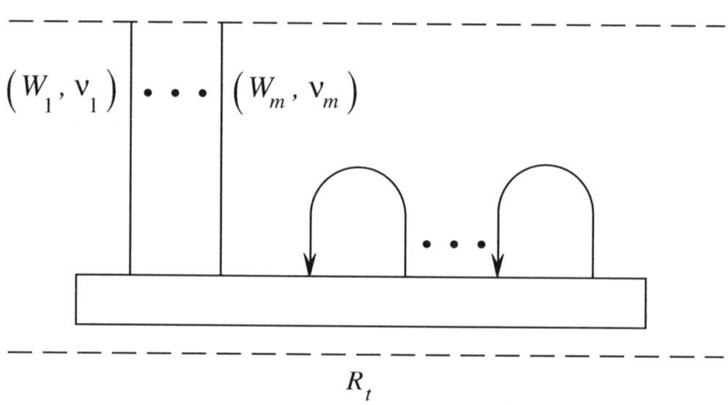

Figure 1.1

Fix a closed regular neighborhood U_t of R_t in the strip $\mathbb{R}^2 \times [0, 1]$. It is clear that U_t is a handlebody of genus g. We provide U_t with right-handed orientation and provide the surface ∂U_t with the induced orientation. The graph R_t lies in $\mathrm{Int}(U_t)$ except for the upper bases of the vertical bands

(1.2.a) $$\{\, [j - (1/10), j + (1/10)] \times 0 \times 1 \mid j = 1, \ldots, m \,\}$$

which lie on the surface ∂U_t. Orient these intervals to the right and provide the set of these intervals with the natural order from left to right. Provide each interval $[j - (1/10), j + (1/10)] \times 0 \times 1$ with the mark (W_j, ν_j). In this way the surface ∂U_t acquires the structure of a d-surface of type t. This d-surface is denoted Σ_t and is called the standard d-surface of type t.

A non-connected standard d-surface is a disjoint union of a finite number of connected standard d-surfaces. The empty set will be considered as the empty standard d-surface.

Since the d-surfaces Σ_{-t} and $-\Sigma_t$ have the same type $-t$, they are d-homeomorphic. For each decorated type t, fix a d-homeomorphism $\text{rev}_t : \Sigma_{-t} \to -\Sigma_t$. We may assume that for any t, we have $\text{rev}_{-t} = (-\text{rev}_t)^{-1} : \Sigma_t \to -\Sigma_{-t}$. Indeed, in the case $m \neq 0$ we choose a representative t from each pair $\{t, -t\}$, take an arbitrary d-homeomorphism $\text{rev}_t : \Sigma_{-t} \to -\Sigma_t$, and define rev_{-t} by the formula $\text{rev}_{-t} = (-\text{rev}_t)^{-1}$. If $m = 0$ then $t = -t$ and we take rev_t to be an orientation-reversing involution of Σ_t.

1.3. Parametrized d-surfaces. A connected d-surface Σ of type t is said to be parametrized if it is equipped with a d-homeomorphism $\Sigma_t \to \Sigma$. This homeomorphism is called a parametrization of Σ. The negation for d-surfaces naturally lifts to a negation for parametrized d-surfaces. Namely, for any parametrized connected d-surface $(\Sigma, f : \Sigma_t \to \Sigma)$, its opposite is defined to be $(-\Sigma, (-f)\text{rev}_t : \Sigma_{-t} \to -\Sigma)$. The equality $(-\text{rev}_t) \circ \text{rev}_{-t} = \text{id}$ ensures involutivity of this negation.

The notion of a parametrization extends to non-connected d-surfaces in the obvious way. Simply instead of connected standard d-surfaces we should use non-connected ones. In other words, a non-connected d-surface is parametrized if and only if all its components are parametrized. The negation extends to non-connected d-surfaces component-wise.

The simplest example of a parametrized d-surface is provided by a standard d-surface with the identity parametrization.

The structure of a parametrized d-surface is an involutive space-structure compatible with disjoint union (see Section III.1.1). Formally speaking we define the space-structure $\mathfrak{A}^d = \mathfrak{A}^d_{\mathcal{V}}$ of parametrized d-surfaces as follows. For any topological space X, the class $\mathfrak{A}^d(X)$ consists of homeomorphisms of standard d-surfaces onto X. The action $\mathfrak{A}^d(X) \to \mathfrak{A}^d(Y)$ of a homeomorphism $X \to Y$ is defined in the obvious way via composition. It is clear that an $\mathfrak{A}^d$-space is nothing but a parametrized d-surface (possibly empty). An $\mathfrak{A}^d$-homeomorphism of $\mathfrak{A}^d$-spaces is a d-homeomorphism of parametrized d-surfaces commuting with the parametrizations.

1.4. Modular functor $\mathcal{T}$. We define a modular functor $\mathcal{T} = \mathcal{T}_{\mathcal{V}}$ based on the space-structure $\mathfrak{A}^d$ of parametrized d-surfaces.

For every decorated type $t = (g; (W_1, \nu_1), ..., (W_m, \nu_m))$, we define a projective K-module Ψ_t as follows. For $i = (i_1, ..., i_g) \in I^g$, set

$$\Phi(t; i) = W_1^{\nu_1} \otimes W_2^{\nu_2} \otimes ... \otimes W_m^{\nu_m} \otimes \bigotimes_{r=1}^{g} (V_{i_r} \otimes V_{i_r}^*)$$

where $W^{+1} = W$ and $W^{-1} = W^*$. (Here $\otimes$ is the tensor product in $\mathcal{V}$ so that $\Phi(t;i)$ is an object of $\mathcal{V}$.) Set

$$\Psi_t = \bigoplus_{i \in I^g} \operatorname{Hom}(\mathbb{1}, \Phi(t;i)). \tag{1.4.a}$$

The module Ψ_t may be interpreted in terms of the ribbon graph R_t and its v-colorings. The choice of $i \in I^g$ determines a coloring of the g uncolored bands of R_t (we count these bands from left to right). Having made such a choice, the module $\operatorname{Hom}(\mathbb{1}, \Phi(t;i))$ is exactly the module of possible colors for the coupon of R_t. We say that Ψ_t is the module generated by v-colorings of R_t. For example, if $g = 1$ and $m = 0$ then $\Psi_t = K^{\operatorname{card}(I)}$.

For a parametrized d-surface Σ, we define the module $\mathcal{T}(\Sigma)$ to be the non-ordered tensor product of the modules Ψ_t where t runs over the types of components of Σ. For instance, if Σ is connected and non-empty then $\mathcal{T}(\Sigma) = \Psi_t$ where $t = t(\Sigma)$ is the type of Σ. If $\Sigma = \emptyset$ then $\mathcal{T}(\Sigma) = K$. By definition $\mathcal{T}(\Sigma) = \mathcal{T}(\Sigma^0)$ where Σ^0 is the standard d-surface parametrizing Σ.

The action $f_\# : \mathcal{T}(\Sigma) \to \mathcal{T}(\Sigma')$ of a d-homeomorphism $f : \Sigma \to \Sigma'$ commuting with parametrizations is defined in a tautological way. If Σ, Σ' are connected then $f_\#$ is the identity in $\mathcal{T}(\Sigma) = \mathcal{T}(\Sigma')$. For non-connected Σ, Σ', we define $f_\#$ to be the tensor product of the identity isomorphisms induced by the restrictions of f to connected components.

1.4.1. Lemma. *The functor* ($\Sigma \mapsto \mathcal{T}(\Sigma)$, $f \mapsto f_\#$) *is a modular functor based on the space-structure* $\mathfrak{A}^d$.

The proof is obvious: axioms (III.1.2.1) - (III.1.2.3) of a modular functor follow directly from the definitions.

The construction of the modular functor $\mathcal{T}$ is rather formal. Parametrized d-surfaces may be regarded as copies of standard d-surfaces so that it is no wonder that we may ascribe to them "standard" modules determined by their decorated types. We confine ourselves to tautological homeomorphisms between these copies and there is nothing spectacular in assigning to such homeomorphisms the identity endomorphisms of modules. The really deep aspect in the definition of $\mathcal{T}$ is the choice of the modules $\{\Psi_t\}_t$. The fact that these modules are generated by colorings of elementary ribbon graphs will eventually ensure non-triviality of the theory and in particular naturality of $\mathcal{T}$ with respect to homeomorphisms of surfaces not commuting with parametrizations.

Example III.1.5.2 suggests that one may regard the module $\mathcal{T}(\Sigma)$ as a quantum version of the group ring $K[H_1(\Sigma)]$. In accordance with this viewpoint, the dimension of this module depends on the genus of Σ in an exponential way, see Section 12 where we compute the dimension of Ψ_t. In the simplest case where

$m = 0$, i.e., $t = (g;)$

$$\mathrm{Dim}(\mathcal{T}(\Sigma_t)) = \mathrm{Dim}(\Psi_t) = \mathcal{D}^{2g-2} \sum_{j \in I} (\dim(j))^{2-2g}.$$

In the sequel we shall need an endomorphism $\eta = \{\eta(\Sigma) : \mathcal{T}(\Sigma) \to \mathcal{T}(\Sigma)\}_\Sigma$ of the modular functor $\mathcal{T}$. For a connected parametrized d-surface Σ of type $t = (g; (W_1, \nu_1), \dots, (W_m, \nu_m))$, the endomorphism $\eta(\Sigma) : \Psi_t \to \Psi_t$ preserves the splitting (1.4.a) and acts in each summand $\mathrm{Hom}(\mathbb{1}, \Phi(t; i))$ as multiplication by $\mathcal{D}^{1-g}\dim(i)$ where for $i = (i_1, \dots, i_g) \in I^g$, we set $\dim(i) = \prod_{n=1}^{g} \dim(i_n)$. For non-connected parametrized d-surfaces, the homomorphism $\eta(\Sigma)$ is defined by multiplicativity so that $\eta(\Sigma_1 \amalg \Sigma_2) = \eta(\Sigma_1) \otimes \eta(\Sigma_2)$. For the empty d-surface $\Sigma = \emptyset$, set $\eta(\Sigma) = \mathrm{id}_K : K \to K$.

1.5. Ribbon graphs in 3-manifolds revisited. In Section II.2.3 we introduced ribbon graphs without "free ends" in 3-manifolds. We define here more general ribbon graphs with free ends which should be attached to certain specified arcs in the boundary of ambient 3-manifolds.

Let M be a 3-manifold whose boundary is endowed with a finite family of disjoint marked arcs. A ribbon graph in M is an oriented surface Ω embedded in M and decomposed as a union of a finite number of directed annuli, directed bands, and coupons such that: Ω meets ∂M transversally along the distinguished arcs in ∂M which are bases of certain bands of Ω; other bases of bands lie on the bases of coupons; otherwise the bands, coupons, and annuli are disjoint; the orientation of Ω induces on each distinguished arc of ∂M the orientation opposite to the given one; near each distinguished arc in ∂M with the sign $+1$ (resp. -1) the incident band is directed inside (resp. outside) M. It is understood that every distinguished arc of ∂M serves as a base of a certain band of Ω.

We define colored and v-colored ribbon graphs in M in the same way as in Section I.2.2 with one additional condition on the colors of bands: for each distinguished arc in ∂M labelled with an object W, the band attached to this arc should be colored with the same object W. Thus, the labels and signs of the distinguished arcs in ∂M determine the colors and directions of incident bands.

1.6. Decorated 3-manifolds. A decorated 3-manifold is a compact oriented 3-manifold with parametrized decorated boundary and with a v-colored ribbon graph sitting in this 3-manifold. We shall regard this v-colored ribbon graph and the structure on the boundary as a kind of additional structure on the underlying 3-manifold. We define the boundary of a decorated 3-manifold to be the usual boundary with its structure of parametrized d-surface.

A d-homeomorphism of decorated 3-manifolds is a degree 1 homeomorphism of 3-manifolds preserving all additional structures in question. A d-homeomorphism of decorated 3-manifolds $M \to M'$ restricts to a d-homeomorphism $\partial M \to \partial M'$ that commutes with the parametrizations.

An example of a decorated 3-manifold is provided by the cylinder $\Sigma \times [0, 1]$ over a parametrized d-surface Σ. We decorate the cylinder as follows. The surfaces $\Sigma \times 0$ and $\Sigma \times 1$ are parametrized and decorated via the identifications $\Sigma \times 0 = -\Sigma$ and $\Sigma \times 1 = \Sigma$. Provide $\Sigma \times [0, 1]$ with the ribbon graph formed by the cylinders over the distinguished arcs in Σ. More exactly, let $\alpha_1, ..., \alpha_m$ be the distinguished arcs in Σ. For each $r = 1, ..., m$, consider the band $\alpha_r \times [0, 1] \subset \Sigma \times [0, 1]$ with the bases $\alpha_r \times 0$, $\alpha_r \times 1$. The color of this band is chosen to be the label of α_r. This band is directed down (towards $\Sigma \times 0$) if the sign of α_r is equal to $+1$ and up in the opposite case. Provide the surface of $\alpha_r \times [0, 1]$ with the product orientation (recall that the arcs $\alpha_1, ..., \alpha_m$ are oriented, the interval $[0,1]$ is right-oriented). The bands $\alpha_1 \times [0, 1], ..., \alpha_m \times [0, 1]$ form a colored ribbon graph in $\Sigma \times [0, 1]$. The cylinder $\Sigma \times [0, 1]$ with this colored ribbon graph is a decorated 3-manifold with the boundary $(-\Sigma \times 0) \amalg (\Sigma \times 1)$. This decorated 3-manifold is called the cylinder over Σ and is denoted $\Sigma \times [0, 1]$.

It is obvious that the structure of decorated 3-manifolds is a space-structure compatible with disjoint union. Formally speaking, the corresponding functor $\mathfrak{B}^d = \mathfrak{B}^d_{\mathcal{V}}$ assigns to any topological space X the class of homeomorphisms of decorated 3-manifolds onto X considered up to composition with d-homeomorphisms of decorated 3-manifolds. It is clear that a $\mathfrak{B}^d$-space is nothing but a decorated 3-manifold. A $\mathfrak{B}^d$-homeomorphism of $\mathfrak{B}^d$-spaces is a d-homeomorphism of decorated 3-manifolds.

1.6.1. Lemma. *The pair $(\mathfrak{B}^d, \mathfrak{A}^d)$ is a cobordism theory in the sense of Section* III.1.3.

The proof is obvious. The gluing involved in (III.1.3.1) is the usual gluing of 3-manifolds along orientation-reversing homeomorphisms of their boundary components. (Note that the free ends of ribbon graphs match well under the gluings in question.)

1.7. Standard decorated handlebodies. Let $t = (g; (W_1, \nu_1), ..., (W_m, \nu_m))$ be a decorated type. As it was explained in Section 1.4 any choice of $x \in \mathrm{Hom}(\mathbb{1}, \Phi(t; i))$ with $i \in I^g$ determines a v-coloring of R_t. The handlebody U_t with this v-colored ribbon graph inside is a decorated 3-manifold. It is bounded by the standard d-surface Σ_t with the identity parametrization. This decorated 3-manifold is called a standard decorated handlebody and is denoted $H(U_t, R_t, i, x)$.

Another standard decorated handlebody can be obtained from the ribbon $(m, 0)$-graph $-R_t$ (see Figure 1.2). We have $-R_t = \mathrm{mir}(R_t)$ where mir is the mirror reflection of $\mathbb{R}^3$ with respect to the plane $\mathbb{R}^2 \times 1/2 \subset \mathbb{R}^3$ (this plane is orthogonal to the plane of Figure 1.1 and intersects it along the horizontal line lying between the broken lines and equidistant from them). The ribbon graph $-R_t$ is formed by one coupon and $m + g$ bands lying in the plane of the picture. As usual, the preferred side of $-R_t$ is the one turned towards the reader and the bottom base of the coupon is its lower horizontal base. The r-th vertical band is colored with W_r and directed down if $\nu_r = 1$ and up otherwise. The g cup-like bands of $-R_t$ are not colored, they are directed from left to right. The coupon is not colored.

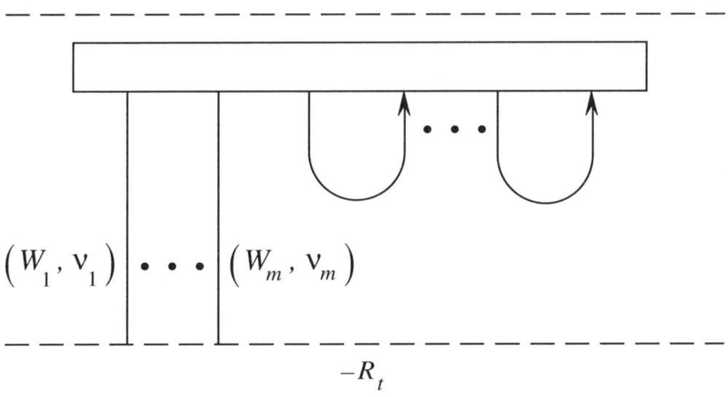

Figure 1.2

Set $U_t^- = \text{mir}(U_t)$. Clearly, U_t^- is a regular neighborhood of $-R_t$ in $\mathbb{R}^2 \times [0, 1]$. We provide U_t^- with right-handed orientation and provide $\partial(U_t^-)$ with the induced orientation. It is clear that the intersection $(-R_t) \cap \partial(U_t^-)$ consists of the intervals $\{ [j - (1/10), j + (1/10)] \times 0 \times 0 \mid j = 1, ..., m \}$. Orient these intervals to the left and provide the set of these intervals with the natural order from left to right. Provide each interval $[j - (1/10), j + (1/10)] \times 0 \times 0$ with the mark $(W_j, -\nu_j)$. In this way the surface $\partial(U_t^-)$ acquires the structure of a d-surface of type $-t$. Note that the mirror reflection mir induces mutually inverse d-homeomorphisms $-\Sigma_t \to \partial(U_t^-)$ and $\partial(U_t^-) \to -\Sigma_t$.

It is clear that any $i \in I^g$ determines a coloring of g uncolored bands of $-R_t$. Any

$$x \in \text{Hom}(\Phi(t; i), \mathbb{1}) = (\text{Hom}(\mathbb{1}, \Phi(t; i)))^\star$$

with $i \in I^g$ determines a v-coloring of $-R_t$. The handlebody U_t^- with this v-colored ribbon graph inside and with the parametrization $\text{mir} \circ \text{rev}_t : \Sigma_{-t} \to \partial(U_t^-)$ of its boundary is a decorated 3-manifold. It is denoted $H(U_t^-, -R_t, i, x)$ and is also called a standard decorated handlebody. We shall regard the module $\Psi_t^\star = (\mathcal{T}(\Sigma_t))^\star$ as the module generated by v-colorings of $-R_t$.

1.8. Operator invariants of decorated 3-cobordisms. Now we are ready to extend the modular functor $\mathcal{T}$ constructed in Section 1.4 to a topological quantum field theory based on decorated 3-cobordisms.

In accordance with the terminology of Section III.1 a decorated 3-cobordism is a triple $(M, \partial_- M, \partial_+ M)$ where $\partial_- M$ and $\partial_+ M$ are parametrized d-surfaces and M is a decorated 3-manifold with $\partial M = (-\partial_- M) \amalg \partial_+ M$. We define a

K-homomorphism

(1.8.a) $$\tau(M) = \tau(M, \partial_- M, \partial_+ M) : \mathcal{T}(\partial_- M) \to \mathcal{T}(\partial_+ M)$$

as follows. For any connected component Σ of $\partial_- M$ of type $t = t(\Sigma)$, glue U_t to M along the given parametrization $\partial U_t = \Sigma_t \to \Sigma$. Perform such gluings corresponding to all components of $\partial_- M$. In a sense we fill in the components of the bottom base of M using the standard handlebodies. Similarly, for every component Σ of $\partial_+ M$ of type $t = t(\Sigma)$, glue U_t^- to M along the d-homeomorphism $(-f) \circ \text{mir} : \partial(U_t^-) \to -\Sigma$ where $f : \partial U_t = \Sigma_t \to \Sigma$ is the given parametrization of Σ.

These gluings result in a *closed* oriented 3-manifold $\tilde{M}$ with a ribbon graph $\tilde{\Omega}$ sitting inside $\tilde{M}$. The graph $\tilde{\Omega}$ is obtained by gluing the given ribbon graph in M and the ribbon graphs $\{R_t, -R_t\}_t$ in the standard handlebodies. The ribbon graph $\tilde{\Omega}$ is in general neither colored nor v-colored. Indeed, the cap-like and cup-like bands and the coupons of $\tilde{\Omega}$ contained in the newly glued handlebodies are not colored. We vary the colors of these bands in the set $\{V_i\}_{i \in I}$ and vary the colors of the coupons in the corresponding K-modules discussed in Sections 1.4 and 1.7. For each such choice, say y, of colors, we get a v-coloring of $\tilde{\Omega}$. Applying the invariant τ of v-colored graphs introduced in Section II.2 we get a certain $\tau(\tilde{M}, \tilde{\Omega}, y) \in K$. It is obvious that the mapping $y \mapsto \tau(\tilde{M}, \tilde{\Omega}, y)$ is polylinear with respect to the colors of coupons. In view of the interpretation of the modules $\mathcal{T}(\partial_- M)$ and $\mathcal{T}(\partial_+ M)$ in terms of the colorings of R_t and $-R_t$, the mapping $y \mapsto \tau(\tilde{M}, \tilde{\Omega}, y)$ induces a K-homomorphism

$$\mathcal{T}(\partial_- M) \otimes_K (\mathcal{T}(\partial_+ M))^\star \to K.$$

We consider the adjoint K-homomorphism $\mathcal{T}(\partial_- M) \to \mathcal{T}(\partial_+ M)$ and define the homomorphism (1.8.a) to be its composition with $\eta(\partial_+ M) : \mathcal{T}(\partial_+ M) \to \mathcal{T}(\partial_+ M)$ (see Section 1.4).

Note that if $\partial_- M$ or $\partial_+ M$ is empty then we just do not perform the corresponding gluings. If M is closed then we retrieve the invariant τ of closed oriented 3-manifolds with embedded v-colored ribbon graphs introduced in Chapter II. Note also that the homomorphism $\tau(M)$ is preserved under isotopy of the parametrization homeomorphisms of the bases of M. (Isotopies should preserve the images of distinguished arcs.)

1.9. Theorem. *The function* $(M, \partial_- M, \partial_+ M) \mapsto \tau(M) : \mathcal{T}(\partial_- M) \to \mathcal{T}(\partial_+ M)$ *extends the modular functor* $\mathcal{T}$ *to a non-degenerate topological quantum field theory.*

Theorem 1.9 is proven in Section 2. The main point is to verify the functoriality axiom which is by no means straightforward. The proof is based on a geometric technique that enables us to present decorated 3-cobordisms by ribbon graphs in $\mathbb{R}^3$ and to express the operator invariants of 3-cobordisms via the operator invariants of ribbon graphs.

Theorem 1.9 yields a TQFT $(\mathcal{T}, \tau)$ based on parametrized d-surfaces and decorated 3-manifolds. As we shall see in Section 4 this TQFT has anomalies which may be computed in terms of Maslov indices of Lagrangian spaces. This computation will lead us to parametrization-free and anomaly-free renormalizations of $(\mathcal{T}, \tau)$.

Up to normalization, the TQFT $(\mathcal{T}, \tau)$ is multiplicative with respect to connected sum. The connected sum of two decorated 3-cobordisms M_1 and M_2 is defined as follows. Take closed 3-balls $B_1 \subset \text{Int}(M_1)$ and $B_2 \subset \text{Int}(M_2)$ disjoint from the given ribbon graphs and glue $M_1 \setminus \text{Int}(B_1)$ to $M_2 \setminus \text{Int}(B_2)$ along a homeomorphism $\partial B_1 \to \partial B_2$. This homeomorphism is chosen so that the orientations in $M_1 \setminus \text{Int}(B_1)$ and $M_2 \setminus \text{Int}(B_2)$ induced by those in M_1, M_2 are compatible. The gluing yields a decorated 3-cobordism $M_1 \# M_2$ between $\partial_-(M_1) \amalg \partial_-(M_2)$ and $\partial_+(M_1) \amalg \partial_+(M_2)$. The construction of connected sum has indeterminacy which amounts to the choice of connected components of M_1, M_2 containing B_1, B_2; for connected M_1, M_2, the sum is defined uniquely. It is easy to check that

$$\tau(M_1 \# M_2) = \mathcal{D}\,\tau(M_1 \amalg M_2) = \mathcal{D}\,\tau(M_1) \otimes \tau(M_2) \tag{1.9.a}$$

(cf. formula (II.2.3.a)). In particular,

$$\tau(M_1 \# (S^1 \times S^2)) = \mathcal{D}\,\tau(M_1) \tag{1.9.b}$$

where we view $S^1 \times S^2$ as a 3-manifold decorated with the empty ribbon graph.

1.10. Remark. The fact that the invariant $\tau(M)$ introduced in Chapter II may be included in a TQFT sheds additional light on the very definition of this invariant. Consider a closed oriented 3-manifold M obtained by surgery on S^3 along a framed m-component link $L \subset S^3$. Let U be a closed regular neighborhood of L in S^3. The surface ∂U splits M into a union of the link exterior $E = S^3 \setminus \text{Int}(U)$ and m copies of $S^1 \times B^2$. The components of $\partial U = \partial E$ are parametrized 2-tori, the parametrization being induced by the framing of L as in Section II.2.1. We have $\mathcal{T}(S^1 \times S^1) = K^{\text{card}(I)}$ with a natural basis numerated by elements of I and $\mathcal{T}(\partial U) = (K^{\text{card}(I)})^{\otimes m}$ with a natural basis numerated by colorings of L. We may compute $\tau(M)$ as the composition of $\tau(E, \emptyset, \partial U) : K \to \mathcal{T}(\partial U)$ with the tensor product of m copies of $\tau(S^1 \times B^2, S^1 \times S^1, \emptyset) : K^{\text{card}(I)} \to K$. (We should take into account the gluing anomaly.) Such a computation of $\tau(M)$ gives a formula similar to (II.2.2.a) where the roles of $F(\Gamma(L, \lambda))$ and $\dim(\lambda(L_n))$ are played by the matrix coefficients of $\tau(E, \emptyset, \partial U)$ and $\tau(S^1 \times B^2, S^1 \times S^1, \emptyset)$ respectively.

1.11. Exercise. Let B^3 be the 3-ball regarded as a decorated 3-manifold with an arbitrary parametrization of the boundary by the standard d-surface $\Sigma_{(0;)}$ and the empty ribbon graph. Show that $\tau(B^3, \emptyset, \partial B^3) : \mathcal{T}(\emptyset) = K \to \mathcal{T}(\partial B^3) = K$ is the identity id_K. In other words, $\tau(B^3) = 1 \in K$. Show that $\tau(B^3, \partial B^3, \emptyset) : \mathcal{T}(\partial B^3) = K \to \mathcal{T}(\emptyset) = K$ is multiplication by $\mathcal{D}^{-1}$.

2. Proof of Theorem 1.9

2.0. Outline. We first reduce Theorem 1.9 to three lemmas formulated in Section 2.1. The proof of these lemmas relies on a technique of presentation of 3-cobordisms by ribbon graphs in $\mathbb{R}^3$. This technique in its full generality is rather heavy. The reader may first confine himself to the case of cobordisms with connected bases which is technically much simpler.

Only a small part of the ideas and results involved in the proof of Theorem 1.9 will be used in further sections. We shall often refer to Lemmas 2.2.2 and 2.2.3 which are quite straightforward and independent of the rest of the section. Presentations of 3-cobordisms by ribbon graphs in $\mathbb{R}^3$ will be used in Sections 4.3, 5.4, 10.4, and in the last two sections of Chapter V.

2.1. Reduction to Lemmas. To prove Theorem 1.9 we have to verify the axioms (III.1.4.1) - (III.1.4.4). The naturality axiom (III.1.4.1) follows from Theorem II.2.3.2. The multiplicativity axiom (III.1.4.2) follows from the multiplicativity of τ with respect to disjoint union of closed 3-manifolds (see Section II.2). The normalization and functoriality axioms are ensured by the following two lemmas.

2.1.1. Lemma. *For any parametrized decorated surface* Σ,

$$\tau(\Sigma \times [0,1]) = \mathrm{id}_{\mathcal{T}(\Sigma)}. \tag{2.1.a}$$

2.1.2. Lemma. *If a decorated 3-cobordism $M = M_2 M_1$ is obtained from decorated 3-cobordisms M_1 and M_2 by gluing along a d-homeomorphism $p : \partial_+(M_1) \to \partial_-(M_2)$ commuting with parametrizations then for some invertible $k \in K$,*

$$\tau(M) = k\, \tau(M_2)\, p_{\#}\, \tau(M_1). \tag{2.1.b}$$

To show that the TQFT $(\mathcal{T}, \tau)$ is non-degenerate we compute the value of τ on the decorated handlebody $H(U_t, R_t, i, x)$ where $x \in \mathrm{Hom}(\mathbb{1}, \Phi(t; i))$ (see Section 1.4). Here we regard $H(U_t, R_t, i, x)$ as a cobordism between the empty surface and $\partial H(U_t, R_t, i, x) = \Sigma_t$ so that $\tau(H(U_t, R_t, i, x)) \in \mathcal{T}(\Sigma_t) = \Psi_t$ (cf. Section III.1.4).

2.1.3. Lemma. *For any decorated type t with the underlying genus g and any $x \in \mathrm{Hom}(\mathbb{1}, \Phi(t; i))$ with $i \in I^g$, we have $\tau(H(U_t, R_t, i, x)) = x$.*

This lemma implies non-degeneracy of $(\mathcal{T}, \tau)$.

The remaining part of Section 2 is devoted to the proof of these lemmas. In Section 2.2 we introduce a functor h_0 used in the proof. In Sections 2.3 - 2.5 we introduce a geometric technique enabling us to present decorated 3-cobordisms by ribbon graphs in $\mathbb{R}^3$ and to compute the operator invariants of 3-cobordisms. Lemmas 2.1.1 - 2.1.3 are proven in Sections 2.6 - 2.9. For the convenience of the reader, we distinguish in the proof of Lemma 2.1.2 and in the relevant preliminary material the (simpler) case of cobordisms with connected bases.

2.2. Algebraic preliminaries. Recall the category $\mathrm{Proj}(K)$ of (finitely generated) projective K-modules and K-linear homomorphisms. Denote by h_0 the covariant functor $\mathcal{V} \to \mathrm{Proj}(K)$ transforming an object V into $\mathrm{Hom}(\mathbb{1}, V)$ and transforming a morphism $f : V \to W$ into the homomorphism $g \mapsto fg : \mathrm{Hom}(\mathbb{1}, V) \to \mathrm{Hom}(\mathbb{1}, W)$. The remaining part of this subsection is concerned with the behavior of the tensor product in $\mathcal{V}$ with respect to h_0. This study will be used in the proof of Lemma 2.1.2 for cobordisms with non-connected bases. The reader pursuing the case of cobordisms with connected bases may skip the rest of this subsection.

It is obvious that the functor h_0 does not preserve tensor product. For example, if V is a simple object of $\mathcal{V}$ non-isomorphic to $\mathbb{1}$ then $h_0(V) = h_0(V^*) = 0$ whilst $h_0(V \otimes V^*) = K$. On the other hand, the tensor multiplication of morphisms induces a K-homomorphism

$$\varphi : h_0(W_1) \otimes_K h_0(W_2) \otimes_K \ldots \otimes_K h_0(W_n) \to h_0(W_1 \otimes \ldots \otimes W_n) \tag{2.2.a}$$

for any objects $W_1, W_2, \ldots, W_n$ of $\mathcal{V}$. The following argument shows that this homomorphism is an embedding onto a direct summand of $h_0(W_1 \otimes \ldots \otimes W_n)$ and that it has a canonical left inverse. Note that tensor multiplication of morphisms induces a K-homomorphism

$$\bigotimes_{r=1}^{n} \mathrm{Hom}(W_r, \mathbb{1}) \to \mathrm{Hom}(W_1 \otimes \ldots \otimes W_n, \mathbb{1}). \tag{2.2.b}$$

By Lemma II.4.2.3 the module $\mathrm{Hom}(W_r, \mathbb{1})$ is dual to $h_0(W_r)$. For projective K-modules, the tensor product commutes with the passage to the dual module (up to a canonical isomorphism). Therefore dualizing the homomorphism (2.2.b) we get a homomorphism

$$\psi : h_0(W_1 \otimes \ldots \otimes W_n) \to h_0(W_1) \otimes_K \ldots \otimes_K h_0(W_n). \tag{2.2.c}$$

It follows from the definitions that $\psi\varphi = \mathrm{id}$. This implies that φ is an embedding onto a direct summand of $h_0(W_1 \otimes \ldots \otimes W_n)$.

It is clear that the endomorphism $\varphi\psi$ of $h_0(W_1 \otimes \ldots \otimes W_n)$ is a projection onto the image of φ. We shall need an explicit formula for $\varphi\psi$. So, consider for any object W of $\mathcal{V}$ and for any $i \in I$ the morphism $p(W, i) : W \to W$ corresponding (under the functor F constructed in Section I.2) to the ribbon tangle presented in

Figure II.1.1 where j should be replaced by W. Set

$$p_W = \mathcal{D}^{-2} \sum_{i \in I} \dim(i)\, p(W, i) : W \to W$$

where $\mathcal{D} \in K$ is the rank of $\mathcal{V}$. It follows from the isotopy invariance of F that p_W is natural with respect to morphisms in $\mathcal{V}$: for any morphism $f : V \to W$, we have $p_W f = f p_V$.

2.2.1. Lemma. *For any objects $W_1, W_2, ..., W_n$ of the category $\mathcal{V}$, we have*

$$\varphi\psi = h_0(p_{W_1} \otimes ... \otimes p_{W_n}) \in \mathrm{End}_K(h_0(W_1 \otimes ... \otimes W_n)). \tag{2.2.d}$$

We shall prove this lemma at the end of Section 2.2 using Lemmas 2.2.2 - 2.2.4.

2.2.2. Lemma. *For any objects V, W of the category $\mathcal{V}$, there is a canonical K-linear splitting*

$$h_0(V \otimes W) = \bigoplus_{i \in I} \left(h_0(V \otimes V_i^*) \otimes_K h_0(V_i \otimes W) \right). \tag{2.2.e}$$

The isomorphism transforming the right-hand side into the left-hand side is given by the formula

$$x \otimes y \mapsto (\mathrm{id}_V \otimes d_{V_i} \otimes \mathrm{id}_W)(x \otimes y), \tag{2.2.f}$$

where $x \in h_0(V \otimes V_i^*)$, $y \in h_0(V_i \otimes W)$, cf. Figure 2.1.

Proof. Recall that for any objects U, V, W of $\mathcal{V}$, there exists a bijection between the sets $h_0(V \otimes W)$ and $\mathrm{Hom}(V^*, W)$ (see Exercise I.1.8.1). Using such bijections it is easy to establish a correspondence between (2.2.e) and the splitting provided by Lemma II.4.2.2 (with V and i replaced by V^* and i^* respectively).

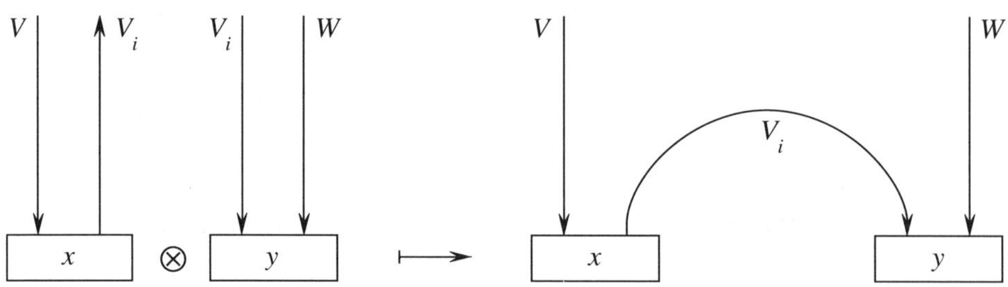

Figure 2.1

2.2.3. Lemma. *For any objects $W_1, W_2, \ldots, W_n$ of $\mathcal{V}$ with $n \geq 2$, there is a canonical K-linear splitting*

$$h_0(W_1 \otimes \ldots \otimes W_n) = \tag{2.2.g}$$

$$= \bigoplus_{i_1,\ldots,i_{n-1}\in I} \left(h_0(W_1 \otimes V_{i_1}^*) \otimes \bigotimes_{r=2}^{n-1} h_0(V_{i_{r-1}} \otimes W_r \otimes V_{i_r}^*) \otimes h_0(V_{i_{n-1}} \otimes W_n) \right).$$

Proof. The case $n = 2$ is contained in the previous lemma. The general case is proven by induction, the inductive assumption should be applied to the objects $W_1, \ldots, W_{n-2}, W_{n-1} \otimes W_n$.

A graphical form for elements of the summand corresponding to a sequence $i_1, \ldots, i_{n-1} \in I$ is given in Figure 2.2. Lemma 2.2.3 yields another proof of the fact that the homomorphism φ considered above is an embedding. The image of this embedding is the summand corresponding to $i_1 = \ldots = i_{n-1} = 0$.

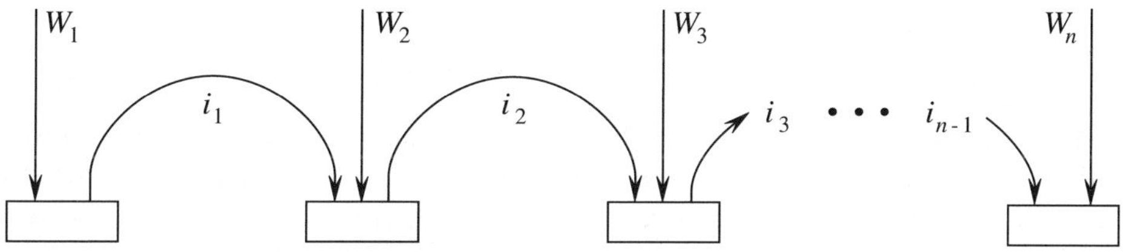

Figure 2.2

2.2.4. Lemma. (i) $p_{\mathbb{1}} = \mathrm{id}_{\mathbb{1}}$.
(ii) *For any $j \in I \setminus \{0\}$, we have $p_{V_j} = 0$.*

Proof. The arguments given in Section II.3.3 show that for any $j \in I$, the morphism $p(V_j, i) : V_j \to V_j$ is multiplication by $S_{i,j}(\dim(j))^{-1}$. Therefore

$$p_{V_j} = \left(\mathcal{D}^{-2}(\dim(j))^{-1} \sum_{i\in I} \dim(i)\, S_{i,j} \right) \mathrm{id}_{V_j}.$$

It follows from formula (II.3.8.b) that the right-hand side equals 0 if $j \neq 0$ and equals $\mathrm{id}_{\mathbb{1}}$ if $j = 0$.

2.2.5. Proof of Lemma 2.2.1. Set $\eta = h_0(p_{W_1} \otimes \ldots \otimes p_{W_n})$. To prove that $\varphi\psi = \eta$ it suffices to show that $\psi = \psi\eta$ and that $\mathrm{Im}(\eta) \subset \mathrm{Im}(\varphi)$. Then, since $\varphi\psi$ is a projection onto $\mathrm{Im}(\varphi)$, we would have $\varphi\psi = \varphi\psi\eta = \eta$.

To prove that $\psi = \psi\eta$ we should show that for any $x \in h_0(W_1 \otimes \ldots \otimes W_n)$ and any $\{y_r \in \mathrm{Hom}(W_r, \mathbb{1})\}_{r=1}^n$, we have

$$(y_1 \otimes \ldots \otimes y_n)\, x = (y_1 \otimes \ldots \otimes y_n)(p_{W_1} \otimes \ldots \otimes p_{W_n})\, x.$$

This follows from the equalities $y_i p_{W_i} = p_{1\!\!1} y_i = y_i$ where $i = 1, ..., n$.

To prove that $\mathrm{Im}(\eta) \subset \mathrm{Im}(\varphi)$ we shall show that each summand of the decomposition (2.2.g) is mapped by η into $\mathrm{Im}(\varphi)$. Fix $i_1, ..., i_{n-1} \in I$ and consider an element of the corresponding summand presented in Figure 2.2. The image of this element under η is presented in Figure 2.3. Here we added n small circles linking the vertical bands. The bold points on the circles indicate that we vary their colors in the set I and sum up the operator invariants of the resulting v-colored ribbon graphs with coefficients $\mathcal{D}^{-2n} \prod_i \dim(i)$ where i runs over the colors of these n circles.

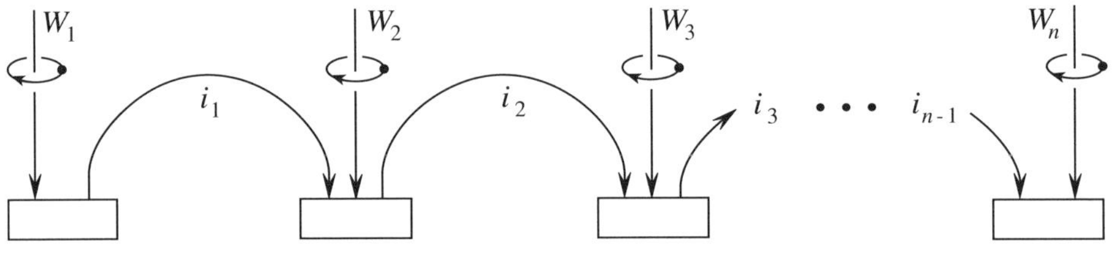

Figure 2.3

If $i_1 = i_2 = ... = i_{n-1} = 0$ then the direct summand in question is just $\mathrm{Im}(\varphi)$. In this case we may efface the arcs marked by $i_1, ..., i_{n-1}$ in Figures 2.2 and 2.3. The n circles linking the vertical arcs may be pushed down so that they become unlinked with the arcs. The operator invariant of each of these circles colored with $i \in I$ equals $\dim(i)$. Taking into account the coefficients mentioned above and summing up over $i \in I$ we obtain that the factor contributed by these circles is equal to 1. This shows that η is the identity on $\mathrm{Im}(\varphi)$.

Let us prove that if at least one term of the sequence $i_1, ..., i_{n-1} \in I$ is non-zero then the corresponding summand on the right-hand side of (2.2.g) is transformed by η into 0. It suffices to show that the element of $h_0(W_1 \otimes ... \otimes W_n)$ shown in Figure 2.3 is equal to zero. We may deform the left circle so that it links the arc colored by i_1. A segment of the last arc linked by this circle together with the circle itself represents the morphism $p_{V_{i_1}}$. Lemma 2.2.4 ensures that if $i_1 \neq 0$ then this morphism is equal to 0 so that Figure 2.3 presents a zero morphism. If $i_1 = 0$ then we may efface the arc colored by i_1 and apply the same argument to the second circle, etc. This implies the inclusion $\mathrm{Im}(\eta) \subset \mathrm{Im}(\varphi)$ and the claim of the lemma.

2.3. Presentation of 3-cobordisms by graphs in $\mathbb{R}^3$: the case of connected bases. As we know, any framed link in $\mathbb{R}^3$ gives rise to a closed oriented 3-manifold via surgery on $S^3 = \mathbb{R}^3 \cup \{\infty\}$. More generally, a pair consisting of a framed link L and a ribbon (0,0)-graph in $\mathbb{R}^3$ gives rise to a ribbon graph in the closed 3-manifold obtained from S^3 by the surgery along L. This construction is universal in the sense that it produces all closed connected oriented 3-manifolds and all ribbon graphs in these manifolds. Here we extend this approach to ribbon graphs in 3-cobordisms with connected bases. More precisely, we describe a geometric construction deriving such decorated 3-cobordisms from certain ribbon graphs in $\mathbb{R}^3$. We compute the operator invariants of these cobordisms via the functor F of Chapter I.

Let m^-, g^-, m^+, g^+ be non-negative integers and let $k^- = m^- + 2g^-$ and $k^+ = m^+ + 2g^+$. Let Ω be a ribbon (k^-, k^+)-graph in $\mathbb{R}^3$ (see Section I.2). We say that Ω is a (partially v-colored) special ribbon graph if it satisifes the following conditions (i) - (iii). Set $\varepsilon = 1/10$.

(i) For all odd i with $1 \leq i \leq 2g^- - 1$, the bottom boundary intervals

$$[m^- + i - \varepsilon, m^- + i + \varepsilon] \times 0 \times 0, \quad [m^- + i + 1 - \varepsilon, m^- + i + 1 + \varepsilon] \times 0 \times 0$$

are the bases of a band e_i^- of Ω directed towards the left base.

(ii) For all odd j with $1 \leq j \leq 2g^+ - 1$, the top boundary intervals

$$[m^+ + j - \varepsilon, m^+ + j + \varepsilon] \times 0 \times 1, \quad [m^+ + j + 1 - \varepsilon, m^+ + j + 1 + \varepsilon] \times 0 \times 1$$

are the bases of a band e_j^+ directed towards the right base.

(iii) These $g^- + g^+$ bands are not colored, all other bands and all coupons of Ω are colored. Certain (but not necessarily all) annuli of Ω may be colored.

For instance, the ribbon tangle presented by the diagram in Figure 2.4 is a special ribbon graph with $m^- = m^+ = m$ and $g^- = g^+ = g$. This ribbon tangle consists of m vertical untwisted unlinked bands and g copies of the same (2,2)-tangle. The vertical bands are colored with objects $W_1, ..., W_m$ of $\mathcal{V}$, other bands and annuli are not colored. The r-th vertical band is directed down if $\nu_r = 1$ and up otherwise, other bands are directed as shown in the figure. Here e_i^- and e_j^+ are the cap-like and cup-like bands respectively.

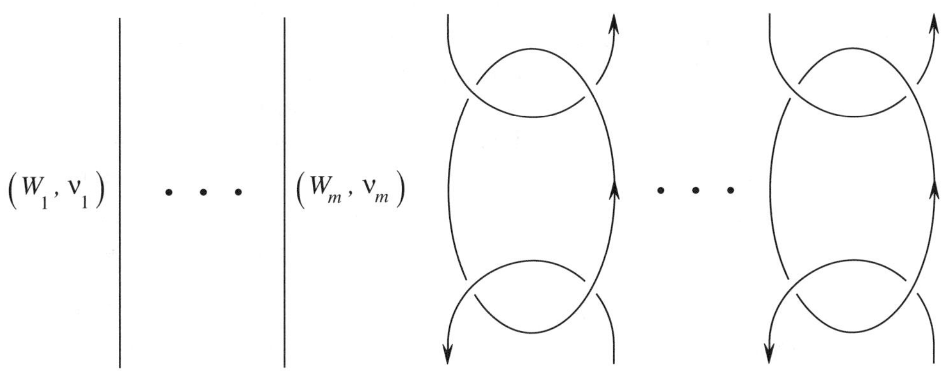

Figure 2.4

To any special ribbon graph Ω we associate two decorated types t^- and t^+ as follows. For $i = 1, ..., m^-$, denote by W_i^- the color of the band of Ω attached to the boundary interval $[i-\varepsilon, i+\varepsilon] \times 0 \times 0$. Set $\nu_i^- = 1$ if this band is directed towards this interval and set $\nu_i^- = -1$ if it is directed away. Similarly, for $j = 1, ..., m^+$, denote by W_j^+ the color of the band of Ω attached to the boundary interval $[j-\varepsilon, j+\varepsilon] \times 0 \times 1$. Set $\nu_j^+ = -1$ if this band is directed towards this interval and set $\nu_j^+ = 1$ if it is directed away. Set

$$t^- = (g^-; (W_1^-, \nu_1^-), ..., (W_{m^-}^-, \nu_{m^-}^-)), \quad t^+ = (g^+; (W_1^+, \nu_1^+), ..., (W_{m^+}^+, \nu_{m^+}^+)).$$

We shall call t^- and t^+ the bottom (decorated) type and the top (decorated) type of Ω.

Each special ribbon graph Ω with the bottom type t^- and the top type t^+ gives rise to a decorated 3-cobordism (M, Σ_-, Σ_+) where the parametrized d-surfaces Σ_-, Σ_+ have types t^-, t^+ respectively. The construction of this cobordism goes as follows. Attach to the bottom of Ω one uncolored coupon $Q^- = [0, k^- + 1] \times 0 \times [-1, 0]$ with the top base $[0, k^- + 1] \times 0 \times 0$. Similarly, attach to the top of Ω one uncolored coupon $Q^+ = [0, k^+ + 1] \times 0 \times [1, 2]$ with the bottom base $[0, k^+ + 1] \times 0 \times 1$. This results in a partially v-colored ribbon graph in $\mathbb{R}^3$ without free ends. Denote this ribbon graph by Ω_0. Recall the standard ribbon graphs $R_t, -R_t$ and the standard handlebodies U_t, U_t^- introduced in Section 1. There is an obvious embedding $f^- : R_{t^-} \to \Omega_0$ whose image comprises the coupon Q^-, the bands $\{e_i^-\}_i$, and small pieces of m^- colored bands of Ω_0 attached to the top base of Q^-. These small pieces are narrow rectangles with lower bases $\{[i-\varepsilon, i+\varepsilon] \times 0 \times 0\}_{i=1,...,m^-}$ lying on the top base of Q^-. The embedding f^- maps the coupon, the cap-like bands, and the vertical bands of R_{t^-} onto Q^-, the bands $\{e_i^-\}_i$, and these narrow rectangles, respectively. The embedding f^- extends to an orientation-preserving embedding $U_{t^-} \to S^3$, also denoted by f^-. We shall assume that the image of U_{t^-} intersects Ω_0 only along $f^-(R_{t^-})$. Similarly, there is an obvious embedding $f^+ : -R_{t^+} \to \Omega_0$ whose image comprises the coupon Q^+, the bands $\{e_j^+\}_j$, and small pieces of the m^+ colored bands of Ω_0 attached to the bottom base of Q^+. These small pieces are narrow rectangles with upper bases $\{[j-\varepsilon, j+\varepsilon] \times 0 \times 1\}_{j=1,...,m^+}$ lying on the bottom base of Q^+. The embedding f^+ maps the coupon, the cap-like bands, and the vertical bands of $-R_{t^+}$ onto Q^+, the bands $\{e_j^+\}_j$, and these narrow rectangles, respectively. The embedding f^+ extends to an orientation-preserving embedding $U_{t^+}^- \to S^3$, also denoted by f^+. We shall assume that the image of f^+ is disjoint from the image of f^- and intersects Ω_0 only along $f^+(-R_{t^+})$.

Cut out the open handlebodies $f^-(\operatorname{Int} U_{t^-})$ and $f^+(\operatorname{Int} U_{t^+}^-)$ from S^3. This results in a compact oriented 3-cobordism E with bottom base $\Sigma_- = f^-(\partial U_{t^-})$ and top base $\Sigma_+ = f^+(\partial U_{t^+}^-)$. We provide E with an orientation induced by right-handed orientation in S^3, and Σ_-, Σ_+ with orientations such that $\partial E = (-\Sigma_-) \cup \Sigma_+$. It is clear that the homeomorphism f^- restricted to $\Sigma_{t^-} = \partial U_{t^-}$ yields an orientation-preserving homeomorphism $\Sigma_{t^-} \to \Sigma_-$. The image of the distinguished marked arcs on Σ_{t^-} under this homeomorphism yields the structure of a parametrized d-surface on Σ_-. The distinguished arcs on Σ_- are the upper horizontal bases of the narrow rectangles attached to Q^- (as mentioned above) with their natural order from left to right, right-handed orientation, and marks $(W_1^-, \nu_1^-), \dots, (W_{m^-}^-, \nu_{m^-}^-)$. Similarly, the homeomorphism f^+ restricted to $\partial U_{t^+}^-$ and composed with the mirror reflection $\Sigma_{t^+} = \partial U_{t^+} \to \partial U_{t^+}^-$ yields an orientation-preserving homeomorphism $\Sigma_{t^+} \to \Sigma_+$. The image of the distinguished marked arcs on Σ_{t^+} under this homeomorphism yields the structure of a parametrized d-surface on Σ_+. The distinguished arcs on Σ_+ are the lower horizontal bases of the rectangles attached to Q^+ (as mentioned above) with their natural order from left to right, right-handed orientation, and marks $(W_1^+, \nu_1^+), \dots, (W_{m^+}^+, \nu_{m^+}^+)$.

It is obvious that the part of Ω_0 lying in E is a ribbon graph Ω_0' in E. The partial coloring of Ω_0 induces a coloring of all bands and coupons of Ω_0'. (Indeed, the only uncolored coupons Q^-, Q^+ and uncolored bands $\{e_i^-\}_i$, $\{e_j^+\}_j$ are cut out from Ω_0.) All annuli of Ω survive in Ω_0' and those which have been colored keep their colors. The uncolored annuli of Ω yield uncolored annuli in Ω_0'. The cores of these uncolored annuli form a framed link in $\mathrm{Int}(E)$, the framing being orthogonal to the annuli. Surgery on E along this framed link produces a compact oriented 3-cobordism M with parametrized decorated bases $\partial_- M = \Sigma_-$ and $\partial_+ M = \Sigma_+$. The ribbon graph Ω_0' with the uncolored annuli removed survives the surgery and gives rise to a v-colored ribbon graph in M. This completes the construction of the decorated 3-cobordism M associated to Ω.

We now give an explicit formula for computing the homomorphism $\tau(M) : \Psi_{t^-} \to \Psi_{t^+}$ from the operator invariants of Ω. With respect to the splittings (1.4.a) of Ψ_{t^-} and Ψ_{t^+} the homomorphism $\tau(M)$ may be presented by a block-matrix τ_i^j where i runs over sequences $i_1, \dots, i_{g^-} \in I$ and j runs over sequences $j_1, \dots, j_{g^+} \in I$. Each such sequence i determines a coloring $e_n^- \mapsto i_n$ of the uncolored cap-like bands of Ω incident to its bottom boundary. Similarly, each sequence $j = (j_1, \dots, j_{g^+}) \in I^{g_+}$ determines a coloring $e_n^+ \mapsto j_n$ of the uncolored cup-like bands of Ω incident to its top boundary. Therefore a pair ($i \in I^{g_-}$, $j \in I^{g_+}$) determines a coloring of uncolored bands of Ω. Let $L = L_1 \cup \dots \cup L_m$ be the framed link formed by the uncolored annuli of Ω. Every $\lambda \in \mathrm{col}(L)$ determines (together with i and j) a v-coloring of Ω. Denote the resulting v-colored ribbon graph in $\mathbb{R}^3$ by (Ω, i, j, λ). Recall its operator invariant $F(\Omega, i, j, \lambda) : \Phi(t^-; i) \to \Phi(t^+; j)$ defined in Chapter I. Recall the notation

$$\dim(\lambda) = \prod_{n=1}^{m} \dim(\lambda(L_n)).$$

The composition of a morphism $\mathbb{1} \to \Phi(t^-; i)$ with $F(\Omega, i, j, \lambda)$ defines a K-linear homomorphism $\mathrm{Hom}(\mathbb{1}, \Phi(t^-; i)) \to \mathrm{Hom}(\mathbb{1}, \Phi(t^+; j))$ denoted by $F_0(\Omega, i, j, \lambda)$.

It follows from the very definition of $\tau(M)$ that

$$\tau_i^j = \Delta^{\sigma(L)} \mathcal{D}^{-g^+ - \sigma(L) - m} \dim(j) \sum_{\lambda \in \mathrm{col}(L)} \dim(\lambda)\, F_0(\Omega, i, j, \lambda). \tag{2.3.a}$$

In the case $L = \emptyset$ the right-hand side reduces to one term $\mathcal{D}^{-g^+} \dim(j)\, F_0(\Omega, i, j)$.

To present 3-cobordisms with empty top base we may use the same constructions as above where Ω is a $(k^-, 0)$-graph and we do not involve Q^+, f^+ and do not cut out $f^+(\mathrm{Int}\, U_{t+}^-)$. The case of empty bottom base is treated similarly. Formula (2.3.a) extends to these cases; in order to avoid using the genus of the empty set we should replace the term $-g^+$ with $(\chi(\Sigma_+)/2) - 1$ where χ is the Euler characteristic. (The Euler characteristic of the empty set is equal to zero.)

Note finally that the directions (of the cores) of uncolored annuli of Ω are irrelevant for the construction of M. In our pictures we shall supress these directions.

2.4. Presentation of 3-cobordisms by graphs in $\mathbb{R}^3$: the general case. The results of Section 2.3 generalize to decorated 3-cobordisms with arbitrary, possibly non-connected bases. The difference is that instead of one coupon, Q^-, glued to the bottom line of Ω and one coupon, Q^+, glued to the top line of Ω, we should glue to these lines several coupons. Let Ω be a ribbon (k^-, k^+)-graph in $\mathbb{R}^3$. Assume that we have positive integers $k_1^-, ..., k_{r-}^-$ and $k_1^+, ..., k_{r+}^+$ such that $k_1^- + ... + k_{r-}^- = k^-$ and $k_1^+ + ... + k_{r+}^+ = k^+$. Let us glue to the bottom of Ω disjoint coupons $Q_1^-, ..., Q_{r-}^- \subset \mathbb{R} \times 0 \times [-1, 0]$ numbered from left to right such that the upper horizontal base of each Q_i^- includes k_i^- consecutive bottom boundary intervals of Ω. Let us glue to the top line of Ω disjoint coupons $Q_1^+, ..., Q_{r+}^+ \subset \mathbb{R} \times 0 \times [1, 2]$ numbered from left to right such that the lower horizontal base of each Q_j^+ includes k_j^+ consecutive top boundary intervals of Ω. We assume that moving from left to right along the top base of any Q_i^- or along the bottom base of any Q_j^+ we encounter first the bases of colored bands of Ω and then several pairs of bases of uncolored bands of Ω. The uncolored bands should have both their bases lying on the same Q_i^- or on the same Q_j^+. (The bases of each such band should lie next to each other and form one of the pairs mentioned above.) Each such band should be directed toward its left (resp. right) base if this base lies in the bottom (resp. top) line of Ω. All other bands and all coupons of Ω are assumed to be colored. Certain but not necessarily all annuli of Ω are colored. A ribbon graph Ω satisfying these assumptions is said to be a (partially v-colored) special ribbon graph. Now we perform the same constructions as in Section 2.3 with the obvious changes. This yields a decorated 3-cobordism M associated to Ω. Here the parametrized d-surfaces $\partial_- M$ and $\partial_+ M$ have r_- and r_+ connected components respectively.

The construction of M depends on the choice of sequences $k_1^-, ..., k_{r-}^-$ and $k_1^+, ..., k_{r+}^+$ determining the splitting of the boundary ends of Ω into consecutive families. By abuse of language we shall say sometimes that Ω presents M. In the case of connected bases, where $r^- = r^+ = 1$, both sequences in question reduce to one-term sequences k^- and k^+.

Formula (2.3.a) for $\tau(M)$ extends to the general case as follows. The coloring and directions of the bands of Ω attached to the coupons $Q_1^-, \ldots, Q_{r^-}^-$ and $Q_1^+, \ldots, Q_{r^+}^+$ determine in the obvious way certain decorated types $t_1^-, \ldots, t_{r^-}^-$ and $t_1^+, \ldots, t_{r^+}^+$ respectively. Denote the numerical genera underlying these types by $g_1^-, \ldots, g_{r^-}^-$ and $g_1^+, \ldots, g_{r^+}^+$ respectively. By definition,

$$\mathcal{T}(\partial_- M) = \bigotimes_{p=1}^{r^-} \Psi_{t_p^-} = \bigoplus_{i^1 \in I^{g_1^-}, \ldots, i^{r^-} \in I^{g_{r^-}^-}} \left(\bigotimes_{p=1}^{r^-} \mathrm{Hom}(\mathbb{1}, \Phi(t_p^-; i^p)) \right).$$

Similarly,

$$\mathcal{T}(\partial_+ M) = \bigoplus_{j^1 \in I^{g_1^+}, \ldots, j^{r^+} \in I^{g_{r^+}^+}} \left(\bigotimes_{q=1}^{r^+} \mathrm{Hom}(\mathbb{1}, \Phi(t_q^+; j^q)) \right).$$

With respect to these splittings the homomorphism $\tau(M) : \mathcal{T}(\partial_- M) \to \mathcal{T}(\partial_+ M)$ may be presented by a block-matrix

$$\{\tau_i^j : \otimes_{p=1}^{r^-} \mathrm{Hom}(\mathbb{1}, \Phi(t_p^-; i^p)) \to \otimes_{q=1}^{r^+} \mathrm{Hom}(\mathbb{1}, \Phi(t_q^+; j^q))\}_{i,j}$$

where i runs over sequences $(i^1 \in I^{g_1^-}, \ldots, i^{r^-} \in I^{g_{r^-}^-})$ and j runs over sequences $(j^1 \in I^{g_1^+}, \ldots, j^{r^+} \in I^{g_{r^+}^+})$. Let L denote the framed m-component link formed by the uncolored annuli of Ω. Every $\lambda \in \mathrm{col}(L)$ determines (together with i and j as above) a v-coloring of Ω. Denote the resulting v-colored ribbon graph in $\mathbb{R}^3$ by (Ω, i, j, λ). The morphism

$$F(\Omega, i, j, \lambda) : \otimes_{p=1}^{r^-} \Phi(t_p^-; i^p) \to \otimes_{q=1}^{r^+} \Phi(t_q^+; j^q)$$

induces a K-homomorphism

$$\mathrm{Hom}(\mathbb{1}, \otimes_{p=1}^{r^-} \Phi(t_p^-; i^p)) \to \mathrm{Hom}(\mathbb{1}, \otimes_{q=1}^{r^+} \Phi(t_q^+; j^q)).$$

Composing this homomorphism on the right with the natural inclusion

$$\otimes_{p=1}^{r^-} \mathrm{Hom}(\mathbb{1}, \Phi(t_p^-; i^p)) \to \mathrm{Hom}(\mathbb{1}, \otimes_{p=1}^{r^-} \Phi(t_p^-; i^p))$$

and on the left with the natural projection

$$\mathrm{Hom}(\mathbb{1}, \otimes_{q=1}^{r^+} \Phi(t_q^+; j^q)) \to \otimes_{q=1}^{r^+} \mathrm{Hom}(\mathbb{1}, \Phi(t_q^+; j^q))$$

(cf. Section 2.2) we get a homomorphism

$$\otimes_{p=1}^{r^-} \mathrm{Hom}(\mathbb{1}, \Phi(t_p^-; i^p)) \to \otimes_{q=1}^{r^+} \mathrm{Hom}(\mathbb{1}, \Phi(t_q^+; j^q)).$$

Denote the last homomorphism by $F_0(\Omega, i, j, \lambda)$. It follows from the definitions that

$$\tau_i^j = \Delta^{\sigma(L)} \mathcal{D}^{-\sigma(L)-m-1+\chi(\partial_+ M)/2} \prod_{q=1}^{r^+} \dim(j^q) \sum_{\lambda \in \operatorname{col}(L)} \dim(\lambda)\, F_0(\Omega, i, j, \lambda)$$

where χ is the Euler characteristic.

2.5. Lemma. *Any decorated 3-cobordism considered up to homeomorphism may be presented by a special ribbon graph in $\mathbb{R}^3$.*

Proof. The proof of this lemma is essentially contained in Sections II.2 and 1.8 which provide a construction inverse to the construction of 3-cobordisms from ribbon graphs in $\mathbb{R}^3$. Consider a decorated 3-cobordism M. As in Section 1.8, glue to M standard handlebodies along the given parametrizations of components of ∂M. This results in a closed oriented 3-manifold $\tilde{M}$ with an embedded ribbon graph $\tilde{\Omega}$. Each component of ∂M contributes one coupon to $\tilde{\Omega}$, other coupons of $\tilde{\Omega}$ correspond to coupons of the original ribbon graph in M. Denote by $Q_1^-, ..., Q_{r^-}^-$ (resp. $Q_1^+, ..., Q_{r^+}^+$) the coupons of $\tilde{\Omega}$ which arise from components of $\partial_- M$ (resp. from components of $\partial_+ M$), the numbering of these coupons is arbitrary. Present $\tilde{M}$ as the result of surgery on the 3-sphere along a framed link L. By slightly deforming $\tilde{\Omega} \subset \tilde{M}$, we may push it into the exterior of L in S^3. Thus we may assume that $\tilde{\Omega} \subset S^3 \setminus L$. Let Ω_0 denote the disjoint union of Ω with uncolored annuli obtained by thickening of the link L orthogonally to its framing. Pulling the coupons $Q_1^-, ..., Q_{r^-}^-$ and $Q_1^+, ..., Q_{r^+}^+$ down and up respectively, we may deform Ω_0 in $S^3 = \mathbb{R}^3 \cup \{\infty\}$ so that it satisfies the following conditions: the coupons $Q_1^-, ..., Q_{r^-}^-$ lie in $\mathbb{R} \times 0 \times [-1, 0]$ with their top bases being the segments (I.2.1.a) for a certain integer k; the coupons $Q_1^+, ..., Q_{r^+}^+$ lie in $\mathbb{R} \times 0 \times [1, 2]$ with their bottom bases being the segments (I.2.1.b) for a certain integer l; the remaining part of Ω_0 lies in the strip $\mathbb{R}^2 \times [0, 1]$. Cutting off the coupons $Q_1^-, ..., Q_{r^-}^-$ and $Q_1^+, ..., Q_{r^+}^+$ we get from Ω_0 a special ribbon graph in $\mathbb{R}^2 \times [0, 1]$. It follows from definitions that this ribbon graph presents M.

2.6. Lemma. *Let $t = (g; (W_1, \nu_1), ..., (W_m, \nu_m))$ be a decorated type. Let Ω be the ribbon tangle in $\mathbb{R}^3$ shown in Figure 2.4. Then Ω (together with the 1-term sequences $m + g$ and $m + g$) presents the cylinder $\Sigma_t \times [0, 1]$.*

Proof. Let M denote the decorated 3-cobordism presented by Ω. To construct M we attach two uncolored coupons Q^- and Q^+ to the bottom and top of Ω and consider their closed regular neighborhoods, T^- and T^+ in $\mathbb{R}^2 \times [-\infty, 0]$ and $\mathbb{R}^2 \times [1, \infty]$ respectively. These neighborhoods are closed disjoint 3-balls in S^3. In the case $g = 0$ we do not perform any surgery but simply cut out the open 3-balls $\operatorname{Int}(T^-)$ and $\operatorname{Int}(T^+)$ from S^3. It is clear that the resulting

3-cobordism M is the cylinder over the 2-sphere with m distinguished arcs marked by $(W_1, \nu_1), \ldots, (W_m, \nu_m)$.

Assume that $g \geq 1$. Denote by T_r^- (resp. by T_r^+) a closed regular neighborhood in $\mathbb{R}^2 \times [0, 1]$ of the r-th cap-like (resp. the r-th cup-like) band of Ω. It is clear that T_r^- and T_r^+ are solid 3-cylinders with bases lying in ∂T^- and ∂T^+ respectively. Set

$$T = T^- \cup (\cup_{r=1}^g T_r^-) \cup T^+ \cup (\cup_{r=1}^g T_r^+) \subset S^3.$$

It is clear that T is a disjoint union of two unknotted unlinked handlebodies of genus g in S^3.

The cobordism M is obtained from $S^3 \setminus \text{Int}(T)$ by surgery along the framed link determined by the annuli of Ω. Let A_r be the r-th annulus of Ω where $r = 1, \ldots, g$. We present this annulus in the form $A_r = D_r \setminus \text{Int}(D'_r)$ where D_r and D'_r are concentric 2-disks in $\mathbb{R}^2 \times [0, 1]$ such that $D'_r \subset \text{Int}(D_r)$ and D'_r transversally intersects Ω along two short intervals lying on two bands of Ω linked by A_r, see Figure 2.5. Consider a regular neighborhood $D_r \times [-1, 1]$ in $\mathbb{R}^2 \times (0, 1)$ of the larger disk $D_r = D_r \times 0$. We assume that there are no redundant crossings, i.e., that $D_r \times [-1, 1]$ is disjoint from $D_s \times [-1, 1], T_s^-, T_s^+$ for $r \neq s$. We also assume that $D_r \times [-1, 1]$ crosses T_r^- (resp. T_r^+) in the simplest possible way, i.e., along a subcylinder $B_r^- \times [-1, 1]$ (resp. $B_r^+ \times [-1, 1]$) concentric to $D_r \times [-1, 1]$. Here B_r^- and B_r^+ are small closed disjoint 2-disks in $\text{Int}(D'_r)$.

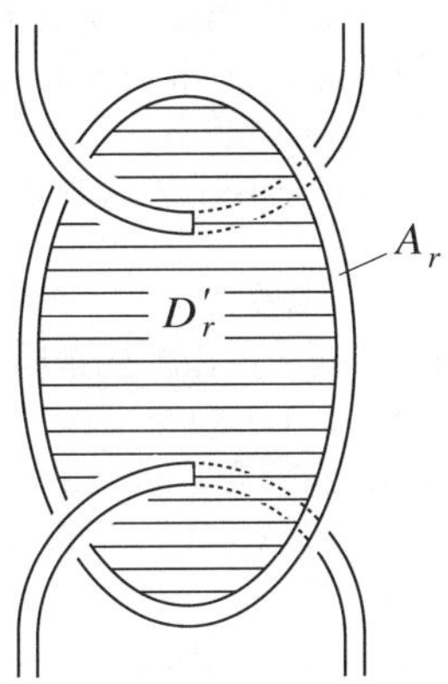

Figure 2.5

The surgery along the framed knot defined by A_r may be described as follows. Consider the solid torus

$$A_r \times [-1, 1] = (D_r \times [-1, 1]) \setminus (\text{Int}(D'_r) \times [-1, 1]) \subset S^3.$$

Its boundary consists of four annuli $A_r \times (-1)$, $A_r \times 1$, $\partial D_r \times [-1, 1]$, $\partial D'_r \times [-1, 1]$. We remove the interior of $A_r \times [-1, 1]$ from $S^3 \setminus \text{Int}(T)$ and glue in its place the standard solid torus $D^2 \times S^1$ where D^2 is a 2-disk. The gluing is performed along a homeomorphism $\partial(A_r \times [-1,1]) \to \partial(D^2 \times S^1)$ carrying each circle $\partial D'_r \times t$ with $t \in [-1,1]$ onto a circle $\partial D^2 \times x$ with $x \in S^1$. Let $E(r)$ denote the solid 3-cylinder formed by the disks $D^2 \times x$ glued to $\partial D'_r \times t$ with $t \in [-1,1]$. Let $F(r)$ denote the complementary solid 3-cylinder $\overline{(D^2 \times S^1) \setminus E(r)}$.

We present M as the union of $g+1$ cobordisms with boundary as follows. For $r = 1, ..., g$, consider the genus 2 handlebody

$$(D'_r \times [-1,1]) \setminus \mathrm{Int}(T) = (D'_r \setminus (\mathrm{Int}(B_r^- \cup B_r^+))) \times [-1,1]$$

and glue $E(r)$ to it as specified above. This gives a 3-cobordism with bases $\partial B_r^- \times [-1,1]$ and $\partial B_r^+ \times [-1,1]$ lying in $\partial_- M$ and $\partial_+ M$ respectively. This cobordism is a cylinder over $\partial B_r^- \times [-1,1]$. Indeed, for $t \in [-1,1]$, the disk $D^2 \times x \subset E(r)$ glued to $\partial D'_r \times t$ and the disk with two holes $(D'_r \setminus \mathrm{Int}(B_r^- \cup B_r^+)) \times t$ form an annulus with bases $\partial B_r^- \times t$ and $\partial B_r^+ \times t$. These annuli corresponding to all $t \in [-1,1]$ form the cylinder in question. When r runs over $1, ..., g$ we get g cylinder cobordisms inside M. The complementary part of M is obtained from $S^3 \setminus \mathrm{Int}(T \cup (D_r \times [-1,1]))$ by gluing the cylinders $F(1), F(2), ..., F(r)$. This gluing may be accomplished inside S^3. More exactly, we may glue each $F(r)$ inside $D_r \times [-1,1]$. The result of these gluings is the complement in S^3 of the union of two open 3-balls $\mathrm{Int}(T^-) \cup \mathrm{Int}(T^+)$ with $m+2g$ unlinked vertical tubes connecting these balls. As it was already noticed in the case $g = 0$, such a space has a natural cylindrical structure. In this way M splits as the union of $g+1$ cylindrical cobordisms with boundary. One may check that the cylindrical structures are compatible on the boundary and yield an identification $M = \partial_- M \times [0,1]$. The homeomorphism $\partial_- M \to \partial_+ M$ induced by this identification commutes with parametrizations. Therefore the decorated 3-cobordism M is homeomorphic to $\Sigma_t \times [0,1]$.

2.7. Proof of Lemma 2.1.1. In view of the multiplicativity of τ with respect to disjoint union it suffices to consider the case when Σ is connected. Without loss of generality we may assume that $\Sigma = \Sigma_t$ where $t = (g; (W_1, \nu_1), ..., (W_m, \nu_m))$ is a decorated type and Σ is endowed with the identity parametrization.

Lemma 2.6 provides a special ribbon graph Ω in $\mathbb{R}^3$ presenting $\Sigma \times [0,1]$. To compute the homomorphism $\tau(\Sigma \times [0,1]) : \Psi_t \to \Psi_t$ we apply (2.3.a). Present this homomorphism by a block-matrix $\{\tau_i^j : \mathrm{Hom}(\mathbb{1}, \Phi(t;i)) \to \mathrm{Hom}(\mathbb{1}, \Phi(t;j))\}_{i,j}$ with $i, j \in I^g$. The sequences i and j determine a coloring of the uncolored bands of Ω as in Section 2.3. Let L be the trivial framed link formed by g uncolored annuli of Ω. Clearly $\sigma(L) = 0$. By (2.3.a),

$$\tau_i^j = \mathcal{D}^{-2g} \dim(j) \sum_{\lambda \in \mathrm{col}(L)} \dim(\lambda) F_0(\Omega, i, j, \lambda). \tag{2.7.a}$$

The arguments of Section II.3.6 (see also Exercise II.3.10.2) shows that the sum on the right-hand side of (2.7.a) is equal to 0 unless $i = j$. (To see this we may consider separately all (2,2)-tangles involved in Ω. The role of the object V_r in Section II.3.6 is played here by $(V_i)^*$.) The same arguments show that if $i = j = (j_1, j_2, ..., j_g)$ then the right-hand side of (2.7.a) is equal to $h_0(F(\Omega_j))$ where Ω_j is the colored ribbon tangle in $\mathbb{R}^3$ consisting of $m + 2g$ vertical unlinked untwisted bands whose colors are $W_1, ..., W_m, V_{j_1}, V_{j_1}, V_{j_2}, V_{j_2} ..., V_{j_g}, V_{j_g}$ respectively and whose directions are determined by the sequence $\nu_1, ..., \nu_m, 1, -1, 1, -1, ..., 1, -1$. It is obvious that $F(\Omega_j) = \text{id}$ and therefore $\tau_j^j = \text{id}$.

2.8. Proof of Lemma 2.1.2. If we replace the cobordism M_1 by the connected sum of its components we change both $\tau(M_1)$ and $\tau(M_2 M_1)$ by the same power of $\mathcal{D}$ (cf. formulas (1.9.a) and (II.2.3.a)). Similar remarks apply to M_2. Therefore, without loss of generality we may assume that both M_1 and M_2 are connected.

Consider first the case when the bases of M_1 and M_2 are connected surfaces. Present M_1 and M_2 by special ribbon graphs $\Omega_1 \subset \mathbb{R}^3$ and $\Omega_2 \subset \mathbb{R}^3$ respectively. The existence of a d-homeomorphism $\partial_+(M_1) \to \partial_-(M_2)$ implies that Ω_1 and Ω_2 are composable: we may put Ω_2 on the top of Ω_1 matching the bands together with their colors and directions. Denote by Ω the (partially v-colored) special ribbon graph in $\mathbb{R}^3$ obtained by putting Ω_2 on the top of Ω_1 and compressing the result into $\mathbb{R}^2 \times [0, 1]$.

We prove that Ω presents M. Denote the genus of $\partial_+(M_1)$ by g. Let $A_1, ..., A_g$ be the annuli of Ω obtained from g uncolored bands of Ω_1 attached to its top boundary intervals and g uncolored bands of Ω_2 attached to its bottom boundary intervals (these bands are glued to each other to form exactly g annuli). Complete the plane $\mathbb{R}^2 \times (1/2)$ to a 2-sphere $S^2 = (\mathbb{R}^2 \times (1/2)) \cup \{\infty\} \subset S^3$. We may assume that every annulus A_r intersects S^2 along two arcs each corresponding to a top boundary interval of Ω_1 and a bottom boundary interval of Ω_2. Let X be a closed regular neighborhood of $S^2 \cup (\cup_{s=1}^{g} A_s)$ in S^3. Clearly, X is a handlebody of genus $2g$. To construct the decorated 3-cobordism presented by Ω we have to apply surgery along all uncolored annuli of Ω. Since the uncolored annuli $A_1, ..., A_g$ lie in X we may first apply surgery to X along $A_1, ..., A_g$. The result of this surgery, say N, is a 3-submanifold of the 3-cobordism presented by Ω. The manifold N may be easily identified to be the cylinder $\partial_+(M_1) \times [-1, 1]$. When we cut out N from the ambient cobordism there remain two connected pieces which may be identified with M_1 and M_2. The cylinder structure in N determines the gluing of M_1 and M_2 along the (only) homeomorphism $\partial_+(M_1) \to \partial_-(M_2)$ commuting with parametrizations. This shows that the decorated 3-cobordism presented by Ω is homeomorphic to M.

Let us prove formula (2.1.b). Denote the numerical genera of the d-surfaces $\partial_-(M_1), \partial_+(M_1) \approx \partial_-(M_2), \partial_+(M_2)$ by f, g, h respectively. Note that the homomorphism $p_\#$ in (2.1.b) is the identity endomorphism of Ψ_t where t is the decorated type of $\partial_+(M_1) \approx \partial_-(M_2)$. We present the homomorphisms $\tau(M), \tau(M_1), \tau(M_2)$ by block-matrices as in Section 2.3. To prove (2.1.b) we should show that for any $i \in I^f$ and $l \in I^h$,

$$\tau^l_i(M) = k \sum_{j \in I^g} \tau^l_j(M_2)\, \tau^j_i(M_1) \tag{2.8.a}$$

where k is an invertible element of the ground ring K independent of the choice of i and l. Fix $i \in I^f$ and $l \in I^h$. Denote the framed links in S^3 formed by the uncolored annuli of $\Omega, \Omega_1, \Omega_2$ by L, L_1, L_2 respectively. It is clear that L is the union of L_1, L_2, and $\{A_1, ..., A_g\}$. For colorings $\lambda_1 \in \mathrm{col}(L_1)$, $\lambda_2 \in \mathrm{col}(L_2)$ and a sequence $j = (j_1, ..., j_g) \in I^g$, denote by $\lambda(j; \lambda_1, \lambda_2)$ the coloring of L extending λ_1 and λ_2 and assigning j_n to A_n for all $n = 1, ..., g$. It is obvious that

$$\dim(\lambda(j; \lambda_1, \lambda_2)) = \dim(j) \dim(\lambda_1) \dim(\lambda_2). \tag{2.8.b}$$

It follows from the functoriality of F that

$$F(\Omega, i, l, \lambda(j; \lambda_1, \lambda_2)) = F(\Omega_2, j, l, \lambda_2)\, F(\Omega_1, i, j, \lambda_1). \tag{2.8.c}$$

Since h_0 is a covariant functor we have

$$F_0(\Omega, i, l, \lambda(j; \lambda_1, \lambda_2)) = F_0(\Omega_2, j, l, \lambda_2)\, F_0(\Omega_1, i, j, \lambda_1). \tag{2.8.d}$$

Multiplying the right-hand side and the left-hand side of this formula by the right-hand side and the left-hand side of (2.8.b) and summing up over all λ_1, λ_2, j as above we get

$$\sum_{\lambda \in \mathrm{col}(L)} \dim(\lambda)\, F_0(\Omega, i, l, \lambda) = \sum_{j \in I^g} \dim(j) \times \tag{2.8.e}$$

$$\times \left(\sum_{\lambda_2 \in \mathrm{col}(L_2)} \dim(\lambda_2) F_0(\Omega_2, j, l, \lambda_2) \sum_{\lambda_1 \in \mathrm{col}(L_1)} \dim(\lambda_1) F_0(\Omega_1, i, j, \lambda_1) \right).$$

Set

$$k = (\mathcal{D}\Delta^{-1})^{\sigma(L_1)+\sigma(L_2)-\sigma(L)}. \tag{2.8.f}$$

It is clear that

$$\Delta^{\sigma(L)} \mathcal{D}^{-h-\sigma(L)-m} = k\, \Delta^{\sigma(L_2)} \mathcal{D}^{-h-\sigma(L_2)-m_2} \Delta^{\sigma(L_1)} \mathcal{D}^{-g-\sigma(L_1)-m_1} \tag{2.8.g}$$

where m, m_1, m_2 denote the number of components of L, L_1, L_2 respectively so that $m = m_1 + m_2 + g$. Multiplying (2.8.e) and (2.8.g) and applying (2.3.a) we get an equality equivalent to (2.8.a).

The case of 3-cobordisms with non-connected bases is considered in a similar way. If the surface $\partial_+(M_1) \approx \partial_-(M_2)$ is connected whilst $\partial_-(M_1)$ and/or $\partial_+(M_2)$ are disconnected then we just repeat the same argument until we get (2.8.c). The homomorphisms F_0 appearing in (2.8.d) are not mere compositions of F and h_0 as we have to involve the injections and projections used in Section 2.4. However, the same injections and projections appear on both sides of (2.8.d). This causes no trouble: we still have (2.8.d) and proceed as above.

In the case when the surface $\partial_+(M_1) \approx \partial_-(M_2)$ is non-connected we need more serious changes. Number the components of this surface by integers $1, 2, ..., r$. Present M_1, M_2 by special ribbon graphs Ω_1, Ω_2 in $\mathbb{R}^3$ equipped with numerical sequences determining a splitting of their boundary intervals into consecutive families. Since we use the same numeration for components of $\partial_+(M_1)$ and $\partial_-(M_2)$ the ribbon graphs Ω_1 and Ω_2 are composable: we may put Ω_2 on the top of Ω_1 matching the bands together with their colors and directions. Moreover, under this matching the r families of consecutive top boundary intervals of Ω_1 correspond to the r families of consecutive bottom boundary intervals of Ω_2. Denote by $\Omega_2\,\Omega_1$ the (partially v-colored) ribbon graph in $\mathbb{R}^3$ obtained by putting Ω_2 on the top of Ω_1 and compressing the result into $\mathbb{R}^2 \times [0, 1]$. We keep the splitting of the top (resp. bottom) boundary intervals of $\Omega_2\,\Omega_1$ into families induced by the one for Ω_2 (resp. Ω_1). It is clear that $\Omega_2\,\Omega_1$ is a special ribbon graph. In general, this graph *does not* present the composition M of the cobordisms M_1 and M_2. To construct a ribbon graph presenting M, we add to Ω several uncolored annuli. Denote by $\alpha_1, ..., \alpha_r$ the untwisted uncolored annuli lying in the plane $\mathbb{R}^2 \times (1/2)$ and encircling the r families of top boundary intervals of Ω_1 glued to bottom boundary intervals of Ω_2. The cores of these annuli bound r disjoint disks in $\mathbb{R}^2 \times (1/2)$ each containing one of these families. Denote by Ω^r the special ribbon graph consisting of $\Omega_2\,\Omega_1$ and $r-1$ annuli $\alpha_1, ..., \alpha_{r-1}$. We claim that Ω^r presents M. Indeed, consider the uncolored annuli $\{A_s\}_{s=1}^g$ of $\Omega_2\,\Omega_1$ obtained by gluing the uncolored bands of Ω_1 attached to its top boundary intervals with the uncolored bands of Ω_2 attached to its bottom boundary intervals. Here g is the sum of genera of components of $\partial_+(M_1)$. Each annulus A_s transversally intersects the 2-sphere $S^2 = (\mathbb{R}^2 \times (1/2)) \cup \{\infty\}$ along two arcs disjoint from $\alpha_1 \cup \alpha_2 \cup ... \cup \alpha_r$. To construct the 3-cobordism presented by Ω^r we should surger S^3 along the cores of $A_1, ..., A_g, \alpha_1, ..., \alpha_{r-1}$ and uncolored annuli of Ω_1, Ω_2.

Let X be a closed regular neighborhood of $S^2 \cup (\cup_{s=1}^g A_s)$ in S^3. We may assume that X consists of the cylinder $S^2 \times [-\delta, \delta]$ for a small positive δ and $2g$ one-handles attached to it. Note that the uncolored annuli $A_1, ..., A_g, \alpha_1, ..., \alpha_{r-1}$ lie inside $X \subset S^3$. Surgering the handlebody X along the cores of $A_1, ..., A_g$ we get a cylinder N over a connected closed surface of genus g. Cut out the cylinders $\alpha_1 \times [-\delta, \delta], ..., \alpha_{r-1} \times [-\delta, \delta]$ from N and glue in $2(r-1)$ copies of the 3-dimensional cylinder $D^2 \times [-\delta, \delta]$ along the product homeomorphisms $\partial D^2 \times [-\delta, \delta] \to S^1 \times [-\delta, \delta]$ where S^1 runs over all $2(r-1)$ components of $\partial\alpha_1, ..., \partial\alpha_{r-1}$. The resulting manifold, say N', has r connected components and may be identified with $\partial_+(M_1) \times [-\delta, \delta]$. The manifold N' naturally embeds in the 3-cobordism presented by Ω^r because the gluing of the 3-dimensional cylinders used above may be performed inside the solid tori $D^2 \times S^1$ attached under the surgery to 2-tori bounding regular neighborhoods of $\alpha_1, ..., \alpha_{r-1}$. When we cut out N' from M there remain two connected pieces which may be identified with M_1 and M_2. This shows that the decorated 3-cobordism presented by Ω^r is homeomorphic to M.

For the sake of computation, it is more convenient to use the special ribbon graph

$$\Omega = \Omega^r \cup \alpha_r = (\Omega_2 \, \Omega_1) \cup \alpha_1 \cup ... \cup \alpha_r$$

instead of Ω^r. It is easy to observe that α_r lies in a 3-ball in $N' \subset M$ as an unknotted untwisted annulus. Therefore Ω presents the decorated 3-cobordism obtained from M by surgery along a trivial knot with zero framing. Hence Ω presents $M\#(S^1 \times S^2)$.

Now we are ready to complete the proof of the lemma. We argue in the same way as in the case of connected bases. A generic coloring $\lambda = \lambda(j; \beta; \lambda_1, \lambda_2)$ of Ω is determined by colorings j, λ_1, λ_2 analogous to those involved in (2.8.b) and by a sequence $\beta \in I^r$ determining the colors of $\alpha_1, ..., \alpha_r$. We may present Ω as a composition of Ω_2, Γ, and Ω_1 where Γ is the ribbon tangle consisting of $\alpha_1, ..., \alpha_r$ and r families of vertical bands linked by these annuli. The sequences j and β determine a coloring of Γ in the obvious way. Functoriality of F implies that

$$F(\Omega, i, l, \lambda(j; \beta; \lambda_1, \lambda_2)) = F(\Omega_2, j, l, \lambda_2) \, F(\Gamma, j, j, \beta) \, F(\Omega_1, i, j, \lambda_1).$$

Lemma 2.2.1 computes

$$\mathcal{D}^{-2r} \sum_{\beta \in I^r} \dim(\beta) \, F(\Gamma; j; \beta)$$

to be the composition of the inclusion and projection that appear in the definitions of $F_0(\Omega_2, j, l, \lambda_2)$ and $F_0(\Omega_1, i, j, \lambda_1)$ respectively. Therefore

$$\sum_{\beta \in I^r} \dim(\beta) \, F_0(\Omega, i, l, \lambda(j; \beta; \lambda_1, \lambda_2)) = \mathcal{D}^{2r} F_0(\Omega_2, j, l, \lambda_2) \, F_0(\Omega_1, i, j, \lambda_1).$$

The remaining part of the proof repeats the arguments given in the cases considered above with the obvious changes. Although the ribbon graph Ω presents $M\#(S^1 \times S^2)$ and not M, this does not spoil the argument because of (1.9.b).

2.9. Proof of Lemma 2.1.3. Let $t = (g; (W_1, \nu_1), ..., (W_m, \nu_m))$ and $i = (i_1, ..., i_g) \in I^g$. Let $\Xi(t, i, x)$ be the ribbon $(0, m + 2g)$-graph in $\mathbb{R}^3$ obtained from the ribbon graph Ω shown in Figure 2.4 by attaching one coupon from below and coloring the cap-like bands with $V_{i_1}, ..., V_{i_g}$. (The annuli and the cup-like bands remain uncolored.) The coupon of $\Xi(t, i, x)$ is colored with $x \in \mathrm{Hom}(\mathbb{1}, \Phi(t; i))$. It is easy to see that $\Xi(t, i, x)$ presents the standard decorated handlebody $H(U_t, R_t, i, x)$. Indeed, the decorated 3-cobordism presented by $\Xi(t, i, x)$ may be obtained from the cobordism presented by Ω by gluing the handlebody $H(U_t, R_t, i, x)$ to the bottom base along the parametrization of this base. Since Ω presents $\Sigma_t \times [0, 1]$ such a gluing yields $H(U_t, R_t, i, x)$. Applying to $\Xi(t, i, x)$ the same arguments as in the proof of Lemma 2.1.1 we get $\tau(H(U_t, R_t, i, x)) = x$.

2.10. Remark. The reader may notice a certain redundancy in our argument: we use the same computation to prove Lemmas 2.1.1 and 2.1.3. In fact, Lemma 2.1.1 follows from Lemmas 2.1.2 and 2.1.3. It suffices to apply Lemma 2.1.2 to the composition of the standard decorated handlebody $H(U_t, R_t, i, x)$ with the cylinder over its boundary. The result of this gluing is homeomorphic to the same handlebody which shows that the homomorphism $\tau(\Sigma_t \times [0, 1])$ acts as the identity on $\tau(H(U_t, R_t, i, x)) = x$ for any x (cf. Remark III.4.5.3).

2.11. Exercise. Show that for any object W of $\mathcal{V}$, the morphism $p_W : W \to W$ defined in Section 2.2 is a projector, i.e., that $(p_W)^2 = p_W$.

3. Lagrangian relations and Maslov indices

3.0. Outline. Lagrangian spaces, Lagrangian relations, and Maslov indices play a crucial role in the study of gluing anomalies of 3-dimensional TQFT's. We discuss here a part of the theory of Lagrangian spaces and Maslov indices necessary for the study of TQFT's. This includes the definition and properties of Lagrangian relations and cocycle identities for Maslov indices (Lemmas 3.6 and 3.7).

3.1. Symplectic and Lagrangian spaces. A symplectic vector space is a finite dimensional real vector space H endowed with an antisymmetric bilinear form $\omega : H \times H \to \mathbb{R}$. The same vector space H with the opposite form $-\omega$ will be denoted by $-H$. If ω is non-degenerate then we say that the symplectic vector space H is non-degenerate. A direct sum of symplectic vector spaces is a symplectic vector space in the obvious way.

For a linear subspace A of a symplectic vector space (H, ω), denote by $\mathrm{Ann}(A)$ the annihilator of A with respect to ω. Thus $\mathrm{Ann}(A) = \{h \in H \mid \omega(h, A) = 0\}$. It is obvious that $\mathrm{Ann}(A) \supset \mathrm{Ann}(H)$ and that for any linear subspaces $A, A' \subset H$,

$$\mathrm{Ann}(A + A') = \mathrm{Ann}(A) \cap \mathrm{Ann}(A'). \tag{3.1.a}$$

If H is non-degenerate then $\dim(\mathrm{Ann}(A)) = \dim(H) - \dim(A)$ where dim denotes the dimension over $\mathbb{R}$.

A linear subspace $A \subset H$ is called isotropic if $A \subset \mathrm{Ann}(A)$, i.e., if $\omega(A, A) = 0$. A Lagrangian subspace of H is a maximal isotropic subspace of H. In other words, a linear subspace $\lambda \subset H$ is Lagrangian if $\lambda = \mathrm{Ann}(\lambda)$. It is clear that any Lagrangian subspace of H contains $\mathrm{Ann}(H)$. The set of Lagrangian subspaces of H is denoted by $\Lambda(H)$.

3.1.1. Lemma. *Let* (H, ω) *be a symplectic vector space. For any linear subspaces* $A, A' \subset H$*, we have* $\operatorname{Ann}(\operatorname{Ann}(A)) = A + \operatorname{Ann}(H)$ *and*

$$\operatorname{Ann}(A \cap A') = \operatorname{Ann}(A) + \operatorname{Ann}(A').$$

Proof. Replacing A, A' with $A + \operatorname{Ann}(H)$, $A' + \operatorname{Ann}(H)$ respectively we do not change the annihilators $\operatorname{Ann}(A)$, $\operatorname{Ann}(A')$, $\operatorname{Ann}(A \cap A')$. Therefore we may assume that $\operatorname{Ann}(H) \subset A$, $\operatorname{Ann}(H) \subset A'$. Factorizing out $\operatorname{Ann}(H)$ we reduce the lemma to the case of non-degenerate H.

Assume that H is non-degenerate. The obvious inclusion $A \subset \operatorname{Ann}(\operatorname{Ann}(A))$ and the equalities

$$\dim(\operatorname{Ann}(\operatorname{Ann}(A))) = \dim(H) - \dim(\operatorname{Ann}(A)) = \dim(A)$$

imply that $A = \operatorname{Ann}(\operatorname{Ann}(A))$. It follows from (3.1.a) that

$$\operatorname{Ann}(\operatorname{Ann}(A) + \operatorname{Ann}(A')) = \operatorname{Ann}(\operatorname{Ann}(A)) \cap \operatorname{Ann}(\operatorname{Ann}(A')) = A \cap A'.$$

Taking the annihilators we obtain $\operatorname{Ann}(A) + \operatorname{Ann}(A') = \operatorname{Ann}(A \cap A')$.

3.2. Lagrangian contractions. Lagrangian contractions will serve as technical tools in the study of Lagrangian relations in the next two subsections.

Let (H, ω) be a symplectic vector space. Let A be an isotropic subspace of H so that $A \subset \operatorname{Ann}(A)$. Denote by $H|A$ the quotient vector space $\operatorname{Ann}(A)/A$ with the antisymmetric bilinear form $(x \bmod A, y \bmod A) = \omega(x, y)$. For a linear space $\lambda \subset H$, set

$$\lambda|A = (\lambda + A) \cap \operatorname{Ann}(A)/A \ \subset H|A.$$

We say that $\lambda|A$ is obtained from λ by contraction along A.

3.2.1. Lemma. *If* λ *is a Lagrangian subspace of* H *then* $\lambda|A$ *is a Lagrangian subspace of* $H|A$.

The formula $\lambda \mapsto \lambda|A$ defines a mapping $\Lambda(H) \to \Lambda(H|A)$ called the Lagrangian contraction along A.

Proof of Lemma. For any $x, y, a, b \in H$,

$$\omega(x + a, y + b) = \omega(x, y) + \omega(x + a, b) + \omega(a, y + b) - \omega(a, b).$$

If $x, y \in \lambda$, $a, b \in A$, and $x + a, y + b \in \operatorname{Ann}(A)$ then all four expressions on the right-hand side are equal to 0 so that $\omega(x + a, y + b) = 0$. This implies that $\lambda|A \subset \operatorname{Ann}(\lambda|A)$. Let us verify the opposite inclusion $\operatorname{Ann}(\lambda|A) \subset \lambda|A$. Let x be an element of $\operatorname{Ann}(A)$ annihilating $(\lambda + A) \cap \operatorname{Ann}(A)$. Then

$$x \in \operatorname{Ann}((\lambda + A) \cap \operatorname{Ann}(A)) = \operatorname{Ann}(\lambda + A) + \operatorname{Ann}(\operatorname{Ann}(A)) =$$

$$= (\operatorname{Ann}(\lambda) \cap \operatorname{Ann}(A)) + A + \operatorname{Ann}(H) \subset \operatorname{Ann}(\lambda) + A = \lambda + A.$$

Therefore $x \in (\lambda + A) \cap \mathrm{Ann}(A)$. This implies that $\mathrm{Ann}(\lambda|A) = \lambda|A$.

3.3. Lagrangian relations. Lagrangian relations mimic on the algebraic level the homological properties of odd-dimensional cobordisms of manifolds. Here we define Lagrangian relations and outline their basic properties.

Let H_1, H_2 be non-degenerate symplectic vector spaces. A Lagrangian relation between H_1 and H_2 is a Lagrangian subspace of $(-H_1) \oplus H_2$. For a Lagrangian relation $N \subset (-H_1) \oplus H_2$, we shall use the notation $N : H_1 \Rightarrow H_2$. For instance, for any symplectic isomorphism $f : H_1 \to H_2$, its graph $\{h \oplus f(h) \mid h \in H_1\}$ is a Lagrangian relation between H_1 and H_2 (here we need the non-degeneracy of H_1, H_2). In this way, Lagrangian relations generalize symplectic isomorphisms. In particular, the graph of the identity endomorphism of a non-degenerate symplectic space H

$$\mathrm{diag}_H = \{h \oplus h \in (-H) \oplus H \mid h \in H\}$$

is a Lagrangian relation $H \Rightarrow H$. It is called the diagonal Lagrangian relation.

Non-degenerate symplectic vector spaces, as objects, and Lagrangian relations, as morphisms, form a category. The composition N_2N_1 of Lagrangian relations $N_1 : H_1 \Rightarrow H_2$ and $N_2 : H_2 \Rightarrow H_3$ is the subspace of $(-H_1) \oplus H_3$ consisting of $h_1 \oplus h_3$ such that for a certain $h_2 \in H_2$, we have $h_1 \oplus h_2 \in N_1$ and $h_2 \oplus h_3 \in N_2$. To verify that N_2N_1 is a Lagrangian subspace of $(-H_1) \oplus H_3$ consider the symplectic space $H = (-H_1) \oplus H_2 \oplus (-H_2) \oplus H_3$ and its isotropic subspace

$$A = 0 \oplus \mathrm{diag} \oplus 0 = \{0 \oplus h \oplus h \oplus 0 \mid h \in H_2\}.$$

It follows from the non-degeneracy of H_2 that $\mathrm{Ann}(A) = (-H_1) \oplus \mathrm{diag} \oplus H_3$. Therefore $H|A = (-H_1) \oplus H_3$. It remains to observe that $N_2N_1 = (N_1 \oplus N_2)|A$. Lemma 3.2.1 implies that N_2N_1 is a Lagrangian subspace of $(-H_1) \oplus H_3$.

It follows directly from definitions that the composition of Lagrangian relations is associative. The diagonal Lagrangian relations play the role of identity endomorphisms of symplectic spaces.

For a Lagrangian relation $N : H_1 \Rightarrow H_2$, we define the symmetric Lagrangian relation $N_s : H_2 \Rightarrow H_1$ to be the subspace of $(-H_2) \oplus H_1$ consisting of (h_2, h_1) such that $(h_1, h_2) \in N$. Clearly, $(N_s)_s = N$.

3.4. Lagrangian actions. Lagrangian relations act on Lagrangian spaces as follows. Let H_1, H_2 be non-degenerate symplectic vector spaces. Each Lagrangian relation $N : H_1 \Rightarrow H_2$ induces two mappings $N_* : \Lambda(H_1) \to \Lambda(H_2)$ and $N^* : \Lambda(H_2) \to \Lambda(H_1)$. The mapping N_* carries $\lambda \in \Lambda(H_1)$ into the linear space $N_*(\lambda) \subset H_2$ that consists of $h_2 \in H_2$ such that for a certain $h_1 \in \lambda$, we have $(h_1, h_2) \in N$. To show that $N_*(\lambda)$ is Lagrangian observe that the annihilator of $\lambda \oplus 0$ in $(-H_1) \oplus H_2$ is equal to $\lambda \oplus H_2$. (We use the symbol λ for a Lagrangian subspace of H_1 and for the same Lagrangian space regarded as a subspace of $-H_1$.) The symplectic space

$$((-H_1) \oplus H_2)|(\lambda \oplus 0) = (\lambda \oplus H_2)/(\lambda \oplus 0)$$

can be identified with H_2. Under this identification, $N|(\lambda \oplus 0) = N_*(\lambda)$. Hence $N_*(\lambda)$ can be obtained from the Lagrangian space $N \subset (-H_1) \oplus H_2$ by contraction along $\lambda \oplus 0$. Lemma 3.2.1 implies that $N_*(\lambda)$ is a Lagrangian subspace of H_2.

Set $N^* = (N_s)_* : \Lambda(H_2) \to \Lambda(H_1)$. Thus, for $\lambda \in \Lambda(H_2)$, the Lagrangian space $N^*(\lambda) \subset H_1$ consists of $h_1 \in H_1$ such that for a certain $h_2 \in \lambda$, we have $(h_1, h_2) \in N$.

For example, if N is the graph of a symplectic isomorphism $f : H_1 \to H_2$ then the mapping $N_* : \Lambda(H_1) \to \Lambda(H_2)$ carries $\lambda \subset H_1$ into $f(\lambda) \subset H_2$ and the mapping $N^* : \Lambda(H_2) \to \Lambda(H_1)$ carries $\lambda \subset H_2$ into $f^{-1}(\lambda) \subset H_1$.

It is left to the reader to verify that the action of Lagrangian relations is compatible with composition: for any Lagrangian relations $N_1 : H_1 \Rightarrow H_2$ and $N_2 : H_2 \Rightarrow H_3$, we have

$$(N_2N_1)_* = (N_2)_*(N_1)_*, \qquad (N_2N_1)^* = (N_1)^*(N_2)^*.$$

3.5. Maslov indices. Let $\lambda_1, \lambda_2, \lambda_3$ be three isotropic subspaces of a symplectic vector space (H, ω). Consider a bilinear form $\langle ., . \rangle$ on $(\lambda_1 + \lambda_2) \cap \lambda_3$ defined as follows. For $a, b \in (\lambda_1 + \lambda_2) \cap \lambda_3$ where $a = a_1 + a_2$ with $a_1 \in \lambda_1, a_2 \in \lambda_2$ set

$$\langle a, b \rangle = \omega(a_2, b).$$

Note that a_2 is determined by a up to addition of elements of $\lambda_1 \cap \lambda_2$. These elements annihilate $b \in \lambda_1 + \lambda_2$ so that $\omega(a_2, b)$ is well-defined.

The bilinear form $\langle ., . \rangle$ on $(\lambda_1 + \lambda_2) \cap \lambda_3$ is symmetric. Indeed, if $b = b_1 + b_2$ with $b_1 \in \lambda_1, b_2 \in \lambda_2, b \in \lambda_3$ then

$$\omega(a, b) = \omega(a_1, b_1) = \omega(a_2, b_2) = 0$$

and therefore

$$\omega(a_2, b) - \omega(b_2, a) = \omega(a_2 - a, b - b_2) = \omega(-a_1, b_1) = 0.$$

The symmetric bilinear form $\langle ., . \rangle$ on $(\lambda_1 + \lambda_2) \cap \lambda_3$ may be degenerate, its annihilator contains $(\lambda_1 \cap \lambda_3) + (\lambda_2 \cap \lambda_3)$.

The Maslov index $\mu(\lambda_1, \lambda_2, \lambda_3) \in \mathbb{Z}$ of the triple $(\lambda_1, \lambda_2, \lambda_3)$ is the signature of the bilinear form $\langle ., . \rangle$ on $(\lambda_1 + \lambda_2) \cap \lambda_3$, i.e., the number of positive entries minus the number of negative entries in its diagonal form.

The Maslov index $\mu(\lambda_1, \lambda_2, \lambda_3)$ is antisymmetric: for any isotropic spaces $\lambda_1, \lambda_2, \lambda_3 \subset H$,

$$\mu(\lambda_1, \lambda_2, \lambda_3) = -\mu(\lambda_2, \lambda_1, \lambda_3) = -\mu(\lambda_1, \lambda_3, \lambda_2). \tag{3.5.a}$$

The first equality follows from the fact that the triples $(\lambda_1, \lambda_2, \lambda_3)$ and $(\lambda_2, \lambda_1, \lambda_3)$ determine opposite symmetric bilinear forms on the vector space $(\lambda_1 + \lambda_2) \cap \lambda_3 =$

$(\lambda_2 + \lambda_1) \cap \lambda_3$. (Indeed, $\omega(a_2, b) = \omega(a - a_1, b) = -\omega(a_1, b)$.) To prove the equality $\mu(\lambda_1, \lambda_2, \lambda_3) = -\mu(\lambda_1, \lambda_3, \lambda_2)$ note that the splitting $a = a_1 + a_2$ with $a_1 \in \lambda_1, a_2 \in \lambda_2, a \in \lambda_3$ may be rewritten as $a_2 = -a_1 + a$. The formula $a \mapsto a_2$ defines an isomorphism of the form induced by $\langle ., . \rangle$ on

$$((\lambda_1 + \lambda_2) \cap \lambda_3) \,/\, (\lambda_1 \cap \lambda_3)$$

onto the negative of the form induced by $\langle ., . \rangle$ on

$$((\lambda_1 + \lambda_3) \cap \lambda_2) \,/\, (\lambda_1 \cap \lambda_2).$$

Hence $\mu(\lambda_1, \lambda_2, \lambda_3) = -\mu(\lambda_1, \lambda_3, \lambda_2)$.

Formula (3.5.a) implies that the Maslov index $\mu(\lambda_1, \lambda_2, \lambda_3)$ is invariant under cyclic permutations of the triple $(\lambda_1, \lambda_2, \lambda_3)$. Another important corollary of (3.5.a): if two of the three Lagrangian spaces coincide then $\mu(\lambda_1, \lambda_2, \lambda_3) = 0$. There is a more general criterion ensuring triviality of the Maslov index: if $\lambda_3 \subset (\lambda_1 \cap \lambda_3) + (\lambda_2 \cap \lambda_3)$ then $\mu(\lambda_1, \lambda_2, \lambda_3) = 0$. Indeed, in this case in the decomposition $a = a_1 + a_2$ used above we may take $a_1 \in \lambda_1 \cap \lambda_3, a_2 \in \lambda_2 \cap \lambda_3$. The inclusions $a_2, b \in \lambda_3$ imply that $\omega(a_2, b) = 0$ so that the form $\langle ., . \rangle$ on $(\lambda_1 + \lambda_2) \cap \lambda_3$ is identically 0.

3.6. Lemma. *Let (H, ω) be a symplectic vector space. For any Lagrangian subspaces $\lambda_1, \lambda_2, \lambda_3, \lambda_4$ of H, we have*

$$\mu(\lambda_1, \lambda_2, \lambda_3) - \mu(\lambda_1, \lambda_2, \lambda_4) + \mu(\lambda_1, \lambda_3, \lambda_4) - \mu(\lambda_2, \lambda_3, \lambda_4) = 0. \tag{3.6.a}$$

Proof. Let E be the linear subspace of $H \oplus H \oplus H \oplus H$ consisting of elements $h_1 \oplus h_2 \oplus h_3 \oplus h_4$ such that $h_i \in \lambda_i$ for $i = 1, 2, 3, 4$ and $h_1 + h_2 + h_3 + h_4 = 0$. (The last summation is performed in H.) Define a bilinear symmetric form $\rho : E \times E \to \mathbb{R}$ by the formula

$$\rho(h_1 \oplus h_2 \oplus h_3 \oplus h_4, h_1' \oplus h_2' \oplus h_3' \oplus h_4') = -\omega(h_1, h_2') - \omega(h_2, h_3')$$

$$-\omega(h_3, h_4') - \omega(h_4, h_1') - \omega(h_1', h_2) - \omega(h_2', h_3) - \omega(h_3', h_4) - \omega(h_4', h_1).$$

Let E_i be the linear subspace of E singled out by the linear equation $h_i = 0$ where $i = 1, 2, 3, 4$. Denote the signature of the form ρ restricted to E_i by σ_i.

Let us show that

$$\mu(\lambda_1, \lambda_2, \lambda_3) = \sigma_4. \tag{3.6.b}$$

Consider the isomorphism $E_4 \to (\lambda_1 + \lambda_2) \cap \lambda_3$ defined by the formula

$$h_1 \oplus h_2 \oplus h_3 \oplus 0 \mapsto h_3 \, (= -h_1 - h_2).$$

If $h = h_1 \oplus h_2 \oplus h_3 \oplus 0 \in E_4$ and $h' = h_1' \oplus h_2' \oplus h_3' \oplus 0 \in E_4$ then

$$\omega(h_1, h_2') = \omega(h_1, h_1' + h_2') = \omega(h_1, -h_3') =$$

$$= \omega(h_1 + h_3, -h'_3) = \omega(-h_2, -h'_3) = \omega(h_2, h'_3).$$

Exchanging h and h' we get $\omega(h'_1, h_2) = \omega(h'_2, h_3)$. Observe that

$$\omega(h'_2, h_3) = -\omega(h_3, h'_2) = -\omega(h_3 + h_2, h'_2) = \omega(h_1, h'_2) = \omega(h_2, h'_3).$$

Therefore

$$\rho(h, h') = -4\,\omega(h_2, h'_3) = 4\,\omega(-h_2, h'_3) = 4\,\langle h_3, h'_3\rangle$$

where $\langle ., .\rangle$ is the symmetric bilinear form on $(\lambda_1 + \lambda_2) \cap \lambda_3$ defined in Section 3.5. This shows that the isomorphism $E_4 \to (\lambda_1 + \lambda_2) \cap \lambda_3$ constructed above transforms $\rho|_{E_4}$ into $4\langle ., .\rangle$. Therefore $\sigma_4 = \mu(\lambda_1, \lambda_2, \lambda_3)$. Applying this equality to permutations of the tuple $\lambda_1, \lambda_2, \lambda_3, \lambda_4$ we get

(3.6.c) $\qquad \mu(\lambda_2, \lambda_3, \lambda_4) = \sigma_1, \quad \mu(\lambda_1, \lambda_3, \lambda_4) = \sigma_2, \quad \mu(\lambda_1, \lambda_2, \lambda_4) = \sigma_3.$

Set

$$G = E_2 \cap E_4 = \{a \oplus 0 \oplus (-a) \oplus 0 \mid a \in \lambda_1 \cap \lambda_3\}.$$

It is easy to see that $\rho(G, E_2) = \rho(G, E_4) = 0$. In other words, the annihilator $\mathrm{Ann}(G) = \mathrm{Ann}_\rho(G) \subset E$ of G with respect to ρ contains the linear space E_2+E_4 generated by E_2 and E_4. Let us show that $\mathrm{Ann}(G) = E_2 + E_4$.

We verify the inclusion $\mathrm{Ann}(G) \subset E_2+E_4$. Let $h = h_1\oplus h_2\oplus h_3\oplus h_4 \in \mathrm{Ann}(G)$. Let $a \in \lambda_1 \cap \lambda_3$. The equality $\rho(h, a \oplus 0 \oplus (-a) \oplus 0) = 0$ implies that $\omega(h_4, a) = \omega(h_2, a)$. On the other hand $\omega(h_2+h_4, a) = -\omega(h_1+h_3, a) = 0$ since $a \in \lambda_1\cap\lambda_3$. This shows that $\omega(h_4, a) = \omega(h_2, a) = 0$. In other words, both h_2 and h_4 lie in the annihilator of $\lambda_1 \cap \lambda_3$ with respect to ω. Since the spaces λ_1, λ_3 are Lagrangian, this annihilator equals $\lambda_1 + \lambda_3$. Therefore $h_2 = h_{21} + h_{23}$ for certain $h_{21} \in \lambda_1$ and $h_{23} \in \lambda_3$. Now we may split h as the sum of $(-h_{21}) \oplus h_2 \oplus (-h_{23}) \oplus 0$ and $(h_1 + h_{21}) \oplus 0 \oplus (h_3 + h_{23}) \oplus h_4$. These elements belong to E_4 and E_2 respectively. Hence $h \in E_2 + E_4$. Thus $\mathrm{Ann}(G) = E_2 + E_4 \supset G$.

By Lemma 3.8 formulated below, the inclusion $G \subset \mathrm{Ann}(G)$ implies that $\sigma(\rho) = \sigma(\rho|_{\mathrm{Ann}(G)})$. The signature of $\rho|_{\mathrm{Ann}(G)}$ is equal to $\sigma(\rho|_{E_2}) + \sigma(\rho|_{E_4}) = \sigma_2 + \sigma_4$ because $E_2 \cap E_4$ annihilates both E_2 and E_4 with respect to ρ. Thus, $\sigma(\rho) = \sigma_2 + \sigma_4$. A similar argument shows that $\sigma(\rho) = \sigma_1 + \sigma_3$. Therefore $\sigma_2 + \sigma_4 = \sigma_1 + \sigma_3$. Combining this equality with (3.6.b) and (3.6.c) we get (3.6.a).

3.7. Lemma. *Let H, H' be non-degenerate symplectic vector spaces and let $N \subset (-H) \oplus H'$ be a Lagrangian relation $H \Rightarrow H'$. Let $\lambda_1, \lambda_2 \subset H$ and $\lambda'_1, \lambda'_2 \subset H'$ be Lagrangian subspaces. Then*

(3.7.a) $$\mu(\lambda_1, \lambda_2, N^*(\lambda'_1)) + \mu(N_*(\lambda_1), \lambda'_1, \lambda'_2) =$$

$$= \mu(\lambda_1, \lambda_2, N^*(\lambda'_2)) + \mu(N_*(\lambda_2), \lambda'_1, \lambda'_2).$$

This lemma generalizes Lemma 3.6 which corresponds to the case where $H = H'$ and N is the graph of the identity endomorphism of H.

Proof of Lemma. Set

$$\lambda_{11} = \lambda_1 \oplus \lambda'_1, \quad \lambda_{21} = \lambda_2 \oplus \lambda'_1, \quad \lambda_{12} = \lambda_1 \oplus \lambda'_2, \quad \lambda_{22} = \lambda_2 \oplus \lambda'_2.$$

It is obvious that $\lambda_{11}, \lambda_{21}, \lambda_{12}, \lambda_{22}$ are Lagrangian subspaces of $(-H) \oplus H'$. Let us show that

$$\mu(\lambda_1, \lambda_2, N^*(\lambda'_1)) = -\mu(\lambda_{11}, \lambda_{21}, N). \tag{3.7.b}$$

The Maslov index $\mu(\lambda_{11}, \lambda_{21}, N)$ is the signature of the symmetric bilinear form $\langle .,. \rangle$ on

$$(\lambda_{11} + \lambda_{21}) \cap N = ((\lambda_1 + \lambda_2) \oplus \lambda'_1)) \cap N$$

induced by the antisymmetric bilinear form on $(-H) \oplus H'$. The projection on the first summand $p_1 : H \oplus H' \to H$ carries $((\lambda_1 + \lambda_2) \oplus \lambda'_1)) \cap N$ onto $(\lambda_1 + \lambda_2) \cap N^*(\lambda'_1)$. It is easy to deduce from the definitions that this projection carries the form $\langle .,. \rangle$ on $((\lambda_1 + \lambda_2) \oplus \lambda'_1)) \cap N$ into the negative of the bilinear form on $(\lambda_1 + \lambda_2) \cap N^*(\lambda'_1)$ induced by the antisymmetric bilinear form on H (and used to define $\mu(\lambda_1, \lambda_2, N^*(\lambda'_1)))$. This implies (3.7.b).

We have

$$\mu(N_*(\lambda_1), \lambda'_1, \lambda'_2) = \mu(\lambda'_1, \lambda'_2, N_*(\lambda_1)) = \mu(\lambda'_1, \lambda'_2, (N_s)^*(\lambda_1)) =$$

$$= -\mu(\lambda'_1 \oplus \lambda_1, \lambda'_2 \oplus \lambda_1, N_s) = \mu(\lambda_{11}, \lambda_{12}, N).$$

Here the third equality follows from (3.7.b) and the fourth equality follows from the fact that the flip $x \oplus y \mapsto y \oplus x$ defines an isomorphism $(-H') \oplus H \to -((-H) \oplus H')$. Therefore the left-hand side of (3.7.a) may be computed as follows:

$$\mu(\lambda_1, \lambda_2, N^*(\lambda'_1)) + \mu(N_*(\lambda_1), \lambda'_1, \lambda'_2) = -\mu(\lambda_{11}, \lambda_{21}, N) + \mu(\lambda_{11}, \lambda_{12}, N) =$$

$$= \mu(\lambda_{21}, \lambda_{12}, N) - \mu(\lambda_{11}, \lambda_{21}, \lambda_{12}).$$

The last equality is obtained by application of (3.6.a) to the 4-tuple $(\lambda_{11}, \lambda_{21}, \lambda_{12}, N)$. The triviality criterion for Maslov indices mentioned in Section 3.5 and the obvious inclusion $\lambda_{11} \subset (\lambda_{11} \cap \lambda_{21}) + (\lambda_{11} \cap \lambda_{12})$ imply that $\mu(\lambda_{11}, \lambda_{21}, \lambda_{12}) = 0$. Thus,

$$\mu(\lambda_1, \lambda_2, N^*(\lambda'_1)) + \mu(N_*(\lambda_1), \lambda'_1, \lambda'_2) = \mu(\lambda_2 \oplus \lambda'_1, \lambda_1 \oplus \lambda'_2, N). \tag{3.7.c}$$

The three Maslov indices in this formula apply to Lagrangian spaces in H, H', and $(-H) \oplus H'$ respectively.

Applying (3.7.c) to the symmetric Lagrangian relation $N_s : H' \Rightarrow H$ and the Lagrangian spaces $\lambda'_2, \lambda'_1, \lambda_2, \lambda_1$ we obtain that the right-hand side of (3.6.a) is equal to the same Maslov index $\mu(\lambda_2 \oplus \lambda'_1, \lambda_1 \oplus \lambda'_2, N)$ as the left-hand side.

3.8. Lemma. *Let E be a finite dimensional real vector space endowed with a symmetric bilinear form $\rho : E \times E \to \mathbb{R}$. Let G be a linear subspace of E such that $G \subset \mathrm{Ann}(G)$. Then $\sigma(\rho) = \sigma(\rho|_{\mathrm{Ann}(G)})$.*

Proof. Replacing G with $G + \mathrm{Ann}(E)$ we do not change the annihilator $\mathrm{Ann}(G)$. Therefore we may assume that $\mathrm{Ann}(E) \subset G$. Factorizing out $\mathrm{Ann}(E)$ we may assume that E is non-degenerate. Note that in this case $G = \mathrm{Ann}(\mathrm{Ann}(G))$ (cf. Lemma 3.1.1).

Denote by A the vector space $\mathrm{Ann}(G)$ with the symmetric bilinear form $-\rho$. It suffices to verify that $\sigma(\rho) + \sigma(A) = \sigma(E \oplus A) = 0$. Choose an orthogonal splitting $E \oplus A = X^+ \oplus X^- \oplus X^0$ with X^+ positive definite, X^- positive definite, and $X^0 = \mathrm{Ann}(E \oplus A) = 0 \oplus G$ self-orthogonal.

Set
$$J = \{(g, h) \in E \oplus A \mid g \in \mathrm{Ann}(G), h \in g + G\}.$$
It is easy to check that J is a self-orthogonal linear space and
$$\dim(J) = \dim(A) + \dim(G) = \dim(E).$$
Since J is self-orthogonal, $J \cap X^+ = 0$ so that
$$\dim(J) \leq \dim(X^-) + \dim(X^0).$$
Similarly, the formula $J \cap X^- = 0$ implies that $\dim(J) \leq \dim(X^+) + \dim(X^0)$. Summing up these inequalities we obtain
$$2\dim(J) \leq \dim(X^+) + \dim(X^-) + 2\dim(X^0) = \dim(E \oplus A) + \dim(X^0) =$$
$$= \dim(E) + \dim(\mathrm{Ann}(G)) + \dim(G) = 2\dim(E) = 2\dim(J).$$
Hence the inequalities above are equalities so that $\dim(X^+) = \dim(X^-)$. Therefore $\sigma(E \oplus A) = \dim(X^+) - \dim(X^-) = 0$.

3.9. Exercise. Verify that for any Lagrangian subspace λ of a symplectic vector space H,
$$\dim(\lambda) = (1/2)\,(\dim(H) + \dim(\mathrm{Ann}(H))).$$

4. Computation of anomalies

4.0. Outline. The TQFT $(\mathcal{T}, \tau)$ constructed in Section 1 has anomalies computed here in terms of Maslov indices of Lagrangian spaces in the homologies of surfaces.

The main result of this section (Theorem 4.3) will be crucial in further computations in Sections 5 - 9. It will be used to define various renormalizations of $(\mathcal{T}, \tau)$ (see Sections 6 and 9).

4.1. Lagrangian spaces in the homologies of surfaces. For any oriented surface Σ, the real vector space $H_1(\Sigma;\mathbb{R})$ supports the intersection pairing

$$H_1(\Sigma;\mathbb{R}) \times H_1(\Sigma;\mathbb{R}) \to \mathbb{R}. \tag{4.1.a}$$

This pairing is antisymmetric so that $H_1(\Sigma;\mathbb{R})$ is a symplectic vector space. For compact Σ, the annihilator of this space is equal to the image of the inclusion homomorphism $H_1(\partial\Sigma;\mathbb{R}) \to H_1(\Sigma;\mathbb{R})$. If Σ is closed then the pairing (4.1.a) is non-degenerate and the dimension of any Lagrangian subspace of $H_1(\Sigma;\mathbb{R})$ equals $(1/2)\dim H_1(\Sigma;\mathbb{R})$. If $\Sigma = \emptyset$ then $H_1(\Sigma;\mathbb{R}) = 0$ contains a zero Lagrangian subspace.

A natural source of Lagrangian spaces in the homologies of surfaces is provided by the theory of 3-manifolds. An oriented compact 3-manifold M gives rise to a Lagrangian subspace in $H_1(\partial M;\mathbb{R})$ which is the kernel of the inclusion homomorphism $H_1(\partial M;\mathbb{R}) \to H_1(M;\mathbb{R})$. The fact that this subspace is Lagrangian is a well known corollary of the Poincaré duality (see Hempel [He]).

We state a theorem due to C.T.C. Wall which elucidates the role of Maslov indices in the framework of low-dimensional topology and gives us a key to the anomalies of 3-dimensional TQFT's.

Let W be an oriented compact 4-manifold. The signature $\sigma(W)$ of W is the signature of the homological intersection pairing $H_2(W;\mathbb{R}) \times H_2(W;\mathbb{R}) \to \mathbb{R}$. Let M_0 be an oriented compact 3-manifold properly embedded into W so that M_0 transversally intersects ∂W along $\partial M_0 = M_0 \cap \partial W$. Assume that M_0 splits W into two 4-manifolds W_1 and W_2 (see Figure 4.1). For $i = 1, 2$, denote by M_i the compact 3-manifold $\partial W_i \setminus \mathrm{Int}(M_0)$. Orient M_1 and M_2 so that $\partial W_1 = M_0 \cup (-M_1)$ and $\partial W_2 = (-M_0) \cup M_2$. The orientations of M_0, M_1, M_2 induce the same orientation in the surface $\Sigma = \partial M_0 = \partial M_1 = \partial M_2$. We shall use this orientation to define the intersection form in $H_1(\Sigma;\mathbb{R})$ and the Maslov indices of Lagrangian subspaces of $H_1(\Sigma;\mathbb{R})$. For $i = 0, 1, 2$, denote by λ_i the kernel of the inclusion homomorphism $H_1(\Sigma;\mathbb{R}) \to H_1(M_i;\mathbb{R})$. As we know, λ_i is a Lagrangian subspace of $H_1(\Sigma;\mathbb{R})$.

4.1.1. Theorem. *Under the conditions above*

$$\sigma(W) = \sigma(W_1) + \sigma(W_2) + \mu(\lambda_1, \lambda_0, \lambda_2).$$

For a proof (and for a generalization to higher dimensions), see [Wall].

4.2. Lagrangian functor. We assign to each parametrized d-surface a Lagrangian subspace of its homologies and to each decorated 3-cobordism a Lagrangian relation between the homologies of the bases. This construction may be regarded as a covariant functor from the category of decorated 3-cobordisms to the category whose objects are non-degenerate symplectic spaces with fixed Lagrangian

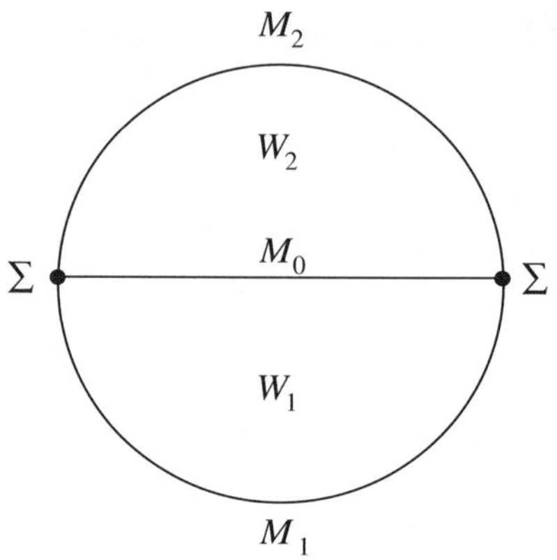

Figure 4.1

subspaces and whose morphisms are Lagrangian relations. We shall not pursue this categorical interpretation but simply describe the construction in question.

For a parametrized d-surface Σ, we define a distinguished Lagrangian space $\lambda(\Sigma) \subset H_1(\Sigma;\mathbb{R})$. (By homologies of decorated surfaces we mean the homologies of the underlying topological surfaces.) For a standard surface $\Sigma = \Sigma_t$, we take $\lambda(\Sigma)$ to be the kernel of the inclusion homomorphism $H_1(\Sigma_t;\mathbb{R}) \to H_1(U_t;\mathbb{R})$. In other words, $\lambda(\Sigma_t)$ is the subspace of $H_1(\Sigma_t;\mathbb{R})$ generated by the homology classes of the meridianal loops of the standard handlebody U_t. For any connected parametrized d-surface Σ, set $\lambda(\Sigma) = f_*(\lambda(\Sigma_t))$ where f is the parametrization homeomorphism $\Sigma_t \to \Sigma$ and f_* is the induced isomorphism of 1-homologies. For a disconnected parametrized d-surface Σ, we define $\lambda(\Sigma)$ to be the subspace of $H_1(\Sigma;\mathbb{R})$ generated by the distinguished Lagrangian subspaces in the 1-homologies of the connected components.

For any decorated 3-cobordism $(M, \partial_- M, \partial_+ M)$, we have

$$H_1(\partial M;\mathbb{R}) = (-H_1(\partial_- M;\mathbb{R})) \oplus H_1(\partial_+ M;\mathbb{R}).$$

The kernel of the inclusion homomorphism $H_1(\partial M;\mathbb{R}) \to H_1(M;\mathbb{R})$ yields a Lagrangian relation $H_1(\partial_- M;\mathbb{R}) \Rightarrow H_1(\partial_+ M;\mathbb{R})$. We denote this Lagrangian relation by $N(M)$. (The construction of $N(M)$ uses neither the parametrizations and marks of $\partial_- M$, $\partial_+ M$ nor the given ribbon graph in M.) If M is the cylinder $\partial_- M \times [0,1]$ then $N(M)$ is the diagonal Lagrangian relation. The construction $M \mapsto N(M)$ is covariant: if a 3-cobordism M is obtained from 3-cobordisms M_1 and M_2 by gluing along a degree 1 homeomorphism $p: \partial_+(M_1) \to \partial_-(M_2)$ then

$$N(M) = N(M_2)\, p_*\, N(M_1)$$

where p_* denotes the Lagrangian relation $H_1(\partial_+(M_1);\mathbb{R}) \Rightarrow H_1(\partial_-(M_2);\mathbb{R})$ induced by the symplectic isomorphism $p_*: H_1(\partial_+(M_1);\mathbb{R}) \to H_1(\partial_-(M_2);\mathbb{R})$.

The action of the Lagrangian relation $N = N(M)$ on Lagrangian spaces may be described as follows. For any $\lambda \in \Lambda(H_1(\partial_- M; \mathbb{R}))$, the Lagrangian space $N_*(\lambda) \subset H_1(\partial_+ M; \mathbb{R})$ consists of the elements of $H_1(\partial_+ M; \mathbb{R})$ homological in M to elements of λ. For any $\lambda \in \Lambda(H_1(\partial_+ M; \mathbb{R}))$, the Lagrangian space $N^*(\lambda) \subset H_1(\partial_- M; \mathbb{R})$ consists of the elements of $H_1(\partial_- M; \mathbb{R})$ homological in M to elements of λ.

For a decorated 3-cobordism $(M, \partial_- M, \partial_+ M)$, denote the Lagrangian spaces $\lambda(\partial_-(M)) \subset H_1(\partial_- M; \mathbb{R})$ and $\lambda(\partial_+(M)) \subset H_1(\partial_+ M; \mathbb{R})$ by $\lambda_-(M)$ and $\lambda_+(M)$ respectively. Warning: these Lagrangian spaces are determined by the parametrizations of the bases of M and not by the intrinsic topology of M.

4.3. Theorem (computation of anomalies). *Let $M = M_2 M_1$ be a decorated 3-cobordism obtained from decorated 3-cobordisms M_1 and M_2 by gluing along a d-homeomorphism $p : \partial_+(M_1) \to \partial_-(M_2)$ commuting with parametrizations. Set*

$$N_r = N(M_r) : H_1(\partial_-(M_r); \mathbb{R}) \Rightarrow H_1(\partial_+(M_r); \mathbb{R})$$

for $r = 1, 2$. Then

$$\tau(M) = (\mathcal{D}\Delta^{-1})^m \tau(M_2)\, p_\# \, \tau(M_1)$$

with $m = \mu(p_*(N_1)_*(\lambda_-(M_1)), \lambda_-(M_2), N_2^*(\lambda_+(M_2)))$.

This theorem shows that the gluing anomaly is completely determined by the topology of cobordisms under gluing and parametrizations of their bases. Note that M_1 and M_2 play symmetric roles in the definition of m. Indeed, since p commutes with parametrizations we have $\lambda_-(M_2) = p_*(\lambda_+(M_1))$. The anomaly does not depend on the distinguished ribbon graphs in the cobordisms.

Proof of Theorem. Let us first consider the case where the cobordisms M_1, M_2 and their bases are connected. Let s, t, u be the decorated types of the d-surfaces $\partial_-(M_1) = \partial_- M$, $\partial_+(M_1) \approx \partial_-(M_2)$, and $\partial_+(M_2) = \partial_+ M$ respectively. We shall use the links L, L_1, L_2 introduced in Section 2.8 in the course of the proof of Lemma 2.1.2. A comparison of formulas (2.1.b) and (2.8.f) shows that it suffices to prove that $\sigma(L_1) + \sigma(L_2) - \sigma(L) = m$.

Recall the 4-manifold W_L obtained from the 4-ball B^4 by attaching 2-handles along the components of the framed link $L \subset S^3 = \partial B^4$. It is clear that the 3-manifold ∂W_L is glued from M and two standard handlebodies U_s and $U_u \approx \overline{U_u}$ where the gluing is performed along the parametrization homeomorphisms $\partial U_s \to \partial_- M$ and $\partial U_u \to \partial_+ M$. Therefore we may view W_L as a 4-dimensional cobordism (W_L, U_s, U_u) with boundary $(M, \partial_- M, \partial_+ M)$. Similar remarks apply to W_{L_1} and W_{L_2}. The arguments of Section 2.8 may be extended to show that the cobordism (W_L, U_s, U_u) is obtained from the cobordisms (W_{L_1}, U_s, U_t) and (W_{L_2}, U_t, U_u) by gluing two copies of U_t. (To see this, take a 3-ball $B^3 \subset B^4$ splitting B^4 into two halves such that ∂B^3 is the two-sphere used in Section 2.8.

Consider the union of 2-handles attached to B^4 along regular neighborhoods of the annuli $A_1, \ldots, A_g \subset \partial B^4$ mentioned in Section 2.8 with a narrow regular neighborhood $B^3 \times [-\delta, \delta]$ of B^3 in B^4 (here $\delta > 0$). This union is a 4-manifold which may be identified with the cylinder $U_t \times [-\delta, \delta]$ embedded in W_L. Its complement in W_L consists of two connected pieces which may be identified with W_{L_1} and W_{L_2}. Note that the manifold N appearing in Section 2.8 is nothing but $\partial U_t \times [-\delta, \delta]$.)

By definition, the numbers $\sigma(L), \sigma(L_1)$, and $\sigma(L_2)$ are the signatures of the 4-manifolds W_L, W_{L_1}, and W_{L_2}. Therefore to compute the expression $\sigma(L_1) + \sigma(L_2) - \sigma(L)$ we may apply Wall's formula. Let $\lambda_0, \lambda_1, \lambda_2 \subset H_1(\Sigma_t; \mathbb{R})$ be the kernels of the homology homomorphisms induced by the inclusions $\Sigma_t = \partial U_t \to U_t$, $\Sigma_t \to \partial W_{L_1} \setminus \operatorname{Int}(U_t)$, and $\Sigma_t \to \partial W_{L_2} \setminus \operatorname{Int}(U_t)$. Theorem 4.1.1 implies that $\sigma(L) = \sigma(L_1) + \sigma(L_2) - \mu(\lambda_1, \lambda_0, \lambda_2)$. Here μ denotes the Maslov index of Lagrangian subspaces of $H_1(\Sigma_t; \mathbb{R})$ with the symplectic form induced by the canonical orientation in the standard surface Σ_t. The minus in front of μ is due to the fact that the orientation in $\Sigma_t = \partial U_t$ used in Wall's formula is opposite to the canonical one. Indeed, the former orientation is induced by the one in M_2 via the parametrization homeomorphism $\Sigma_t \to \partial_-(M_2)$ whereas, as we know, $\partial(M_2) = (-\partial_-(M_2)) \amalg \partial_+(M_2)$. Under the isomorphism $H_1(\Sigma_t; \mathbb{R}) = H_1(\partial_-(M_2); \mathbb{R})$ induced by the parametrization $\Sigma_t \to \partial_-(M_2)$ the Lagrangian spaces $\lambda_0, \lambda_1, \lambda_2$ correspond to $\lambda_-(M_2)$, $p_*(N_1)_*(\lambda_-(M_1))$, $N_2^*(\lambda_+(M_2))$ respectively. Therefore $\mu(\lambda_1, \lambda_0, \lambda_2) = m$. This implies the claim of the theorem. The case of non-connected cobordisms or non-connected bases is considered along the same lines, using the observations made in Section 2.8.

5. Action of the modular groupoid

5.0. Outline. We define a projective linear action of the modular groups of surfaces in the modules of states. This action relates the 3-dimensional TQFT's to the theory of modular groups, Teichmüller spaces, etc. It is this relationship which suggested the terms modular functor and modular category.

We also define a (more general) projective linear action of the so-called modular groupoid. The 2-cocycle determined by this action is computed in terms of Maslov indices.

The action of the modular groupoid will be instrumental up to the end of the chapter. In this section we use it to describe the dependency of the operator invariant of 3-cobordisms on the choice of parametrizations of bases.

5.1. Action of the modular group. The modular group Mod_n of genus n is the group of isotopy classes of degree 1 self-homeomorphisms of a closed oriented surface of genus n. This group is also called the mapping class group or the homeotopy group of genus n. A similar definition applies to d-homeomorphisms of d-surfaces. Two d-homeomorphisms $\Sigma \to \Sigma'$ are isotopic if they may be related by an isotopy preserving the images of the distinguished arcs. For a decorated type $t = (n; (W_1, \nu_1), ..., (W_m, \nu_m))$, we define the modular group Mod_t to be the group of the isotopy classes of d-homeomorphisms $\Sigma_t \to \Sigma_t$. (We do not assume that these homeomorphisms commute with the identity parametrization of Σ_t.) If $m = 0$ then $\mathrm{Mod}_t = \mathrm{Mod}_n$.

Set $\Sigma = \Sigma_t$ and $\lambda_t = \lambda(\Sigma_t) \subset H_1(\Sigma; \mathbb{R})$. We construct a projective linear action of Mod_t on the module $\mathcal{T}(\Sigma) = \Psi_t$. For a d-homeomorphism $g : \Sigma \to \Sigma$, consider the cylinder $\Sigma \times [0, 1]$ decorated as in Section 1.6 with the only difference that its bottom and top bases are parametrized by g and id_Σ respectively. This cylinder is a decorated 3-cobordism between (Σ, g) and $(\Sigma, \mathrm{id}_\Sigma)$. Denote this cobordism by $M(g)$. By definition, $\lambda_-(M(g)) = g_*(\lambda_t)$ and $\lambda_+(M(g)) = \lambda_t$. Set

$$\varepsilon(g) = \tau(M(g)) : \Psi_t \to \Psi_t.$$

Observe that for isotopic d-homeomorphisms $g, g' : \Sigma \to \Sigma$, we have $\varepsilon(g) = \varepsilon(g')$. Indeed, let $g_u : \Sigma \to \Sigma$, $u \in [0, 1]$ be an isotopy between $g_0 = g$ and $g_1 = g'$. The homeomorphism

$$(x, u) \mapsto (g' g_u^{-1}(x), u) : \Sigma \times [0, 1] \to \Sigma \times [0, 1]$$

(where $x \in \Sigma$, $u \in [0, 1]$) regarded as a homeomorphism $M(g) \to M(g')$ commutes with the parametrizations of the bases and preserves the distinguished colored ribbon graphs. Hence, this is a d-homeomorphism $M(g) \to M(g')$. The invariance of the operator invariant of decorated 3-cobordisms under d-homeomorphisms implies that $\varepsilon(g) = \varepsilon(g')$.

By Theorem 1.9 (or Lemma 2.1.1), $\varepsilon(\mathrm{id}_\Sigma) = \mathrm{id}$. It is easy to see that for any d-homeomorphisms $g, h : \Sigma_t \to \Sigma_t$, the cobordism $M(gh)$ is obtained from $M(g) \amalg M(h)$ by gluing the top base of $M(h)$ to the bottom base of $M(g)$ along $g : \Sigma \to \Sigma$. The Maslov index corresponding to this gluing via Theorem 4.3 is equal to

$$\mu((gh)_*(\lambda_t), g_*(\lambda_t), \lambda_t) = \mu(h_*(\lambda_t), \lambda_t, g_*^{-1}(\lambda_t)).$$

Theorem 4.3 implies that

$$\varepsilon(gh) = (\mathcal{D}\Delta^{-1})^{\mu(h_*(\lambda_t), \lambda_t, g_*^{-1}(\lambda_t))}\, \varepsilon(g)\, \varepsilon(h). \tag{5.1.a}$$

This shows that $g \mapsto \varepsilon(g)$ is a projective linear action of Mod_t on Ψ_t.

Although we have not used formula (3.6.a) in these computations, it is this formula that makes (5.1.a) consistent. Indeed, in order to compute ε for a composition of three d-homeomorphisms f, g, h we may use the equality $fgh = (fg)h$ or the equality $fgh = f(gh)$. By (3.6.a), the powers of $\mathcal{D}\Delta^{-1}$ in the corresponding expressions for $\varepsilon(fgh)$ are equal.

5.2. Modular groupoid. The action of modular groups extends to an action of the groupoid formed by homeomorphisms of d-surfaces which may not preserve the order of distinguished arcs. The idea is that we may consider cylinder cobordisms between d-surfaces with different orders of distinguished arcs.

A weak d-homeomorphism of d-surfaces is a degree 1 homeomorphism of the underlying surfaces preserving the distinguished arcs together with their orientations and marks. A weak d-homeomorphism does not need to preserve the order of the arcs. Two weak d-homeomorphisms $\Sigma \to \Sigma'$ are isotopic if they may be related by an isotopy preserving the images of distinguished arcs.

To each pair of standard d-surfaces (Σ, Σ') we assign the set (possibly empty) $\mathcal{M}(\Sigma, \Sigma')$ consisting of the isotopy classes of weak d-homeomorphisms $\Sigma \to \Sigma'$. The modular groupoid is the family of sets $\{\mathcal{M}(\Sigma, \Sigma')\}_{\Sigma,\Sigma'}$ provided with associative multiplication $\mathcal{M}(\Sigma, \Sigma') \times \mathcal{M}(\Sigma', \Sigma'') \to \mathcal{M}(\Sigma, \Sigma'')$ induced by composition of homeomorphisms. Here Σ and Σ' run independently over all standard d-surfaces. (Note that connected standard d-surfaces are weakly d-homeomorphic if and only if their types are equal up to the order of the marks.)

For any decorated type t, we have $\mathcal{M}(\Sigma_t, \Sigma_t) = \mathrm{Mod}_t$ so that the modular groupoid contains the modular groups. We construct a projective action of the modular groupoid extending the action ε of modular groups. For a weak d-homeomorphism of standard d-surfaces $g : \Sigma \to \Sigma'$, consider the cylinder $\Sigma' \times [0, 1]$ decorated as in Section 1.6 with the only difference that its top base $\Sigma' \times 1$ is decorated and parametrized via $\mathrm{id}_{\Sigma'}$ whereas the bottom base $\Sigma' \times 0$ is decorated and parametrized via g. Denote this cobordism by $M(g)$. Set $\varepsilon(g) = \tau(M(g)) : \mathcal{T}(\Sigma) \to \mathcal{T}(\Sigma')$. As in Section 5.1, the actions of isotopic weak d-homeomorphisms coincide. Theorem 1.9 ensures that $\varepsilon(\mathrm{id}_\Sigma) = \mathrm{id}_{\mathcal{T}(\Sigma)}$. The same argument as in Section 5.1 shows that for any weak d-homeomorphisms $h : \Sigma \to \Sigma'$ and $g : \Sigma' \to \Sigma''$, we have

$$\varepsilon(gh) = (\mathcal{D}\Delta^{-1})^{\mu(h_*(\lambda(\Sigma)),\lambda(\Sigma'),g_*^{-1}(\lambda(\Sigma'')))}\, \varepsilon(g)\, \varepsilon(h). \tag{5.2.a}$$

5.3. Applications to decorated 3-cobordisms. We employ the action ε of the modular groupoid to study the operator invariants of decorated 3-cobordisms. More precisely, we describe in terms of this action the dependency of the operator invariant on the choice of parametrizations of bases. This description will be used in Section 6.

Let $(M, \partial_- M, \partial_+ M)$ be a decorated 3-cobordism with the parametrizations $f_- : \Sigma_- \to \partial_- M$ and $f_+ : \Sigma_+ \to \partial_+ M$. Let $g_- : \Sigma_- \to \Sigma'_-$ and $g_+ : \Sigma_+ \to \Sigma'_+$ be weak d-homeomorphisms of standard d-surfaces. Provide $\partial_- M$ and $\partial_+ M$ with the structure of parametrized d-surfaces via $f'_- = f_-(g_-)^{-1} : \Sigma'_- \to \partial_- M$ and $f'_+ = f_+(g_+)^{-1} : \Sigma'_+ \to \partial_+ M$. Denote the resulting parametrized d-surfaces by $\partial'_- M$ and $\partial'_+ M$ respectively. These are the same surfaces as $\partial_- M, \partial_+ M$ with the same distinguished marked arcs but (possibly) different orders in the set of distinguished arcs and different parametrizations. The cobordism M with the newly parametrized bases is a decorated 3-cobordism, say M', between $\partial'_- M$ and $\partial'_+ M$.

By definition of $\mathcal{T}$, we have $\mathcal{T}(\partial_- M) = \mathcal{T}(\Sigma_-)$, $\mathcal{T}(\partial'_- M) = \mathcal{T}(\Sigma'_-)$, and similar equalities with - replaced by $+$. Consider the diagram

$$(5.3.a)\qquad \begin{array}{ccccccc} \mathcal{T}(\partial_- M) & = & \mathcal{T}(\Sigma_-) & \xrightarrow{\varepsilon(g_-)} & \mathcal{T}(\Sigma'_-) & = & \mathcal{T}(\partial'_- M) \\ \downarrow{\scriptstyle \tau(M)} & & \downarrow{\scriptstyle \tau(M)} & & \downarrow{\scriptstyle \tau(M')} & & \downarrow{\scriptstyle \tau(M')} \\ \mathcal{T}(\partial_+ M) & = & \mathcal{T}(\Sigma_+) & \xrightarrow{\varepsilon(g_+)} & \mathcal{T}(\Sigma'_+) & = & \mathcal{T}(\partial'_+ M). \end{array}$$

We claim that this diagram is commutative up to a numerical factor computed as follows. Note that the Lagrangian relations $N(M) : H_1(\partial_- M; \mathbb{R}) \Rightarrow H_1(\partial_+ M; \mathbb{R})$ and $N(M') : H_1(\partial'_- M; \mathbb{R}) \Rightarrow H_1(\partial'_+ M; \mathbb{R})$ coincide. Set $N = N(M) = N(M')$. Set

$$\mu_+ = \mu(N_*(\lambda_-(M)), \lambda_+(M), \lambda_+(M')), \ \ \mu_- = \mu(\lambda_-(M), \lambda_-(M'), N^*(\lambda_+(M'))).$$

(For the definition of λ_- and λ_+, see Section 4.2.) Then

$$(5.3.b)\qquad \tau(M')\,\varepsilon(g_-) = (\mathcal{D}\Delta^{-1})^{\mu_+ - \mu_-}\,\varepsilon(g_+)\,\tau(M).$$

To prove this formula denote by X the decorated 3-cobordism obtained from M and the cylinder $M(g_+)$ by gluing of the bottom base of $M(g_+)$ to the top base of M along $f'_+ : \Sigma'_+ \times 0 \to \partial_+ M$. Considered up to d-homeomorphism, X is the same cobordism as M with the only difference that the top base of X is parametrized via f'_+ rather than f_+ (the bottom base of X is parametrized via f_-). The same decorated 3-cobordism X may be obtained from M' and $M(g_-)$ by gluing the top base of $M(g_-)$ to the bottom base of M' along $f'_- : \Sigma'_- \times 1 \to \partial_-(M')$. Note that the gluing homeomorphisms in question commute with parametrizations. Therefore we may apply Theorem 4.3 to these two splittings of X which gives

$$\tau(X) = (\mathcal{D}\Delta^{-1})^{\mu_+}\,\varepsilon(g_+)\,\tau(M) = (\mathcal{D}\Delta^{-1})^{\mu_-}\,\tau(M')\,\varepsilon(g_-).$$

This implies (5.3.b).

5.4. Computations on the torus. We may explicitly compute the action of the modular group $\mathrm{Mod}_1 = SL(2, \mathbb{Z})$ of the 2-torus $\Sigma_{(1;)} = \partial U_{(1;)} = S^1 \times S^1$ with an empty set of distinguished arcs. This turns out to be the projective linear action of $SL(2, \mathbb{Z})$ defined in Section II.3.9 in terms of matrices S and T. This fact places the constructions of Section II.3.9 in a proper perspective and emphasizes once more the role of the non-degeneracy axiom in the definition of modular categories.

The genus 1 handlebody $U_{(1;)}$ is a closed regular neighborhood of a ribbon graph in $\mathbb{R}^3$ consisting of a coupon with one cap-like band attached from above, as in Figure 5.1. Consider two loops α, β in $\Sigma_{(1;)} = \partial U_{(1;)}$ shown in Figure 5.1. Their homological classes form a basis of $H_1(\Sigma_{(1;)}; \mathbb{Z}) = \mathbb{Z} \oplus \mathbb{Z}$. We identify elements of this group with 2-columns via $m\alpha + n\beta \longleftrightarrow \begin{bmatrix} m \\ n \end{bmatrix}$. The endomorphisms of $H_1(\Sigma_{(1;)}; \mathbb{Z})$ are identified with integer 2×2-matrices acting on the 2-columns by multiplication on the left.

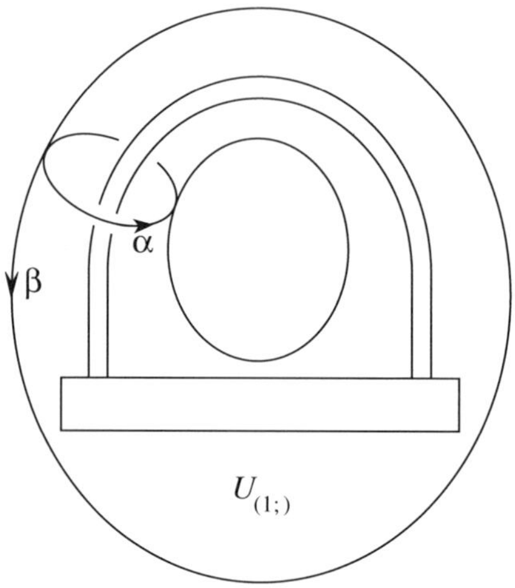

Figure 5.1

It is well known that every automorphism of $H_1(\Sigma_{(1;)}; \mathbb{Z})$ is induced by an orientation-preserving self-homeomorphism of $\Sigma_{(1;)}$ unique up to isotopy. Therefore $\text{Mod}_1 = SL(2, \mathbb{Z})$. The construction of Section 5.1 yields a projective linear action of $SL(2, \mathbb{Z})$ on the module $\Psi_{(1;)} = \oplus_{i \in I} \text{Hom}(\mathbb{1}, V_i \otimes V_i^*)$. This is a free K-module of rank card(I) with the basis $\{b_i = b_{V_i} : \mathbb{1} \to V_i \otimes V_i^*\}_{i \in I}$. We compute the action of the generators $s, t \in SL(2, \mathbb{Z})$ specified in Section II.3.9.

Consider the special ribbon graph presented by the left diagram in Figure 5.2. Let M denote the decorated 3-cobordism determined by this graph. Since the complement of the Hopf link in S^3 is the cylinder over the 2-torus, the cobordism M is homeomorphic to $S^1 \times S^1 \times [0, 1]$ with a certain parametrization of the bases. To verify that M is homeomorphic to the cobordism $M(s)$ as defined in Section 5.1, we compute the composition

$$H_1(\Sigma_{(1;)}; \mathbb{Z}) \xrightarrow{\approx} H_1(\partial_- M; \mathbb{Z}) \xrightarrow{\approx} H_1(\partial_+ M; \mathbb{Z}) \xrightarrow{\approx} H_1(\Sigma_{(1;)}; \mathbb{Z})$$

where the first and third isomorphisms are induced by the parametrizations of the bases of M and the second isomorphism is obtained by pushing loops in the bottom base of M to the top base using the cylindrical structure on M. A direct computation shows that this composition carries α, β into $\beta, -\alpha$ respectively. Therefore this endomorphism of $H_1(\Sigma_{(1;)}; \mathbb{Z})$ equals s. Hence, $M = M(s)$. Now we may compute $\varepsilon(s) : K^{\text{card}(I)} \to K^{\text{card}(I)}$. We present $\varepsilon(s)$ by a matrix $[\tau_i^j : Kb_i \to Kb_j]_{i,j \in I}$ and compute τ_i^j using formula (2.3.a). (Here $L = \emptyset$ so that the sum on the right-hand side of this formula degenerates to one term.) The computation given in Figure 5.3 shows that $\varepsilon(s) = \mathcal{D}^{-1}S$. It is easy to deduce from Lemma 2.6 that the right ribbon graph in Figure 5.2 also presents the cylinder $S^1 \times S^1 \times [0, 1]$ with a certain parametrization of the bases. As above, homological computations show that this is $M(t)$. We compute $\varepsilon(t)$ following the lines of Section 2.7 or simply using the result of this section. There is one additional left-hand twist here so that instead of the identity operator we obtain $\varepsilon(t) = T^{-1}$.

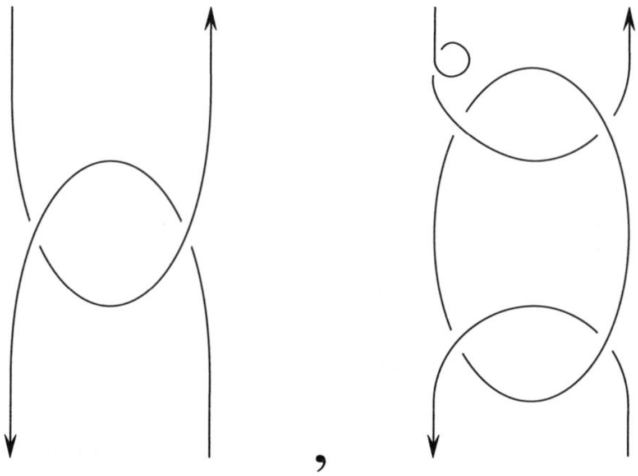

Figure 5.2

$\tau_i^j(b_i) \doteq \mathcal{D}^{-1}\dim(j)$ 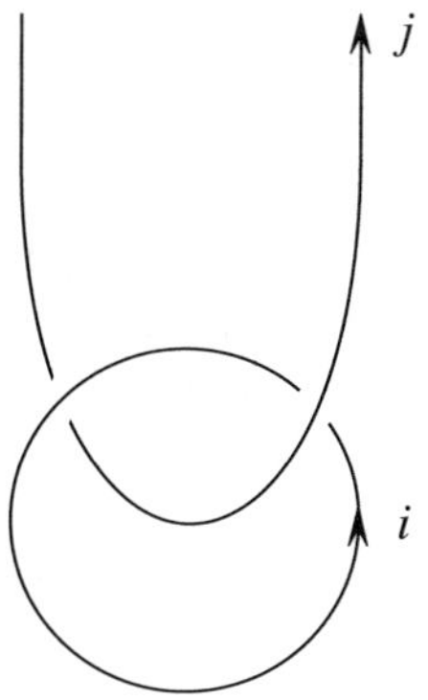

$\doteq \mathcal{D}^{-1} S_{i,j}\, b_j$

Figure 5.3

5.5. Remark. Systematic appearance of Lagrangian spaces and Maslov indices suggests an intimate relationship between 3-dimensional TQFT's and the Segal-Shale-Weil representations of the metaplectic groups (cf. [LV]).

5.6. Exercises. 1. Denote the integer $\mu_+ - \mu_-$ used in Section 5.3 by $\mu(M, M')$. Show that $\mu(M', M) = -\mu(M, M')$ and $\mu(M, M') + \mu(M', M'') = \mu(M, M'')$.

2. Let $g : \Sigma \to \Sigma'$ be a weak d-homeomorphism of standard d-surfaces such that $g_*(\lambda(\Sigma)) = \lambda(\Sigma')$. Show that the homomorphism $\varepsilon(g) : \mathcal{T}(\Sigma) \to \mathcal{T}(\Sigma')$ does not depend on the choice of $\mathcal{D}$. (Hint: the first Betti number of the manifold $\tilde{M}(g)$, derived as in Section 1.8 from the cylinder $M(g)$, is equal to the genus of Σ; cf. also Exercise II.2.5 and the definition of η in Section 1.4.)

6. Renormalized 3-dimensional TQFT

6.0. Outline. In order to apply to a surface, Σ, the modular functor constructed in Section 1 this surface should be parametrized by a standard one. The goal of this section is to relax this underlying structure. We show how to replace the parametrization of Σ with a much weaker structure which amounts to a choice of Lagrangian subspace in $H_1(\Sigma;\mathbb{R})$. We also show how to dispense with the order of distinguished arcs on Σ. This leads to a simplified version $(\mathcal{T}^e, \tau^e)$ of the TQFT $(\mathcal{T}, \tau)$ which applies to so-called extended surfaces and extended 3-cobordisms.

The TQFT $(\mathcal{T}^e, \tau^e)$ has gluing anomalies which will be explicitly computed in Section 7. This TQFT will be further improved in Section 9 where we kill the anomalies. However, the constructions of Section 9 are rather formal: in a sense we transform the problem of anomalies into a problem of gluing. From the computational point of view this does not change anything at all. The author's experience is that the TQFT $(\mathcal{T}^e, \tau^e)$ is quite suitable for computations.

The reader should pay attention to the notion of a weak e-homeomorphism introduced in Section 6.3. These homeomorphisms will be instrumental in Sections 7 and 8.

6.1. Extended surfaces. The theory of extended surfaces runs parallel to the theory of parametrized decorated surfaces. In the "extended" setting we involve neither parametrizations of surfaces nor orders in the sets of arcs. Instead we use Lagrangian spaces in real 1-homologies.

An extended surface or, briefly, an e-surface is a closed oriented surface Υ endowed with a finite (possibly empty) family of disjoint marked arcs and a Lagrangian space $\lambda(\Upsilon) \subset H_1(\Upsilon;\mathbb{R})$. (For the definition of a marked arc, see Section 1.1.) The disjoint union of a finite number of e-surfaces is an e-surface in the obvious way. The empty set will be treated as the empty e-surface. Warning: in general the connected components of an e-surface do not inherit the structure of an e-surface. The point is that the Lagrangian space $\lambda(\Upsilon)$ is not obliged to be decomposable. (A linear subspace of $H_1(\Upsilon;\mathbb{R})$ is decomposable if it splits as the direct sum of its intersections with the homology groups of the components of Υ.)

An e-homeomorphism of extended surfaces is a degree 1 homeomorphism of the underlying surfaces preserving distinguished arcs together with their orientations and marks and inducing an isomorphism of 1-homologies which preserves the distinguished Lagrangian subspace.

The structure of an e-surface is subject to negation as follows. For any e-surface Υ, the opposite e-surface $-\Upsilon$ is obtained from Υ by reversing the orientation of Υ, reversing the orientation of distinguished arcs, and multiplying the signs of these arcs by -1 while keeping the labels and the distinguished Lagrangian space. Clearly, $-(-\Upsilon) = \Upsilon$. The transformation $\Upsilon \mapsto -\Upsilon$ is natural in the sense that any e-homeomorphism $f : \Upsilon \to \Upsilon'$ gives rise to an e-homeomorphism $-f : -\Upsilon \to -\Upsilon'$ which coincides with f as a mapping.

It is clear that the structure of an e-surface is an involutive space-structure compatible with disjoint union (see Section III.1.1). Formally speaking, the space-structure $\mathfrak{A}^e = \mathfrak{A}^e_{\mathcal{V}}$ of extended surfaces is defined as follows. For any topological space X, the class $\mathfrak{A}^e(X)$ consists of homeomorphisms of e-surfaces onto X considered up to composition with e-homeomorphisms of e-surfaces. The involution in $\mathfrak{A}^e(X)$ transforms any homeomorphism $\Upsilon \to X$ into the same homeomorphism viewed as a mapping $-\Upsilon \to X$. (In contrast to the decorated setting we do not need the mappings $\{\mathrm{rev}_t\}_t$ used in Section 1.) It is clear that an $\mathfrak{A}^e$-space is nothing but an e-surface. An $\mathfrak{A}^e$-homeomorphism of $\mathfrak{A}^e$-spaces is just an e-homeomorphism of e-surfaces.

6.2. Extended 3-manifolds. An extended 3-manifold is a compact oriented 3-manifold with the structure of an e-surface on its boundary and with a v-colored ribbon graph sitting in this 3-manifold. (For the definition of a ribbon graph in a 3-manifold, see Section 1.5, a ribbon graph will always meet the boundary of the ambient 3-manifold along the distinguished arcs.) We define the boundary of an extended 3-manifold to be the usual boundary with its structure of e-surface. Note that for closed 3-manifolds, there is no difference between extended and decorated manifolds.

An e-homeomorphism of extended 3-manifolds is a degree 1 homeomorphism of 3-manifolds preserving all additional structures in question (including the v-colored ribbon graph). An e-homeomorphism of extended 3-manifolds $M \to M'$ restricts to an e-homeomorphism $\partial M \to \partial M'$.

For any e-surface Υ, the cylinder $\Upsilon \times [0, 1]$ is oriented and provided with a v-colored ribbon graph in the same way as in Section 1.6. The boundary of this cylinder is regarded as an e-surface via the identifications $\Upsilon \times 0 = -\Upsilon$ and $\Upsilon \times 1 = \Upsilon$. In this way $\Upsilon \times [0, 1]$ acquires the structure of an extended 3-manifold.

It is obvious that the structure of extended 3-manifolds is a space-structure in the sense of Section III.1.1. The corresponding functor $\mathfrak{B}^e = \mathfrak{B}^e_{\mathcal{V}}$ assigns to any topological space X the class of homeomorphisms of extended 3-manifolds onto X considered up to composition with e-homeomorphisms of extended 3-manifolds. It is clear that a $\mathfrak{B}^e$-space is just an extended 3-manifold. A $\mathfrak{B}^e$-homeomorphism of $\mathfrak{B}^e$-spaces is an e-homeomorphism of extended 3-manifolds. The space-structure $\mathfrak{B}^e$ is compatible with disjoint union in the obvious way.

6.2.1. Lemma. *The pair* $(\mathfrak{B}^e, \mathfrak{A}^e)$ *is a cobordism theory in the sense of Section* III.1.3.

This lemma follows from the definitions. The gluing involved in (III.1.3.1) is the usual gluing along homeomorphisms.

6.3. Modular functor $\mathcal{T}^e$. We define a modular functor $\mathcal{T}^e = \mathcal{T}^e_{\mathcal{V}}$ based on $\mathfrak{A}^e$.

First of all we assign to every e-surface Υ a K-module $\mathcal{T}^e(\Upsilon)$. The idea is to consider all parametrizations of Υ by standard d-surfaces and to identify the modules of states of these surfaces using the action of the modular groupoid. By a parametrization of Υ we mean a pair (a standard d-surface Σ, a degree 1 homeomorphism $f : \Sigma \to \Upsilon$ which transforms the distinguished arcs of Σ onto those of Υ preserving orientations and marks). We do not impose any compatibility conditions on the distinguished Lagrangian space $\lambda(\Upsilon)$ and the Lagrangian space $\lambda(\Sigma)$ defined in Section 4.2. It is obvious that the class of parametrizations of Υ is non-empty. There are only a finite number of standard d-surfaces which may parametrize Υ. If Υ is connected then this number is equal to $m!$ where m is the number of distinguished arcs in Υ. On the other hand, the parametrization homeomorphisms $f : \Sigma \to \Upsilon$ in general are infinite in number (even considered up to isotopy) since we may always compose any such homeomorphism with a self-homeomorphism of Σ preserving the distinguished arcs.

For parametrizations $f_0 : \Sigma_0 \to \Upsilon$, $f_1 : \Sigma_1 \to \Upsilon$, the composition $f_1^{-1} f_0 : \Sigma_0 \to \Sigma_1$ is a weak d-homeomorphism. Set

$$\varphi(f_0, f_1) = (\mathcal{D}\Delta^{-1})^{-\mu((f_0)_*(\lambda(\Sigma_0)), \lambda(\Upsilon), (f_1)_*(\lambda(\Sigma_1)))}\, \varepsilon(f_1^{-1} f_0) : \mathcal{T}(\Sigma_0) \to \mathcal{T}(\Sigma_1).$$

For any parametrizations f_0, f_1, f_2 of Υ, we have

$$\varphi(f_1, f_2)\, \varphi(f_0, f_1) = \varphi(f_0, f_2). \tag{6.3.a}$$

This follows from formula (5.2.a), formula (3.6.a) applied to the Lagrangian spaces $(f_r)_*(\lambda(\Sigma_r))$ with $r = 0, 1, 2$ and $\lambda(\Upsilon)$, and the invariance of the Maslov index under symplectic isomorphisms. It is clear that $\varphi(f_0, f_0) = \text{id}$. Applying (6.3.a) to $f_2 = f_0$ we conclude that $\varphi(f_0, f_1)$ and $\varphi(f_1, f_0)$ are mutually inverse isomorphisms.

Identify the modules $\{\mathcal{T}(\Sigma)\}_{(\Sigma, f)}$ corresponding to all parametrizations of Υ along the intertwining isomorphisms $\{\varphi(f_0, f_1)\}_{(f_0, f_1)}$. This yields a projective K-module $\mathcal{T}^e(\Upsilon)$ depending solely on Υ. Elements of this module are compatible families $\{x(\Sigma, f) \in \mathcal{T}(\Sigma)\}_{(\Sigma, f)}$ where (Σ, f) runs over all parametrizations of Υ and compatibility means that for any parametrizations (Σ_0, f_0), (Σ_1, f_1) of Υ, we have $x(\Sigma_1, f_1) = \varphi(f_0, f_1)(x(\Sigma_0, f_0))$. It is clear that such a family is completely determined by any of its representatives. Hence, for any parametrization $f_0 : \Sigma_0 \to \Upsilon$, the formula $\{x(\Sigma, f)\}_{(\Sigma, f)} \mapsto x(\Sigma_0, f_0)$ defines an isomorphism $\mathcal{T}^e(\Upsilon) \to \mathcal{T}(\Sigma_0)$. The inverse isomorphism $\mathcal{T}(\Sigma_0) \to \mathcal{T}^e(\Upsilon)$ is denoted by $(f_0)_\#$.

In order to extend the construction $\Upsilon \mapsto \mathcal{T}^e(\Upsilon)$ to a modular functor we need to define the action of e-homeomorphisms. We shall give a more general definition applying to weak e-homeomorphisms. A weak e-homeomorphism of e-surfaces is a degree 1 homeomorphism of the underlying surfaces preserving distinguished arcs with their orientations and marks. A weak e-homeomorphism does not need to preserve the distinguished Lagrangian spaces.

Let $g : \Upsilon \to \Upsilon'$ be a weak e-homeomorphism. For any parametrization $f : \Sigma \to \Upsilon$, the pair (Σ, gf) is a parametrization of Υ'. Set

$$g_\# = (\mathcal{D}\Delta^{-1})^{-\mu(g_*(\lambda(\Upsilon)),\lambda(\Upsilon'),(gf)_*(\lambda(\Sigma)))}(gf)_\#(f_\#)^{-1} : \mathcal{T}^e(\Upsilon) \to \mathcal{T}^e(\Upsilon'). \tag{6.3.b}$$

For instance, if $g = \mathrm{id} : \Upsilon \to \Upsilon$ then $g_\# = \mathrm{id}_{\mathcal{T}^e(\Upsilon)}$.

6.3.1. Lemma. *For any weak e-homeomorphism $g : \Upsilon \to \Upsilon'$, the homomorphism $g_\# : \mathcal{T}^e(\Upsilon) \to \mathcal{T}^e(\Upsilon')$ does not depend on the choice of parametrization of Υ.*

Proof. Let $f_0 : \Sigma_0 \to \Upsilon$ and $f_1 : \Sigma_1 \to \Upsilon$ be two parametrizations of Υ. To show that the homomorphisms $g_\# : \mathcal{T}^e(\Upsilon) \to \mathcal{T}^e(\Upsilon')$ derived from f_0 and f_1 coincide we have to show that

$$(\mathcal{D}\Delta^{-1})^{-\mu(g_*(\lambda(\Upsilon)),\lambda(\Upsilon'),(gf_1)_*(\lambda(\Sigma_1)))}\,\varphi(f_0, f_1) =$$
$$= (\mathcal{D}\Delta^{-1})^{-\mu(g_*(\lambda(\Upsilon)),\lambda(\Upsilon'),(gf_0)_*(\lambda(\Sigma_0)))}\,\varphi(gf_0, gf_1).$$

By definition of the intertwining isomorphisms φ, it suffices to establish the identity

$$-\mu(g_*(\lambda(\Upsilon)), \lambda(\Upsilon'), (gf_1)_*(\lambda(\Sigma_1))) - \mu((f_0)_*(\lambda(\Sigma_0)), \lambda(\Upsilon), (f_1)_*(\lambda(\Sigma_1))) =$$
$$-\mu(g_*(\lambda(\Upsilon)), \lambda(\Upsilon'), (gf_0)_*(\lambda(\Sigma_0))) - \mu((gf_0)_*(\lambda(\Sigma_0)), \lambda(\Upsilon'), (gf_1)_*(\lambda(\Sigma_1))).$$

This identity follows from Lemma 3.6, the skew-symmetry of the Maslov index, and the invariance of the Maslov index under isomorphisms of symplectic linear spaces.

6.3.2. Lemma. *For any weak e-homeomorphisms $g : \Upsilon \to \Upsilon'$ and $g' : \Upsilon' \to \Upsilon''$, we have*

$$(g'g)_\# = (\mathcal{D}\Delta^{-1})^{\mu(g_*(\lambda(\Upsilon)),\lambda(\Upsilon'),(g')_*^{-1}(\lambda(\Upsilon'')))}\,g'_\# g_\#. \tag{6.3.c}$$

Proof. To compute $g_\#$, $g'_\#$, and $(g'g)_\#$ we use an arbitrary parametrization $f : \Sigma \to \Upsilon$ and the parametrizations $f' = gf$, $f'' = g'gf$ of Υ' and Υ''. To establish (6.3.c) it suffices to show that

$$-\mu((g'g)_*(\lambda(\Upsilon)), \lambda(\Upsilon''), f''_*(\lambda(\Sigma))) - \mu(g_*(\lambda(\Upsilon)), \lambda(\Upsilon'), (g')_*^{-1}(\lambda(\Upsilon''))) =$$
$$= -\mu(g_*(\lambda(\Upsilon)), \lambda(\Upsilon'), f'_*(\lambda(\Sigma))) - \mu(g'_*(\lambda(\Upsilon')), \lambda(\Upsilon''), f''_*(\lambda(\Sigma))).$$

This equality follows from Lemma 3.6 and the invariance of the Maslov index under isomorphisms of symplectic linear spaces.

6.3.3. Lemma. *Assigning to each e-surface* Υ *the module* $\mathcal{T}^e(\Upsilon)$ *and to each e-homeomorphism g the homomorphism* $g_\#$ *we get a modular functor* $\mathcal{T}^e$ *based on* $\mathfrak{A}^e$.

Proof. We should verify axioms (III.1.2.1) - (III.1.2.3) with the words "$\mathfrak{A}$-space", "$\mathfrak{A}$-homeomorphism" replaced with "e-surface", "e-homeomorphism" respectively. If g is an e-homeomorphism then the Maslov index in formula (6.3.b) vanishes so that $g_\# = (gf)_\#(f_\#)^{-1}$. This implies (III.1.2.1). (This axiom follows also from Lemma 6.3.2 because of the vanishing of the corresponding Maslov index.) Axioms (III.1.2.2) and (III.1.2.3) follow directly from definitions and the properties of $\mathcal{T}$.

6.4. Remarks. 1. The naturality condition (iii) in (III.1.2.3) holds also in the case where f and g are weak e-homeomorphisms. This follows from definitions and the additivity of the Maslov index with respect to direct sums of symplectic spaces and their Lagrangian subspaces.

2. There is a simpler construction of the modular functor $\mathcal{T}^e$ which applies to those e-surfaces Υ whose distinguished Lagrangian space is integral and decomposable. (A subspace of $H_1(\Upsilon;\mathbb{R})$ is integral if it is generated by its intersection with the lattice of integer homologies.) If $\lambda(\Upsilon)$ is integral and decomposable we may restrict ourselves to parametrizations $f:\Sigma\to\Upsilon$ such that $f_*(\lambda(\Sigma))=\lambda(\Upsilon)$. The integrality and decomposability of $\lambda(\Upsilon)$ is a necessary and sufficient condition for the existence of such parametrizations. (This follows from two facts: (i) symplectic endomorphisms of $H_1(\Upsilon;\mathbb{Z})$ act transitively on the set of integral Lagrangian subspaces of $H_1(\Upsilon;\mathbb{R})$, (ii) if Υ is connected then any symplectic endomorphism of $H_1(\Upsilon;\mathbb{Z})$ is induced by an orientation-preserving self-homeomorphism of Υ.) In this restricted class of parametrizations the Maslov index in the definition of $\varphi(f_0, f_1)$ vanishes. If we confine ourselves to such parametrizations then the definition of the homomorphism τ^e in the next subsection and the verification of its properties become considerably simpler due to the vanishing of almost all Maslov indices. The reader may pursue this simplified approach keeping in mind that it leads to a slightly less general TQFT.

6.5. Homomorphism τ^e. We define an operator invariant $\tau^e=\tau^e_{\mathcal{V}}$ of extended 3-cobordisms. In accordance with the terminology of Section III.1 an extended 3-cobordism is a triple $(M,\partial_-M,\partial_+M)$ where ∂_-M and ∂_+M are e-surfaces and M is an extended 3-manifold with $\partial M=(-\partial_-M)\amalg\partial_+M$. This equality implies that $\lambda(\partial M)=\lambda(\partial_-M)\oplus\lambda(\partial_+M)$.

For any extended 3-cobordism M, we define a K-homomorphism

$$\tau^e(M) : \mathcal{T}^e(\partial_- M) \to \mathcal{T}^e(\partial_+ M)$$

as follows. Choose arbitrary parametrizations $f_- : \Sigma_- \to \partial_- M$ and $f_+ : \Sigma_+ \to \partial_+ M$ where Σ_- and Σ_+ are standard d-surfaces. These parametrizations provide $\partial_- M$ and $\partial_+ M$ with the structure of a parametrized d-surface. In this way the cobordism M acquires the structure of a decorated 3-cobordism which we denote by $\check{M}$. Set $N = N(M) : H_1(\partial_- M; \mathbb{R}) \Rightarrow H_1(\partial_+ M; \mathbb{R})$. Set

$$m_-(\check{M}) = \mu(N^*(\lambda_+(M)), \lambda_-(M), \lambda_-(\check{M}))$$

and

$$m_+(\check{M}) = \mu(N_*(\lambda_-(\check{M})), \lambda_+(M), \lambda_+(\check{M}))$$

where $\lambda_+(M) = \lambda(\partial_+ M)$ is the Lagrangian space in $H_1(\partial_+ M; \mathbb{R})$ determined by the structure of the e-surface on $\partial_+ M$ whilst $\lambda_+(\check{M})$ is the Lagrangian space in $H_1(\partial_+ M; \mathbb{R})$ determined by the parametrization, i.e., $\lambda_+(\check{M}) = (f_+)_*(\lambda(\Sigma_+))$. Similar notation applies to the bottom base. Set

$$k(\check{M}) = (\mathcal{D}\Delta^{-1})^{m_-(\check{M}) - m_+(\check{M})}.$$

The modular functor $\mathcal{T}$ assigns to the bases of $\check{M}$ the modules $\mathcal{T}(\partial_- \check{M}) = \mathcal{T}(\Sigma_-)$ and $\mathcal{T}(\partial_+ \check{M}) = \mathcal{T}(\Sigma_+)$ respectively. The TQFT $(\mathcal{T}, \tau)$ yields a linear operator $\tau(\check{M}) : \mathcal{T}(\Sigma_-) \to \mathcal{T}(\Sigma_+)$. We define $\tau^e(M) : \mathcal{T}^e(\partial_- M) \to \mathcal{T}^e(\partial_+ M)$ to be the composition of the following three homomorphisms:

$$\mathcal{T}^e(\partial_- M) \xrightarrow{(f_-)_\#^{-1}} \mathcal{T}(\Sigma_-) \xrightarrow{k(\check{M})\,\tau(\check{M})} \mathcal{T}(\Sigma_+) \xrightarrow{(f_+)_\#} \mathcal{T}^e(\partial_+ M).$$

(Recall that $(f_-)_\# : \mathcal{T}(\Sigma_-) \to \mathcal{T}^e(\partial_- M)$ and $(f_+)_\# : \mathcal{T}(\Sigma_+) \to \mathcal{T}^e(\partial_+ M)$ are isomorphisms.)

6.5.1. Lemma. *The homomorphism $\tau^e(M)$ does not depend on the choice of parametrizations of $\partial_- M$ and $\partial_+ M$.*

Proof. Let $f'_- : \Sigma'_- \to \partial_- M$ and $f'_+ : \Sigma'_+ \to \partial_+ M$ be another pair of parametrizations of $\partial_- M$ and $\partial_+ M$ where Σ'_-, Σ'_+ are standard d-surfaces. Using f'_- and f'_+ we regard M as a decorated 3-cobordism which we denote by $\check{M}'$. Set $\varphi_- = \varphi(f_-, f'_-)$ and $\varphi_+ = \varphi(f_+, f'_+)$. Consider the diagram

$$\begin{array}{ccccccc}
\mathcal{T}^e(\partial_- M) & \xrightarrow{(f_-)_\#^{-1}} & \mathcal{T}(\Sigma_-) & \xrightarrow{k(\check{M})\,\tau(\check{M})} & \mathcal{T}(\Sigma_+) & \xrightarrow{(f_+)_\#} & \mathcal{T}^e(\partial_+ M) \\
{\scriptstyle \mathrm{id}}\downarrow & & {\scriptstyle \varphi_-}\downarrow & & \downarrow{\scriptstyle \varphi_+} & & \downarrow{\scriptstyle \mathrm{id}} \\
\mathcal{T}^e(\partial_- M) & \xrightarrow{(f'_-)_\#^{-1}} & \mathcal{T}(\Sigma'_-) & \xrightarrow{k(\check{M}')\,\tau(\check{M}')} & \mathcal{T}(\Sigma'_+) & \xrightarrow{(f'_+)_\#} & \mathcal{T}^e(\partial_+ M).
\end{array}$$

By definition of $\mathcal{T}^e(\partial_- M)$ and $\mathcal{T}^e(\partial_+ M)$, the left and right squares are commutative. Let us prove that the middle square is also commutative. Commutativity of this diagram would imply the claim of the lemma.

We have

$$\varphi_+ = (\mathcal{D}\Delta^{-1})^{-\mu(\lambda_+(\check{M}),\lambda_+(M),\lambda_+(\check{M}'))}\,\varepsilon((f'_+)^{-1}f_+)$$

and similarly with the subscript $+$ replaced by $-$. This and formula (5.3.b) applied to $g_+ = (f'_+)^{-1}f_+$ and $g_- = (f'_-)^{-1}f_-$ imply that

$$\varphi_+\tau(\check{M}) = (\mathcal{D}\Delta^{-1})^m\tau(\check{M}')\varphi_-$$

where

$$m = \mu(\lambda_-(\check{M}), \lambda_-(M), \lambda_-(\check{M}')) - \mu(\lambda_+(\check{M}), \lambda_+(M), \lambda_+(\check{M}'))+$$

$$+\mu(\lambda_-(\check{M}), \lambda_-(\check{M}'), N^*(\lambda_+(\check{M}'))) - \mu(N_*(\lambda_-(\check{M})), \lambda_+(\check{M}), \lambda_+(\check{M}')).$$

To complete the proof it remains to show that

$$m + m_-(\check{M}) - m_+(\check{M}) - m_-(\check{M}') + m_+(\check{M}') = 0. \tag{6.5.a}$$

We shall deduce this equality from Lemmas 3.6 and 3.7. Applying the formula (3.6.a) to the 4-tuple of Lagrangian spaces

$$\lambda_-(\check{M}),\ \lambda_-(M),\ \lambda_-(\check{M}'),\ N^*(\lambda_+(M))$$

we obtain that the sum of the first summand of the expression for m with $m_-(\check{M})-m_-(\check{M}')$ is equal to $-\mu(\lambda_-(\check{M}), \lambda_-(\check{M}'), N^*(\lambda_+(M)))$. Applying (3.6.a) to the 4-tuple of Lagrangian spaces

$$\lambda_+(\check{M}),\ \lambda_+(M),\ \lambda_+(\check{M}'),\ N_*(\lambda_-(\check{M}))$$

we get that the (algebraic) sum of the second and forth summands of m with $-m_+(\check{M})$ is equal to $-\mu(\lambda_+(M), \lambda_+(\check{M}'), N_*(\lambda_-(\check{M})))$. Therefore the left-hand side of (6.5.a) is equal to an algebraic sum of four Maslov indices. It is straightforward to check that this sum equals the difference between the left-hand side and the right-hand side of formula (3.7.a) where $\lambda_1 = \lambda_-(\check{M})$, $\lambda_2 = \lambda_-(\check{M}')$, $\lambda'_1 = \lambda_+(\check{M}')$, and $\lambda'_2 = \lambda_+(M)$. This completes the proof of the lemma.

6.6. Theorem. *The function $(M, \partial_- M, \partial_+ M) \mapsto \tau^e(M)$ extends the modular functor $\mathcal{T}^e$ to a non-degenerate topological quantum field theory based on extended surfaces and extended 3-manifolds.*

This theorem is essentially obvious: all axioms of TQFT's and non-degeneracy of $(\mathcal{T}^e, \tau^e)$ follow from the corresponding properties of $(\mathcal{T}, \tau)$. We shall compute the anomalies of $(\mathcal{T}^e, \tau^e)$ in Section 7.

The TQFT $(\mathcal{T}^e, \tau^e) = (\mathcal{T}^e_{\mathcal{V}}, \tau^e_{\mathcal{V}})$ should be viewed as a reparametrized version of $(\mathcal{T}, \tau)$ adapted to the "extended" setting. This setting is considerably more natural and simpler than the "decorated" one. The parametrizations of surfaces and orderings of marked arcs involved in the decorated setting are quite clumsy, and the appearance of Lagrangian spaces seems to be a rather mild price to pay in order to get rid of them. The values of τ^e on closed oriented 3-manifolds with v-colored ribbon graphs sitting inside coincide with the values of τ and thus with the invariant defined in Chapter II.

The TQFT $(\mathcal{T}^e, \tau^e)$ behaves nicely with respect to connected sum. For any extended 3-cobordisms M_1, M_2 we have

$$\tau^e(M_1\#M_2) = \mathcal{D}\,\tau^e(M_1 \amalg M_2) = \mathcal{D}\,\tau^e(M_1) \otimes \tau^e(M_2).$$

6.7. Forgetful functor $(\mathcal{T}, \tau) \to (\mathcal{T}^e, \tau^e)$. There is a natural forgetful functor relating the TQFT's $(\mathcal{T}, \tau)$ and $(\mathcal{T}^e, \tau^e)$. This functor is not a homomorphism of TQFT's in the sense of Section III.1.4 because these TQFT's are based on different cobordism theories. We shall not develop a general theory of functors relating TQFT's based on different cobordism theories. Instead we simply define the functor in question.

Let Σ be a d-surface with a parametrization $f : \Sigma^0 \to \Sigma$ where Σ^0 is a standard d-surface. Denote by $\underline{\Sigma}$ the underlying e-surface of Σ obtained by forgetting the order of distinguished arcs and the parametrization while fixing the Lagrangian space $f_*(\lambda(\Sigma^0)) \subset H_1(\Sigma; \mathbb{R})$. The construction $\Sigma \mapsto \underline{\Sigma}$ is natural with respect to d-homeomorphisms: any d-homeomorphism $g : \Sigma \to \Sigma'$ of parametrized d-surfaces induces a weak e-homeomorphism $\underline{g}$ of the underlying e-surfaces coinciding with g as a mapping. If, in addition, g commutes with parametrizations then $\underline{g}$ is an e-homeomorphism. Similarly, each decorated 3-manifold M gives rise to an underlying extended 3-manifold $\underline{M}$. We should just replace the structure of parametrized d-surface on ∂M with the underlying structure of e-surface.

For any parametrized d-surface Σ, we define a canonical isomorphism $\theta(\Sigma) : \mathcal{T}(\Sigma) \to \mathcal{T}^e(\underline{\Sigma})$. Consider the given parametrization $f : \Sigma^0 \to \Sigma$ by a standard d-surface Σ^0. Viewed as a mapping $\Sigma^0 \to \underline{\Sigma}$ it yields a parametrization $\underline{f}$ of the e-surface $\underline{\Sigma}$. We define $\theta(\Sigma)$ to be the composition of the identification isomorphism $\mathcal{T}(\Sigma) = \mathcal{T}(\Sigma^0)$ and the isomorphism $(\underline{f})_\# : \mathcal{T}(\Sigma^0) \to \mathcal{T}^e(\underline{\Sigma})$.

The isomorphisms $\{\theta(\Sigma)\}_\Sigma$ form a forgetful functor $(\mathcal{T}, \tau) \to (\mathcal{T}^e, \tau^e)$ in the following sense.

6.7.1. Theorem. *For any d-homeomorphism $g : \Sigma \to \Sigma'$ of parametrized d-surfaces, the following diagram is commutative:*

$$\begin{array}{ccc} \mathcal{T}(\Sigma) & \xrightarrow{\varepsilon(g)} & \mathcal{T}(\Sigma') \\ {\scriptstyle\theta(\Sigma)}\downarrow & & \downarrow{\scriptstyle\theta(\Sigma')} \\ \mathcal{T}^e(\underline{\Sigma}) & \xrightarrow{(\underline{g})_\#} & \mathcal{T}^e(\underline{\Sigma'}). \end{array} \tag{6.7.a}$$

For any decorated 3-cobordism M, the following diagram is commutative:

$$\begin{array}{ccc} \mathcal{T}(\partial_- M) & \xrightarrow{\tau(M)} & \mathcal{T}(\partial_+ M) \\ {\scriptstyle \theta(\partial_- M)}\downarrow & & \downarrow{\scriptstyle \theta(\partial_+ M)} \\ \mathcal{T}^e(\partial_-(\underline{M})) & \xrightarrow{\tau^e(\underline{M})} & \mathcal{T}^e(\partial_+(\underline{M})). \end{array} \tag{6.7.b}$$

Commutativity of (6.7.a) implies that the forgetful functor transforms the action ε of the modular groupoid into an action of the groupoid formed by the isotopy classes of weak e-homeomorphisms. In this way we may reformulate ε completely in terms of weak e-homeomorphisms.

To prove Theorem 6.7.1 we need the following lemma.

6.7.2. Lemma. *Let Σ be a closed oriented surface with a finite family of disjoint marked arcs. Let λ, λ' be Lagrangian subspaces of $H_1(\Sigma;\mathbb{R})$. Denote by C the cylinder $\Sigma\times[0,1]$ viewed as an extended 3-cobordism between (Σ,λ) and (Σ,λ'). Let $\mathrm{id}^{\lambda,\lambda'}$ denote the identity mapping $(\Sigma,\lambda)\to(\Sigma,\lambda')$ viewed as a weak e-homeomorphism. Then*

$$\tau^e(C) = (\mathrm{id}^{\lambda,\lambda'})_\# : \mathcal{T}^e(\Sigma,\lambda) \to \mathcal{T}^e(\Sigma,\lambda').$$

Proof. Choose a standard d-surface Σ^0 and a degree 1-homeomorphism $f:\Sigma^0\to\Sigma$ preserving the distinguished arcs with their orientations and marks. We regard f as a parametrization of the e-surface (Σ,λ). Denote by f' the same mapping f viewed as a parametrization of the e-surface (Σ,λ'). In other words, $f' = \mathrm{id}^{\lambda,\lambda'}\circ f:\Sigma^0\to(\Sigma,\lambda')$. By definition,

$$\tau^e(C) = (\mathcal{D}\Delta^{-1})^{m_-(\check{C})-m_+(\check{C})}\, f'_\#\,\tau(\check{C})\, f_\#^{-1}$$

where $\check{C}$ is the cylinder $\Sigma\times[0,1]$ whose bases are parametrized and decorated via f. By Theorem 1.9 (or Lemma 2.1.1), $\tau(\check{C}) = \mathrm{id}:\mathcal{T}(\Sigma^0)\to\mathcal{T}(\Sigma^0)$. It is easy to compute that $m_-(\check{C}) = \mu(\lambda',\lambda,f_*(\lambda(\Sigma^0)))$ and $m_+(\check{C})=0$. Therefore

$$\tau^e(C) = (\mathcal{D}\Delta^{-1})^{-\mu(\lambda,\lambda',f_*(\lambda(\Sigma^0)))}\, f'_\#\, f_\#^{-1}.$$

Formula (6.3.b) implies that the last expression is equal to $(\mathrm{id}^{\lambda,\lambda'})_\#$.

6.7.3. Proof of Theorem 6.7.1. The commutativity of (6.7.b) is essentially tautological. To compute $\tau^e(\underline{M})$ in (6.7.b) we should choose parametrizations of the e-surfaces $\partial_-(\underline{M})$, $\partial_+(\underline{M})$ and transform in this way the extended 3-cobordism $\underline{M}$ into a decorated 3-cobordism $\check{\underline{M}}$. We choose the parametrizations to be those induced by the parametrizations of $\partial_- M$ and $\partial_+ M$. This transforms $\underline{M}$ into the original decorated 3-cobordism M. The corresponding numbers m_- and m_+ as defined in Section 6.5 are both equal to 0. Now, the commutativity of (6.7.b) follows directly from the definition of τ^e given in Section 6.5 because the homomorphisms $(f_-)_\#, (f_+)_\#$ used there coincide with the homomorphisms $\theta(\partial_- M), \theta(\partial_+ M)$.

We first show commutativity of (6.7.a) in the case where g commutes with parametrizations. We may assume that Σ and Σ' are connected d-surfaces of a certain type t. Let f and f' denote the given parametrizations $\Sigma_t \to \Sigma$ and $\Sigma_t \to \Sigma'$. Since $f' = gf$, the homomorphism $\varepsilon(g) : \mathcal{T}(\Sigma) \to \mathcal{T}(\Sigma')$ is the composition of the identification isomorphisms $\mathcal{T}(\Sigma) = \mathcal{T}(\Sigma_t) = \mathcal{T}(\Sigma')$ provided by the very definition of the modular functor $\mathcal{T}$ on parametrized d-surfaces. The Maslov index which enters (6.3.b) is equal to 0 because

$$g_*(\lambda(\underline{\Sigma})) = g_* f_*(\lambda(\Sigma_t)) = f'_*(\lambda(\Sigma_t)) = \lambda(\underline{\Sigma}').$$

Formula (6.3.b) implies that $(\underline{g})_\# = (\underline{f}')_\# (\underline{f})_\#^{-1}$. Comparing this expression for $(\underline{g})_\#$ with the definitions of homomorphisms $\theta(\Sigma)$ and $\theta(\Sigma')$, we obtain commutativity of (6.7.a).

Let us prove commutativity of (6.7.a) for an arbitrary weak d-homeomorphism $g : \Sigma \to \Sigma'$. Let f and f' denote the given parametrizations of Σ and Σ'. Denote by Σ'' the surface Σ parametrized and decorated via $g^{-1} f'$. Present g as the composition of the identity mapping $g' = \mathrm{id} : \Sigma \to \Sigma''$ and a d-homeomorphism $g'' : \Sigma'' \to \Sigma'$ commuting with parametrizations and coinciding with g as a mapping. (Note that $g' = (g'')^{-1} g$ is a weak d-homeomorphism.) Using formula (5.2.a) and Lemma 6.3.2 we may reduce commutativity of (6.7.a) to commutativity of two diagrams obtained from (6.7.a) by the substitutions $g = g'$ and $g = g''$. (The Maslov indices appearing in (5.2.a) and Lemma 6.3.2 both vanish.) The case $g = g''$ was treated above. Consider the case $g = g'$. It follows from the definition of ε that $\varepsilon(g') = \tau(C)$ where C is the cylinder $\Sigma \times [0, 1]$ whose bottom and top bases are parametrized and decorated via $g' f$ and f respectively. It follows from Lemma 6.7.2 that $(\underline{g}')_\# = \tau^e(\underline{C})$. Therefore diagram (6.7.a) with $g = g'$ coincides with (6.7.b) for $M = C$. Commutativity of (6.7.b) implies commutativity of (6.7.a).

6.8. Bundle of states. Let Σ be a closed oriented surface with a finite family of disjoint marked arcs. For any Lagrangian space $\lambda \in \Lambda = \Lambda(H_1(\Sigma; \mathbb{R}))$, the pair (Σ, λ) is an e-surface so that we may consider the module $\mathcal{T}^e(\Sigma, \lambda)$. The modules corresponding to all λ form a flat bundle over Λ called the bundle of states of Σ. Strictly speaking, we can not use here the notion of a topological bundle because the fibers in question are projective modules over a commutative ring K without any topology. Instead of giving appropriate general definitions, we simply define the bundle we have in mind.

Denote by $p : \tilde{\Lambda} \to \Lambda$ the universal covering of Λ (the space Λ is connected and has infinite cyclic fundamental group, see [Ar]). By [Ler], [Tu1], there exists a function $\ell : \tilde{\Lambda} \times \tilde{\Lambda} \to \mathbb{Z}$ satisfying the following two properties: (a) it is locally constant on the set of pairs $x, y \in \tilde{\Lambda}$ such that the Lagrangian spaces $p(x)$ and

$p(y)$ are transversal to each other, (b) for any $x, y, z \in \tilde{\Lambda}$,

$$\ell(x, y) - \ell(x, z) + \ell(y, z) = \mu(p(x), p(y), p(z)).$$

Moreover, the function ℓ is uniquely characterized by these properties. This implies that ℓ is invariant under covering transformations and $\ell(x, x) = \mu(p(x), p(x), p(x)) = 0$ for any $x \in \tilde{\Lambda}$.

For $x, y \in \tilde{\Lambda}$, consider the isomorphism

$$L(x, y) = (\mathcal{D}\Delta^{-1})^{-\ell(x,y)}(\mathrm{id}^{p(x),p(y)})_{\#} : \mathcal{T}^e(\Sigma, p(x)) \to \mathcal{T}^e(\Sigma, p(y))$$

(cf. Lemma 6.7.2). It follows from Lemma 6.3.2 and the properties of ℓ that for any $x, y, z \in \tilde{\Lambda}$,

(i) $L(x, x) = \mathrm{id}$,

(ii) $L(y, z)\, L(x, y) = L(x, z)$,

(iii) the isomorphisms $\{L(x, y)\}_{x,y}$ commute with the covering transformations of p.

Using the first two properties we can regard the union $\cup_{x\in\tilde{\Lambda}} \mathcal{T}^e(\Sigma, p(x))$ as a trivial bundle over $\tilde{\Lambda}$. The isomorphisms $\{L(x, y)\}_{x,y}$ provide this bundle with a flat connection. The property (iii) shows that this flat bundle may be pushed down to Λ.

6.9. Exercises. 1. Let Υ' be an e-surface obtained from an e-surface Υ by adding an arc marked with $(\mathbb{1}, 1)$ or $(\mathbb{1}, -1)$ to the set of distinguished arcs. Show that $\mathcal{T}^e(\Upsilon') = \mathcal{T}^e(\Upsilon)$.

2. Let Υ' be an e-surface obtained from an e-surface Υ by replacing a distinguished arc marked with $(W, -1)$ (where W is an object of $\mathcal{V}$) by the same arc marked with $(W^*, 1)$. Show that $\mathcal{T}^e(\Upsilon') = \mathcal{T}^e(\Upsilon)$.

3. Let Υ' be an e-surface obtained from an e-surface Υ by replacing two distinguished arcs marked with $(V, 1)$ and $(W, 1)$ (where V, W are objects of $\mathcal{V}$) by one arc marked with $(V \otimes W, 1)$. Show that $\mathcal{T}^e(\Upsilon') = \mathcal{T}^e(\Upsilon)$.

4. A weak e-homeomorphism of extended 3-manifolds is a degree 1 homeomorphism of 3-manifolds preserving their additional structures except possibly the Lagrangian space associated to the boundary. A weak e-homeomorphism of extended 3-manifolds $M_1 \to M_2$ restricts to a weak e-homeomorphism $\partial M_1 \to \partial M_2$. Let $g : (M_1, \partial_-(M_1), \partial_+(M_1)) \to (M_2, \partial_-(M_2), \partial_+(M_2))$ be a weak e-homeomorphism of extended 3-cobordisms. Set $N = N(M_2) : H_1(\partial_-(M_2);\mathbb{R}) \Rightarrow H_1(\partial_+(M_2);\mathbb{R})$ and

$$m = \mu(N_*(g|_{\partial_-(M_1)})_*(\lambda_-(M_1)), (g|_{\partial_+(M_1)})_*(\lambda_+(M_1)), \lambda_+(M_2)) - \mu((g|_{\partial_-(M_1)})_*(\lambda_-(M_1)), \lambda_-(M_2), N^*(\lambda_+(M_2))). \tag{6.9.a}$$

Show that the following diagram is commutative:

$$\begin{array}{ccc} \mathcal{T}^e(\partial_-(M_1)) & \xrightarrow{(g|_{\partial_-(M_1)})_\#} & \mathcal{T}^e(\partial_-(M_2)) \\ {\scriptstyle (\mathfrak{D}\Delta^{-1})^m \tau^e(M_1)} \downarrow & & \downarrow {\scriptstyle \tau^e(M_2)} \\ \mathcal{T}^e(\partial_+(M_1)) & \xrightarrow{(g|_{\partial_+(M_1)})_\#} & \mathcal{T}^e(\partial_+(M_2)). \end{array}$$

7. Computations in the renormalized TQFT

7.0. Outline. We compute the gluing anomalies of $(\mathcal{T}^e, \tau^e)$. In fact, the gluing of extended 3-cobordisms may be performed not only along e-homeomorphisms but, more generally, along weak e-homeomorphisms. We compute the anomalies of $(\mathcal{T}^e, \tau^e)$ for these generalized gluings. As an application, we compute the invariant τ for closed oriented 3-manifolds fibered over a circle.

7.1. Theorem. *Let $M = M_2 M_1$ be an extended 3-cobordism obtained from extended 3-cobordisms M_1 and M_2 by gluing along a weak e-homeomorphism $p : \partial_+(M_1) \to \partial_-(M_2)$. For $r = 1, 2$, set $N_r = N(M_r) : H_1(\partial_-(M_r); \mathbb{R}) \Rightarrow H_1(\partial_+(M_r); \mathbb{R})$. Set*

$$n = \mu(p_*(N_1)_*(\lambda_-(M_1)), p_*(\lambda_+(M_1)), N_2^*(\lambda_+(M_2)))$$

and

$$n' = \mu(p_*(\lambda_+(M_1)), \lambda_-(M_2), N_2^*(\lambda_+(M_2))).$$

Then

$$\tau^e(M) = (\mathfrak{D}\Delta^{-1})^{n+n'} \tau^e(M_2)\, p_\#\, \tau^e(M_1).$$

Note that if $p : \partial_+(M_1) \to \partial_-(M_2)$ is an e-homeomorphism then $n' = 0$.

Proof of Theorem. Consider the e-surfaces $\partial_-(M_1), \partial_+(M_1), \partial_+(M_2)$ and fix their parametrizations

$$f_- : \Sigma_- \to \partial_-(M_1), \quad f_0 : \Sigma_0 \to \partial_+(M_1), \quad f_+ : \Sigma_+ \to \partial_+(M_2)$$

by certain standard d-surfaces $\Sigma_-, \Sigma_0, \Sigma_+$. Clearly, $pf_0 : \Sigma_0 \to \partial_-(M_2)$ is a parametrization of $\partial_-(M_2)$. These parametrizations provide M_1, M_2, M with the structure of decorated 3-cobordisms denoted by $\check{M}_1, \check{M}_2, \check{M}$ respectively. Theorem 4.3 implies that $\tau(\check{M}) = (\mathfrak{D}\Delta^{-1})^m \tau(\check{M}_2)\, p_\#\, \tau(\check{M}_1)$ where

$$m = \mu(p_*(N_1)_*(\lambda_-(\check{M}_1)), \lambda_-(\check{M}_2), N_2^*(\lambda_+(\check{M}_2))).$$

The Lagrangian spaces in the last formula are determined by the parametrizations f_-, pf_0, f_+, i.e., $\lambda_-(\check{M}_1) = (f_-)_*(\lambda(\Sigma_-))$, etc. Note that under the identifications $\mathcal{T}(\partial_+(\check{M}_1)) = \mathcal{T}(\Sigma_0)$ and $\mathcal{T}(\partial_-(M_2)) = \mathcal{T}(\Sigma_0)$ determined by f_0 and pf_0 the isomorphism $p_\# : \mathcal{T}(\partial_+(\check{M}_1)) \to \mathcal{T}(\partial_-(M_2))$ corresponds to the identity endomorphism of $\mathcal{T}(\Sigma_0)$.

By definition, $\tau^e(M) = k(\check{M})\,(f_+)_\#\,\tau(\check{M})\,(f_-)_\#^{-1}$. Similarly

$$\tau^e(M_1) = k(\check{M}_1)\,(f_0)_\#\,\tau(\check{M}_1)\,(f_-)_\#^{-1}, \quad \tau^e(M_2) = k(\check{M}_2)\,(f_+)_\#\,\tau(\check{M}_2)\,(pf_0)_\#^{-1}.$$

By (6.3.b), the homomorphism $p_\# : \mathcal{T}^e(\partial_+(M_1)) \to \mathcal{T}^e(\partial_-(M_2))$ is equal to $(\mathcal{D}\Delta^{-1})^{-q}(pf_0)_\#(f_0)_\#^{-1}$ where $q = \mu(p_*(\lambda_+(M_1)), \lambda_-(M_2), \lambda_-(\check{M}_2))$. Therefore

$$\tau^e(M_2)\,p_\#\,\tau^e(M_1) = k(\check{M}_1)\,k(\check{M}_2)\,(\mathcal{D}\Delta^{-1})^{-q}\,(f_+)_\#\,\tau(\check{M}_2)\,\tau(\check{M}_1)\,(f_-)_\#^{-1} =$$

$$= k(\check{M}_1)\,k(\check{M}_2)\,(\mathcal{D}\Delta^{-1})^{-q-m}\,(f_+)_\#\,\tau(\check{M})\,(f_-)_\#^{-1} = (\mathcal{D}\Delta^{-1})^{l}\,\tau^e(M)$$

where

$$l = m_-(\check{M}_1) - m_+(\check{M}_1) + m_-(\check{M}_2) - m_+(\check{M}_2) - m_-(\check{M}) + m_+(\check{M}) - q - m.$$

It remains to show that $l + n + n' = 0$.

Let us first compute the numbers

$$x = m_-(\check{M}_1) - m_-(\check{M}) + n \quad \text{and} \quad y = -m_+(\check{M}_2) + m_+(\check{M}) - m.$$

It is obvious that $\lambda_-(\check{M}) = \lambda_-(\check{M}_1)$ and $\lambda_-(M) = \lambda_-(M_1)$. Applying Lemma 3.7 to the Lagrangian relation N_1 and the Lagrangian spaces

$$\lambda_1 = \lambda_-(M_1),\ \lambda_2 = \lambda_-(\check{M}_1),\ \lambda_1' = \lambda_+(M_1),\ \lambda_2' = p_*^{-1}N_2^*(\lambda_+(M_2))$$

we get

$$x = \mu((N_1)_*(\lambda_-(\check{M}_1)), \lambda_+(M_1), p_*^{-1}N_2^*(\lambda_+(M_2))).$$

Here we use the equalities $N(M) = N_2 p_* N_1$ and $(N(M))^* = N_1^* p^* N_2^*$. Similarly, using the equalities $\lambda_+(\check{M}) = \lambda_+(\check{M}_2)$ and $\lambda_+(M) = \lambda_+(M_2)$, we compute that

$$y = \mu(N_2^*(\lambda_+(M_2)), \lambda_-(\check{M}_2), p_*(N_1)_*(\lambda_-(\check{M}_1))) =$$

$$= \mu(p_*^{-1}N_2^*(\lambda_+(M_2)), \lambda_+(\check{M}_1), (N_1)_*(\lambda_-(\check{M}_1))).$$

The second equality follows from the formula $\lambda_-(\check{M}_2) = p_*(\lambda_+(\check{M}_1))$. Thus, x and y are the Maslov indices of certain Lagrangian spaces in $H_1(\partial_+(M_1);\mathbb{R})$. Applying Lemma 3.6 to the 4-tuple of Lagrangian spaces

$$p_*^{-1}N_2^*(\lambda_+(M_2)),\ \lambda_+(\check{M}_1),\ (N_1)_*(\lambda_-(\check{M}_1)),\ \lambda_+(M_1) \subset H_1(\partial_+(M_1);\mathbb{R})$$

we get

$$x + y - m_+(\check{M}_1) = \mu(p_*^{-1}N_2^*(\lambda_+(M_2)), \lambda_+(\check{M}_1), \lambda_+(M_1)).$$

Applying Lemma 3.6 to the 4-tuple of Lagrangian spaces

$$p_*^{-1}N_2^*(\lambda_+(M_2)),\ \lambda_+(\check{M}_1),\ \lambda_+(M_1),\ p_*^{-1}(\lambda_-(M_2)) \subset H_1(\partial_+(M_1);\mathbb{R})$$

we get that the sum of the last Maslov index with $m_-(\check{M}_2) - q + n'$ is equal to zero. This implies the equality $l + n + n' = 0$.

7.2. Applications to mapping tori. Let Υ be an extended surface and let $g : \Upsilon \to \Upsilon$ be a weak e-homeomorphism. The mapping torus M_g of g is the closed extended 3-manifold obtained from the cylinder $\Upsilon \times [0, 1]$ by gluing the top base to the bottom base along $g : \Upsilon \times 1 \to \Upsilon \times 0$. (The distinguished ribbon graph in M_g consists of colored annuli whose number is equal to the number of orbits of the permutation induced by g in the set of distinguished arcs of Υ.) It is clear that M_g is a fiber bundle over a circle. The following theorem computes $\tau(M_g) = \tau^e(M_g)$ in terms of traces and Maslov indices.

7.2.1. Theorem. *For any weak e-homeomorphism* $g : \Upsilon \to \Upsilon$,

$$\tau(M_g) = (\mathcal{D}\Delta^{-1})^r \operatorname{Tr}(g_\# : \mathcal{T}^e(\Upsilon) \to \mathcal{T}^e(\Upsilon))$$

with $r = \mu(\lambda(g), \lambda(\Upsilon) \oplus g_*(\lambda(\Upsilon)), \operatorname{diag}_H)$ *where* $H = H_1(\Upsilon; \mathbb{R})$, g_* *is the endomorphism of* H *induced by* g, μ *is the Maslov index of the Lagrangian spaces in* $-H \oplus H$, *and* $\lambda(g)$ *is the graph of* g_*, *i.e., the subspace of* $(-H) \oplus H$ *consisting of vectors* $x \oplus g_*(x)$ *with* $x \in H$.

For the definition of diag_H, see Section 3.3.

Note that the extended 3-manifold M_g and its invariant $\tau(M_g)$ do not depend on the Lagrangian space $\lambda(\Upsilon)$. However, the module $\mathcal{T}^e(\Upsilon)$ and thus the homomorphism $g_\#$ depend on the choice of $\lambda(\Upsilon)$. Different choices lead to homomorphisms which are conjugate up to a power of $\mathcal{D}\Delta^{-1}$. This explains the appearance of the factor $(\mathcal{D}\Delta^{-1})^r$ in Theorem 7.2.1.

In the case when g acts as the identity in $H_1(\Upsilon; \mathbb{R})$ we have $\lambda(g) = \operatorname{diag}_H$ so that $r = 0$ and $\tau(M_g) = \operatorname{Tr}(g_\#)$. Applying this formula to $g = \operatorname{id}_\Upsilon$ we get $\tau(\Upsilon \times S^1) = \operatorname{Dim}(\mathcal{T}^e(\Upsilon))$. In Section 12 we shall explicitly compute $\operatorname{Dim}(\mathcal{T}^e(\Upsilon))$ in terms of the genus of Υ and labels of the distinguished arcs.

Proof of Theorem. Set $P = \mathcal{T}^e(\Upsilon)$ and $Q = \mathcal{T}^e(-\Upsilon)$. The cylinder $\Upsilon \times [0, 1]$ gives rise to two extended 3-cobordisms: $M_1^\Upsilon = (\Upsilon \times [0, 1], \emptyset, (-\Upsilon \times 0) \amalg (\Upsilon \times 1))$ and $M_2^\Upsilon = (\Upsilon \times [0, 1], (\Upsilon \times 0) \amalg (-\Upsilon \times 1), \emptyset)$. Denote the operators $\tau^e(M_1^\Upsilon) : K \to Q \otimes_K P$ and $\tau^e(M_2^\Upsilon) : P \otimes_K Q \to K$ by b^Υ and d^Υ respectively. The same argument as in Section III.2.3 shows that these two operators satisfy the conditions of Lemma III.2.2. Moreover, in the notation of that lemma $k = k' = 1$. For, consider the Maslov indices n and n' from Theorem 7.1 corresponding to the gluing in Figure III.2.1. Since this gluing is performed along an (identity) e-homeomorphism, $n' = 0$. By Theorem 7.1, we have $n = \mu(\lambda_1, \lambda_2, \lambda_3)$ where

$$\lambda_1 = \lambda(\Upsilon) \oplus \operatorname{diag}_H, \ \lambda_2 = \lambda(\Upsilon) \oplus \lambda(\Upsilon) \oplus \lambda(\Upsilon), \ \lambda_3 = \operatorname{diag}_{-H} \oplus \lambda(\Upsilon)$$

are Lagrangian spaces in $H \oplus (-H) \oplus H$. These Lagrangian spaces satisfy the formula $\lambda_2 \subset (\lambda_1 \cap \lambda_2) + (\lambda_3 \cap \lambda_2)$ which ensures vanishing of the Maslov index. Therefore $n = n' = 0$. Thus, b^{Υ} and d^{Υ} satisfy the conditions of Lemma III.2.2 with $k = k' = 1$. This allows us to use these operators to compute the trace of any K-linear homomorphism $G : P \to P$. Namely,

$$\operatorname{Tr}(G) = d^{\Upsilon} \operatorname{Perm}_{Q,P} (\mathrm{id}_Q \otimes G)\, b^{\Upsilon} \in \operatorname{End}_K(K) = K$$

where $\operatorname{Perm}_{Q,P}$ is the flip $q \otimes p \to p \otimes q : Q \otimes_K P \to P \otimes_K Q$ (cf. Lemma III.2.6). As in Section III.2.3 we have $d^{\Upsilon} \operatorname{Perm}_{Q,P} = d^{-\Upsilon}$. Applying these formulas to $G = g_{\#}$ we get $\operatorname{Tr}(g_{\#}) = d^{-\Upsilon} (\mathrm{id}_Q \otimes g_{\#})\, b^{\Upsilon}$.

Now we may compute $\tau(M_g) = \tau^e(M_g)$. The mapping torus M_g may be obtained by gluing M_1^{Υ} to $M_2^{-\Upsilon}$ along the weak e-homeomorphism of their bases

$$\mathrm{id}_{-\Upsilon} \amalg g : (-\Upsilon \times 0) \amalg (\Upsilon \times 1) \to (-\Upsilon \times 0) \amalg (\Upsilon \times 1).$$

By Theorem 7.1,

$$\tau^e(M_g) = (\mathcal{D}\Delta^{-1})^r\, \tau^e(M_2^{-\Upsilon})\, (\mathrm{id}_{-\Upsilon} \amalg g)_{\#}\, \tau^e(M_1^{\Upsilon}) =$$

$$= (\mathcal{D}\Delta^{-1})^r\, d^{-\Upsilon}\, (\mathrm{id}_Q \otimes g_{\#})\, b^{\Upsilon} = (\mathcal{D}\Delta^{-1})^r \operatorname{Tr}(g_{\#})$$

where

$$r = \mu(\lambda(g), \lambda(\Upsilon) \oplus g_*(\lambda(\Upsilon)), \mathrm{diag}_H) + \mu(\lambda(\Upsilon) \oplus g_*(\lambda(\Upsilon)), \lambda(\Upsilon) \oplus \lambda(\Upsilon), \mathrm{diag}_H).$$

The same criterion for the vanishing of Maslov indices as above ensures that the second Maslov index is equal to 0. This gives the formula for r specified in the statement of the theorem.

7.3. Exercise. Show that the self-duality pairing $d_{\Upsilon} : \mathcal{T}^e(\Upsilon) \otimes_K \mathcal{T}^e(-\Upsilon) \to K$ defined as in Section III.2 is natural with respect to weak e-homeomorphisms of e-surfaces. (Hint: use Lemma 6.7.2 and Theorem 7.1.)

8. Absolute anomaly-free TQFT

8.0. Outline. In this section, we turn aside from our quest for a 3-dimensional anomaly-free TQFT derived from $\mathcal{V}$ and instead consider an anomaly-free TQFT obtained by tensor multiplication of $(\mathcal{T}^e_{\mathcal{V}}, \tau^e_{\mathcal{V}})$ and $(\mathcal{T}^e_{\overline{\mathcal{V}}}, \tau^e_{\overline{\mathcal{V}}})$. The anomalies of these two TQFT's cancel each other which gives an anomaly-free TQFT. Moreover, using weak e-homeomorphisms we can get rid of distinguished Lagrangian spaces in homologies. This yields an anomaly-free TQFT $(\widehat{\mathcal{T}}, \widehat{\tau})$ assigning projective K-modules to closed oriented surfaces with marked arcs and assigning homomorphisms of these modules to compact oriented 3-cobordisms with embedded v-colored ribbon graphs. In contrast to previously considered 3-dimensional

TQFT's this one involves neither parametrizations of surfaces nor Lagrangian spaces in homologies.

8.1. Absolute square. Recall the mirror modular category $\overline{\mathcal{V}}$ defined in Sections I.1.4 and II.1.9.2. As the rank of $\overline{\mathcal{V}}$, we take the rank $\mathcal{D}$ of $\mathcal{V}$. Since the underlying categories of $\overline{\mathcal{V}}$ and $\mathcal{V}$ are the same, the structures of extended surfaces and extended 3-manifolds over $\overline{\mathcal{V}}$ and $\mathcal{V}$ coincide. Therefore the TQFT's $(\mathcal{T}^e_{\mathcal{V}}, \tau^e_{\mathcal{V}})$ and $(\mathcal{T}^e_{\overline{\mathcal{V}}}, \tau^e_{\overline{\mathcal{V}}})$ are based on the same cobordism theory $(\mathfrak{B}^e, \mathfrak{A}^e)$. Equality (II.2.4.a) implies that

$$\mathcal{D}\Delta_{\mathcal{V}}^{-1} = (\mathcal{D}\Delta_{\overline{\mathcal{V}}}^{-1})^{-1}.$$

Therefore the anomalies of these two TQFT's are mutually inverse. Hence, their tensor product (in the sense of Section III.1.4)

$$(\mathcal{T}^s, \tau^s) = (\mathcal{T}^e_{\mathcal{V}}, \tau^e_{\mathcal{V}}) \otimes (\mathcal{T}^e_{\overline{\mathcal{V}}}, \tau^e_{\overline{\mathcal{V}}})$$

is an anomaly-free TQFT based on $(\mathfrak{B}^e, \mathfrak{A}^e)$. It may be viewed as a kind of absolute square of $(\mathcal{T}^e_{\mathcal{V}}, \tau^e_{\mathcal{V}})$.

Weak e-homeomorphisms act in $(\mathcal{T}^s, \tau^s)$ as follows. For any weak e-homeomorphism of e-surfaces $g : \mathrm{Y}_1 \to \mathrm{Y}_2$, we define its action $\tilde{g} : \mathcal{T}^s(\mathrm{Y}_1) \to \mathcal{T}^s(\mathrm{Y}_2)$ to be the tensor product of $g_\# : \mathcal{T}^e_{\mathcal{V}}(\mathrm{Y}_1) \to \mathcal{T}^e_{\mathcal{V}}(\mathrm{Y}_2)$ and $g_\# : \mathcal{T}^e_{\overline{\mathcal{V}}}(\mathrm{Y}_1) \to \mathcal{T}^e_{\overline{\mathcal{V}}}(\mathrm{Y}_2)$, for the definition of the last homomorphisms, see Section 6.3. It follows from Lemma 6.3.2 that the formula $g \mapsto \tilde{g}$ transforms composition into composition without any additional factors. Similarly, Theorem 7.1 implies that the TQFT $(\mathcal{T}^s, \tau^s)$ produces no anomalies when we glue extended 3-manifolds along weak e-homeomorphisms.

8.2. Elimination of Lagrangian spaces. The TQFT $(\mathcal{T}^s, \tau^s)$ defined in Section 8.1 is based on the cobordism theory $(\mathfrak{B}^e, \mathfrak{A}^e)$ formed by e-surfaces and extended 3-manifolds. It turns out that we may forget about Lagrangian spaces in homologies of surfaces. More precisely, the TQFT $(\mathcal{T}^s, \tau^s)$ induces another TQFT $(\widehat{\mathcal{T}}, \widehat{\tau}) = (\widehat{\mathcal{T}}_{\mathcal{V}}, \widehat{\tau}_{\mathcal{V}})$ based on space-structures which differ from the structures of e-surfaces and extended 3-manifolds in that we omit all mention to Lagrangian spaces.

We first define the cobordism theory $(\widehat{\mathfrak{B}}, \widehat{\mathfrak{A}})$ underlying $(\widehat{\mathcal{T}}, \widehat{\tau})$. The space-structures $\widehat{\mathfrak{A}}$ and $\widehat{\mathfrak{B}}$ are defined in the same way as $\mathfrak{A}^e$, $\mathfrak{B}^e$ (see Sections 6.1, 6.2) with the difference that all references to Lagrangian spaces should be omitted. Thus, an $\widehat{\mathfrak{A}}$-space is a closed oriented surface equipped with a finite family of disjoint marked arcs. A $\widehat{\mathfrak{B}}$-space is a compact oriented 3-manifold equipped with a v-colored ribbon graph over $\mathcal{V}$. (For the definition of a ribbon graph in a 3-manifold, see Section 1.5. Note that the families of arcs on surfaces as well as ribbon graphs in 3-manifolds may be empty.)

Now we define the modular functor $\widehat{\mathcal{T}}$ based on $\widehat{\mathfrak{A}}$. Let Σ be a closed oriented surface equipped with a finite family of disjoint marked arcs. For any Lagrangian spaces $\lambda, \lambda' \in \Lambda(H_1(\Sigma;\mathbb{R}))$, the identity mapping $(\Sigma, \lambda) \to (\Sigma, \lambda')$ is a weak e-homeomorphism of e-surfaces inducing an isomorphism $\mathcal{T}^s(\Sigma, \lambda) \to \mathcal{T}^s(\Sigma, \lambda')$. These isomorphisms form a commuting system. Identifying the modules $\{\mathcal{T}^s(\Sigma, \lambda) \mid \lambda \in \Lambda(H_1(\Sigma;\mathbb{R}))\}$ along these isomorphisms we get a K-module $\widehat{\mathcal{T}}(\Sigma)$ independent of the choice of Lagrangian space but canonically isomorphic to any one of the modules $\mathcal{T}^s(\Sigma, \lambda)$. The action of an $\widehat{\mathfrak{A}}$-homeomorphism $g : \Sigma_1 \to \Sigma_2$ is defined in the obvious way: choose arbitrary $\lambda_1 \in \Lambda(H_1(\Sigma_1;\mathbb{R}))$, $\lambda_2 \in \Lambda(H_1(\Sigma_2;\mathbb{R}))$ and define $g_\# : \widehat{\mathcal{T}}(\Sigma_1) \to \widehat{\mathcal{T}}(\Sigma_2)$ to be the composition

$$\widehat{\mathcal{T}}(\Sigma_1) \xrightarrow{\approx} \mathcal{T}^s(\Sigma_1, \lambda_1) \xrightarrow{\approx} \mathcal{T}^s(\Sigma_2, \lambda_2) \xrightarrow{\approx} \widehat{\mathcal{T}}(\Sigma_2)$$

where the first and third mappings are the canonical isomorphisms and the second mapping is the isomorphism induced by the weak e-homeomorphism $g : (\Sigma_1, \lambda_1) \to (\Sigma_2, \lambda_2)$. This composition does not depend on the choice of the Lagrangian spaces λ_1, λ_2. It is easy to check that $\widehat{\mathcal{T}}$ is a modular functor based on closed oriented surfaces with marked arcs.

The operator invariant $\widehat{\tau}$ of a 3-cobordism M is defined in a similar way: choose Lagrangian spaces in the 1-homologies of the bases, consider the operator invariant $\tau^s(M) : \mathcal{T}^s(\partial_- M) \to \mathcal{T}^s(\partial_+ M)$, and consider the induced homomorphism $\widehat{\mathcal{T}}(\partial_- M) \to \widehat{\mathcal{T}}(\partial_+ M)$. It follows from Lemma 6.7.2 and Theorem 7.1 that this homomorphism does not depend on the choice of Lagrangian spaces.

8.3. Theorem. *$(\widehat{\mathcal{T}}, \widehat{\tau})$ is an anomaly-free TQFT based on closed oriented surfaces with marked arcs and compact oriented 3-cobordisms with $\mathcal{V}$-colored ribbon graphs.*

This theorem is essentially obvious. The axioms of TQFT's and anomaly-freeness follow directly from the corresponding properties of $(\mathcal{T}^s, \tau^s)$ and Theorem 7.1. We call $(\widehat{\mathcal{T}}, \widehat{\tau})$ the absolute anomaly-free TQFT derived from $\mathcal{V}$.

The passage from $(\mathcal{T}^e, \tau^e)$ to $(\widehat{\mathcal{T}}, \widehat{\tau})$ simplifies the theory by eliminating Lagrangian spaces and anomalies. On the other hand, we loose some information when we take the tensor product of two TQFT's. For instance, the invariant $\widehat{\tau}$ does not distinguish a closed oriented 3-manifold M from the same manifold with the opposite orientation $-M$ whilst τ^e may distinguish these manifolds. Nevertheless, the TQFT $(\widehat{\mathcal{T}}, \widehat{\tau})$ is non-trivial even if we restrict it to closed oriented surfaces with empty systems of arcs and compact oriented 3-cobordisms with empty ribbon graphs. This restricted TQFT will be studied further in Chapter VII. It will be computed in terms of state models on triangulations of 3-manifolds. For the more powerful TQFT $(\mathcal{T}^e, \tau^e)$, such a computation is not known.

Note finally that for any closed oriented 3-manifold M with a v-colored ribbon graph Ω inside, we have

$$\widehat{\tau}(M,\Omega)=\tau^{s}(M,\Omega)=\tau_{\mathcal{V}}(M,\Omega)\,\tau_{\overline{\mathcal{V}}}(M,\Omega)=\tau_{\mathcal{V}}(M,\Omega)\,\tau_{\mathcal{V}}(-M,\Omega).$$

9. Anomaly-free TQFT

9.0. Outline. We show how to kill the gluing anomalies of $(\mathcal{T}^e,\tau^e)$. The idea is to incorporate the anomalies into an additional structure on the 3-manifolds. This additional structure is used to renormalize the TQFT via multiplication of the operator invariants of cobordisms by certain scalar factors. (We do not change the modular functor.) This approach follows the general lines of Section III.6; for the convenience of the reader, we reproduce the necessary definitions adapted to the 3-dimensional setting. The resulting 3-dimensional TQFT $(\mathcal{T}^e,\tau^w)$ is anomaly-free and non-degenerate. This TQFT is more fundamental than the one constructed in Section 8, but we pay the price of a slightly more complicated ground structure on the 3-manifolds.

9.1. Weighted extended 3-manifolds. A weighted extended 3-manifold is a pair (M,m) where M is an extended 3-manifold in the sense of Section 6.2 and m is an integer satisfying the only condition that if $M=\emptyset$ then $m=0$. We say that M is the underlying extended 3-manifold of (M,m) and m is its (additive) weight. The disjoint union of weighted extended 3-manifolds is defined in the obvious way, the weight being additive under disjoint union. By an e-homeomorphism of weighted extended 3-manifolds with equal weights we mean an e-homeomorphism of the underlying extended 3-manifolds.

For any e-surface Υ, the cylinder $\Upsilon\times[0,1]$ acquires the structure of an extended 3-manifold as described in Section 6.2. Ascribing to this cylinder the zero weight we get the weighted extended cylinder over Υ.

It is obvious that the structure of weighted extended 3-manifolds is a space-structure compatible with disjoint union. Formally speaking, the corresponding functor $\mathfrak{B}^w$ is defined by the formula $\mathfrak{B}^w(X)=\mathfrak{B}^e(X)\times\mathbb{Z}$ if $X\neq\emptyset$ and $\mathfrak{B}^w(\emptyset)$ is a one-element set. It is clear that a $\mathfrak{B}^w$-space is just a weighted extended 3-manifold. A $\mathfrak{B}^w$-homeomorphism of $\mathfrak{B}^w$-spaces is an e-homeomorphism of weighted extended 3-manifolds with equal weights.

The space-structures $\mathfrak{B}^w$ and $\mathfrak{A}^e$ form a cobordism theory. The boundary of a weighted extended 3-manifold is the usual boundary with its structure of an e-surface. The gluing of weighted extended 3-manifolds is defined as follows. Let (M,m) be a weighted extended 3-manifold whose boundary is the disjoint union of e-surfaces X,Y,Z such that X is e-homeomorphic to $-Y$. Denote by R

the Lagrangian relation

$$-H_1(Z;\mathbb{R}) \Rightarrow H_1(X \amalg Y;\mathbb{R}) = H_1(X;\mathbb{R}) \oplus H_1(Y;\mathbb{R})$$

determined by the cobordism $(M, -Z, X \amalg Y)$. Denote by M' the extended 3-manifold obtained from M by gluing X to Y along an e-homeomorphism $f : X \to -Y$. We define the weighted extended 3-manifold obtained from (M, m) by gluing X to Y along f to be the pair

$$(M',\ m - \mu(\lambda(-f), \lambda(X) \oplus \lambda(Y), R_*(\lambda(-Z))))$$

where $\lambda(-f) \subset H_1(X;\mathbb{R}) \oplus H_1(Y;\mathbb{R})$ is the graph of the symplectic isomorphism $(-f)_* : -H_1(X;\mathbb{R}) \to H_1(Y;\mathbb{R})$ and

$$\lambda(X) \subset H_1(X;\mathbb{R}),\ \lambda(Y) \subset H_1(Y;\mathbb{R}),\ \lambda(-Z) \subset -H_1(Z;\mathbb{R})$$

are the Lagrangian spaces determined by the structure of the e-surfaces on $X, Y, -Z$. (For the definition of the graph of a symplectic isomorphism, see Section 3.3.) Here μ denotes the Maslov index of Lagrangian spaces in $H_1(X;\mathbb{R}) \oplus H_1(Y;\mathbb{R})$.

9.1.1. Lemma. *The space-structures $\mathfrak{B}^w$ and $\mathfrak{A}^e$ form a cobordism theory in the sense of Section* III.1.3.

Proof. The only non-trivial point is that the successive gluing along disjoint closed subsurfaces of ∂M gives rise to the same weight as the gluing along their union. This follows from the formula (3.7.c). More exactly, let M, m, X, Y, Z, f, R be the same objects as above. Assume that X is a disjoint union of e-surfaces X_1 and X_2. For $r = 1, 2$, set $Y_r = f(X_r)$ and $f_r = f|_{X_r} : X_r \to -Y_r$. The desired equality of weights follows from elementary homological computations and formula (3.7.c) applied to the non-degenerate symplectic spaces

$$H = -H_1(X_1 \amalg Y_1;\mathbb{R}),\quad H' = H_1(X_2 \amalg Y_2;\mathbb{R}),$$

the Lagrangian relation $N = R_*(\lambda(-Z)) : H \Rightarrow H'$, and the Lagrangian spaces

$$\lambda_1 = \lambda(f_1) \subset H,\quad \lambda_2 = \lambda(-X_1) \oplus \lambda(-Y_1) \subset H,$$

$$\lambda_1' = \lambda(X_2) \oplus \lambda(Y_2) \subset H',\quad \lambda_2' = \lambda(-f_2) \subset H'.$$

Using (3.7.c) we should keep in mind that the three Maslov indices in this formula apply to Lagrangian spaces in H, H', and $(-H) \oplus H'$ respectively.

9.2. Renormalized TQFT. We define a TQFT $(\mathcal{T}^e, \tau^w) = (\mathcal{T}^e_{\mathcal{V}}, \tau^w_{\mathcal{V}})$ based on the cobordism theory $(\mathfrak{B}^w, \mathfrak{A}^e)$. Here $\mathcal{T}^e$ is the modular functor introduced in Section 6 based on e-surfaces. The operator invariant τ^w of a weighted extended 3-cobordism $(M, m) = (M, \partial_- M, \partial_+ M, m)$ is defined by the formula

$$\tau^w(M, m) = (\mathcal{D}\Delta^{-1})^{-m} \tau^e(M) : \mathcal{T}^e(\partial_- M) \to \mathcal{T}^e(\partial_+ M).$$

9.2.1. Theorem. *$(\mathcal{T}^e, \tau^w)$ is an anomaly-free non-degenerate TQFT based on extended surfaces and weighted extended 3-manifolds.*

Proof. We already know that $\mathcal{T}^e$ is a modular functor, so we only have to check axioms (III.1.4.1) - (III.1.4.4). Verification of the naturality, multiplicativity, and normalization axioms is straightforward. Let us check the functoriality axiom. Let (M, m) be a weighted extended 3-cobordism obtained from weighted extended 3-cobordisms (M_1, m_1) and (M_2, m_2) by gluing along an e-homeomorphism $p: \partial_+(M_1) \to \partial_-(M_2)$. Here the role of M, X, and Y in the definition of gluing given above is played by $M_1 \amalg M_2$, $\partial_+(M_1)$, and $-\partial_-(M_2)$ respectively. The definition of τ^w and Theorem 7.1 imply that

$$\tau^w(M, m) = (\mathcal{D}\Delta^{-1})^{-m}\tau^e(M) = (\mathcal{D}\Delta^{-1})^{n+n'-m}\tau^e(M_2)\, p_\#\, \tau^e(M_1) =$$

$$= (\mathcal{D}\Delta^{-1})^{n+n'+m_1+m_2-m}\tau^w(M_2, m_2)\, p_\#\, \tau^w(M_1, m_1)$$

where n, n' are the integers specified in the statement of Theorem 7.1. To complete the proof it remains to show that $m = n + n' + m_1 + m_2$.

For $r = 1, 2$, set $N_r = N(M_r) : H_1(\partial_-(M_r); \mathbb{R}) \Rightarrow H_1(\partial_+(M_r); \mathbb{R})$. Set

$$H = -H_1(\partial_+(M_1); \mathbb{R}), \quad H' = -H_1(\partial_-(M_2); \mathbb{R}).$$

From the definition of gluing of weighted 3-manifolds we have $m = m_1 + m_2 - m'$ where

$$m' = \mu(\lambda(-p),\ \lambda_+(M_1) \oplus \lambda_-(M_2),\ (N_1)_*(\lambda_-(M_1)) \oplus N_2^*(\lambda_+(M_2))).$$

Here μ denotes the Maslov index of Lagrangian spaces in the symplectic space $(-H) \oplus H'$ and $\lambda(-p)$ denotes the graph of the symplectic isomorphism $(-p)_* : H \to H'$. Warning: we use here (as in Section 3) the same symbol for a Lagrangian subspace of a symplectic vector space and for the same Lagrangian space regarded as a subspace of the opposite symplectic vector space. In the expression for m' the Lagrangian spaces $\lambda_-(M_2), N_2^*(\lambda_+(M_2))$ are regarded as subspaces of $H' = -H_1(\partial_-(M_2); \mathbb{R})$.

To compute m' we use formula (3.7.c) with

$$N = \lambda(-p) \subset (-H) \oplus H',\ \lambda_1 = (N_1)_*(\lambda_-(M_1)) \subset H,$$

$$\lambda_2 = \lambda_+(M_1) \subset H,\ \lambda_1' = \lambda_-(M_2) \subset H',\ \lambda_2' = N_2^*(\lambda_+(M_2)) \subset H'.$$

We obtain

$$m' = \mu(\lambda_2 \oplus \lambda_1', \lambda_1 \oplus \lambda_2', N) = \mu(\lambda_1, \lambda_2, (-p)_*^{-1}(\lambda_1')) + \mu((-p)_*(\lambda_1), \lambda_1', \lambda_2') =$$

$$= \mu((-p)_*(\lambda_1), (-p)_*(\lambda_2), \lambda_1') + \mu((-p)_*(\lambda_1), \lambda_1', \lambda_2') =$$

$$= \mu((-p)_*(\lambda_1), (-p)_*(\lambda_2), \lambda_2') + \mu((-p)_*(\lambda_2), \lambda_1', \lambda_2').$$

(The last equality follows from Lemma 3.6.) The last two Maslov indices are exactly the Maslov indices used in Theorem 7.1 to define n and n'. However, we work here in the symplectic space $H' = -H_1(\partial_-(M_2); \mathbb{R})$ opposite to the one used in Theorem 7.1. Therefore $m' = -n - n'$, so that $m = m_1 + m_2 + n + n'$.

9.2.2. Corollary. *Considered up to isomorphism, the TQFT $(\mathcal{T}^e, \tau^w)$ does not depend on the choice of d-homeomorphisms $\{\mathrm{rev}_t : \Sigma_{-t} \to -\Sigma_t\}_t$ (see Section 1.2).*

Indeed, different choices lead to TQFT's based on the same cobordism theory and coinciding on closed oriented 3-manifolds with v-colored ribbon graphs. Theorems 9.2.1 and III.3.3 imply that these TQFT's are isomorphic. As an exercise the reader may give a direct construction of an isomorphism between these TQFT's.

9.3. Summary. The construction of the TQFT $(\mathcal{T}^e, \tau^w)$ is one of the major achievements of this monograph. It crowns our efforts towards a 3-dimensional topological quantum field theory. From any modular category $\mathcal{V}$ we derive a non-degenerate anomaly-free 3-dimensional TQFT. Our construction involves four major steps: (a) a construction of isotopy invariants of ribbon graphs in $\mathbb{R}^3$ (see Chapter I), (b) a definition of the topological invariant τ of closed 3-manifolds with embedded v-colored ribbon graphs (see Chapter II), (c) a definition of the TQFT $(\mathcal{T}, \tau)$ (see Sections 1 and 2), (d) a renormalization of this TQFT. In contrast to the constructions of Section 8 we loose no information under the passage from $(\mathcal{T}, \tau)$ to $(\mathcal{T}^e, \tau^w)$: up to technical though important modifications these TQFT's are equivalent.

The category $\mathcal{V}$ is used in the construction of $(\mathcal{T}^e, \tau^w)$ at every stage. This category also appears in the very definition of extended surfaces and weighted extended 3-manifolds. The objects of $\mathcal{V}$ are used as labels (or colors) of distinguished arcs on surfaces and bands and annuli of ribbon graphs. The morphisms of $\mathcal{V}$ are used as colors of coupons of ribbon graphs.

We may restrict $(\mathcal{T}^e, \tau^w)$ to surfaces with an empty set of distinguished arcs and to 3-cobordisms with empty ribbon graphs. This gives an anomaly-free TQFT whose domain of definition does not depend on $\mathcal{V}$. Generally speaking, though, even this restricted TQFT depends on the choice of $\mathcal{V}$. For instance, the dimension of the module of states of a closed surface strongly depends on this choice (see Section 12).

We may apply the results of Chapter III to $(\mathcal{T}^e, \tau^w)$. In particular, the invariant τ^w of weighted closed oriented 3-manifolds with v-colored ribbon graphs is a quantum invariant in the sense of Section III.4. This is essentially the K-valued invariant defined in Chapter II. Namely, if (M, m) is a weighted extended 3-manifold with empty boundary then $\tau^w(M, m) = (\mathcal{D}\Delta^{-1})^{-m}\tau(M)$. By Theorem III.4.4, this invariant determines the TQFT $(\mathcal{T}^e, \tau^w)$ up to isomorphism.

9.4. Remark. The definition of gluing of 3-manifolds given in Section 9.1 and its properties established in Lemma 9.1.1 extend to weak e-homeomorphisms word for word. The computation of gluing anomalies given in the proof of Theorem 9.2.1 also extends to weak e-homeomorphisms without any changes. These observations may be useful for concrete computations.

10. Hermitian TQFT

Throughout this section, $\mathcal{V}$ is a Hermitian modular category with conjugation $f \to \overline{f}$, ground ring K, and rank $\mathcal{D} \in K$ such that $\overline{\mathcal{D}} = \mathcal{D}$ (see Section II.5).

10.0. Outline. The TQFT $(\mathcal{T}^e, \tau^w)$ derived from the Hermitian modular category $\mathcal{V}$ is shown to be Hermitian. We compute explicitly the Hermitian form on the module of states of a surface and show that it is invariant under the action of the modular groupoid defined in Section 5.2.

10.1. Negation. In order to speak of Hermitian structures on $(\mathcal{T}^e, \tau^w)$ we need a negation for weighted extended 3-manifolds. Let (M, Ω, m) be a weighed extended 3-manifold where M is a compact oriented 3-manifold with extended boundary, Ω is a v-colored ribbon graph in M, and $m \in \mathbb{Z}$. We define the opposite weighted extended 3-manifold to be $(-M, -\Omega, -m)$ where $-M$ is the manifold M with the opposite orientation and the opposite extended structure on the boundary, $-\Omega$ is obtained from Ω by the transformations (T1) and (T2) introduced in Section II.5.1. It is easy to see that $-\Omega$ is a v-colored ribbon graph in $-M$ so that the triple $(-M, -\Omega, -m)$ is a weighted extended 3-manifold. In this way the structure of weighted extended 3-manifolds $\mathfrak{B}^w = \mathfrak{B}^w_{\mathcal{V}}$ becomes an involutive space-structure. Since $\partial(-M) = -\partial M$, the cobordism theory $(\mathfrak{B}^w, \mathfrak{A}^e)$ is involutive. Note that the conjugation $f \to \overline{f}$ in $\mathcal{V}$ is used in the definition of the transformation (T2).

10.2. Theorem. *The TQFT $(\mathcal{T}^e, \tau^w) = (\mathcal{T}^e_{\mathcal{V}}, \tau^w_{\mathcal{V}})$ derived from the Hermitian modular category $\mathcal{V}$ admits a unique Hermitian structure.*

Proof. As we know, the TQFT $(\mathcal{T}^e, \tau^w)$ is non-degenerate and anomaly-free so that we may apply Theorem III.5.3. This theorem guarantees that the Hermitian structure in question is unique and that its existence would follow from the identity $\tau^w(-M, -\Omega, -m) = \overline{\tau^w(M, \Omega, m)}$ where (M, Ω, m) runs over weighted extended 3-manifolds with empty boundary. It follows from formula (II.5.6.a) that

(10.2.a) $$\overline{\mathcal{D}\Delta^{-1}} = (\mathcal{D}\Delta^{-1})^{-1}.$$

Since $\tau^w(M, \Omega, m) = (\mathcal{D}\Delta^{-1})^{-m}\tau(M, \Omega)$, the desired identity follows from Theorem II.5.4.

10.3. Naturality of the Hermitian form. For every e-surface Υ, Theorem 10.2 provides the module $\mathcal{T}^e(\Upsilon)$ with a Hermitian form that is invariant under e-homeomorphisms of e-surfaces. The next theorem asserts that, more generally, this form is invariant under weak e-homeomorphisms.

10.3.1. Theorem. *For any weak e-homeomorphism of e-surfaces $g : \Upsilon \to \Upsilon'$, the induced isomorphism $g_\# : \mathcal{T}^e(\Upsilon) \to \mathcal{T}^e(\Upsilon')$ preserves the Hermitian form.*

Proof. Every weak e-homeomorphism may be presented as a composition of an e-homeomorphism and a weak e-homeomorphism which is the identity as a mapping. Since e-homeomorphisms induce isometries of the Hermitian forms, it is enough to consider the case where Υ and Υ' have the same underlying topological surface with marked arcs, say Σ, and $g = \mathrm{id}_\Sigma : \Upsilon \to \Upsilon'$. (The e-surfaces Υ, Υ' may differ by the distinguished Lagrangian spaces $\lambda(\Upsilon), \lambda(\Upsilon') \subset H_1(\Sigma; \mathbb{R})$.)

Denote by $C(\Upsilon, \Upsilon')$ the cylinder $\Sigma \times [0, 1]$ whose boundary is provided with the structure of an e-surface via the identifications $\Sigma \times 0 = -\Upsilon$ and $\Sigma \times 1 = \Upsilon'$. It is clear that $(C(\Upsilon, \Upsilon'), \Upsilon, \Upsilon')$ is an extended 3-cobordism between Υ and Υ'. By Lemma 6.7.2,

$$\tau(C(\Upsilon, \Upsilon'), \Upsilon, \Upsilon') = g_\# : \mathcal{T}^e(\Upsilon) \to \mathcal{T}^e(\Upsilon').$$

Exchanging the roles of Υ and Υ' we get

$$\tau(C(\Upsilon', \Upsilon), \Upsilon', \Upsilon) = (g^{-1})_\# = (g_\#)^{-1} : \mathcal{T}^e(\Upsilon') \to \mathcal{T}^e(\Upsilon)$$

where the equality $(g^{-1})_\# = (g_\#)^{-1}$ follows from Lemma 6.3.2.

It is straightforward to verify that the mapping $(a, t) \mapsto (a, 1-t) : \Sigma \times [0, 1] \to \Sigma \times [0, 1]$ induces an e-homeomorphism

$$(-C(\Upsilon, \Upsilon'), \Upsilon', \Upsilon) \approx (C(\Upsilon', \Upsilon), \Upsilon', \Upsilon)$$

that extends the identity homeomorphisms of the bases. Therefore

$$\tau(-C(\Upsilon, \Upsilon'), \Upsilon', \Upsilon) = \tau(C(\Upsilon', \Upsilon), \Upsilon', \Upsilon) = (g_\#)^{-1}.$$

This formula together with (III.5.2.a) implies that for any $x, y \in \mathcal{T}^e(\Upsilon)$,

$$\langle g_\#(x), g_\#(y)\rangle = \langle \tau(C(\Upsilon, \Upsilon'), \Upsilon, \Upsilon')(x), g_\#(y)\rangle =$$

$$= \langle x, \tau(-C(\Upsilon, \Upsilon'), \Upsilon', \Upsilon) g_\#(y)\rangle = \langle x, y\rangle.$$

10.3.2. Corollary. *The action of the modular groupoid defined in Section 5.2 preserves the Hermitian structure on $\mathcal{T}$ induced by the Hermitian structure on $\mathcal{T}^e$ via the forgetful functor.*

This follows from the previous theorem and Theorem 6.7.1.

10.4. Explicit description of the Hermitian structure. For an e-surface Υ, Theorem 10.2 yields a Hermitian form $\langle .,. \rangle_\Upsilon : \mathcal{T}^e(\Upsilon) \otimes_K \mathcal{T}^e(\Upsilon) \to K$. We compute this Hermitian form explicitly. Since this form is multiplicative with respect to disjoint union we may restrict ourselves to the case of connected Υ. It is clear that Υ is weakly e-homeomorphic to the underlying e-surface of a standard d-surface Σ_t for a certain decorated type $t = (g; (W_1, \nu_1), ..., (W_m, \nu_m))$. Therefore it is enough to consider the case where Υ is the underlying e-surface of Σ_t with the Lagrangian space in 1-homologies $\lambda(\Sigma_t) = \mathrm{Ker}(\mathrm{incl}_* : H_1(\Sigma_t; \mathbb{R}) \to H_1(U_t; \mathbb{R}))$. In this case

$$\mathcal{T}^e(\Upsilon) = \mathcal{T}(\Sigma_t) = \Psi_t = \bigoplus_{i \in I^g} \mathrm{Hom}(\mathbb{1}, \Phi(t; i))$$

where g is the genus of Υ.

10.4.1. Theorem. *The direct summands of Ψ_t are orthogonal with respect to the Hermitian form $\langle .,. \rangle_\Upsilon$. For elements of the same summand $x, y \in Hom(\mathbb{1}, \Phi(t; i))$, we have*

$$\langle x, y \rangle_\Upsilon = \mathcal{D}^{g-1}(\dim(i))^{-1}\mathrm{tr}(x\overline{y}) = \mathcal{D}^{g-1}(\dim(i))^{-1}\mathrm{tr}(\overline{y}x).$$

Proof. Let

$$x \in \mathrm{Hom}(\mathbb{1}, \Phi(t; i)) \subset \mathcal{T}^e(\Upsilon), \quad y \in \mathrm{Hom}(\mathbb{1}, \Phi(t; j)) \subset \mathcal{T}^e(\Upsilon)$$

where $i, j \in I^g$. To compute $\langle x, y \rangle_\Upsilon$ we shall use formula (III.5.3.a). Consider the standard handlebodies $H(U_t, R_t, i, x)$ and $H(U_t, R_t, j, y)$. Endow each of them with the zero numerical weight and the Lagrangian space $\lambda(\Sigma_t)$. Denote the resulting weighted extended 3-manifolds by M and N respectively. It is clear that $\partial M = \partial N = \Upsilon$. Gluing of M to $-N$ along the identity homeomorphism of their boundaries yields a closed weighted extended 3-manifold $M \cup_{\mathrm{id}} -N$. By (III.5.3.a) (cf. also Exercise III.5.4), $\langle x, y \rangle_\Upsilon = \tau^w(M \cup_{\mathrm{id}} -N)$. The numerical weight of $M \cup_{\mathrm{id}} -N$ computed according to Section 9.1 is equal to zero because the corresponding Maslov index vanishes. Therefore

$$\langle x, y \rangle_\Upsilon = \tau^w(M \cup_{\mathrm{id}} -N) = \tau(M \cup_{\mathrm{id}} -N).$$

It should be stressed that $M \cup_{\mathrm{id}} -N$ is a pair (a closed oriented 3-manifold, a distinguished v-colored ribbon graph in this manifold). In particular, by $\tau(M \cup_{\mathrm{id}} -N)$ we mean the invariant τ of this pair. The 3-manifold in question is obtained by gluing two copies of the standard handlebody of genus g along the identity homeomorphism of the boundaries. This 3-manifold splits as a connected sum of g copies of $S^1 \times S^2$. The distinguished ribbon graph in $M \cup_{\mathrm{id}} -N$ is obtained by gluing two copies of R_t. The coloring of its bands is determined by the sequences t, i, j, the two coupons of this ribbon graph are colored with x and $\overline{y}$.

To compute $\tau(M \cup_{\mathrm{id}} -N)$ we shall present $M \cup_{\mathrm{id}} -N$ as the result of surgery on the 3-sphere. Consider the special ribbon (0,0)-graph $\Omega \subset \mathbb{R}^3$ shown in Figure 10.1. This ribbon graph consists of two coupons, m vertical bands, and g fragments each formed by three annuli, a cup-like band, and a cap-like band. The vertical bands are directed and colored, their directions and colors are determined by t in the usual way. The cap-like bands are directed to the left and colored with $i_1, ..., i_g$ where $i = (i_1, ..., i_g)$. The cup-like bands are directed to the right and colored with $j_1, ..., j_g$ where $j = (j_1, ..., j_g)$. The annuli are neither directed nor colored. These annuli form a $3g$-component framed link $L \subset S^3$ which splits as a disjoint union of g three-component framed links.

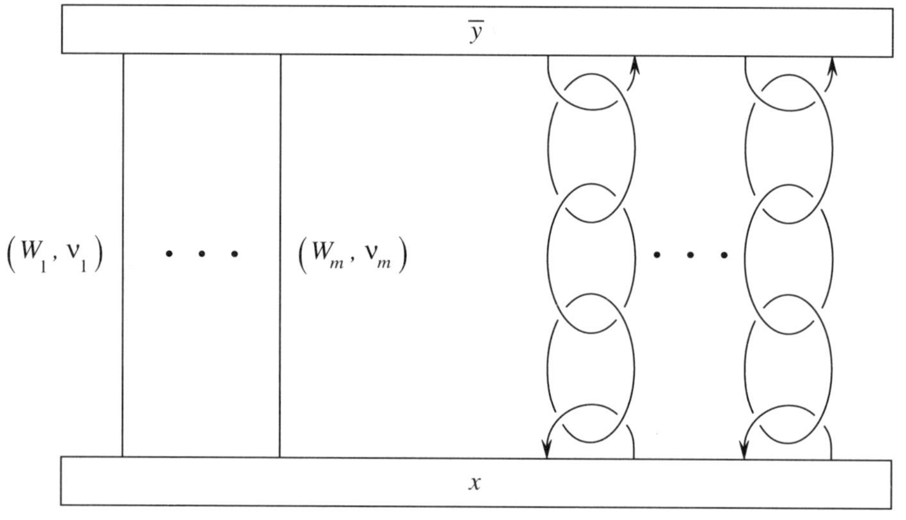

Figure 10.1

We shall show that the ribbon graph Ω presents the closed extended 3-manifold $M \cup_{\mathrm{id}} -N$. This means that surgering S^3 along L while keeping the remaining part of Ω we get $M \cup_{\mathrm{id}} -N$ with its distinguished ribbon graph. This allows us to compute $\tau(M \cup_{\mathrm{id}} -N)$ from the definition of τ given in Section II.2.3. It is easy to check that $\sigma(L) = 0$ so that the normalizing factor in the definition of $\tau(M \cup_{\mathrm{id}} -N)$ is equal to $\mathcal{D}^{-3g-1}$. To compute $\tau(M \cup_{\mathrm{id}} -N)$ we vary the colors of $3g$ components of L and sum up the operator invariants of the resulting v-colored ribbon graphs with the appropriate coefficients. We apply the result of Exercise II.3.10.2 two times to the k-th fragment of Ω where $k = 1, ..., g$. This yields that if $i_k \neq j_k$ then the summation mentioned above gives 0, whereas if $i_k = j_k$ then up to multiplication by $\mathcal{D}^4 \dim(i_k)^{-1}$ we may replace the k-th fragment with a pair of vertical untwisted unlinked bands colored with i_k. Therefore if $i \neq j$ then the summation gives 0 so that $\langle x, y\rangle_Y = \tau(M \cup_{\mathrm{id}} -N) = 0$. If $i = j$ then up to multiplication by $\mathcal{D}^{4g}(\dim(i))^{-1}$ we may replace all g fragments of Ω with vertical untwisted unlinked bands. The operator invariant of the resulting v-colored ribbon graph is equal to $\overline{y}(x)$. Therefore

$$\langle x, y\rangle_\Upsilon = \tau(M \cup_{\mathrm{id}} -N) = \mathcal{D}^{g-1}(\dim(i))^{-1}\overline{y}(x) = \mathcal{D}^{g-1}(\dim(i))^{-1}\mathrm{tr}(\overline{y}x).$$

To prove that Ω presents $M \cup_{\mathrm{id}} -N$ we use the technique of decorated 3-manifolds. We can lift the negation of extended 3-manifolds to a negation of decorated 3-manifolds as follows. For a decorated 3-manifold N, the opposite decorated 3-manifold $-N$ is obtained from N by reversing the orientation in N, negating the structure of parametrized d-surface on ∂N, and applying the transformations (T1) and (T2) to the given v-colored ribbon graph in N. Let us view $M = H(U_t, R_t, i, x)$ and $N = H(U_t, R_t, j, y)$ as decorated 3-manifolds whose boundary is decorated and parametrized via the identity $\mathrm{id} : \Sigma_t \to \Sigma_t$. Recall the special ribbon graph $\Xi(t, i, x)$ presenting the decorated 3-cobordism $(M, \emptyset, \partial M)$ (see Section 2.9). Recall the negation of v-colored ribbon graphs in $\mathbb{R}^3$ defined in Section II.5.1. This negation extends to partially colored ribbon graphs in $\mathbb{R}^3$ word for word. It is straightforward to verify that the ribbon graph $-\Xi(t, j, y)$ is special and presents $(-N, \partial N, \emptyset)$. (More exactly, the special ribbon graph $-\Xi(t, j, y)$ presents a decorated 3-cobordism d-homeomorphic to $(-N, \partial N, \emptyset)$, the d-homeomorphism in question is induced by the mirror reflection of $\mathbb{R}^3$ with respect to the plane $\mathbb{R}^2 \times 1/2 \subset \mathbb{R}^3$.) Putting $-\Xi(t, j, y)$ on the top of $\Xi(t, i, x)$ we get Ω. By the results of Section 2.8 (the case of connected bases), Ω presents the composition of decorated 3-cobordisms $(M, \emptyset, \partial M)$ and $(-N, \partial N, \emptyset)$, i.e., $M \cup_{\mathrm{id}} -N$.

10.5. Example. We may explicitly compute the Hermitian pairing in the module $\mathcal{T}^e(S^1 \times S^1) = \Psi_{(1;)}$. This is a free K-module of rank card(I) with basis $\{b_i = b_{V_i} : \mathbb{1} \to V_i \otimes V_i^*\}_{i \in I}$ (see Section 5.4). Theorem 10.4.1 implies that $\langle b_i, b_j\rangle = 0$ for $i \neq j$ and $\langle b_i, b_i\rangle = (\dim(i))^{-1}\mathrm{tr}(\overline{b_i}b_i)$. By (II.5.1.a), $\langle b_i, b_i\rangle = 1$. Therefore $\{b_i\}_{i\in I}$ is an orthonormal basis of $\mathcal{T}^e(S^1 \times S^1)$.

It follows from the computations of Section 5.4 and Corollary 10.3.2 that the action of the matrices $\mathcal{D}^{-1}S$ and T^{-1} on $\mathcal{T}^e(S^1 \times S^1)$ preserves the Hermitian form. That T is an isometry means that $v_i\overline{v_i} = 1$, which we already know. That $\mathcal{D}^{-1}S$ is an isometry (together with (II.3.8.a)) implies that for any $i, j \in I$,

$$\overline{S_{i,j}} = S_{i,j^*}.$$

10.6. Computation of Hermitian forms. Theorem 10.4.1 suggests that we should study in greater detail the Hermitian form $(x, y) \mapsto \mathrm{tr}(x\overline{y})$ on $\mathrm{Hom}(\mathbb{1}, W)$, where W is an object of $\mathcal{V}$. Here we discuss the structure of this form in the case where W decomposes into a tensor product.

Assume that $W = W_1 \otimes ... \otimes W_n$ where $W_1, ..., W_n$ are objects of $\mathcal{V}$. Lemma 2.2.3 gives us a splitting of $\mathrm{Hom}(\mathbb{1}, W) = h_0(W)$ into a direct sum of K-modules, each of the summands being a tensor product of n modules.

10.6.1. Lemma. *The direct summands of* $\mathrm{Hom}(\mathbb{1}, W) = h_0(W)$ *are orthogonal to each other with respect to the form* $(x, y) \mapsto \mathrm{tr}(x\overline{y})$.

Proof. Let x and y be elements of the summands corresponding to the sequences $i_1, ..., i_{n-1}$ and $j_1, ..., j_{n-1}$ respectively. To compute $\mathrm{tr}(x\overline{y}) = \mathrm{tr}(\overline{y}x)$ present x and y as in Figure 2.2, apply the transformation $D \mapsto -D$ described in Section II.5.1 to the picture of y, and attach the resulting picture of $\overline{y}$ on the top of the picture of x (here we use Lemma II.5.1.3). This gives a diagram of a ribbon (0,0)-graph with $2n$ coupons presenting the composition $\overline{y}x$. For each $r = 1, ..., n-1$, we can split this diagram by a vertical line into two smaller diagrams comprising $2r$ coupons and $2(n-r)$ coupons respectively. This line crosses the diagram in two points lying on the bent arcs colored with i_r and j_r. Now, if $i_r \neq j_r$ then the morphism $V_{i_r} \to V_{j_r}$ represented by the left diagram is equal to zero so that $\overline{y}x = 0$. This shows that if the sequences $i_1, ..., i_{n-1}$ and $j_1, ..., j_{n-1}$ are distinct then the corresponding summands of $\mathrm{Hom}(\mathbb{1}, W)$ are orthogonal.

10.6.2. Lemma. *The canonical Hermitian form in the direct summand of* $h_0(W)$ *corresponding to* $i = (i_1, ..., i_{n-1}) \in I^{n-1}$ *is equal to the tensor product of the Hermitian forms* $(x, y) \mapsto \mathrm{tr}(x\overline{y})$ *on the tensor factors of this summand multiplied by* $(\dim(i))^{-1} = \prod_{r=1}^{n-1}(\dim(i_r))^{-1}$.

Proof. Use the same pictures as in the proof of the previous lemma. Apply induction on k and the fact that any endomorphism z of V_{i_r} is multiplication by $(\dim(i_r))^{-1}\mathrm{tr}(z)$.

10.7. Signatures. The results of the previous subsection allow us to compute the canonical Hermitian form in $\mathrm{Hom}(\mathbb{1}, V_{j_1} \otimes ... \otimes V_{j_n})$ in terms of the Hermitian forms in $\mathrm{Hom}(\mathbb{1}, V_i \otimes V_j \otimes V_k)$ where i, j, k run over I. If $K = \mathbb{C}$ with the usual complex conjugation this allows us to compute the signature $\sigma^{j_1 \dots j_n}$ of the Hermitian form in $\mathrm{Hom}(\mathbb{1}, V_{j_1} \otimes ... \otimes V_{j_n})$. For $i \in I$, denote by ε_i the sign of $\dim(i) \in \mathbb{R}$. Then

$$\sigma^{j_1 \dots j_n} = \sum_{i_2, \dots, i_{n-2} \in I} \left(\prod_{r=2}^{n-2} \varepsilon_{i_r} \prod_{r=2}^{n-1} \sigma^{i_{r-1} j_r i_r^*} \right)$$

where $i_1 = j_1$ and $i_{n-1} = j_n^*$. Combining this result with Theorem 10.4.1 we get a computation of the signature of the Hermitian form in $\mathcal{T}^e(\mathrm{Y})$.

10.8. Exercise. Let Ω be a special ribbon graph in $\mathbb{R}^3$ presenting a decorated 3-cobordism $(N, \partial_- N, \partial_+ N)$. Then the special ribbon graph $-\Omega$ presents a decorated 3-cobordism d-homeomorphic to $(-N, \partial_+ N, \partial_- N)$.

11. Unitary TQFT

Throughout this section, $\mathcal{V}$ is a unitary modular category over $K = \mathbb{C}$ with conjugation $f \to \overline{f}$ and rank $\mathcal{D} > 0$.

11.0. Outline. We show that the TQFT $(\mathcal{T}^e, \tau^w)$ derived from the unitary modular category $\mathcal{V}$ is itself unitary in the sense of Section III.5.2. We use this fact to establish two estimates for the absolute value of $\tau(M)$ where τ is the invariant of closed oriented 3-manifolds defined in Chapter II. At the end of the section we state a theorem presenting $\tau(M)$ as a limit. The results of this section demonstrate that unitary TQFT's are considerably more sensitive to the topology of manifolds than general TQFT's.

11.1. Theorem. *The TQFT* $(\mathcal{T}^e, \tau^w) = (\mathcal{T}^e_{\mathcal{V}}, \tau^w_{\mathcal{V}})$ *derived from the unitary modular category* $\mathcal{V}$ *is a unitary TQFT.*

Proof. Theorem 10.2 implies that the TQFT $(\mathcal{T}^e, \tau^w)$ is Hermitian. By a remark made in the first paragraph of Section II.5.5, we have $\dim(i) > 0$ for any $i \in I$. The inequalities $\mathcal{D} > 0$, $\dim(i) > 0$, $\operatorname{tr}(x\overline{x}) > 0$ and Theorem 10.4.1 imply that the Hermitian form $\langle .,. \rangle_\Upsilon$ is positive definite for every e-surface Υ. Hence $(\mathcal{T}^e, \tau^w)$ is a unitary TQFT.

11.2. Theorem. *Let* Υ *be an e-surface and* M_g *be the mapping torus of a weak e-homeomorphism* $g : \Upsilon \to \Upsilon$. *Then*

$$|\tau(M_g)| \leq \dim_{\mathbb{C}}(\mathcal{T}^e(\Upsilon)).$$

Here $\dim_{\mathbb{C}}$ is the usual dimension of vector spaces over $\mathbb{C}$.

Proof of Theorem. It follows from (10.2.a) that $|\mathcal{D}\Delta^{-1}| = 1$. Theorem 7.2.1 implies that $|\tau(M_g)| = |\operatorname{Tr}(g_\#)|$. Theorem 10.3.1 ensures that $g_\# : \mathcal{T}^e(\Upsilon) \to \mathcal{T}^e(\Upsilon)$ preserves the Hermitian form. By Theorem 11.1, this form is unitary. Therefore $|\operatorname{Tr}(g_\#)| \leq \dim_{\mathbb{C}}(\mathcal{T}^e(\Upsilon))$.

11.3. Corollary. *If a closed oriented* 3*-manifold* M *(with the empty ribbon graph inside) fibers over* S^1 *with fiber a surface of genus* $\leq g$ *then*

$$|\tau(M)| \leq \mathcal{D}^{2g-2} \sum_{j \in I} (\dim(j))^{2-2g}.$$

This follows from Theorem 11.2 and the formula for the dimension of $\mathcal{T}^e(\Upsilon)$ established in Section 12.

11.4. The Heegaard genus. The Heegaard genus is a classical numerical invariant of closed oriented 3-manifolds, see, for instance, [He]. Its definition is based on the notion of a Heegaard decomposition. A Heegaard decomposition of a closed 3-manifold M is a splitting $M = U \cup U'$ where U, U' are two 3-dimensional handlebodies embedded in M such that $U \cap U' = \partial U = \partial U'$. Note that 3-dimensional handlebodies with homeomorphic boundaries are themselves homeomorphic. Hence we may view U' as a copy of U glued to U along a homeomorphism of the boundaries. It is well known that every closed 3-manifold M admits a Heegaard decomposition (cf. Exercise 11.7).

In the case of orientable M the handlebodies U, U' and the surface $\partial U = \partial U'$ are orientable. The genus of this surface is called the genus of the Heegaard decomposition $M = U \cup U'$. The genus $g(M)$ of M is the minimal integer $g \geq 0$ such that M admits a Heegaard decomposition of genus g. The following theorem estimates the genus from below in terms of the invariant τ derived from the unitary modular category $\mathcal{V}$.

11.5. Theorem. *If a closed oriented 3-manifold M admits a Heegaard decomposition of genus g then $|\tau(M)| \leq \mathcal{D}^{g-1}$.*

Proof. Let U be an orientable 3-dimensional handlebody with g handles. We provide U with an empty ribbon graph, zero weight, and the distinguished Lagrangian space in $H_1(\partial U; \mathbb{R})$ which is the kernel of the inclusion homomorphism $H_1(\partial U; \mathbb{R}) \to H_1(U; \mathbb{R})$. In this way U acquires the structure of a weighted extended 3-manifold. Since M admits a Heegaard decomposition of genus g, it may be obtained by gluing U to $-U$ along an orientation-preserving homeomorphism $f : \partial U \to \partial U$. Set $x = \tau^w(U) \in \mathcal{T}^e(\partial U)$. We have $\tau^w(M) = \langle f_\#(x), x\rangle$ where $f_\#$ is the unitary endomorphism of $\mathcal{T}^e(\partial U)$ induced by f. Therefore

$$|\tau(M)| = |\tau^w(M)| = |\langle f_\#(x), x\rangle| \leq \langle f_\#(x), f_\#(x)\rangle^{1/2} \langle x, x\rangle^{1/2} = \langle x, x\rangle.$$

Note that gluing U to $-U$ along the identity $\mathrm{id} : U \to U$ we get the connected sum $g(S^1 \times S^2)$ of g copies of $S^1 \times S^2$ provided with zero weight. (The Maslov indices corresponding to this gluing from Theorem 7.1 vanish.) Hence

$$\langle x, x\rangle = \tau^w(g(S^1 \times S^2)) = \tau(g(S^1 \times S^2)) = \mathcal{D}^{g-1}.$$

Therefore $|\tau(M)| \leq \mathcal{D}^{g-1}$.

11.5.1. Corollary. *For any closed oriented 3-manifold M, $|\tau(M)| \leq \mathcal{D}^{g(M)-1}$.*

11.5.2. Corollary. *For each $j \in I$, we have* $\dim(j) \geq 1$.

Proof. Let Υ_g denote a closed connected oriented surface of genus $g \geq 0$. We regard Υ_g as an e-surface with empty family of distinguished arcs and an arbitrary distinguished Lagrangian space. By Theorem 7.2.1 and the computations of

Section 12,

$$\tau(\Upsilon_g \times S^1) = \dim_{\mathbb{C}}(\mathcal{T}^e(\Upsilon_g)) = \mathcal{D}^{2g-2} \sum_{j\in I} (\dim(j))^{2-2g}.$$

We construct a Heegaard decomposition of $\Upsilon_g \times S^1$ of genus $2g+1$. Choose two small disjoint closed disks $D_0, D_1 \subset \Upsilon_g$. Present S^1 as a union of two semi-circles α_0, α_1 with common ends. For $r = 0, 1$, set

$$U_r = ((\Upsilon \setminus \mathrm{Int}(D_r)) \times \alpha_r) \cup (D_{1-r} \times \alpha_{1-r}) \subset \Upsilon_g \times S^1.$$

It is easy to observe that U_0, U_1 are handlebodies of genus $2g + 1$ forming a Heegaard decomposition of $\Upsilon_g \times S^1$. By Theorem 11.5,

$$\mathcal{D}^{2g-2} \sum_{j\in I} (\dim(j))^{2-2g} \leq \mathcal{D}^{2g}.$$

Therefore the sum $\sum_{j\in I}(\dim(j))^{2-2g}$ is bounded from above by $\mathcal{D}^2$ for any $g \geq 0$. Note that $\{\dim(j)\}_{j\in I}$ are positive real numbers. Hence for each $j \in I$, we have $\dim(j) \geq 1$.

11.5.3. Corollary. $|\Delta| = \mathcal{D} \geq (\mathrm{card}(I))^{1/2}$.

11.6. Wenzl's limit. Under favorable algebraic conditions we compute the invariant $\tau = \tau_{\mathcal{V}}$ of a closed 3-manifold as a limit of certain expressions. (This limit was first considered by Wenzl [We5].) The expressions in question involve cabled links in S^3 whose components are colored with a "fundamental" object of $\mathcal{V}$. The algebraic conditions that allow us to consider the Wenzl limit amount to the unitarity of $\mathcal{V}$ and the existence of a fundamental simple object of $\mathcal{V}$. It is this specific algebraic setting which makes the Wenzl limit especially intriguing.

An object V of a modular category is said to be fundamental if the family $\mathbb{1}, V, V^{\otimes 2} = V \otimes V, V^{\otimes 3}, \ldots$ dominates this category in the sense of Section II.1.3. For instance, the category $\mathcal{V}(G, K, c, \varphi)$ introduced in Section I.1.7.2 has fundamental objects if and only if the group G is cyclic. Every generator of G is a fundamental object of this category.

We describe the cabling of links involved in the Wenzl limit. Let L be a framed oriented link in S^3 with m components $L_1, \ldots, L_m$. Let $z = (z_1, \ldots, z_m)$ be a sequence of non-negative integers. As usual, we may thicken $L_1, \ldots, L_m$ into annuli orthogonal to the framing. For $i = 1, \ldots, m$, choose inside the i-th annulus z_i disjoint concentric subannuli. (They are isotopic to each other and to the original annulus.) The orientation of L induces directions of the resulting $|z| = z_1 + z_2 + \ldots + z_m$ annuli. This yields a ribbon graph in $\mathbb{R}^3$ consisting of $|z|$ annuli. It is denoted by L^z. We say that L^z is obtained from L by cabling.

Fix $l \in I$ and assign to each component of L^z the object V_l of $\mathcal{V}$. This gives a colored ribbon graph in $\mathbb{R}^3$ denoted by L_l^z.

11.6.1. Theorem. *Let V_l with $l \in I$ be a fundamental object of a unitary modular category $(\mathcal{V}, \{V_i\}_{i\in I})$. Let M be the closed oriented 3-manifold obtained by surgery on S^3 along a framed oriented link $L \subset S^3$. Then*

$$\tau_{\mathcal{V}}(M) = \Delta^{\sigma(L)} \mathcal{D}^{-\sigma(L)+m-1} \lim_{N\to\infty} \frac{\sum_{z\in\{1,2,\dots,N\}^m} (\dim(l))^{-|z|} F(L_l^z)}{N^m}. \tag{11.6.a}$$

For the definition of $\sigma(L)$, see Section II.2.1. Note that the ground ring of $\mathcal{V}$ is $\mathbb{C}$ and, in particular, $F(L_l^z) \in \mathbb{C}$. Recall that $\dim(l) \in \mathbb{R}$. This allows us to consider the limit on the right-hand side of (11.6.a). Note also that the left-hand side and, due to this theorem, the right-hand side of (11.6.a) do not depend on the choice of orientation of L.

The proof of Theorem 11.6.1 is given in Section 12.7. It uses the Verlinde algebra of $\mathcal{V}$ discussed in Sections 12.2 and 12.4.

11.7. Exercise. Let M be a triangulated closed 3-manifold. Let U denote a closed regular neighborhood of (the union of) edges and vertices of the triangulation of M. Show that $M = U \cup \overline{(M \setminus U)}$ is a Heegaard decomposition of M. (Hint: notice that $\overline{M \setminus U}$ is a closed regular neighborhood of (the union of) edges and vertices of the cell subdivision of M dual to the given triangulation, cf. Section IX.2.2). Show that in the case of orientable M the genus of this decomposition equals $\mu_3 + 1$ where μ_3 is the number of 3-simplices in the triangulation of M. Hence $\mu_3 + 1 \geq g(M)$ and $|\tau(M)| \leq \mathcal{D}^{\mu_3}$. (This estimate is very rough.)

12. Verlinde algebra

In this section $(\mathcal{V}, \{V_i\}_{i\in I})$ is a strict modular category with ground ring K and rank $\mathcal{D} \in K$. (In Section 12.7 $\mathcal{V}$ is unitary).

12.0. Outline. Our goal is to compute the dimension of the module $\mathcal{T}(\Sigma_t) = \Psi_t$ introduced in Section 1. The main role in this computation is played by a commutative associative algebra $\mathbb{V}(\mathcal{V})$ called the Verlinde algebra (or the fusion algebra) of $\mathcal{V}$. At the end of the section we use this algebra to prove Theorem 11.6.1.

This section is essentially independent of the constructions of Sections 1 - 11. We need only the definition of the module Ψ_t given in Section 1 and Lemmas 2.2.2 and 2.2.3.

12.1. Formula for Dim(Ψ_t). Let $t = (g; (W_1, \nu_1), \dots, (W_m, \nu_m))$ be a decorated type where g is a non-negative integer, $W_1, \dots, W_m$ are objects of $\mathcal{V}$, and $\nu_1, \dots, \nu_m = \pm 1$. We give an explicit formula computing $\operatorname{Dim}(\Psi_t) \in K$ in terms of the matrix $S = [S_{i,j}]_{i,j\in I}$ used in the definition of modular categories. Recall that $S_{i,j} = \mathrm{tr}(c_{V_j,V_i} \circ c_{V_i,V_j}) \in K$ where c is the braiding in $\mathscr{V}$.

Consider first the case when $\nu_1 = \ldots = \nu_m = 1$ and all objects $W_1, \ldots, W_m$ belong to the family $\{V_i\}_{i\in I}$.

12.1.1. Theorem. *Let $t = (g; (V_{i_1}, 1), \ldots, (V_{i_m}, 1))$ with $i_1, \ldots, i_m \in I$. Then*

$$\mathrm{Dim}(\Psi_t) = \mathscr{D}^{2g-2} \sum_{j\in I} \left((\dim(j))^{2-2g-m} \prod_{n=1}^{m} S_{i_n,j} \right).$$

Recall that $\mathscr{D}^2 = \sum_{j\in I} (\dim(j))^2$. Theorem 12.1.1 implies that $\mathrm{Dim}(\Psi_t)$ does not depend on the choice of rank $\mathscr{D}$ of $\mathscr{V}$. In the case $m = 0$ we get the following formula.

12.1.2. Corollary. *If $t = (g;)$ then*

$$\mathrm{Dim}(\Psi_t) = \mathscr{D}^{2g-2} \sum_{j\in I} (\dim(j))^{2-2g}.$$

The reader may directly verify this formula for $g = 0$ and $g = 1$.

To generalize Theorem 12.1.1 to arbitrary marks we need the following notation. For an object W of $\mathscr{V}$ and $j \in I$, set $\mathrm{Dim}_j(W) = \mathrm{Dim}(\mathrm{Hom}(V_j, W))$. For $\nu = \pm 1$, set

$$S_{(W,\nu),j} = \sum_{i\in I} \mathrm{Dim}_i(W^\nu)\, S_{i,j} \in K \tag{12.1.a}$$

where $W^1 = W$ and $W^{-1} = W^*$. For instance, if $W = V_r$ and $\nu = 1$ then $S_{(W,\nu),j} = S_{r,j}$.

12.1.3. Theorem. *For any decorated type $t = (g; (W_1, \nu_1), \ldots, (W_m, \nu_m))$,*

$$\mathrm{Dim}(\Psi_t) = \mathscr{D}^{2g-2} \sum_{j\in I} \left((\dim(j))^{2-2g-m} \prod_{n=1}^{m} S_{(W_n,\nu_n),j} \right). \tag{12.1.b}$$

It is clear that Theorem 12.1.3 contains Theorem 12.1.1 as a special case. To prove Theorem 12.1.3 we introduce the Verlinde algebra of $\mathscr{V}$ (Sections 12.2 - 12.4). Theorem 12.1.3 is proven in Sections 12.5 - 12.6.

12.2. Verlinde's algebra $\mathbb{V}$. Consider the K-module $\mathbb{V}$ freely generated by abstract generators $\{b_i\}_{i\in I}$ numerated by the set I. The elements of $\mathbb{V}$ are formal linear combinations $\sum_{i\in I} k_i b_i$ where $k_i \in K$ for all $i \in I$. Addition, subtraction, and the action of K are defined coordinate-wise. We define multiplication in $\mathbb{V}$ by the formula

$$b_i b_j = \sum_{r \in I} h_r^{i,j} b_r$$

where

$$h_r^{i,j} = \mathrm{Dim}_r(V_i \otimes V_j) = \mathrm{Dim}(\mathrm{Hom}(V_r, V_i \otimes V_j)) \in K.$$

12.2.1. Lemma. $\mathbb{V}$ *is a commutative associative algebra with unit* b_0.

Proof. Commutativity follows from the identity $h_r^{i,j} = h_r^{j,i}$. The equality $b_i b_0 = b_i$ for all i follows from the equality $V_0 = \mathbb{1}$ and definitions. Associativity is equivalent to the following identity:

$$\sum_{r \in I} h_r^{i,j} h_q^{r,p} = \sum_{r \in I} h_q^{i,r} h_r^{j,p}$$

for any $i, j, p, q \in I$. This identity follows from the equalities

$$\mathrm{Hom}(V_q, V_i \otimes V_j \otimes V_p) = \bigoplus_{r \in I} \left(\mathrm{Hom}(V_q, V_r \otimes V_p) \otimes_K \mathrm{Hom}(V_r, V_i \otimes V_j)\right),$$

$$\mathrm{Hom}(V_q, V_i \otimes V_j \otimes V_p) = \bigoplus_{r \in I} \left(\mathrm{Hom}(V_q, V_i \otimes V_r) \otimes_K \mathrm{Hom}(V_r, V_j \otimes V_p)\right).$$

These equalities follow from Lemma II.4.2.2 (cf. below Lemmas VI.1.1.1 and VI.1.1.2).

12.2.2. Remark. The algebra $\mathbb{V}$ may be regarded as a version of the Grothendieck ring $K_0(\mathcal{V})$ of the category $\mathcal{V}$. This ring is defined in the case where $\mathcal{V}$ is an additive category so that we may speak of direct sums of objects. The Grothendieck ring $K_0(\mathcal{V})$ is the K-module generated by the isomorphism classes of the objects of $\mathcal{V}$ subject to the relation $[V \oplus W] = [V] + [W]$ where the square brackets denote the isomorphism class of the module. Multiplication in $K_0(\mathcal{V})$ is defined by the formula $[V]\,[W] = [V \otimes W]$. Assume that every object of $\mathcal{V}$ may be decomposed into a direct sum of objects belonging to the family $\{V_i\}_{i \in I}$ (possibly with multiplicities). Then the ring $K_0(\mathcal{V})$ is canonically isomorphic to $\mathbb{V}$. The isomorphism $K_0(\mathcal{V}) \to \mathbb{V}$ carries $[V]$ into $\sum_{i \in I} \mathrm{Dim}_i(V)\, b_i$. The inverse isomorphism is defined by the formula $b_i \mapsto [V_i]$.

12.3. Trace tr : $\mathbb{V} \to K$. Consider the algebra $\mathrm{End}(\mathbb{V})$ of K-linear endomorphisms of $\mathbb{V}$. Choosing a basis of $\mathbb{V}$ over K (for instance, $\{b_i\}_{i \in I}$) we may represent any K-endomorphism of $\mathbb{V}$ 57 over K. The standard trace of matrices yields a function $\mathrm{tr} : \mathrm{End}(\mathbb{V}) \to K$. As usual, the trace does not depend on the choice of basis of $\mathbb{V}$.

The action of $\mathbb{V}$ on itself via left multiplication yields a linear representation $\rho : \mathbb{V} \to \mathrm{End}(\mathbb{V})$. Thus, $\rho(v)w = vw$ for $v, w \in \mathbb{V}$. For any $v \in \mathbb{V}$, set $\mathrm{tr}(v) =$

$\mathrm{tr}(\rho(v)) \in K$. Choosing $\{b_i\}_{i\in I}$ as the basis for $\mathbb{V}$ we obtain

$$\mathrm{tr}(\sum_{i\in I} k_i b_i) = \sum_{i\in I} k_i \,\mathrm{tr}(b_i) = \sum_{i,j\in I} k_i h_j^{i,j}.$$

The trace $\mathrm{tr} : \mathbb{V} \to K$ is K-linear and satisfies the identity $\mathrm{tr}(vw) = \mathrm{tr}(wv)$ for any $v, w \in \mathbb{V}$.

12.4. Algebraic structure of $\mathbb{V}$. The algebra $\mathbb{V}$ can be decomposed into a direct product of algebras of rank 1. To this end consider the matrix $S^{-1} = [s'_{i,j}]_{i,j\in I}$ inverse to the matrix S. (By (II.3.8.a), we have $s'_{i,j} = \mathcal{D}^{-2} S_{i,j^*}$). For $i \in I$, set

$$b^i = \sum_{k\in I} s'_{i,k} b_k = \mathcal{D}^{-2} \sum_{k\in I} S_{i,k^*} b_k.$$

It is clear that $\{b^i\}_{i\in I}$ is a basis for $\mathbb{V}$ and that $b_k = \sum_{i\in I} S_{k,i} b^i$ for any $k \in I$. The next lemma shows that multiplication in $\mathbb{V}$ is diagonal in the basis $\{b^i\}_{i\in I}$. Invertibility of $\dim(i) \in K$ used in the statement of this lemma follows from Lemma II.4.2.4.

12.4.1. Lemma. *For any $i, j \in I$, we have $b^i b^j = \delta^i_j (\dim(i))^{-1} b^i$ where δ^i_j is the Kronecker symbol.*

Proof. We have

$$b^i b^j = \sum_{k,l\in I} s'_{i,k} s'_{j,l} b_k b_l = \sum_{k,l,r\in I} s'_{i,k} s'_{j,l} h_r^{k,l} b_r = \sum_{k,l,r,p\in I} s'_{i,k} s'_{j,l} h_r^{k,l} S_{r,p} b^p.$$

By Theorem II.4.5.1, the last expression is equal to

$$\sum_{k,l,p\in I} s'_{i,k} s'_{j,l} S_{k,p} S_{l,p} (\dim(p))^{-1} b^p = \sum_{p\in I} \delta^i_p \delta^j_p (\dim(p))^{-1} b^p = \delta^i_j (\dim(i))^{-1} b^i.$$

12.5. Computation of $\mathrm{Dim}(\Psi_t)$. We compute the dimension of Ψ_t in terms of the representation $\rho : \mathbb{V} \to \mathrm{End}(\mathbb{V})$. In the case of non-zero genus this computation involves the trace $\mathrm{tr} : \mathbb{V} \to K$. In the case of zero genus it involves a certain matrix coefficient of ρ.

We need the following notation. For an object W of $\mathcal{V}$ and $\nu = \pm 1$ set

$$[W, \nu] = \sum_{i\in I} \mathrm{Dim}_i(W^\nu)\, b_i \in \mathbb{V}. \tag{12.5.a}$$

12.5.1. Lemma. *For a decorated type $t = (g; (W_1, \nu_1), \ldots, (W_m, \nu_m))$ with $g > 0$, set*

$$v^t = \prod_{n=1}^{m} [W_n, \nu_n] \, (\sum_{i\in I} b_i b_{i^*})^{g-1} \in \mathbb{V}. \tag{12.5.b}$$

Then $\mathrm{Dim}(\Psi_t) = \mathrm{tr}(v^t) \in K$.

Proof. Replacing $(W_n, -1)$ by $(W_n^*, 1)$ we change neither Ψ_t nor v^t. Therefore it suffices to consider the case $\nu_1 = \nu_2 = ... = \nu_m = 1$.

Lemma 2.2.2 implies that

$$\Psi_t = \bigoplus_{r\in I} \big(\mathrm{Hom}(\mathbb{1}, W_1 \otimes V_r^*) \otimes_K \Psi_{t^r}\big) = \bigoplus_{r\in I} (\mathrm{Hom}(V_r, W_1) \otimes_K \Psi_{t^r})$$

where t^r is the decorated type obtained from t by replacing W_1 with V_r. Therefore

$$\mathrm{Dim}(\Psi_t) = \sum_{r\in I} \mathrm{Dim}_r(W_1)\,\mathrm{Dim}(\Psi_{t^r}). \tag{12.5.c}$$

Comparing this equality with (12.5.a) and (12.5.b) and taking into account the additivity of the trace we conclude that it suffices to prove this lemma in the case where $W_1 = V_r$ with $r \in I$. Since the tensor product in $\mathcal{V}$ is commutative up to braiding isomorphisms we may apply similar arguments to $W_2, ..., W_m$. Therefore it suffices to consider the case where $W_1 = V_{r_1}, W_2 = V_{r_2}, ..., W_m = V_{r_m}$ for certain $r_1, ..., r_m \in I$.

Let $i = (i_0, i_1, ..., i_{g-1}) \in I^g$. Recall the object $\Phi(t;i)$ of $\mathcal{V}$ introduced in Section 1.4. It is convenient to replace $\Phi(t;i)$ with an isomorphic object:

$$\Phi(t;i) \approx V_{i_0} \otimes V_{r_1} \otimes V_{r_2} \otimes ... \otimes V_{r_m} \otimes \bigotimes_{n=1}^{g-1} (V_{i_n} \otimes V_{i_n}^*) \otimes V_{i_0}^*. \tag{12.5.d}$$

It follows from Lemma 2.2.3 that

$$\mathrm{Dim}_0(\Phi(t;i)) = \mathrm{Dim}(\mathrm{Hom}(\mathbb{1}, \Phi(t;i))) = \tag{12.5.e}$$

$$= \sum_{j_1,...,j_{m+2g-3}\in I} \left(\prod_{n=1}^{m} h_{j_n}^{r_n, j_{n-1}} \prod_{n=1}^{g-1} (h_{j_{m+2n-1}}^{i_n, j_{m+2n-2}} h_{j_{m+2n}}^{i_n^*, j_{m+2n-1}}) \right)$$

where $j_0 = j_{m+2g-2} = i_0$.

To compute $\mathrm{Dim}(\Psi_t)$ we have to sum up the right-hand side of (12.5.e) over all $i = (i_0, i_1, ..., i_{g-1}) \in I^g$. We first freeze i_0 and sum up the right-hand sides of (12.5.e) over all $i_1, ..., i_{g-1}$. Note that $(h_k^{qj})_{j,k\in I}$ is the matrix of the operator $\rho(b_q)$ with respect to the basis $\{b_i\}_{i\in I}$. Therefore the resulting sum equals the i_0-th diagonal matrix coefficient of the operator

$$\rho\left(\prod_{n=1}^{m} b_{r_n} (\sum_{i\in I} b_i b_{i^*})^{g-1}\right) : \mathbb{V} \to \mathbb{V}$$

with respect to the basis $\{b_i\}_{i\in I}$. Summing up over all $i_0 \in I$ we get $\mathrm{Dim}(\Psi_t) = \mathrm{tr}(v^t)$.

12.5.2. Lemma. *For any decorated type* $t = (0; (W_1, \nu_1), \ldots, (W_m, \nu_m))$, *we have* $\mathrm{Dim}(\Psi_t) = a_{0,0}$ *where* $(a_{i,j})_{i,j \in I}$ *is the matrix of* $\rho(\prod_{n=1}^m [W_n, \nu_n])$ *with respect to the basis* $\{b_i\}_{i \in I}$ *of* $\mathbb{V}$.

Proof. The proof is analogous to the proof of Lemma 12.5.1. Instead of the isomorphism (12.5.d) we should use the isomorphism

$$W_1 \otimes W_2 \otimes \ldots \otimes W_m = V_0 \otimes W_1 \otimes W_2 \otimes \ldots \otimes W_m \otimes V_0$$

induced by the equality $V_0 = \mathbb{1}$.

12.6. Proof of Theorem 12.1.3. Let us first compute the product $v^t b^j$ where v^t is defined by (12.5.b) and $j \in I$. It follows from Lemma 12.4.1 that

$$b_i b^j = \sum_{k \in I} S_{i,k} b^k b^j = S_{i,j} (\dim(j))^{-1} b^j.$$

Therefore

$$v^t b^j = \left\{ (\dim(j))^{2-2g-m} \prod_{n=1}^m \Big(\sum_{i \in I} \mathrm{Dim}_i(W_n^{\nu_n}) S_{i,j} \Big) \Big(\sum_{i \in I} S_{i,j} S_{i^*,j} \Big)^{g-1} \right\} b^j.$$

Since $S_{i^*,j} = S_{i,j^*}$, it follows from (II.3.8.a) that $\sum_{i \in I} S_{i,j} S_{i^*,j} = \mathscr{D}^2$. Using (12.1.a) we get

$$v^t b^j = \left\{ \mathscr{D}^{2g-2} (\dim(j))^{2-2g-m} \prod_{n=1}^m S_{(W_n,\nu_n),j} \right\} b^j. \tag{12.6.a}$$

If $g > 0$ then Lemma 12.5.1 implies that $\mathrm{Dim}(\Psi_t) = \mathrm{tr}(v^t)$. Thus, (12.6.a) implies (12.1.b).

Consider the case $g = 0$. Set

$$\gamma = \mathscr{D}^{-2} \sum_{i \in I} (\dim(i))^3 \, b^i \in \mathbb{V}.$$

For any $v \in \mathbb{V}$, denote the matrix of the operator $\rho(v) : \mathbb{V} \to \mathbb{V}$ with respect to the basis $\{b_i\}_i$ by $(a_{i,j}^v)_{i,j}$. Let us show that for any $v \in \mathbb{V}$,

$$a_{0,0}^v = \mathrm{tr}(\gamma v).$$

Since both sides are K-linear functions of v, it is enough to consider the case $v = b_j$ with $j \in I$. We have $b_j b_0 = b_j$ and therefore $a_{0,0}^v = \delta_0^j$. On the other hand,

$$\gamma v = \gamma b_j = \mathscr{D}^{-2} \sum_{i,k \in I} (\dim(i))^3 S_{j,k} b^i b^k = \mathscr{D}^{-2} \Big(\sum_{i \in I} (\dim(i))^2 S_{j,i} \Big) b^i.$$

Since $\mathrm{tr}(b^i) = (\dim(i))^{-1}$,

$$\mathrm{tr}(\gamma v) = \mathcal{D}^{-2} \sum_{i \in I} \dim(i) S_{j,i}.$$

It follows from (II.3.8.b) that the last expression equals $\delta_0^j = a_{0,0}^v$.

By Lemma 12.5.2, we have $\mathrm{Dim}(\Psi_t) = a_{0,0}^v = \mathrm{tr}(\gamma v)$ where

$$v = \prod_{n=1}^{m} [W_n, \nu_n] = \prod_{n=1}^{m} \left(\sum_{i \in I} \mathrm{Dim}_i(W_n^{\nu_n}) b_i \right).$$

It is easy to compute that

$$\gamma v b^j = v \gamma b^j = \mathcal{D}^{-2} (\dim(j))^2 v b^j = \left\{ \mathcal{D}^{-2} (\dim(j))^{2-m} \prod_{n=1}^{m} S_{(W_n,\nu_n),j} \right\} b^j.$$

These observations imply (12.1.b) in the case $g = 0$.

12.7. Proof of Theorem 11.6.1. Set $V = V_l$. Let $z = (z_1, ..., z_m)$ be a sequence of non-negative integers. Denote by $\tilde{L}^z$ the colored framed oriented link obtained from L by assigning to each component L_n the tensor product $V^{\otimes z_n} = V \otimes ... \otimes V$ of z_n copies of V. Corollary I.2.8.3 and Lemma II.4.4 imply that

$$F(L_l^z) = F(\tilde{L}^z) = \sum_{\lambda \in \mathrm{col}(L)} \left(F(L, \lambda) \prod_{n=1}^{m} \mathrm{Dim}_{\lambda(L_n)}(V^{\otimes z_n}) \right).$$

Substituting this expression in the formula (11.6.a) and comparing with the definition of $\tau(M)$ we observe that to prove (11.6.a) it suffices to show that for any $k \in I$,

$$\lim_{N \to \infty} N^{-1} \left(\sum_{z=1}^{N} (\dim(l))^{-z} \mathrm{Dim}_k(V^{\otimes z}) \right) = \mathcal{D}^{-2} \dim(k). \tag{12.7.a}$$

To compute $\mathrm{Dim}_k(V^{\otimes z})$ we may inductively apply (II.4.5.c) or use Lemma 2.2.3. Little inspection shows that

$$\mathrm{Dim}_k(V^{\otimes z}) = \sum_{l=r_0, r_1, ..., r_{z-2}, r_{z-1}=k \in I} \left(\prod_{s=1}^{z-1} h_{r_s}^{l r_{s-1}} \right). \tag{12.7.b}$$

We rewrite the last formula in a more concise form as follows. Denote by f the $\mathbb{C}$-linear endomorphism $(\dim(l))^{-1} \rho(b_l)$ of $\mathbb{V}$. By definition,

$$f(b_k) = (\dim(l))^{-1} \sum_{j \in I} h_j^{lk} b_j.$$

It follows from (12.7.b) that for any integer $z \geq 1$,

$$f^z(b_0) = \sum_{k \in I} (\dim(l))^{-z} \mathrm{Dim}_k(V^{\otimes z})\, b_k.$$

Therefore to prove (12.7.a) it suffices to show that

$$\lim_{N\to\infty} N^{-1} \sum_{z=1}^{N} f^z(b_0) = \mathcal{D}^{-2} \sum_{k\in I} \dim(k)\, b_k. \tag{12.7.c}$$

Here the limit is taken with respect to the standard topology in $\mathbb{V} = \mathbb{C}^{\mathrm{card}(I)}$.

To prove (12.7.c) we compute the eigenvalues of $f : \mathbb{V} \to \mathbb{V}$. It follows from Lemma 12.4.1 that f is diagonal with respect to the basis $\{b^i\}_{i\in I}$ of $\mathbb{V}$. We have

$$f(b^i) = (\dim(l))^{-1} b_l b^i = \alpha_i b^i$$

where $\alpha_i = (\dim(l))^{-1}(\dim(i))^{-1} S_{l,i}$. Note that $\alpha_0 = 1$. Corollary II.5.5.2 implies that $|\alpha_i| \leq 1$ for any $i \in I \setminus \{0\}$. (It is here that we use the unitarity of $\mathcal{V}$.) We shall show that $\alpha_i \neq 1$ for $i \neq 0$. Indeed, assume that $\alpha_i = 1$ for a certain $i \in I$. An analysis of the proof of the inequality $|\alpha_i| \leq 1$ shows that the equality $\alpha_i = 1$ implies that for any $j \in I$, the composition of a morphism $V_j \to V_l \otimes V_i$ with

$$c_{V_i,V_l}\, c_{V_l,V_i} : V_l \otimes V_i \to V_l \otimes V_i$$

induces the identity endomorphism of $\mathrm{Hom}(V_j, V_l \otimes V_i)$. Using the fact that the family $\{V_j\}_{j\in I}$ dominates $V_l \otimes V_i$, we deduce that $c_{V_i,V_l}\, c_{V_l,V_i}$ is the identity endomorphism of $V_l \otimes V_i$. In other words, $c_{V_i,V_l} = c^{-1}_{V_l,V_i}$. Hence, we may push bands colored with V_l across bands colored with V_i without changing the operator invariant of v-colored ribbon graphs. Since the object V_l is fundamental we may decompose the identity endomorphism of any object W of $\mathcal{V}$ in a finite sum of compositions $W \to V_l^{\otimes n} \to W$, cf. Figure 12.1. Therefore we may push a band colored with W across a band colored with V_i without changing the operator invariant. Consider the Hopf link whose components are colored with V_i and V_k and push the component colored with V_k far away from the second component. This gives a trivial link. Since this transformation preserves the operator invariant, we get $S_{i,k} = \dim(i)\dim(k) = \dim(i) S_{0,k}$ for all $k \in I$. Hence, the i-th row of the matrix S is proportional to its 0-th row. Since S is invertible, this is possible only for $i = 0$. Therefore $\alpha_i \neq 1$ for all $i \in I \setminus \{0\}$.

Since $f(b^0) = b^0$,

$$\lim_{N\to\infty} N^{-1} \left(\sum_{z=1}^{N} f^z(b^0) \right) = b^0.$$

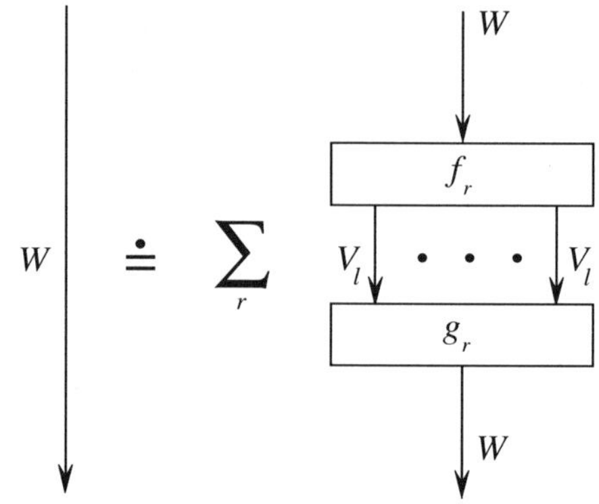

Figure 12.1

For $i \neq 0$, we have

$$\lim_{N\to\infty} N^{-1}\left(\sum_{z=1}^{N} f^z(b^i)\right) = \left(\lim_{N\to\infty} N^{-1}(\alpha_i^{N+1}-1)/(\alpha_i-1)\right) b^i = 0.$$

This implies that for any $x = \sum_{k\in I} x_k b^k \in \mathbb{V}$, the sequence $\{N^{-1}\sum_{z=1}^{N} f^z(x)\}_N$ converges to $x_0 b^0 = x b^0$. Therefore for $x = b_0$, this sequence converges to

$$b_0 b^0 = b^0 = \mathcal{D}^{-2}\sum_{k\in I}\dim(k)\, b_k.$$

This implies (12.7.c) and the claim of the theorem.

Notes

Section 1. A mathematical construction of a non-trivial 3-dimensional TQFT was first outlined by Reshetikhin and Turaev at the end of 1988, see [RT1], [RT2]. We derived 3-dimensional TQFT's from quasitriangular Hopf algebras and, in particular, from the quantum group $U_q(sl_2\mathbb{C})$ at roots of unity.

Here we follow the lines of [RT2, Section 4] using modular categories instead of Hopf algebras. Note that in [RT2] we restricted ourselves to a simplified setting where surfaces have no distinguished arcs, ribbon graphs have no free ends, and cobordisms have connected bases. The general case is treated here for the first time.

There is another, essentially equivalent, approach to the construction of 3-dimensional TQFT's due to Kontsevich [Kon], Crane [Cr1], Kohno [Ko1], [Ko2]. These authors start with the 2-dimensional modular functor arising in 2-dimensional conformal field theory. They appeal to Heegaard decompositions or handle decompositions of 3-manifolds to show that this modular functor extends to a 3-dimensional TQFT.

Section 2. The technique of presentation of 3-cobordisms by ribbon graphs in $\mathbb{R}^3$ is introduced in [RT2]. Lemma 2.1.2 is contained in [RT2] in the case of cobordisms with connected bases. Lemmas 2.1.1 and 2.1.3 are new.

Section 3. The definitions and results of Section 3 are standard except Lemma 3.7. (The proof of Lemma 3.6 seems also to be new.) For more information on Maslov indices, the reader is referred to [Ar], [GS], [Ler], [LV].

Section 4. The anomalies of the 3-dimensional TQFT were computed in [CLM] and [Tu12].

Section 5. The action of the modular group was defined in [RT2] using presentations of homeomorphisms of surfaces by tangles in $\mathbb{R}^3$. A counterpart of this action is well known in 2-dimensional conformal field theory. The action of the modular groupoid is introduced here for the first time.

Section 6. The TQFT $(\mathcal{T}^e, \tau^e)$ is new. The idea to involve Lagrangian spaces in 1-homologies of surfaces in order to renormalize the 3-dimensional TQFT constructed from quantum groups was put forward by Walker [Wa], although he did not fully develop this approach.

The function $\ell(x, y)$ used in Section 6.8 was introduced by Leray [Ler] for pairs $x, y \in \tilde{\Lambda}$ such that $p(x)$ and $p(y)$ are transversal. For the extension to arbitrary $x, y \in \tilde{\Lambda}$, see [Tu1], [Tu2].

Sections 7 and 8. The material of these sections is new.

Section 9. Theorem 9.2.1 is new. The idea to use numerical weights in order to eliminate anomalies was suggested by Walker [Wa].

Section 10. The material of this section is new.

Section 11. Estimates of genera of surfaces via quantum invariants were studied by Walker in 1990 (unfortunately his results are not available to the author). See also [Ga], [Ko3].

The limit in Theorem 11.6 was first considered by Wenzl [We5] in the setting of quantum groups at the roots of unity of type $e^{2\pi i/r}$. He showed that the limit exists and yields an invariant of closed oriented 3-manifolds. In the first versions of his paper, Wenzl conjectures that this limit yields the invariants of 3-manifolds defined in [RT2] and [TW]. Here we prove this conjecture. It has been also proven by Wenzl himself in a different way, see [We5].

Section 12. The definition of fusion algebra and the formula for the dimension of the module of states reproduce in our setting the definitions and results of Verlinde [Ver] in 2-dimensional conformal field theory.

Chapter V
Two-dimensional modular functors

1. Axioms for a 2-dimensional modular functor

1.0. Outline. After a few preliminaries we give an axiomatic definition of a 2-dimensional modular functor. This is a version of the definition of a modular functor given in Section III.1 with a few additional axioms specific to the dimension 2.

The exposition in this section is independent of the material of Chapters I - IV.

1.1. Monoidal classes. A monoidal class $\mathcal{C}$ is a class with a preferred element $\mathbb{1} \in \mathcal{C}$ and a rule assigning to any $V, W \in \mathcal{C}$ an element $VW \in \mathcal{C}$ such that $\mathbb{1}V = V\mathbb{1} = V$ for any $V \in \mathcal{C}$ and $(UV)W = U(VW)$ for any $U, V, W \in \mathcal{C}$. For example, the objects of a strict monoidal category form a monoidal class with multiplication $VW = V \otimes W$.

1.2. Marked surfaces. Fix a monoidal class $\mathcal{C}$. A marked surface or, briefly, m-surface (over $\mathcal{C}$) is a compact oriented surface Σ endowed with a Lagrangian subspace of $H_1(\Sigma; \mathbb{R})$ and such that each connected component of $\partial\Sigma$ is provided with a base point, sign ± 1, and an element of $\mathcal{C}$. The element of $\mathcal{C}$ assigned to a component X of $\partial\Sigma$ is called the label of X. The pair consisting of the label of X and the sign of X is called the mark of X. We provide $\partial\Sigma$ with the orientation induced by the one in Σ.

For example, a closed m-surface is a closed oriented surface Σ with a distinguished Lagrangian subspace of $H_1(\Sigma; \mathbb{R})$. We consider the empty set as an empty marked surface.

An m-homeomorphism of m-surfaces is a homeomorphism of the underlying surfaces preserving the orientation, the distinguished Lagrangian space in 1-homologies, and the base points, labels, and signs of the boundary components. Two m-homeomorphisms $\Sigma_1 \to \Sigma_2$ are said to be isotopic if they can be related by an isotopy in the class of m-homeomorphisms.

A disjoint union of a finite family of m-surfaces is an m-surface in the obvious way. The distinguished Lagrangian space in its 1-homology is the direct sum of the Lagrangian spaces assigned to the surfaces of this family. The operation of disjoint union for m-surfaces will be denoted by the symbol $\amalg$. Warning: the Lagrangian space assigned to a non-connected m-surface is not obliged to split as a direct sum of Lagrangian spaces in the 1-homologies of the components. This shows that in general there is no natural way to regard the connected components of an m-surface as m-surfaces.

As we shall explain in Section 4.2, the notion of an m-surface is essentially equivalent to the notion of an extended surface introduced in Section IV.6. However, m-surfaces are more convenient for an axiomatic study of 2-dimensional modular functors.

1.3. Gluing of boundary components. Let Σ be an m-surface over a monoidal class $\mathscr{C}$. Two distinct components of $\partial\Sigma$ are said to be subject to gluing if they are labelled with the same element of $\mathscr{C}$ and their signs are opposite. Let X, Y be components of $\partial\Sigma$ subject to gluing. There is a (unique up to isotopy) orientation-reversing homeomorphism $X \to Y$ carrying the base point of X into the base point of Y. Identifying $X \subset \Sigma$ with $Y \subset \Sigma$ along such a homeomorphism we obtain an orientable surface Σ' and a projection $\Sigma \to \Sigma'$ that carries X and Y onto the same simple loop $p(X) = p(Y) \subset \mathrm{Int}(\Sigma')$ and restricts to a homeomorphism $\Sigma \setminus (X \cup Y) \to \Sigma \setminus p(X)$.

We provide Σ' with the structure of an m-surface as follows. Endow the components of $\partial\Sigma' = \partial\Sigma \setminus (X \cup Y)$ with the base points, labels, and signs determined by the corresponding data for Σ. Orient Σ' so that p preserves orientation. Provide $H_1(\Sigma'; \mathbb{R})$ with the linear subspace which is the image of the given Lagrangian subspace of $H_1(\Sigma; \mathbb{R})$ under $p_* : H_1(\Sigma; \mathbb{R}) \to H_1(\Sigma'; \mathbb{R})$. (This image is a Lagrangian subspace of $H_1(\Sigma'; \mathbb{R})$.) The resulting m-surface will be denoted by $\Sigma/[X = Y]$. We shall say that it is obtained from Σ by gluing X and Y.

The construction of the m-surface Σ' involves a choice of a homeomorphism, $X \to Y$, in the isotopy class of base point preserving orientation-reversing homeomorphisms. In Appendix III, we show that different choices lead to essentially the same m-surfaces. In the sequel, we shall ignore the dependence of Σ' on the choice of this homeomorphism.

The gluing is natural with respect to m-homeomorphisms. Indeed, let $f : \Sigma_1 \to \Sigma_2$ be an m-homeomorphism of m-surfaces and X, Y be components of $\partial(\Sigma_1)$ subject to gluing. Then $f(X), f(Y) \subset \partial(\Sigma_2)$ are subject to gluing and f induces an m-homeomorphism $\Sigma_1/[X = Y] \to \Sigma_2/[f(X) = f(Y)]$.

1.4. Planar m-surfaces. We introduce a class of m-surfaces lying in the plane $\mathbb{R}^2$. Note first that for any subsurface E of the plane, the intersection form in $H_1(E; \mathbb{R})$ is zero. Therefore $H_1(E; \mathbb{R})$ contains only one Lagrangian subspace, coinciding with $H_1(E; \mathbb{R})$.

For finite sequences $W, W_1, W_2, ..., W_n \in \mathscr{C}$ and $\nu, \nu_1, ..., \nu_n \in \{+1, -1\}$, we define a planar m-surface as follows. Let $B^2 \subset \mathbb{R}^2$ be a closed Euclidean 2-disk. Let $b_1, b_2, ..., b_n$ be disjoint closed Euclidean 2-disks contained in B^2 such that their centers lie on the horizontal line passing through the center of B^2. It is understood that moving along this line from left to right we meet the centers of

$b_1, b_2, \ldots, b_n$ in this order. Set

$$E = B^2 \setminus \left(\bigcup_{r=1}^{n} \mathrm{Int}(b_r) \right).$$

It is clear that E is a compact surface. We provide E with counterclockwise orientation and with the Lagrangian space $H_1(E, \mathbb{R})$. The boundary of E consists of $n+1$ Euclidean circles: $\partial E = \partial B^2 \cup \left(\cup_{r=1}^{n} \partial b_r\right)$. Provide these circles with base points which are their lowest points with respect to the second (vertical) coordinate in $\mathbb{R}^2$. Mark ∂b_r with (W_r, ν_r) for $r = 1, \ldots, n$ and mark ∂B^2 with (W, ν). The resulting m-surface is denoted by $E((W, \nu); (W_1, \nu_1), \ldots, (W_n, \nu_n))$ or by $E(\nu W; \nu_1 W_1, \ldots, \nu_n W_n)$, see Figure 1.1. The circle ∂b_r is called the r-th internal boundary component of this m-surface, the circle ∂B^2 is called the external boundary component. The counterclockwise orientation of E induces clockwise orientations in $\partial b_1, \ldots, \partial b_n$ and counterclockwise orientation in ∂B^2.

We shall ignore the dependence of the m-surface $E(\nu W; \nu_1 W_1, \ldots, \nu_n W_n)$ on the choice of centers and radii of $B^2, b_1, b_2, \ldots, b_n$. The reason is that the m-surfaces corresponding to different choices are related by canonical m-homeomorphisms commuting up to isotopy.

For $W \in \mathscr{C}$, we shall denote a pair $(W, \nu \in \{+1, -1\})$ by νW. The pair $(W, -1)$ will be denoted by $-W$. The pair $(W, +1)$ will be denoted by W or $+W$.

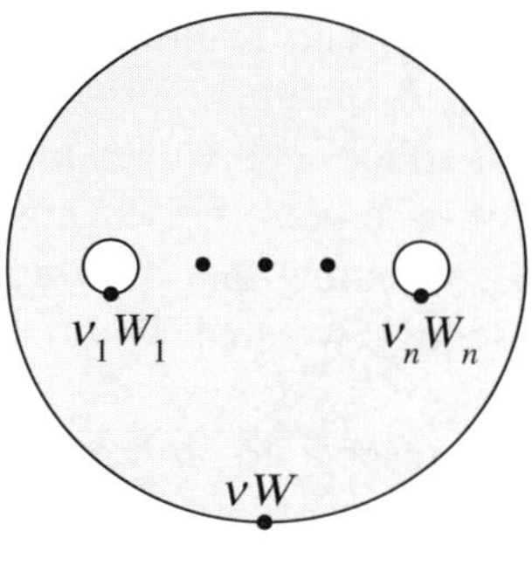

$E(\nu W;\ \nu_1 W_1, \ldots, \nu_n W_n)$

Figure 1.1

In the sequel, we shall often consider the m-surface $E(\nu W; \nu_1 W_1, \ldots, \nu_n W_n)$ for $n = 0, 1, 2$. For $n = 0$, this surface is a marked disk $E(\nu W)$, see Figure 1.2. For $n = 1$, this is an annulus. For $n = 2$, this is a disk with two holes. We shall call a disk with two holes a trinion.

1.5. Modular functors. Fix a commutative ring with unit, K, and a monoidal class $\mathscr{C}$. By m-surfaces, we shall mean m-surfaces over $\mathscr{C}$.

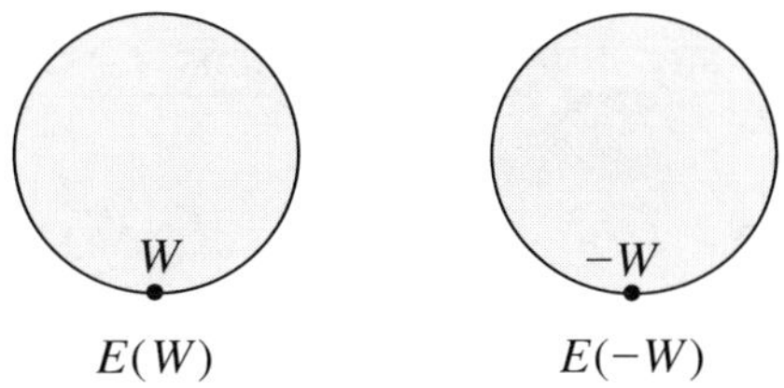

Figure 1.2

A two-dimensional modular functor over $(\mathcal{C}, K)$, briefly 2-DMF, assigns to every m-surface Σ a projective K-module (of finite type) $\mathcal{H}(\Sigma)$ and to every m-homeomorphism of m-surfaces $f : \Sigma_1 \to \Sigma_2$ a K-isomorphism $f_\# : \mathcal{H}(\Sigma_1) \to \mathcal{H}(\Sigma_2)$ depending only on the isotopy class of f. The module $\mathcal{H}(\Sigma)$ is called the module of states of Σ, its elements are called states on Σ. Further data and conditions are formulated as axioms (1.5.1) - (1.5.8).

(1.5.1) (Functoriality axiom). For arbitrary m-homeomorphisms of m-surfaces $g : \Sigma_1 \to \Sigma_2$ and $f : \Sigma_2 \to \Sigma_3$, we have $(fg)_\# = f_\# g_\# : \mathcal{H}(\Sigma_1) \to \mathcal{H}(\Sigma_3)$.

Thus, $\mathcal{H}$ is a covariant functor from the category of m-surfaces and m-homeomorphisms into the category of projective K-modules and K-isomorphisms. For the identity homeomorphism of an m-surface $\mathrm{id}_\Sigma : \Sigma \to \Sigma$, we have $(\mathrm{id}_\Sigma)_\# = \mathrm{id}_{\mathcal{H}(\Sigma)}$. (Proof: apply (1.5.1) to $f = g = \mathrm{id}_\Sigma$.) For any m-homeomorphism $f : \Sigma_1 \to \Sigma_2$, we have $f_\#^{-1} = (f_\#)^{-1}$. (Proof: apply (1.5.1) to $g = f^{-1}$.)

(1.5.2) (Disjoint union axiom). For disjoint m-surfaces Σ_1, Σ_2, there is an identification isomorphism

$$\mathcal{H}(\Sigma_1 \amalg \Sigma_2) = \mathcal{H}(\Sigma_1) \otimes_K \mathcal{H}(\Sigma_2) \tag{1.5.a}$$

satisfying the following three conditions.

(i) (Commutativity). The diagram

$$\begin{array}{ccc} \mathcal{H}(\Sigma_1 \amalg \Sigma_2) & = & \mathcal{H}(\Sigma_1) \otimes_K \mathcal{H}(\Sigma_2) \\ {\scriptstyle =}\downarrow & & \downarrow{\scriptstyle \mathrm{Perm}} \\ \mathcal{H}(\Sigma_2 \amalg \Sigma_1) & = & \mathcal{H}(\Sigma_2) \otimes_K \mathcal{H}(\Sigma_1) \end{array}$$

is commutative. (Here Perm is the flip $x \otimes y \mapsto y \otimes x$.)

(ii) (Associativity). For any m-surfaces $\Sigma_1, \Sigma_2, \Sigma_3$, the composition of identifications

$$(\mathcal{H}(\Sigma_1) \otimes_K \mathcal{H}(\Sigma_2)) \otimes_K \mathcal{H}(\Sigma_3) = \mathcal{H}(\Sigma_1 \amalg \Sigma_2) \otimes_K \mathcal{H}(\Sigma_3) = \mathcal{H}(\Sigma_1 \amalg \Sigma_2 \amalg \Sigma_3) =$$

$$= \mathcal{H}(\Sigma_1) \otimes_K \mathcal{H}(\Sigma_2 \amalg \Sigma_3) = \mathcal{H}(\Sigma_1) \otimes_K (\mathcal{H}(\Sigma_2) \otimes_K \mathcal{H}(\Sigma_3))$$

is the standard isomorphism $(x \otimes y) \otimes z \mapsto x \otimes (y \otimes z)$.

(iii) (Naturality). For any m-homeomorphisms $f : \Sigma_1 \to \Sigma_1'$, $g : \Sigma_2 \to \Sigma_2'$, the diagram

$$\text{(1.5.b)} \qquad \begin{array}{ccc} \mathcal{H}(\Sigma_1 \amalg \Sigma_2) & \xrightarrow{(f \amalg g)_\#} & \mathcal{H}(\Sigma_1' \amalg \Sigma_2') \\ {\scriptstyle =}\downarrow & & \downarrow{\scriptstyle =} \\ \mathcal{H}(\Sigma_1) \otimes_K \mathcal{H}(\Sigma_2) & \xrightarrow{f_\# \otimes g_\#} & \mathcal{H}(\Sigma_1') \otimes_K \mathcal{H}(\Sigma_2') \end{array}$$

is commutative.

(1.5.3) (Normalization axiom). $\mathcal{H}(\emptyset) = K$ and for any m-surface Σ, the identification $\mathcal{H}(\Sigma) = \mathcal{H}(\emptyset \amalg \Sigma) = \mathcal{H}(\emptyset) \otimes_K \mathcal{H}(\Sigma)$ is induced by $\mathcal{H}(\emptyset) = K$.

Axioms (1.5.1) - (1.5.3) reproduce in the present setting, the definition of a modular functor given in Section III.1. Thus, every 2-DMF is a modular functor in the sense of Section III.1. The following axioms (1.5.4) - (1.5.8) are specific to the dimension 2.

(1.5.4) (Gluing axiom). Let Σ' be the m-surface obtained from an m-surface Σ by identification of two components of $\partial\Sigma$ subject to gluing (see Section 1.3). Then there is a distinguished K-homomorphism $\mathcal{H}(\Sigma) \to \mathcal{H}(\Sigma')$ called the gluing homomorphism, that satisfies the following conditions:

(i) (Naturality). If an m-homeomorphism of m-surfaces $f : \Sigma_1 \to \Sigma_2$ induces an m-homeomorphism of m-surfaces $f' : \Sigma_1' \to \Sigma_2'$ where Σ_1', Σ_2' are obtained by gluing of boundary components from Σ_1, Σ_2 respectively, then the following diagram of induced isomorphisms and gluing homomorphisms is commutative:

$$\text{(1.5.c)} \qquad \begin{array}{ccc} \mathcal{H}(\Sigma_1) & \xrightarrow{f_\#} & \mathcal{H}(\Sigma_2) \\ \downarrow & & \downarrow \\ \mathcal{H}(\Sigma_1') & \xrightarrow{f'_\#} & \mathcal{H}(\Sigma_2'). \end{array}$$

(ii) (Commutativity). The gluing homomorphisms corresponding to gluings along disjoint pairs of boundary components commute as follows. For distinct boundary components X, Y, X', Y' of any m-surface Σ such that the pairs (X, Y) and (X', Y') are subject to gluing, the following diagram of gluing homomorphisms is commutative:

$$\begin{array}{ccc} \mathcal{H}(\Sigma) & \longrightarrow & \mathcal{H}(\Sigma/[X = Y]) \\ \downarrow & & \downarrow \\ \mathcal{H}(\Sigma/[X' = Y']) & \longrightarrow & \mathcal{H}(\Sigma/[X = Y, X' = Y']). \end{array}$$

(iii) (Compatibility with disjoint union). For the gluing homomorphism $q : \mathcal{H}(\Sigma_1) \to \mathcal{H}(\Sigma_1')$ corresponding to a gluing $\Sigma_1 \to \Sigma_1'$ and for any m-surface Σ_2, the following diagram of gluing homomorphisms and identification isomorphisms is commutative:

$$\begin{array}{ccc} \mathcal{H}(\Sigma_1 \amalg \Sigma_2) & = & \mathcal{H}(\Sigma_1) \otimes_K \mathcal{H}(\Sigma_2) \\ \downarrow & & \downarrow{\scriptstyle q\otimes \mathrm{id}} \\ \mathcal{H}(\Sigma_1' \amalg \Sigma_2) & = & \mathcal{H}(\Sigma_1') \otimes_K \mathcal{H}(\Sigma_2). \end{array}$$

Condition (ii) allows us to define the gluing homomorphism for a gluing of several disjoint pairs of boundary components of an m-surface. It is defined as the composition of the gluing homomorphisms corresponding to consecutive gluings of these pairs.

Note that we do not require the gluing homomorphisms to be surjective (or injective).

We shall often consider the gluing homomorphisms in the following special case. Assume that we have two m-surfaces Σ_1, Σ_2 and that certain components $X_1 \subset \partial\Sigma_1$, $X_2 \subset \partial\Sigma_2$ are subject to gluing. We form the disjoint union $\Sigma_1 \amalg \Sigma_2$ and glue X_1 to X_2. This results in an m-surface Σ. We say that Σ is obtained by gluing Σ_1 and Σ_2 along X_1, X_2. Composing the identification isomorphism (1.5.a) with the gluing homomorphism $\mathcal{H}(\Sigma_1 \amalg \Sigma_2) \to \mathcal{H}(\Sigma)$, we get a K-linear homomorphism $\mathcal{H}(\Sigma_1) \otimes_K \mathcal{H}(\Sigma_2) \to \mathcal{H}(\Sigma)$. For $h_1 \in \mathcal{H}(\Sigma_1), h_2 \in \mathcal{H}(\Sigma_2)$, the image of $h_1 \otimes h_2$ under this homomorphism is denoted by $h_1 \diamond h_2$ and called the result of gluing h_1 and h_2 along X_1, X_2. The pairing

$$(h_1, h_2) \mapsto h_1 \diamond h_2 : \mathcal{H}(\Sigma_1) \times \mathcal{H}(\Sigma_2) \to \mathcal{H}(\Sigma)$$

is bilinear. Note that gluing Σ_2, Σ_1 along X_2, X_1 gives the same m-surface Σ. It follows from (1.5.2) (i) that the corresponding pairing $\mathcal{H}(\Sigma_2) \times \mathcal{H}(\Sigma_1) \to \mathcal{H}(\Sigma)$ is obtained from the previous one via the permutation $(h_2, h_1) \mapsto (h_1, h_2)$.

The gluing of states on surfaces is associative in the following sense. Assume that we have three m-surfaces $\Sigma_1, \Sigma_2, \Sigma_3$ such that Σ_1 may be glued to Σ_2 along boundary components $X_1 \subset \partial\Sigma_1, X_2 \subset \partial\Sigma_2$ and Σ_2 may be glued to Σ_3 along boundary components $X_2' \subset \partial\Sigma_2, X_3 \subset \partial\Sigma_3$ where $X_2 \neq X_2'$. It follows from definitions and axioms (1.5.1) - (1.5.4) that for any $h_1 \in \mathcal{H}(\Sigma_1), h_2 \in \mathcal{H}(\Sigma_2), h_3 \in \mathcal{H}(\Sigma_3)$,

$$(h_1 \diamond h_2) \diamond h_3 = h_1 \diamond (h_2 \diamond h_3). \tag{1.5.d}$$

On the left-hand side $h_1 \diamond h_2$ is the result of gluing h_1 and h_2 along X_1, X_2, and $(h_1 \diamond h_2) \diamond h_3$ is the result of gluing $h_1 \diamond h_2$ and h_3 along X_2', X_3. Similarly, on the right-hand side $h_2 \diamond h_3$ is the result of gluing h_2 and h_3 along X_2', X_3, and $h_1 \diamond (h_2 \diamond h_3)$ is the result of gluing h_1 and $h_2 \diamond h_3$ along X_1, X_2.

Thus, similarly to puzzle games where various pictures are glued from small pieces of colored paper, we may glue the states on surfaces from smaller pieces. This may lead to interesting identities between these pieces, cf. Figure 1.4.

(1.5.5) (Disk axiom). The K-modules $\mathcal{H}(E(+\mathbb{1})), \mathcal{H}(E(-\mathbb{1}))$ are isomorphic to K. For any $\nu \in \{+1, -1\}$ and any (free) generator $e_\nu \in \mathcal{H}(E(\nu\mathbb{1})) = K \cdot e_\nu$, we have the following: if an m-surface Σ' is obtained from an m-surface Σ by gluing $E(\nu\mathbb{1})$ along a boundary component of Σ marked with $(\mathbb{1}, -\nu)$, then the gluing homomorphism

$$h \mapsto h \diamond e_\nu : \mathcal{H}(\Sigma) \to \mathcal{H}(\Sigma')$$

is an isomorphism.

The inverse isomorphism $\mathcal{H}(\Sigma') \to \mathcal{H}(\Sigma)$ is called the excision of e_ν. Note that the transformation $\Sigma' \mapsto \Sigma$ cuts out $E(\nu\mathbb{1})$ from the interior of Σ' and creates an additional boundary component marked with $(\mathbb{1}, -\nu)$. By the disk axiom, this transformation does not change the module of states up to isomorphism.

For $V, W \in \mathcal{C}$, consider the trinion $E(-VW; V, W)$ whose internal boundary components are marked with $(V, +1)$, $(W, +1)$ and whose external boundary component is marked with $(VW, -1)$, see Figure 1.3.

(1.5.6a) (First excision axiom). For any $V, W \in \mathcal{C}$, there is a distinguished state $e_{V,W} \in \mathcal{H}(E(-VW; V, W))$. For any m-surface Σ' obtained by gluing an m-surface Σ and $E(-VW; V, W)$ along a component of $\partial\Sigma$ marked with $+VW$ and the external component of $\partial E(-VW; V, W)$, the gluing homomorphism

$$h \mapsto h \diamond e_{V,W} : \mathcal{H}(\Sigma) \to \mathcal{H}(\Sigma') \tag{1.5.e}$$

is an isomorphism.

The isomorphism inverse to (1.5.e) is called the excision of $e_{V,W}$.

The axiom (1.5.6a) relates the gluing homomorphisms to multiplication in $\mathcal{C}$. This axiom shows that we may cut off the distinguished states on trinions from arbitrary states on m-surfaces. Note that the surface Σ has fewer boundary components than Σ'. Iterating the excision isomorphisms, we may reduce the computation of the module of states to the case where $\partial\Sigma = S^1$ or $\partial\Sigma = \emptyset$.

We shall present $e_{V,W} \in \mathcal{H}(E(-VW; V, W))$ graphically by drawing the symbol $e_{V,W}$ on the trinion $E(-VW; V, W)$, see Figure 1.3.

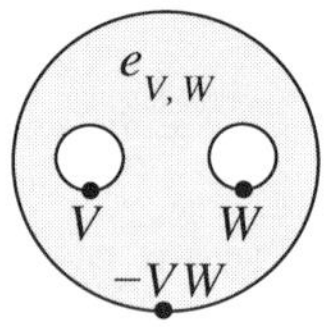

Figure 1.3

(1.5.6b) (Second excision axiom). For any $W \in \mathcal{C}$ and any m-surface Σ' obtained by gluing an m-surface Σ and $E(-W; \mathbb{1}, W)$ along a component of $\partial\Sigma$

marked with $-W$ and the second internal component of $\partial E(-W;\mathbb{1}, W)$, the gluing homomorphism

$$h \to h \diamond e_{\mathbb{1},W} : \mathcal{H}(\Sigma) \to \mathcal{H}(\Sigma')$$

is an isomorphism.

(1.5.7) (Basic identity). For any $U, V, W \in \mathcal{C}$, we have

$$e_{U,V} \diamond e_{UV,W} = e_{V,W} \diamond e_{U,VW}. \tag{1.5.f}$$

On the left-hand side, the pairing $\diamond$ corresponds to the gluing of m-surfaces $E(-UV;U,V)$ and $E(-UVW;UV,W)$ along the boundary components labelled with UV. On the right-hand side, the pairing $\diamond$ corresponds to the gluing of m-surfaces $E(-VW;V,W)$ to $E(-UVW;U,VW)$ along the boundary components labelled with VW. Both gluings give $E(-UVW;U,V,W)$ so that equality (1.5.f) makes sense. For the graphical form of (1.5.f), see Figure 1.4, where the equality means the equality of the elements of $\mathcal{H}(E(-UVW;U,V,W))$ obtained by gluing the pieces on the right-hand and left-hand sides.

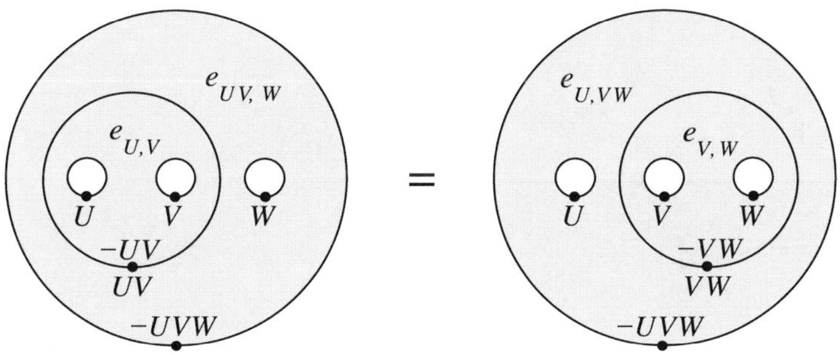

Figure 1.4

Note that on both sides of (1.5.f) we glue planar disks with holes, E_1, E_2, along the external boundary component of E_1 and an internal boundary component of E_2. In a sense, we fill one of the holes of E_2 with E_1. This gives another planar disk with holes, E, and a pairing $(h_1, h_2) \mapsto h_1 \diamond h_2 : \mathcal{H}(E_1) \times \mathcal{H}(E_2) \to \mathcal{H}(E)$. We shall often use such pairings in the formulas to follow.

Axiom (1.5.7) may be viewed as a consistency condition for the excision isomorphisms. This axiom will play a fundamental role in the study of 2-DMF's in Section 2.

To formulate the next axiom we need a few preliminary constructions. Fix generators $e_+ \in \mathcal{H}(E(\mathbb{1})) = \mathcal{H}(E(+\mathbb{1}))$ and $e_- \in \mathcal{H}(E(-\mathbb{1}))$. For any $V, W \in \mathcal{C}$, we shall define an element

$$x(V, W) \in \mathcal{H}(E(-W; WV, -V)).$$

Using the disk axiom with $\nu = 1$ and $\nu = -1$ consecutively, we obtain isomorphisms

$$\mathcal{H}(E(-V;\mathbb{1},V)) \to \mathcal{H}(E(-V;V)) \to \mathcal{H}(E(-V;-\mathbb{1},V)).$$

(We first glue in e_- and then cut out e_+.) Let $z_V \in \mathcal{H}(E(-V;-\mathbb{1},V))$ denote the image of $e_{\mathbb{1},V} \in \mathcal{H}(E(-V;\mathbb{1},V))$ under these isomorphisms.

It is obvious that the trinion $E(-V;-\mathbb{1},V)$ is homeomorphic to $E(-\mathbb{1};V,-V)$ via an m-homeomorphism ξ that carries the broken intervals drawn in Figure 1.5 onto the broken intervals. The m-homeomorphism ξ is unique up to isotopy. Consider the induced isomorphism

$$\xi_\# : \mathcal{H}(E(-V;-\mathbb{1},V)) \to \mathcal{H}(E(-\mathbb{1};V,-V)). \tag{1.5.g}$$

Gluing $\xi_\#(z_V)$ to $e_{W,\mathbb{1}}$ along the boundary components labelled by $\mathbb{1}$ yields

$$\xi_\#(z_V) \diamond e_{W,\mathbb{1}} \in \mathcal{H}(E(-W;W,V,-V)).$$

Cutting out $e_{W,V}$, we get $x(V,W)$, see Figure 1.6.

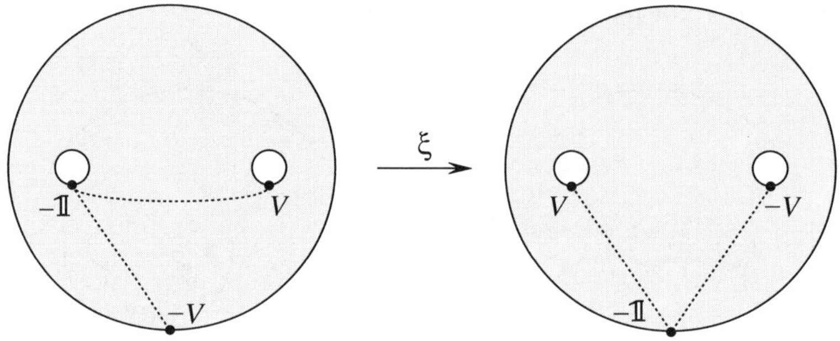

Figure 1.5

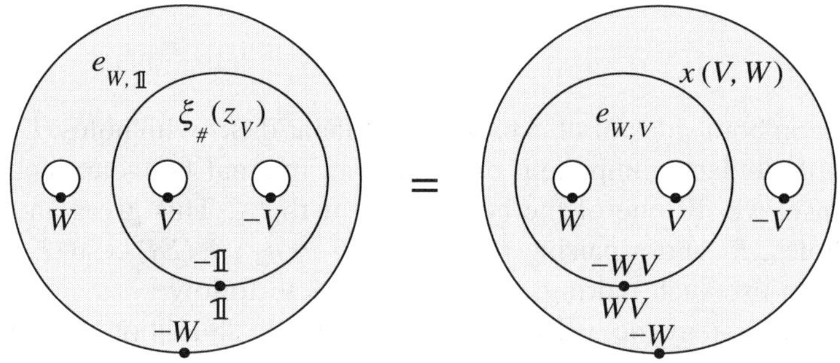

Figure 1.6

Note that gluing $E(-WV)$ to $E(-W;WV,-V)$ along the boundary components labelled by WV yields $E(-W;-V)$. For $D \in \mathcal{H}(E(-WV))$, set

$$\tilde{D} = D \diamond x(V,W) \in \mathcal{H}(E(-W;-V)).$$

(1.5.8) (Duality axiom). To each $V \in \mathscr{C}$ there are assigned certain $V^* \in \mathscr{C}$ and $D = D_V \in \mathscr{H}(E(-V^*V))$ such that $\tilde{D} \in \mathscr{H}(E(-V^*; -V))$ has the following non-degeneracy property. If an m-surface Σ' is obtained by gluing an m-surface Σ and $E(-V^*; -V)$ along a component of $\partial\Sigma$ marked with $+V^*$ and the external component of $\partial E(-V^*; -V)$ then the homomorphism

$$h \to h \diamond \tilde{D} : \mathscr{H}(\Sigma) \to \mathscr{H}(\Sigma') \tag{1.5.h}$$

is an isomorphism.

The element $V^* \in \mathscr{C}$ is called the dual of V. We do NOT require that $V^{**} = V$, $(VW)^* = W^*V^*$, or $\mathbb{1}^* = \mathbb{1}$.

Note that the state $x(V, V^*)$ on $E(-V^*; V^*V, -V)$ depends on the choice of generators $e_+ \in \mathscr{H}(E(\mathbb{1}))$ and $e_- \in \mathscr{H}(E(-\mathbb{1}))$. Therefore $x(V, V^*)$ and $\tilde{D}$ are defined up to multiplication by invertible elements of K. This does not spoil the formulation of the duality axiom.

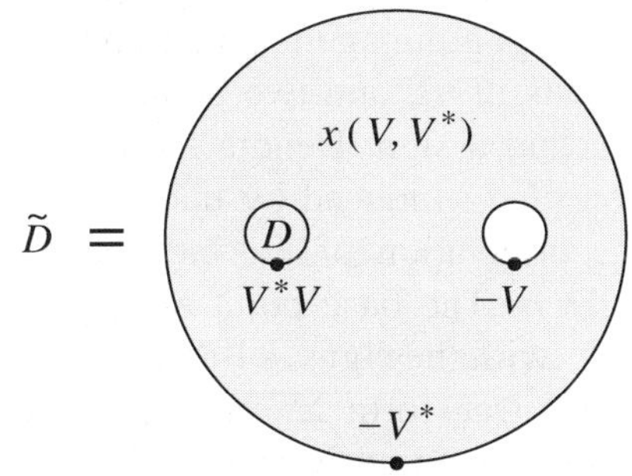

Figure 1.7

This completes the list of axioms for a 2-dimensional modular functor. The simplest example of a 2-DMF is obtained as follows. Let $\mathscr{C}$ be the class of all projective K-modules. Assign the module K to all marked surfaces over $\mathscr{C}$ and the identity $\mathrm{id}_K : K \to K$ to all m-homeomorphisms and gluings. Take the distinguished states mentioned in (1.5.6a), (1.5.8) to be $1 \in K$. This example shows that the axioms of a 2-DMF are consistent.

In Section 2 we shall show that 2-DMF's are closely related to ribbon categories. We shall derive from any 2-DMF $\mathscr{H}$ over $(\mathscr{C}, K)$ a ribbon Ab-category $\mathscr{V}_{\mathscr{H}}$ with ground ring K. The objects of this category are the elements of $\mathscr{C}$. The ribbon category $\mathscr{V}_{\mathscr{H}}$ is called the underlying category of $\mathscr{H}$.

1.6. Rational modular functors. An important class of 2-DMF's is formed by so-called rational modular functors. A rational 2-DMF is a pair $(\mathscr{H}, \{V_i\}_{i \in I})$ where $\mathscr{H}$ is a 2-DMF over a monoidal class $\mathscr{C}$ and a commutative ring with unit K and $\{V_i\}_{i \in I}$ is a family of elements of $\mathscr{C}$ numerated by a *finite* set I such that the following four axioms are satisfied.

(1.6.1) (Second normalization axiom). There exists a unique $0 \in I$ such that $V_0 = \mathbb{1}$.

(1.6.2) (Third normalization axiom). For each $i \in I$, the module of states $\mathcal{H}(E(-V_i; V_i))$ is isomorphic to K. For each $i \in I$, there exists a unique $i^* \in I$ such that $\mathcal{H}(E(V_i; V_{i^*}))$ is isomorphic to K. If $i, j \in I$ and $j \neq i^*$ then $\mathcal{H}(E(V_i; V_j)) = 0$.

To formulate the next axiom, we need more notation. Let Σ be an m-surface. A simple loop on Σ is a closed curve in $\text{Int}(\Sigma)$ which has no self-intersections and which is provided with an orientation and a base point. (We do not assume that the loop is parametrized.) By an s-loop on Σ we mean a simple loop on Σ such that its homology class in $H_1(\Sigma; \mathbb{R})$ belongs to the distinguished Lagrangian subspace of $H_1(\Sigma; \mathbb{R})$. Cutting out Σ along an s-loop a we get a compact oriented surface Σ^a. We shall transform this surface into an m-surface. We assign to Σ^a the preimage of the given Lagrangian subspace of $H_1(\Sigma; \mathbb{R})$ under the natural homomorphism $H_1(\Sigma^a; \mathbb{R}) \to H_1(\Sigma; \mathbb{R})$. (This preimage is a Lagrangian subspace of $H_1(\Sigma^a; \mathbb{R})$.) The boundary of Σ^a consists of $\partial\Sigma$ and two copies of a. Each component of $\partial\Sigma$ has a base point, a label, and a sign. Denote the copies of a in $\partial\Sigma^a$ by a_+ and a_- so that the orientation of a_+ induced by that of Σ^a coincides with the given orientation in a and the orientation of a_- induced by that of Σ^a is opposite to the given orientation in a. The base point on a gives rise to base points on a_+, a_-. We provide a_+, a_- with the signs $+1$ and -1 respectively. Labelling both a_+ and a_- with $V \in \mathcal{C}$ we transform Σ^a into an m-surface denoted by $\Sigma^a(V)$. We have the obvious gluing mapping $\Sigma^a(V) \to \Sigma$ induced by the identification $a_+ = a_-$. By the gluing axiom, this mapping gives rise to a gluing homomorphism $\mathcal{H}(\Sigma^a(V)) \to \mathcal{H}(\Sigma)$.

(1.6.3) (Splitting axiom). For any s-loop a on an m-surface Σ, the sum of gluing homomorphisms

$$\bigoplus_{i \in I} \mathcal{H}(\Sigma^a(V_i)) \to \mathcal{H}(\Sigma) \tag{1.6.a}$$

is an isomorphism.

It should be emphasized that the family $\{V_i\}_{i \in I}$ does not depend on the choice of Σ and a. The isomorphism (1.6.a) gives a splitting

$$\mathcal{H}(\Sigma) = \bigoplus_{i \in I} \mathcal{H}(\Sigma^a(V_i)). \tag{1.6.b}$$

Note that this splitting is available only under the assumption that a is a simple loop in $\text{Int}(\Sigma)$ whose homology class in $H_1(\Sigma; \mathbb{R})$ belongs to the distinguished Lagrangian space.

To state the last axiom, we need to involve the underlying ribbon category of $\mathcal{H}$. It would be nice to reformulate this axiom in terms of m-surfaces avoiding the underlying category.

(1.6.4) (Non-degeneracy axiom). The pair consisting of the underlying ribbon category of $\mathcal{H}$ and the family $\{V_i\}_{i\in I}$ satisfies the non-degeneracy axiom (II.1.4.4).

The axioms of 2-DMF's and rational 2-DMF's look pretty complicated. It is by no means clear that they define interesting objects worth any serious attention. A most striking phenomenon is the existence of non-trivial rational 2-DMF's and their deep connections with the theory of representations of Lie algebras and topology of 3-manifolds. In Section 4 we shall show that any modular category $\mathcal{V}$ with rank $\mathcal{D}$ gives rise to a rational 2-DMF whose underlying ribbon category is isomorphic to $\mathcal{V}$. Combining this construction with those of Chapter XI we obtain that semisimple complex Lie algebras endowed with roots of unity give rise to rational 2-DMF's. The relationships between modular categories, rational 2-DMF's, and invariants of 3-manifolds will be discussed in Section 4.1 in more detail.

1.7. Remark. The mapping $i \mapsto i^* : I \to I$ given in (1.6.2) is an involution. Indeed, the marked annulus $E(V_i; V_j)$ is m-homeomorphic to $E(V_j; V_i)$ and therefore $i^{**} = i$ for any $i \in I$. Note also that $0^* = 0$ because, by the disk axiom,

$$\mathcal{H}(E(V_0; V_0)) = \mathcal{H}(E(\mathbb{1}; \mathbb{1})) = \mathcal{H}(E(\mathbb{1})) = K.$$

2. Underlying ribbon category

Fix up to the end of this section a monoidal class $\mathcal{C}$, a commutative ring with unit, K, and a 2-DMF $\mathcal{H}$ over $(\mathcal{C}, K)$.

2.0. Outline. We derive from the 2-DMF $\mathcal{H}$ a ribbon Ab-category $\mathcal{V}_{\mathcal{H}}$.

2.1. Objects and morphisms of $\mathcal{V}_{\mathcal{H}}$. The objects of $\mathcal{V}_{\mathcal{H}}$ are the elements of $\mathcal{C}$. For $V, W \in \mathcal{C}$, consider the marked annulus $E(-V; W) = E(-V; +W)$ (see Section 1.4). Set

$$\mathrm{Hom}(V, W) = \mathcal{H}(E(-V; W)).$$

Thus, morphisms $V \to W$ are the states on $E(-V; W)$.

Let us define the composition rule

$$\mathrm{Hom}(U, V) \times \mathrm{Hom}(V, W) \to \mathrm{Hom}(U, W) \tag{2.1.a}$$

where $U, V, W \in \mathcal{C}$. Gluing the annuli $E(-U; V)$ and $E(-V; W)$ along their boundary components labelled with V yields the annulus $E(-U; W)$. The gluing homomorphism

$$\mathcal{H}(E(-U;V)) \otimes_K \mathcal{H}(E(-V;W)) \to \mathcal{H}(E(-U;W))$$

induces a bilinear pairing (2.1.a), see Figure 2.1 where $g \in \operatorname{Hom}(U,V)$, $f \in \operatorname{Hom}(V,W)$. We use this pairing as a composition rule. Associativity of composition follows from (1.5.d).

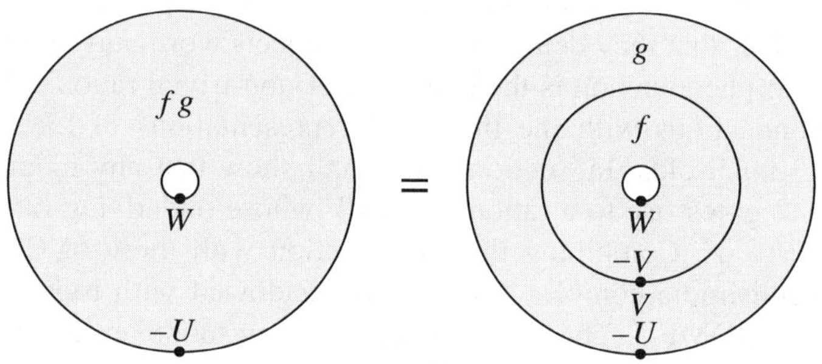

Figure 2.1

2.2. Lemma. *There exists a unique state e_- on the disk $E(-\mathbb{1})$ such that*

$$e_- \diamond (e_- \diamond e_{\mathbb{1},\mathbb{1}}) = e_-. \tag{2.2.a}$$

The state e_- is a free generator of $\mathcal{H}(E(-\mathbb{1}))$, i.e., $\mathcal{H}(E(-\mathbb{1})) = K \cdot e_-$.

Proof. Consider a generator e of $\mathcal{H}(E(-\mathbb{1})) = K \cdot e$. Gluing $E(-\mathbb{1})$ to $E(-\mathbb{1};\mathbb{1},\mathbb{1})$ along the second internal component of $\partial E(-\mathbb{1};\mathbb{1},\mathbb{1})$ gives $E(-\mathbb{1};\mathbb{1})$. By the second excision axiom, the gluing homomorphism

$$h \mapsto h \diamond e_{\mathbb{1},\mathbb{1}} : \mathcal{H}(E(-\mathbb{1})) \to \mathcal{H}(E(-\mathbb{1};\mathbb{1}))$$

is an isomorphism. Gluing a copy of $E(-\mathbb{1})$ to $E(-\mathbb{1};\mathbb{1})$ gives $E(-\mathbb{1})$. By the disk axiom, the gluing homomorphism

$$g \mapsto e \diamond g : \mathcal{H}(E(-\mathbb{1};\mathbb{1})) \to \mathcal{H}(E(-\mathbb{1}))$$

is an isomorphism. We conclude that $e \diamond (e \diamond e_{\mathbb{1},\mathbb{1}})$ is a generator of $\mathcal{H}(E(-\mathbb{1})) = K \cdot e$. Therefore $e \diamond (e \diamond e_{\mathbb{1},\mathbb{1}}) = xe$ where x is an invertible element of K. Set $e_- = x^{-1}e \in \mathcal{H}(E(-\mathbb{1}))$. It is clear that e_- is a generator of $\mathcal{H}(E(-\mathbb{1}))$ and is the unique element of $\mathcal{H}(E(-\mathbb{1}))$ satisfying (2.2.a).

2.3. Identity morphisms. Let $V \in \mathcal{C}$. Gluing e_- to $e_{\mathbb{1},V}$ along the first internal component of $\partial E(-V;\mathbb{1},V)$ gives the state $e_- \diamond e_{\mathbb{1},V}$ on $E(-V;V)$. We denote this state as well as the corresponding morphism $V \to V$ by id_V, see Figure 2.2. The following sequence of lemmas results in Lemma 2.3.4 which shows that id_V is the identity endomorphism of V.

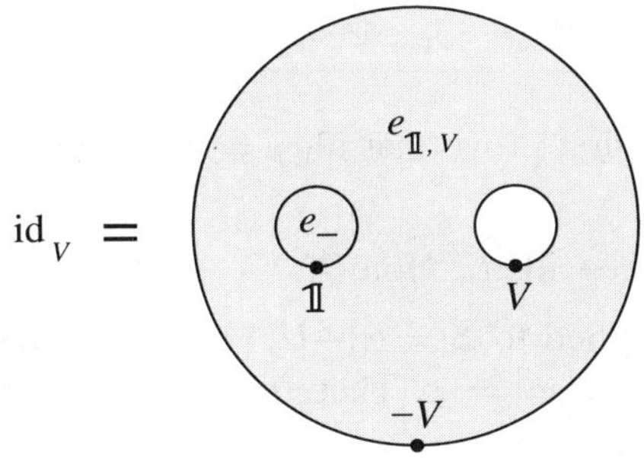

Figure 2.2

2.3.1. Lemma. *The gluing of $E(-V;V)$ to an m-surface Σ along the external component of $\partial E(-V;V)$ and a component of $\partial\Sigma$ marked with $+V$ yields an m-surface Σ' such that the gluing homomorphism* $h \to \mathrm{id}_V \diamond h : \mathcal{H}(\Sigma) \mapsto \mathcal{H}(\Sigma')$ *is an isomorphism.*

Proof. This follows from the first excision axiom and the disk axiom.

2.3.2. Lemma. *The gluing of $E(-V;V)$ to an m-surface Σ along the internal component of $\partial E(-V;V)$ and a component of $\partial\Sigma$ marked with $-V$ yields an m-surface Σ' such that the gluing homomorphism* $h \to \mathrm{id}_V \diamond h : \mathcal{H}(\Sigma) \mapsto \mathcal{H}(\Sigma')$ *is an isomorphism.*

Proof. This follows from the second excision axiom and the disk axiom.

2.3.3. Lemma. $(\mathrm{id}_V)^2 = \mathrm{id}_V$.

Proof. The formula $(\mathrm{id}_V)^2 = \mathrm{id}_V$ is proven in Figure 2.3. The left picture presents $(\mathrm{id}_V)^2$. The equality follows from the basic identity (1.5.f). By (2.2.a), the right picture presents $e_- \diamond e_{\mathbb{1},V} = \mathrm{id}_V$.

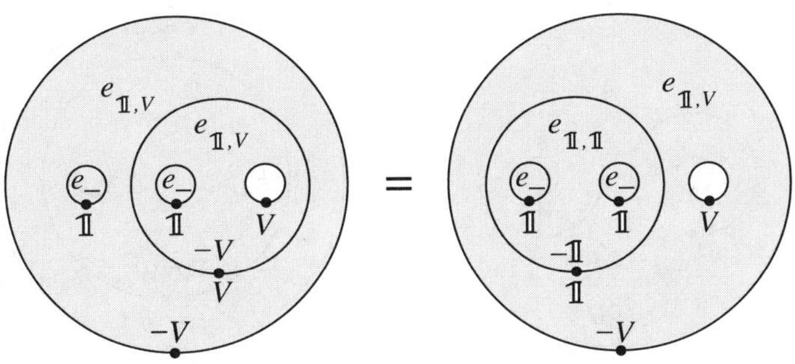

Figure 2.3

2.3.4. Lemma. *The morphism* $\mathrm{id}_V : V \to V$ *is the identity endomorphism of* V.

Proof. Let $U \in \mathscr{C}$. We first prove that $\mathrm{id}_V f = f$ for any $f \in \mathrm{Hom}(U, V)$. Denote the mapping

$$f \mapsto \mathrm{id}_V f : \mathrm{Hom}(U, V) \to \mathrm{Hom}(U, V)$$

by φ. Lemma 2.3.1 applied to $\Sigma = E(-U; V)$ implies that φ is an isomorphism. Lemma 2.3.3 implies that $\varphi^2 = \varphi$. Therefore φ is the identity so that $\mathrm{id}_V f = f$ for any $f \in \mathrm{Hom}(U, V)$.

Similar arguments, using Lemma 2.3.2 instead of Lemma 2.3.1, show that $f\,\mathrm{id}_V = f$ for any $f \in \mathrm{Hom}(V, U)$. Hence id_V is the identity endomorphism of V.

2.3.5. Lemma. *Let* $\mathrm{id}'_V : V \to V$ *be the morphism corresponding to the state on* $E(-V; V)$ *obtained by gluing* e_- *to* $e_{V,\mathbb{1}}$ *along the second internal component of* $\partial E(-V; V, \mathbb{1})$. *Then* $\mathrm{id}_V = \mathrm{id}'_V$.

Proof. Arguments analogous to those used in the proof of Lemma 2.3.3 show that $(\mathrm{id}'_V)^2 = \mathrm{id}'_V$. Similar to Lemmas 2.3.1 and 2.3.4, $f = \mathrm{id}'_V f$ for any $f \in \mathrm{Hom}(U, V)$. In particular, $\mathrm{id}_V = \mathrm{id}'_V \mathrm{id}_V$. By Lemma 2.3.4, $\mathrm{id}'_V \mathrm{id}_V = \mathrm{id}'_V$.

2.4. Tensor product in $\mathscr{V}_{\mathscr{H}}$. The tensor product of two objects V, W of $\mathscr{V}_{\mathscr{H}}$ is defined to be their product in $\mathscr{C}$. Thus, $V \otimes W = VW \in \mathscr{C}$. By the definition of a monoidal class, this tensor product is strictly associative and $V \otimes \mathbb{1} = \mathbb{1} \otimes V = V$ for any V.

Let $f : V \to V'$ and $g : W \to W'$ be two morphisms in $\mathscr{V}_{\mathscr{H}}$. Denote the corresponding states on $E(-V; V')$ and $E(-W; W')$ by the same symbols f, g. Glue these states to $e_{V,W}$ along the boundary components labelled by V, W. This gives $f \diamond (g \diamond e_{V,W}) \in \mathscr{H}(E(-VW; V', W'))$. The first excision axiom implies that $f \diamond (g \diamond e_{V,W}) = e_{V',W'} \diamond h$ for a unique $h \in \mathscr{H}(E(-VW; V'W'))$. Set $f \otimes g = h$ (see Figure 2.4). Thus, $f \diamond (g \diamond e_{V,W}) = e_{V',W'} \diamond (f \otimes g)$.

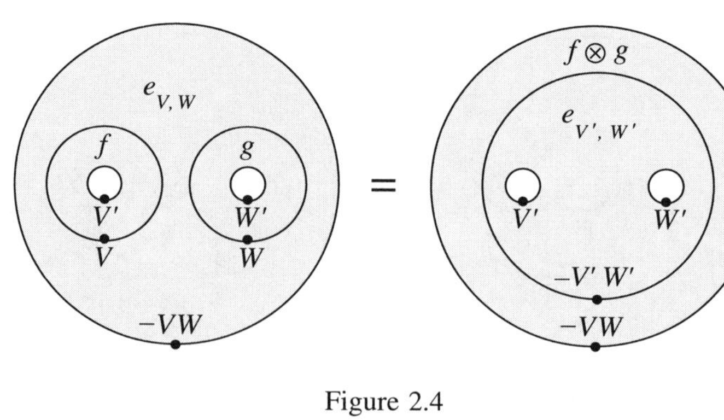

Figure 2.4

2.4.1. Theorem. *The category $\mathcal{V}_{\mathcal{H}}$ with the tensor product defined above is a strict monoidal category.*

Proof. Let us verify the axioms of a monoidal category formulated in Section I.1.1. Let $f : U \to U', g : V \to V', h : W \to W'$ be three morphisms in the category $\mathcal{V}_{\mathcal{H}}$. In Figure 2.5 we show that $(f \otimes g) \otimes h = f \otimes (g \otimes h)$. To simplify the pictures, we omit in Figure 2.5 (and in certain figures to follow) the labels of boundary circles. They can be reconstructed uniquely from the states specified in the pictures. Comparing the first and last pictures in Figure 2.5 and applying the first excision axiom we get $(f \otimes g) \otimes h = f \otimes (g \otimes h)$.

The same argument as in the proof of the equality $\mathrm{id}_V f = f$ in Section 2.3.4 proves the equalities in Figure 2.6. Therefore $\mathrm{id}_V \otimes \mathrm{id}_W = \mathrm{id}_{V \otimes W}$. In Figure 2.7 we show that $f' f \otimes g' g = (f' \otimes g')(f \otimes g)$ for any $f : U \to U'$, $f' : U' \to U''$, $g : V \to V'$, $g' : V' \to V''$. For a morphism $f : U \to U'$, the formula $\mathrm{id}_{\mathbb{1}} \otimes f = f$ is equivalent to the formula in Figure 2.8. Gluing e_- to both sides we observe that this formula is equivalent to the equality $\mathrm{id}_{U'} f = f \mathrm{id}_U$. This equality is obvious since both sides are equal to f. The equality $f \otimes \mathrm{id}_{\mathbb{1}} = f$ is proven similarly, although instead of id we should consider the morphism id' defined in Lemma 2.3.5 and use the result of that Lemma.

2.5. Braiding in $\mathcal{V}_{\mathcal{H}}$. In this subsection we provide $\mathcal{V}_{\mathcal{H}}$ with a braiding. Consider the trinion $\mathbb{E}$ obtained by cutting out two small disjoint open disks $\mathrm{Int}(b_1)$, $\mathrm{Int}(b_2)$ from a larger Euclidean 2-disk B^2. Assume that the centers of these three disks lie on a horizontal line in $\mathbb{R}^2$ and that the radii of b_1, b_2 are equal. Let γ be a homeomorphism $\mathbb{E} \to \mathbb{E}$ characterized (up to isotopy) by the following properties: γ is the identity on ∂B^2, γ permutes ∂b_1 and ∂b_2 and transforms the intervals x_1, x_2 connecting the base point of ∂B^2 with the base points of $\partial b_1, \partial b_2$ in the way shown in Figure 2.9. It is instructive to think of γ as the final stage of an isotopy of B^2 in itself. This isotopy slowly pushes the disks b_1, b_2 (via parallel translations) in the counterclockwise direction. When b_1 reaches the position of b_2 and b_2 reaches the position of b_1 we get $\gamma : \mathbb{E} \to \mathbb{E}$. We shall call γ the positive permutation of ∂b_1 and ∂b_2.

For $V, W \in \mathcal{C}$, the homeomorphism $\gamma : \mathbb{E} \to \mathbb{E}$ induces an m-homeomorphism $E(-VW; V, W) \to E(-VW; W, V)$ (for notation, see Section 1.5). This m-homeomorphism is denoted by $\gamma_{V,W}$ or simply by γ.

Consider the homomorphism $\gamma_\# : \mathcal{H}(E(-VW; V, W)) \to \mathcal{H}(E(-VW; W, V))$ and the state $\gamma_\#(e_{V,W})$ on $E(-VW; W, V)$. Cutting off $e_{W,V}$, we get a certain state on $E(-VW; WV)$. Denote this state and the corresponding morphism $VW \to WV$ by $c_{V,W}$ (see Figure 2.10). Thus

$$\gamma_\#(e_{V,W}) = e_{W,V} \diamond c_{V,W}.$$

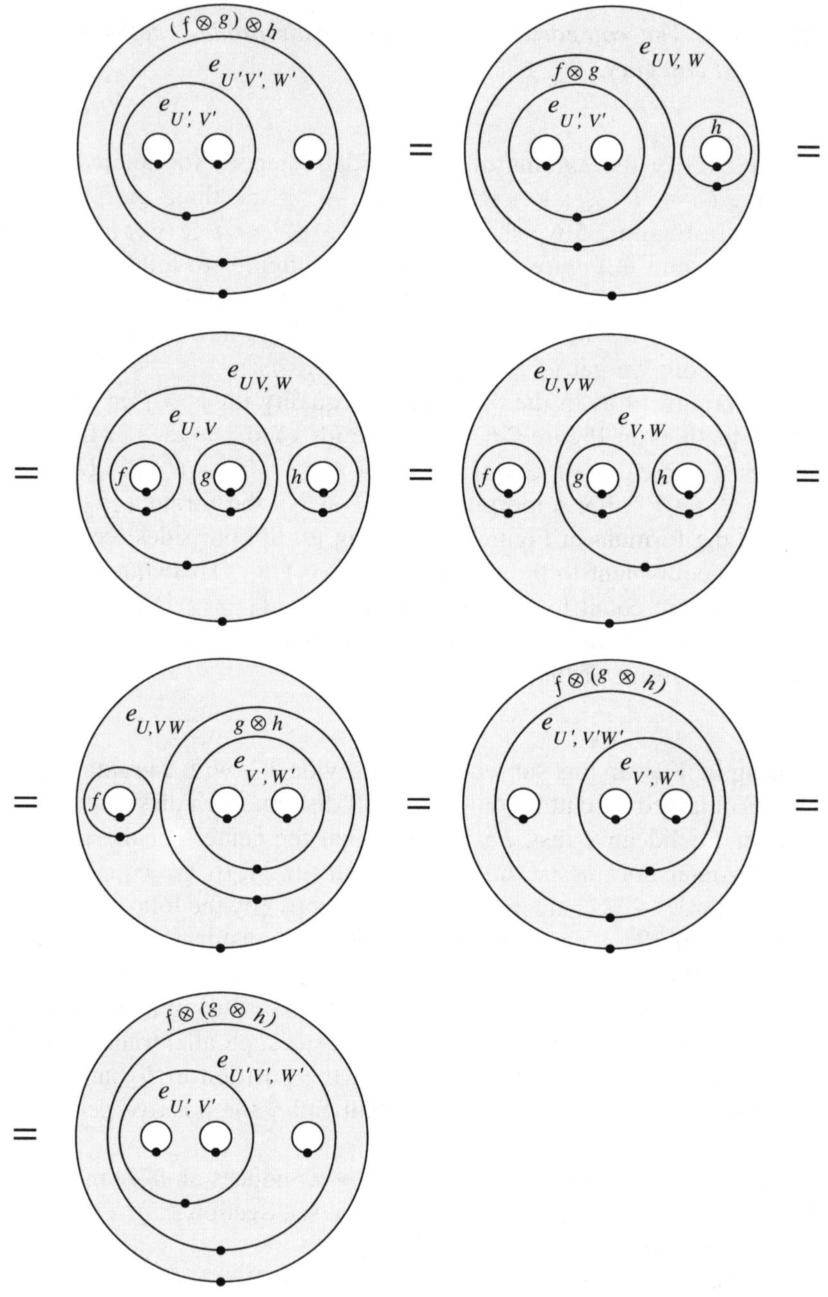
(f ⊗ g) ⊗ h
e_{U'V', W'}
e_{U', V'}
=
f ⊗ g
e_{UV, W}
e_{U', V'}
h
=
e_{UV, W}
e_{U, V}
f
g
h
=
e_{U, VW}
e_{V, W}
f
g
h
=
e_{U, VW}
g ⊗ h
e_{V', W'}
f
=
f ⊗ (g ⊗ h)
e_{U', V'W'}
e_{V', W'}
=
f ⊗ (g ⊗ h)
e_{U'V', W'}
e_{U', V'}

Figure 2.5

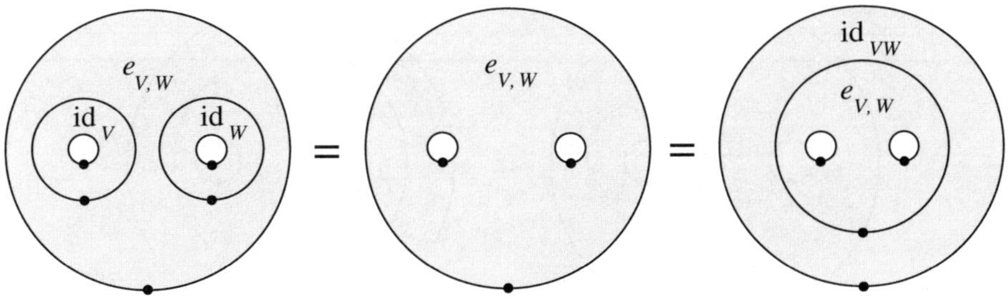

Figure 2.6

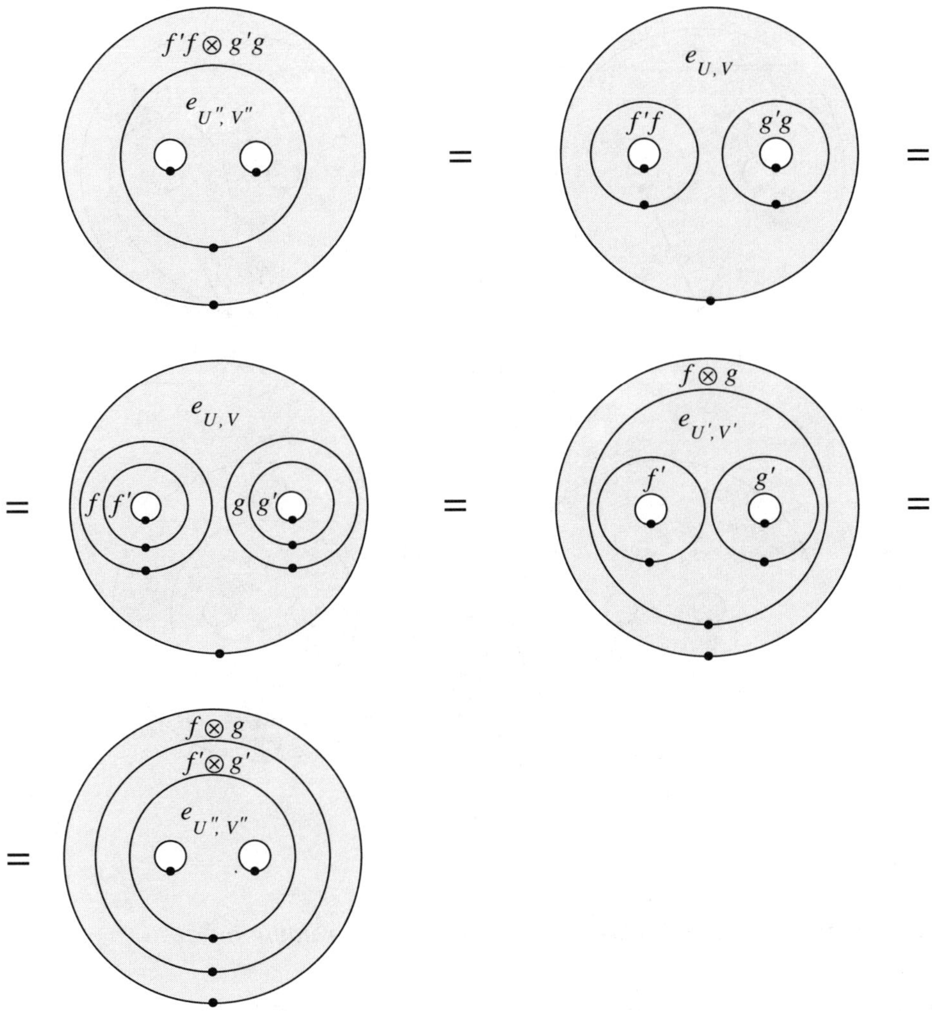

Figure 2.7

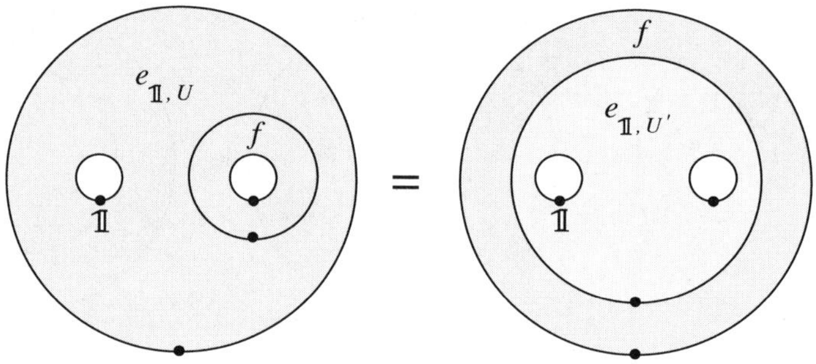

Figure 2.8

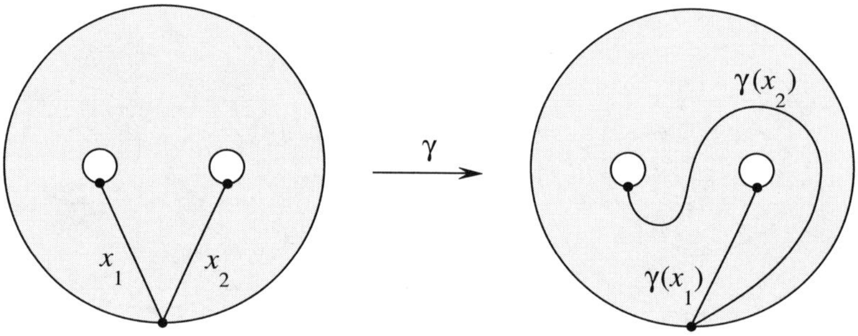

Figure 2.9

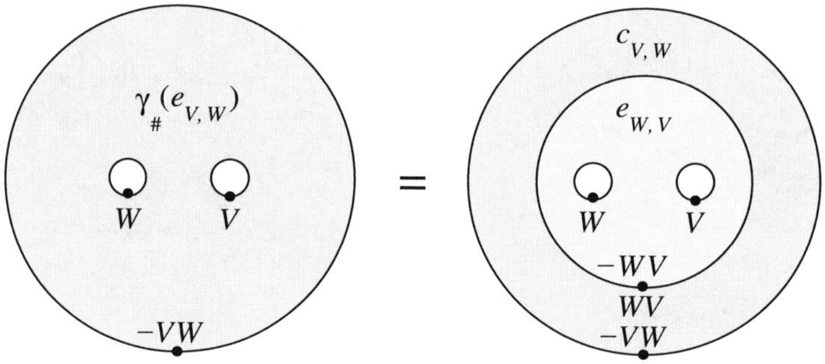

Figure 2.10

2.5.1. Lemma. *The morphisms $\{c_{V,W}\}_{V,W}$ form a braiding in $\mathcal{V}_{\mathcal{H}}$.*

Proof. The invertibility of $c_{V,W}$ will be proven in Section 2.6. Here we verify formulas (I.1.2.b), (I.1.2.c), and (I.1.2.d).

Formula (I.1.2.b) in our present notation may be rewritten as follows

$$c_{U,VW} = (\mathrm{id}_V \otimes c_{U,W})(c_{U,V} \otimes \mathrm{id}_W). \tag{2.5.a}$$

The disk with three holes $E(-UVW; U, V, W)$ can be obtained by gluing the trinions $E(-UVW; U, VW)$ and $E(-VW; V, W)$. Similarly, $E(-UVW; V, W, U)$ can be obtained by gluing $E(-UVW; VW, U)$ and $E(-VW; V, W)$. Extending the homeomorphism

$$\gamma_{U,VW} : E(-UVW; U, VW) \to E(-UVW; VW, U)$$

by the identity self-homeomorphism of $E(-VW; V, W)$ we obtain an m-homeomorphism $E(-UVW; U, V, W) \to E(-UVW; V, W, U)$. Denote this m-homeomorphism by Γ. It follows from the basic identity (1.5.f) and definitions that

$$\Gamma_\#(e_{U,V} \diamond e_{UV,W}) = \Gamma_\#(e_{V,W} \diamond e_{U,VW}) = e_{V,W} \diamond e_{VW,U} \diamond c_{U,VW}. \tag{2.5.b}$$

We shall compute the state $\Gamma_\#(e_{U,V} \diamond e_{UV,W})$ in another way. Present Γ as the composition $\Gamma^2\Gamma^1$ of homeomorphisms

$$\Gamma^1 : E(-UVW; U, V, W) \to E(-UVW; V, U, W)$$

and

$$\Gamma^2 : E(-UVW; V, U, W) \to E(-UVW; V, W, U)$$

where Γ^1 is the positive permutation of the U-labelled and V-labelled circles and Γ^2 is the positive permutation of the U-labelled and W-labelled circles. It follows from definitions that

$$\Gamma^1_\#(e_{U,V} \diamond e_{UV,W}) = (\gamma_{U,V})_\#(e_{U,V}) \diamond e_{UV,W} = e_{V,U} \diamond c_{U,V} \diamond e_{UV,W}.$$

This state on $E(-UVW; V, U, W)$ is shown in the top left part of Figure 2.11. As in Section 2.3, we do not change this state if we glue to it the state id_W on the annulus $E(-W; W)$; the gluing in question proceeds along the component of $\partial E(-UVW; V, U, W)$ labelled with W. Thus

$$e_{V,U} \diamond c_{U,V} \diamond e_{UV,W} = e_{V,U} \diamond c_{U,V} \diamond (\mathrm{id}_W \diamond e_{UV,W}).$$

By the definition of the tensor product in $\mathcal{V}_{\mathcal{H}}$, the last expression equals

$$e_{V,U} \diamond e_{VU,W} \diamond (c_{U,V} \otimes \mathrm{id}_W) = e_{U,W} \diamond e_{V,UW} \diamond (c_{U,V} \otimes \mathrm{id}_W)$$

(see Figure 2.11). Thus

$$\Gamma^1_\#(e_{U,V} \diamond e_{UV,W}) = e_{U,W} \diamond e_{V,UW} \diamond (c_{U,V} \otimes \mathrm{id}_W).$$

To compute the image of the last state on $E(-UVW; V, U, W)$ under $\Gamma^2_\#$ note that Γ^2 acts inside $E(-UW; U, W) \subset E(-UVW; V, U, W)$ and keeps $c_{U,V} \otimes \mathrm{id}_W$ intact.

A direct computation, similar to the one in Figure 2.11, shows that

$$\begin{aligned}\Gamma^2_{\#}(e_{U,W} \diamond e_{V,UW} \diamond (c_{U,V} \otimes \mathrm{id}_W)) &= e_{W,U} \diamond c_{U,W} \diamond e_{V,UW} \diamond (c_{U,V} \otimes \mathrm{id}_W) = \\ &= e_{W,U} \diamond c_{U,W} \diamond (\mathrm{id}_V \diamond e_{V,UW}) \diamond (c_{U,V} \otimes \mathrm{id}_W) = \\ &= e_{W,U} \diamond e_{V,WU} \diamond (\mathrm{id}_V \otimes c_{U,W}) \diamond (c_{U,V} \otimes \mathrm{id}_W) = \\ &= e_{V,W} \diamond e_{VW,U} \diamond (\mathrm{id}_V \otimes c_{U,W}) \diamond (c_{U,V} \otimes \mathrm{id}_W).\end{aligned}$$

Since $\Gamma = \Gamma^2\,\Gamma^1$, we have $\Gamma_{\#} = \Gamma^2_{\#}\Gamma^1_{\#}$ and therefore

$$\Gamma_{\#}(e_{U,V} \diamond e_{UV,W}) = e_{V,W} \diamond e_{VW,U} \diamond (\mathrm{id}_V \otimes c_{U,W}) \diamond (c_{U,V} \otimes \mathrm{id}_W).$$

Comparing this formula with (2.5.b) and applying excision homomorphisms we get (2.5.a). The proof of the second braiding identity (I.1.2.c) is similar.

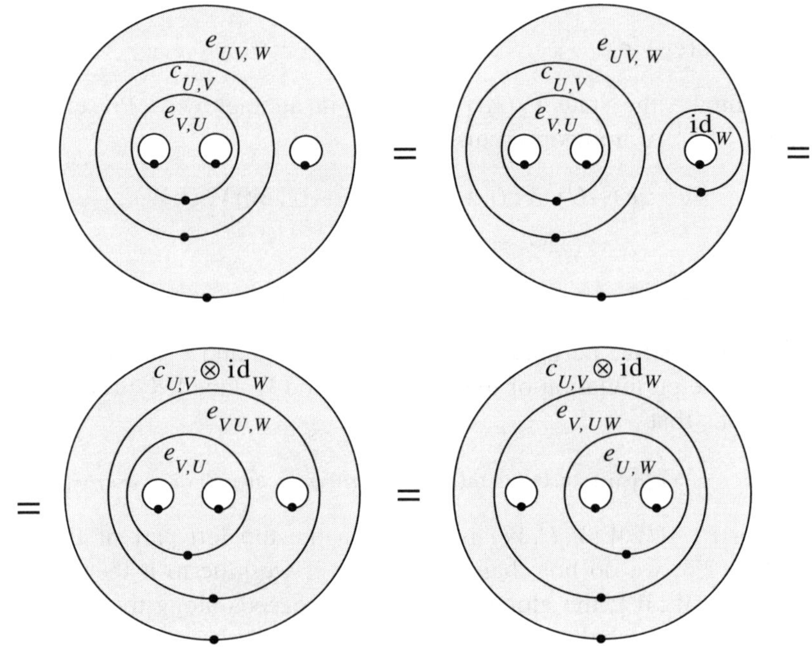

Figure 2.11

We may rewrite (I.1.2.d) in the following form:

$$(g \otimes f) \diamond c_{V,W} = c_{V',W'} \diamond (f \otimes g)$$

where $f \in \mathcal{H}(E(-V;V'))$, $g \in \mathcal{H}(E(-W;W'))$, and

$$f \otimes g \in \mathcal{H}(E(-VW;V'W')), \quad g \otimes f \in \mathcal{H}(E(-WV;W'V')).$$

We have

$$e_{W',V'} \diamond c_{V',W'} \diamond (f \otimes g) = (\gamma_{V',W'})_{\#}(e_{V',W'}) \diamond (f \otimes g) =$$

$$= \gamma_\#(e_{V',W'} \diamond (f \otimes g)) = \gamma_\#(f \diamond (g \diamond e_{V,W})) = g \diamond (f \diamond (\gamma_{V,W})_\#(e_{V,W})) =$$

$$= g \diamond (f \diamond e_{W,V} \diamond c_{V,W}) = e_{W',V'} \diamond (g \otimes f) \diamond c_{V,W}$$

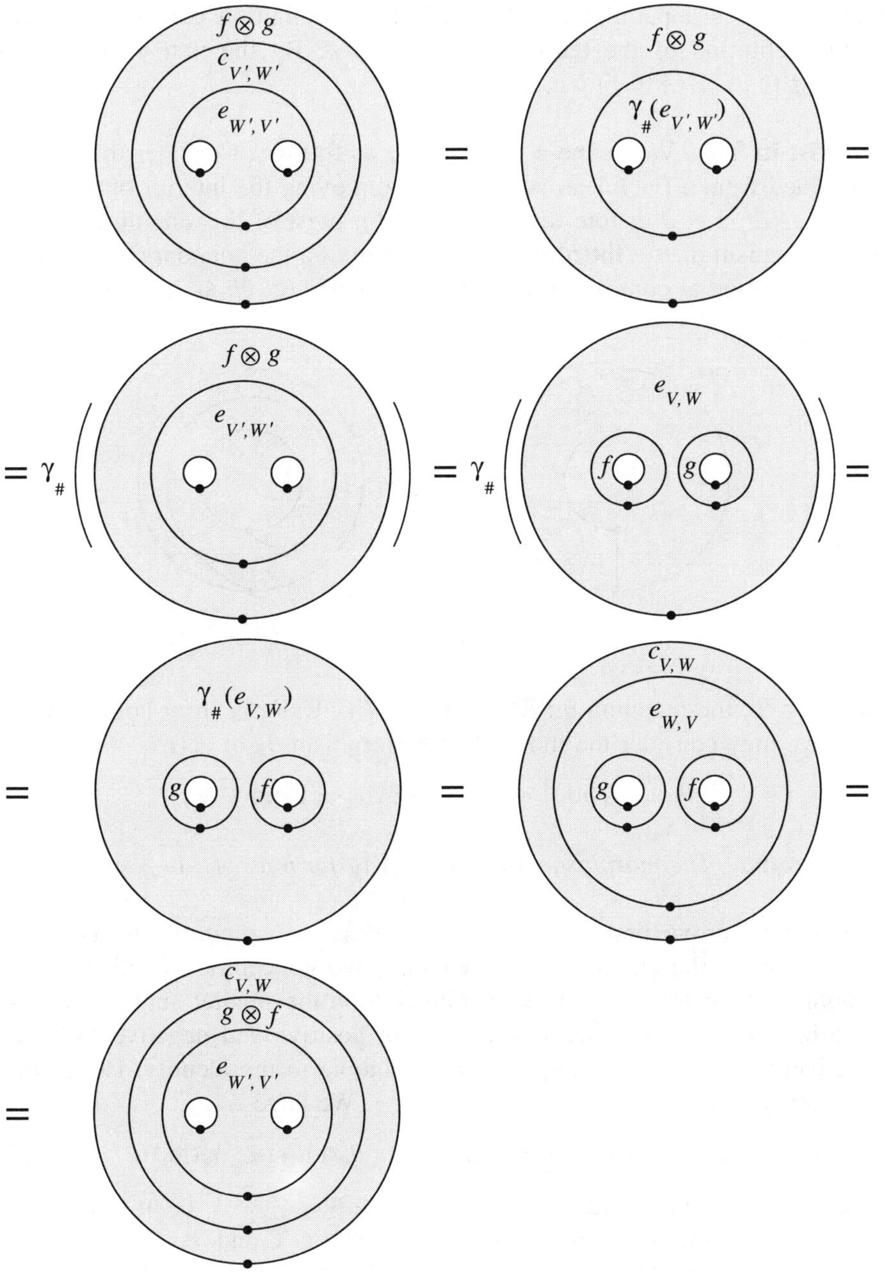

Figure 2.12

(see Figure 2.12). These equalities follow from: (i) the definition of the braiding in $\mathcal{V}_{\mathcal{H}}$, (ii) the fact that γ considered up to isotopy is the identity near the external boundary, (iii) the definition of the tensor product in $\mathcal{V}_{\mathcal{H}}$, (iv) the fact that near each internal boundary component the homeomorphism γ considered up to isotopy is a parallel translation, (v) the definition of the braiding, and (vi) the definition of the tensor product in $\mathcal{V}_{\mathcal{H}}$. By the first excision axiom, $c_{V',W'} \diamond (f \otimes g) = (g \otimes f) \diamond c_{V,W}$.

2.6. Twist in $\mathcal{V}_{\mathcal{H}}$. We define a twist in $\mathcal{V}_{\mathcal{H}}$ as follows. Consider the annulus in $\mathbb{R}^2$ obtained from a Euclidean 2-disk B^2 by removing the interior of a concentric 2-disk $b \subset B^2$. Let θ denote the positive Dehn twist of this annulus, i.e., a self-homeomorphism of $B^2 \setminus \text{Int}(b)$ that is the identity on the boundary and transforms the vertical interval connecting the base points of ∂B^2, ∂b as indicated in Figure 2.13.

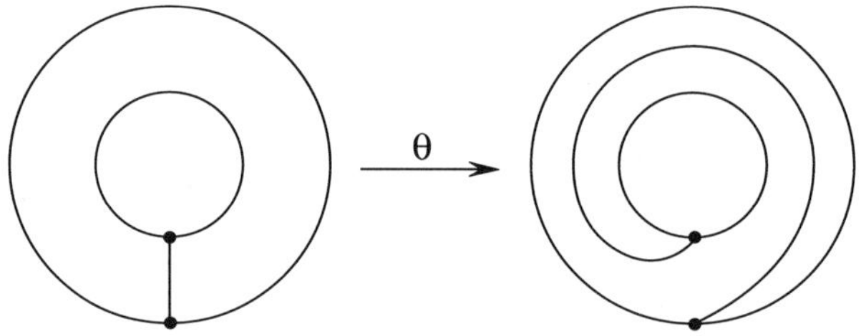

Figure 2.13

For $V \in \mathcal{C}$, the mapping $\theta : E(-V; V) \to E(-V; V)$ is an m-homeomorphism so that we may consider the induced endomorphism $\theta_\#$ of $\mathcal{H}(E(-V; V))$. Set

$$\theta_V = \theta_\#(\text{id}_V) \in \mathcal{H}(E(-V; V)) = \text{Hom}(V, V).$$

2.6.1. Lemma. *The morphisms $\{\theta_V : V \to V\}_V$ form a twist in $\mathcal{V}_{\mathcal{H}}$.*

Proof. Let us prove first that $\theta_V : V \to V$ is an isomorphism. The core of $E(-V; V)$ splits the annulus $E(-V; V)$ into two concentric annuli. Let η be a self-homeomorphism of $E(-V; V)$ acting as θ on the smaller annulus and as θ^{-1} on the bigger annulus. Since θ and θ^{-1} are positive and negative Dehn twists respectively, the m-homeomorphism η is isotopic to the identity. Hence $\eta_\#$ acts as the identity in $\mathcal{H}(E(-V; V)) = \text{Hom}(V, V)$. We have

$$\text{id}_V = \eta_\#(\text{id}_V) = \eta_\#(\text{id}_V \text{id}_V) = \theta_\#(\text{id}_V)\,(\theta^{-1})_\#(\text{id}_V).$$

Similarly, $(\theta^{-1})_\#(\text{id}_V)\,\theta_\#(\text{id}_V) = \text{id}_V$. Therefore $\theta_V : V \to V$ is an isomorphism.

Let us check the naturality of $\{\theta_V\}_V$. Let $U, V \in \mathcal{C}$ and $f \in \mathcal{H}(E(-U; V)) = \text{Hom}(U, V)$. Denote the identity mapping $\text{id} : E(-U; V) \to E(-U; V)$ by j. Clearly, $j_\#(f) = f$. Note that the gluing of j to $\theta : E(-V; V) \to E(-V; V)$ along the internal component of $\partial E(-U; V)$ gives a self-homeomorphism of $E(-U; V)$ isotopic to θ. Therefore

$$\theta_V f = \theta_\#(\mathrm{id}_V)\, j_\#(f) = \theta_\#(\mathrm{id}_V f) = \theta_\#(f).$$

Similarly,

$$f\,\theta_U = j_\#(f)\,\theta_\#(\mathrm{id}_U) = \theta_\#(f\,\mathrm{id}_U) = \theta_\#(f).$$

Therefore $\theta_V f = f\,\theta_U$.

Let us prove that

$$(\theta_V \otimes \theta_W)\, c_{W,V}\, c_{V,W} = \theta_{VW}. \tag{2.6.a}$$

We shall need an elementary equality in the mapping class group of the trinion $\mathbb{E}$ considered in Section 2.5. This group consists of the isotopy classes of homeomorphisms $\mathbb{E} \to \mathbb{E}$ that are equal to the identity on $\partial\mathbb{E}$. The equality in question relates the homeomorphism $\gamma : \mathbb{E} \to \mathbb{E}$ defined in Section 2.5 with the Dehn twists acting on $\mathbb{E}$ near $\partial\mathbb{E}$. Denote the positive Dehn twist in a narrow annulus neighborhood of the first (resp. second) internal component of $\partial\mathbb{E}$ by Θ_1 (resp. Θ_2). Denote the positive Dehn twist in an annulus neighborhood of the external component of $\partial\mathbb{E}$ by Θ. Then

$$\Theta = \Theta_1\Theta_2\gamma^2.$$

One way to prove this equality is to rewrite it in the form $\Theta\gamma^{-1} = \Theta_1\Theta_2\gamma$ and to compare the images under $\Theta\gamma^{-1}$, $\Theta_1\Theta_2\gamma$ of the intervals x_1, x_2 drawn in Figure 2.9. (Alternatively one may compute the action of $\Theta\gamma^{-1}$ and $\Theta_1\Theta_2\gamma$ on the fundamental group.)

Let us glue $e_{V,W}$ to both sides of (2.6.a). The gluing of $e_{V,W}$ to the left-hand side is computed in Figure 2.14. The equalities in Figure 2.14 follow from: (i) the definition of the tensor product and the definition of $c_{W,V}$, (ii) the naturality of the gluing homomorphisms and the fact that the homeomorphism γ, considered up to isotopy, is the identity near the external component of $\partial\mathbb{E}$ and a parallel translation near the internal components of $\partial\mathbb{E}$, (iii) the definition of $c_{V,W}$ and the properties of the identity morphisms, (iv) the formula $\gamma^2 = \Theta_1^{-1}\Theta_2^{-1}\Theta$ and the definition of θ_V, θ_W, and (v) the properties of the identity morphisms. Hence the left-hand and right-hand sides of (2.6.a) yield the same under gluing of $e_{V,W}$. This and the first excision axiom imply (2.6.a).

2.6.2. Corollary. *For any $V, W \in \mathscr{C}$, the morphism $c_{V,W} : V \otimes W \to W \otimes V$ is invertible.*

This follows from (2.6.a) and the invertibility of $\theta_V, \theta_W, \theta_{VW}$.

2.7. Lemma. *Let $\mathscr{V}$ be a monoidal category. Assume that to each object V of $\mathscr{V}$ there are assigned an object V^* of $\mathscr{V}$ and two morphisms $b_V : \mathbb{1} \to V \otimes V^*$,*

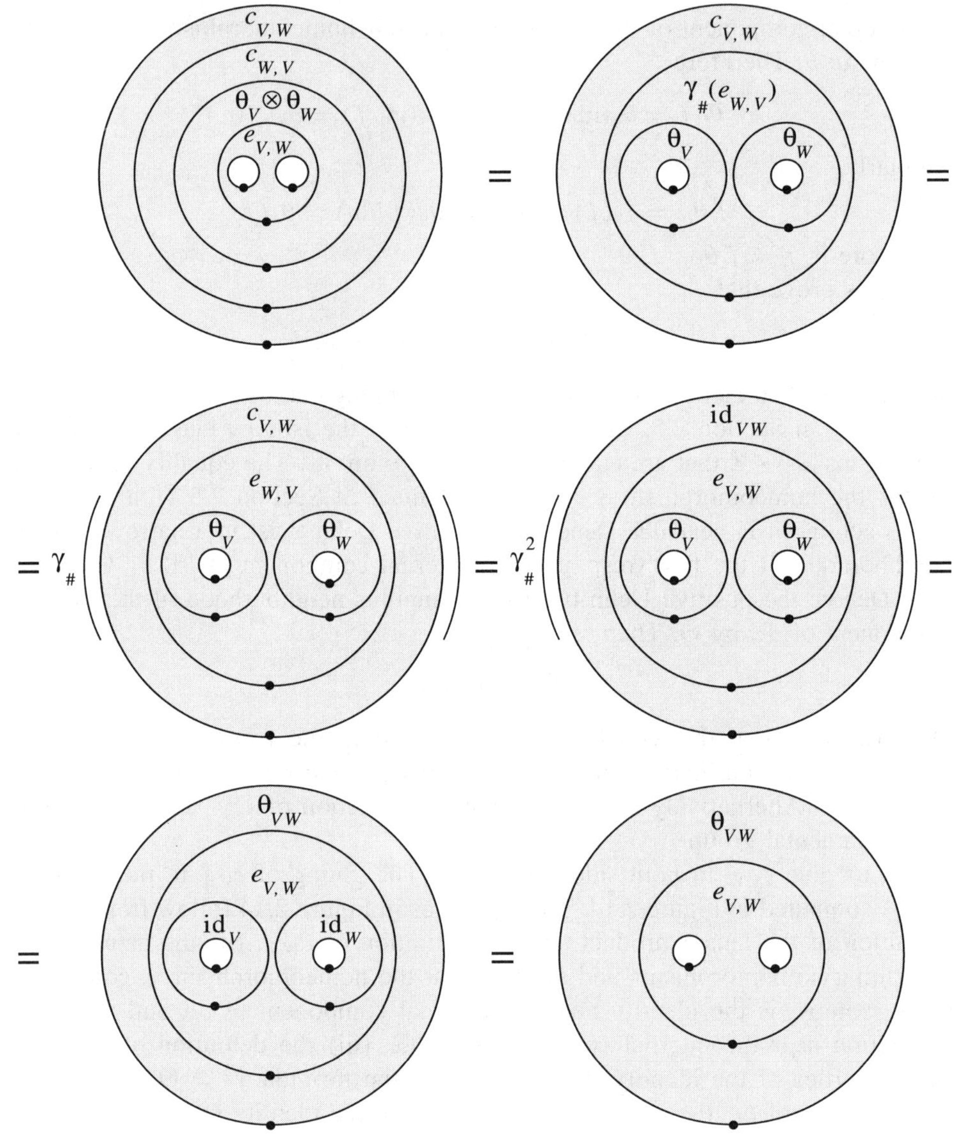

Figure 2.14

$d_V : V^* \otimes V \to \mathbb{1}$ such that

$$(2.7.a) \qquad (d_V \otimes \mathrm{id}_{V^*})(\mathrm{id}_{V^*} \otimes b_V) = \mathrm{id}_{V^*}$$

and the homomorphism

$$(2.7.b) \qquad f \mapsto (f \otimes \mathrm{id}_{V^*})\, b_V : \mathrm{Hom}(V, V) \to \mathrm{Hom}(\mathbb{1}, V \otimes V^*)$$

is injective. Then the family $\{b_V, d_V\}_V$ is a duality in $\mathcal{V}$.

Proof. For an object V of $\mathcal{V}$, set $f_V = (\mathrm{id}_V \otimes d_V)(b_V \otimes \mathrm{id}_V)$. We should check that $f_V = \mathrm{id}_V$. We have

$$(f_V \otimes \mathrm{id}_{V^*})b_V = (\mathrm{id}_V \otimes d_V \otimes \mathrm{id}_{V^*})(b_V \otimes \mathrm{id}_V \otimes \mathrm{id}_{V^*})b_V =$$

$$(\mathrm{id}_V \otimes d_V \otimes \mathrm{id}_{V^*})(b_V \otimes b_V) == (\mathrm{id}_V \otimes d_V \otimes \mathrm{id}_{V^*})(\mathrm{id}_V \otimes \mathrm{id}_{V^*} \otimes b_V)b_V = (\mathrm{id}_V \otimes \mathrm{id}_{V^*})b_V$$

where the last equality follows from (2.7.a). The injectivity of the homomorphism (2.7.b) implies that $f_V = \mathrm{id}_V$.

2.8. Duality in $\mathcal{V}_{\mathcal{H}}$. Let $D = D_V \in \mathcal{H}(E(-V^*V))$ be the distinguished state given in the duality axiom. Recall that

$$\tilde{D} = D \diamond x(V, V^*) \in \mathcal{H}(E(-V^*; -V))$$

(see Figure 1.7). It is understood that in the definition of $x(V, V^*)$ we use the generator $e_- \in \mathcal{H}(E(-\mathbb{1}))$ given by Lemma 2.2 and an arbitrary generator $e_+ \in \mathcal{H}(E(+\mathbb{1}))$. It follows from the disk axiom that $D_V = e_- \diamond d_V$ for a unique $d_V \in \mathcal{H}(E(-VV^*; \mathbb{1}))$ (see Figure 2.15). The morphism $V^* \otimes V \to \mathbb{1}$ determined by d_V is also denoted d_V.

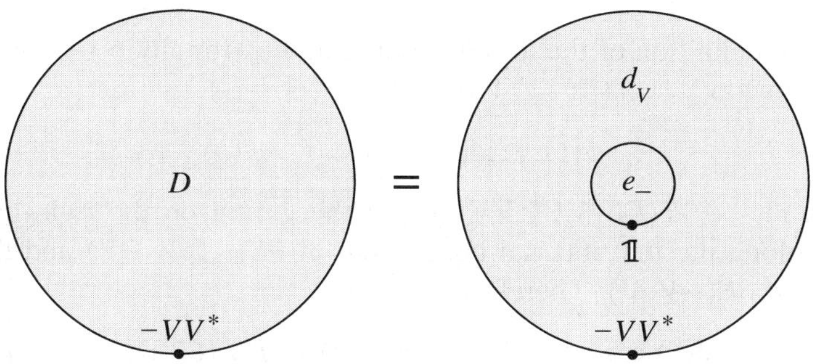

Figure 2.15

Let us construct $b_V : \mathbb{1} \to V \otimes V^*$. For any $b \in \mathcal{H}(E(-\mathbb{1}; VV^*))$, set

$$\psi(b) = \tilde{D} \diamond e_{V,V^*} \diamond b \in \mathcal{H}(E(-\mathbb{1}; V, -V)),$$

see Figure 2.16. This yields a K-homomorphism

$$\psi : \mathcal{H}(E(-\mathbb{1}; VV^*)) \to \mathcal{H}(E(-\mathbb{1}; V, -V)).$$

The non-degeneracy of $\tilde{D}$ and the first excision axiom imply that ψ is an isomorphism.

Recall the state z_V on $E(-V; -\mathbb{1}, V)$ and the isomorphism

$$\xi_\# : \mathcal{H}(E(-V; -\mathbb{1}, V)) \to \mathcal{H}(E(-\mathbb{1}; V, -V))$$

used to define $x(V, V^*)$. Set

$$b_V = \psi^{-1}(\xi_\#(z_V)) \in \mathcal{H}(E(-\mathbb{1}; VV^*)).$$

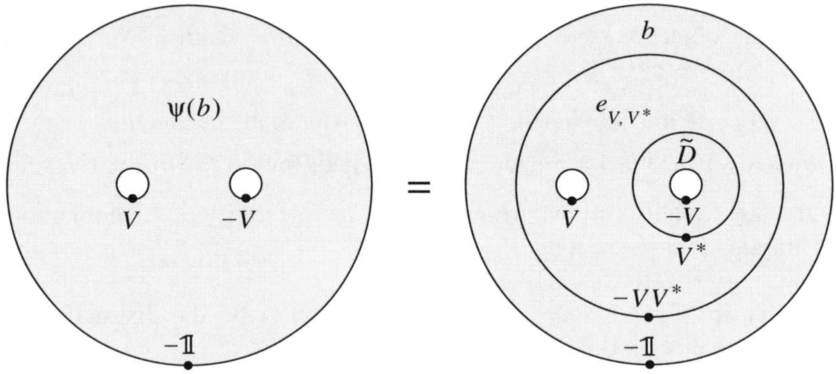

Figure 2.16

Denote the corresponding morphism $\mathbb{1} \to V \otimes V^*$ by the same symbol b_V.

2.8.1. Lemma. *The morphisms* $\{b_V, d_V\}_V$ *satisfy the conditions of Lemma* 2.7 *and therefore form a duality in* $\mathcal{V}_{\mathcal{H}}$.

Proof. By the definition of the tensor product in $\mathcal{V}_{\mathcal{H}}$, for any $b \in \mathcal{H}(E(-\mathbb{1}; VV^*))$ and $f \in \operatorname{Hom}(V, V) = \mathcal{H}(E(-V; V))$,

$$\psi((f \otimes \mathrm{id}_{V^*}) \diamond b) = f \diamond \psi(b)$$

where $f \otimes \mathrm{id}_{V^*} \in \mathcal{H}(E(-VV^*; VV^*))$ and the gluing on the right-hand side is performed along the first internal component of $\partial E(-\mathbb{1}; V, -V)$ and the external component of $\partial E(-V; V)$. Therefore

$$\xi_{\#}^{-1}\psi((f \otimes \mathrm{id}_{V^*}) \diamond b_V) = \xi_{\#}^{-1}(f \diamond \psi(b_V)) = f \diamond (\xi_{\#}^{-1}\psi)(b_V) = f \diamond z_V.$$

By the definition of z_V, we have $e_+ \diamond z_V = \mathrm{id}_V \in \mathcal{H}(E(-V; V))$. Therefore

$$e_+ \diamond \xi_{\#}^{-1}\psi((f \otimes \mathrm{id}_{V^*})\, b_V) = e_+ \diamond (f \diamond z_V) = f \diamond (e_+ \diamond z_V) = f \diamond \mathrm{id}_V = f. \tag{2.8.a}$$

Hence the mapping $f \mapsto (f \otimes \mathrm{id}_{V^*})\, b_V$ has a left inverse. Therefore this mapping is injective.

Let us prove (2.7.a). By the definition of the tensor product of morphisms, we have the equality in Figure 2.17. Consider the state $\tilde{D} \diamond e_{V,V^*}$ on $E(-VV^*; V, -V)$. Let us glue this state to both diagrams in Figure 2.17 along the second internal boundary component. Applying the basic identity on the left-hand side and the definition of ψ on the right-hand side we get the first equality in Figure 2.18. The second equality in Figure 2.18 follows from the formula $\psi(b_V) = \xi_{\#}(z_V)$ and the equality in Figure 1.6 with $W = V^*$. Let us cut out $e_{V^*,V}$ from the first and last diagram in Figure 2.18. Glue $D \in \mathcal{H}(E(-V^*V))$ in the resulting hole. On the right-hand side we obtain $\tilde{D} \in \mathcal{H}(E(-V^*; -V))$, this follows directly from the definition of $\tilde{D}$. To compute the left-hand side replace D with $e_- \diamond d_V$

(cf. Figure 2.15) and replace $\tilde{D}$ with $\tilde{D} \diamond \mathrm{id}_{V^*}$. By the definition of the tensor product of morphisms, this gives the left diagram in Figure 2.19. Now, replace $\tilde{D}$ with $\tilde{D} \diamond \mathrm{id}_{V^*}$ in the right diagram in Figure 2.19. The non-degeneracy of D allows us to cut off $\tilde{D}$ on both sides. By the definition of the identity morphism, $e_- \diamond e_{\mathbb{1},V^*} = \mathrm{id}_{V^*}$. Thus, the equality in Figure 2.19 implies (2.7.a).

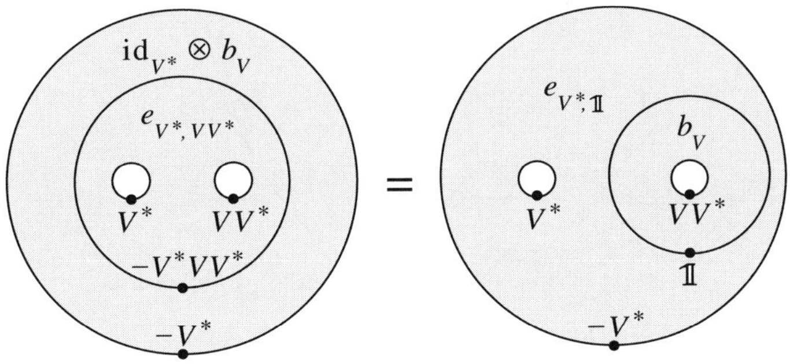

Figure 2.17

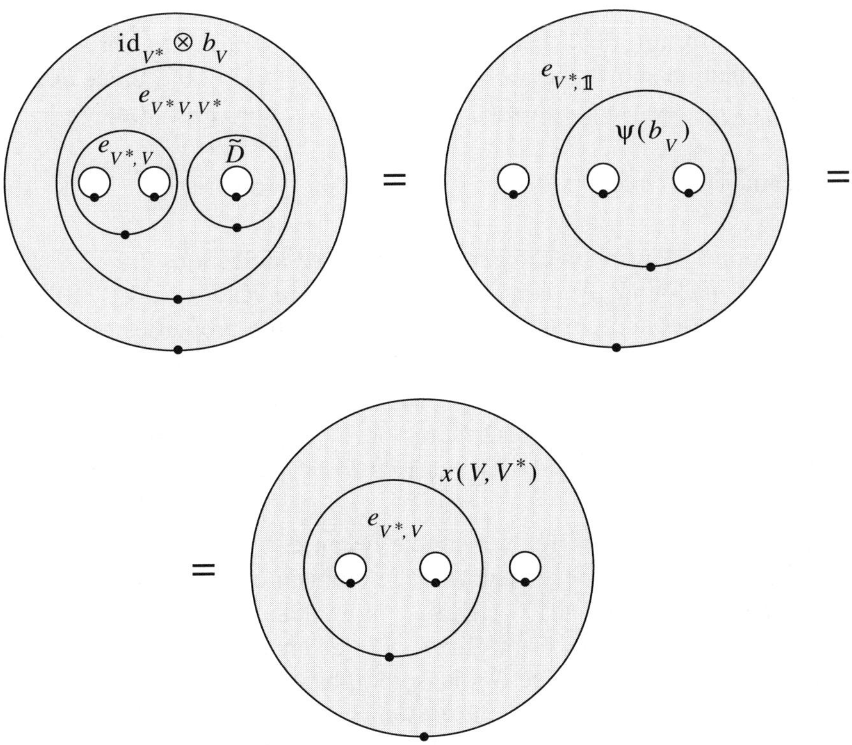

Figure 2.18

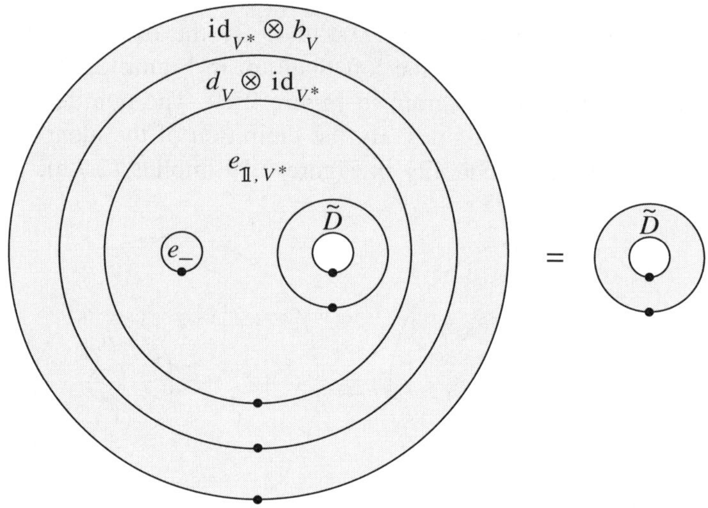

Figure 2.19

2.8.2. Lemma. *The duality in $\mathcal{V}_{\mathcal{H}}$ is compatible with twist.*

Proof. It follows from (2.8.a) that $e_+ \diamond \xi_\#^{-1}\psi((\theta_V \otimes \mathrm{id}_{V^*})\, b_V) = \theta_V$ for any $V \in \mathcal{C}$. A similar calculation shows that $e_+ \diamond \xi_\#^{-1}\psi((\mathrm{id}_V \otimes \theta_{V^*})\, b_V) = \theta_V$. Since ψ, $\xi_\#$, and the excision of e_+ are isomorphisms, we have $(\theta_V \otimes \mathrm{id}_{V^*})\, b_V = (\mathrm{id}_V \otimes \theta_{V^*})\, b_V$.

2.9. Theorem. *The category $\mathcal{V}_{\mathcal{H}}$ is a strict ribbon* Ab*-category.*

Proof. The axioms of a ribbon category were verified in Sections 2.1 - 2.8. The K-linear structure in $\mathrm{Hom}(V, W)$ is induced by the one in $\mathcal{H}(E(-V; W))$. Bilinearity of the tensor product and composition follows from the properties of the gluing homomorphisms.

2.10. Theorem. *For any rational 2-DMF $(\mathcal{H}, \{V_i\}_{i\in I})$, the ribbon* Ab*-category $\mathcal{V}_{\mathcal{H}}$ with the distinguished objects $\{V_i\}_{i\in I}$ is a strict modular category.*

Proof. The objects $\{V_i\}_{i\in I}$ of $\mathcal{V}_{\mathcal{H}}$ are simple because, by (1.6.2), $\mathrm{Hom}(V_i, V_i) = \mathcal{H}(E(-V_i; V_i)) \simeq K$ where $\simeq$ denotes isomorphism of K-modules.

Let us verify axioms (II.1.4.1) - (II.1.4.4) of modular categories. The existence of $0 \in I$ with $V_0 = \mathbb{1}$ follows from (1.6.1). Let us check (II.1.4.2). It suffices to show that for each $i \in I$, the object V_{i^*} is isomorphic to $V_i^* = (V_i)^*$ where $i^* \in I$ is provided by (1.6.2). We shall show the existence of morphisms $g \in \mathrm{Hom}(V_{i^*}, V_i^*)$ and $f \in \mathrm{Hom}(V_i^*, V_{i^*})$ such that $\mathrm{id}_{V_{i^*}} = fg$. Such morphisms are mutually inverse isomorphisms. Indeed, since V_i^* is a simple object, $gf = k\, \mathrm{id}_{V_i^*}$ with $k \in K$. The equalities

$$\mathrm{id}_{V_{i^*}} = \mathrm{id}_{V_{i^*}} \mathrm{id}_{V_{i^*}} = (fg)(fg) = f(gf)g = kfg = k\,\mathrm{id}_{V_{i^*}}$$

imply that $k = 1$.

Let us construct f, g. By (1.6.2), $\mathcal{H}(E(-V_{i^*}; V_{i^*})) = K$. Applying (1.6.3) to the core of the annulus $E(-V_{i^*}; V_{i^*})$ (oriented counterclockwise), we get

$$\mathcal{H}(E(-V_{i^*}; V_{i^*})) = \bigoplus_{j \in I} \mathcal{H}(E(-V_{i^*}; -V_j)) \otimes_K \mathcal{H}(E(V_j; V_{i^*})).$$

By (1.6.2), the j-th summand on the right-hand side is equal to 0 unless $j = i$ and $\mathcal{H}(E(V_i; V_{i^*})) = K$. Therefore $\mathrm{id}_{V_{i^*}} \in \mathcal{H}(E(-V_{i^*}; V_{i^*}))$ can be obtained by gluing certain $g' \in \mathcal{H}(E(-V_{i^*}; -V_i))$ and $f' \in \mathcal{H}(E(V_i; V_{i^*}))$ along the boundary component labelled with V_i. The annulus $E(-V_{i^*}; -V_i)$ can be obtained by gluing two annuli $E(-V_{i^*}; V_i^*)$ and $E(-V_i^*; -V_i)$ along the boundary component labelled with V_i^*. The duality axiom implies that $g' = \tilde{D} \diamond g$ where $g \in \mathcal{H}(E(-V_{i^*}; V_i^*))$ and $\tilde{D} \in \mathcal{H}(E(-V_i^*; -V_i))$. Hence

$$\mathrm{id}_{V_{i^*}} = f' \diamond g' = f' \diamond (\tilde{D} \diamond g) = (f' \diamond \tilde{D}) \diamond g.$$

Therefore $\mathrm{id}_{V_{i^*}}$ is the composition of $g \in \mathrm{Hom}(V_{i^*}, V_i^*)$ with a morphism $f : V_i^* \to V_{i^*}$ represented by $f' \diamond \tilde{D} \in \mathcal{H}(E(-V_i^*; V_{i^*}))$.

Let us prove the domination axiom (II.1.4.3). Let $V \in \mathcal{C}$. Applying axiom (1.6.3) to the core of the annulus $E(-V; V)$ (equipped with clockwise orientation) we get

$$\mathcal{H}(E(-V; V)) = \bigoplus_{i \in I} \mathcal{H}(E(-V; V_i)) \otimes_K \mathcal{H}(E(-V_i; V)).$$

We rewrite this formula as

$$\mathrm{Hom}(V, V) = \bigoplus_{i \in I} \mathrm{Hom}(V, V_i) \otimes_K \mathrm{Hom}(V_i, V)$$

where each inclusion $\mathrm{Hom}(V, V_i) \otimes_K \mathrm{Hom}(V_i, V) \to \mathrm{Hom}(V, V)$ is induced by the composition of morphisms. This implies that the family $\{V_i\}_{i \in I}$ dominates V.

The non-degeneracy axiom (II.1.4.4) for $\mathcal{V}_{\mathcal{H}}$ follows directly from axiom (1.6.4). This completes the proof of the theorem.

2.11. Remark. We may generalize the gluing homomorphisms provided by axiom (1.5.4). Instead of boundary components of an m-surface labelled with the same object of $\mathcal{V}_{\mathcal{H}}$ (and opposite signs) we may consider boundary components labelled with isomorphic objects of $\mathcal{V}_{\mathcal{H}}$ (and opposite signs). Moreover, it is enough to have a morphism between these objects in $\mathcal{V}_{\mathcal{H}}$. Let X, Y be boundary components of an m-surface Σ marked by $+V$ and $-W$, respectively. Let $f \in \mathcal{H}(E(-V; +W))$ be a morphism $V \to W$ in the category $\mathcal{V}_{\mathcal{H}}$. Gluing the annulus $E(-V; +W)$ to Σ along $X \cup Y$ we obtain an m-surface Σ'. The homomorphism $h \mapsto h \diamond f : \mathcal{H}(\Sigma) \to \mathcal{H}(\Sigma')$ can be viewed as a generalized gluing homomorphism determined by f. In the case $V = W$ and $f = \mathrm{id}_V$ we essentially recover the homomorphism provided by (1.5.4). (Note that although the m-surface Σ' differs from $\Sigma/[X = Y]$, they are canonically m-homeomorphic, see Section 4.4.)

2.12. Exercises. 1. Show that for any $V \in \mathcal{C}$, we have $\mathcal{H}(E(-V)) = \mathrm{Hom}(V, \mathbb{1})$ and $\mathcal{H}(E(+V)) = \mathrm{Hom}(\mathbb{1}, V)$. Show that $\mathcal{H}(S^2) = K$.

2. Give a proof of the invertibility of the braiding morphisms constructed in Section 2.5 without using the twist.

3. Let Σ' be an m-surface obtained from an m-surface Σ by replacing the label, $V \in \mathcal{C}$, of a distinguished arc with another label $W \in \mathcal{C}$ such that V and W are isomorphic in the category $\mathcal{V}_{\mathcal{H}}$. Show that the modules $\mathcal{H}(\Sigma)$ and $\mathcal{H}(\Sigma')$ are isomorphic.

3. Weak and mirror modular functors

3.0. Outline. In Section 3.1 we introduce a version of 2-DMF's which applies to a larger class of homeomorphisms of m-surfaces.

In Section 3.2 we briefly discuss a mirror modular functor.

3.1. Weak 2-DMF's. For practical computations, it is convenient to involve the homeomorphisms of m-surfaces that do not preserve the distinguished Lagrangian spaces in 1-homologies. We introduce here a version of 2-DMF's that satisfies the functoriality axiom (1.5.1) in a weakened form (up to a factor).

A weak m-homeomorphism of m-surfaces is a homeomorphism of the underlying surfaces preserving the orientation and the base points, labels, and signs of the boundary components. A weak m-homeomorphism does not need to preserve the distinguished Lagrangian space in 1-homologies.

A weak 2-DMF $\mathcal{H}$ is defined in the same way as a 2-DMF with the following changes: to any weak m-homeomorphism $f : \Sigma_1 \to \Sigma_2$ the functor $\mathcal{H}$ assigns a K-isomorphism $f_\# : \mathcal{H}(\Sigma_1) \to \mathcal{H}(\Sigma_2)$ depending only on the isotopy class of f; in the naturality conditions in axioms (1.5.2) and (1.5.4) we replace "m-homeomorphism" with "weak m-homeomorphism"; instead of (1.5.1) we have

(3.1.1) There is an invertible element $\nabla \in K$ such that for any weak m-homeomorphisms of m-surfaces $g : \Sigma_1 \to \Sigma_2$ and $f : \Sigma_2 \to \Sigma_3$,

$$(fg)_\# = \nabla^{\mu(g_*(\lambda_1), \lambda_2, (f_*)^{-1}(\lambda_3))} f_\# g_\# \tag{3.1.a}$$

where λ_i is the distinguished Lagrangian subspace of $H_1(\Sigma_i; \mathbb{R})$, the star $*$ denotes the induced homomorphism in the real 1-homologies, and μ denotes the Maslov index of the Lagrangian subspaces of $H_1(\Sigma_2; \mathbb{R})$.

Note that a weak 2-DMF is not a modular functor in the sense of Chapter III, unless $\nabla = 1$. The element $\nabla \in K$ is called the charge of $\mathcal{H}$.

If the homeomorphisms f, g in (3.1.1) preserve the distinguished Lagrangian spaces in 1-homologies then $\mu(g_*(\lambda_1), \lambda_2, (f_*)^{-1}(\lambda_3)) = 0$, and we recover axiom (1.5.1). Therefore, restricting a weak 2-DMF $\mathcal{H}$ to m-homeomorphisms, we obtain a 2-DMF in the sense of Section 1. The underlying ribbon category of this 2-DMF is called the underlying ribbon category of $\mathcal{H}$ and denoted $\mathcal{V}_{\mathcal{H}}$.

The definitions of Section 1.6 apply to weak 2-DMF's word for word and give a weak rational 2-DMF.

3.2. Mirror modular functor. Consider the following "negation" of the structure of an m-surface. For an m-surface Σ, the opposite m-surface $\neg\Sigma$ is obtained from Σ by reversing the orientation of Σ while keeping the Lagrangian space in the 1-homologies and the labels, signs, and base points of the components of $\partial\Sigma$. Clearly $\neg(\neg\Sigma) = \Sigma$. The transformation $\Sigma \mapsto \neg\Sigma$ is natural in the sense that any (weak) m-homeomorphism $f : \Sigma_1 \to \Sigma_2$ gives rise to a (weak) m-homeomorphism $\neg f : \neg\Sigma_1 \to \neg\Sigma_2$ which coincides with f as a mapping. This transformation commutes with disjoint union.

For a 2-DMF $\mathcal{H}$ over $(\mathcal{C}, K)$, we define a mirror modular functor $\overline{\mathcal{H}}$ over $(\mathcal{C}, K)$. For an m-surface Σ, set $\overline{\mathcal{H}}(\Sigma) = \mathcal{H}(\neg\Sigma)$. For an m-homeomorphism f, set $f_\# = (\neg f)_\#$. Similarly, the gluing homomorphism $\overline{\mathcal{H}}(\Sigma) \to \overline{\mathcal{H}}(\Sigma')$ is defined to be the gluing homomorphism $\mathcal{H}(\neg\Sigma) \to \mathcal{H}(\neg\Sigma')$. We keep the family $\{V_i\}_{i\in I}$ and the rule $V \mapsto V^* : \mathcal{C} \to \mathcal{C}$. The surfaces $\neg E(-VW; V, W)$ and $\neg E(\pm V)$ can be identified with $E(-VW; W, V)$ and $E(\pm V)$ via the symmetry in $\mathbb{R}^2$ with respect to the vertical axis. The state

$$\gamma_\#(e_{V,W}) \in \mathcal{H}(E(-VW; W, V)) = \mathcal{H}(\neg E(-VW; V, W)) = \overline{\mathcal{H}}(E(-VW; V, W)),$$

defined in Section 2.5, is chosen as the preferred element of $\overline{\mathcal{H}}(E(-VW; V, W))$. Take

$$D_V \in \mathcal{H}(E(-V^*V)) = \mathcal{H}(\neg E(-V^*V)) = \overline{\mathcal{H}}(E(-V^*V))$$

as the preferred element of $\overline{\mathcal{H}}(E(-V^*V))$. It is straightforward to verify that $\overline{\mathcal{H}}$ satisfies all axioms of a 2-DMF except possibly the duality axiom. Conjecturally, the duality axiom is also satisfied.

For a weak 2-DMF $\mathcal{H}$ with charge $\nabla \in K$, we similarly define a weak 2-DMF $\overline{\mathcal{H}}$ with charge $\nabla^{-1} \in K$ (modulo a verification of the duality axiom).

3.3. Exercise. Verify the basic identity (1.5.7) in $\overline{\mathcal{H}}$. Show that the underlying ribbon category of $\overline{\mathcal{H}}$ is isomorphic to the mirror ribbon category $\overline{\mathcal{V}_{\mathcal{H}}}$.

4. Construction of modular functors

4.0. Outline. We show that every modular category $\mathcal{V}$ gives rise in a canonical way to a weak rational 2-DMF $\mathcal{H}(\mathcal{V})$. The underlying ribbon category of $\mathcal{H}(\mathcal{V})$ is isomorphic to $\mathcal{V}$.

4.1. Theorem. *Let $\mathcal{V}$ be a modular category with ground ring K and rank $\mathcal{D} \in K$. Let $\mathcal{C}$ be the class of objects of $\mathcal{V}$. There exists a weak rational 2-DMF $\mathcal{H} = \mathcal{H}(\mathcal{V})$ over $(\mathcal{C}, K)$ with charge $\nabla = \mathcal{D}\Delta_{\mathcal{V}}^{-1} \in K$ such that the underlying ribbon* Ab*-category of $\mathcal{H}$ is isomorphic to $\mathcal{V}$.*

Restricting $\mathcal{H}$ to m-homeomorphisms, we obtain a rational 2-DMF with underlying ribbon category $\mathcal{V}$.

The relationships between modular categories and 2-DMF's may be summarized as follows. There are two arrows

$$\text{modular category } \mathcal{V} \text{ with rank } \mathcal{D} \mapsto$$

$$\text{weak rational 2-DMF } \mathcal{H}(\mathcal{V}) \text{ with charge } \mathcal{D}\Delta_{\mathcal{V}}^{-1}$$

and

$$\text{weak rational 2-DMF} \mapsto \text{underlying modular category.}$$

The composition of these arrows (in this order) is the identity. It seems natural to conjecture that the composition of these arrows in the opposite order is also the identity so that any weak rational 2-DMF $\mathcal{H}$ is isomorphic to the weak 2-DMF $\mathcal{H}(\mathcal{V}_{\mathcal{H}})$. We conjecture also that if ∇ is the charge of a weak rational 2-DMF $\mathcal{H}$ then $\nabla\Delta_{\mathcal{V}}$ is a rank of the underlying modular category $\mathcal{V}$.

To prove Theorem 4.1, we shall construct $\mathcal{H}$ from the 3-dimensional TQFT $(\mathcal{T}^e, \tau^e)$ defined in Section IV.6. The 2-DMF $\mathcal{H}$ is essentially a disguised version of the modular functor $\mathcal{T}^e$, although to define the gluing homomorphisms we need τ^e. To sum up, our construction of a weak rational 2-DMF involves four major steps: (a) a construction of isotopy invariants of ribbon graphs in $\mathbb{R}^3$ (see Chapter I), (b) a definition of the topological invariant τ of closed 3-manifolds with embedded v-colored ribbon graphs (see Chapter II), (c) a definition of the TQFT $(\mathcal{T}^e, \tau^e)$ (see Sections IV.1 - IV.6), (d) a definition of the gluing homomorphisms and a verification of the axioms of 2-DMF's. This shows that the study of 2-DMF's is closely related to the study of 3-manifold invariants and 3-dimensional TQFT's.

The proof of Theorem 4.1 occupies the remaining part of this section and Section 5. In this section we define the modules of states of m-surfaces, the action of weak m-homeomorphisms, and the gluing homomorphisms. We also verify axioms (1.5.1) - (1.5.4). In Section 5 we shall verify the remaining axioms.

4.2. Marked surfaces versus extended surfaces. Recall that an extended surface or, briefly, an e-surface is a closed oriented surface Υ endowed with a finite (possibly empty) family of disjoint marked arcs and a Lagrangian space $\lambda(\Upsilon) \subset H_1(\Upsilon; \mathbb{R})$ (cf. Section IV.6.1, for the definition of a marked arc, see Section IV.1.1). The notions of marked and extended surfaces are essentially equivalent; one may switch from the language of marked surfaces to the language of extended surfaces and back. We shall use only one direction of this equivalence, namely, that every marked surface naturally determines an extended surface. More exactly, as we shall see below, every m-surface Σ embeds into an e-surface $\overline{\Sigma}$ such that the complement $\overline{\Sigma} \setminus \Sigma$ consists of 2-disks.

Let B^2 denote the unit disk $\{z \in \mathbb{C} \mid |z| \leq 1\}$. Set $S^1 = \partial B^2 = \{z \in \mathbb{C} \mid |z| = 1\}$. We provide S^1 with clockwise orientation and base point $-\sqrt{-1}$. Endow B^2 with the horizontal diameter connecting the points $-1, +1 \in S^1$ and oriented from -1 to $+1$.

Let Σ be an m-surface with distinguished Lagrangian space $\lambda(\Sigma) \subset H_1(\Sigma; \mathbb{R})$. For a component X of $\partial\Sigma$ with base point $x \in X$, a parametrization of X is an orientation-preserving homeomorphism $(S^1, -\sqrt{-1}) \to (X, x)$. (Recall that the orientation of X is induced by the orientation of Σ.) Choose parametrizations of all components of $\partial\Sigma$ and glue a copy of B^2 along each of these parametrizations. This yields a closed surface denoted by $\overline{\Sigma}$. Clearly, $\Sigma \subset \overline{\Sigma}$ and $\overline{\Sigma} \backslash \Sigma$ consists of m copies of $\mathrm{Int}(B^2)$ where m is the number of components of $\partial\Sigma$. We provide $\overline{\Sigma}$ with the structure of an e-surface as follows. It is clear that the orientation of Σ extends to an orientation of $\overline{\Sigma}$. The inclusion homomorphism $\mathrm{incl}_* : H_1(\Sigma; \mathbb{R}) \to H_1(\overline{\Sigma}; \mathbb{R})$ preserves the homological intersection form. The kernel of this homomorphism is the annihilator of the homological intersection form in $H_1(\Sigma; \mathbb{R})$. Therefore $\mathrm{incl}_*(\lambda(\Sigma))$ is a Lagrangian subspace of $H_1(\overline{\Sigma}; \mathbb{R})$. Set $\lambda(\overline{\Sigma}) = \mathrm{incl}_*(\lambda(\Sigma))$. The role of distinguished arcs on $\overline{\Sigma}$ is played by the preferred diameters of the m copies of B^2 in $\overline{\Sigma}$. Mark each of these m diameters with the same object of $\mathcal{V}$ and the same sign ± 1 as the corresponding component of $\partial\Sigma$. In this way, $\overline{\Sigma}$ acquires the structure of an e-surface. By definition, if $\Sigma = \emptyset$ then $\overline{\Sigma} = \emptyset$.

Strictly speaking, the e-surface $\overline{\Sigma}$ depends on the choice of parametrizations of the components of $\partial\Sigma$. However, the e-surfaces corresponding to different parametrizations are e-homeomorphic via e-homeomorphisms defined in a canonical way (up to isotopy). This follows from the fact that any two parametrizations of a component of $\partial\Sigma$ are isotopic (cf. Appendix III). Therefore we can regard the e-surfaces corresponding to different parametrizations of $\partial\Sigma$ as copies of the same e-surface.

A weak m-homeomorphism of marked surfaces $f : \Sigma_1 \to \Sigma_2$ extends uniquely (up to isotopy) to a weak e-homeomorphism $\overline{f} : \overline{\Sigma}_1 \to \overline{\Sigma}_2$. It is obvious that the formulas $(\Sigma \mapsto \overline{\Sigma}, f \mapsto \overline{f})$ define a covariant functor from the category of m-surfaces and (weak) m-homeomorphisms into the category of e-surfaces and (weak) e-homeomorphisms. Although we shall not need it, note that this is an equivalence of categories.

4.3. The module $\mathcal{H}(\Sigma)$. For a marked surface Σ, set

$$\mathcal{H}(\Sigma) = \mathcal{T}^e(\overline{\Sigma})$$

where $\mathcal{T}^e$ is the modular functor defined in Section IV.6.

For a weak m-homeomorphism of marked surfaces $f : \Sigma_1 \to \Sigma_2$, set

$$f_\# = \overline{f}_\# : \mathcal{H}(\Sigma_1) = \mathcal{T}^e(\overline{\Sigma_1}) \to \mathcal{T}^e(\overline{\Sigma_2}) = \mathcal{H}(\Sigma_2). \tag{4.3.a}$$

Axioms (1.5.1) - (1.5.3) follow from Lemma IV.6.3.3. The commutativity of diagram (1.5.b) for weak m-homeomorphisms f, g follows from Remark IV.6.4.1. Axiom (3.1.1) with $\nabla = \mathcal{D}\Delta_{\mathcal{V}}^{-1} \in K$ follows from Lemma IV.6.3.2.

4.4. A model for gluing. To define the gluing homomorphisms, we need a slightly different definition of the gluing of m-surfaces. Let Σ be an m-surface. Let X, Y be components of $\partial\Sigma$ marked with $+W$ and $-W$ respectively, where W is an object of $\mathcal{V}$. Consider the m-surface Σ' obtained by attaching the cylinder $X \times [0, 1]$ to Σ along its bases: the bottom base $X \times 0 = X$ is identified with $X \subset \Sigma$ and the top base $X \times 1$ is identified with $Y \subset \Sigma$ via an orientation-reversing base point preserving homeomorphism $X \times 1 \to Y$. Extend the orientation of Σ to an orientation of Σ' and endow the components of $\partial\Sigma' = \partial\Sigma \setminus (X \cup Y)$ with the base points, labels, and signs determined by the corresponding data for Σ. Finally, provide $H_1(\Sigma'; \mathbb{R})$ with the Lagrangian subspace which is the image of the given Lagrangian subspace of $H_1(\Sigma; \mathbb{R})$ under the inclusion homomorphism $H_1(\Sigma; \mathbb{R}) \to H_1(\Sigma'; \mathbb{R})$. It follows from the next lemma that the m-surface Σ' is m-homeomorphic to the m-surface $\Sigma/[X = Y]$ via an m-homeomorphism defined in a canonical way up to isotopy (cf. Section 1.3). For technical reasons, attaching a cylinder as above is more convenient than the identification $X = Y$. In Sections 4 and 5, we shall use Σ' as a model for $\Sigma/[X = Y]$. We shall denote the m-surface Σ' by $\Sigma/[X \approx Y]$.

4.4.1. Lemma. *Let $\tilde{\Sigma}$ be an m-surface obtained by attaching the cylinder $X \times [0, 1]$ to Σ along $X \times 0 = X$. Let $U = X \times [-1, 0]$ be a cylindrical neighborhood of $X = X \times 0$ in Σ. Let $j : \Sigma \to \tilde{\Sigma}$ be a mapping that is the identity outside U and carries each point $x \times t \in U$ with $x \in X, t \in [-1, 0]$ into the point*

$$x \times (2t + 1) \in X \times [-1, 1] = U \cup_{X \times 0} (X \times [0, 1]) \subset \tilde{\Sigma}.$$

Then j is an m-homeomorphism and the isotopy class of j does not depend on the choice of U.

Proof. It is obvious that j is an m-homeomorphism. All cylinder neighborhoods of X in Σ are isotopic and therefore the m-homeomorphisms obtained from different neighborhoods of X are isotopic in the class of m-homeomorphisms.

4.5. Elementary cobordisms. Let Σ, Σ', X, Y, W be the same objects as in Section 4.4. To define the gluing homomorphism $\mathcal{H}(\Sigma) \to \mathcal{H}(\Sigma')$ we shall construct an extended 3-cobordism M between the e-surfaces $\overline{\Sigma}$ and $\overline{\Sigma'}$. (For the definition of extended 3-cobordisms, see Section IV.6.)

Let $B_X \subset \overline{\Sigma}, B_Y \subset \overline{\Sigma}$ be the copies of the unit 2-disk B^2 bounded by X, Y in $\overline{\Sigma}$. Let $\alpha_X \subset B_X, \alpha_Y \subset B_Y$ be the distinguished diameters of these disks marked with $+W$ and $-W$ respectively. Consider the cylinder $\overline{\Sigma} \times [0,1]$ regarded as an extended 3-cobordism with bottom base $\overline{\Sigma} \times 0$ and top base $\overline{\Sigma} \times 1$. Recall that the distinguished ribbon graph in $\overline{\Sigma} \times [0,1]$ consists of the vertical bands $\alpha \times [0,1]$ where α runs over the distinguished arcs in $\overline{\Sigma}$ (cf. Sections IV.1.6 and IV.6.2). The extended 3-cobordism M between $\overline{\Sigma}$ and $\overline{\Sigma'}$ will be obtained by attaching to $\overline{\Sigma} \times [0,1]$ a 1-handle connecting $B_X \times 1$ to $B_Y \times 1$ (cf. Figure 4.1).

First, attach the 1-handle $C = B_X \times [0,1]$ to the top base of $\overline{\Sigma} \times [0,1]$ as follows. Glue $B_X \times 0 \subset C$ to $B_X \times 1 \subset \overline{\Sigma} \times 1$ along $B_X \times 0 = B_X = B_X \times 1$. Glue $B_X \times 1 \subset C$ to $B_Y \times 1 \subset \overline{\Sigma} \times 1$ along a homeomorphism induced by the homeomorphism $z \mapsto -\mathrm{Re}(z) + \sqrt{-1}\,\mathrm{Im}(z) : B^2 \to B^2$. Note that the arc $\alpha_X \times 1 \subset C$ is identified with $\alpha_Y \times 1$ via an orientation-reversing homeomorphism. This gluing yields a 3-cobordism $M = (\overline{\Sigma} \times [0,1]) \cup C$ with bottom base $\overline{\Sigma} \times 0$ and top base

$$(\overline{\Sigma} \setminus \mathrm{Int}(B_X \cup B_Y)) \cup_{X \cup Y} (\partial B_X \times [0,1]) = \overline{\Sigma'}.$$

The bands $\alpha_X \times [0,1], \alpha_Y \times [0,1] \subset \overline{\Sigma} \times [0,1]$, and $\alpha_X \times [0,1] \subset C$ form a band, say β, in M. The directions of the bands $\alpha_X \times [0,1], \alpha_Y \times [0,1]$ (down and up, respectively) are compatible and induce a direction of β. Similarly, the product orientations of the rectangles $\alpha_X \times [0,1], \alpha_Y \times [0,1]$ (induced by the given orientations in α_X, α_Y and right-hand orientation in [0,1]) are compatible and induce an orientation of β. We color β with W. The band β meets the boundary of M along its bases $\alpha_X \times 0, \alpha_Y \times 0 \subset \overline{\Sigma} \times 0$ and the intervals $x \times [0,1], x' \times [0,1] \subset C$ where x, x' are the endpoints of α_X. To eliminate the last intersection, push β slightly inside M keeping its bases fixed. This gives a colored oriented directed band $\beta' \subset M$. We endow M with the colored ribbon graph formed by β' and the vertical bands $\alpha \times [0,1] \subset \overline{\Sigma} \times [0,1] \subset M$, where α runs over the distinguished arcs in $\overline{\Sigma}$ distinct from α_X, α_Y. In this way we transform M into an extended 3-cobordism between $\overline{\Sigma}$ and $\overline{\Sigma'}$. This cobordism is denoted by the same symbol M. It is schematically shown in Figure 4.1.

The dual handle decomposition of M consists of one 2-handle (with a standard band inside) attached to $\overline{\Sigma'} \times [0,1]$ from below. By a 2-handle with standard band we mean the product $B^2 \times [0,1]$ with the band $\alpha \times [0,1]$ where α is the distinguished diameter of B^2. The intervals $\alpha \times 0, \alpha \times 1$ are the bases of this band. The band is colored with W, directed towards $\alpha \times 0$ and endowed with the product of right-hand orientations in α and [0, 1]. The 2-handle $B^2 \times [0,1]$ is glued to the base $\overline{\Sigma'} \times 0$ of $\overline{\Sigma'} \times [0,1]$ along a homeomorphism $f \times \mathrm{id} : S^1 \times [0,1] \to X \times [0,1]$

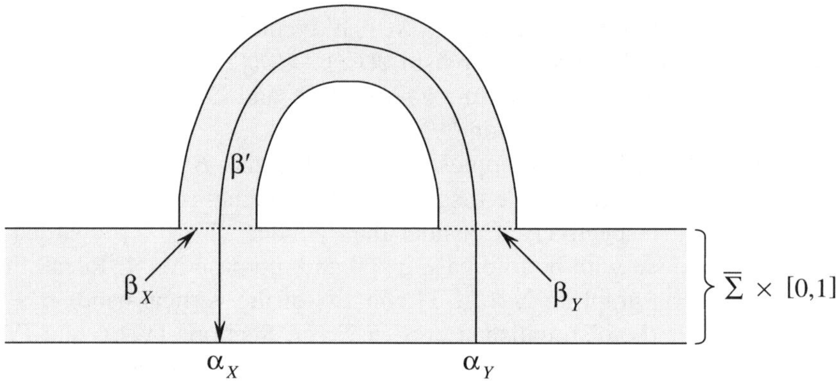

Figure 4.1

where $f : S^1 \to X$ is a parametrization of X and $X \times [0, 1]$ is the cylinder glued to Σ to form $\Sigma' = \Sigma' \times 0$.

Consider the Lagrangian relation $N = N(M) : H_1(\overline{\Sigma}; \mathbb{R}) \Rightarrow H_1(\overline{\Sigma'}; \mathbb{R})$, see Section IV.4.2. We shall need the following property of N:

$$N_*(\lambda(\overline{\Sigma})) = \lambda(\overline{\Sigma'}), \quad N^*(\lambda(\overline{\Sigma'})) = \lambda(\overline{\Sigma}). \tag{4.5.a}$$

Indeed, if X and Y lie on different components of Σ, then the Lagrangian relation N is induced by an isomorphism $H_1(\overline{\Sigma}; \mathbb{R}) \approx H_1(\overline{\Sigma'}; \mathbb{R})$ which transforms $\lambda(\overline{\Sigma})$ onto $\lambda(\overline{\Sigma'})$. If X and Y lie on the same component of Σ, then the symplectic space $H_1(\overline{\Sigma'}; \mathbb{R})$ is a direct sum of $H_1(\overline{\Sigma}; \mathbb{R})$ and a 2-dimensional non-degenerate symplectic space. We have $\lambda(\overline{\Sigma'}) = \lambda(\overline{\Sigma}) \oplus \mathbb{R}m$ where m is the vector of that 2-dimensional symplectic space represented by a meridian of the 1-handle C. It is easy to check that for any Lagrangian space $\lambda \subset H_1(\overline{\Sigma}; \mathbb{R})$, we have $N_*(\lambda) = \lambda \oplus \mathbb{R}m$ and $N^*(\lambda \oplus \mathbb{R}m) = \lambda$. This implies (4.5.a).

4.6. Gluing homomorphisms. Let Σ' be an m-surface obtained from an m-surface Σ by gluing, as described in Section 4.4. Consider the extended 3-cobordism M between the e-surfaces $\overline{\Sigma}$ and $\overline{\Sigma'}$ constructed in Section 4.5. We define the gluing homomorphism $\mathcal{H}(\Sigma) \to \mathcal{H}(\Sigma')$ to be the composition of the identification isomorphisms $\mathcal{H}(\Sigma) = \mathcal{T}^e(\overline{\Sigma})$, $\mathcal{H}(\Sigma') = \mathcal{T}^e(\overline{\Sigma'})$ and the homomorphism

$$\tau^e(M, \overline{\Sigma}, \overline{\Sigma'}) : \mathcal{T}^e(\overline{\Sigma}) \to \mathcal{T}^e(\overline{\Sigma'}).$$

We verify axiom (1.5.4). The naturality of gluing homomorphisms with respect to m-homeomorphisms follows directly from the naturality of τ^e with respect to e-homeomorphisms of extended 3-cobordisms. To prove Theorem 4.1 we also have to check the naturality of gluing homomorphisms with respect to weak m-homeomorphisms. Thus, we should verify that diagram (1.5.c) is commutative for any weak m-homeomorphism $f : \Sigma_1 \to \Sigma_2$ inducing a weak m-homeomorphism

$f' : \Sigma'_1 \to \Sigma'_2$ where the m-surfaces Σ'_1, Σ'_2 are obtained from Σ_1, Σ_2 by the gluing of boundary components. Consider the extended 3-cobordisms $(M_1, \overline{\Sigma_1}, \overline{\Sigma'_1})$ and $(M_2, \overline{\Sigma_2}, \overline{\Sigma'_2})$ determined by gluings $\Sigma_1 \mapsto \Sigma'_1, \Sigma_2 \mapsto \Sigma'_2$ as above. The homeomorphisms f, f' extend to a weak e-homeomorphism $g : M_1 \to M_2$. Therefore, by the result of Exercise IV.6.9.4, diagram (1.5.c) is commutative up to multiplication by $(\mathcal{D}\Delta^{-1})^{\mu_1 - \mu_2}$ where μ_1, μ_2 are the Maslov indices in (IV.6.9.a). We claim that $\mu_1 = \mu_2 = 0$. By (4.5.a), the second and third Lagrangian spaces in the formula for μ_2 coincide. Therefore $\mu_2 = 0$. Under the isomorphisms

$$\overline{f}_* : H_1(\overline{\Sigma_1}; \mathbb{R}) \to H_1(\overline{\Sigma_2}; \mathbb{R}), \quad \overline{f'}_* : H_1(\overline{\Sigma'_1}; \mathbb{R}) \to H_1(\overline{\Sigma'_2}; \mathbb{R})$$

the Lagrangian relation $N(M_1)$ corresponds to the Lagrangian relation $N = N(M_2)$. By definition, $g|_{\partial_-(M_1)} = \overline{f}$ and $g|_{\partial_+(M_1)} = \overline{f'}$. Therefore

$$N_*(g|_{\partial_-(M_1)})_*(\lambda_-(M_1)) = N_* \overline{f}_*(\lambda(\overline{\Sigma_1})) =$$

$$= \overline{f'}_*(N(M_2))_*(\lambda(\overline{\Sigma_1})) = \overline{f'}_*(\lambda(\overline{\Sigma'_1})) = (g|_{\partial_+(M_1)})_*(\lambda_+(M_1)).$$

Here the third equality follows from (4.5.a). Hence the first and second Lagrangian spaces in the formula for μ_1 coincide. Therefore $\mu_1 = \mu_2 = 0$, and diagram (1.5.c) is commutative.

Let us verify condition (ii) of axiom (1.5.4). Let $B_X, B_Y, B_{X'}, B_{Y'}$ be copies of the unit 2-disk bounded by X, Y, X', Y' in $\overline{\Sigma}$. Set $\Sigma' = \Sigma/[X \approx Y]$ and $\Sigma'' = \Sigma/[X \approx Y, X' \approx Y']$ (cf. Section 4.4). Let $(M_1, \overline{\Sigma}, \overline{\Sigma'})$ be the extended 3-cobordism obtained from $\overline{\Sigma} \times [0, 1]$ by attaching a 1-handle connecting $B_X \times 1$ to $B_Y \times 1$. Similarly, let $(M_2, \overline{\Sigma'}, \overline{\Sigma''})$ be the extended 3-cobordism obtained from $\overline{\Sigma'} \times [0, 1]$ by attaching a 1-handle connecting $B_{X'} \times 1$ to $B_{Y'} \times 1$. Gluing M_1 to M_2 along the identity $\mathrm{id} : \overline{\Sigma'} \to \overline{\Sigma'}$ gives an extended 3-cobordism $(M, \overline{\Sigma}, \overline{\Sigma''})$. The cobordism M is obtained from $\overline{\Sigma} \times [0, 1]$ by attaching two 1-handles: one connecting $B_X \times 1$ to $B_Y \times 1$ and one connecting $B_{X'} \times 1$ to $B_{Y'} \times 1$. By definition, the gluing homomorphisms $\mathcal{H}(\Sigma) \to \mathcal{H}(\Sigma')$ and $\mathcal{H}(\Sigma') \to \mathcal{H}(\Sigma'')$ are equal to $\tau^e(M_1)$ and $\tau^e(M_2)$ respectively. To compute the composition $\tau^e(M_2)\,\tau^e(M_1)$ we use Theorem IV.7.1. In the notation of that theorem, $p = \mathrm{id}$, $\lambda_-(M_1) = \lambda(\overline{\Sigma})$, and $\lambda_+(M_1) = \lambda_-(M_2) = \lambda(\overline{\Sigma'})$. We have $p_*(\lambda_+(M_1)) = \lambda_-(M_2)$ and therefore $n' = 0$. It follows from (4.5.a) that $(N(M_1))_*(\lambda_-(M_1)) = \lambda_+(M_1)$ and therefore $n = 0$. By Theorem IV.7.1, $\tau^e(M_2)\,\tau^e(M_1) = \tau^e(M)$. A similar computation shows that the composition of gluing homomorphisms $\mathcal{H}(\Sigma) \to \mathcal{H}(\Sigma/[X' \approx Y'])$ and $\mathcal{H}(\Sigma/[X' \approx Y']) \to \mathcal{H}(\Sigma'')$ is also equal to $\tau^e(M)$. This implies condition (ii) of (1.5.4).

Condition (iii) of (1.5.4) follows directly from the multiplicativity of τ^e with respect to disjoint union of extended 3-cobordisms and the fact that for any e-surface Υ, we have $\tau^e(\Upsilon \times [0, 1]) = \mathrm{id}_{\mathcal{T}^e(\Upsilon)}$.

5. Construction of modular functors continued

5.0. Outline. We finish the proof of Theorem 4.1.

5.1. Digression on the TQFT $(\mathcal{T}^e, \tau^e)$. We analyze the modules of states of extended 2-spheres and develop a technique which allows us to present elements of these modules by v-colored ribbon graphs in $\mathbb{R}^3$.

Let $V_1, ... V_m, W_1, ..., W_n$ be objects of $\mathcal{V}$ and let $\varepsilon_1, ..., \varepsilon_m, \nu_1, ..., \nu_n \in \{+1, -1\}$. Denote the sequences $(V_1, \varepsilon_1), ..., (V_m, \varepsilon_m)$ and $(W_1, \nu_1), ..., (W_n, \nu_n)$ by η and η' respectively. Let Ω be a v-colored ribbon graph in $\mathbb{R}^3$ such that η (resp. η') is the sequence of colors and directions of those bands of Ω which hit the bottom (resp. top) boundary intervals. As usual, the sign $+1$ corresponds to the downward direction near the corresponding boundary interval and the sign -1 corresponds to the upward direction. In the notation of Chapter I, $\eta = \text{source}(\Omega)$, $\eta' = \text{target}(\Omega)$, and Ω represents a morphism $\eta \to \eta'$ in the category of ribbon graphs $\text{Rib}_{\mathcal{V}}$. We inscribe Ω in a 3-ball B^3 so that the free bands of Ω (slightly extended and rounded) are attached to a big circle in $S^2 = \partial B^3$, see Figure 5.1 where this circle lies in the plane of the picture. Provide the bases of the bands lying on this circle with clockwise orientation. We view these $m+n$ bases as arcs on $S^2 = \partial B^3$. Mark the bottom and top arcs with $(V_1, -\varepsilon_1), ..., (V_m, -\varepsilon_m)$ and $(W_1, \nu_1), ..., (W_n, \nu_n)$ respectively. Provide B^3 with right-handed orientation.

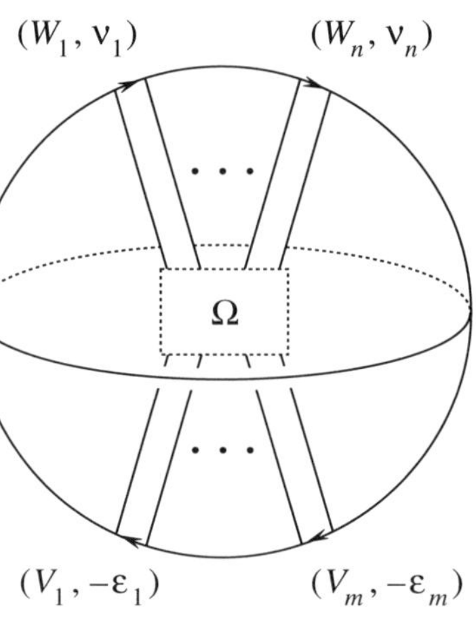

Figure 5.1

It is clear that the pair (B^3, Ω) is an extended 3-manifold. It is bounded by an extended 2-sphere with $m+n$ marked arcs (and with the zero Lagrangian space in 1-homologies). Denote this extended 2-sphere by $\Upsilon = \Upsilon(\eta; \eta')$. The TQFT $(\mathcal{T}^e, \tau^e)$, constructed in Section IV.6, yields an element $\tau^e(B^3, \Omega) \in \mathcal{T}^e(\Upsilon)$. The following lemma shows that $\tau^e(B^3, \Omega)$ is completely determined by the operator

$$F(\Omega) \in \operatorname{Hom}(V_1^{\varepsilon_1} \otimes ... \otimes V_m^{\varepsilon_m}, W_1^{\nu_1} \otimes ... W_n^{\nu_n}).$$

(Here $\operatorname{Hom} = \operatorname{Hom}_{\mathcal{V}}$, $W^{+1} = W$, and $W^{-1} = W^*$.)

5.1.1. Lemma. *Let Ω_1, Ω_2 be v-colored ribbon graphs in $\mathbb{R}^3$ such that* $\operatorname{source}(\Omega_1) = \operatorname{source}(\Omega_2) = \eta$, $\operatorname{target}(\Omega_1) = \operatorname{target}(\Omega_2) = \eta'$, *and* $F(\Omega_1) = F(\Omega_2)$. *Then* $\tau^e(B^3, \Omega_1) = \tau^e(B^3, \Omega_2) \in \mathcal{T}^e(\Upsilon(\eta; \eta'))$.

Proof. Consider arbitrary v-colored ribbon $(0,0)$-graphs Ω'_1, Ω'_2 in $\mathbb{R}^3$ which coincide outside a 3-ball and are equal to Ω_1, Ω_2 inside this ball. We claim that $F(\Omega'_1) = F(\Omega'_2)$. Indeed, we may present Ω'_1 as the closure of the composition $\Omega\,\Omega_1$ where Ω is a v-colored ribbon graph with $\operatorname{target}(\Omega) = \eta$ and $\operatorname{source}(\Omega) = \eta'$. Similarly, Ω'_2 is the closure of the composition $\Omega\,\Omega_2$ with the same Ω. By Corollary I.2.7.2,

$$F(\Omega'_1) = \operatorname{tr} F(\Omega\,\Omega_1) = \operatorname{tr}\,(F(\Omega)F(\Omega_1)) = \operatorname{tr}\,(F(\Omega)F(\Omega_2)) = \operatorname{tr} F(\Omega\,\Omega_2) = F(\Omega'_2).$$

To compute $\tau^e(B^3, \Omega_1)$, $\tau^e(B^3, \Omega_2)$ we glue to (B^3, Ω_1), (B^3, Ω_2) a 3-ball with a standard v-colored ribbon graph inside and evaluate the operator invariant F of the resulting ribbon graphs in S^3 (see Section IV.1). These ribbon graphs in S^3 satisfy the conditions of the previous paragraph. Therefore they have equal operator invariants. Hence $\tau^e(B^3, \Omega_1) = \tau^e(B^3, \Omega_2)$.

5.1.2. Lemma. *Let Ω_f be the elementary v-colored ribbon graph with* $\operatorname{source}(\Omega) = ((V_1, \varepsilon_1), ..., (V_m, \varepsilon_m))$, $\operatorname{target}(\Omega) = ((W_1, \nu_1), ..., (W_n, \nu_n))$, *and*

$$f \in \operatorname{Hom}(V_1^{\varepsilon_1} \otimes ... \otimes V_m^{\varepsilon_m}, W_1^{\nu_1} \otimes ... W_n^{\nu_n}).$$

(see Figure I.2.4*). Let* $\Upsilon = \Upsilon((V_1, \varepsilon_1), ..., (V_m, \varepsilon_m); (W_1, \nu_1), ..., (W_n, \nu_n))$. *The homomorphism*

$$f \mapsto \tau^e(B^3, \Omega_f) : \operatorname{Hom}(V_1^{\varepsilon_1} \otimes ... \otimes V_m^{\varepsilon_m}, W_1^{\nu_1} \otimes ... \otimes W_n^{\nu_n}) \to \mathcal{T}^e(\Upsilon) \tag{5.1.a}$$

is an isomorphism.

Recall that Ω_f consists of one f-colored coupon with m untwisted unlinked bands attached from below and n untwisted unlinked bands attached from above. The i-th band attached from below is colored with V_i and directed down if $\varepsilon_i = 1$ and up otherwise. The i-th band attached from above is colored with W_i and directed down if $\nu_i = 1$ and up otherwise. The surface of this ribbon graph is oriented counterclockwise.

Proof of Lemma. In the case $m = 0$ the surface Υ coincides with the standard decorated surface Σ_t where $t = (0; (W_1, \nu_1), ..., (W_n, \nu_n))$, see Section IV.1. We have

$$\mathcal{T}^e(\Upsilon) = \mathcal{T}(\Sigma_t) = \mathrm{Hom}(\mathbb{1}, W_1^{\nu_1} \otimes ... W_n^{\nu_n}).$$

It follows from definitions that $\tau^e(B^3, \Omega_f) = f$. This implies the claim of the lemma for $m = 0$.

The case $m \neq 0$ may be reduced to the case $m = 0$ pushing the bottom bands of Ω_f upwards along the big circle. Indeed, the extended 3-ball (B^3, Ω_f) is e-homeomorphic to the extended 3-ball (B^3, Ω') bounded by the extended 2-sphere

$$\Upsilon' = \Upsilon(\emptyset; (W_1, \nu_1), ..., (W_n, \nu_n), (V_m, -\varepsilon_m), (V_{m-1}, -\varepsilon_{m-1}), ..., (V_1, -\varepsilon_1)),$$

the ribbon graph Ω' is schematically shown in Figure 5.2. (In contrast to Figure 5.1, in Figure 5.2 and in the figures to follow we represent bands by their cores; as usual, the bands themselves are parallel to the plane of the picture.) Consider an e-homeomorphism $j : \Upsilon \to \Upsilon'$ that extends to an e-homeomorphism $(B^3, \Omega_f) \to (B^3, \Omega')$. We have $j_\#(\tau^e(B^3, \Omega_f)) = \tau^e(B^3, \Omega')$. By Lemma 5.1.1, $\tau^e(B^3, \Omega') = \tau^e(B^3, \Omega_g)$ where g is the image of f under the isomorphism

$$\mathrm{Hom}(V_1^{\varepsilon_1} \otimes ... \otimes V_m^{\varepsilon_m}, W_1^{\nu_1} \otimes ... W_n^{\nu_n}) = \mathrm{Hom}(\mathbb{1}, W_1^{\nu_1} \otimes ... \otimes W_n^{\nu_n} \otimes V_m^{-\varepsilon_m} \otimes ... \otimes V_1^{-\varepsilon_1})$$

(cf. Exercises I.1.8.1, I.2.9.2, and Figure 5.2; it is understood that in Figure 5.2 the coupon of Ω_g is incident to $m + n$ bands attached from above). Therefore the homomorphism (5.1.a) is the composition of the isomorphism $f \mapsto g = g(f)$, the isomorphism $g \mapsto \tau^e(B^3, \Omega_g) \in \mathcal{T}^e(\Upsilon')$, and the isomorphism $j_\#^{-1} : \mathcal{T}^e(\Upsilon') \to \mathcal{T}^e(\Upsilon)$.

5.2. A computation of $\mathcal{H}$. In this subsection we consider the modules of states of planar m-surfaces introduced in Section 1.4. We adjust the technique of Section 5.1 in order to present the elements of these modules by v-colored ribbon graphs in $\mathbb{R}^3$. Such presentations will be used systematically up to the end of Section 5.

Consider the marked disk with holes $E = E((W, -\nu); (W_1, \nu_1), ..., (W_n, \nu_n))$, where $W, W_1, W_2, ..., W_n$ are objects of $\mathcal{V}$ and $\nu, \nu_1, ..., \nu_n \in \{+1, -1\}$. It is clear that $\overline{E}$ is an extended 2-sphere (for the definition of $\overline{E}$, see Section 4.2). We can describe $\overline{E}$ more precisely. Recall that each component X of ∂E is a metric circle in $\mathbb{R}^2$. We parametrize X by the homeomorphism $S^1 \to X$, which stretches the unit circle uniformly over X and carries $-\sqrt{-1} \in S^1$ into the base point of X. It is clear that the images of the points $-1, +1 \in S^1$ lie on the horizontal line dividing E into two equal parts (and used in the definition of E in Section 1.4). This line intersects E in $n + 1$ intervals whose endpoints are exactly the images of $-1, +1$ on the components of ∂E. These $n + 1$ intervals together with the distinguished arcs in $\overline{E}$ (i.e., the distinguished diameters of the 2-disks glued to E in $\overline{E}$) form a circle in $\overline{E}$.

We shall identify the extended 2-sphere $\overline{E}$ with the extended 2-sphere $\Upsilon = \Upsilon((W, \nu); (W_1, \nu_1), ..., (W_n, \nu_n))$ introduced in Section 5.1. (Both extended spheres contain $n + 1$ distinguished arcs marked by $(W, -\nu)$, (W_1, ν_1), ...,

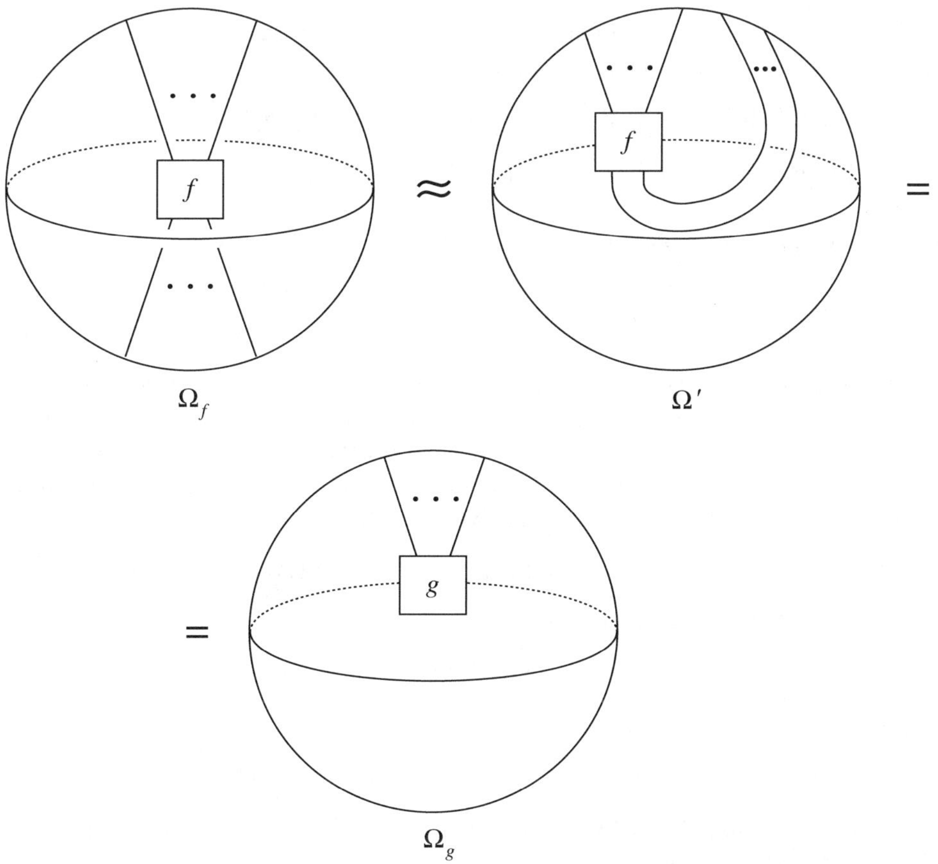

Figure 5.2

(W_n, ν_n).) The identification $\Upsilon = \overline{E}$ is determined by an m-homeomorphism $\Upsilon \to \overline{E}$, which carries the big circle in Υ mentioned in Section 5.1 onto the circle in $\overline{E}$ constructed above. Such an m-homeomorphism is well-defined up to isotopy. Thus,

$$\mathcal{T}^e(\Upsilon) = \mathcal{T}^e(\overline{E}) = \mathcal{H}(E).$$

Combining these equalities with the technique of Section 5.1 we obtain that every v-colored ribbon $(1, n)$-graph Ω in $\mathbb{R}^3$ with source$(\Omega) = (W, \nu)$ and target$(\Omega) = ((W_1, \nu_1), \ldots, (W_n, \nu_n))$ gives rise to an element $\tau^e(B^3, \Omega) \in \mathcal{H}(E)$. In our figures we shall represent this element by the graph Ω inscribed in a circle.

5.2.1. Lemma. *The state $\tau^e(B^3, \Omega) \in \mathcal{H}(E)$ depends only on $F(\Omega)$.*

This follows from Lemma 5.1.1.

Taking Ω to be the elementary ribbon graph used in Lemma 5.1.2 (with $m = 1, V_1 = W, \varepsilon_1 = \nu$) we obtain a homomorphism

$$\mathrm{Hom}(W^{\nu}, W_1^{\nu_1} \otimes \ldots \otimes W_n^{\nu_n}) \to \mathcal{H}(E((W, -\nu); (W_1, \nu_1), \ldots, (W_n, \nu_n)))$$

defined by the rule $f \mapsto \tau^e(B^3, \Omega_f)$ and denoted by $\rho_{\nu W}^{\nu_1 W_1, \ldots, \nu_n W_n}$.

5.2.2. Lemma. *The homomorphism $\rho_{\nu W}^{\nu_1 W_1, \ldots, \nu_n W_n}$ is an isomorphism.*

This follows from Lemma 5.1.2.

5.2.3. Lemma. *Let Ω_1 be a v-colored ribbon $(1, n)$-graph in $\mathbb{R}^3$ such that* $\mathrm{source}(\Omega_1) = (W, \nu)$ *and* $\mathrm{target}(\Omega_1) = ((W_1, \nu_1), \ldots, (W_n, \nu_n))$. *Let Ω_0 be a v-colored ribbon $(1, m)$-graph in $\mathbb{R}^3$ with* $\mathrm{source}(\Omega_0) = (W_i, \nu_i)$ *for a certain $i \in \{1, \ldots, n\}$. Let $\Omega \subset \mathbb{R}^3$ be the v-colored ribbon $(1, m+n-1)$-graph obtained by attaching Ω_0 to the i-th upper free band of Ω_1. Set*

$$E_1 = E((W, \nu); (W_1, \nu_1), \ldots, (W_n, \nu_n)), \qquad E_0 = E((W_i, \nu_i); \mathrm{target}(\Omega_0)),$$

and $E = E((W, \nu); \mathrm{target}(\Omega))$. *Then*

$$\tau^e(B^3, \Omega) = \tau^e(B^3, \Omega_0) \diamond \tau^e(B^3, \Omega_1) \tag{5.2.a}$$

where

$$\tau^e(B^3, \Omega) \in \mathcal{H}(E), \quad \tau^e(B^3, \Omega_0) \in \mathcal{H}(E_0), \quad \tau^e(B^3, \Omega_1) \in \mathcal{H}(E_1).$$

The gluing of E_0 and E_1 in (5.2.a) *proceeds along the external component of ∂E_0 and the i-th internal component of ∂E_1.*

Note that

$$\mathrm{target}(\Omega) = ((W_1, \nu_1), \ldots, (W_{i-1}, \nu_{i-1}), \mathrm{target}(\Omega_0), (W_{i+1}, \nu_{i+1}), \ldots, (W_n, \nu_n)).$$

Proof of Lemma. Let M_1 denote the disjoint union of the extended 3-balls (B^3, Ω_0), (B^3, Ω_1). We regard M_1 as a cobordism between $\partial_-(M_1) = \emptyset$ and $\partial_+(M_1) = \overline{E_0} \amalg \overline{E_1}$. It follows from the properties of TQFT's that $\tau^e(M_1) : K = \mathcal{T}^e(\emptyset) \to \mathcal{T}^e(\overline{E_0}) \otimes_K \mathcal{T}^e(\overline{E_1})$ carries $1 \in K$ into $\tau^e(B^3, \Omega_0) \otimes \tau^e(B^3, \Omega_1)$.

To compute the state $\tau^e(B^3, \Omega_0) \diamond \tau^e(B^3, \Omega_1)$ on E we should consider the cylinder over $\overline{E_0 \amalg E_1} = \overline{E_0} \amalg \overline{E_1}$ and attach a 1-handle connecting the 2-disks in $\overline{E_0}, \overline{E_1}$ glued to the external component of ∂E_0 and the i-th internal component of ∂E_1. This gives an extended 3-cobordism, M_2, between $\overline{E_0} \amalg \overline{E_1}$ and $\overline{E}$. By definition,

$$\tau^e(B^3, \Omega_0) \diamond \tau^e(B^3, \Omega_1) = \tau^e(M_2)(\tau^e(B^3, \Omega_0) \otimes \tau^e(B^3, \Omega_1)) = \tau^e(M_2)\, \tau^e(M_1)(1).$$

To compute the right-hand side, we glue M_1 to M_2 from below along the identity $\mathrm{id} : \partial_+(M_1) \to \partial_-(M_2)$. This gives an extended 3-cobordism $M = M_2 M_1$

between $\emptyset$ and $\overline{E}$. The cobordism M is schematically shown in Figure 5.3 where, in order to distinguish M_1 from M_2, we have shaded M_2. By Theorem IV.7.1, we have $\tau^e(M) = \tau^e(M_2)\,\tau^e(M_1)$. (The corresponding gluing anomaly is equal to 1 because $H_1(S^2;\mathbb{R}) = 0$.) Hence

$$\tau^e(B^3, \Omega_0) \diamond \tau^e(B^3, \Omega_1) = \tau^e(M)(1).$$

It is easy to observe that the extended 3-manifold M is e-homeomorphic to the extended 3-ball (B^3, Ω). Moreover, the e-homeomorphism in question extends the identification of the boundaries $\partial M = \overline{E} = \partial(B^3, \Omega)$. Therefore $\tau^e(M)(1) = \tau^e(B^3, \Omega)$.

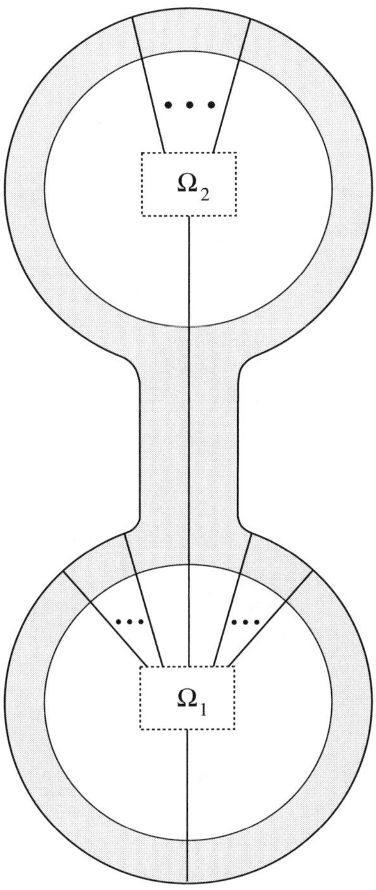

Figure 5.3

5.3. The disk axiom. We verify (1.5.5). By the results of Section 5.2, we have an isomorphism

$$\rho^{\emptyset}_{\nu 1\!\!1} : \mathrm{Hom}(1\!\!1^{\nu}, 1\!\!1) \to \mathcal{H}(E((-\nu)1\!\!1)).$$

Clearly, $\operatorname{Hom}(\mathbb{1}^+, \mathbb{1}) = \operatorname{Hom}(\mathbb{1}, \mathbb{1}) = K \cdot \mathrm{id}_{\mathbb{1}}$. By Corollary I.2.6.2, the morphisms $b_{\mathbb{1}} : \mathbb{1} \to \mathbb{1}^*$ and $d_{\mathbb{1}} : \mathbb{1}^* \to \mathbb{1}$ are mutually inverse isomorphisms. Therefore $\operatorname{Hom}(\mathbb{1}^-, \mathbb{1}) = \operatorname{Hom}(\mathbb{1}^*, \mathbb{1}) = K \cdot d_{\mathbb{1}}$. The generators

$$e_- = \rho^{\emptyset}_{+\mathbb{1}}(\mathrm{id}_{\mathbb{1}}) \in \mathcal{H}(E(-\mathbb{1})) \simeq K, \qquad e_+ = \rho^{\emptyset}_{-\mathbb{1}}(d_{\mathbb{1}}) \in \mathcal{H}(E(+\mathbb{1})) \simeq K$$

are exhibited in Figure 5.4. More precisely, denote by Ω_-, Ω_+ the elementary v-colored ribbon (1,0)-graphs contained in the circles in Figure 5.4. Then $e_- = \tau^e(B^3, \Omega_-)$ and $e_+ = \tau^e(B^3, \Omega_+)$.

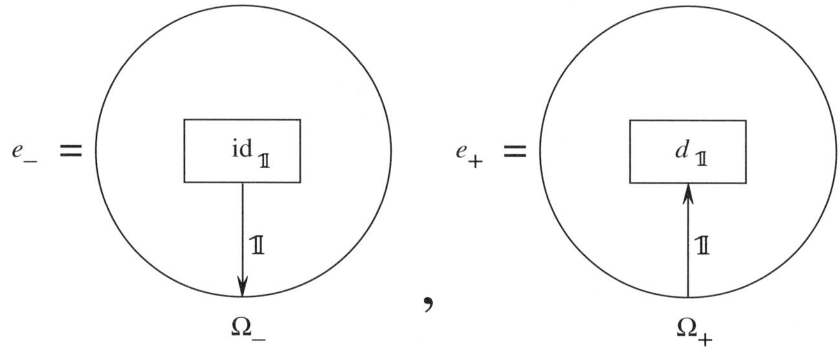

Figure 5.4

It is enough to verify the second part of axiom (1.5.5) for the generators e_-, e_+ introduced above. We need the following lemma.

5.3.1. Lemma. *Let* Υ *and* Υ' *be e-surfaces such that* Υ' *is obtained from* Υ *by forgetting a distinguished arc* α *marked with* $(\mathbb{1}, \nu)$ *where* $\nu = \pm 1$. *Let* (M, Υ, Υ') *be the extended 3-cobordism obtained from the cylinder* $\Upsilon \times [0, 1]$ *by replacing the band* $\alpha \times [0, 1]$ *with the ribbon* $(1, 0)$*-graph* Ω_ν *defined above. Then the homomorphism* $\tau^e(M) : \mathcal{T}^e(\Upsilon) \to \mathcal{T}^e(\Upsilon')$ *is an isomorphism.*

Proof. By replacing $\alpha \times [0, 1]$ with Ω_ν, we mean drawing Ω_ν on $\alpha \times [0, 1]$ and forgetting $\alpha \times [0, 1]$. In other words, we replace the band $\alpha \times [0, 1]$ with a shorter band, say $\alpha \times [0, 1/2]$, colored with $\mathbb{1}$, directed downwards if $\nu = -1$ and upwards if $\nu = 1$, and one coupon lying in $\Upsilon \times [1/2, 1)$ such that its bottom base contains $\alpha \times 1/2$. The coupon is colored with $\mathrm{id}_{\mathbb{1}}$ if $\nu = -1$ and with $d_{\mathbb{1}} : \mathbb{1}^* \to \mathbb{1}$ if $\nu = 1$. The distinguished ribbon graph in M consists of this copy of Ω_ν and the bands $\beta \times [0, 1]$ where β runs over the distingiushed arcs in Υ distinct from α. Forgetting Ω_ν transforms M into $\Upsilon' \times [0, 1]$.

Consider the case $\nu = -1$. Here is the key observation underlying the proof of the lemma. If a v-colored ribbon graph $\Omega' \subset \mathbb{R}^3$ is obtained from a v-colored ribbon graph $\Omega \subset \mathbb{R}^3$ by adding a coupon colored with $\mathrm{id}_{\mathbb{1}}$ and a band connecting the bottom base of this coupon to the top base of a coupon of Ω then $F(\Omega) =$

$F(\Omega')$. This follows from definitions and the isotopy invariance of F, see Section I.2. Thus, we may safely add or remove such a coupon and a band without changing the operator invariants of ribbon graphs in $\mathbb{R}^3$.

It is enough to consider the case of connected Υ. To compute $\tau^e(M)$, we need to transform M into a decorated cobordism, i.e., to parametrize Υ and Υ' by standard decorated surfaces (see Section IV.1). Parametrize Υ' by a standard decorated surface Σ' of type $t' = (g; (W_1, \nu_1), ..., (W_m, \nu_m))$ where $(W_1, \nu_1), ..., (W_m, \nu_m)$ are the marks of the distinguished arcs on Υ' and g is the genus of Υ'. Let Σ be the standard decorated surface of type $t = (g; (\mathbb{1}, +1), (W_1, \nu_1), ..., (W_m, \nu_m))$. Recall that $\Sigma = \Sigma_t$ is the boundary of a regular neighborhood U_t of the ribbon graph $R_t \subset \mathbb{R}^3$ (see Section IV.1.2). There is a homeomorphism $U_t \to U_{t'}$ obtained by pushing a cylindrical neighborhood of the leftmost band inside $U_{t'}$. This homeomorphism restricts to a homeomorphism $\Sigma \to \Sigma'$. The composition of the last homeomorphism and the parametrization $\Sigma' \to \Upsilon'$ yields a parametrization of Υ. In this way the cobordism (M, Υ, Υ') acquires the structure of a decorated 3-cobordism. Denote this decorated cobordism by $\check{M}$.

In the notation of Section IV.1.4, $\Phi(t'; i) = \mathbb{1} \otimes \Phi(t; i) = \Phi(t; i)$ for any $i \in I^g$. Therefore

$$\mathcal{T}(\Sigma') = \Psi_{t'} = \Psi_t = \mathcal{T}(\Sigma).$$

The definition of the homomorphism $\tau(\check{M}) : \mathcal{T}(\Sigma) \to \mathcal{T}(\Sigma')$ given in Section IV.1.8 is based on a reduction to v-colored ribbon graphs in $\mathbb{R}^3$ and their operator invariants. The observation made at the beginning of the proof implies that to compute $\tau(\check{M})$ we may forget about the additional band and coupon. Therefore $\tau(\check{M}) = \tau(\Upsilon' \times [0, 1]) = \mathrm{id}_{\Psi_t}$. By definition (see Section IV.6.5), the homomorphism $\tau^e(M) : \mathcal{T}^e(\Upsilon) \to \mathcal{T}^e(\Upsilon')$ equals (up to an invertible factor) the composition of an isomorphism $\mathcal{T}^e(\Upsilon) \approx \mathcal{T}(\Sigma)$, the homomorphism $\tau(\check{M}) = \mathrm{id}_{\Psi_t}$, and an isomorphism $\mathcal{T}^e(\Sigma') \approx \mathcal{T}(\Upsilon')$. Therefore $\tau^e(M)$ is an isomorphism.

The case $\nu = +1$ is treated similarly with obvious changes.

5.3.2. Verification of the second part of (1.5.5). Let Ω_+, Ω_- denote the v-colored ribbon (1,0)-graphs inscribed in the circles in Figure 5.4. Let Υ_ν denote the extended 2-sphere $\overline{E(\nu\mathbb{1})}$. (It contains one distinguished arc marked by $(\mathbb{1}, \nu)$.) As explained in Section 5.2, the boundary of the extended 3-ball (B^3, Ω_ν) can be identified with Υ_ν. Let M_1 be the disjoint union of the cylinder $\overline{\Sigma} \times [0, 1]$ and the extended 3-ball (B^3, Ω_ν). We regard M_1 as a cobordism between $\partial_-(M_1) = \overline{\Sigma}$ and $\partial_+(M_1) = \overline{\Sigma} \amalg \Upsilon_\nu$. It follows from the properties of TQFT's and definitions that the homomorphism $\tau^e(M_1) : \mathcal{T}^e(\overline{\Sigma}) \to \mathcal{T}^e(\overline{\Sigma}) \otimes_K \mathcal{T}^e(\Upsilon_\nu)$ is the tensor product of the identity acting on $\mathcal{T}^e(\overline{\Sigma})$ and the homomorphism $K = \mathcal{T}^e(\emptyset) \to \mathcal{T}^e(\Upsilon_\nu)$ that carries $1 \in K$ into $e_\nu = \tau^e(B^3, \Omega_\nu) \in \mathcal{T}^e(\Upsilon_\nu)$. Thus, $\tau^e(M_1)$ carries any $h \in \mathcal{T}^e(\overline{\Sigma})$ into $h \otimes e_\nu \in \mathcal{T}^e(\overline{\Sigma}) \otimes_K \mathcal{T}^e(\Upsilon_\nu)$. By Lemma 5.1.2, $\mathcal{T}^e(\Upsilon_\nu) = K \cdot e_\nu$. Therefore the homomorphism $\tau^e(M_1)$ is an isomorphism.

To compute the homomorphism $h \mapsto h \diamond e_\nu : \mathcal{H}(\Sigma) \to \mathcal{H}(\Sigma')$ we should consider the cylinder over the e-surface $\overline{\Sigma} \amalg \Upsilon_\nu$ and attach a 1-handle from above as specified in Section 4.5. This gives an extended 3-cobordism, M_2, between $\overline{\Sigma} \amalg \Upsilon_\nu$ and $\overline{\Sigma'}$. For any $h \in \mathcal{H}(\Sigma) = \mathcal{T}^e(\overline{\Sigma})$,

$$h \diamond e_\nu = \tau^e(M_2)(h \otimes e_\nu) = \tau^e(M_2)\,\tau^e(M_1)(h).$$

To compute $\tau^e(M_2)\,\tau^e(M_1)$ we glue M_1 to M_2 from below along $\partial_+(M_1) = \partial_-(M_2)$. This gives an extended 3-cobordism $M = M_2 M_1$ between $\overline{\Sigma}$ and $\overline{\Sigma'}$. Up to an invertible factor, $\tau^e(M) = \tau^e(M_2)\,\tau^e(M_1)$. It is easy to observe that the extended 3-cobordism M is exactly of the type studied in Lemma 5.3.1. (The corresponding picture is similar to Figure 5.3 where Ω_0 is replaced with Ω_ν and (B^3, Ω_1) is replaced with the cylinder over $\overline{\Sigma}$, cf. Figure 5.7 where M_2 is shaded.) Therefore $\tau^e(M)$ is an isomorphism. Hence $\tau^e(M_2) = \tau^e(M)(\tau^e(M_1))^{-1}$ is an isomorphism.

5.4. The first excision axiom. For objects V, W of $\mathcal{V}$, set $E = E(-VW; V, W)$ where $VW = V \otimes W$. Set

$$e_{V,W} = \rho_{+V\otimes W}^{+V,+W}(\mathrm{id}_{V\otimes W}) \in \mathcal{H}(E)$$

where

$$\rho_{+V\otimes W}^{+V,+W} : \mathrm{Hom}(V \otimes W, V \otimes W) \to \mathcal{H}(E)$$

is the isomorphism constructed in Section 5.2. The definition of $e_{V,W}$ may be reformulated as follows. Let $\Omega_{V,W}$ denote the elementary v-colored ribbon graph Ω_f used in Lemma 5.1.2 where $m = 1$, $n = 2$, $V_1 = V \otimes W$, $W_1 = V$, $W_2 = W$, $\nu_1 = \nu_2 = \varepsilon_1 = 1$, and $f = \mathrm{id}_{V\otimes W}$. Let $B^3_{V,W}$ denote the extended 3-ball $(B^3, \Omega_{V,W})$. As explained in Section 5.2, the boundary of $B^3_{V,W}$ may be identified with the extended 2-sphere $\overline{E}$. By definition,

$$e_{V,W} = \tau^e(B^3_{V,W}) \in \mathcal{T}^e(\partial B^3_{V,W}) = \mathcal{T}^e(\overline{E}) = \mathcal{H}(E).$$

For a graphical form of this equality, see Figure 5.5.

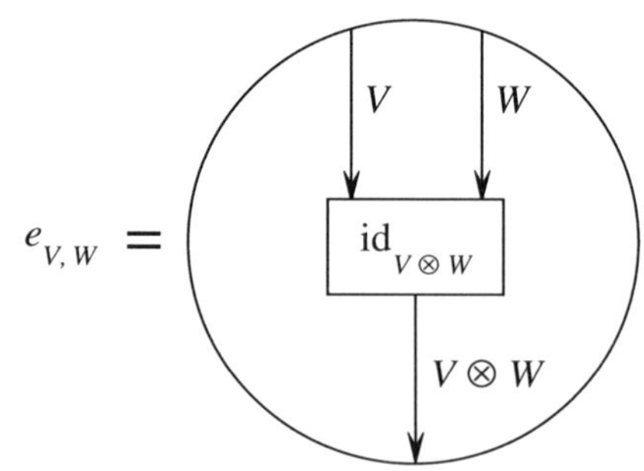

Figure 5.5

To verify the second part of axiom (1.5.6a) we need the following lemma.

5.4.1. Lemma. *Let* Υ *and* Υ' *be e-surfaces such that* Υ' *is obtained from* Υ *by replacing a distinguished arc* α *marked by* $+V \otimes W$ *with two disjoint subarcs* $\alpha', \alpha'' \subset \alpha$ *marked by* $+V$ *and* $+W$*, respectively. (The orientations of* α', α'' *are induced by the one in* α *and* α' *preceeds* α'' *on* α*.) Let* (M, Υ, Υ') *be the extended 3-cobordism obtained from the cylinder* $\Upsilon \times [0, 1]$ *by replacing the band* $\alpha \times [0, 1]$ *with* $\Omega_{V,W}$*. Then the homomorphism* $\tau^e(M) : \mathcal{T}^e(\Upsilon) \to \mathcal{T}^e(\Upsilon')$ *is an isomorphism.*

Proof. By replacing the band $\alpha \times [0, 1]$ with $\Omega_{V,W}$, we mean replacing $\alpha \times [0, 1]$ with three shorter bands $\alpha \times [0, 1/3]$, $\alpha' \times [2/3, 1]$, $\alpha'' \times [2/3, 1]$ directed towards $\Upsilon \times 0$ and colored with $V \otimes W, V, W$, respectively, and the coupon $\alpha \times [1/3, 2/3]$ with bottom base $\alpha \times 1/3$ and color $\mathrm{id}_{V\otimes W}$. The distinguished ribbon graph in M consists of this copy of $\Omega_{V,W}$ and the bands $\beta \times [0, 1]$ where β runs over the distinguished arcs in Υ distinct from α.

Besides the elementary v-colored ribbon graph $\Omega_{V,W}$ in $\mathbb{R}^3$ introduced above, we need the elementary v-colored ribbon graph $\Omega^{V,W}$ in $\mathbb{R}^3$. It is defined as Ω_f in Lemma 5.1.2 where $m = 2$, $n = 1$, $V_1 = V$, $V_2 = W$, $W_1 = V \otimes W$, $\varepsilon_1 = \varepsilon_2 = \nu_1 = 1$, and $f = \mathrm{id}_{V\otimes W}$. Clearly, $F(\Omega_{V,W}) = F(\Omega^{V,W}) = \mathrm{id}_{V\otimes W}$.

Let $(M', \Upsilon', \Upsilon)$ be the extended 3-cobordism obtained from the cylinder $\Upsilon \times [0, 1]$ by replacing the band $\alpha \times [0, 1]$ with $\Omega^{V,W}$. It is clear that the composition $M'M$ is obtained from $\Upsilon \times [0, 1]$ by replacing the band $\alpha \times [0, 1]$ with $\Omega^{V,W}\Omega_{V,W}$. The equality $F(\Omega^{V,W}\Omega_{V,W}) = \mathrm{id}_{V\otimes W}$ implies that this does not change the operator invariant. Hence $\tau^e(M'M) = \tau^e(\Upsilon \times [0, 1]) = \mathrm{id}_{\mathcal{T}^e(\Upsilon)}$. Similarly, $\tau^e(MM') = \mathrm{id}_{\mathcal{T}^e(\Upsilon')}$. On the other hand, $\tau^e(M'M) = k\, \tau^e(M')\, \tau^e(M)$ and $\tau^e(MM') = k' \tau^e(M)\, \tau^e(M')$ with invertible $k, k' \in K$. Therefore $\tau^e(M)$ is an isomorphism.

5.4.2. Verification of the second part of (1.5.6a). Our arguments are similar to those used in Section 5.3.2. Let M_1 denote the disjoint union of the cylinder $\overline{\Sigma} \times [0, 1]$ and the extended 3-ball $B^3_{V,W}$. We regard M_1 as a cobordism between $\partial_-(M_1) = \overline{\Sigma}$ and $\partial_+(M_1) = \overline{\Sigma} \amalg \overline{E}$ where $E = E(-VW; V, W)$. It follows from the properties of TQFT's that $\tau^e(M_1)$ is the tensor product of the identity acting on $\mathcal{T}^e(\overline{\Sigma})$ and the homomorphism

$$\tau^e(B^3_{V,W}, \emptyset, \overline{E}) : K = \mathcal{T}^e(\emptyset) \to \mathcal{T}^e(\overline{E})$$

that carries $1 \in K$ into $e_{V,W} = \tau^e(B^3_{V,W}) \in \mathcal{T}^e(\overline{E})$. Hence, $\tau^e(M_1)(h) = h \otimes e_{V,W}$ for any $h \in \mathcal{T}^e(\overline{\Sigma})$.

To compute the gluing homomorphism (1.5.e) we should consider the cylinder over $\overline{\Sigma \amalg E} = \overline{\Sigma} \amalg \overline{E}$ and attach a 1-handle from above as specified in Section 4.5. This gives an extended 3-cobordism M_2 between $\overline{\Sigma} \amalg \overline{E}$ and $\overline{\Sigma'}$. For any

$h \in \mathcal{H}(\Sigma) = \mathcal{T}^e(\overline{\Sigma})$,

$$h \diamond e_{V,W} = \tau^e(M_2)(h \otimes e_{V,W}) = \tau^e(M_2)\,\tau^e(M_1)(h) \in \mathcal{H}(\Sigma') = \mathcal{T}^e(\overline{\Sigma'}).$$

To compute $\tau^e(M_2)\,\tau^e(M_1)$, we glue M_1 to M_2 from below along the identity $\mathrm{id} : \partial_+(M_1) \to \partial_-(M_2)$. This gives an extended 3-cobordism $M = M_2 M_1$ between $\overline{\Sigma}$ and $\overline{\Sigma'}$. We have $\tau^e(M) = k\,\tau^e(M_2)\,\tau^e(M_1)$ with invertible $k \in K$. (In fact, $k = 1$ but we do not need this.) Thus, $h \diamond e_{V,W} = k^{-1}\tau^e(M)(h)$ for any $h \in \mathcal{T}^e(\overline{\Sigma})$. It is easy to observe that the extended 3-cobordism M is exactly of the type studied in Lemma 5.4.1. (The corresponding picture is similar to Figure 5.3 where Ω_0 is replaced with $\Omega_{V,W}$ and (B^3, Ω_1) is replaced with the cylinder over $\overline{\Sigma}$, cf. Figure 5.7.) Therefore $\tau^e(M) : \mathcal{T}^e(\overline{\Sigma}) \to \mathcal{T}^e(\overline{\Sigma'})$ is an isomorphism. Hence the homomorphism $h \mapsto h \diamond e_{V,W} : \mathcal{H}(\Sigma) \to \mathcal{H}(\Sigma')$ is an isomorphism.

5.5. The second excision axiom

5.5.1. Lemma. *Let Υ and Υ' be e-surfaces such that Υ has a distinguished arc α marked with $-W$ and Υ' is obtained from Υ by adding a distinguished arc marked with $+\mathbb{1}$. Let (M, Υ, Υ') be the extended 3-cobordism obtained from the extended cylinder $\Upsilon \times [0,1]$ by replacing the band $\alpha \times [0,1]$ with the v-colored ribbon graph Ω^W shown in Figure 5.6. Then the homomorphism $\tau^e(M) : \mathcal{T}^e(\Upsilon) \to \mathcal{T}^e(\Upsilon')$ is an isomorphism.*

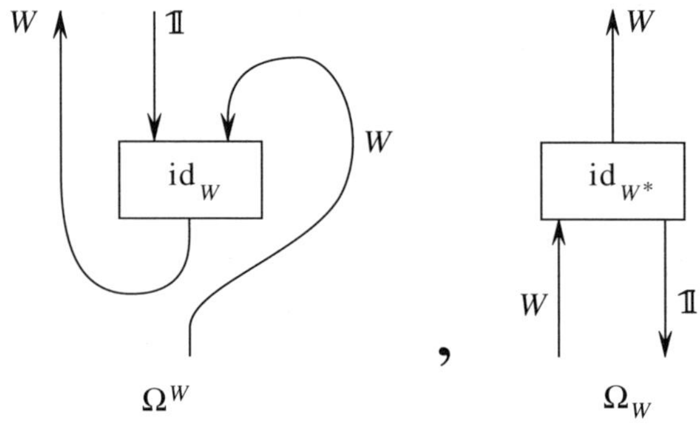

Figure 5.6

The proof of Lemma 5.5.1 is similar to the proof of Lemma 5.4.1. The key observation is that the v-colored ribbon graphs $\Omega^W, \Omega_W \subset \mathbb{R}^3$ shown in Figure 5.6 satisfy the identites $F(\Omega^W) = F(\Omega_W) = \mathrm{id}_{W^*}$ and therefore $F(\Omega^W \Omega_W) = F(\Omega_W \Omega^W) = \mathrm{id}_{W^*}$.

Axiom (1.5.6b) is verified in the same way as (1.5.6a). Instead of $B^3_{V,W}$, we use $B^3_{\mathbb{1},W}$, and instead of Lemma 5.4.1 refer to Lemma 5.5.1. The corresponding

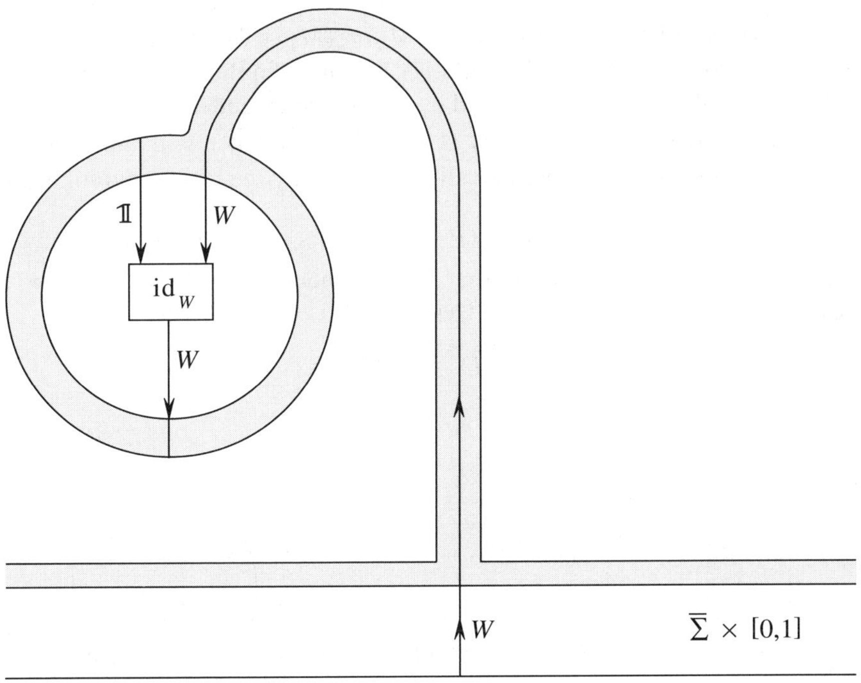

Figure 5.7

cobordism $M = M_2 M_1$ is schematically shown in Figure 5.7 (the part representing M_2 is shaded).

5.6. Verification of (1.5.7). Set $E_{U,V} = E(-UV; U, V)$. Set

$$E_{UV,W} = E(-UVW; UV, W), \quad E = E(-UVW; U, V, W).$$

Let M_1 denote the disjoint union of the extended 3-balls $B^3_{U,V}$ and $B^3_{UV,W}$ (for notation, see Section 5.4). Regard M_1 as a cobordism between the empty set and $\overline{E_{U,V}} \amalg \overline{E_{UV,W}}$. We have $\tau^e(M_1)(1) = e_{U,V} \otimes e_{UV,W}$ where $1 \in K = \mathcal{T}^e(\emptyset)$, $e_{U,V} \in \mathcal{H}(E_{U,V}) = \mathcal{T}^e(\overline{E_{U,V}})$ and $e_{UV,W} \in \mathcal{H}(E_{UV,W}) = \mathcal{T}^e(\overline{E_{UV,W}})$.

To compute the left-hand side of the basic identity (1.5.f) we should consider the cylinder over $\overline{E_{U,V}} \amalg \overline{E_{UV,W}}$ and attach a 1-handle from above as specified in Section 4.5. This gives an extended 3-cobordism, M_2, between $\overline{E_{U,V}} \amalg \overline{E_{UV,W}}$ and $\overline{E}$. We have

$$e_{U,V} \diamond e_{UV,W} = \tau^e(M_2)(e_{U,V} \otimes e_{UV,W}) = \tau^e(M_2)\,\tau^e(M_1)(1) \in \mathcal{T}^e(\overline{E}) = \mathcal{H}(E).$$

To compute the composition $\tau^e(M_2)\,\tau^e(M_1)$ we glue M_1 to M_2 from below along $\overline{E_{U,V}} \amalg \overline{E_{UV,W}}$. This gives an extended 3-cobordism $M = M_2 M_1$ between $\emptyset$ and $\overline{E}$. The manifold M is schematically presented by the left diagram in Figure 5.8 where M_2 is shaded. It follows from Theorem IV.7.1 and the remarks at the end of Section 4.5 that $\tau^e(M) = \tau^e(M_2)\,\tau^e(M_1)$. Therefore $e_{U,V} \diamond e_{UV,W} = \tau^e(M)(1)$. The right-hand side of (1.5.f) is computed similarly to be $\tau^e(M')$ where M' is the extended 3-manifold obtained by gluing the extended 3-balls $B^3_{V,W}$, $B^3_{U,VW}$ to the cylinder over $\overline{E(-VW;V,W)} \amalg \overline{E(-UVW;U,VW)}$ with an attached 1-handle. The manifold M' is schematically presented by the right diagram in Figure 5.8.

The boundaries of both M and M' are identified with $\overline{E}$. This identification extends to a homeomorphism of the 3-balls underlying M, M'. Lemma 5.2.1 implies the equalities in Figure 5.8. Hence $\tau^e(M) = \tau^e(M')$ and

$$e_{U,V} \diamond e_{UV,W} = \tau^e(M) = \tau^e(M') = e_{V,W} \diamond e_{U,VW}.$$

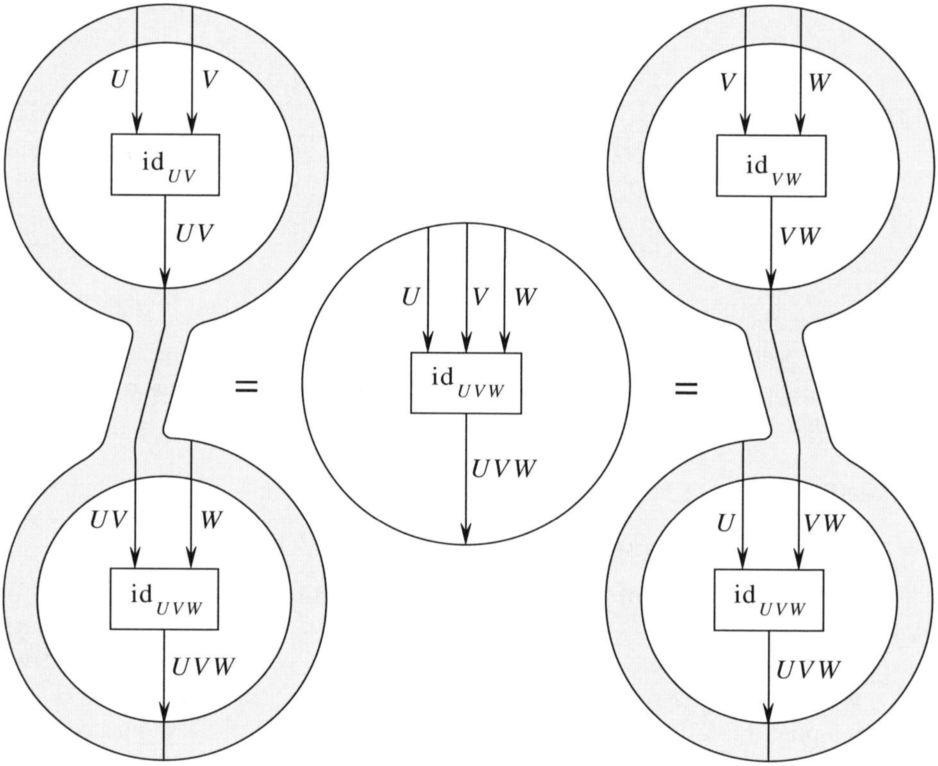

Figure 5.8

5.7. The normalization axioms. Axiom (1.6.1) follows directly from axiom (II.1.4.1) in the definition of modular categories. Let us verify axiom (1.6.2). For $i \in I$, the e-surface $\overline{E(-V_i;V_i)}$ is a 2-sphere with two distinguished arcs marked by $(V_i, -1)$ and $(V_i, +1)$. Hence

$$\mathcal{H}(E(-V_i;V_i)) = \mathcal{T}^e(\overline{E(-V_i;V_i)}) \simeq \mathrm{Hom}(\mathbb{1}, V_i^* \otimes V_i) = \mathrm{Hom}(V_i, V_i) = K.$$

Similarly, for $i, j \in I$,

$$\mathcal{H}(E(V_i; V_j)) \simeq \mathrm{Hom}(\mathbb{1}, V_i \otimes V_j) = \mathrm{Hom}(V_i^*, V_j) = \mathrm{Hom}(V_{i^*}, V_j).$$

The last module equals K if $j = i^*$ and 0 otherwise.

5.8. The duality axiom. The object V^* is the dual object to V provided by duality in the category $\mathcal{V}$. By the results of Section 5.2, we have an isomorphism

$$\rho^{\emptyset}_{+V^*V} : \mathrm{Hom}(V^*V, \mathbb{1}) \to \mathcal{H}(E(-V^*V)).$$

Set $D_V = \rho^{\emptyset}_{+V^*V}(d_V)$ where $d_V \in \mathrm{Hom}(V^* \otimes V, \mathbb{1})$ is the duality morphism in $\mathcal{V}$. We may present D_V graphically by the same diagram as e_- in Figure 5.4 although the band should be colored with V^*V and the coupon should be colored with d_V (cf. Figure 5.20). To verify axiom (1.5.8) we present z_V and $x(V, W)$ by ribbon graphs where W is an arbitrary object of $\mathcal{V}$. We shall use the generators $e_+ \in \mathcal{H}(E(\mathbb{1}))$, $e_- \in \mathcal{H}(E(-\mathbb{1}))$ shown in Figure 5.4. By Lemma 5.2.3, the left diagram in Figure 5.9 presents the state $e_- \diamond e_{\mathbb{1},V}$ on $E(-V; V)$. The first equality

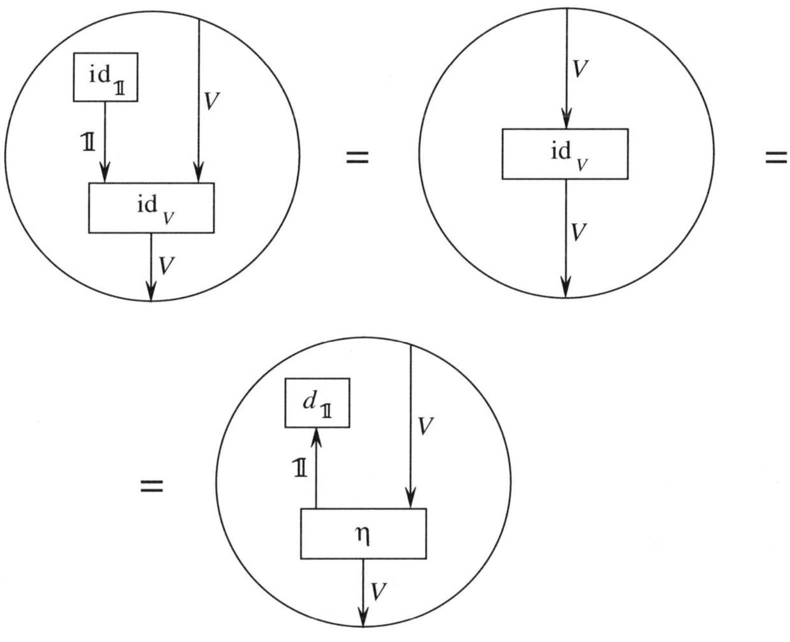

Figure 5.9

in Figure 5.9 follows from Lemma 5.2.1. Set

$$\eta = b_{\mathbb{1}} \otimes \mathrm{id}_V : V = \mathbb{1} \otimes V \to \mathbb{1}^* \otimes V.$$

Note that $(d_{\mathbb{1}} \otimes \mathrm{id}_V)\,\eta = \mathrm{id}_V$. Therefore Lemma 5.2.1 implies also the second equality in Figure 5.9. Recall that $z_V \in \mathcal{H}(E(-V; -\mathbb{1}, V))$ is the image of $e_- \diamond e_{\mathbb{1},V}$

under the excision of e_+. It follows from Lemma 5.2.3 and the equalities in Figures 5.4, 5.9 that the element $z \in \mathcal{H}(E(-V; -\mathbb{1}, V))$ shown in Figure 5.10 has the property $e_+ \diamond z = e_- \diamond e_{\mathbb{1},V} \in \mathcal{H}(E(-V; V))$. Hence $z = z_V$. Using the definition of the m-homeomorphism $\xi : E(-V; -\mathbb{1}, V) \to E(-\mathbb{1}; V, -V)$, we can give a graphical formula for $\xi_\#(z_V) \in \mathcal{H}(E(-\mathbb{1}; V, -V))$, see Figure 5.11 where the second and third equalities follow from Lemma 5.2.1. (We use the equality $F(\cap_{\mathbb{1}}^-)(\mathrm{id}_{\mathbb{1}} \otimes b_{\mathbb{1}}) = \mathrm{id}_{\mathbb{1}}$ which follows from the equalities $F(\cap_{\mathbb{1}}^-) = F(\cap_{\mathbb{1}}) = d_{\mathbb{1}}$ and $\mathrm{id}_{\mathbb{1}} \otimes b_{\mathbb{1}} = b_{\mathbb{1}}$.) Now we compute $\xi_\#(z_V) \diamond e_{W,\mathbb{1}}$, see Figure 5.12 where $\mathrm{id} = \mathrm{id}_{W \otimes V}$ and the second equality follows from Lemma 5.2.1. It follows from Lemma 5.1.2 that the element $x \in \mathcal{H}(E(-W; WV, -V))$ shown in Figure 5.13 has the property $e_{W,V} \diamond x = \xi_\#(z_V) \diamond e_{W,\mathbb{1}}$. Hence $x = x(V, W)$. This gives a graphical expression for $\tilde{D}_V = D_V \diamond x(V, V^*)$, see Figure 5.14 where the second equality follows from Lemma 5.2.1 and the formula $(d_V \otimes \mathrm{id}_{V^*})(\mathrm{id}_{V^*} \otimes b_V) = \mathrm{id}_{V^*}$. Now, to prove that the gluing homomorphism (1.5.h) is an isomorphism, it suffices to repeat the arguments used in Section 5.4 using the following lemma instead of Lemma 5.4.1.

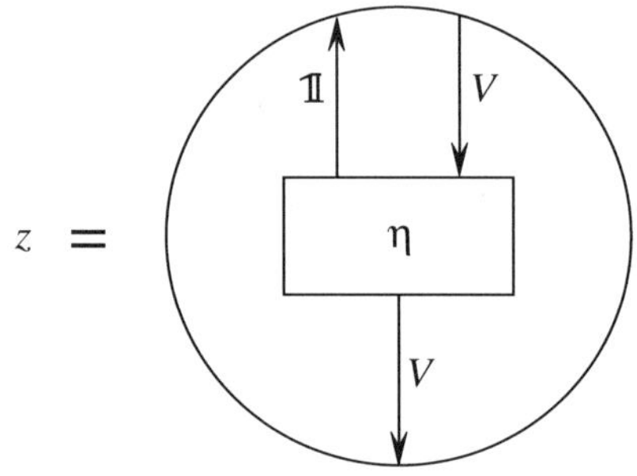

Figure 5.10

5.8.1. Lemma. *Let* Υ *and* Υ' *be e-surfaces such that* Υ' *is obtained from* Υ *by replacing the mark* $+V^*$ *of a distinguished arc* α *with* $-V$. *Let* (M, Υ, Υ') *be the extended* 3*-cobordism obtained from the cylinder* $\Upsilon \times [0, 1]$ *by replacing the band* $\alpha \times [0, 1]$ *with the v-colored ribbon graph* Ω_1 *shown in Figure* 5.15. *Then the homomorphism* $\tau^e(M) : \mathcal{T}^e(\Upsilon) \to \mathcal{T}^e(\Upsilon')$ *is an isomorphism.*

The proof of this lemma is analogous to the proof of Lemma 5.4.1, the role of $\Omega_{V,W}$ and $\Omega^{V,W}$ is played by the ribbon graphs shown in Figure 5.15.

5.9. The splitting axiom: the case of a non-separating loop. Here we verify axiom (1.6.3) in the case where a_+ and a_- lie in the same connected component of Σ^a.

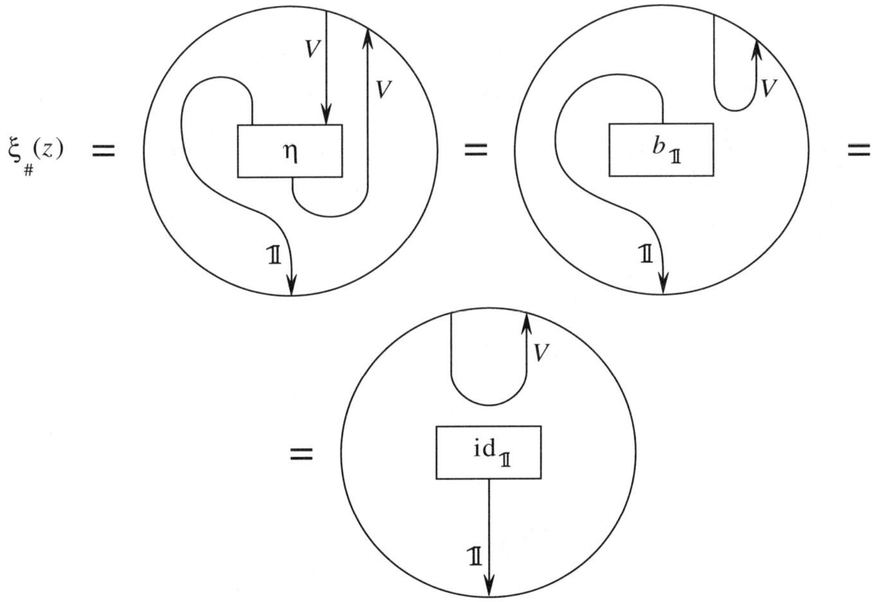

Figure 5.11

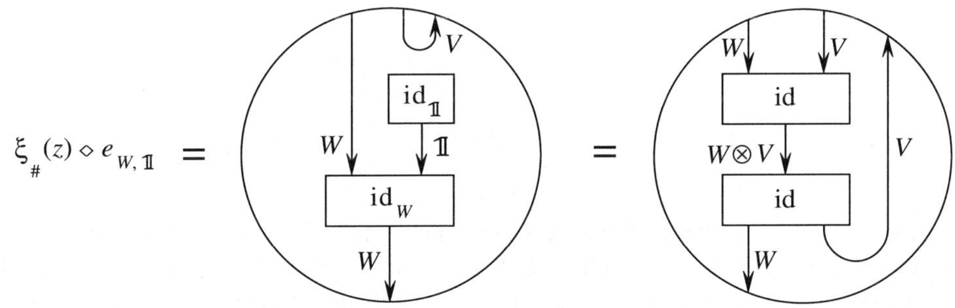

Figure 5.12

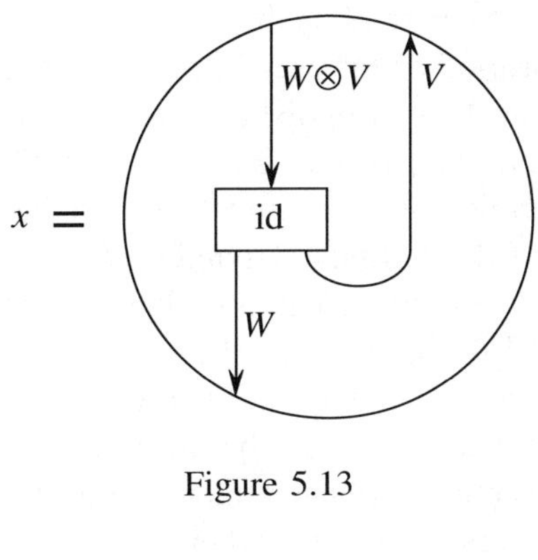

Figure 5.13

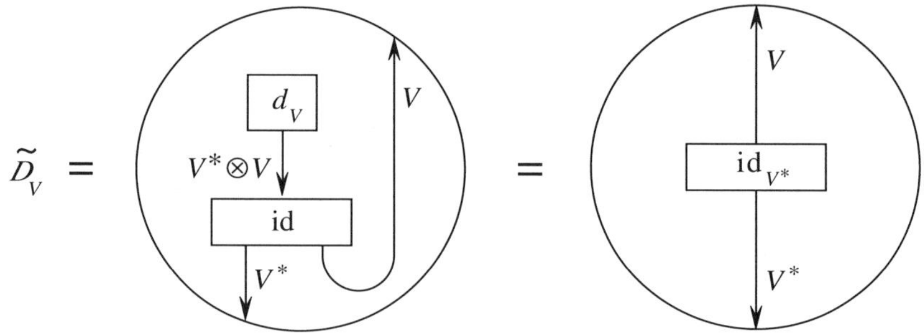

Figure 5.14

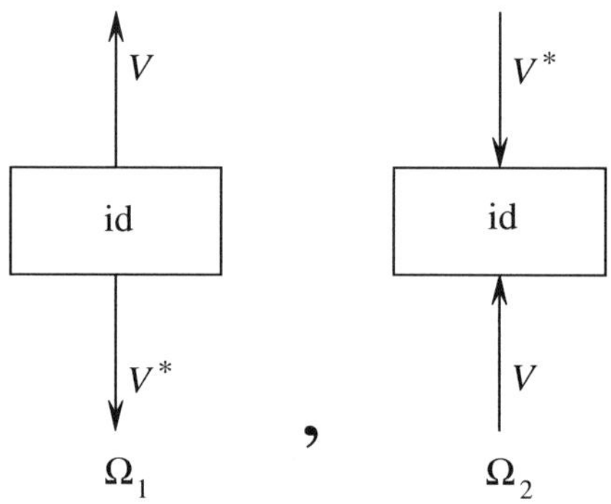

Figure 5.15

It is enough to consider the case where the surface Σ is connected. Moreover, we can restrict ourselves to standard $\overline{\Sigma}$ and a as follows. Let g be the genus of Σ and $(W_1, \nu_1), \ldots, (W_m, \nu_m)$ be the marks of the distinguished arcs on Σ (enumerated in an arbitrary way). Since there is a non-separating loop on Σ, we have $g \geq 1$. Denote the tuple $(g; (W_1, \nu_1), \ldots, (W_m, \nu_m))$ by t. Consider the standard decorated surface $\Sigma_t = \partial U_t$ where U_t is the standard handlebody in $\mathbb{R}^3$ obtained as a closed regular neighborhood of the standard ribbon graph $R_t \subset \mathbb{R}^2 \times [0, 1]$ (see Figure IV.1.1). Recall that the distinguished Lagrangian space of Σ_t is the kernel of the inclusion homomorphism $H_1(\Sigma_t; \mathbb{R}) \to H_1(U_t; \mathbb{R})$. Denote by a_t a simple loop on Σ_t which is a meridian of the leftmost cap-like band of R_t. (In other words, $a_t \subset \partial U_t$ is a meridian of the leftmost 1-handle of the handlebody U_t.) The orientation of a_t is chosen so that its linking number with the leftmost cap-like band of R_t directed to the left is equal to $+1$. The base point of a is its highest point over the plane of R_t. It is obvious that there is a weak e-homeomorphism $\overline{\Sigma} \to \Sigma_t$ carrying a onto a_t. (Note that there may be no e-homeomorphism $\overline{\Sigma} \to \Sigma_t$.) The invariance of the gluing homomorphisms under weak e-homeomorphisms implies that we may restict ourselves to the case where $\overline{\Sigma} = \Sigma_t$ and $a = a_t$. For $k \in I$, the surface $\overline{\Sigma^a(V_k)}$ may be identified with the standard decorated surface $\Sigma_{t_k} = \partial U_{t_k}$ where t_k is the tuple $(g-1; (W_1, \nu_1), ..., (W_m, \nu_m), (V_k, 1), (V_k, -1))$.

We have $\mathcal{H}(\Sigma) = \mathcal{T}^e(\overline{\Sigma}) = \mathcal{T}^e(\Sigma_t)$. By definition (see Section IV.1.4), $\mathcal{T}^e(\Sigma_t) = \mathcal{T}(\Sigma_t) = \Psi_t$ where

$$\Psi_t = \bigoplus_{(i_1,...,i_g)\in I^g} \operatorname{Hom}(1\!\!1, W_1^{\nu_1} \otimes ... \otimes W_m^{\nu_m} \otimes \bigotimes_{r=1}^{g} (V_{i_r} \otimes V_{i_r}^*)).$$

Similarly, $\mathcal{H}(\Sigma^a(V_k)) = \mathcal{T}^e(\Sigma_{t_k}) = \Psi_{t_k}$. It follows from definitions that

$$\Psi_t = \bigoplus_{k\in I} \Psi_{t_k}.$$

Therefore

$$\mathcal{T}^e(\Sigma_t) = \bigoplus_{k\in I} \mathcal{T}^e(\Sigma_{t_k}).$$

Consider the gluing homomorphism

$$\Psi_{t_k} = \mathcal{T}^e(\Sigma_{t_k}) \to \mathcal{T}^e(\Sigma_t) = \Psi_t. \tag{5.9.a}$$

We claim that this is just the inclusion $\Psi_{t_k} \hookrightarrow \Psi_t$. This claim implies the splitting axiom.

By definition, the gluing homomorphism (5.9.a) is equal to $\tau^e(M_k)$ where M_k is an extended 3-cobordism obtained from the cylinder $\Sigma_t \times [0, 1]$ by attaching a 2-handle $B^2 \times [-1, 1]$ along an embedding $\partial B^2 \times [-1, 1] \hookrightarrow \Sigma_t \times 0$ onto a regular neighborhood of the loop $a_t \times 0 \subset \Sigma_t \times 0$, cf. Section 4.5. Since both bases of M_k are standard decorated surfaces we may regard M_k as a decorated 3-cobordism and use the technique of Section IV.2 to compute $\tau^e(M_k) = \tau(M_k)$.

We shall present decorated 3-cobordisms by ribbon graphs, as in Section IV.2.3. By Lemma IV.2.6, the cylinder $C = \Sigma_t \times [0, 1]$ is presented by the colored ribbon tangle Ω_t drawn in Figure IV.2.4. In order to obtain M_k we should add the 2-handle mentioned above. A direct analysis of definitions shows that M_k may be presented by the colored ribbon tangle Ω_{t_k} obtained from Ω_t by coloring its leftmost bottom cap-like band with k. (Geometrically, this is the same ribbon tangle, but since the band in question is colored we no longer cut out its neighborhood from S^3 in order to obtain the corresponding 3-cobordism. This amounts to attaching one 2-handle to C.) We may apply formula (IV.2.3.a) to compute $\tau(C), \tau(M_k)$. The only difference between the resulting expressions is that in the case of C all bottom cap-like bands of Ω_t are uncolored and we can color them with arbitrary $i_1, ..., i_g \in I^g$. In the case of M_k, we can color the $g-1$ uncolored bottom cap-like bands of Ω_{t_k} with arbitrary $i_2, ..., i_g \in I^g$. This shows that the operator $\tau(M_k)$ is the restriction of $\tau(C)$ to Ψ_{t_k}. By Lemma IV.2.1.1, $\tau(C) = \mathrm{id}$. Therefore $\tau(M_k)$ is the inclusion $\Psi_{t_k} \hookrightarrow \Psi_t$.

5.10. The splitting axiom: the case of a separating loop. We verify axiom (1.6.3) in the case where a_+ and a_- lie in different connected components of Σ^a. It is enough to consider the case where the surface Σ is connected. Then the surface $\Sigma^a(V_k)$ has two connected components $\Sigma_+(V_k)$ and $\Sigma_-(V_k)$ containing a_+ and a_- respectively. Hence

$$\mathcal{H}(\Sigma^a(V_k)) = \mathcal{H}(\Sigma_+(V_k)) \otimes_K \mathcal{H}(\Sigma_-(V_k)).$$

Note that any non-separating s-loop on $\Sigma_+(V_k)$ is simultaneously a non-separating s-loop on Σ. The results of the previous subsection allow us to split both $\mathcal{H}(\Sigma)$ and $\mathcal{H}(\Sigma_+(V_k))$ along such loops. Similar arguments apply to $\Sigma_-(V_k)$. Therefore, splitting along non-separating s-loops on $\Sigma_+(V_k), \Sigma_-(V_k)$, we can reduce the verification of (1.6.3) to the case where the genera of Σ, $\Sigma_+(V_k)$, and $\Sigma_-(V_k)$ are equal to 0.

Let $(U_1, \varepsilon_1), \ldots, (U_m, \varepsilon_m), (V_k, -1)$ be the marks of the distinguished arcs on $\Sigma_-(V_k)$. Denote the tuple $(0; (U_1, \varepsilon_1), \ldots, (U_m, \varepsilon_m), (V_k, -1))$ by t_-. As in the previous subsection, we may restrict ourselves to the case where $\Sigma_-(V_k)$ is the standard decorated surface Σ_{t_-} of type t_-. Similarly, we can assume that $\Sigma_+(V_k)$ is the standard decorated surface Σ_{t_+} of a certain type $t_+ = (0; (V_k, 1), (W_1, \nu_1), \ldots, (W_n, \nu_n))$. Then the surface $\overline{\Sigma}$ may be identified with Σ_t where

$$t = (0; (U_1, \varepsilon_1), \ldots, (U_m, \varepsilon_m), (V_k, -1), (V_k, +1), (W_1, \nu_1), \ldots, (W_n, \nu_n)).$$

Set $U = U_1^{\varepsilon_1} \otimes \ldots \otimes U_m^{\varepsilon_m}$ and $W = W_1^{\nu_1} \otimes \ldots \otimes W_n^{\nu_n}$. We have

$$\mathcal{T}^e(\Sigma_{t_-}) = \Psi_{t_-} = \mathrm{Hom}(\mathbb{1}, U \otimes V_k^*),$$

$$\mathcal{T}^e(\Sigma_{t_+}) = \Psi_{t_+} = \mathrm{Hom}(\mathbb{1}, V_k \otimes W),$$

$$\mathcal{T}^e(\Sigma_t) = \Psi_t = \mathrm{Hom}(\mathbb{1}, U \otimes W).$$

Consider the gluing homomorphism

$$\mathcal{T}^e(\Sigma_{t_-}) \otimes_K \mathcal{T}^e(\Sigma_{t_+}) = \mathcal{T}^e(\Sigma_{t_-} \amalg \Sigma_{t_+}) \to \mathcal{T}^e(\Sigma_t) \tag{5.10.a}$$

defined in Section 4.6. We claim that this homomorphism is equal to the homomorphism

$$\mathrm{Hom}(\mathbb{1}, U \otimes V_k^*) \otimes_K \mathrm{Hom}(\mathbb{1}, V_k \otimes W) \to \mathrm{Hom}(\mathbb{1}, U \otimes W)$$

defined by the formula

$$x \otimes y \mapsto (\mathrm{id}_U \otimes d_{V_k} \otimes \mathrm{id}_W)(x \otimes y) \tag{5.10.b}$$

where $x \in \mathrm{Hom}(\mathbb{1}, U \otimes V_k^*)$, $y \in \mathrm{Hom}(\mathbb{1}, V_k \otimes W)$. This claim and Lemma IV.2.2.2 would imply the splitting axiom.

The homomorphism (5.10.a) is equal to $\tau^e(M_k)$ where M_k is an extended 3-cobordism between $\Sigma_{t_-} \amalg \Sigma_{t_+}$ and Σ_t obtained by attaching a 2-handle $B^2 \times [-1, 1]$ to the cylinder $\Sigma_t \times [0, 1]$ along an embedding $\partial B^2 \times [-1, 1] \hookrightarrow \Sigma_t \times 0$ onto a regular neighborhood of the loop $a_t \times 0 \subset \Sigma_t \times 0$, cf. Section 4.5. Since the bases of M_k are standard surfaces we may regard M_k as a decorated 3-cobordism and use the technique of Section IV.2.4 to compute $\tau^e(M_k) = \tau(M_k)$. Let Ω_t denote the colored ribbon tangle in $\mathbb{R}^3$ consisting of $m+n$ vertical untwisted unlinked bands colored with $U_1, ..., U_m, W_1, ..., W_n$. The directions of these bands are determined by the signs $\varepsilon_1, ..., \varepsilon_m, \nu_1, ..., \nu_n$ in the usual way. The colored ribbon tangle Ω_t presents C in the sense specified in Section IV.2. Indeed, attaching to Ω_t a coupon from below and a coupon from above and removing from S^3 regular neighborhoods of these two coupons we obtain $\Sigma_t \times [0, 1]$. (No surgery is involved.) In order to obtain M_k, we should paste the 2-handle (with a V_k-colored band inside) mentioned above. This can be done inside a regular neighborhood of the lower coupon. A direct analysis of definitions shows that M_k may be presented by the colored ribbon tangle Ω shown in Figure 5.16 together with the integers $k_1^- = m+1, k_2^- = n+1, k_1^+ = m+n$ (see Section IV.2.4, here $r^- = 2, r^+ = 1$). Note that the cap-like band in Figure 5.16 is the band contained in the 2-handle in question. It follows from definitions (or from the results of Section IV.2.4) that the homomorphism $\tau(M_k)$ is induced by the morphism

$$F(\Omega) : U \otimes V_k^* \otimes V_k \otimes W \to U \otimes W.$$

Therefore the gluing homomorphism (5.10.a) is computed by formula (5.10.b).

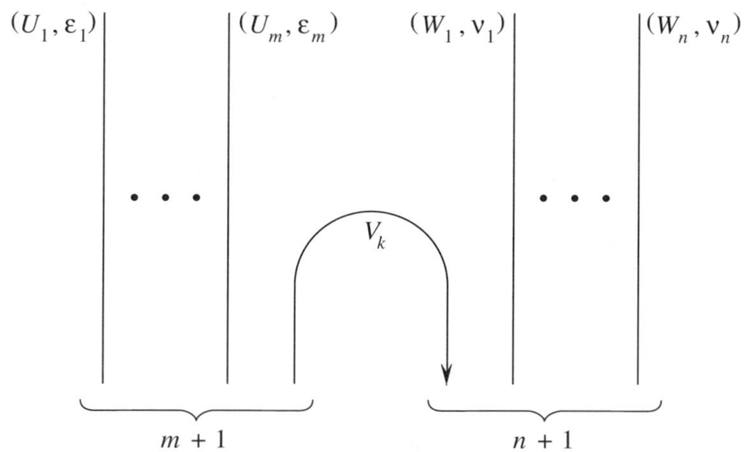

Figure 5.16

5.11. The underlying ribbon category. Let $\mathcal{W}$ denote the underlying ribbon category of the weak 2-DMF $\mathcal{H}$. We shall construct an isomorphism of ribbon categories $\mathcal{R} : \mathcal{V} \to \mathcal{W}$. The categories $\mathcal{W}$ and $\mathcal{V}$ have the same class of objects and we define $\mathcal{R}$ to be the identity on this class. For any morphism $f : V \to W$

in $\mathcal{V}$, define $\mathcal{R}(f) : V \to W$ as follows. Consider the additive isomorphism

$$\rho_{+V}^{+W} : \mathrm{Hom}_{\mathcal{V}}(V, W) \to \mathcal{H}(E(-V; W)) = \mathrm{Hom}_{\mathcal{W}}(V, W)$$

defined in Section 5.2. Set $\mathcal{R}(f) = \rho_{+V}^{+W}(f)$. It is easy to deduce from Lemmas 5.2.1, 5.2.3, and the definition of the composition in $\mathcal{W}$ that $\mathcal{R}$ carries the composition of morphisms in $\mathcal{V}$ into the composition in $\mathcal{W}$.

The identity endomorphism of an object V of $\mathcal{W}$ is equal to $e_- \diamond e_{\mathbb{1},V} \in \mathcal{H}(E(-V; V))$. It follows from the computation of $e_- \diamond e_{\mathbb{1},V}$ given in Section 5.8, that $\mathcal{R}$ carries the identity of V in $\mathcal{V}$ into the identity of V in $\mathcal{W}$. Thus, $\mathcal{R} : \mathcal{V} \to \mathcal{W}$ an isomorphism of Ab-categories. We claim that $\mathcal{R}$ carries the tensor product, braiding, twist, and duality in $\mathcal{V}$ into the tensor product, braiding, twist, and duality in $\mathcal{W}$. This will imply that $\mathcal{R}$ is an isomorphism of ribbon categories.

The fact that $\mathcal{R}$ commutes with tensor product is proven in Figure 5.17. Here $f : V \to V'$ and $g : W \to W'$ are morphisms in $\mathcal{V}$. Both diagrams in Figure 5.17 represent elements of $\mathcal{H}(E(-VW; V', W'))$. The left diagram represents $\mathcal{R}(f) \diamond (\mathcal{R}(g) \diamond e_{V,W})$ where $\mathcal{R}(f) \in \mathrm{Hom}_{\mathcal{W}}(V, V') = \mathcal{H}(E(-V; V'))$ and $\mathcal{R}(g) \in \mathrm{Hom}_{\mathcal{W}}(W, W') = \mathcal{H}(E(-W; W'))$. The right diagram represents $e_{V',W'} \diamond \mathcal{R}(f \otimes g)$ where

$$\mathcal{R}(f \otimes g) \in \mathrm{Hom}_{\mathcal{W}}(V \otimes W, V' \otimes W') = \mathcal{H}(E(-VW; V'W')).$$

The equality in Figure 5.17 follows from Lemma 5.2.1. This equality and the definition of the tensor product in $\mathcal{W}$ imply that $\mathcal{R}(f \otimes g) = \mathcal{R}(f) \otimes \mathcal{R}(g)$.

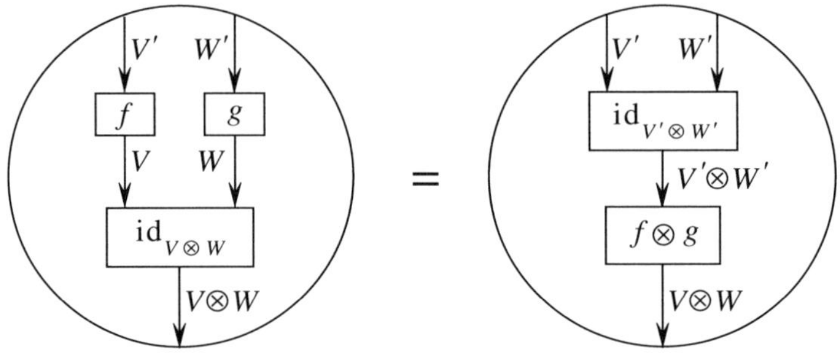

Figure 5.17

The fact that $\mathcal{R}$ carries the braiding $c_{V,W} : V \otimes W \to W \otimes V$ in $\mathcal{V}$ into the braiding in $\mathcal{W}$ is proven in Figure 5.18. Both diagrams in Figure 5.18 represent elements of $\mathcal{H}(E(-VW; W, V))$. The left diagram represents $e_{W,V} \diamond \mathcal{R}(c_{V,W})$ where

$$\mathcal{R}(c_{V,W}) \in \mathrm{Hom}_{\mathcal{W}}(V \otimes W, W \otimes V) = \mathcal{H}(E(-VW; WV)).$$

There is an obvious e-homeomorphism of the extended 3-balls presented schematically in Figure 5.5 and on the right-hand side of Figure 5.18. This homeomorphism restricts to the e-homeomorphism of the boundaries induced by the homeomorphism $\gamma : E(-VW; V, W) \to E(-VW; W, V)$ defined in Section 2.5. Therefore the right diagram in Figure 5.18 represents the state $\gamma_\#(e_{V,W}) \in \mathcal{H}(E(-VW; W, V))$. The equality in Figure 5.18 follows from Lemma 5.2.1. Therefore $e_{W,V} \diamond \mathcal{R}(c_{V,W}) = \gamma_\#(e_{V,W})$. This equality and the definition of the braiding in $\mathcal{W}$ (see Section 2.5) imply that $\mathcal{R}(c_{V,W})$ is the braiding in $\mathcal{W}$.

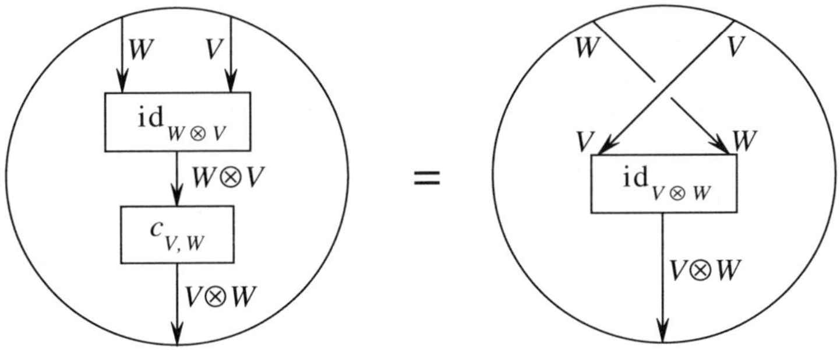

Figure 5.18

The fact that $\mathcal{R}$ carries the twist $\theta_V : V \to V$ in $\mathcal{V}$ into the twist in $\mathcal{W}$ is proven similarly, see Figure 5.19. The diagrams in Figure 5.19 represent elements of $\mathcal{H}(E(-V; V)) = \mathrm{Hom}_{\mathcal{W}}(V, V)$. The left diagram represents $\mathcal{R}(\theta_V)$. The right diagram represents $\theta_\#(\mathcal{R}(\mathrm{id}_V))$ where $\theta : E(-V; V) \to E(-V; V)$ is the m-homeomorphism considered in Section 2.6. As we know, $\mathcal{R}$ carries the identity in $\mathcal{V}$ into the identity in $\mathcal{W}$. Therefore the second diagram represents $\theta_\#(\mathrm{id}_V)$, that is the twist in $\mathcal{W}$.

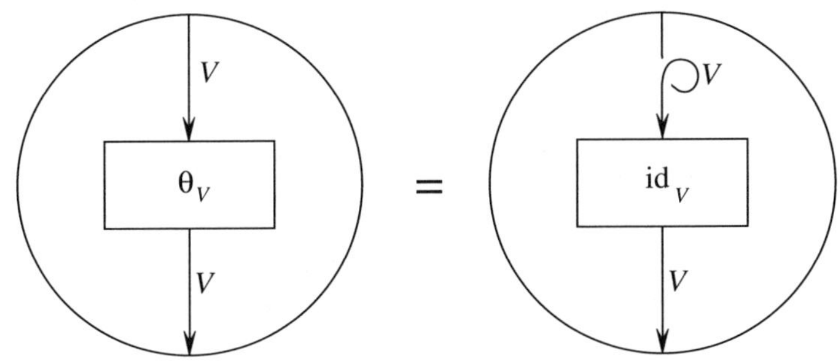

Figure 5.19

The fact that $\mathcal{R}$ carries the duality morphism $V^* \otimes V \to \mathbb{1}$ in $\mathcal{V}$ into the duality morphism $V^* \otimes V \to \mathbb{1}$ in $\mathcal{W}$ is almost tautological. Denote these two morphisms by d_V and d'_V respectively. The equalities in Figure 5.20 follow from the definition of D_V and Lemma 5.2.1. The $\mathrm{id}_{\mathbb{1}}$-colored coupon with the $\mathbb{1}$-colored band attached from below represents e_-. By Lemma 5.2.3, cutting off this coupon is equivalent to the excision of e_-. Therefore, the duality morphism d'_V can be computed as in Figure 5.21. This implies that $d'_V = \mathcal{R}(d_V)$.

It remains to show that $\mathcal{R}$ carries the duality morphism $\mathbb{1} \to V \otimes V^*$ in $\mathcal{V}$ into the duality morphism $\mathbb{1} \to V \otimes V^*$ in $\mathcal{W}$. Note that in any ribbon category, any morphism $x \in \mathrm{Hom}(\mathbb{1}, V \otimes V^*)$ can be written in the form

$$x = ((\mathrm{id}_V \otimes d_V)(x \otimes \mathrm{id}_V) \otimes \mathrm{id}_{V^*})\, b_V.$$

(To obtain this equality, use the graphical calculus of Section I.1.1 and the identity $(d_V \otimes \mathrm{id}_{V^*})(\mathrm{id}_{V^*} \otimes b_V) = \mathrm{id}_{V^*}$.) Hence there exists at most one morphism $x : \mathbb{1} \to V \otimes V^*$ such that $(\mathrm{id}_V \otimes d_V)(x \otimes \mathrm{id}_V) = \mathrm{id}_V$. It is clear that both the duality morphism $\mathbb{1} \to V \otimes V^*$ in $\mathcal{W}$, and the image under $\mathcal{R}$ of the duality morphism $b_V : \mathbb{1} \to V \otimes V^*$ in $\mathcal{V}$, satisfy the last condition. Therefore they are equal. Thus, $\mathcal{R}$ is an isomorphism of ribbon categories.

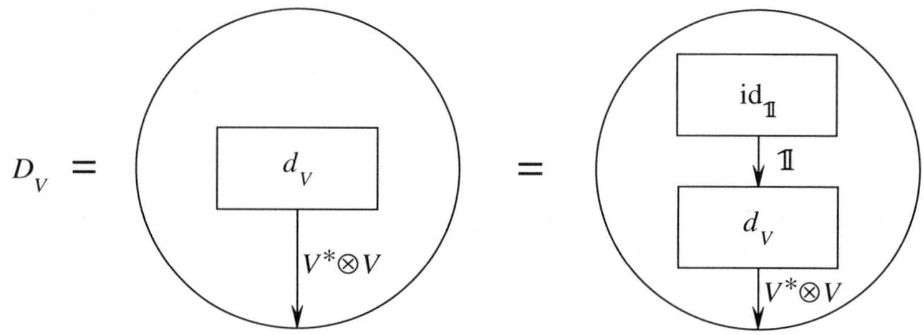

Figure 5.20

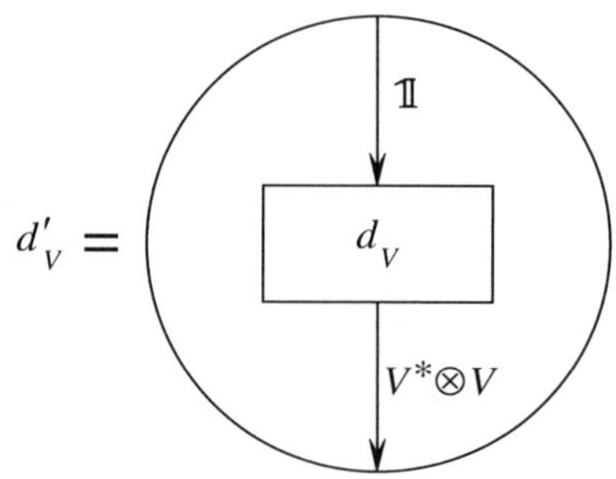

Figure 5.21

5.12. The non-degeneracy axiom. The isomorphism of ribbon categories $\mathcal{R} : \mathcal{V} \to \mathcal{W} = \mathcal{V}_{\mathcal{H}}$ constructed in Section 5.11 carries the distinguished objects $\{V_i\}_{i \in I}$ of $\mathcal{V}$ into the distinguished objects $\{V_i\}_{i \in I}$ of $\mathcal{W}$. Since axiom (1.6.4) holds for $\mathcal{V}$, it holds also for $\mathcal{W}$. This finishes the construction of a weak rational 2-DMF and the proof of Theorem 4.1.

Notes

Section 1. The concept of a rational modular functor was conceived in the framework of 2-dimensional conformal field theory (CFT), see [MS1], [MS2], [Se]. Our exposition differs in a few essential points. As marks we use pairs (an element of a fixed monoidal class, a sign ± 1) rather than elements of a finite set I, as it is common in CFT. This enables us to introduce axioms (1.5.6), (1.5.7), (1.5.8) crucial for our study. We consider gluing homomorphisms rather than just the splitting (1.6.b). Also, we pay proper attention to Lagrangian spaces in 1-homologies which were missing in previous works.

Axioms (1.5.1) - (1.5.5), (1.6.1) - (1.6.3) are known in 2-dimensional CFT (modulo the remarks above). Axioms (1.5.6) - (1.5.8), (1.6.4) are new.

In CFT the module $\mathscr{H}(\Sigma)$ assigned to a surface Σ is called the module of conformal blocks of Σ. This term apparently has no convincing justification in physics; we prefer the more neutral term "the module of states".

Note also that in CFT one usually applies modular functors to closed surfaces with distinguished points provided with tangent vectors and labels. These are smooth analogues of the e-surfaces introduced in Chapter IV; we use oriented arcs rather than points with tangent vectors in order to avoid smooth structures on surfaces. As was already mentioned above, the notions of an e-surface and an m-surface are essentially equivalent.

Sections 2 - 5. The material of these sections is new.

Part II

The Shadow World

Chapter VI
$6j$-symbols

Fix up to the end of this chapter a strict semisimple ribbon Ab-category ($\mathcal{V}$, $\{V_i\}_{i\in I}$) with ground ring K.

1. Algebraic approach to $6j$-symbols

1.0. Outline. We derive from the category $\mathcal{V}$ a system of tensors, called $6j$-symbols. They are numerated by ordered 6-tuples of elements of the set I. Each $6j$-symbol is a tensor of 4 variables running over certain multiplicity modules. As it will be clear from definitions, the system of $6j$-symbols describes associativity of the tensor product in $\mathcal{V}$ in terms of multiplicity modules.

We study in this section simple algebraic properties of $6j$-symbols. Their geometric nature will be revealed in Section 5.

1.1. Lemmas. The following two lemmas will be instrumental in the definition of $6j$-symbols.

1.1.1. Lemma. *Let U, V, W be objects of $\mathcal{V}$. For each $i \in I$, the formula $(f, g) \mapsto (\mathrm{id}_V \otimes g)f$ defines a K-homomorphism*

$$\mathrm{Hom}(U, V \otimes V_i) \otimes_K \mathrm{Hom}(V_i, W) \to \mathrm{Hom}(U, V \otimes W). \tag{1.1.a}$$

The direct sum of these homomorphisms is an isomorphism

$$\bigoplus_{i\in I} \mathrm{Hom}(U, V \otimes V_i) \otimes_K \mathrm{Hom}(V_i, W) \to \mathrm{Hom}(U, V \otimes W). \tag{1.1.b}$$

Proof. Consider the isomorphisms $\mathrm{Hom}(U, V \otimes W) \to \mathrm{Hom}(V^* \otimes U, W)$ and $\mathrm{Hom}(U, V \otimes V_i) \to \mathrm{Hom}(V^* \otimes U, V_i)$ constructed in Exercise I.1.8.1. Under these isomorphisms the homomorphism (1.1.a) corresponds to the homomorphism

$$\mathrm{Hom}(V^* \otimes U, V_i) \otimes_K \mathrm{Hom}(V_i, W) \to \mathrm{Hom}(V^* \otimes U, W)$$

induced by composition. (To see this use the graphical notation for morphisms.) Therefore the lemma follows from Lemma II.4.2.2.

1.1.2. Lemma. *Let U, V, W be objects of $\mathcal{V}$. For each $i \in I$, the formula $(f, g) \mapsto (g \otimes \mathrm{id}_W)f$ defines a K-homomorphism*

$$\mathrm{Hom}(U, V_i \otimes W) \otimes_K \mathrm{Hom}(V_i, V) \to \mathrm{Hom}(U, V \otimes W). \tag{1.1.c}$$

The direct sum of these homomorphisms is an isomorphism

$$\bigoplus_{i \in I} \mathrm{Hom}(U, V_i \otimes W) \otimes_K \mathrm{Hom}(V_i, V) \to \mathrm{Hom}(U, V \otimes W). \tag{1.1.d}$$

The proof of this lemma repeats the proof of the previous lemma with the isomorphisms of Exercise I.1.8.1 replaced by the isomorphisms of Exercise I.2.9.3.

Note that the direct sums which appear in (1.1.b) and (1.1.d) are finite even if the set I is infinite. Indeed, it follows from Lemma II.4.2.1 that only a finite number of summands are non-trivial.

1.2. Multiplicity modules. For any $i, j, k \in I$ consider the K-modules

$$H_k^{ij} = \mathrm{Hom}(V_k, V_i \otimes V_j) \quad \text{and} \quad H_{ij}^k = \mathrm{Hom}(V_i \otimes V_j, V_k).$$

These are "modules of multiplicities". By Lemma II.4.2.1, these modules are projective. By Lemma II.4.2.3, the bilinear pairing

$$(x, y) \mapsto \mathrm{tr}(yx) : H_k^{ij} \otimes_K H_{ij}^k \to K$$

is non-degenerate. We shall identify the dual modules $(H_k^{ij})^\star = \mathrm{Hom}_K(H_k^{ij}, K)$ and $(H_{ij}^k)^\star = \mathrm{Hom}_K(H_{ij}^k, K)$ with H_{ij}^k and H_k^{ij} respectively (via this pairing).

In the graphical notation elements $x \in H_k^{ij}$, $y \in H_{ij}^k$ are presented in Figure 1.1. Here we have diagrams of two v-colored elementary ribbon graphs in $\mathbb{R}^3$ whose operator invariants are equal to x and y respectively. As usual, the symbols $i, j, k, \ldots$ attached to strings signify that these strings are colored with $V_i, V_j, V_k, \ldots$ respectively. The trace $\mathrm{tr}(yx) = \mathrm{tr}(xy)$ is presented in Figure 1.2. Here, as in Chapters I and II, the symbol $\doteq$ denotes equality of the operator invariant of a v-colored ribbon graph and a morphism in the category $\mathcal{V}$. Note that the composition $yx : V_k \to V_k$ is multiplication by $(\dim(k))^{-1}\mathrm{tr}(yx)$.

1.3. Definition of $6j$-symbols. For each (ordered) 6-tuple of elements of the set I, we define a tensor of 4 variables called the $6j$-symbol of this tuple. The $6j$-symbol of a 6-tuple $(i, j, k, l, m, n) \in I^6$ is a K-linear homomorphism

$$\begin{Bmatrix} i & j & k \\ l & m & n \end{Bmatrix} : H_m^{kl} \otimes_K H_k^{ij} \to H_m^{in} \otimes_K H_n^{jl}. \tag{1.3.a}$$

It is defined as follows. Lemma 1.1.1 yields an isomorphism

$$\bigoplus_{n \in I} H_m^{in} \otimes_K H_n^{jl} \to \mathrm{Hom}(V_m, V_i \otimes V_j \otimes V_l). \tag{1.3.b}$$

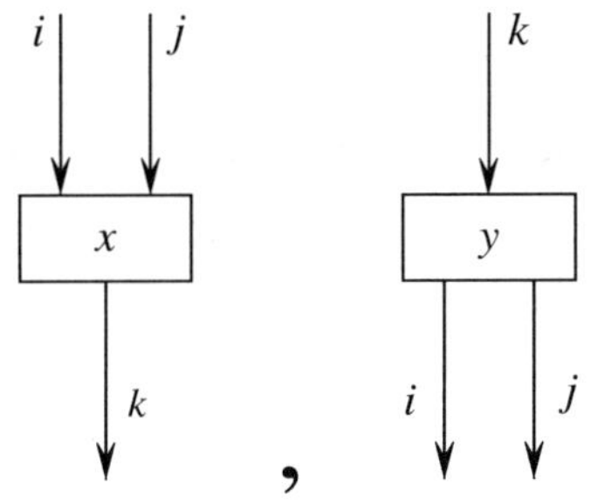

Figure 1.1

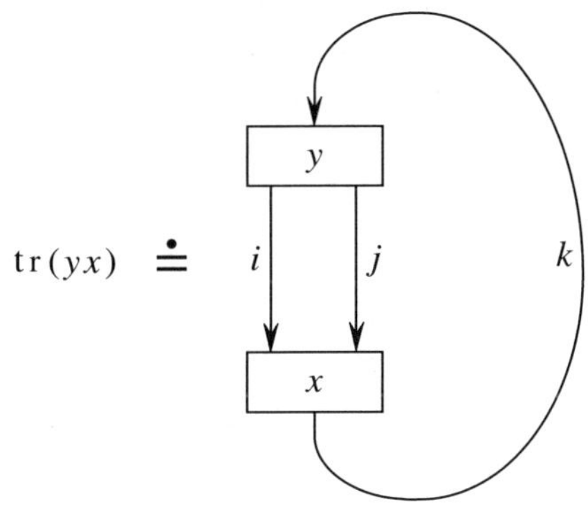

Figure 1.2

Lemma 1.1.2 yields an isomorphism

$$\bigoplus_{k \in I} H_m^{kl} \otimes_K H_k^{ij} \to \mathrm{Hom}(V_m, V_i \otimes V_j \otimes V_l). \tag{1.3.c}$$

Composing the second isomorphism with the inverse of the first one we get an isomorphism

$$\bigoplus_{k \in I} H_m^{kl} \otimes_K H_k^{ij} \to \bigoplus_{n \in I} H_m^{in} \otimes_K H_n^{jl}. \tag{1.3.d}$$

Restricting this isomorphism to the summand on the left-hand side corresponding to a given $k \in I$ and projecting into the summand on the right-hand side corresponding to a given $n \in I$ we get the homomorphism (1.3.a). Thus, the homomorphisms (1.3.a) corresponding to fixed i, j, l, m and arbitrary k, n form a block-matrix of the isomorphism (1.3.d).

The $6j$-symbol (1.3.a) has the following finiteness property: for any i, j, l, m, there is only a finite number of pairs (k, n) such that (1.3.a) is non-zero. This follows from Lemma II.4.2.1.

In the graphical notation the definition of 6j-symbol is presented in Figure 1.3. The left graph in Figure 1.3 has two empty boxes which may be filled in with arbitrary elements of H_m^{kl} and H_k^{ij}. When we put in these boxes elements, say $x \in H_m^{kl}$ and $y \in H_k^{ij}$, we get a diagram of a v-colored ribbon graph. Its operator invariant is equal to $(y \otimes \mathrm{id}_{V_l})x : V_m \to V_i \otimes V_j \otimes V_l$. Thus the left graph represents the K-homomorphism

(1.3.e) $$x \otimes y \mapsto (y \otimes \mathrm{id}_{V_l})x : H_m^{kl} \otimes_K H_k^{ij} \to \mathrm{Hom}(V_m, V_i \otimes V_j \otimes V_l).$$

Similarly, the graph in the right part of Figure 1.3 represents the K-homomorphism

$$a \otimes b \mapsto (\mathrm{id}_{V_i} \otimes b)a : H_m^{in} \otimes_K H_n^{jl} \to \mathrm{Hom}(V_m, V_i \otimes V_j \otimes V_l).$$

The equality in Figure 1.3 indicates that the composition of the last homomorphism with homomorphism (1.3.a) summed up over all $n \in I$ is equal to (1.3.e). This equality is a straightforward corollary of definitions.

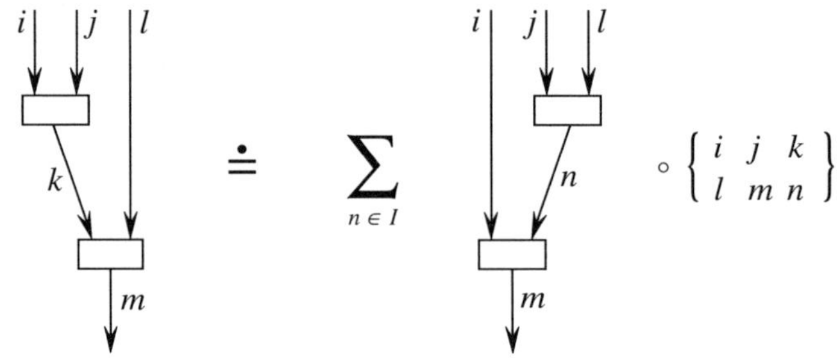

Figure 1.3

1.4. Reformulation of the 6*j*-symbols. Projectivity of the K-modules H_m^{kl} and H_k^{ij} implies that the 6j-symbol (1.3.a) corresponds to an element of the K-module

(1.4.a) $$(H_m^{kl} \otimes_K H_k^{ij})^\star \otimes_K H_m^{in} \otimes_K H_n^{jl} = H_{kl}^m \otimes_K H_{ij}^k \otimes_K H_m^{in} \otimes_K H_n^{jl}.$$

We shall denote this element by

(1.4.b) $$\left\{ \begin{matrix} i & j & k \\ l & m & n \end{matrix} \right\}'.$$

This is an equivalent version of the 6j-symbol (1.3.a). The relationship between these two forms of 6j-symbol may be more exactly described as follows. Each element x of the module (1.4.a) determines an adjoint homomorphism $\tilde{x} : H_m^{kl} \otimes_K H_k^{ij} \to H_m^{in} \otimes_K H_n^{jl}$ as follows. Present x as a sum $\sum_p x_1^p \otimes x_2^p \otimes x_3^p \otimes x_4^p$ where p runs over a certain finite set of indices and $x_1^p \in H_{kl}^m$, $x_2^p \in H_{ij}^k$, $x_3^p \in H_m^{in}$, $x_4^p \in H_n^{jl}$. The homomorphism $\tilde{x}$ is defined by the formula

$$\tilde{x}(y_1 \otimes y_2) = \sum_p \mathrm{tr}(x_1^p y_1)\, \mathrm{tr}(x_2^p y_2)\, x_3^p \otimes x_4^p$$

for any $y_1 \in H^{kl}_m, y_2 \in H^{ij}_k$. The $6j$-symbol (1.4.b) is the unique element of the module (1.4.a) such that its adjoint homomorphism is equal to (1.3.a). The $6j$-symbol (1.4.b) admits the following geometric interpretation.

1.4.1. Lemma. *We have the equality in Figure* 1.4.

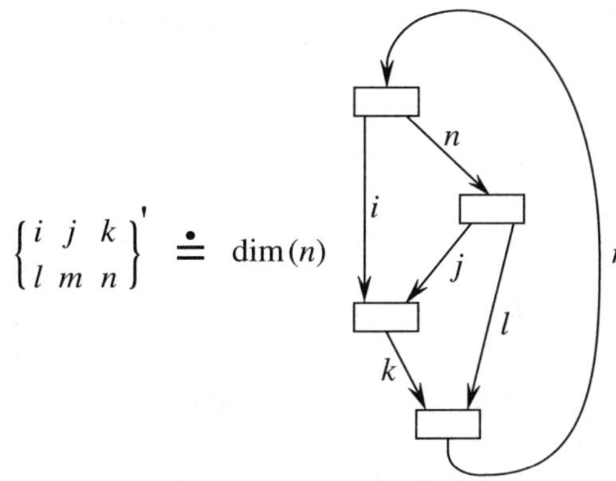

Figure 1.4

Both sides of the formula in Figure 1.4 should be treated as K-linear functionals

$$H^{kl}_m \otimes_K H^{ij}_k \otimes_K H^m_{in} \otimes_K H^n_{jl} \to K.$$

Proof of Lemma. Denote by α the K-homomorphism

$$H^{kl}_m \otimes_K H^{ij}_k \otimes_K H^n_{jl} \to H^{in}_m$$

adjoint to the $6j$-symbol (1.3.a). We first prove that for any $x \in H^{kl}_m, y \in H^{ij}_k$, $z \in H^n_{jl}$,

$$\alpha(x \otimes y \otimes z) = \dim(n)(\mathrm{id}_{V_i} \otimes z)(y \otimes \mathrm{id}_{V_l})x. \tag{1.4.c}$$

By Lemma 1.1.1

$$(y \otimes \mathrm{id}_{V_l})x = \sum_{q \in I, r \in R_q} (\mathrm{id}_{V_i} \otimes b_{r,q})a_{r,q} \tag{1.4.d}$$

where R_q is a finite set of indices depending on $q \in I$ and $a_{r,q} \in H^{iq}_m$, $b_{r,q} \in H^{jl}_q$. By the very definition of $6j$-symbols, we have for each $q \in I$,

$$\begin{Bmatrix} i & j & k \\ l & m & q \end{Bmatrix}(x \otimes y) = \sum_{r \in R_q} a_{r,q} \otimes b_{r,q}.$$

Composing (1.4.d) with $\mathrm{id}_{V_i} \otimes z$ we get

$$(\mathrm{id}_{V_i} \otimes z)(y \otimes \mathrm{id}_{V_l})x = \sum_{q \in I, r \in R_q} (\mathrm{id}_{V_i} \otimes z\, b_{r,q})a_{r,q}.$$

The homomorphism $z\, b_{r,q} : V_q \to V_n$ is multiplication by $(\dim(n))^{-1}\mathrm{tr}(zb_{r,n})$ if $q = n$ and equals zero if $q \neq n$. Hence

$$(\mathrm{id}_{V_i} \otimes z)(y \otimes \mathrm{id}_{V_l})x = \sum_{r \in R_n} (\dim(n))^{-1}\mathrm{tr}(zb_{r,n})a_{r,n}.$$

This implies (1.4.c).

Now, for any $t \in H^m_{in}$, we have

(1.4.e) $$t\,\alpha(x \otimes y \otimes z) = \dim(n)\, t\, (\mathrm{id}_{V_i} \otimes z)(y \otimes \mathrm{id}_{V_l})x \in \mathrm{Hom}(V_m, V_m).$$

Applying the trace to the left-hand side we get the canonical coupling between the $6j$-symbol (1.4.b) and $x \otimes y \otimes t \otimes z$. On the other hand, if we assign x, y, z, t to the boxes in Figure 1.4 then the operator invariant of the resulting v-colored ribbon graph will be equal to the trace of $t(\mathrm{id}_{V_i} \otimes z)(y \otimes \mathrm{id}_{V_l})x$. Therefore the equality in Figure 1.4 follows directly from (1.4.e).

1.5. The Biedenharn-Elliott identity. One of the main properties of the $6j$-symbols is a far-reaching generalization of the Biedenharn-Elliott identity well known for $6j$-symbols associated to the Lie algebra $sl_2(\mathbb{C})$ (see [BL1]). We give here two forms of this identity formulated in terms of the $6j$-symbol (1.3.a) and the modified $6j$-symbol (1.4.b). We shall write id^{ij}_k for the identity endomorphism of H^{ij}_k.

1.5.1. Theorem. *For any $j_0, j_1, ..., j_8 \in I$, we have*

$$\sum_{j \in I} (\mathrm{id}^{j_1 j_7}_{j_0} \otimes \left\{\begin{matrix} j_2 & j_3 & j \\ j_4 & j_7 & j_8 \end{matrix}\right\})(\left\{\begin{matrix} j_1 & j & j_6 \\ j_4 & j_0 & j_7 \end{matrix}\right\} \otimes \mathrm{id}^{j_2 j_3}_{j})(\mathrm{id}^{j_6 j_4}_{j_0} \otimes \left\{\begin{matrix} j_1 & j_2 & j_5 \\ j_3 & j_6 & j \end{matrix}\right\}) =$$

$$= (\left\{\begin{matrix} j_1 & j_2 & j_5 \\ j_8 & j_0 & j_7 \end{matrix}\right\} \otimes \mathrm{id}^{j_3 j_4}_{j_8})\, P_{23}\, (\left\{\begin{matrix} j_5 & j_3 & j_6 \\ j_4 & j_0 & j_8 \end{matrix}\right\} \otimes \mathrm{id}^{j_1 j_2}_{j_5})$$

where P_{23} is the permutation (the flip) of the second and third factors in the tensor product.

Both sides of this equality are homomorphisms $Y \to X$ where

(1.5.a) $$Y = H^{j_6 j_4}_{j_0} \otimes_K H^{j_5 j_3}_{j_6} \otimes_K H^{j_1 j_2}_{j_5} \quad \text{and} \quad X = H^{j_1 j_7}_{j_0} \otimes_K H^{j_2 j_8}_{j_7} \otimes_K H^{j_3 j_4}_{j_8}.$$

The finiteness of $6j$-symbols implies that the sum in Theorem 1.5.1 is finite.

Proof of Theorem. Consider the module

$$Z = \mathrm{Hom}(V_{j_0}, V_{j_1} \otimes V_{j_2} \otimes V_{j_3} \otimes V_{j_4}).$$

Computing here the tensor product in the order 1(2(34)) and using Lemma 1.1.1 we may identify Z with the direct sum of the modules $\{X = X(j_7, j_8)\}$ corresponding to arbitrary $j_7, j_8 \in I$. Computing the same tensor product in the order ((12)3)4 and using Lemma 1.1.2 we may identify Z with the direct sum of the modules $\{Y = Y(j_5, j_6)\}$ corresponding to arbitrary $j_5, j_6 \in I$. This gives an isomorphism

$$\bigoplus_{j_7, j_8 \in I} X(j_7, j_8) \approx \bigoplus_{j_5, j_6 \in I} Y(j_5, j_6). \tag{1.5.b}$$

This isomorphism may be presented as a composition of isomorphisms induced by rearrangements of brackets. There are two different ways leading from 1(2(34)) to ((12)3)4, namely 1(2(34)) $\mapsto$ 1((23)4) $\mapsto$ (1(23))4 $\mapsto$ ((12)3)4 and 1(2(34)) $\mapsto$ (12)(34) $\mapsto$ ((12)3)4. Each arrow here corresponds to a tensor product of a matrix of $6j$-symbols and a certain identity isomorphism id_k^{ij}. The compositions of these two sequences of arrows give the same isomorphism (1.5.b). This yields the claim of the theorem.

The same argument may be given in a graphical form, see Figures 1.5 and 1.6. To simplify these figures we have shrunk all coupons into points. In Figure 1.6 the symbol $L(j, j_0, j_1, ..., j_8)$ denotes the summand on the left-hand side of the equality in Theorem 1.5.1. Each of these two figures provides a decomposition of the embedding $Y(j_5, j_6) \hookrightarrow Z$ into a linear combination of compositions $Y(j_5, j_6) \to X(j_7, j_8) \hookrightarrow Z$. Since $Z = \oplus_{j_7, j_8} X(j_7, j_8)$ we have the equality of the corresponding homomorphisms $Y(j_5, j_6) \to X(j_7, j_8)$. In other words, for each pair $j_7, j_8 \in I$, the coefficients in the final expressions in Figures 1.5 and 1.6 are equal.

1.5.2. Tensor contractions. In order to reformulate Theorem 1.5.1 in terms of the symbol (1.4.b) we pass to non-ordered tensor products of multiplicity modules (cf. Section III.1.2) and replace composition of homomorphisms with contraction of tensors. When there is a tensor product (ordered or non-ordered) of several K-modules so that among the factors there is a matched pair H_k^{ij}, H_{ij}^k we may contract any element of this tensor product using the canonical duality between H_k^{ij} and H_{ij}^k. For example, any element $x \otimes y \otimes z \in H \otimes H_k^{ij} \otimes H_{ij}^k$ contracts into $\mathrm{tr}(zy)x \in H$. This operation is called the contraction along H_{ij}^k (or along H_k^{ij}) and denoted by $*_{ij}^k$. The contraction is well-defined only when there is a decomposition into a tensor product and the factors H_k^{ij}, H_{ij}^k to be contracted are matched. It is clear that contractions along disjoint pairs of factors commute.

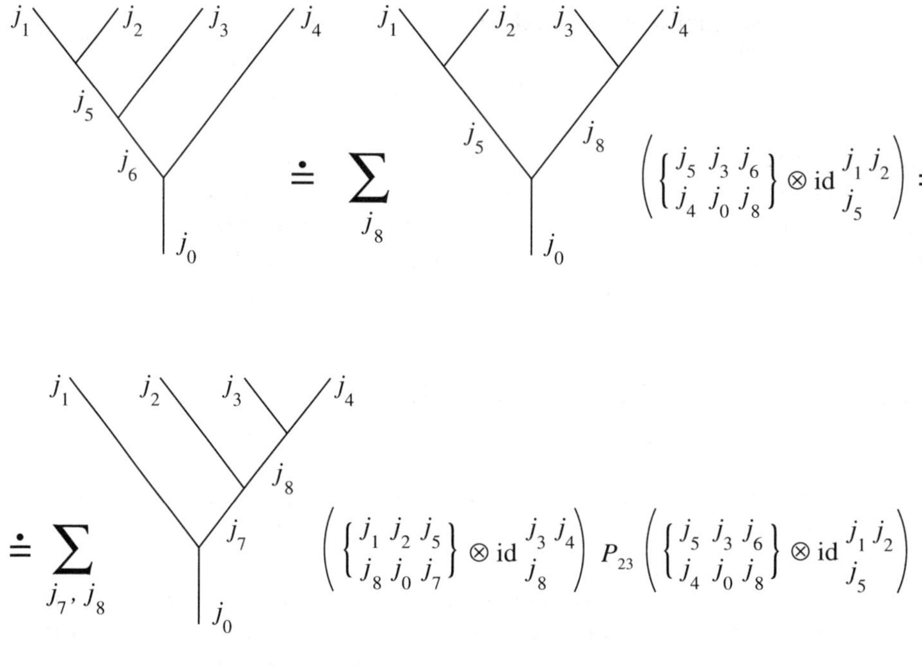

Figure 1.5

1.5.3. Corollary of Theorem 1.5.1. *For any $j_0, j_1, \ldots, j_8 \in I$, we have*

$$(1.5.c)\quad \sum_{j\in I} *^{j}_{j_2 j_3} *^{j_7}_{j j_4} *^{j_6}_{j_1 j}\left(\left\{\begin{matrix} j_1 & j_2 & j_5\\ j_3 & j_6 & j\end{matrix}\right\}' \otimes \left\{\begin{matrix} j_1 & j & j_6\\ j_4 & j_0 & j_7\end{matrix}\right\}' \otimes \left\{\begin{matrix} j_2 & j_3 & j\\ j_4 & j_7 & j_8\end{matrix}\right\}'\right) =$$

$$= *^{j_0}_{j_5 j_8}\left(\left\{\begin{matrix} j_5 & j_3 & j_6\\ j_4 & j_0 & j_8\end{matrix}\right\}' \otimes \left\{\begin{matrix} j_1 & j_2 & j_5\\ j_8 & j_0 & j_7\end{matrix}\right\}'\right).$$

Here both sides lie in the non-ordered tensor product of six K-modules

$$H^{j_6}_{j_5 j_3},\ H^{j_5}_{j_1 j_2},\ H^{j_0}_{j_6 j_4},\ H^{j_1 j_7}_{j_0},\ H^{j_2 j_8}_{j_7},\ H^{j_3 j_4}_{j_8}.$$

The permutation P_{23} used in Theorem 1.5.1 does not appear in this context because we are dealing here with non-ordered tensor products.

Proof of Corollary. Note first that the homomorphism (1.3.a) may be expressed in terms of the modified $6j$-symbol (1.4.b) as follows: for any $x \in H^{kl}_m \otimes_K H^{ij}_k$, we have

$$\left\{\begin{matrix} i & j & k\\ l & m & n\end{matrix}\right\}(x) = *^{m}_{kl} *^{k}_{ij}\left(\left\{\begin{matrix} i & j & k\\ l & m & n\end{matrix}\right\}' \otimes x\right).$$

Now we can prove the Corollary. Denote by A the tensor product of the three $6j$-symbols on the left-hand side of (1.5.c). Denote by B the tensor product of the two $6j$-symbols on the right-hand side of (1.5.c). Recall the modules X, Y

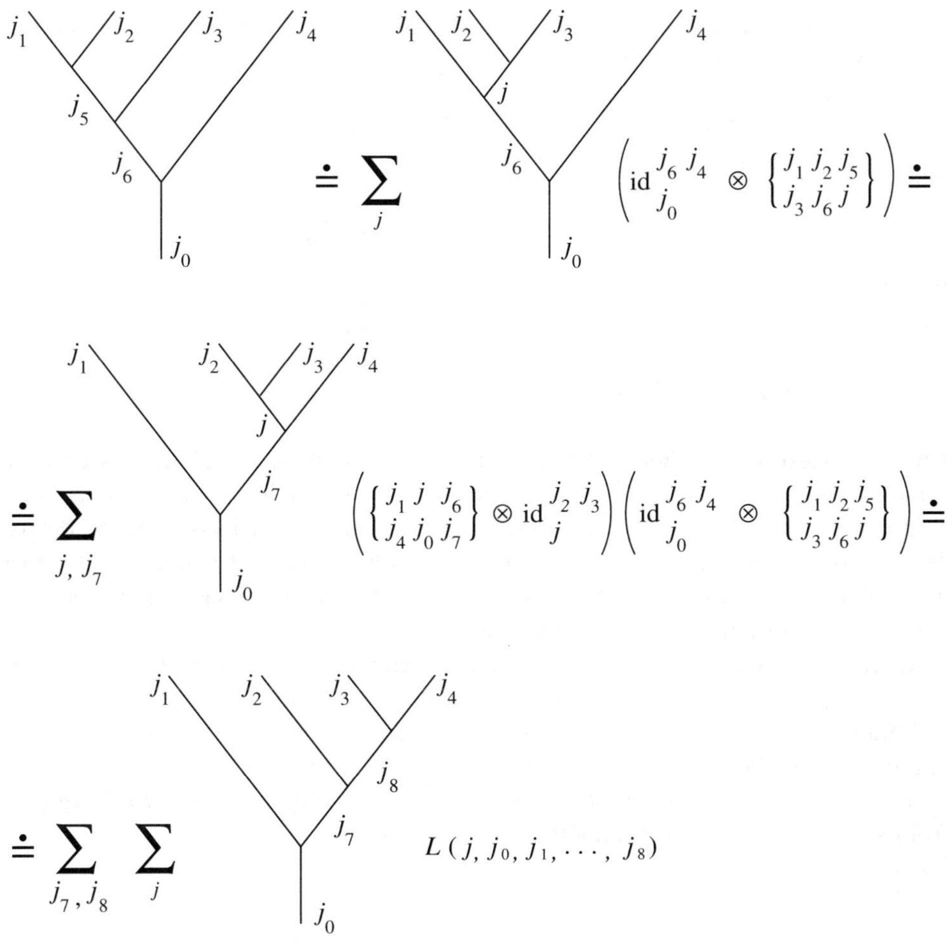

Figure 1.6

given by (1.5.a). Denote by f, g the homomorphisms $Y \to X$ provided by the left-hand and right-hand sides of the equality in Theorem 1.5.1. By this theorem, $f = g$. It follows from the remarks above that for any $y \in Y$,

$$f(y) = \sum_{j \in I} *^{j_7}_{j j_4} *^{j}_{j_2 j_3} *^{j_0}_{j_6 j_4} *^{j_6}_{j_1 j} *^{j_6}_{j_5 j_3} *^{j_5}_{j_1 j_2} (A \otimes y) =$$

$$= *^{j_0}_{j_6 j_4} *^{j_6}_{j_5 j_3} *^{j_5}_{j_1 j_2} \left(\sum_{j \in I} *^{j}_{j_2 j_3} *^{j_7}_{j j_4} *^{j_6}_{j_1 j}(A) \otimes y \right).$$

Note that contractions involved in this formula commute and may be applied in an arbitrary order. Similarly,

$$g(y) = *^{j_0}_{j_5 j_8} *^{j_5}_{j_1 j_2} *^{j_0}_{j_6 j_4} *^{j_6}_{j_5 j_3} (B \otimes y) = *^{j_0}_{j_6 j_4} *^{j_6}_{j_5 j_3} *^{j_5}_{j_1 j_2} \left(*^{j_0}_{j_5 j_8}(B) \otimes y \right).$$

Since $f(y) = g(y)$ for any

$$y \in Y = H^{j_6 j_4}_{j_0} \otimes_K H^{j_5 j_3}_{j_6} \otimes_K H^{j_1 j_2}_{j_5},$$

we have

$$\sum_{j \in I} *^{j}_{j_2 j_3} *^{j_7}_{j j_4} *^{j_6}_{j_1 j}(A) = *^{j_0}_{j_5 j_8}(B).$$

This is the claim of the Corollary.

1.6. Overview of further sections. Until now we have developed essentially a purely algebraic approach to $6j$-symbols, the figures have served as a mere graphical notation for algebraic formulas. This technique is inadequate to derive deep properties of $6j$-symbols, in particular their tetrahedral symmetry. One of the problems encountered here is the rigid structure of coupons which does not allow us to pull bands from one base to another.

In the next sections we develop another (though related) technique which gives more insight in the properties of $6j$-symbols. The essential ingredients of this technique are unimodal categories, symmetrized multiplicity modules, and framed graphs. These three notions will be discussed in the three following sections. Armed with the results of these sections we shall be able to consider $6j$-symbols from a more flexible and productive viewpoint.

2. Unimodal categories

2.0. Outline. Unimodality is a minor technical condition on semisimple (and modular) ribbon categories. It is under this condition that we shall give a geometric formulation of $6j$-symbols.

2.1. Unimodal categories. An element $i \in I$ is said to be self-dual if $i^* = i$. For each self-dual $i \in I$, we have

$$\mathrm{Hom}(V \otimes V, \mathbb{1}) = \mathrm{Hom}(V, V^*) \approx \mathrm{Hom}(V, V) = K$$

where $V = V_i$ and the symbol $\approx$ indicates the existence of a K-isomorphism. The formula $x \mapsto x(\mathrm{id}_V \otimes \theta_V) c_{V,V}$ defines a K-endomorphism of $\mathrm{Hom}(V \otimes V, \mathbb{1})$ which must be multiplication by a certain $\varepsilon_i \in K$. In view of (I.1.2.d), (I.1.2.h) the square of this endomorphism is the mapping $x \mapsto x\theta_{V \otimes V}$. Since $x\theta_{V \otimes V} = \theta_{\mathbb{1}} x = x$ we have $(\varepsilon_i)^2 = 1$. In particular, if K is a field then $\varepsilon_i = \pm 1$.

The semisimple category $(\mathcal{V}, \{V_i\}_{i\in I})$ is called unimodal if for every self-dual $i \in I$, we have $\varepsilon_i = 1$. In other words, the semisimple category $(\mathcal{V}, \{V_i\}_{i\in I})$ is unimodal if for any $i \in I$ with $i^* = i$ and any $x \in \mathrm{Hom}(V \otimes V, \mathbb{1})$, we have $x(\mathrm{id}_V \otimes \theta_V)c_{V,V} = x$. A unimodal modular category will be called unimodular category.

For example, the category $\mathcal{V}(G, K, c, \varphi)$ introduced in Sections I.1.7.2 and II.1.7.2 is unimodal if and only if $\varphi(g) = 1$ for any element $g \in G$ of order 2. In particular, if $\varphi = 1$ then this category is unimodal.

2.2. Lemma. *If $\mathcal{V}$ is unimodal then there exists a family of isomorphisms $\{w_i : V_i \to (V_{i^*})^*\}_{i\in I}$ such that for every $i \in I$, we have*

$$d_{V_i}(w_{i^*} \otimes \mathrm{id}_{V_i}) = F(\cap^-_{V_{i^*}})(\mathrm{id}_{V_{i^*}} \otimes w_i). \tag{2.2.a}$$

The isomorphism $w_0 : V_0 \to (V_0)^*$ may be chosen so that both sides of (2.2.a) *with $i = 0$ are equal to $id_{\mathbb{1}} : \mathbb{1} \to \mathbb{1}$.*

The graphical form of (2.2.a) is given in Figure 2.1.

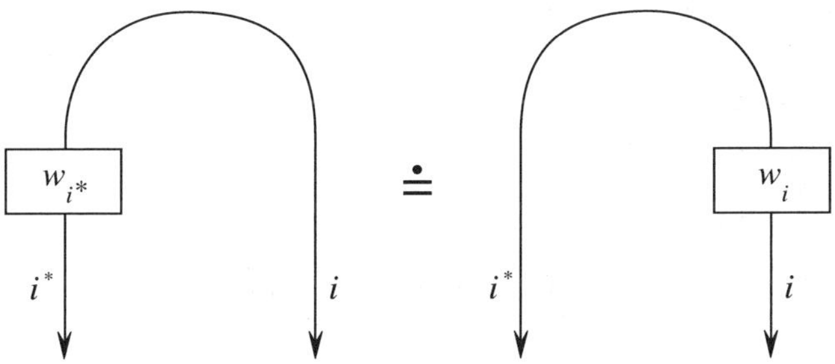

Figure 2.1

Proof of Lemma. Consider first the case $i^* \neq i$. Take as w_i an arbitrary isomorphism $V_i \to (V_{i^*})^*$ and choose w_{i^*} so that it satisfies (2.2.a). There is only one such w_{i^*} defined by the formula

$$w_{i^*} = (F(\cap^-_{V_{i^*}}) \otimes \mathrm{id}_{V_i^*})(\mathrm{id}_{V_{i^*}} \otimes w_i \otimes \mathrm{id}_{V_i^*})(\mathrm{id}_{V_{i^*}} \otimes b_{V_i}).$$

(The reader may draw the corresponding picture.)

Let $i^* = i$. We shall show that an arbitrary isomorphism $w = w_i = w_{i^*} : V_i \to (V_i)^*$ satisfies (2.2.a). Set $V = V_i$. Let $x \in \mathrm{Hom}(V \otimes V, \mathbb{1})$ be the morphism presented by the left diagram in Figure 2.1. It is easy to express $w = w_i$ via x. In the algebraic notation $w = (x \otimes \mathrm{id}_{V^*})(\mathrm{id}_V \otimes b_V)$, while drawing the corresponding picture is left to the reader. Substituting this expression in place of w_i in the right diagram in Figure 2.1 we get the left diagram in Figure 2.2. Using the isotopy in Figure 2.2 we observe that these diagrams represent $x(\mathrm{id}_V \otimes \theta_V)c_{V,V} = x$. Hence the equality in Figure 2.1 is true.

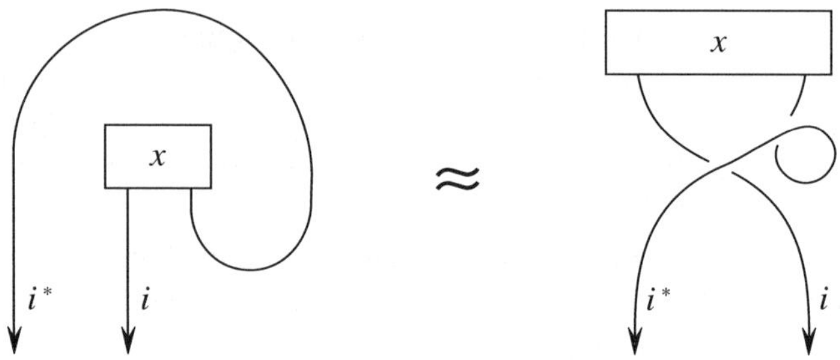

Figure 2.2

Let us show that $w_0 = b_{\mathbb{1}} : \mathbb{1} \to \mathbb{1}^*$ satisfies the last assertion of the lemma. By Corollary I.2.6.2, w_0 is an isomorphism. The results of the previous paragraph show that equality (2.2.a) is true. The morphism presented by the right diagram in Figure 2.1 with $i = 0$ equals

$$F(\cap_{\mathbb{1}}^-)(\mathrm{id}_{\mathbb{1}} \otimes b_{\mathbb{1}}) = F(\cap_{\mathbb{1}}^-)b_{\mathbb{1}} = F(\cap_{\mathbb{1}}^-)F(\cup_{\mathbb{1}}) = F(\cap_{\mathbb{1}}^-\cup_{\mathbb{1}}) = \mathrm{tr}(\mathrm{id}_{\mathbb{1}}) = \mathrm{id}_{\mathbb{1}}.$$

2.3. Exercises. 1. Show that if $\mathcal{V}$ is unimodal then the mirror category $(\overline{\mathcal{V}}, \{V_i\}_{i\in I})$ is also unimodal.

2. Show that any family of isomorphisms $\{w_i : V_i \to (V_{i^*})^*\}_{i\in I}$ satisfying (2.2.a) also satisfies the equalities

$$(w_i \otimes \mathrm{id}_{V_i^*})b_{V_i} = (\mathrm{id}_{V_{i^*}^*} \otimes w_{i^*})F(\cup_{V_{i^*}}^-), \quad (\mathrm{id}_{V_i^*} \otimes w_i^{-1})b_{V_{i^*}} = (w_{i^*}^{-1} \otimes \mathrm{id}_{V_i})F(\cup_{V_i}^-).$$

3. Show that the existence of a family of isomorphisms satisfying Lemma 2.2 is equivalent to unimodality of $\mathcal{V}$.

4. For each self-dual $i \in I$, the formula $x \mapsto (\theta_V \otimes \mathrm{id}_V)c_{V,V}x$ (where $V = V_i$) defines an endomorphism of $\mathrm{Hom}(\mathbb{1}, V_i \otimes V_i) \approx K$ which is multiplication by a certain $\varepsilon^i \in K$. Show that $(\varepsilon^i)^2 = 1$. Show that $\varepsilon^i = \varepsilon_i$.

5. The element 0 of I is self-dual because the object $V_0^* = \mathbb{1}^*$ is isomorphic to $V_0 = \mathbb{1}$. Show that $\varepsilon_0 = 1$.

3. Symmetrized multiplicity modules

3.0. Outline. By definition, each multiplicity module H_k^{ij} or H_{ij}^k is determined by an ordered triple (i, j, k) of elements of the set I. We define here a symmetric version of multiplicity modules. The symmetrized multiplicity modules are canonically isomorphic to the original non-symmetric ones but depend on *non-ordered* triples of indices. This will allow us to rewrite $6j$-symbols in a more convenient form, see Section 5. Note that whereas the definitions of Section 1 do not use the ribbon structure in the category $\mathcal{V}$, the definition of symmetrized multiplicity modules uses both the braiding and the twist in $\mathcal{V}$.

3.1. Conventions. From now on we assume (unless explicitly stated otherwise) that the fixed semisimple category $(\mathcal{V}, \{V_i\}_{i\in I})$ is unimodal. In order to develop a geometric approach to $6j$-symbols and in particular in order to define a symmetric version of multiplicity modules we need to make certain choices. Namely, we fix a family of isomorphisms $\{w_i : V_i \to (V_{i^*})^*\}_{i\in I}$ and two families $\{\dim'(i)\}_{i\in I}$, $\{v'_i\}_{i\in I}$ of elements of K. The isomorphisms $\{w_i\}_{i\in I}$ are assumed to satisfy the conditions of Lemma 2.2. The set $\{\dim'(i)\}_{i\in I}$ is just a family of square roots of the dimensions $\{\dim(i)\}_{i\in I}$ such that $\dim'(0) = 1$ and for every $i \in I$,

$$(\dim'(i))^2 = \dim(i), \quad \dim'(i^*) = \dim'(i).$$

Recall that the twist θ acts in each simple object V_i via multiplication by an invertible element $v_i \in K$. Equality (I.1.8.a) implies that $v_{i^*} = v_i$ for all $i \in I$. The equality $\theta_{\mathbb{1}} = \mathrm{id}_{\mathbb{1}}$ implies that $v_0 = 1$. We assume that $v'_i \in K$ is a "half-twist" such that $v'_0 = 1$ and for every $i \in I$,

$$(v'_i)^2 = v_i, \quad v'_{i^*} = v'_i.$$

The existence of the family $\{w_i : V_i \to (V_{i^*})^*\}_{i\in I}$ is ensured by Lemma 2.2. The existence of the families $\{\dim'(i)\}_{i\in I}$, $\{v'_i\}_{i\in I}$ is a quite mild restriction on the category $\mathcal{V}$. Although in general we can not extract square roots in K, we may overcome this problem by formally adding to K elements $\{\dim'(i)\}_{i\in I}$, $\{v'_i\}_{i\in I}$ with the desired properties and extending $\mathcal{V}$ to include these elements (cf. Section II.1.6).

Neither the family $\{w_i\}_{i\in I}$, nor the families $\{\dim'(i)\}_{i\in I}$, $\{v'_i\}_{i\in I}$ are unique. It is important that we fix this data once for all. Further constructions will depend on this data, although a different choice would lead to equivalent results.

3.2. Symmetrization of multiplicity modules. Let $i, j, k \in I$. Set

$$H^{ijk} = \mathrm{Hom}(\mathbb{1}, V_i \otimes V_j \otimes V_k).$$

The modules $\{H^{ijk}\}_{i,j,k\in I}$ present an equivalent version of the multiplicity modules considered in Section 1. Using the braiding in $\mathcal{V}$ it is easy to show that the six modules corresponding to the permutations of the triple $\{i, j, k\}$ (i.e., the modules $H^{ijk}, H^{jki}, H^{kij}, H^{jik}, H^{ikj}, H^{kji}$) are mutually isomorphic. Using the unimodality of $\mathcal{V}$ we construct a commuting system of isomorphisms between these modules. In other words, we construct an action of the symmetric group S_3 in this system of modules. Note that the braiding directly induces an action of the group of braids on 3 strings in this system of modules. In order to factorize down to S_3 we involve the twists. Identifying the modules $H^{ijk}, H^{jki}, H^{kij}, H^{jik}, H^{ikj}, H^{kji}$ along the resulting action of S_3 we shall obtain the symmetrized multiplicity module associated to the non-ordered triple $\{i, j, k\}$.

It suffices to construct isomorphisms $\sigma_1(ijk) : H^{ijk} \to H^{jik}$ and $\sigma_2(ijk) : H^{ijk} \to H^{ikj}$ such that

$$\sigma_1(jik)\,\sigma_1(ijk) = \mathrm{id}, \tag{3.2.a}$$

$$\sigma_2(ikj)\,\sigma_2(ijk) = \mathrm{id}, \tag{3.2.b}$$

$$\sigma_1(jki)\,\sigma_2(jik)\,\sigma_1(ijk) = \sigma_2(kij)\,\sigma_1(ikj)\,\sigma_2(ijk). \tag{3.2.c}$$

For $x \in H^{ijk} = \mathrm{Hom}(\mathbb{1}, V_i \otimes V_j \otimes V_k)$, set

$$\sigma_1(ijk)(x) = v_i' v_j' (v_k')^{-1} (c_{V_i,V_j} \otimes \mathrm{id}_{V_k}) x$$

where $c_{V_i,V_j} : V_i \otimes V_j \to V_j \otimes V_i$ is the braiding isomorphism. In the graphical notation the element x and its image under $\sigma_1(ijk)(x)$ are represented by (the v-colored ribbon graphs presented by) the diagrams in Figure 3.1.

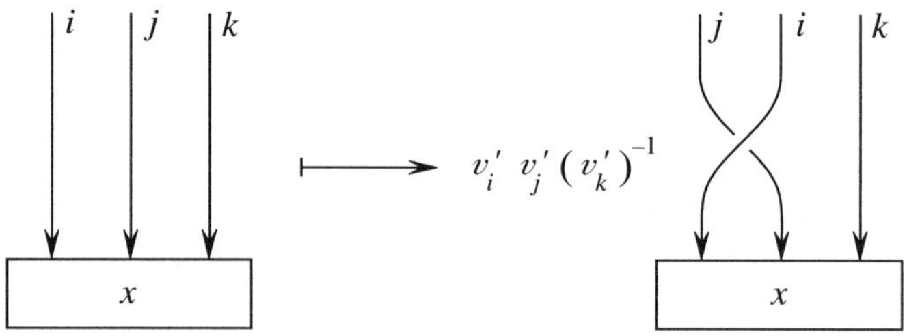

Figure 3.1

Similarly, set

$$\sigma_2(ijk)(x) = v_j' v_k' (v_i')^{-1} (\mathrm{id}_{V_i} \otimes c_{V_j,V_k}) x.$$

It is a nice geometric exercise to see that the ribbon graphs presented by the diagrams in Figure 3.2 are isotopic (with the top ends fixed !). Therefore formula (3.2.a) follows from definitions and the fact that insertion of a positive (resp. negative) twist in a band of color V_l leads to multiplication of the operator invariant by $v_l = (v_l')^2$ (resp. $v_l^{-1} = (v_l')^{-2}$). The proof of (3.2.b) is similar and (3.2.c) follows from Yang-Baxter identity (I.1.2.f).

The K-module obtained by identification of the modules $H^{ijk}, H^{jki}, H^{kij}, H^{jik}, H^{ikj}, H^{kji}$ along the isomorphisms constructed above is denoted by $H(i, j, k)$. An element x of $H(i, j, k)$ is a function assigning to every ordering i_1, i_2, i_3 of the set $\{i, j, k\}$ an element $x^{i_1 i_2 i_3}$ of $H^{i_1 i_2 i_3}$ so that the elements corresponding to different orderings are related by these isomorphisms. It is obvious that the formula $x \mapsto x^{i_1 i_2 i_3}$ defines a canonical isomorphism $H(i, j, k) \to H^{i_1 i_2 i_3}$. The module $H(i, j, k)$ is called the symmetrized multiplicity module associated to the

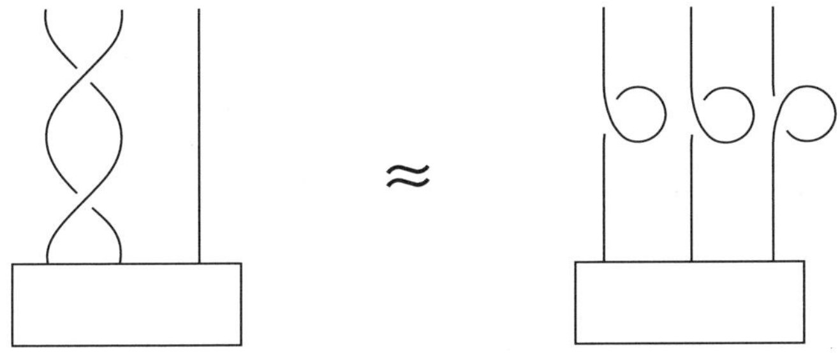

Figure 3.2

triple $\{i, j, k\}$. It is important to emphasize that this module is determined by the non-ordered triple $\{i, j, k\}$: permutations of this triple do not change $H(i, j, k)$. For instance, $H(i, j, k)$ and $H(j, i, k)$ are actually the same module (and not merely isomorphic modules).

The symmetrized multiplicity modules are closely related to the multiplicity modules introduced in Section 1. For any triple $i, j, k \in I$, we define an isomorphism $H_k^{ij} \to H^{ijk^*}$ by the formula

$$x \mapsto (x \otimes w_{k^*}^{-1}) b_{V_k}$$

(see Figure 3.3). Composing this isomorphism with the canonical isomorphism $H^{ijk^*} \to H(i, j, k^*)$ (constructed in the previous paragraph) we get an isomorphism $H_k^{ij} \to H(i, j, k^*)$. Similarly, we define an isomorphism $H_{ij}^k \to H^{kj^*i^*}$ by the formula

$$y \mapsto (y \otimes w_{j^*}^{-1} \otimes w_{i^*}^{-1})(\mathrm{id}_{V_i} \otimes b_{V_j} \otimes \mathrm{id}_{V_i^*}) b_{V_i}$$

(see Figure 3.4). Composing this isomorphism $H_{ij}^k \to H^{kj^*i^*}$ with the canonical isomorphism $H^{kj^*i^*} \to H(k, j^*, i^*)$ we get an isomorphism $H_{ij}^k \to H(k, j^*, i^*)$.

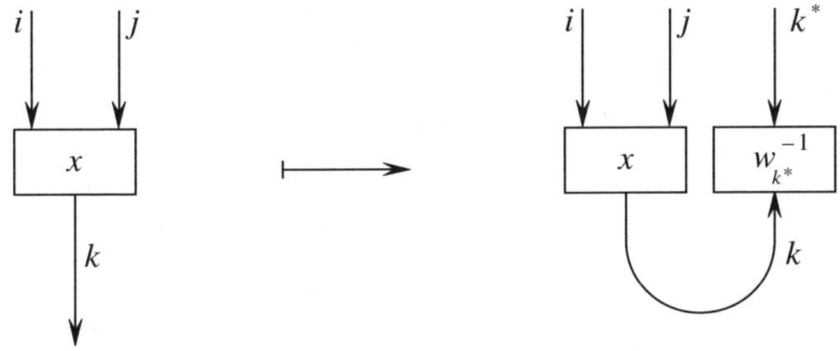

Figure 3.3

3.3. Duality for symmetrized multiplicity modules. Let A and B be two non-ordered triples of elements of I. We say that these triples are dual if there is a fixed bijection between the elements of A and the elements of B so that the

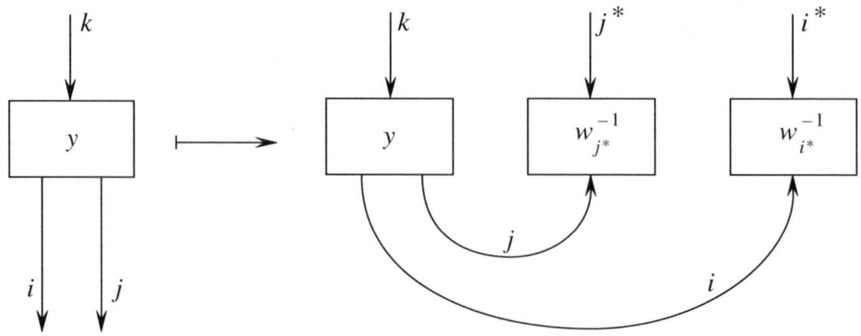

Figure 3.4

corresponding terms are related by the involution $i \mapsto i^* : I \to I$. For any dual triples A and B, we construct a non-degenerate pairing $H(A) \otimes_K H(B) \to K$ as follows. Let i, j, k be elements of A written down in a certain order and let i^*, j^*, k^* be the corresponding elements of B. There is a bilinear pairing $a^{ijk} : H^{ijk} \times H^{k^*j^*i^*} \to K$ assigning to any pair $(x \in H^{ijk}, y \in H^{k^*j^*i^*})$ the operator invariant of the v-colored ribbon (0,0)-graph shown in Figure 3.5.

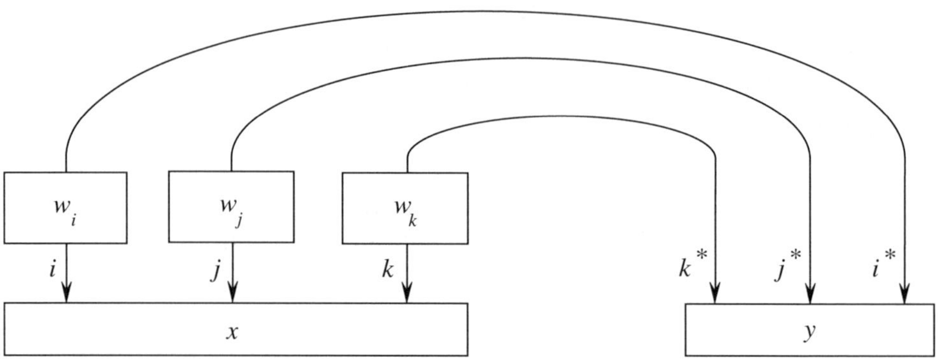

Figure 3.5

It is easy to deduce from formula (2.2.a) (cf. Figure 2.1) that

$$a^{ijk}(x, y) = a^{k^*j^*i^*}(y, x).$$

The pairing a^{ijk} is compatible with the identification isomorphisms between the modules $H^{ijk}, H^{jki}, H^{kij}, \; H^{jik}, H^{ikj}, H^{kji}$ induced by permutations of indices. Specifically, for any $x \in H^{ijk}, y \in H^{k^*j^*i^*}$ we have

$$a^{ijk}(x, y) = a^{jik}(\sigma_1(ijk)(x), \sigma_2(k^*j^*i^*)(y)) = a^{ikj}(\sigma_2(ijk)(x), \sigma_1(k^*j^*i^*)(y)).$$

(The reader is advised to draw the corresponding pictures.) These equalities follow from definitions and the formulas $\sigma_1(ijk) = (\sigma_1(jik))^{-1}$, $\sigma_2(ijk) = (\sigma_2(ikj))^{-1}$

proven above. Thus we have a correctly defined bilinear pairing

$$H(A) \otimes_K H(B) = H(i, j, k) \otimes_K H(i^*, j^*, k^*) \to K. \tag{3.3.a}$$

It is easy to see that under the isomorphisms $H^{ij}_{k^*} \to H(i, j, k)$, $H^{k^*}_{ij} \to H(k^*, j^*, i^*) = H(i^*, j^*, k^*)$ the pairing considered in Section 1.2 corresponds to the pairing (3.3.a). Therefore the pairing (3.3.a) is non-degenerate.

For a pair of dual unordered triples $A, B \subset I$, we shall identify the dual of $H(A)$ with $H(B)$ using the pairing (3.3.a).

3.4. Multiplicity modules for the mirror category. We extend Conventions 3.1 to the mirror category $(\overline{\mathcal{V}}, \{V_i\}_{i\in I})$ keeping the families $\{w_i : V_i \to (V_{i^*})^*\}_{i\in I}$, $\{\dim'(i)\}_{i\in I}$ and choosing as half-twists the inverses of the fixed half-twists $\{v'_i\}_i$. The multiplicity modules H^{ijk} corresponding to $\mathcal{V}$ and $\overline{\mathcal{V}}$ are the same because the underlying monoidal categories of $\mathcal{V}$ and $\overline{\mathcal{V}}$ are the same. It is easy to deduce from definitions and formulas (3.2.a), (3.2.b) that the identification isomorphisms $\sigma_1(ijk), \sigma_2(ijk)$ corresponding to $\mathcal{V}$ and $\overline{\mathcal{V}}$ coincide. Therefore the symmetrized multiplicity modules associated to $\mathcal{V}$ are identical with the symmetrized multiplicity modules associated to $\overline{\mathcal{V}}$.

3.5. The color 0. When one of the indices i, j, or k is equal to $0 \in I$ the module $H(i, j, k)$ may be non-trivial only if the other two indices are dual to each other. If this is the case then the module $H(i, j, k)$ is isomorphic to K. We specify here a canonical isomorphism $H(i, i^*, 0) \to K$.

Consider the homomorphism $z : K \to \mathrm{Hom}(\mathbb{1}, V_i \otimes V_{i^*})$ such that

$$z(1) = (\dim'(i))^{-1}(\mathrm{id}_{V_i} \otimes w_{i^*}^{-1})b_{V_i}.$$

(Invertibility of $\dim'(i) \in K$ is ensured by Lemma II.4.2.4.) The homomorphism z is an isomorphism because under the identification $\mathrm{Hom}(\mathbb{1}, V_i \otimes V_{i^*}) = \mathrm{Hom}(V_i^*, V_{i^*})$ introduced in Exercise I.1.8.1 the morphism $(\mathrm{id}_{V_i} \otimes w_{i^*}^{-1})b_{V_i}$ corresponds to the isomorphism $w_{i^*}^{-1}$ which freely generates $\mathrm{Hom}(V_i^*, V_{i^*})$. We identify $H^{ii^*0} = \mathrm{Hom}(\mathbb{1}, V_i \otimes V_{i^*})$ with K via z. It is easy to check that the identification isomorphisms $H^{ii^*0} = K$, $H^{i^*i0} = K$ and the isomorphism $\sigma_1(ii^*0) : H^{ii^*0} \to H^{i^*i0}$ form a commutative diagram. (One should use the result of Exercise 2.3.2.) In this way we obtain an identification $H(i, i^*, 0) = K$.

In the case $i = 0$ we obtain an identification $H^{000} = K$ which is nothing but the obvious equality

$$\mathrm{Hom}(\mathbb{1}, \mathbb{1} \otimes \mathbb{1} \otimes \mathbb{1}) = \mathrm{Hom}(\mathbb{1}, \mathbb{1}) = K.$$

The identification $H(0, 0, 0) = K$ is also induced by this equality.

The duality pairing introduced in Section 3.3 yields a duality between two copies of $H(i, i^*, 0)$. It is easy to verify that under the identification $H(i, i^*, 0) = K$ this pairing is presented by the unit 1×1-matrix over K. It is here that the factor $(\dim'(i))^{-1}$ in the definition of z plays its role.

4. Framed graphs

Everywhere in this section we adhere to Conventions 3.1.

4.0. Outline. We introduce a geometric technique of framed graphs. Framed graphs are versions of ribbon graphs with a relaxed structure in the vertices: instead of coupons we involve 2-disks, i.e., we omit all references to bases of coupons. The distinction between framed and ribbon graphs mirrors on the geometric level the distinction between symmetrized and non-symmetrized multiplicity modules.

We show how to pass from ribbon graphs to framed graphs and back. To this end we introduce two transformations: (i) ribboning, which transforms any framed graph into a ribbon graph, and (ii) rounding of corners, which acts in the opposite direction. We caution that these geometric transformations are not mutually inverse. We define an invariant of colored framed graphs corresponding to the operator invariant F of colored ribbon graphs under both (i) and (ii). Thus, from the viewpoint of operator invariants, the transformations (i) and (ii) may be regarded as mutually inverse. This shows that the languages of ribbon and framed graphs are essentially equivalent. The choice of language is mainly a matter of convenience. As we shall see in Section 5, framed graphs perfectly suit the theory of $6j$-symbols. The technique of framed graphs will be used also in Chapter X.

4.1. Trivalent graphs and their colorings. By a graph we mean a finite 1-dimensional cell space. Each graph consists of a finite number of vertices, edges, and circle components. Every edge connects two (possibly coinciding) vertices called the end points of this edge. Note that we deal with open edges, i.e., we do not include end points in the edges. Every vertex x of a graph has a neighborhood consisting of several half-open intervals with common end point x. The number of these intervals is called the valency of x. We shall never consider graphs with 2-valent vertices. By circle components we mean connected components homeomorphic to S^1. These components do not contain vertices. Edges and circle components of a graph are called its 1-strata.

Let γ be a graph. By an oriented 1-stratum of γ we mean a 1-stratum of γ (i.e., an edge or a circle component) equipped with an orientation. By a coloring of γ we mean a function assigning to every oriented 1-stratum of γ an element of the set I so that the elements assigned to opposite orientations of the same 1-stratum are related by the involution $i \mapsto i^* : I \to I$. It is obvious that every graph admits $(\mathrm{card}(I))^n$ colorings where n is the number of its 1-strata. A graph endowed with a coloring is said to be colored.

By a trivalent graph we mean a graph whose vertices have valency 3. For instance, a circle is a trivalent graph because it has no vertices. We shall treat the empty space $\emptyset$ as a trivalent graph.

For each colored trivalent graph γ, we define a K-module $H(\gamma)$. For a vertex x of γ, consider the module $H_x(\gamma) = H(i, j, k)$ where i, j, k are the colors of three 1-strata of γ incident to x and oriented towards x. (Since the triple i, j, k has no natural order, it is important here that symmetrized multiplicity modules are determined by non-ordered triples.) Set $H(\gamma) = \otimes_x H_x(\gamma)$ where x runs over all vertices of γ. It follows from Lemma II.4.2.1 that $H(\gamma)$ is a projective K-module. Set

$$H^\star(\gamma) = (H(\gamma))^\star = \mathrm{Hom}_K(H(\gamma), K).$$

For instance, if γ is a circle or a disjoint union of circles then $H^\star(\gamma) = H(\gamma) = K$. Set $H(\emptyset) = K$.

By a v-colored trivalent graph we mean a colored trivalent graph γ with a preferred element chosen in each module $H_x(\gamma)$ where x runs over all vertices of γ.

Warning: by their very definition the modules $H_x(\gamma)$ and $H(\gamma)$ depend on the data fixed in Section 3.1.

4.2. Framed graphs. The notion of a framed graph is a modification of the notion of a ribbon graph. The main difference is that the role of vertices is no longer played by boxes with distinguished bases but rather by 2-disks without any additional geometric structure. This renders framed graphs much more flexible in comparison to ribbon graphs. We restrict ourselves here to framed graphs with 3-valent vertices and without free ends.

Before giving a definition of framed graphs we need to adjust our terminology. In order to avoid confusion between bands (topological rectangles with a distinguished pair of opposite bases) and Möbius bands we shall use for Möbius bands the term twisted annuli. The untwisted annulus is of course the annulus $S^1 \times [0, 1]$.

A framed graph is a compact surface decomposed into a union of a finite number of annuli (twisted or untwisted), bands, and 2-disks such that the following conditions hold: the bases of bands lie on the boundary of the 2-disks, otherwise the bands, 2-disks, and annuli are disjoint, and each of the 2-disks is incident to exactly 3 bands (counted with multiplicities). The surface in question is said to be the surface of the framed graph. In contrast to ribbon graphs, the surfaces of framed graphs are not assumed to be oriented or even orientable, the bands and annuli of framed graphs are not assumed to be directed.

Each framed graph Γ has a core γ which is a trivalent graph in the sense of Section 4.1. The vertices of γ are the centers of the 2-disks of Γ, the 1-strata of γ are the cores of the annuli and the cores of the bands extended slightly to reach vertices of γ. It is clear that γ lies in the surface of Γ as a deformation retract. Vertices, 1-strata, and colorings of γ will be called vertices, 1-strata, and colorings of Γ respectively. Thus, a coloring of Γ assigns elements of I to oriented cores of bands and annuli of Γ so that the elements assigned to opposite orientations of the same core are related by the involution $i \mapsto i^*$. The set of all colorings of Γ will be denoted by $\mathrm{col}(\Gamma)$. A framed graph endowed with a coloring is said to be colored. Note an important distinction between colorings of ribbon and framed graphs: to color bands and annuli of ribbon graphs we use objects of $\mathcal{V}$ whereas to color bands and annuli of framed graphs we use elements of the set I.

For each 2-disk D of a colored framed graph Γ, we define the module $H_D(\Gamma)$ to be $H_x(\gamma)$ where x is the corresponding vertex of γ. In other words, $H_D = H(i, j, k)$ where i, j, k are the colors of the (cores of the) bands of Γ attached to D and directed towards D. Set $H(\Gamma) = H(\gamma) = \otimes_D H_D(\Gamma)$ so that $H(\Gamma)$ is the tensor product over K of the modules assigned to the 2-disks of Γ. Set $H^\star(\Gamma) = (H(\Gamma))^\star$. When it is necessary to stress the role of the category $\mathcal{V}$ we denote the modules $H(\Gamma), H^\star(\Gamma)$ by $H_{\mathcal{V}}(\Gamma), H^\star_{\mathcal{V}}(\Gamma)$ respectively.

By a v-colored framed graph we mean a colored framed graph endowed with a distinguished element in each module H_D where D runs over all 2-disks of the graph. This element is called the color of D.

By a framed graph in a 3-manifold M we mean a framed graph embedded in $\mathrm{Int}(M)$. Two framed graphs in M are isotopic if they can be smoothly deformed into each other in the class of framed graphs in M preserving the splitting into the annuli, bands, and 2-disks during the deformation. By an isotopy of colored (resp. v-colored) graphs we mean a color-preserving isotopy.

We shall present framed graphs in $\mathbb{R}^3$ by plane diagrams in the way similar to the one used for ribbon graphs. Namely, we shall draw 2-disks and cores of the bands and annuli. Unless specified otherwise, the bands and annuli themselves are assumed to go close and parallel to the plane of the picture. In our pictures of colored framed graphs we shall indicate the colors corresponding to certain directions of the cores; it is understood that the colors corresponding to opposite directions are obtained via the involution $i \mapsto i^*$. Note finally that the notion of a framed graph generalizes the notion of a framed link used in Chapters I and II.

4.3. Invariants of framed graphs and ribboning. For each v-colored framed graph $\Gamma \subset \mathbb{R}^3$, we define an isotopy invariant $\mathbb{F}(\Gamma) = \mathbb{F}_{\mathcal{V}}(\Gamma) \in K$. This invariant should be viewed as a version of the invariant F of v-colored ribbon graphs studied in Chapter I.

We first assume that the surface of Γ is oriented. We shall use the given orientation, say α, of this surface to transform Γ into a v-colored ribbon (0,0)-graph Ω_Γ. (Recall that ribbon (0,0)-graphs are just ribbon graphs without free ends.) The transformation $\Gamma \mapsto \Omega_\Gamma$ is called ribboning. The v-colored ribbon (0,0)-graph Ω_Γ is not uniquely determined by (Γ, α) but its invariant $F(\Omega_\Gamma) \in K$ depends solely on Γ. This will allow us to define $\mathbb{F}(\Gamma) = F(\Omega_\Gamma)$. A posteriori we shall see that $\mathbb{F}(\Gamma)$ does not depend on the choice of α.

The ribboning modifies annuli, bands, and 2-disks of Γ. Roughly speaking, annuli and 2-disks of Γ give rise to annuli and coupons of Ω_Γ, while each band of Γ gives rise to two shorter bands and one coupon between them as in Figure

4.1. Here are the details. We direct the annuli of Γ (i.e., orient their cores) in an arbitrary way and attach to each annulus the object V_i where $i \in I$ is the color of this directed annulus provided by the given coloring of Γ. The transformation of bands is slightly more complicated because we want all bands of Ω_Γ to be directed towards the coupons obtained from the 2-disks of Γ. With this in mind, we add in the middle of each band b of Γ a coupon B, see Figure 4.1. (In contrast to the previous figures, in Figures 4.1 and 4.3 we draw bands and not just their cores; the band b is shaded in Figure 4.1.) Note that the given orientation α of the surface of Γ enables us to choose a preferred side of this surface. In Figure 4.1 and in the next figures the preferred side is the one turned towards the reader. In other words, the band b regarded as a surface is oriented counterclockwise. The arrows in the left part of Figure 4.1 indicate opposite orientations of the core of b and the labels i, i^* specify the corresponding colors. The coupon B cuts b into two shorter bands. We equip these bands with directions leading out of B and with colors V_i, V_{i^*}, as in Figure 4.1. Each base of B is incident to one of these two bands. An arbitrary base of B is chosen to be the bottom base, the opposite base being the top one. (In Figure 4.1 the bottom base of B is the lower horizontal base.) Equip B with the color $w_i : V_i \to V_{i^*}^*$ where V_i is the color of the band incident to the bottom base of B. This procedure replaces each band of Γ with one colored coupon and two colored directed bands.

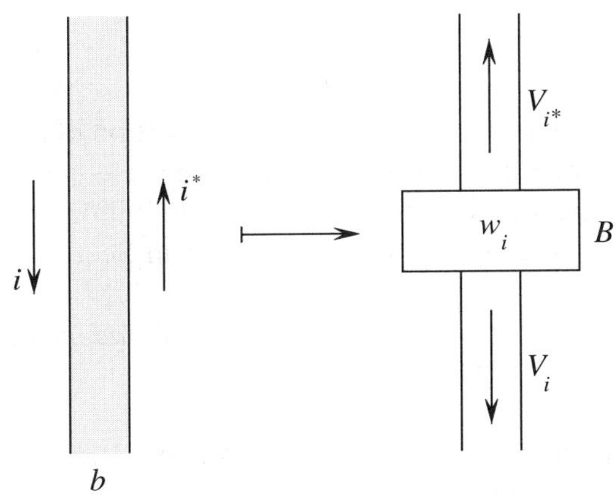

Figure 4.1

It remains to transform the 2-disks of Γ into coupons which is the heart of the construction. Let D be a 2-disk of Γ. Choose a small arc a in ∂D disjoint from the bases of bands attached to D. Split this arc into three consecutive subarcs which will play the role of sides and bottom base of the coupon. The complement of a in ∂D will play the role of the top base. Thus, bases of bands attached to D lie in the top base of the resulting coupon. The orientation α in the surface of Γ restricted to D induces an orientation in ∂D. Consider the three bands of Γ

attached to D. Let i, j, k be the colors of their cores oriented towards D and taken in the order opposite to the one induced by the orientation in ∂D restricted to $\partial D \setminus a$ (see Figure 4.2 where D is oriented counterclockwise). The given element of $H_D = H(i, j, k)$ (the color of D) canonically determines an element of H^{ijk} which we choose to be the color of this coupon.

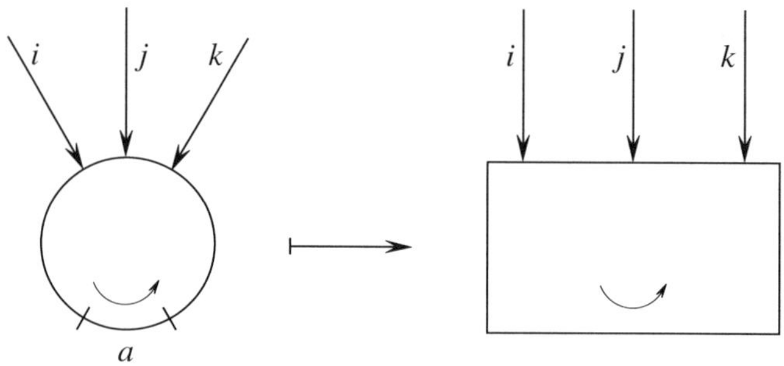

Figure 4.2

Applying this procedure to all bands, annuli, and 2-disks of Γ we get a v-colored ribbon (0,0)-graph $\Omega = \Omega_\Gamma$ in $\mathbb{R}^3$. The orientation in the surface of Ω is induced by α in the obvious way. We say that Ω is obtained by ribboning of Γ. Set

$$\mathbb{F}(\Gamma) = F(\Omega) \in K.$$

The construction of Ω_Γ and, in particular, the insertion of additional coupons in the middle of bands may look artificial. In a sense these coupons are intended to compensate for the coupons which appear in the definitions at the end of Section 3.2, cf. Figures 3.3, 3.4. This will become clearer in Section 4.5.

4.3.1. Lemma. *$\mathbb{F}(\Gamma)$ is an isotopy invariant of Γ independent of the choices made in the construction of Ω.*

Lemma 4.3.1 ensures that $\mathbb{F}(\Gamma)$ is correctly defined. Before proving this lemma we state a few properties of $\mathbb{F}(\Gamma)$.

4.3.2. Lemma. *If a v-colored framed graph Γ' is obtained from Γ by one positive twist of a band of Γ then*

$$\mathbb{F}(\Gamma') = v_i\, \mathbb{F}(\Gamma) = v_{i^*}\, \mathbb{F}(\Gamma) \tag{4.3.a}$$

where i, i^* are the colors corresponding to two directions of this band. If a v-colored framed graph Γ' is obtained from Γ by positive half-twists applied to the three bands incident to a 2-disk D of Γ as in Figure 4.3 then

$$\mathbb{F}(\Gamma') = v'_i v'_j v'_k\, \mathbb{F}(\Gamma) \tag{4.3.b}$$

where i, j, k are the colors of these bands directed towards D.

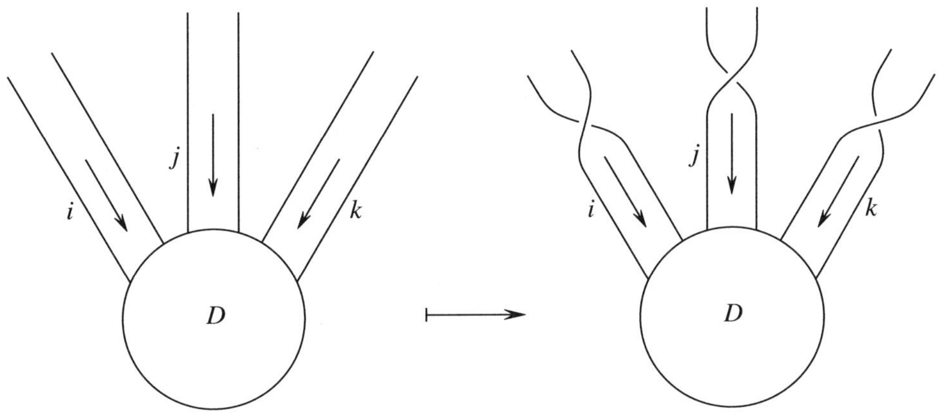

Figure 4.3

In the second claim the framed graphs Γ' and Γ differ in the neighborhood of D shown in Figure 4.3 and coincide otherwise. It is understood that the given orientations in the surfaces of Γ' and Γ coincide outside the neighborhood of D shown in Figure 4.3.

4.3.3. Lemma. $\mathbb{F}(\Gamma)$ *does not depend on the choice of orientation in the surface of* Γ.

This assertion should be contrasted with the fact that the orientation in the surface of Γ has been heavily used in the definition of ribboning of Γ. Different choices of orientation may lead to different v-colored ribbon graphs. By Lemma 4.3.3, all these ribbon graphs have the same operator invariant. This lemma also allows us to view $\mathbb{F}$ as an isotopy invariant of v-colored framed graphs with orientable (but not necessarily oriented) surface.

4.3.4. Proof of Lemma 4.3.1. In the construction of Ω we have made the following choices. We chose directions of annuli, bottom bases of the coupons inserted in the bands of Γ, and small arcs in the boundaries of 2-disks. We shall show that these choices do not affect $F(\Omega)$.

Independence of $F(\Omega)$ of the choice of directions of annuli follows from Lemma I.2.8.1. Independence of the choice of bottom bases of coupons inserted in bands follows from the equality in Figure 4.4 where the bottom bases of the coupons are their lower horizontal bases. This equality in its turn follows from the formula in Figure 2.1.

To show independence of $F(\Omega)$ of the choice of $a \subset \partial D$ it suffices to show that $F(\Omega)$ is not changed when we replace a with an arc a' as in Figure 4.5. When

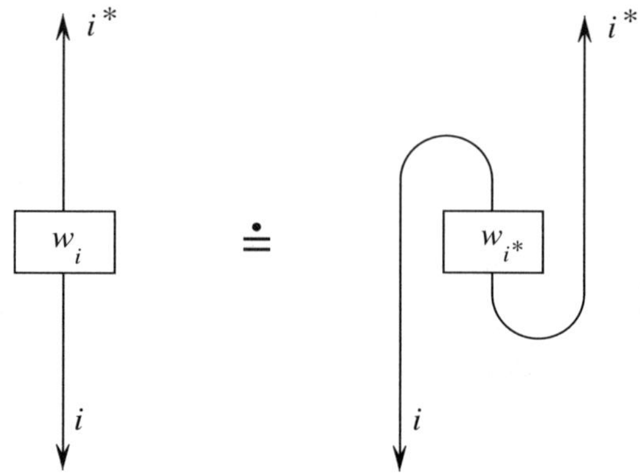

Figure 4.4

we use a (resp. a') to effect the ribboning of Γ we obtain the pictures in the left (resp. right) part of Figure 4.6. We have to prove the equality in this figure where $X \in H^{ijk}$ and $X' \in H^{kij}$ are elements representing the color of D. By definition of H_D, we have $X = \sigma_2(ikj)\,\sigma_1(kij)(X')$. Substituting here the expressions for σ_1, σ_2 (see Section 3.2) we get the desired equality.

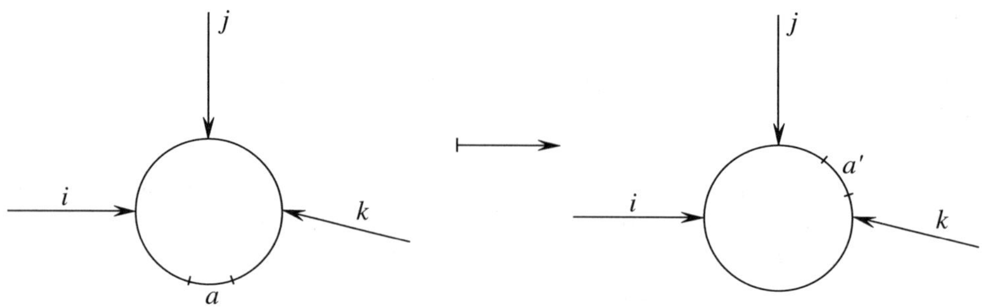

Figure 4.5

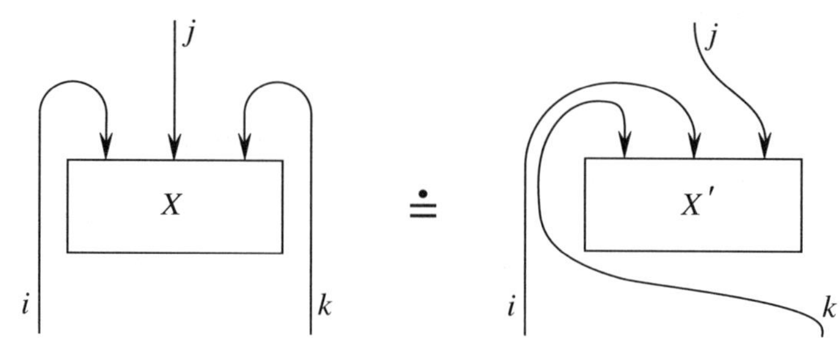

Figure 4.6

4.3.5. Proof of Lemma 4.3.2. The first claim follows directly from the definition of v_i. Let us prove the second claim. The ribboning transforms Γ and Γ' into v-colored ribbon graphs Ω and Ω' respectively. We shall prove that $F(\Omega') = v'_i v'_j v'_k F(\Omega)$. This would imply that $\mathbb{F}(\Gamma') = v'_i v'_j v'_k \mathbb{F}(\Gamma)$.

Let Q and Q' be the coupons of Ω and Ω' obtained from D. (It is understood that we use the same arc $a \subset \partial D$ to construct Ω and Ω'.) Because of the additional half-twists, the orientations in the surfaces of Γ and Γ' induce opposite orientations, say α and $\alpha' = -\alpha$, in D. For concreteness, we assume that α is the counterclockwise orientation and α' is the clockwise orientation in Figure 4.3. To transform D into Q, Q' we have to use opposite orientations both in D and in ∂D.

Let i, j, k be the colors of three bands of Γ attached to D, directed towards D and taken in the order determined by α, as in Figure 4.3. Let $y \in H^{ijk}$ be the color of Q determined by the given color $x_D \in H_D$ of D. Using $\alpha' = -\alpha$ we encounter the same three bands but in the opposite order. Therefore the color of Q' is a certain $y' \in H^{kji}$ determined by the same $x_D \in H_D$. By definition of $H_D = H(i, j, k)$ we have

$$y' = \sigma_1(jki)\,\sigma_2(jik)\,\sigma_1(ijk)(y). \tag{4.3.c}$$

In a neighborhood of Q, Q' the graphs Ω and Ω' look as in Figure 4.7 where each boldface point stands for one positive half-twist of the band around its core. Here we have drawn mirror images of the letters y' and Q' to stress that the coupon Q' is oriented clockwise. In other words, we look at this coupon from the wrong side, the natural view would be from below the page. Keeping three top ends fixed and rotating to the right we deform the right-hand graph into the graph drawn in the left part of Figure 4.8. In particular, this deformation undoes the half-twists of the bands and yields a coupon oriented counterclockwise, i.e., in the same way as Q. Substituting the graphical expressions for σ_1, σ_2 in (4.3.c), and substituting the resulting expression for y' in Figure 4.8 we get (up to isotopy) the ribbon graph in the right part of Figure 4.8. Thus, $F(\Omega') = v'_i v'_j v'_k F(\Omega)$.

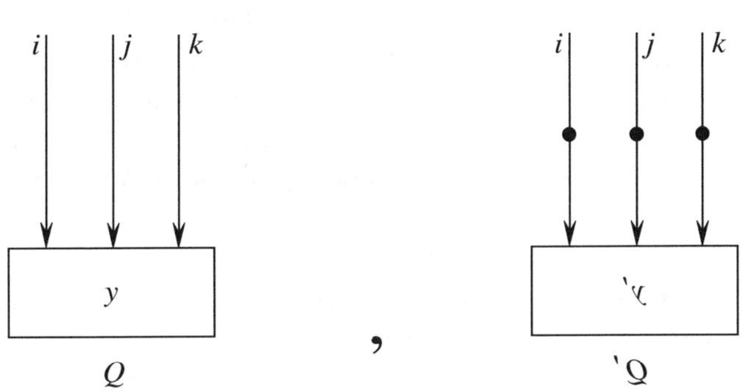

Figure 4.7

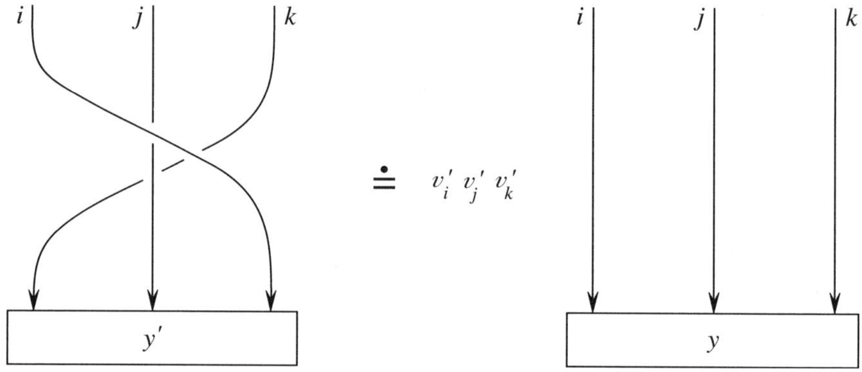

Figure 4.8

4.3.6. Proof of Lemma 4.3.3. Let Γ be a colored framed graph and let α, α' be different orientations of its surface. We prove that $\mathbb{F}(\Gamma, \alpha) = \mathbb{F}(\Gamma, \alpha')$. Consider first the case where Γ is connected and $\alpha' = -\alpha$. If Γ is an annulus then rotation of Γ in $\mathbb{R}^3$ around its core by an angle of π yields the same annulus with the opposite orientation. Since $\mathbb{F}$ is invariant under isotopies, $\mathbb{F}(\Gamma, \alpha) = \mathbb{F}(\Gamma, -\alpha)$. Assume that Γ is not an annulus. Apply to each 2-disk of Γ the transformation shown in Figure 4.3. This yields a new framed graph Γ_0 with the same 2-disks as in Γ. We provide the surface of Γ_0 with an orientation α_0 which coincides on these 2-disks with $-\alpha$. By Lemma 4.3.2 we have $\mathbb{F}(\Gamma_0, \alpha_0) = v\,\mathbb{F}(\Gamma, \alpha)$ with

$$v = \Big(\prod_i v_i'\Big)^2 = \prod_i v_i$$

where i runs over the colors of the bands of Γ. Next, apply to each band of Γ_0 one negative twist. This yields a new framed graph Γ_1 with the same 2-disks as in Γ. We provide the surface of Γ_1 with an orientation α_1 which coincides on these 2-disks with $-\alpha$. By Lemma 4.3.2 we have $\mathbb{F}(\Gamma_1, \alpha_1) = v^{-1}\mathbb{F}(\Gamma_0, \alpha_0)$. It is easy to observe that Γ_1 is isotopic to Γ via an isotopy identical on the 2-disks and transforming α_1 into $-\alpha$. The isotopy invariance of $\mathbb{F}$ implies that $\mathbb{F}(\Gamma_1, \alpha_1) = \mathbb{F}(\Gamma, -\alpha)$. Combining these equalities we get $\mathbb{F}(\Gamma, \alpha) = \mathbb{F}(\Gamma, -\alpha)$.

If Γ is not connected we apply the same argument to those connected components of Γ where $\alpha' = -\alpha$.

4.4. Invariants of framed graphs continued. The invariant $\mathbb{F}(\Gamma)$ defined in Section 4.3 readily extends to the case when Γ is a v-colored framed graph in $\mathbb{R}^3$ with non-orientable surface. Let $b_1, \ldots, b_n$ be the bands and annuli of Γ. Let $(i_1, i_1^*), \ldots, (i_n, i_n^*)$ be their colors. For any sequence of integers $m = (m_1, \ldots, m_n)$, apply to $b_1, \ldots, b_n$ respectively $m_1, \ldots, m_n$ half-twists in the positive direction. Denote the resulting v-colored framed graph by Γ_m. We choose m so that the surface of Γ_m is orientable. Then the invariant $\mathbb{F}(\Gamma_m) \in K$ is well-defined. Set

$$\mathbb{F}(\Gamma) = (v'_{i_1})^{-m_1}(v'_{i_2})^{-m_2}...(v'_{i_n})^{-m_n}\mathbb{F}(\Gamma_m). \tag{4.4.a}$$

A simple cohomological argument shows that different sequences m (such that the surface of Γ_m is orientable) can be related by the following two modifications and their inverses. The first modification increases one term of m by 2. The second modification increases by 1 the three terms of m corresponding to three bands attached to a 2-disk of Γ. By Lemma 4.3.2, the right-hand side of formula (4.4.a) is invariant under these modifications of m. Therefore $\mathbb{F}(\Gamma)$ does not depend on the choice of m. Lemmas 4.3.1 and 4.3.3 imply that $\mathbb{F}(\Gamma)$ is correctly defined and is an isotopy invariant of Γ. It is obvious that Lemma 4.3.2 applies to framed graphs with non-orientable surfaces without any changes.

Using the same idea as at the end of Section 1.3 (cf. Figures 1.3 and 1.4) we can derive from $\mathbb{F}$ an isotopy invariant of colored framed graphs in $\mathbb{R}^3$ with non-colored 2-disks. This invariant is also denoted by $\mathbb{F} = \mathbb{F}_{\mathcal{V}}$ and takes values in certain tensor products of multiplicity modules. In this sense it is a tensor invariant. More exactly, its value on a colored framed graph $\Gamma \subset \mathbb{R}^3$ is an element of $H^\star(\Gamma)$ defined as follows. Take for each 2-disk D of Γ, a certain $x_D \in H_D$ and denote the family $\{x_D\}_D$ by x. The pair (Γ, x) is a v-colored framed graph in $\mathbb{R}^3$ so that we may consider $\mathbb{F}(\Gamma, x) \in K$. It is obvious that $\mathbb{F}(\Gamma, x)$ is linear with respect to each x_D. Therefore the formula $\otimes_D x_D \mapsto \mathbb{F}(\Gamma, x)$ defines a K-homomorphism $H(\Gamma) \to K$. This homomorphism regarded as an element of $H^\star(\Gamma)$ yields $\mathbb{F}(\Gamma)$.

4.5. Rounding the corners. Essentially no information is lost under the passage from the operator invariant F of ribbon graphs to the invariant $\mathbb{F}$ of framed graphs (at least in the case of trivalent graphs without free ends). Indeed, for each v-colored ribbon graph Ω with trivalent coupons and colors of bands belonging to the set $\{V_i\}_{i\in I}$, we can define in a canonical way a v-colored framed graph Γ such that $\mathbb{F}(\Gamma) = F(\Omega)$. We shall need this construction only in the special case when each coupon of Ω has the following property: either there are 2 bands attached to the bottom base and 1 band attached to the top base or the other way round, the bands incident to the top base are incoming whereas the bands incident to the bottom base are outgoing. The graph Γ is obtained from Ω by replacing all coupons with 2-disks while keeping the annuli and bands. It is understood that we forget directions of bands and annuli and the orientation of the surface of Ω. The coloring of bands of Ω induces a coloring of bands of Γ: if a band of Ω has a color V_i then the corresponding band of Γ, being considered with the same direction as in Ω, has the color i and being considered with the opposite direction has the color i^*. Analogous conventions apply to the annuli of Γ. The colors of 2-disks of Γ are obtained from the colors of coupons of Ω via the isomorphisms constructed at the end of Section 3.2. We say that Γ is obtained from Ω by rounding the corners. The equality $\mathbb{F}(\Gamma) = F(\Omega)$ is a nice exercise on definitions. (Note that the ribbon graph Ω_Γ obtained by ribboning of Γ does not coincide with

Ω. Still, $F(\Omega_\Gamma) = F(\Omega)$. To see this, look at Figures 3.3, 3.4 and notice that the isomorphisms constructed at the end of Section 3.2 involve insertion of a coupon (colored with w_i^{-1}) in every outgoing band. This is equivalent to insertion of one coupon in every band. This coupon is compensated by the coupon inserted under ribboning and colored with w_i.)

The equality $\mathbb{F}(\Gamma) = F(\Omega)$ implies a similar equality for the derived tensor invariants of colored (but not v-colored) graphs. The only subtle point is that these invariants are not equal but rather correspond to each other under the isomorphisms defined at the end of Section 3.2.

The operations of ribboning and rounding the corners act in opposite directions: from framed graphs to ribbon graphs and back. On the geometric level these operations are not mutually inverse. However, on the level of invariants F and $\mathbb{F}$ they are indeed mutually inverse. This fact will allow us to pass from framed graphs to ribbon graphs and back without changing the invariants in question. We shall extensively use this in Section 5 in order to deduce identities for invariants of framed graphs from identities involving ribbon graphs.

4.6. Effacing of 0-colored bands. Let Γ be a colored framed graph in $\mathbb{R}^3$ containing a few annuli and/or bands colored with $0 \in I$. Assume that for every 2-disk D of Γ incident to a 0-colored band, the colors of (the cores of) two other bands incident to D and oriented towards D are dual to each other. Let Γ' be the colored framed graph in $\mathbb{R}^3$ obtained from Γ by effacing all 0-colored bands and annuli (and keeping the colors of the remaining bands and annuli). We shall show that $H(\Gamma') = H(\Gamma)$ and we shall give an explicit formula computing $\mathbb{F}(\Gamma') \in H(\Gamma')$ from $\mathbb{F}(\Gamma) \in H(\Gamma)$.

Here is a more detailed description of Γ'. By a 0-annulus (respectively 0-band) we mean an annulus (respectively a band) such that the given color of its core with a certain (and hence with any) orientation is $0 \in I$. Remove from Γ all 0-annuli (twisted and untwisted) and all 0-bands. Efface any 2-disk of Γ incident to three such bands. By assumption there are no 2-disks of Γ incident to exactly two 0-bands. Each 2-disk D of Γ incident to one 0-band together with two other bands attached to D forms a new, longer band of Γ' or an annulus of Γ'. According to our assumptions the colors of these two bands of Γ are compatible and induce a coloring of the new band (or annulus) of Γ'. This completes the description of the colored framed graph Γ'. Because of the identification $H(i, i^*, 0) = K$ specified in Section 3.5 we have $H_D(\Gamma) = K$ for any 2-disk D of Γ incident to exactly one 0-band. Therefore when we remove such a disk, the tensor product of the modules $H_D(\Gamma)$ corresponding to all 2-disks of Γ does not change. Hence $H(\Gamma') = H(\Gamma)$.

For any 2-disk D of Γ incident to one 0-band, set $d'(D) = \dim'(i) = \dim'(i^*) \in K$ where i and i^* are the colors of two other bands incident to D. Set $d'(\Gamma) = \prod_D d'(D) \in K$ where D runs over all 2-disks of Γ incident to exactly one 0-band.

4.6.1. Lemma. $\mathbb{F}(\Gamma') = d'(\Gamma)\,\mathbb{F}(\Gamma)$.

Proof. It follows from definitions, Exercise I.2.9.2, and the equality $v'_0 = 1$ that effacing of a 0-annulus does not change $\mathbb{F}(\Gamma)$. Similar arguments show that effacing a 0-band which connects a 2-disk to itself also does not change $\mathbb{F}(\Gamma)$. Denote by Γ_0 the colored framed graph obtained from Γ by effacing one 0-band connecting different 2-disks D_1, D_2. Let us show that $\mathbb{F}(\Gamma) = (d'(D_1)d'(D_2))^{-1}\mathbb{F}(\Gamma_0)$.

To visualize the passage from Γ to Γ_0 we first deform Γ so that this band looks as in the left part of Figure 4.9. By definition,

$$(d'(D_1)d'(D_2))^{-1} = (\dim'(i)\dim'(j))^{-1}$$

where i, j are the colors of the bands attached to D_1, D_2, see Figure 4.9. Provide $H_{D_1}(\Gamma) = K$ and $H_{D_2}(\Gamma) = K$ with preferred elements corresponding to $1 \in K$. Set

$$f = (\dim'(i))^{-1} w_{i^*}^{-1} \in \mathrm{Hom}(V_i^*, V_{i^*}), \quad g = (\dim'(j))^{-1} w_{j^*}^{-1} \in \mathrm{Hom}(V_j^*, V_{j^*}).$$

The ribboning of the fragment of Γ shown in the left part of Figure 4.9 yields the ribbon graph in Figure 4.9. Five coupons colored with $w_0, w_i, w_j, w_{i^*}, w_{j^*}$ arise in accordance with the instructions for ribboning. Two coupons colored with f and g are due to the definition of the identification isomorphisms $H_{D_1}(\Gamma) = K$, $H_{D_2}(\Gamma) = K$. It follows from (I.1.2.e) that the coupon colored with w_0 and two incident 0-bands may be removed without changing the operator invariant. The coupons colored with f and w_{i^*} cancel each other and contribute the scalar factor $(\dim'(i))^{-1}$. Similarly, the coupons colored with g and w_{j^*} contribute the scalar factor $(\dim'(j))^{-1}$. This yields the equality in Figure 4.9. Therefore $\mathbb{F}(\Gamma) = (\dim'(i)\dim'(j))^{-1}\mathbb{F}(\Gamma_0)$. Applying this formula inductively to all 0-bands of Γ connecting different 2-disks we get $\mathbb{F}(\Gamma') = d'(\Gamma)\,\mathbb{F}(\Gamma)$.

4.7. Invariants of framed graphs in 3-manifolds. In the case where $\mathcal{V}$ is a modular category, the definitions and results of Sections 4.3 - 4.6 extend to framed graphs in any closed oriented 3-manifold M. The only difference is that instead of $F(\Omega)$ we should use the invariant $\tau(M, \Omega) = \tau_{(\mathcal{V},\mathcal{D})}(M, \Omega)$ defined in Section II.2.3. In this way we get for any v-colored framed graph $\Gamma \subset M$, an invariant $\tau(M, \Gamma) \in K$. Lemmas 4.3.1 - 4.3.3 extend to this setting word for word. We may obtain the same invariant $\tau(M, \Gamma)$ using directly the formulas of Section II.2.3 but replacing ribbon graphs in $\mathbb{R}^3$ and their invariant F with framed graphs in $\mathbb{R}^3$ and their invariant $\mathbb{F}$.

As in Section 4.4, for any colored (but not v-colored) framed graph $\Gamma \subset M$, we get an invariant $\tau(M, \Gamma) \in H^\star(\Gamma)$. If $M = S^3$ then $\tau(M, \Gamma) = \mathcal{D}^{-1}\mathbb{F}(\Gamma)$.

4.8. Exercise. Prove that for any colored framed graph $\Gamma \subset \mathbb{R}^3$,

$$\mathbb{F}_{\overline{\mathcal{V}}}(\overline{\Gamma}) = \mathbb{F}_{\mathcal{V}}(\Gamma) \in H^\star_{\mathcal{V}}(\Gamma) = H^\star_{\overline{\mathcal{V}}}(\Gamma)$$

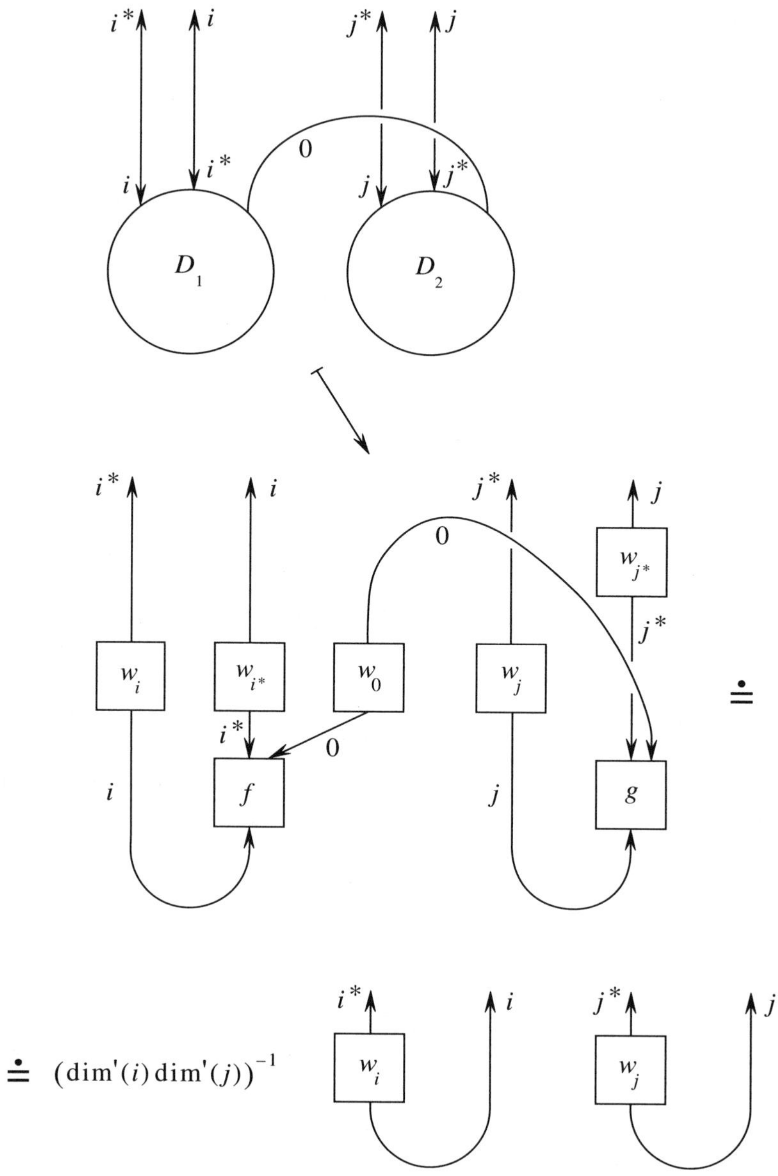

Figure 4.9

where $\overline{\Gamma}$ is the mirror image of Γ with respect to a plane in $\mathbb{R}^3$. If $\mathcal{V}$ is a modular category then for any colored framed graph Γ in a closed oriented 3-manifold M,

$$\tau_{(\mathcal{V},\mathcal{D})}(-M,\Gamma)=\tau_{(\overline{\mathcal{V}},\mathcal{D})}(M,\Gamma).$$

5. Geometric approach to $6j$-symbols

We continue to adhere to Conventions 3.1.

5.0. Outline. We use the technique of framed graphs to define a more geometric and more symmetric version of $6j$-symbols. Its definition proceeds roughly as follows. Consider a 3-dimensional simplex $T \subset \mathbb{R}^3$. We regard the 1-skeleton of T as a trivalent graph lying in the topological 2-sphere ∂T. A regular neighborhood of this graph in ∂T may be viewed as a framed graph in $\mathbb{R}^3$ with 6 edges and 4 vertices. We color the edges with six elements of the set I. The invariant $\mathbb{F}$ of this colored framed graph is a tensor of 4 variables running over certain symmetrized multiplicity modules. We call this tensor the normalized $6j$-symbol corresponding to the chosen 6 elements of I.

We show how to relate the geometric approach of this section with the algebraic approach of Section 1. We also establish a few fundamental properties of normalized $6j$-symbols including the tetrahedral symmetry, the Biedenharn-Elliott identity, the orthonormality identity, and the Racah identity. For completeness, we establish also an analogue of the Yang-Baxter braiding equality (I.1.2.f) for normalized $6j$-symbols, it will not be used in what follows.

We shall use the normalized $6j$-symbols in order to construct state sum invariants of 3-manifolds and shadows, see Chapters VII and X. The tetrahedral symmetry and the identities mentioned above form the background for our study of these invariants.

5.1. Definition of normalized $6j$-symbols. Let $i, j, k, l, m, n \in I$. Consider the colored framed graph $\Gamma = \Gamma(i, j, k, l, m, n) \subset \mathbb{R}^3$ shown in Figure 5.1. The framed graph Γ is formed by 4 disks and 6 bands lying in the 2-sphere $S^2 \subset \mathbb{R}^3$. The symbols $i, j, k, l, m, n \in I$ in Figure 5.1 are the colors assigned to the indicated directions of the bands of Γ. As usual, to the opposite directions of these bands we assign the dual colors $i^*, j^*, k^*, l^*, m^*, n^* \in I$.

By definition, the K-module $H(\Gamma)$ is the non-ordered tensor product over K of the four K-modules $H(i, j, k^*), H(i^*, m, n^*), H(j^*, l^*, n), H(k, l, m^*)$. We define the normalized $6j$-symbol of the tuple (i,j,k, l,m,n) to be

$$\begin{vmatrix} i & j & k \\ l & m & n \end{vmatrix} = \mathbb{F}(\Gamma) \in H^\star(\Gamma). \tag{5.1.a}$$

Here $H^\star(\Gamma) = \mathrm{Hom}_K(H(\Gamma), K)$ is the non-ordered tensor product over K of the modules $H(i^*, j^*, k), H(i, m^*, n), H(j, l, n^*), H(k^*, l^*, m)$.

It follows directly from definitions that the normalized $6j$-symbol has the symmetries of a tetrahedron. In particular,

$$\begin{vmatrix} i & j & k \\ l & m & n \end{vmatrix} = \begin{vmatrix} j & k^* & i^* \\ m & n & l \end{vmatrix} = \begin{vmatrix} k & l & m \\ n^* & i & j^* \end{vmatrix} = \begin{vmatrix} j & i & k \\ m^* & l^* & n^* \end{vmatrix}. \tag{5.1.b}$$

$\Gamma(i,j,k,l,m,n) =$

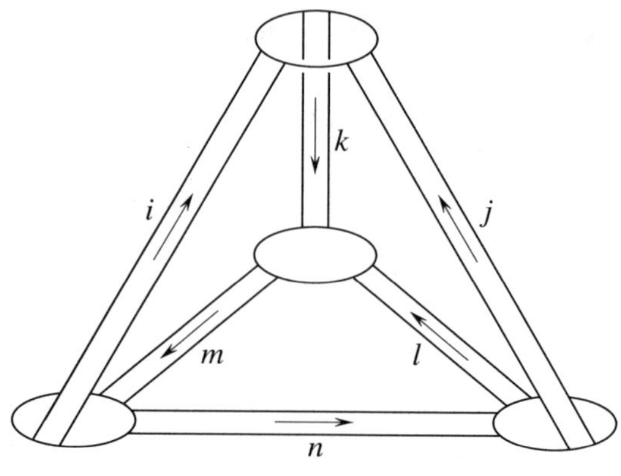

Figure 5.1

These equalities hold because the corresponding four colored framed graphs are isotopic in $\mathbb{R}^3$. For example, the framed graph $\Gamma(i, j, k, l, m, n)$ is isotopic to $\Gamma(j, k^*, i^*, m, n, l)$ which implies the first equality. (It is curious to note that the first three framed graphs are isotopic in S^2 while to pass to the fourth graph we have to get out to $\mathbb{R}^3$.)

More generally, consider any permutation $(a, b, c, d, e, f, a^*, b^*, c^*, d^*, e^*, f^*)$ of the tuple $(i, j, k, l, m, n, i^*, j^*, k^*, l^*, m^*, n^*)$ preserving the set of (non-ordered) triples $(i, j, k^*), (i^*, m, n^*), (j^*, l^*, n), (k, l, m^*)$. Then the normalized $6j$-symbols of the tuples (i, j, k, l, m, n) and (a, b, c, d, e, f) are equal. This follows from the fact that the three permutations corresponding to the equalities (5.1.b) generate the whole group of such permutations.

The framed graph $\Gamma(i, j, k, l, m, n)$ may be easily deformed into the plane $\mathbb{R}^2 \subset \mathbb{R}^3$. This fact and the result of Exercise 4.8 imply that the $6j$-symbols corresponding to the semisimple category $(\mathcal{V}, \{V_i\}_{i\in I})$ and to the mirror category $(\overline{\mathcal{V}}, \{V_i\}_{i\in I})$ are equal.

5.2. Comparison with the previously defined $6j$-symbols. Let $i, j, k, l, m, n \in I$. We compare the normalized $6j$-symbol (5.1.a) with the $6j$-symbol

$$\begin{Bmatrix} i & j & k \\ l & m & n \end{Bmatrix}' \in H^m_{kl} \otimes_K H^k_{ij} \otimes_K H^{in}_m \otimes_K H^{jl}_n \tag{5.2.a}$$

defined in Section 1. A tensor product of the isomorphisms constructed at the end of Section 3.2 yields an isomorphism

$$H^m_{kl} \otimes_K H^k_{ij} \otimes_K H^{in}_m \otimes_K H^{jl}_n \to H^\star(\Gamma)$$

where $\Gamma = \Gamma(i, j, k, l, m, n)$. Pushing forward the element (5.2.a) along the last isomorphism we get an element of $H^\star(\Gamma)$ denoted by

$$\begin{Bmatrix} i & j & k \\ l & m & n \end{Bmatrix}''.$$

It is obvious that the framed graph $\Gamma(i,j,k,l,m,n)$ can be obtained from the ribbon graph in the right part of Figure 1.4 by rounding the corners and isotopy in $\mathbb{R}^3$. Therefore the formula in Figure 1.4 and the results of Section 4.5 imply that

$$\left\{\begin{matrix} i & j & k \\ l & m & n \end{matrix}\right\}'' = \dim(n) \begin{vmatrix} i & j & k \\ l & m & n \end{vmatrix}. \tag{5.2.b}$$

5.3. Computation for the color 0. Let us compute the $6j$-symbol (5.1.a) for the case $n = 0$. This $6j$-symbol is an element of the tensor product of four modules $H(i^*, j^*, k)$, $H(i, m^*, 0)$, $H(j, l, 0)$, $H(k^*, l^*, m)$. If $m \neq i$ or $l \neq j^*$ then this tensor product is zero and the $6j$-symbol (5.1.a) is equal to 0. If $m = i$ and $l = j^*$ then $H(i, m^*, 0) = K$, $H(j, l, 0) = K$ and the tensor product in question is equal to the tensor product of two dual modules $H(i^*, j^*, k) \otimes_K H(i, j, k^*)$.

In general, the tensor product of dual projective K-modules H_1, H_2 contains a canonical element determined by the duality. Indeed, a (non-degenerate) duality pairing $H_1 \otimes_K H_2 \to K$ may be considered as an element of $(H_1 \otimes_K H_2)^\star = H_2^\star \otimes_K H_1^\star = H_1 \otimes_K H_2$. Here the first equality is obvious and the second equality is induced by the adjoint isomorphisms $H_2^\star \to H_1$, $H_1^\star \to H_2$. For example, if H_1, H_2 are free K-modules of finite rank and $\{e_r\}_r$, $\{e^r\}_r$ are dual bases in H_1, H_2 then this canonical element of $H_1 \otimes_K H_2$ is equal to $\sum_r e_r \otimes e^r$.

Denote by $\mathrm{Id}(i, j, k)$ the canonical element of $H(i, j, k) \otimes_K H(i^*, j^*, k^*)$ determined by the duality defined in Section 3.3.

5.3.1. Lemma. *For any $i, j, k, l, m \in I$, we have*

$$\begin{vmatrix} i & j & k \\ l & m & 0 \end{vmatrix} = \delta_{m,i}\, \delta_{l,j^*}\, (\dim'(i))^{-1} (\dim'(j))^{-1} \mathrm{Id}(i, j, k^*),$$

where δ is the Kronecker delta.

Proof. In the case $m \neq i$ or $l \neq j^*$ both sides of the formula are equal to zero. Assume that $m = i$ and $l = j^*$. Denote by Γ_0 the colored framed graph in Figure 5.2 formed by two disks and three bands. It follows from Lemma 4.6.1 that

$$\begin{vmatrix} i & j & k \\ j^* & i & 0 \end{vmatrix} = (\dim'(i))^{-1} (\dim'(j))^{-1} \mathbb{F}(\Gamma_0).$$

Therefore it suffices to note that $\mathbb{F}(\Gamma_0) = \mathrm{Id}(i, j, k^*)$. This follows from the results of Section 3.3 and definitions.

5.4. Fundamental identities between $6j$-symbols. We establish fundamental identities between normalized $6j$-symbols. To this end we need a contraction of tensors parallel to the one introduced in Section 1.5.2. If we have a tensor product of several K-modules so that among the factors there is a matched pair $H(i, j, k)$, $H(i^*, j^*, k^*)$ then we may contract this tensor product using the duality

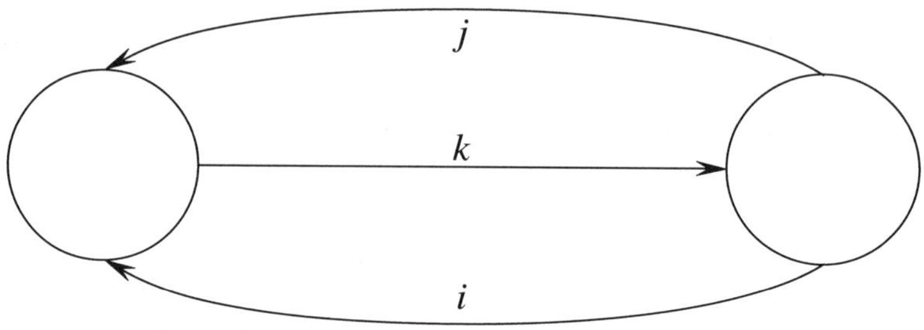

Figure 5.2

between $H(i, j, k)$ and $H(i^*, j^*, k^*)$. This transformation is called the contraction along $H(i, j, k)$ and denoted by $*_{ijk}$. Note that contractions along $H(i, j, k)$ and $H(i^*, j^*, k^*)$ coincide. The remarks made in Section 1.5.2 apply in this setting with the obvious changes.

The contraction $*_{ij}^k$ introduced in Section 1.5.2 is compatible with $*_{ijk^*}$ under the isomorphisms $H_k^{ij} \to H(i, j, k^*)$ and $H_{ij}^k \to H(i^*, j^*, k)$ defined in Figures 3.3 and 3.4. This follows from the discussion in Section 3.3.

The following assertion follows from Corollary 1.5.3, formula (5.2.b), and the invertibility of $\dim(j_7), \dim(j_8)$.

5.4.1. Theorem (the Biedenharn-Elliott identity). *For any* $j_0, j_1, \ldots, j_8 \in I$,

$$\sum_{j\in I} \dim(j) *_{j^* j_2 j_3} *_{j j_4 j_7^*} *_{j j_1 j_6^*} \left(\begin{vmatrix} j_1 & j_2 & j_5 \\ j_3 & j_6 & j \end{vmatrix} \otimes \begin{vmatrix} j_1 & j & j_6 \\ j_4 & j_0 & j_7 \end{vmatrix} \otimes \begin{vmatrix} j_2 & j_3 & j \\ j_4 & j_7 & j_8 \end{vmatrix} \right)$$

$$= *_{j_0^* j_5 j_8} \left(\begin{vmatrix} j_5 & j_3 & j_6 \\ j_4 & j_0 & j_8 \end{vmatrix} \otimes \begin{vmatrix} j_1 & j_2 & j_5 \\ j_8 & j_0 & j_7 \end{vmatrix} \right).$$

Here both sides lie in the non-ordered tensor product of six K-modules

$$H(j_3^*, j_5^*, j_6), H(j_1^*, j_2^*, j_5), H(j_0, j_4^*, j_6^*),$$

$$H(j_0^*, j_1, j_7), H(j_2, j_7^*, j_8), H(j_3, j_4, j_8^*).$$

5.4.2. Theorem (the orthonormality relation). *For any* $i, j, k, k', l, m \in I$,

$$\dim(k) \sum_{n\in I} \dim(n) *_{im^*n} *_{jln^*} \left(\begin{vmatrix} i^* & j^* & k^* \\ l^* & m^* & n^* \end{vmatrix} \otimes \begin{vmatrix} i & j & k' \\ l & m & n \end{vmatrix} \right) = \tag{5.4.a}$$

$$= \delta_{k,k'} \operatorname{Id}(i, j, k^*) \otimes \operatorname{Id}(k, l, m^*).$$

Theorem 5.4.2 is proven in Section 5.5.

5.4.3. Theorem (the Racah identity). *For any* $j_1, j_2, j_3, j_4, j_5, j_6 \in I$,

$$\sum_{j\in I} (v'_j)^{-1} \dim(j) *_{j^* j_1 j_4} *_{j j_2 j_5^*} \left(\begin{vmatrix} j_1 & j_4 & j \\ j_2 & j_5 & j_6 \end{vmatrix} \otimes \begin{vmatrix} j_2 & j_1 & j_3 \\ j_4 & j_5 & j \end{vmatrix} \right) = \tag{5.4.b}$$

$$= v'_{j_3} v'_{j_6} (v'_{j_1} v'_{j_2} v'_{j_4} v'_{j_5})^{-1} \begin{vmatrix} j_1 & j_2 & j_3 \\ j_4 & j_5 & j_6 \end{vmatrix}.$$

This theorem is proven in Section 5.6. Applying this theorem to the mirror semisimple category we obtain the following corollary.

5.4.4. Corollary. *For any* $j_1, j_2, j_3, j_4, j_5, j_6 \in I$,

$$\sum_{j\in I} v'_j \dim(j) *_{j^* j_1 j_4} *_{j j_2 j_5^*} \left(\begin{vmatrix} j_1 & j_4 & j \\ j_2 & j_5 & j_6 \end{vmatrix} \otimes \begin{vmatrix} j_2 & j_1 & j_3 \\ j_4 & j_5 & j \end{vmatrix} \right) = \tag{5.4.c}$$

$$= (v'_{j_3} v'_{j_6})^{-1} v'_{j_1} v'_{j_2} v'_{j_4} v'_{j_5} \begin{vmatrix} j_1 & j_2 & j_3 \\ j_4 & j_5 & j_6 \end{vmatrix}.$$

The Biedenharn-Elliott identity, the orthonormality relation, and the Racah identity form the basis for the study of $6j$-symbols. We shall heavily rely on these identities in our proof of the topological invariance of partition functions introduced in Chapters VII and X. These fundamental identities imply quite a number of other identities between $6j$-symbols. We present one such identity in Section 5.7 and another one in Exercise 5.8.1. (These identities will not be used in further chapters.)

5.5. Proof of Theorem 5.4.2. The easiest way to prove the orthonormality relation is to note that it follows from the Biedenharn-Elliott identity, the tetrahedral symmetry of the normalized $6j$-symbols, and Lemma 5.3.1. To see this it suffices to substitute $j_8 = 0, j_7 = j_2, j_4 = j_3^*$ in the Biedenharn-Elliott identity. (The details are left to the reader as an exercise.)

We give here another, more geometric proof of the orthonormality relation. Consider the isomorphism inverse to (1.3.d). Restricting this isomorphism to the summand in the source corresponding to $n \in I$ and projecting into the summand in the target corresponding to $k \in I$ we get a homomorphism

$$H_m^{in} \otimes_K H_n^{jl} \to H_m^{kl} \otimes_K H_k^{ij}. \tag{5.5.a}$$

Denote this homomorphism by

$$\begin{Bmatrix} i & j & k \\ l & m & n \end{Bmatrix}_{\mathrm{inv}}. \tag{5.5.b}$$

The homomorphisms (5.5.b) corresponding to fixed i, j, l, m and varying k, n form a block-matrix of the isomorphism inverse to (1.3.d). Therefore

$$\sum_{n\in I} \left\{\begin{matrix} i & j & k \\ l & m & n \end{matrix}\right\}_{\mathrm{inv}} \circ \left\{\begin{matrix} i & j & k' \\ l & m & n \end{matrix}\right\} = \delta(k, k' \mid i, j, l, m) \tag{5.5.c}$$

where

$$\delta(k, k' \mid i, j, l, m) : H_m^{k'l} \otimes_K H_{k'}^{ij} \to H_m^{kl} \otimes_K H_k^{ij}$$

is zero if $k \neq k'$ and is the identity homomorphism if $k = k'$.

Denote by

$$\left\{\begin{matrix} i & j & k \\ l & m & n \end{matrix}\right\}'_{\mathrm{inv}} \tag{5.5.d}$$

the element of the module

$$(H_m^{in} \otimes_K H_n^{jl})^\star \otimes_K H_m^{kl} \otimes_K H_k^{ij} = H_{in}^m \otimes_K H_{jl}^n \otimes_K H_m^{kl} \otimes_K H_k^{ij} \tag{5.5.e}$$

corresponding to (5.5.b). In terms of this element and the modified $6j$-symbol (1.4.b) we may rewrite formula (5.5.c) as follows. Consider the tensor

$$\delta_{k,k'}(i, j, l, m) \in H_m^{kl} \otimes_K H_k^{ij} \otimes_K H_{k'l}^m \otimes_K H_{ij}^{k'}$$

equal to 0 if $k \neq k'$ and otherwise corresponding to the tensor product of the elements of $H_m^{kl} \otimes_K H_{kl}^m$ and $H_k^{ij} \otimes_K H_{ij}^k$ determined by duality. Then

$$\sum_{n\in I} *_{in}^m *_{jl}^n \left(\left\{\begin{matrix} i & j & k \\ l & m & n \end{matrix}\right\}'_{\mathrm{inv}} \otimes \left\{\begin{matrix} i & j & k' \\ l & m & n \end{matrix}\right\}' \right) = \delta_{k,k'}(i, j, l, m). \tag{5.5.f}$$

The $6j$-symbol (5.5.d) may be interpreted as a normalized $6j$-symbol along the lines of Section 5.2. To this end we apply the pictorial formalism to (5.5.a) and (5.5.d) reproducing the pictures of Sections 1.3 and 1.4 with the obvious changes. An analogue of Lemma 1.4.1 for the symbol (5.5.d) is given in Figure 5.3. Rounding the corners of the ribbon graph in Figure 5.3 we get a colored framed graph isotopic to $\Gamma(i^*, j^*, k^*, l^*, m^*, n^*)$. A tensor product of isomorphisms constructed at the end of Section 3.2 yields an isomorphism of the module (5.5.e) onto $H^\star(\Gamma(i^*, j^*, k^*, l^*, m^*, n^*))$. The formula in Figure 5.3 and the results of Section 4.5 imply that pushing forward (5.5.d) along the last isomorphism we obtain

$$\dim(k) \left|\begin{matrix} i^* & j^* & k^* \\ l^* & m^* & n^* \end{matrix}\right| \in H^\star(\Gamma(i^*, j^*, k^*, l^*, m^*, n^*)).$$

This observation together with similar observations made in Section 5.2 shows that equality (5.5.f) implies the orthonormality relation (5.4.a).

5.6. Proof of Theorem 5.4.3. Consider the action of the braiding in $\mathcal{V}$ with respect to the decomposition (1.3.c). More exactly, consider the homomorphism

$$\mathrm{Hom}(V_m, V_i \otimes V_j \otimes V_l) \to \mathrm{Hom}(V_m, V_i \otimes V_l \otimes V_j)$$

$$\begin{Bmatrix} i & j & k \\ l & m & n \end{Bmatrix}'_{\text{inv}} \doteq \dim(k)$$

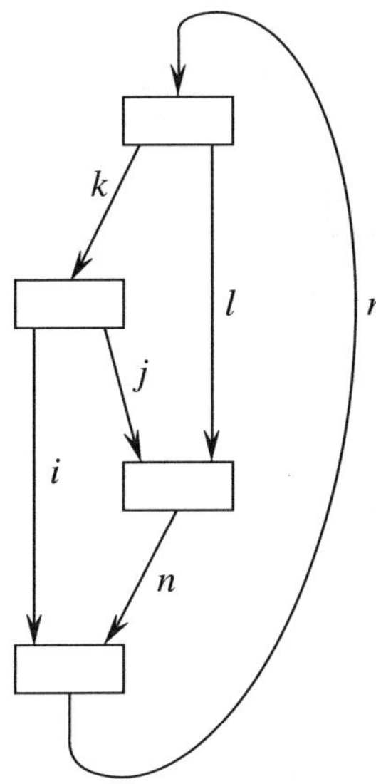

Figure 5.3

obtained by composing with $\mathrm{id}_{V_i} \otimes c_{V_j, V_l}$. Using the decomposition (1.3.c) and the same decomposition with exchanged j and l we present this homomorphism by a block-matrix of homomorphisms

(5.6.a) $$H^{kl}_m \otimes_K H^{ij}_k \to H^{nj}_m \otimes_K H^{il}_n$$

where k, n independently run over I. Denote the resulting homomorphism (5.6.a) by

(5.6.b) $$\begin{Bmatrix} i & j & k \\ l & m & n \end{Bmatrix}_{\text{br}}.$$

Denote the associated element of the module

(5.6.c) $$(H^{kl}_m \otimes_K H^{ij}_k)^\star \otimes_K H^{nj}_m \otimes_K H^{il}_n = H^m_{kl} \otimes_K H^k_{ij} \otimes_K H^{nj}_m \otimes_K H^{il}_n$$

by

(5.6.d) $$\begin{Bmatrix} i & j & k \\ l & m & n \end{Bmatrix}'_{\text{br}}.$$

We shall use the technique of framed graphs to compute the $6j$-symbol (5.6.d) in terms of normalized $6j$-symbols. The definition of the $6j$-symbol (5.6.a) is rewritten in a graphical form in Figure 5.4. For an analogue of Lemma 1.4.1, see Figure 5.5. Note that both sides of the equality in Figure 5.5 should be treated as K-linear functionals

$$H^{kl}_m \otimes_K H^{ij}_k \otimes_K H^m_{nj} \otimes_K H^n_{il} \to K$$

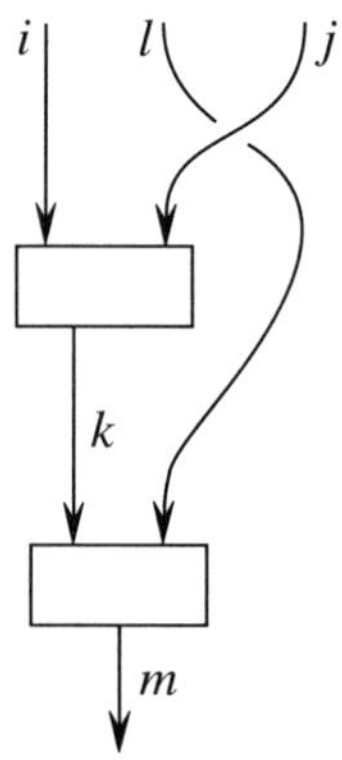

$$\doteq \sum_{n \in I}$$

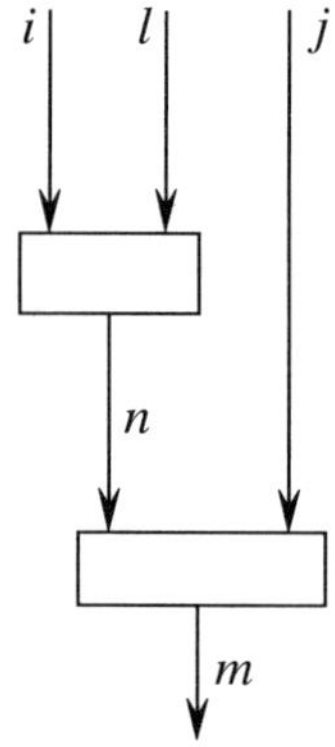

$$\circ \begin{Bmatrix} i & j & k \\ l & m & n \end{Bmatrix}_{\mathrm{br}}$$

Figure 5.4

$$\begin{Bmatrix} i & j & k \\ l & m & n \end{Bmatrix}'_{\mathrm{br}} \doteq \dim(n)$$

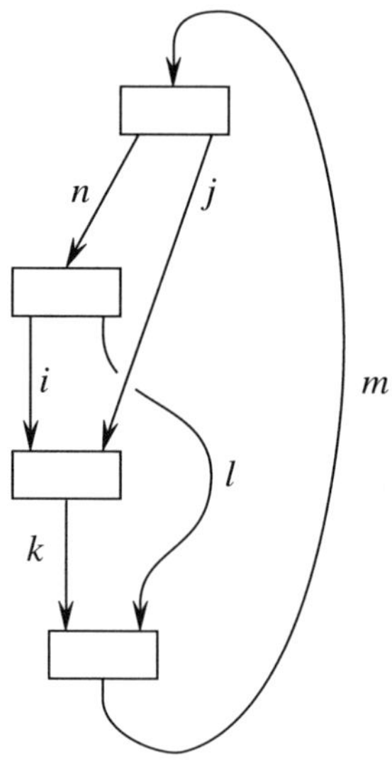

Figure 5.5

or, equivalently, as elements of the module (5.6.c). The proof of the equality in Figure 5.5 repeats the proof of Lemma 1.4.1 with the only difference that instead of $(y \otimes \mathrm{id}_{V_l})x$ we should consider $(\mathrm{id}_{V_i} \otimes c_{V_j,V_l})(y \otimes \mathrm{id}_{V_l})x$.

Rounding the corners of the ribbon graph in Figure 5.5 we obtain a colored framed graph Ψ, see Figure 5.6. A tensor product of isomorphisms constructed in Section 3.2 yields an isomorphism of the module (5.6.c) onto

$$H^\star(\Psi) = H(m, k^*, l^*) \otimes_K H(k, i^*, j^*) \otimes_K H(n, j, m^*) \otimes_K H(i, l, n^*).$$

The equality in Figure 5.5 and results of Section 4.5 imply that pushing forward the 6j-symbol (5.6.d) along these isomorphisms we get $\dim(n)\, \mathbb{F}(\Psi) \in H^\star(\Psi)$. Therefore in order to compute the 6j-symbol (5.6.d) it suffices to compute $\mathbb{F}(\Psi)$.

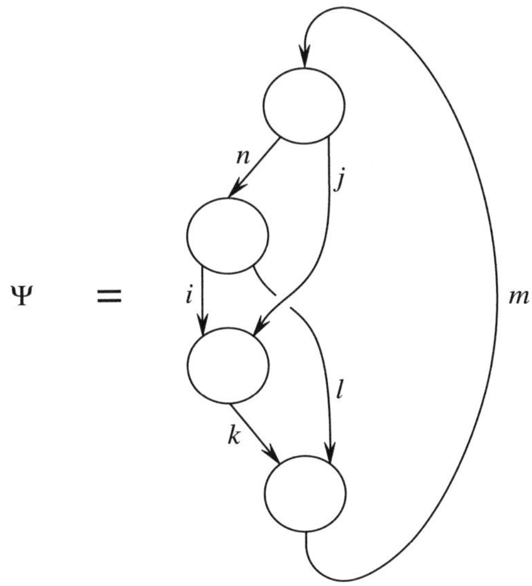

Figure 5.6

The colored framed graph Ψ is isotopic to the colored framed graph obtained from $\Gamma(j, i, k, l, m, n)$ by positive half-twists of the bands marked by i, m and negative half-twists of the bands marked by k, n. (To see this we may deform Ψ in $\mathbb{R}^3$ by pushing the j-band to the left and simultaneously rotating the two 2-disks attached to this band around their vertical axis by an angle of π.) Therefore

$$\mathbb{F}(\Psi) = v'_i v'_m (v'_k v'_n)^{-1}\, \mathbb{F}(\Gamma(j, i, k, l, m, n)) = v'_i v'_m (v'_k v'_n)^{-1} \begin{vmatrix} j & i & k \\ l & m & n \end{vmatrix}.$$

This completes the computation of (5.6.d).

Combining the equalities in Figures 1.3 and 5.4 we get the equality in Figure 5.7. Attaching to both sides from above the same ribbon graph and taking the trace we get for any $i, j, k, l, m, n_0 \in I$ the equality in Figure 5.8 where $M(n, p)$ denotes the K-homomorphism

$$\mathrm{id}_{H^m_{in_0}} \otimes \mathrm{id}_{H^{n_0}_{lj}} \otimes \begin{Bmatrix} i & l & p \\ j & m & n \end{Bmatrix} \begin{Bmatrix} i & j & k \\ l & m & p \end{Bmatrix}_{\mathrm{br}} :$$

$$H^m_{in_0} \otimes_K H^{n_0}_{lj} \otimes_K H^{kl}_m \otimes_K H^{ij}_k \to H^m_{in_0} \otimes_K H^{n_0}_{lj} \otimes_K H^{in}_m \otimes_K H^{lj}_n.$$

The colored ribbon graph in the right part of Figure 5.8 represents a linear functional

$$H^m_{in_0} \otimes_K H^{n_0}_{lj} \otimes_K H^{in}_m \otimes_K H^{lj}_n \to K. \qquad (5.6.e)$$

The right-hand side of the equality in Figure 5.8 is obtained by composing this functional with $M(n, p)$ and summing up over all $n, p \in I$. This yields an element

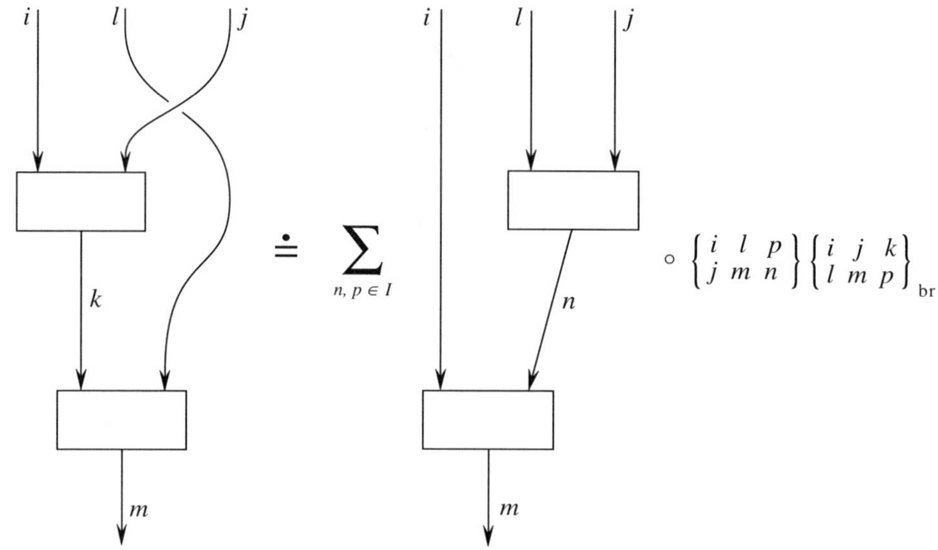

Figure 5.7

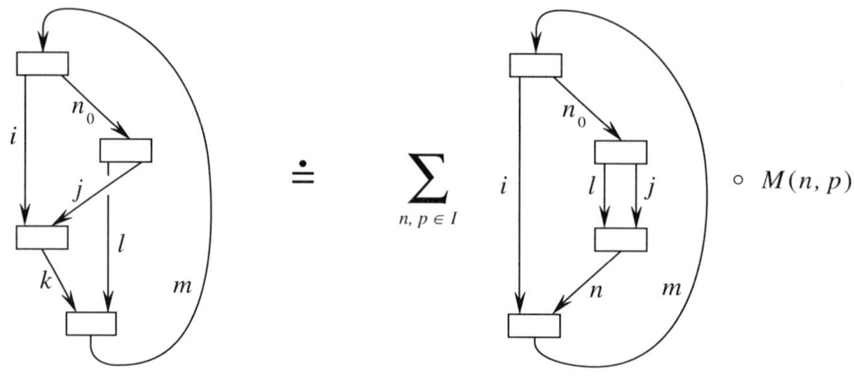

Figure 5.8

of the module

(5.6.f) $\operatorname{Hom}(H^m_{in_0} \otimes_K H^{n_0}_{lj} \otimes_K H^{kl}_m \otimes_K H^{ij}_k, K) = H^{in_0}_m \otimes_K H^{lj}_{n_0} \otimes_K H^m_{kl} \otimes_K H^k_{ij}.$

The equality in Figure 5.8 means that the colored ribbon graph in the left part of this figure represents the same element of this module.

A tensor product of isomorphisms of Section 3.2 yields an isomorphism of the module (5.6.f) onto

(5.6.g) $H(i, n_0, m^*) \otimes_K H(l, j, n_0^*) \otimes_K H(m, k^*, l^*) \otimes_K H(k, i^*, j^*).$

Our next aim is to compute the elements of the last module determined by the left-hand and right-hand sides of the equality in Figure 5.8.

The functional (5.6.e) may be easily computed as the tensor product of the duality pairings

$$H^m_{in_0} \otimes_K H^{in_0}_m \to K, \quad H^{n_0}_{lj} \otimes_K H^{lj}_{n_0} \to K$$

multiplied by $(\dim(n_0))^{-1}\delta^{n_0}_n$. Therefore the element of the module (5.6.f) determined by the right-hand side of the equality in Figure 5.8 is equal to

$$(\dim(n_0))^{-1} \sum_{p\in I} *^p_{il} *^m_{pj} \left(\left\{ \begin{matrix} i & l & p \\ j & m & n_0 \end{matrix} \right\}' \otimes \left\{ \begin{matrix} i & j & k \\ l & m & p \end{matrix} \right\}'_{\mathrm{br}} \right).$$

The computations of the $6j$-symbols (1.4.b), (5.6.d) made above show that passing to the symmetrized multiplicity modules we transform the last expression into

$$\text{(5.6.h)} \qquad \sum_{p\in I} \dim(p) *_{p^*il} *_{m^*pj} v'_i v'_m (v'_k v'_p)^{-1} \left(\left| \begin{matrix} i & l & p \\ j & m & n_0 \end{matrix} \right| \otimes \left| \begin{matrix} j & i & k \\ l & m & p \end{matrix} \right| \right).$$

This is the element of the module (5.6.g) determined by the right-hand side of Figure 5.8.

Now we compute the element of the module (5.6.g) represented by the colored ribbon graph in the left part of Figure 5.8. To this end it suffices to round the corners of this ribbon graph and compute the invariant $\mathbb{F}$ of the resulting colored framed graph, say Γ. The colored framed graph Γ is isotopic to the framed graph obtained from $\Gamma(i, j, k, l, m, n)$ by a positive half-twist of the band marked by n_0 and negative half-twists of the bands marked by j and l. The easiest way to see this is to rotate the coupon of Γ incident to these three bands to the left by an angle of π. Therefore

$$\mathbb{F}(\Gamma) = v'_{n_0}(v'_j v'_l)^{-1}\mathbb{F}(\Gamma(i, j, k, l, m, n_0)) = v'_{n_0}(v'_j v'_l)^{-1} \left| \begin{matrix} i & j & k \\ l & m & n_0 \end{matrix} \right|.$$

The equality in Figure 5.8 implies that the last expression is equal to the expression (5.6.h). Substituting $i = j_1, j = j_2, k = j_3, l = j_4, m = j_5, n_0 = j_6, p = j$ we get the claim of the theorem.

5.7. Theorem (the Yang-Baxter identity). *For any $j_1, j_2, ..., j_9 \in I$ and any $\varepsilon = \pm 1$,*

$$\sum_{j\in I} (v'_{j_5} v'_{j_7} v'_{j_9} v'_j)^{\varepsilon} \dim(j) *_{j^*j_2j_4} *_{jj_1j_6^*} *_{jj_3j_8^*} \left(\left| \begin{matrix} j_2 & j_4 & j \\ j_1 & j_6 & j_5 \end{matrix} \right| \otimes \left| \begin{matrix} j_3 & j & j_8 \\ j_1 & j_7 & j_6 \end{matrix} \right| \otimes \right.$$

$$\left. \otimes \left| \begin{matrix} j_3 & j_4 & j_9 \\ j_2 & j_8 & j \end{matrix} \right| \right) = \sum_{j\in I} (v'_{j_4} v'_{j_6} v'_{j_8} v'_j)^{\varepsilon} \dim(j) *_{j^*j_3j_5} *_{jj_2j_7^*} *_{j^*j_1j_9} \left(\left| \begin{matrix} j_3 & j_5 & j \\ j_2 & j_7 & j_6 \end{matrix} \right| \otimes \right.$$

$$\otimes \begin{vmatrix} j_3 & j_4 & j_9 \\ j_1 & j & j_5 \end{vmatrix} \otimes \begin{vmatrix} j_2 & j_9 & j_8 \\ j_1 & j_7 & j \end{vmatrix} \Big).$$

Proof. Let us transform the second $6j$-symbol from LHS as follows

$$\begin{vmatrix} j_3 & j & j_8 \\ j_1 & j_7 & j_6 \end{vmatrix} = \begin{vmatrix} j_1^* & j_6 & j \\ j_3 & j_8 & j_7 \end{vmatrix} = \tag{5.7.a}$$

$$\sum_{q\in I} (v'_{j_1} v'_{j_3} v'_{j_6} v'_{j_8} (v'_q v'_{j_7} v'_j)^{-1})^{\varepsilon} \dim(q) *_{qj_1j_3^*} *_{qj_6j_8^*} \Big(\begin{vmatrix} j_1^* & j_3 & q \\ j_6 & j_8 & j_7 \end{vmatrix} \otimes \begin{vmatrix} j_6 & j_1^* & j \\ j_3 & j_8 & q \end{vmatrix} \Big).$$

Here the first equality follows from the tetrahedral symmetry, and the second equality follows from Theorems 5.4.3 and 5.4.4. Substitute (5.7.a) into LHS and sum up the terms of the resulting expression involving j over all $j \in I$. This yields

$$\sum_{j\in I} \dim(j) *_{j^*j_2j_4} *_{jj_1j_6^*} *_{jj_3j_8^*} \Big(\begin{vmatrix} j_2 & j_4 & j \\ j_1 & j_6 & j_5 \end{vmatrix} \otimes \begin{vmatrix} j_6 & j_1^* & j \\ j_3 & j_8 & q \end{vmatrix} \otimes \begin{vmatrix} j_3 & j_4 & j_9 \\ j_2 & j_8 & j \end{vmatrix} \Big) =$$

$$= \sum_{j\in I} \dim(j) *_{j^*j_2j_4} *_{jj_1j_6^*} *_{jj_3j_8^*} \Big(\begin{vmatrix} j_2 & j_4 & j \\ j_1 & j_6 & j_5 \end{vmatrix} \otimes \begin{vmatrix} j_8^* & j & j_3^* \\ j_1 & q^* & j_6 \end{vmatrix} \otimes \begin{vmatrix} j_8^* & j_2 & j_9^* \\ j_4 & j_3^* & j \end{vmatrix} \Big) =$$

$$= \sum_{j\in I} \dim(j) *_{j^*j_2j_4} *_{jj_1j_6^*} *_{jj_3j_8^*} \Big(\begin{vmatrix} j_8^* & j_2 & j_9^* \\ j_4 & j_3^* & j \end{vmatrix} \otimes \begin{vmatrix} j_8^* & j & j_3^* \\ j_1 & q^* & j_6 \end{vmatrix} \otimes \begin{vmatrix} j_2 & j_4 & j \\ j_1 & j_6 & j_5 \end{vmatrix} \Big) =$$

$$*_{qj_5j_9^*} \Big(\begin{vmatrix} j_9^* & j_4 & j_3^* \\ j_1 & q^* & j_5 \end{vmatrix} \otimes \begin{vmatrix} j_8^* & j_2 & j_9^* \\ j_5 & q^* & j_6 \end{vmatrix} \Big) = *_{qj_5j_9^*} \Big(\begin{vmatrix} j_5 & j_1^* & j_4 \\ j_3 & j_9 & q \end{vmatrix} \otimes \begin{vmatrix} q & j_5 & j_9 \\ j_2 & j_8 & j_6 \end{vmatrix} \Big).$$

These equalities follow from the tetrahedral symmetry, the fact that we deal with non-ordered tensor products of modules, and Theorem 5.4.1. Therefore LHS is equal to

$$\sum_{q\in I} \dim(q)(v'_q)^{-\varepsilon} (v'_{j_1} v'_{j_3} v'_{j_5} v'_{j_6} v'_{j_8} v'_{j_9})^{\varepsilon} *_{qj_1j_3^*} *_{qj_6j_8^*} *_{qj_5j_9^*} \Big(\begin{vmatrix} j_1^* & j_3 & q \\ j_6 & j_8 & j_7 \end{vmatrix} \otimes$$

$$\otimes \begin{vmatrix} j_5 & j_1^* & j_4 \\ j_3 & j_9 & q \end{vmatrix} \otimes \begin{vmatrix} q & j_5 & j_9 \\ j_2 & j_8 & j_6 \end{vmatrix} \Big).$$

Similarly, substituting in RHS the expression

$$\begin{vmatrix} j_3 & j_4 & j_9 \\ j_1 & j & j_5 \end{vmatrix} = \begin{vmatrix} j_1^* & j_5 & j_4 \\ j_3 & j_9 & j \end{vmatrix} =$$

$$\sum_{q\in I} \dim(q)(v'_{j_1} v'_{j_3} v'_{j_5} v'_{j_9} (v'_q v'_j v'_{j_4})^{-1})^{\varepsilon} *_{qj_1j_3^*} *_{qj_5j_9^*} \Big(\begin{vmatrix} j_1^* & j_3 & q \\ j_5 & j_9 & j \end{vmatrix} \otimes \begin{vmatrix} j_5 & j_1^* & j_4 \\ j_3 & j_9 & q \end{vmatrix} \Big)$$

and applying the formula

$$\sum_{j\in I}\dim(j)*_{j^*j_3j_5}*_{jj_2j_7^*}*_{j^*j_1j_9}\left(\begin{vmatrix} j_3 & j_5 & j\\ j_2 & j_7 & j_6\end{vmatrix}\otimes\begin{vmatrix} j_1^* & j_3 & q\\ j_5 & j_9 & j\end{vmatrix}\otimes\begin{vmatrix} j_2 & j_9 & j_8\\ j_1 & j_7 & j\end{vmatrix}\right)=$$

$$=\sum_{j\in I}\dim(j)*_{j^*j_3j_5}*_{jj_2j_7^*}*_{j^*j_1j_9}\left(\begin{vmatrix} j_1^* & j_3 & q\\ j_5 & j_9 & j\end{vmatrix}\otimes\begin{vmatrix} j_1^* & j & j_9\\ j_2 & j_8 & j_7\end{vmatrix}\otimes\begin{vmatrix} j_3 & j_5 & j\\ j_2 & j_7 & j_6\end{vmatrix}\right)=$$

$$=*_{qj_6j_8^*}\left(\begin{vmatrix} q & j_5 & j_9\\ j_2 & j_8 & j_6\end{vmatrix}\otimes\begin{vmatrix} j_1^* & j_3 & q\\ j_6 & j_8 & j_7\end{vmatrix}\right)$$

we easily obtain that RHS is equal to the same expression as LHS. This completes the proof of the theorem.

5.8. Remark. In the quantum theory of angular momentum one considers also $3j$-symbols, $9j$-symbols, $12j$-symbols, etc., see [BL1], [BL2] and references therein. Although we shall not need them, it is instructive to discuss their possible analogues in our setting. To define analogues of $3j$-symbols we should assume that the objects $\{V_i\}_{i\in I}$ of $\mathcal{V}$ are vector spaces. Fix a basis for each V_i, $i \in I$. Matrix coefficients of homomorphisms $V_k \to V_i \otimes V_j$ may be viewed as linear functionals on H_k^{ij}, i.e., as elements of H_{ij}^k. These elements are analogues of the classical $3j$-symbols. Of course they depend on the choice of the bases of V_i, $i \in I$. Analogues of $9j$-symbols, $12j$-symbols, etc. may be defined in an invariant way without additional assumptions. The standard definitions of these symbols as polynomial expressions of $6j$-symbols directly extend to our setting.

5.9. Exercises. 1. Show that for any $i, k, l \in I$ and any $\varepsilon = \pm 1$,

$$h_i^{kl}\sum_{j\in I}(v_j')^{\varepsilon}\dim(j)*_{ij^*l}\left(\begin{vmatrix} i & k & l\\ i & j & l\end{vmatrix}\right)=h_i^{kl}((v_k')^{-1}v_iv_l)^{\varepsilon}\mathrm{Id}(i,k,l^*).$$

(Hint: substitute $j_6 = 0$ in the Racah identity.)

2. Deduce the Racah identities (5.4.b), (5.4.c) from the Yang-Baxter identity. (Hint: substitute $j_5 = 0$.)

3. Give a proof of Corollary 5.4.4 along the lines of the proof of Theorem 5.7, i.e., without appealing to mirror categories. (Hint: see Chapter I of [Tu13].)

Notes

Section 1. Numerical $6j$-symbols associated to the Lie algebra $sl_2(\mathbb{C})$ are used in mathematical physics in the framework of the quantum theory of angular momentum. They were introduced by Racah [Ra] and Wigner [Wi] in the early 1940's. Both Racah and Wigner realized that the $6j$-symbols reflect associativity of the tensor product of $sl_2(\mathbb{C})$-modules in connection with decompositions into direct sums of irreducible modules. Although there are no fundamental obstructions to dealing with $6j$-symbols associated to arbitrary semisimple Lie algebras, the study of $6j$-symbols has been traditionally limited to the case of $sl_2(\mathbb{C})$. This considerably simplifies definitions and computations because the representation theory of $sl_2(\mathbb{C})$ is exceptionally simple. In particular the corresponding multiplicity modules are either zero or may be identified with $\mathbb{C}$. Therefore the $6j$-symbols associated to $sl_2(\mathbb{C})$ are complex numbers rather than tensors. There are explicit formulas for these numbers as functions of their 6 parameters (see [BL1, page 114]). For more information on the $6j$-symbols associated to $sl_2(\mathbb{C})$ and extensive bibliography, see [BL1], [BL2].

Askey and Wilson [AsW] introduced q-analogues of classical $6j$-symbols in the framework of the theory of q-orthogonal polynomials. It was established by Kirillov and Reshetikhin [KR1] that these (numerical) $6j$-symbols naturally arise in the theory of quantum groups. They appear in the representation theory of the Hopf algebra $U_q(sl_2(\mathbb{C}))$ and play in this setting the same role as the Racah-Wigner $6j$-symbols in the representation theory of $sl_2(\mathbb{C})$.

The definition of $6j$-symbols given in Section 1 directly generalizes the definition of $6j$-symbols associated to the classical and quantum $sl_2(\mathbb{C})$. The Biedenharn-Elliott identity (Theorem 1.5.1) generalizes the identity for $6j$-symbols associated to $sl_2(\mathbb{C})$ discovered independently by L. Biedenharn and J. Elliott in the early 1950's. For "quantum" $6j$-symbols associated to $sl_2(\mathbb{C})$, this identity was established in [KR1].

Section 2. The notion of a unimodal category is new. For modules over quantum groups, the existence of isomorphisms satisfying Lemma 2.2 was established in [RT2] in the case of $U_q(sl_2(\mathbb{C}))$ and in [TW] in the case of quantum groups associated to other classical semisimple Lie algebras.

Section 3. The material of this section is new.

Section 4. The material of this section is new. The definition of $\mathbb{F}$ was inspired by the constructions of [KR1].

Section 5. The geometric definition of normalized $6j$-symbols is new. The orthonormality relation, the Racah identity, and the Yang-Baxter identity generalize identities known for the classical and quantum $6j$-symbols associated to $sl_2(\mathbb{C})$ (see [BL1], [KR1]).

Note that the normalized $6j$-symbols may be defined without appealing to framed graphs. Namely, formula (5.2.b) may be used as the definition of the normalized $6j$-symbol that appears on the right-hand side of (5.2.b).

Chapter VII
Simplicial state sums on 3-manifolds

Fix up to the end of this chapter a strict unimodular category $(\mathcal{V}, \{V_i\}_{i \in I})$ with ground ring K and rank $\mathcal{D} \in K$.

1. State sum models on triangulated 3-manifolds

1.0. Outline. We define a state sum model on any triangulated closed 3-manifold M. The definition goes roughly as follows. The states of the model are colorings of the edges of the triangulation of M with elements of the set I. Having such a state φ, we associate to every tetrahedron T of this triangulation a normalized $6j$-symbol $|T^\varphi|$ determined by the φ-colors of the six edges of T. Multiplying these $6j$-symbols over all tetrahedra T (and further multiplying by a simple numerical factor) we obtain the term $|M|_\varphi \in K$ of the state sum corresponding to φ. To be more precise we should speak of tensor contraction rather than multiplication. The $6j$-symbol $|T^\varphi|$ lies in the tensor product of 4 symmetrized multiplicity modules determined by the 2-faces of T. Every 2-face of (the triangulation of) M is adjacent to exactly two tetrahedra and gives rise to dual factors of the corresponding tensor products. To obtain $|M|_\varphi$ we consider the product $\otimes_T |T^\varphi|$ over all tetrahedra T and contract it along the dual pairs of factors determined by the 2-faces of M.

The main result is independence of the sum $\sum_\varphi |M|_\varphi \in K$ of the choice of triangulation. This sum yields a homeomorphism invariant of M. A similar invariant is defined for compact 3-manifolds with triangulated boundary.

1.1. Normal orientations and colorings. Let M be a triangulated manifold, possibly, with boundary. By a normal orientation of an edge e of M we mean an orientation in the normal vector bundle of e in M. Since this bundle is trivial, e has two (opposite) normal orientations. To specify a normal orientation of e it suffices to specify an orientation in a small disk in M transversally intersecting e in one point.

Denote by $\mathrm{Edg}(M)$ the set of normally oriented edges of M. Inversion of orientation defines an involution $e \mapsto e^*$ in $\mathrm{Edg}(M)$. By a coloring of M we mean a mapping $\varphi : \mathrm{Edg}(M) \to I$ such that $\varphi(e^*) = (\varphi(e))^*$ for any $e \in \mathrm{Edg}(M)$. The set of colorings of M is denoted by $\mathrm{col}(M)$. If M has n non-oriented edges then

$\mathrm{Edg}(M)$ has $2n$ elements and $\mathrm{col}(M)$ has $(\mathrm{card}(I))^n$ elements. Each coloring φ of M gives rise to a dual coloring φ^* of M defined by the formula $\varphi^*(e) = \varphi(e^*)$. Clearly, $\varphi^{**} = \varphi$. The manifold M equipped with a coloring is said to be colored.

It is obvious that the boundary of M inherits a triangulation from M. Each coloring φ of M induces a coloring $\partial\varphi$ of ∂M in the following way. For an edge e of ∂M, every normal orientation α of e in ∂M induces a normal orientation of e in M determined by the pair (the tangent vector directed outside M, α). This gives an imbedding $\mathrm{Edg}(\partial M) \to \mathrm{Edg}(M)$ equivariant under inversion. Composing this imbedding with any coloring $\varphi : \mathrm{Edg}(M) \to I$ we get the coloring $\partial\varphi$ of ∂M.

1.2. Colored tetrahedra and their symbols. Let T be a 3-dimensional tetrahedron (with its standard triangulation consisting of one 3-simplex, four faces, six edges, and four vertices). Suppose that T is colored in the sense of the previous subsection. For each face t of T, we define a K-module H_t as follows. Note first that every edge of t has a natural normal orientation determined by t and T. Namely, if A, B, C are the vertices of t and D is the fourth vertex of T then the normal orientation of the edge AB is determined by the pair of vectors $(\vec{AC}, \vec{AD})$ (or equivalently $(\vec{BC}, \vec{BD})$). Let i, j, k be the (given) colors of the three edges of t with these normal orientations. Set $H_t = H(i, j, k)$. This definition is consistent because the module $H(i, j, k)$ does not depend on the order in the set $\{i, j, k\}$. Denote by $H(T)$ the non-ordered tensor product (over K) of four K-modules associated in this way to the faces of T.

To each colored tetrahedron T we assign an element $|T| \in H(T)$ called the symbol of T. Let γ be the 1-skeleton of the tetrahedron dual to T. Vertices of γ are centers of the faces of T, edges of γ are straight intervals connecting its vertices inside T. The graph γ has 4 vertices and 6 edges. Oriented edges of γ canonically correspond to the normally oriented edges of T. For example, if A, B, C, D are the vertices of T then the oriented edge of γ leading from the center of the face ABD to the center of ABC corresponds to the edge AB with the normal orientation presented by the pair of vectors $(\vec{AD}, \vec{AC})$. In this way the given coloring of T induces a coloring of γ. The graph γ lies in T in a topologically trivial way in the sense that it lies on a 2-sphere embedded in T. Let Γ be the framed graph obtained by thickening γ in this 2-sphere. It is clear that Γ consists of four 2-disks lying in the faces of T and of six untwisted unlinked bands. We may view Γ as the thickened 1-skeleton of the tetrahedron dual to T. Since the core γ of Γ is colored we may consider the modules $H(\Gamma) = H(\gamma)$ and $H^\star(\Gamma) = H^\star(\gamma)$ defined in Sections VI.4.1 and VI.4.2. It follows from definitions that $H(T) = H^\star(\Gamma)$.

Let us embed T in $\mathbb{R}^3$ so that $\Gamma \subset T$ makes a colored framed graph in $\mathbb{R}^3$. Note that there are two isotopy classes of embeddings $T \to \mathbb{R}^3$ obtained from each other by a mirror reflection in $\mathbb{R}^3$. However, the resulted embeddings $\Gamma \to \mathbb{R}^3$ are isotopic because of the planarity of Γ. Therefore we are safe to define

$$|T| = \mathbb{F}(\Gamma) \in H^\star(\Gamma) = H(T).$$

(For the definition of $\mathbb{F}$, see Section VI.4. Note that we continue to adhere to Conventions VI.3.1.)

To have an explicit expression for $|T|$ it is convenient to enumerate the vertices of T. Let i be the color of the edge AB equipped with the normal orientation determined by the pair of vectors $(\vec{AD}, \vec{AC})$. Similarly, let j, k, l, m, n be the colors of the edges BC, AC, CD, AD, BD with the normal orientations determined by the pairs of vectors $(\vec{BD}, \vec{BA})$, $(\vec{AB}, \vec{AD})$, $(\vec{CB}, \vec{CA})$, $(\vec{AC}, \vec{AB})$, $(\vec{BA}, \vec{BC})$ respectively. Then it follows from definitions that

$$|T| = \begin{vmatrix} i & j & k \\ l & m & n \end{vmatrix} \in H(T). \tag{1.2.a}$$

(The reader may draw the corresponding colored framed graph $\Gamma \subset T$ and visualize that it is isotopic to the colored framed graph in Figure VI.5.1.)

In order to apply formula (1.2.a) it is convenient to provide T with the orientation determined by the triple of vectors $(\vec{AB}, \vec{AC}, \vec{AD})$. (Or the other way round, if T is oriented then we choose its vertices A, B, C, D so that the given orientation of T is determined by this triple of vectors.) Now each orientation α of an edge of T gives rise to a normal orientation α' of this edge such that the pair (α, α') determines the given orientation of T. Therefore we may use orientations of the edges of T instead of their normal orientations. It is easy to check that the normal orientations used above correspond to the oriented edges $\vec{BA}, \vec{CB}, \vec{CA}, \vec{DC}, \vec{DA}, \vec{DB}$. This gives the equality in Figure 1.1. Orientations of edges are more handy than normal orientations and will be often used below. However, the language of edge orientations is applicable only when the tetrahedra at hand are oriented.

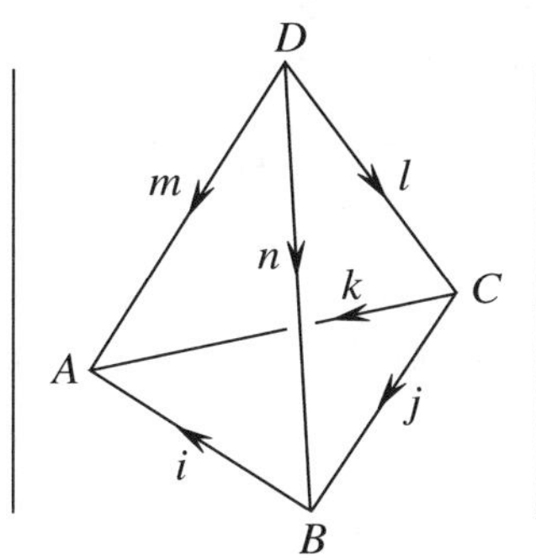

$$= \begin{vmatrix} i & j & k \\ l & m & n \end{vmatrix}$$

Figure 1.1

1.3. Simplicial state sum model on closed 3-manifolds. Let M be a compact triangulated 3-manifold. Consider a coloring φ of M. For each edge e of M, set $\dim_\varphi(e) = \dim(\varphi(\vec{e}))$ and $\dim'_\varphi(e) = \dim'(\varphi(\vec{e}))$ where $\vec{e}$ is the edge e with a certain normal orientation. The equalities $\dim(i) = \dim(i^*)$, $\dim'(i) = \dim'(i^*)$

ensure independence of $\dim_\varphi(e), \dim'_\varphi(e)$ of the choice of normal orientation. Clearly $\dim_\varphi(e) = (\dim'_\varphi(e))^2$.

For each 3-simplex T of M, the coloring φ restricts to a coloring of T. Denote the resulting colored tetrahedra by T^φ. Set

$$H_\varphi = H_\varphi(M) = \bigotimes_T H(T^\varphi)$$

where T runs over all 3-simplices of the given triangulation of M. According to definitions, the module $H(T^\varphi)$ is the non-ordered tensor product of the symmetrized multiplicity modules corresponding to four faces of T^φ.

Assume that $\partial M = \emptyset$. Then each face of (the given triangulation of) M is adjacent to exactly two 3-simplices and contributes two factors to H_φ. These factors correspond to dual triples of colors and therefore they are dual to each other (see Section VI.3.3). Contracting H_φ along all such pairs of factors yields a homomorphism $H_\varphi \to K$. Denote this homomorphism by cntr. Note that the tensor product of $6j$-symbols $|T^\varphi| \in H(T^\varphi)$ over all 3-simplices T of the triangulation of M yields an element $\otimes_T |T^\varphi| \in H_\varphi$. Denote by a the number of vertices of the triangulation of M. Set

$$|M|_\varphi = \mathcal{D}^{-2a}(\prod_e \dim_\varphi(e)) \operatorname{cntr}(\otimes_T |T^\varphi|) \in K$$

where e runs over all (non-oriented) edges of M and T runs over all 3-simplices of M. It is instructive to note that each simplex of M contributes exactly one term in the expression for $|M|_\varphi$. Namely, each vertex contributes $\mathcal{D}^{-2}$, each edge e contributes the factor $\dim_\varphi(e)$, each face gives rise to a contraction along two dual factors of H_φ, and each 3-simplex contributes its symbol.

Consider the state sum (also called the partition function)

$$|M| = |M|_{\mathcal{V}} = \sum_{\varphi \in \operatorname{col}(M)} |M|_\varphi \in K.$$

1.4. Theorem. *For any closed* 3*-manifold M, the state sum $|M|$ does not depend on the choice of triangulation of M.*

This theorem will be proven in Section 2.

Theorem 1.4 yields a K-valued invariant of closed 3-manifolds. Note that the manifolds are not assumed to be oriented or even orientable. It is clear that $|M|$ does not depend on the choice of rank $\mathcal{D}$ of $\mathcal{V}$.

The main advantage of this approach over the one of Chapter II is its intrinsic character: the computation of $|M|$ proceeds inside M and does not appeal to surgery. As we shall see in Section 4, the invariant $|M|$ is closely related to the 3-manifold invariant τ defined in Chapter II.

In Section 1.6 we shall generalize the invariant $|M|$ to compact 3-manifolds with triangulated boundary. The generalized invariant takes value in a certain module associated to the boundary of the manifold at hand. This module is defined in the next subsection.

1.5. Digression on triangulated surfaces. For any triangulated compact surface Σ we define here a projective K-module $\tilde{\mathscr{E}}(\Sigma)$. It will serve as the ambient module of invariants of compact 3-manifolds bounded by Σ.

For each coloring ψ of Σ, we define a K-module $H_\psi(\Sigma)$ as follows. Let t be a 2-simplex of the given triangulation of Σ. Let i, j, k be the ψ-colors of the edges of t with the normal orientation directed "inside t". Set $H_{t,\psi} = H(i, j, k)$ and $H_\psi(\Sigma) = \otimes_t H_{t,\psi}$ where t runs over all faces of Σ. Set

$$\tilde{\mathscr{E}}(\Sigma) = \bigoplus_{\psi \in \mathrm{col}(\Sigma)} H_\psi(\Sigma). \tag{1.5.a}$$

It follows from Lemma II.4.2.1 that $\tilde{\mathscr{E}}(\Sigma)$ is a projective K-module. This module of course depends on the choice of triangulation in Σ.

We shall formally consider the empty set $\emptyset$ as a closed surface with unique triangulation and unique coloring. Set $\tilde{\mathscr{E}}(\emptyset) = K$.

The module $H_{\psi^*}(\Sigma)$ is dual to $H_\psi(\Sigma)$, the duality pairing

$$\langle .,. \rangle_\psi : H_{\psi^*}(\Sigma) \otimes_K H_\psi(\Sigma) \to K$$

being induced by the duality $H(i^*, j^*, k^*) \otimes_K H(i, j, k) \to K$ with $i, j, k \in I$. The duality pairings corresponding to all colorings of Σ determine a bilinear form $\langle .,. \rangle : \tilde{\mathscr{E}}(\Sigma) \otimes_K \tilde{\mathscr{E}}(\Sigma) \to K$ by the formula

$$\langle \bigoplus_{\psi \in \mathrm{col}(\Sigma)} x_\psi, \bigoplus_{\psi \in \mathrm{col}(\Sigma)} y_\psi \rangle = \sum_{\psi \in \mathrm{col}(\Sigma)} \langle x_\psi, y_{\psi^*} \rangle_\psi$$

where $x_\psi, y_\psi \in H_\psi(\Sigma)$ for all $\psi \in \mathrm{col}(\Sigma)$. It is obvious that the form $\langle .,. \rangle$ in $\tilde{\mathscr{E}}(\Sigma)$ is symmetric and non-degenerate. The direct summands $H_\psi(\Sigma)$, $H_{\psi'}(\Sigma)$ of $\tilde{\mathscr{E}}(\Sigma)$ are orthogonal unless $\psi' = \psi^*$ in which case they are coupled via $\langle .,. \rangle_\psi$.

1.6. Simplicial state model on 3-manifolds with boundary. Let M be a triangulated compact 3-manifold with boundary. Let φ be a coloring of M. Consider the tensor product $H_\varphi = \otimes_T H(T^\varphi)$ where T runs over all 3-simplices of the given triangulation of M. Every face of M not lying in ∂M is adjacent to exactly two of these 3-simplices and gives rise to two dual factors in H_φ. Every face t of ∂M gives rise to one factor in H_φ which is nothing but the module $H_{t,\partial\varphi}$ defined in the previous subsection. (Recall that $\partial\varphi$ is the coloring of ∂M induced by φ.) Contracting H_φ along pairs of factors corresponding to the faces of M not lying in ∂M we obtain a homomorphism $H_\varphi \to H_{\partial\varphi}(\partial M) = \otimes_t H_{t,\partial\varphi}$ where t runs over all 2-simplices of ∂M. Denote this homomorphism by cntr. Set

$$|M|_\varphi = \mathcal{D}^{-2a-c}(\prod_e \dim_\varphi(e))(\prod_{e'} \dim'_\varphi(e'))\, \mathrm{cntr}(\otimes_T |T^\varphi|) \in H_{\partial\varphi}(\partial M)$$

where a is the number of vertices of the triangulation of $M \setminus \partial M$, c is the number of vertices of the triangulation of ∂M, e runs over all edges of $M \setminus \partial M$, e' runs over all edges of ∂M, and T runs over all 3-simplices of M. Set

$$|M| = |M|_{\mathcal{V}} = \sum_{\varphi \in \mathrm{col}(M)} |M|_\varphi \in \widetilde{\mathcal{E}}(\partial M).$$

It is obvious that

$$|M| = \oplus_{\psi \in \mathrm{col}(\partial M)} |M, \psi| \tag{1.6.a}$$

where

$$|M, \psi| = \sum_{\varphi \in \mathrm{col}(M),\, \partial\varphi = \psi} |M|_\varphi \in H_\psi(\partial M).$$

1.7. Theorem. *Let M be a compact 3-manifold with triangulated boundary. The state sum $|M| \in \widetilde{\mathcal{E}}(\partial M)$ does not depend on the choice of triangulation of M extending the given triangulation of ∂M.*

This theorem will be proven in Section 2. In view of formula (1.6.a), Theorem 1.7 is equivalent to the assertion that for any $\psi \in \mathrm{col}(\partial M)$, the state sum $|M, \psi|$ does not depend on the choice of triangulation of M extending the given triangulation of ∂M.

Theorem 1.7 yields a topological invariant of compact 3-manifolds with triangulated boundary. One of the most important properties of this invariant is the following multiplicativity theorem. In Section 3 this multiplicativity will be interpreted in the framework of a topological field theory.

1.8. Theorem. *Let M be a compact 3-manifold with triangulated boundary. Let Σ be a closed triangulated surface lying inside M and splitting M into 3-manifolds M_1, M_2 so that $\partial(M_1) = \Sigma \amalg \Sigma_1$, $\partial(M_2) = \Sigma \amalg \Sigma_2$, and $\partial M = \Sigma_1 \amalg \Sigma_2$. Then*

$$|M| = \mathrm{cntr}_\Sigma(|M_1| \otimes |M_2|)$$

where cntr_Σ is the contraction

$$\widetilde{\mathcal{E}}(\Sigma_1) \otimes_K \widetilde{\mathcal{E}}(\Sigma) \otimes_K \widetilde{\mathcal{E}}(\Sigma_2) \otimes_K \widetilde{\mathcal{E}}(\Sigma) \to \widetilde{\mathcal{E}}(\Sigma_1) \otimes_K \widetilde{\mathcal{E}}(\Sigma_2)$$

induced by the form $\langle .,. \rangle$ in $\widetilde{\mathcal{E}}(\Sigma)$.

Theorem 1.8 follows directly from definitions.

1.9. Remarks. 1. It is well known that any triangulation of the boundary of a 3-manifold M extends to a triangulation of M. This extension is non-unique since we may always subdivide it in $\text{Int}(M)$.

2. The definition of $|M|$ seems to be ready for practical computations. Unfortunately, triangulations of 3-manifolds usually involve quite a large number of 3-simplices. For sufficiently complicated 3-manifolds it is rather hard to give an explicit description of a triangulation. From the computational point of view these technical difficulties overshadow the beauty and simplicity of the definition of $|M|$. In the next chapters we shall develop another approach to this invariant based on the theory of skeletons of 3-manifolds. This approach is better suited for computational purposes.

1.10. Exercise. Show that for any closed 3-manifold M, we have $|M|_{\mathcal{V}} = |M|_{\overline{\mathcal{V}}}$.

2. Proof of Theorems 1.4 and 1.7

2.0. Outline. The proof of Theorems 1.4 and 1.7 relies on the theory of bistellar moves on triangulations due to U. Pachner. This theory allows us to connect any two triangulations of the same 3-manifold by a finite sequence of local transformations. We verify that our state sum is preserved under these transformations.

2.1. Bistellar moves. We recall the results of Pachner [Pa1] relating different triangulations of a manifold via so-called bistellar moves.

Let M be a triangulated 3-manifold. For each simplex of the triangulation of M not lying in ∂M, we define the bistellar move along this simplex. This move produces a new triangulation of M. Let T be a 3-dimensional simplex of the triangulation of M with vertices A, B, C, D. The bistellar move along T adds one new vertex O in the center of T and replaces T with four tetrahedra $OABC$, $OABD$, $OACD$, $OBCD$. Let f be a face of the triangulation of M not lying in ∂M. The face f is adjacent to two tetrahedra of M, say $ABCD$ and $A'BCD$, where B, C, D are the vertices of f. The bistellar move along f replaces these two tetrahedra with three tetrahedra $AA'BC$, $AA'BD$, $AA'CD$. This transformation does not add new vertices but adds the edge AA'. The transformations inverse to the bistellar moves along tetrahedra and faces are called bistellar moves along vertices and edges respectively. Thus, there are four types of bistellar moves in dimension 3. By their very definition the bistellar moves do not change the triangulation of the boundary of the 3-manifold.

2.1.1. Theorem (Pachner). *Any two triangulations of a closed 3-manifold can be transformed into one another by a finite sequence of bistellar moves and an ambient isotopy.*

For a proof and generalizations to higher dimensions, see [Pa1]. The proof relies on classical ideas and results due to J. Alexander and M.H.A. Newman. More exactly, Alexander [Al] showed that any two combinatorially equivalent triangulations of a closed 3-manifold may be related by a finite sequence of so-called stellar subdivisions and inverse transformations. Pachner showed that every transformation used by Alexander may be decomposed into a finite sequence of bistellar moves. Note also that according to Moise [Moi] any two triangulations of a 3-manifold (considered up to ambient isotopy) are combinatorially equivalent.

Here is a relative version of Pachner's theorem.

2.1.2. Theorem. *Any two triangulations of a compact 3-manifold M coinciding in ∂M can be transformed into one another by a finite sequence of bistellar moves and an ambient isotopy in M identical on ∂M.*

The proof is obtained by a direct application of Pachner's technique. Instead of the Alexander result mentioned above we should use its relative version established in [TV, Appendix I]. Instead of the Moise theorem mentioned above we should use its relative version also due to Moise: any two triangulations of a 3-manifold M coinciding on ∂M are ambiently isotopic (via isotopies identical on ∂M) to triangulations of M which are combinatorially equivalent to each other.

2.2. Algebraic lemmas

2.2.1. Lemma. *Let Z be a projective K-module and let $\delta_Z \in Z^* \otimes_K Z$ be the element determined by duality (see Section* VI.5.3*). Then the evaluation pairing $y \otimes z \mapsto y(z) : Z^* \otimes_K Z \to K$ transforms δ_Z into* $\mathrm{Dim}(Z) \in K$.

Proof. In the notation of Section I.1.7.1 we have $\mathrm{Dim}(Z) = d_Z \mathrm{Perm}_{Z,Z^*} b_Z(1) \in K$ where $d_Z : Z^* \otimes_K Z \to K$ is the evaluation pairing. Therefore the lemma follows from the obvious equality $\delta_Z = \mathrm{Perm}_{Z,Z^*} b_Z(1)$. (Here is an alternative proof applicable to free modules. Choose a basis $\{z_j\}_j$ of Z and the dual basis $\{z^j\}_j$ of Z^*. It is obvious that the evaluation pairing carries $\delta_Z = \sum_j z^j \otimes z_j$ into $\sum_j z^j(z_j) = \sum_j 1 = \mathrm{Dim}(Z)$.)

2.2.2. Lemma. *Let Z be a projective K-module and $\delta_Z \in Z^* \otimes_K Z$ be the element determined by duality. For any K-module V and any $a \in V \otimes_K Z$, the evaluation pairing $Z \otimes_K Z^* \to K$ transforms $a \otimes \delta \in V \otimes_K Z \otimes_K Z^* \otimes_K Z$ into a.*

A graphical proof along the lines of Section I.1 is given in Figure 2.1.

2.3. Proof of Theorem 1.4. Let M_0 be the manifold M endowed with a triangulation. It is obvious that the state sum $|M_0|$ is invariant under ambient isotopies of the triangulation. In view of Theorem 2.1.1 it remains to check that $|M_0|$ is

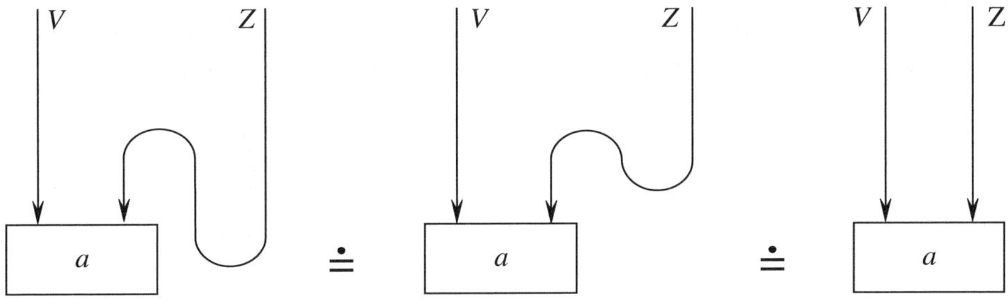

Figure 2.1

invariant under bistellar moves on M_0. Consider first the bistellar move along a 3-simplex T of M_0. Let A, B, C, D be the vertices of T. The bistellar move replaces T with 4 subtetrahedra

$$T_1 = OABC,\ T_2 = OABD,\ T_3 = OBCD,\ T_4 = OACD$$

where O is the center of T. Denote the manifold M endowed with the resulting triangulation by M_1. Fix a coloring φ of M_0. It is clear that φ induces a coloring of all edges of M_1 except OA, OB, OC, OD. Let Φ be the set of colorings of M_1 extending φ. We shall show that

$$|M_0|_\varphi = \sum_{\psi \in \Phi} |M_1|_\psi. \tag{2.3.a}$$

This would imply that $|M_0| = |M_1|$.

We compare different factors on the right-hand and left-hand sides of (2.3.a). It suffices to prove that

$$|T^\varphi| = \mathcal{D}^{-2} \sum_{\psi \in \Phi|_T} \big(\dim_\psi(OA) \dim_\psi(OB) \dim_\psi(OC) \dim_\psi(OD) \times \tag{2.3.b}$$

$$\times \operatorname{cntr}(|T_1^\psi| \otimes |T_2^\psi| \otimes |T_3^\psi| \otimes |T_4^\psi|)\big)$$

where cntr denotes the tensor contraction along the factors corresponding to the faces $OAB, OAC, OAD, OBC, OBD, OCD$ and $\Phi|_T$ denotes the set of colorings of $T_1 \cup T_2 \cup T_3 \cup T_4$ extending φ. Since this equality is a local one we may pass from the language of normal edge orientations to the language of edge orientations as described in Section 1.2. With this view we provide T and its subtetrahedra $T_1 - T_4$ with the orientation determined by the triple of vectors $\vec{AB}, \vec{AC}, \vec{AD}$. To compute $|T^\varphi|$ we shall use formula (1.2.a). Let $j_1, j_2, j_5, j_8, j_0, j_7$ be the φ-colors of the oriented edges $\vec{BA}, \vec{CB}, \vec{CA}, \vec{DC}, \vec{DA}, \vec{DB}$. (The choice of notation for the colors will become clear below.) It follows from (1.2.a) that

$$|T^\varphi| = \begin{vmatrix} j_1 & j_2 & j_5 \\ j_8 & j_0 & j_7 \end{vmatrix}. \tag{2.3.c}$$

Similarly, let $j = j(\psi)$, $j_3 = j_3(\psi)$, $j_4 = j_4(\psi)$, $j_6 = j_6(\psi)$ be the ψ-colors of the edges $\vec{OB}, \vec{OC}, \vec{DO}, \vec{OA}$ respectively. It follows from (1.2.a) that

$$|T_1^\psi| = \begin{vmatrix} j_1 & j_2 & j_5 \\ j_3 & j_6 & j \end{vmatrix}, \quad |T_2^\psi| = \begin{vmatrix} j_1 & j & j_6 \\ j_4 & j_0 & j_7 \end{vmatrix},$$

$$|T_3^\psi| = \begin{vmatrix} j_2 & j_3 & j \\ j_4 & j_7 & j_8 \end{vmatrix}, \quad |T_4^\psi| = \begin{vmatrix} j_6 & j_3^* & j_5 \\ j_8 & j_0 & j_4 \end{vmatrix}.$$

Set

$$\xi(j_0, \dots, j_8) = \sum_{j\in I} \dim(j) *_{j^* j_2 j_3} *_{j j_4 j_7^*} *_{j j_1 j_6^*} \big(|T_1^\psi| \otimes |T_2^\psi| \otimes |T_3^\psi|\big).$$

Substituting here the expressions for $|T_1^\psi|, |T_2^\psi|, |T_3^\psi|$ given above and applying Theorem VI.5.4.1, we get

$$\xi(j_0, \dots, j_8) = *_{j_0^* j_5 j_8}\Big(\begin{vmatrix} j_5 & j_3 & j_6 \\ j_4 & j_0 & j_8 \end{vmatrix} \otimes \begin{vmatrix} j_1 & j_2 & j_5 \\ j_8 & j_0 & j_7 \end{vmatrix} \Big).$$

Therefore the sum on the right-hand side of (2.3.b) is equal to

$$\sum_{j_3, j_4, j_6 \in I} \dim(j_3)\dim(j_4)\dim(j_6) *_{j_3 j_4 j_8^*} *_{j_3 j_5 j_6^*} *_{j_0^* j_4 j_6} \Big(\xi(j_0, \dots, j_8) \otimes \begin{vmatrix} j_6 & j_3^* & j_5 \\ j_8 & j_0 & j_4 \end{vmatrix}\Big)$$

$$= \sum_{j_3, j_4, j_6 \in I} \dim(j_3)\dim(j_4)\dim(j_6) *_{j_3 j_4 j_8^*} *_{j_3 j_5 j_6^*} *_{j_0^* j_4 j_6} *_{j_0^* j_5 j_8}\Big(\begin{vmatrix} j_5 & j_3 & j_6 \\ j_4 & j_0 & j_8 \end{vmatrix} \otimes$$

$$\otimes \begin{vmatrix} j_1 & j_2 & j_5 \\ j_8 & j_0 & j_7 \end{vmatrix} \otimes \begin{vmatrix} j_6 & j_3^* & j_5 \\ j_8 & j_0 & j_4 \end{vmatrix} \Big).$$

Using the tetrahedral symmetry we may rewrite the first and third $6j$-symbols in the last expression as follows:

$$\begin{vmatrix} j_5 & j_3 & j_6 \\ j_4 & j_0 & j_8 \end{vmatrix} = \begin{vmatrix} j_5^* & j_0 & j_8 \\ j_4^* & j_3 & j_6 \end{vmatrix}, \quad \begin{vmatrix} j_6 & j_3^* & j_5 \\ j_8 & j_0 & j_4 \end{vmatrix} = \begin{vmatrix} j_5 & j_0^* & j_8^* \\ j_4 & j_3^* & j_6^* \end{vmatrix}.$$

Now we are in position to apply the orthonormality relation involving the first and third $6j$-symbols and the contractions $*_{j_3 j_5 j_6^*}$, $*_{j_0^* j_4 j_6}$. This relation shows that the last sum is equal to

$$\text{(2.3.d)} \qquad \sum_{j_3, j_4 \in I} \dim(j_3)\dim(j_4)(\dim(j_8))^{-1} *_{j_3 j_4 j_8^*} *_{j_0^* j_5 j_8}\Big(\begin{vmatrix} j_1 & j_2 & j_5 \\ j_8 & j_0 & j_7 \end{vmatrix} \otimes$$

$$\otimes \mathrm{Id}(j_0^*, j_5, j_8) \otimes \mathrm{Id}(j_3, j_4, j_8^*)\Big).$$

The contraction $*_{j_3 j_4 j_8^*}$ in this expression applies to $\mathrm{Id}(j_3, j_4, j_8^*)$. Lemma 2.2.1 shows that

$$*_{j_3 j_4 j_8^*}(\mathrm{Id}(j_3, j_4, j_8^*)) = \mathrm{Dim}(H(j_3, j_4, j_8^*)).$$

The contraction $*_{j_0^* j_5 j_8}$ in (2.3.d) matches the factor $H(j_0^*, j_5, j_8)$ involved in the ambient module of the $6j$-symbol (2.3.c) and the factor $H(j_0, j_5^*, j_8^*)$ involved in $\mathrm{Id}(j_0^*, j_5, j_8)$. Lemma 2.2.2 implies that

$$*_{j_0^* j_5 j_8}\left(\begin{vmatrix} j_1 & j_2 & j_5 \\ j_8 & j_0 & j_7 \end{vmatrix} \otimes \mathrm{Id}(j_0^*, j_5, j_8)\right) = \begin{vmatrix} j_1 & j_2 & j_5 \\ j_8 & j_0 & j_7 \end{vmatrix}.$$

We conclude that the right-hand side of (2.3.b) is equal to

$$\mathscr{D}^{-2}\left(\sum_{j_3, j_4 \in I} \dim(j_3)\dim(j_4)(\dim(j_8))^{-1}\mathrm{Dim}(H(j_3, j_4, j_8^*)) \right) \begin{vmatrix} j_1 & j_2 & j_5 \\ j_8 & j_0 & j_7 \end{vmatrix}.$$

Formula (II.4.5.f) shows that this expression is equal to the $6j$-symbol (2.3.c). This implies (2.3.b).

Let us prove the invariance of $|M_0|$ under the bistellar move along a face BCD. Let $ABCD$ and $A'BCD$ be the tetrahedra of M_0 adjacent to BCD. (For the sake of visual convenience, we may think that these two tetrahedra form a convex body in M containing the interval AA'.) Replacing $ABCD$ and $A'BCD$ with tetrahedra $AA'BC$, $AA'CD$, $AA'BD$ we get a new triangulation M_1 which is the result of the bistellar move on M_0 along BCD. We should prove that $|M_1| = |M_0|$. This formula follows from an equality of local contributions which corresponds to the generalized Biedenharn-Elliott identity (Theorem VI.5.4.1). To establish these correspondence it suffices to denote by $j, j_0, j_1, j_2, j_3, j_4, j_5, j_6, j_7, j_8$ the colors of the oriented edges

$$\vec{AA'}, \vec{DC}, \vec{A'C}, \vec{BA'}, \vec{AB}, \vec{DA}, \vec{BC}, \vec{AC}, \vec{DA'}, \vec{DB}.$$

This completes the proof of the theorem.

2.4. Proof of Theorem 1.7. Proof of Theorem 1.7 is analogous to the proof of Theorem 1.4. The only difference is that instead of Theorem 2.1.1, we should use Theorem 2.1.2.

2.5. Remark. The proof of Theorems 1.4 and 1.7 only uses a few properties of $6j$-symbols namely the tetrahedral symmetry, the generalized Biedenharn-Elliott identity, and the orthonormality. Any system of tensors with these properties would give invariants of 3-manifolds. Moreover, if we apply this approach to oriented 3-manifolds then the tetrahedral symmetry may be restricted to those permutations which preserve orientation of the tetrahedron. Systems of $6j$-symbols with such properties may be extracted from the theory of subfactors (see [Oc], [DJN], [EK]).

3. Simplicial 3-dimensional TQFT

3.0. Outline. We reformulate the simplicial state sum invariant introduced in Section 1 as an operator invariant of 3-cobordisms. This reformulation allows us to construct a TQFT $(\mathscr{E}, |...|)$ based on closed surfaces and compact 3-manifolds (without any additional structure). The construction proceeds in several steps starting off with 3-cobordisms with triangulated bases. As we shall see in Section 4 the TQFT $(\mathscr{E}, |...|)$ is closely related to 3-dimensional TQFT's introduced in Chapter IV.

3.1. Reformulation of the invariant $|M|$. Let $M = (M, \partial_- M, \partial_+ M)$ be a 3-dimensional cobordism with triangulated bases. Thus, M is a compact 3-manifold and $\partial_- M, \partial_+ M$ are triangulated disjoint closed subsurfaces of ∂M whose union equals ∂M. Consider the state sum invariant

$$|M| \in \tilde{\mathscr{E}}(\partial M) = \tilde{\mathscr{E}}(\partial_- M) \otimes_K \tilde{\mathscr{E}}(\partial_+ M) = \mathrm{Hom}_K(\tilde{\mathscr{E}}(\partial_- M), \tilde{\mathscr{E}}(\partial_+ M)).$$

The last isomorphism is induced by the non-degenerate form $\langle .,. \rangle$ in $\tilde{\mathscr{E}}(\partial_- M)$. Therefore the invariant $|M|$ gives rise to a K-homomorphism $\tilde{\mathscr{E}}(\partial_- M) \to \tilde{\mathscr{E}}(\partial_+ M)$. This homomorphism is denoted by $\tilde{e}(M) = \tilde{e}(M, \partial_- M, \partial_+ M)$. It is clear that $\tilde{e}(M)$ is natural with respect to homeomorphisms of 3-cobordisms preserving the triangulation of the boundary. It follows from definitions that the homomorphism $\tilde{e}$ corresponding to the same cobordism M regarded upside down is dual to $\tilde{e}(M)$:

$$\tilde{e}(M, \partial_+ M, \partial_- M) = (\tilde{e}(M, \partial_- M, \partial_+ M))^\star.$$

It follows from Theorem 1.8 that the homomorphism $\tilde{e}$ is multiplicative with respect to gluing of cobordisms: if a triangulated closed surface $\Sigma \subset \mathrm{Int}(M)$ splits M into two cobordisms $(M_1, \partial_- M, \Sigma)$ and $(M_2, \Sigma, \partial_+ M)$ then $\tilde{e}(M) = \tilde{e}(M_2)\, \tilde{e}(M_1)$.

3.2. The module $\mathscr{E}$ for triangulated surfaces. The functor assigning the module $\tilde{\mathscr{E}}(\Sigma)$ to any triangulated closed surface Σ and assigning the homomorphism $\tilde{e}(M)$ to any 3-cobordism M looks like a topological quantum field theory. However this functor does not satisfy the normalization axiom which requires the homomorphism assigned to the cylinder over a surface to be the identity. In order to fulfill this axiom we introduce a submodule of $\tilde{\mathscr{E}}(\Sigma)$ preserved by $\tilde{e}$.

Let Σ be a closed surface with a triangulation α. Consider the cylinder cobordism $\mathrm{Id}_{(\Sigma,\alpha)} = (\Sigma \times [0,1], \Sigma \times 0, \Sigma \times 1)$ with the triangulation of both $\Sigma \times 0$ and $\Sigma \times 1$ induced by α. Consider the endomorphism $\tilde{e}(\mathrm{Id}_{(\Sigma,\alpha)})$ of $\tilde{\mathscr{E}}(\Sigma, \alpha)$ determined by this cobordism. Set

(3.2.a) $$\mathscr{E}(\Sigma, \alpha) = \mathrm{Im}(\tilde{e}(\mathrm{Id}_{(\Sigma,\alpha)})) \subset \tilde{\mathscr{E}}(\Sigma, \alpha).$$

It is obvious that the module $\mathscr{E}(\Sigma, \alpha)$ is natural with respect to triangulation-preserving homeomorphisms of surfaces. This module is multiplicative with respect to disjoint union of surfaces. Indeed, it follows from definitions that for any disjoint closed surfaces Σ_1, Σ_2 with triangulations α_1, α_2, we have

$$\tilde{\mathscr{E}}((\Sigma_1, \alpha_1) \amalg (\Sigma_2, \alpha_2)) = \tilde{\mathscr{E}}(\Sigma_1, \alpha_1) \otimes_K \tilde{\mathscr{E}}(\Sigma_2, \alpha_2)$$

and

$$\tilde{e}(\mathrm{Id}_{(\Sigma_1 \amalg \Sigma_2, \alpha_1 \amalg \alpha_2)}) = \tilde{e}(\mathrm{Id}_{(\Sigma_1, \alpha_1)}) \otimes_K \tilde{e}(\mathrm{Id}_{(\Sigma_2, \alpha_2)}).$$

Therefore

(3.2.b) $$\mathscr{E}((\Sigma_1, \alpha_1) \amalg (\Sigma_2, \alpha_2)) = \mathscr{E}(\Sigma_1, \alpha_1) \otimes_K \mathscr{E}(\Sigma_2, \alpha_2).$$

3.2.1. Lemma. *For any 3-dimensional cobordism M with triangulated bases,*

$$\tilde{e}(M)(\mathscr{E}(\partial_- M)) \subset \mathscr{E}(\partial_+ M).$$

Proof. Denote the homomorphism $\tilde{e}(M) : \tilde{\mathscr{E}}(\partial_- M) \to \tilde{\mathscr{E}}(\partial_+ M)$ by f. Glue the cylinder $\partial_+ M \times [0, 1]$ to M along $\partial_+ M = \partial_+ M \times 0$ to get a new cobordism between $\partial_- M$ and $\partial_+ M$. It is obvious that this cobordism is homeomorphic to M via a homeomorphism identical on the bases. Therefore $\tilde{e}(\mathrm{Id}_{\partial_+ M}) f = f$. Hence $\mathrm{Im}(f) \subset \mathscr{E}(\partial_+ M)$.

Lemma 3.2.1 implies that restricting the homomorphism $\tilde{e}(M)$ to $\mathscr{E}(\partial_- M) \subset \tilde{\mathscr{E}}(\partial_- M)$ we get a K-homomorphism $\mathscr{E}(\partial_- M) \to \mathscr{E}(\partial_+ M)$. It is denoted by $e(M) = e(M, \partial_- M, \partial_+ M)$. Multiplicativity of $\tilde{e}(M)$ with respect to gluing of cobordisms implies a similar multiplicativity of $e(M)$. The next lemma shows that the module $\mathscr{E}(\Sigma, \alpha)$ is projective and that the endomorphism of this module induced by the cylinder cobordism is the identity.

3.2.2. Lemma. *For any closed surface Σ with triangulation α, the module $\mathscr{E}(\Sigma, \alpha)$ is projective and $e(\mathrm{Id}_{(\Sigma, \alpha)}) = \mathrm{id}_{\mathscr{E}(\Sigma, \alpha)}$.*

Proof. Set $E = \tilde{\mathscr{E}}(\Sigma, \alpha)$ and $p = \tilde{e}(\mathrm{Id}_{(\Sigma, \alpha)}) : E \to E$. Since the gluing of the cylinder $\mathrm{Id}_{(\Sigma, \alpha)}$ to itself along a base produces another copy of the same cylinder, the results of Section 3.1 imply that $p^2 = p$. Therefore p acts as the identity in $\mathscr{E}(\Sigma, \alpha) = \mathrm{Im}(p)$ which proves the second claim of the lemma.

It follows from definitions and Lemma II.4.2.1 that the module E is projective. Therefore projectivity of $\mathscr{E}(\Sigma, \alpha) = \mathrm{Im}(p)$ follows from the equality $E = \mathrm{Ker}(p) \oplus \mathrm{Im}(p)$.

3.2.3. Remark. It is instructive to note that the splitting $\tilde{\mathscr{E}}(\Sigma, \alpha) = \mathscr{E}(\Sigma, \alpha) \oplus \mathrm{Ker}(p)$ constructed in the proof of the previous lemma is orthogonal with respect to the non-degenerate bilinear form $\langle ., . \rangle$ in $\tilde{\mathscr{E}}(\Sigma, \alpha)$. Indeed, the cylinder cobordism $\mathrm{Id}_{(\Sigma,\alpha)}$ turned upside down yields a homeomorphic cobordism. Therefore the operator p is self-dual, $p^\star = p$. In other words, $\langle x, p(y)\rangle = \langle p(x), y\rangle$ for any $x, y \in \tilde{\mathscr{E}}(\Sigma, \alpha)$. This implies our claim. Restricting the form $\langle .,.\rangle$ to $\mathscr{E}(\Sigma, \alpha)$ we get a K-valued non-degenerate symmetric bilinear form in $\mathscr{E}(\Sigma, \alpha)$.

3.3. The module $\mathscr{E}$ for non-triangulated surfaces. The functors $(\Sigma \mapsto \mathscr{E}(\Sigma), M \mapsto e(M))$ form a TQFT based on triangulated surfaces and 3-cobordisms with triangulated boundary. In order to get rid of triangulations we employ essentially the same idea as in Section IV.6: we identify the modules of states corresponding to different triangulations of the same surface. This will allow us to assign a projective module $\mathscr{E}(\Sigma)$ to any closed topological surface Σ (without any additional stucture). We construct $\mathscr{E}(\Sigma)$ as follows. For each triangulation α of Σ consider the module $\mathscr{E}(\Sigma, \alpha)$ defined in Section 3.2. For any triangulations α_0, α_1 of Σ, consider the 3-cobordism with triangulated bases $(\Sigma \times [0, 1], \Sigma \times 0, \Sigma \times 1)$ where the triangulations of $\Sigma \times 0$ and $\Sigma \times 1$ are induced by α_0 and α_1 via the identifications $\Sigma \times 0 = \Sigma$, $\Sigma \times 1 = \Sigma$. Denote by $\varphi(\alpha_0, \alpha_1)$ the homomorphism $\mathscr{E}(\Sigma, \alpha_0) \to \mathscr{E}(\Sigma, \alpha_1)$ determined by this cobordism. The results of Sections 3.1 and 3.2 imply the following:

(i) for any triangulation α of Σ, we have $\varphi(\alpha, \alpha) = \mathrm{id}_{\mathscr{E}(\Sigma,\alpha)}$,

(ii) for any triangulations $\alpha_0, \alpha_1, \alpha_2$ of Σ, we have

$$\varphi(\alpha_0, \alpha_2) = \varphi(\alpha_1, \alpha_2)\, \varphi(\alpha_0, \alpha_1).$$

Applying the property (ii) to $\alpha_2 = \alpha_0$ we obtain that

(iii) $\varphi(\alpha_0, \alpha_1)$ and $\varphi(\alpha_1, \alpha_0)$ are mutually inverse isomorphisms.

Identify the modules $\{\mathscr{E}(\Sigma, \alpha)\}_\alpha$ corresponding to all triangulations of Σ along the intertwining isomorphisms $\varphi(\alpha_0, \alpha_1)$ corresponding to all pairs of triangulations. This yields a module $\mathscr{E}(\Sigma)$ independent of the choice of triangulation. Elements of these modules are families $\{x(\alpha) \in \mathscr{E}(\Sigma, \alpha)\}_\alpha$ such that for any two triangulations α_0, α_1 of Σ, we have $x(\alpha_1) = \varphi(\alpha_0, \alpha_1)(x(\alpha_0))$. The formula $\{x(\alpha)\}_\alpha \mapsto x(\alpha_0)$ defines a canonical isomorphism $\mathscr{E}(\Sigma) \to \mathscr{E}(\Sigma, \alpha_0)$ for every triangulation α_0 of Σ. Lemma 3.2.2 implies that the module $\mathscr{E}(\Sigma)$ is projective.

The construction $\Sigma \mapsto \mathscr{E}(\Sigma)$ is natural with respect to homeomorphisms of surfaces. Indeed, let $g : \Sigma \to \Sigma'$ be a homeomorphism of (closed) surfaces. For any triangulation α of Σ, we may consider its direct image $g(\alpha)$ on Σ' which is a triangulation of Σ'. The homeomorphism g induces in the obvious way an isomorphism $\mathscr{E}(\Sigma, \alpha) \to \mathscr{E}(\Sigma', g(\alpha))$. These isomorphisms corresponding to all α commute with the intertwining isomorphisms constructed above. (Indeed, any triangulation A of $\Sigma \times [0, 1]$ induces a triangulation $A' = (g \times \mathrm{id}_{[0,1]})(A)$ of $\Sigma' \times [0, 1]$. If A relates triangulations α_0, α_1 of Σ then A' relates triangulations $g(\alpha_0), g(\alpha_1)$ of Σ'. This implies commutativity of the corresponding diagram of isomorphisms.) Therefore the isomorphisms $\{\mathscr{E}(\Sigma, \alpha) \to \mathscr{E}(\Sigma', g(\alpha))\}_\alpha$ induce an isomorphism $\mathscr{E}(\Sigma) \to \mathscr{E}(\Sigma')$. It is denoted by $g_\#$. It is clear that $(\mathrm{id}_\Sigma)_\# = \mathrm{id}_{\mathscr{E}(\Sigma)}$

and that $(gg')_\# = g_\# g'_\#$ whenever g, g' are composable homeomorphisms of closed surfaces.

The isomorphisms (3.2.b) corresponding to different triangulations obviously commute with the intertwining isomorphisms. Therefore they induce a natural isomorphism

$$\mathscr{E}(\Sigma_1 \amalg \Sigma_2) = \mathscr{E}(\Sigma_1) \otimes_K \mathscr{E}(\Sigma_2). \tag{3.3.a}$$

Summing up, we conclude that the correspondence $(\Sigma \mapsto \mathscr{E}(\Sigma), g \mapsto g_\#)$ is a modular functor in the sense of Section III.1. Here the underlying space-structure $\mathfrak{A}$ assigns a one-point set to closed topological surfaces (in particular to $\emptyset$) and the empty set to other topological spaces.

The next two assertions show that the action $g_\#$ of a homeomorphism g depends only on its isotopy class.

3.3.1. Lemma. *Let Σ be a closed topological surface endowed with a triangulation α. Let h be a self-homeomorphism of Σ isotopic to the identity. Then the homomorphism $\mathscr{E}(\Sigma, \alpha) \to \mathscr{E}(\Sigma, h(\alpha))$ induced by h is equal to $\varphi(\alpha, h(\alpha))$.*

Proof. Let α_0, α_1 be arbitrary triangulations of Σ and let A be a triangulation of the cylinder $\Sigma \times [0,1]$ extending $(\alpha_0 \times 0) \amalg (\alpha_1 \times 1)$. Since h is isotopic to identity there exists a homeomorphism $H : \Sigma \times [0,1] \to \Sigma \times [0,1]$ which is the identity on $\Sigma \times 0$ and such that $H(x \times 1) = h(x) \times 1$ for any $x \in \Sigma$. Using A to compute $\varphi(\alpha_0, \alpha_1)$ and using $H(A)$ to compute $\varphi(\alpha_0, h(\alpha_1))$ we easily conclude that $\varphi(\alpha_0, h(\alpha_1)) = \tilde{h} \circ \varphi(\alpha_0, \alpha_1)$ where $\tilde{h}$ is the homomorphism $\mathscr{E}(\Sigma, \alpha_1) \to \mathscr{E}(\Sigma, h(\alpha_1))$ induced by h. Applying this formula to $\alpha_0 = \alpha_1 = \alpha$ and using the equality $\varphi(\alpha, \alpha) = \mathrm{id}$ we get the claim of the lemma.

3.3.2. Corollary. *The action $g_\# : \mathscr{E}(\Sigma) \to \mathscr{E}(\Sigma')$ of a homeomorphism of closed surfaces $g : \Sigma \to \Sigma'$ depends only on the isotopy class of g.*

Proof. Lemma 3.3.1 implies that any self-homeomorphism of a closed surface Σ isotopic to the identity id_Σ acts in $\mathscr{E}(\Sigma)$ as the identity. This implies the corollary.

3.4. TQFT ($\mathscr{E}$, |...|). We can now construct a topological quantum field theory for compact topological 3-manifolds. The modular functor of this theory is the modular functor $\mathscr{E}$ constructed in Section 3.3. For any compact 3-dimensional cobordism $M = (M, \partial_- M, \partial_+ M)$ we define a K-homomorphism

$$|M| = |M, \partial_- M, \partial_+ M| : \mathscr{E}(\partial_- M) \to \mathscr{E}(\partial_+ M)$$

as follows. Triangulate the bases of this cobordism via certain triangulations α_-, α_+, consider the homomorphism

$$e(M, (\partial_- M, \alpha_-), (\partial_+ M, \alpha_+)) : \mathscr{E}(\partial_- M, \alpha_-) \to \mathscr{E}(\partial_+ M, \alpha_+),$$

and compose it with the canonical isomorphisms

$$\mathscr{E}(\partial_- M, \alpha_-) = \mathscr{E}(\partial_- M), \qquad \mathscr{E}(\partial_+ M, \alpha_+) = \mathscr{E}(\partial_+ M).$$

The resulting homomorphism $\mathscr{E}(\partial_- M) \to \mathscr{E}(\partial_+ M)$ does not depend on the choice of α_-, α_+. This follows from the definition of the intertwining isomorphisms $\varphi(\alpha_0, \alpha_1)$ and the multiplicativity of e with respect to gluing of cobordisms.

It is straightforward to verify that the pair $(\mathscr{E}, |...|)$ satisfies all axioms of TQFT's formulated in Section III.1. Here the role of the space-structure $\mathfrak{B}$ (resp. $\mathfrak{A}$) is played by the functor assigning a one-point set to compact topological 3-manifolds (resp. to closed topological surfaces) and the empty set to other topological spaces. This TQFT does not have anomalies.

Although the operator $|M|$ does not depend on the choice of triangulation in M, we need to triangulate M in order to compute this operator. This is why we call this TQFT a simplicial TQFT (it is also called a lattice TQFT). It would be interesting to give a direct definition of $|M|$ without appealing to triangulations.

For any closed 3-manifold M considered as a cobordism between two copies of the empty space, the homomorphism $|M|$ is multiplication in $\mathscr{E}(\emptyset) = K$ by the invariant $|M| \in K$ defined in Section 1.3.

The TQFT $(\mathscr{E}, |...|)$ may be extended to surfaces with marked arcs and 3-cobordisms with framed graphs inside. For a simplicial construction of such an extension, see [Tu13]; cf. Section X.7.

3.5. Remark. Elements of the module $\mathscr{E}(\Sigma)$ associated to a closed surface Σ may be represented by geometric objects in Σ. Fix a triangulation ζ of Σ. Consider a coloring ψ of (Σ, ζ) and a function h assigning to each 2-face t of ζ a certain $h_t \in H_{t,\psi}$, cf. Section 1.5. Such a pair (ψ, h) determines an element $\otimes_t h_t \in H_\psi(\Sigma) \subset \tilde{\mathscr{E}}(\Sigma)$. The image of $\otimes_t h_t$ under the projection $\tilde{\mathscr{E}}(\Sigma) \to \mathscr{E}(\Sigma)$ should be viewed as the element of $\mathscr{E}(\Sigma)$ represented by (ζ, ψ, h). Of course, there are non-trivial relations between the elements determined by different triples (ζ, ψ, h).

Instead of triangulations we may use the dual language of trivalent graphs (see Section VI.4.1). Each triangulation ζ of Σ gives rise to a trivalent graph $\gamma = \gamma_\zeta \subset \Sigma$ dual to the 1-skeleton of ζ. The vertices of γ are the barycenters of the 2-faces of ζ and the edges of γ connect the barycenters of adjacent 2-faces. (The reader unfamiliar with this construction should draw a few pictures.) Each coloring ψ of ζ induces a dual coloring $\psi^\perp$ of γ by the formula $\psi^\perp(e) = \psi(e^\perp)$ where e is an oriented edge of γ and $e^\perp$ is the dual normally oriented edge of ζ. The formula $\psi \mapsto \psi^\perp$ establishes a bijective correspondence between the colorings of ζ and the colorings of γ. It is obvious that $H_\psi(\Sigma) = H(\gamma, \psi^\perp)$. Therefore instead of the pairs (ψ, h) as above we may simply speak of v-colorings of γ. Thus, every v-coloring of γ determines an element of $\mathscr{E}(\Sigma)$. More generally, any v-colored trivalent graph in Σ (not necessarily dual to a triangulation of Σ) determines an element of $\mathscr{E}(\Sigma)$ (see [Tu13]).

This description of the module $\mathscr{E}(\Sigma)$ resembles the well known description of simplicial homologies in terms of cycles and boundaries. The triples (ζ, ψ, h) as above (or more generally the v-colored trivalent graphs in Σ) are the analogues of cycles, the roles of the boundary operator and homologies are played by the projector $\tilde{\mathscr{E}}(\Sigma) \to \tilde{\mathscr{E}}(\Sigma)$ and its image.

4. Comparison of two approaches

4.0. Outline. We have developed two different approaches to 3-dimensional TQFT's: one based on surgery (see Chapters II and IV) and one based on $6j$-symbols and simplicial state sums (see Chapters VI and VII). Here we discuss relationships between these two approaches.

The theorems stated in this section will be proven in Chapter X.

4.1. The case of closed oriented 3-manifolds. For any closed oriented 3-manifold M, we have two invariants $|M| = |M|_{\mathscr{V}}$ (see Section 1) and $\tau(M) = \tau_{\mathscr{V}}(M)$ (see Chapter II). Both invariants lie in the ground ring K of $\mathscr{V}$. The invariant $|M|$ is definitely weaker because it does not distinguish M from the same manifold with opposite orientation $-M$. The following theorem relates these invariants.

4.1.1. Theorem. *For any closed oriented* 3*-manifold M, we have*

$$|M| = \tau(M)\,\tau(-M).$$

Theorem 4.1.1 shows that the simplicial state sum based on $6j$-symbols computes $\tau(M)\,\tau(-M)$. When $\mathscr{V}$ is a Hermitian modular category we have

$$|M| = \tau(M)\,\tau(-M) = \tau(M)\,\overline{\tau(M)}.$$

When $\mathscr{V}$ is a unitary modular category, we have

$$|M| = \tau(M)\,\tau(-M) = \tau(M)\,\overline{\tau(M)} = |\tau(M)|^2$$

so that $|M|$ computes the square of the absolute value of $\tau(M)$. Theorem 4.1.1 implies that for any unitary unimodular category $\mathscr{V}$, we have $|M|_{\mathscr{V}} \geq 0$.

Theorem 4.1.1 will be proven in Sections X.3 - X.6. The ideas involved in the proof may be outlined as follows. The heart of the proof is a computation of $\tau(M)$ in terms of an arbitrary compact oriented piecewise-linear 4-manifold W bounded by M. We use 2-dimensional skeletons of W, i.e., simple 2-polyhedra in W such that W may be obtained from their regular neighborhoods by attaching 3-handles and 4-handles. Each such 2-polyhedron $X \subset W$ may be naturally decomposed as a disjoint union of 0-strata (vertices), 1-strata, and 2-strata. We endow all 2-strata of X with their self-intersection numbers in W. These numbers are called gleams of the 2-strata, the polyhedron X whose 2-strata are endowed with the gleams is said to be shadowed. We introduce a state sum model on X based on the normalized $6j$-symbols associated to $\mathcal{V}$. This model incorporates the gleams and yields $\tau(M)$. This gives a computation of $\tau(M)$ in terms of W.

The subtle point is the independence of the state sum on X of the choices of X and W with a given $\partial W = M$. We first fix the 4-manifold W and describe local transformations relating arbitrary 2-skeletons of W. This description uses the Matveev-Piergallini moves on simple 2-polyhedra. We define slightly more elaborate shadow moves on shadowed 2-polyhedra. It turns out that the shadowed 2-skeleton $X \subset W$ is determined by W up to shadow moves and a so-called stabilization. To describe possible changes in W with $\partial W = M$ we involve 5-dimensional cobordisms between 4-manifolds. This leads us to the notion of cobordism for shadowed 2-polyhedra. Finally, we show that the shadowed 2-polyhedron X is determined by M up to shadow moves, stabilization, and cobordism. Then we prove that the state sum on X is invariant under these transformations. This implies independence of this state sum of the choices of X and W. To compute this state sum as a function of M we choose W to be the 4-manifold W_L obtained as in Section II.2 from a surgery presentation $L \subset S^3$ of M. Every diagram of L gives rise to a 2-skeleton $X \subset W_L$. The state sum model on X may be computed in terms of the diagram. This allows us to relate this model to the operator invariants introduced in Chapter I and to prove that the state sum on $X \subset W_L$ equals $\tau(M)$.

To prove Theorem 4.1.1 we apply this approach to the 4-manifold $W = M \times [-1, 1]$. We choose as a skeleton of W the 2-skeleton $X \subset M = M \times 0$ of the cell decomposition of M dual to a triangulation of M. The state sum model on X is closely related to the state sum model considered in Section 1 and yields $|M|$. The results stated above imply that this state sum model gives $\tau(\partial W) = \tau(M \amalg (-M)) = \tau(M)\,\tau(-M)$. This implies Theorem 4.1.1.

The proof of Theorem 4.1.1 is based on the notion of a shadow introduced and studied in Chapters VIII - X. The reader willing to simplify his way towards the proof of Theorem 4.1.1 may read Sections VIII.1, VIII.2.1, VIII.2.2, VIII.6, IX.1 and then proceed to Chapter X coming back to Chapters VIII and IX when necessary.

4.2. Theorem. *Suppose that the ground ring K of $\mathcal{V}$ is a field of zero characteristic. Then the simplicial TQFT $(\mathcal{E}, |\ldots|)$ restricted to oriented closed surfaces and oriented compact 3-cobordisms is isomorphic to the absolute anomaly-free TQFT $(\widehat{\mathcal{T}}, \widehat{\tau})$ restricted to oriented closed surfaces (with void families of distinguished arcs) and oriented compact 3-cobordisms (with empty ribbon graphs).*

This theorem is one of the main results of this monograph relating the surgery and simplicial approaches to 3-dimensional TQFT's. The main advantage of the TQFT $(\mathscr{E}, |...|)$ lies in its intrinsic definition. The construction of the module $\mathscr{E}(\Sigma)$ proceeds entirely inside $\Sigma \times [0, 1]$ and the homomorphism $|M|$ is computed entirely inside M. Another strong point of the simplicial approach is its applicability to non-orientable 3-manifolds. On the other hand, the simplicial approach produces a weaker invariant and applies only under the unimodality condition on the modular category.

It is tempting to deduce Theorem 4.2 from Theorem 4.1.1 using the abstract nonsense of Section III.3. Such a deduction requires the non-degeneracy of at least one of the TQFT's in question. The author does not know if these TQFT's (restricted as in the statement of the theorem) are non-degenerate. To circumvent this problem we shall generalize Theorems 4.1.1 and 4.2 to the setting of framed graphs in 3-manifolds. This will eventually allow us to use the non-degeneracy of the TQFT $(\mathscr{T}^e, \tau^e)$ and to apply Theorem III.3.7.

It seems plausible that the condition on the ground ring in Theorem 4.2 may be relaxed.

4.3. Comparison of modular functors. Theorem 4.2 implies that for oriented closed surfaces (with void families of distinguished arcs), the simplicial modular functor $\mathscr{E} = \mathscr{E}_{\mathcal{V}}$ defined in Section 3 is isomorphic to the modular functor $\widehat{\mathscr{T}} = \widehat{\mathscr{T}}_{\mathcal{V}}$ defined in Section IV.8. In particular, for any oriented closed surface Σ, we may compute the module $\mathscr{E}(\Sigma)$ in terms of the modules of states defined in Chapter IV. Namely, if Σ is endowed with a Lagrangian space $\lambda \subset H_1(\Sigma; \mathbb{R})$ then

$$\mathscr{E}_{\mathcal{V}}(\Sigma) = \widehat{\mathscr{T}}_{\mathcal{V}}(\Sigma) = \mathscr{T}^e_{\mathcal{V}}(\Sigma, \lambda) \otimes_K \mathscr{T}^e_{\overline{\mathcal{V}}}(\Sigma, \lambda). \tag{4.3.a}$$

This decomposition is natural with respect to orientation-preserving homeomorphisms of surfaces. (We do not assume that the homeomorphisms preserve λ, for the definition of the action of homeomorphisms in $\mathscr{T}^e$, see Section IV.6.3.) We may combine (4.3.a) with the following lemma.

4.3.1. Lemma. *For an oriented closed surface Σ with distinguished Lagrangian space $\lambda \subset H_1(\Sigma; \mathbb{R})$ (and with void family of distinguished arcs), there is a natural K-isomorphism*

$$\mathscr{T}^e_{\overline{\mathcal{V}}}(\Sigma, \lambda) = (\mathscr{T}^e_{\mathcal{V}}(\Sigma, \lambda))^\star = \mathrm{Hom}_K(\mathscr{T}^e_{\mathcal{V}}(\Sigma, \lambda), K). \tag{4.3.b}$$

Lemma 4.3.1 implies that

$$\mathscr{E}_{\mathcal{V}}(\Sigma) = \mathscr{T}^e_{\mathcal{V}}(\Sigma, \lambda) \otimes_K (\mathscr{T}^e_{\mathcal{V}}(\Sigma, \lambda))^\star = \mathrm{End}(\mathscr{T}^e_{\mathcal{V}}(\Sigma, \lambda)). \tag{4.3.c}$$

Note that the algebra $\mathrm{End}(\mathscr{T}^e_{\mathcal{V}}(\Sigma, \lambda))$ does not depend on the choice of λ. This follows from (4.3.c) or from the fact that the modules $\mathscr{T}^e_{\mathcal{V}}(\Sigma, \lambda)$ corresponding to different Lagrangian spaces are canonically isomorphic (see Section IV.6). The equality $\mathscr{E}_{\mathscr{V}}(\Sigma) = \mathrm{End}(\mathscr{T}^e_{\mathscr{V}}(\Sigma, \lambda))$ gives a geometric interpretation of the algebra $\mathrm{End}(\mathscr{T}^e_{\mathscr{V}}(\Sigma, \lambda))$.

Similarly, if Σ is endowed with a parametrization by a standard surface (with a void family of distinguished arcs) then

$$\mathscr{E}_{\mathscr{V}}(\Sigma) = \mathscr{T}_{\mathscr{V}}(\Sigma) \otimes_K \mathscr{T}_{\overline{\mathscr{V}}}(\Sigma) = \mathscr{T}_{\mathscr{V}}(\Sigma) \otimes_K (\mathscr{T}_{\mathscr{V}}(\Sigma))^\star = \mathrm{End}(\mathscr{T}_{\mathscr{V}}(\Sigma)).$$

The isomorphism of TQFT's provided by Theorem 4.2 and hence the isomorphism $\mathscr{E}_{\mathscr{V}}(\Sigma) = \mathrm{End}(\mathscr{T}^e_{\mathscr{V}}(\Sigma, \lambda))$ are not explicit. Recall the additive generators of $\mathscr{E}_{\mathscr{V}}(\Sigma)$ represented by v-colored triangulations of the surface Σ or, in the dual language, by v-colored trivalent graphs in Σ (cf. Remark 3.5). Conjecturally, the endomorphism of $\mathscr{T}^e_{\mathscr{V}}(\Sigma, \lambda)$ determined by a v-colored trivalent graph $\gamma \subset \Sigma$ coincides (up to a scalar factor) with the endomorphism of $\mathscr{T}^e_{\mathscr{V}}(\Sigma, \lambda)$ defined as follows. Regard a narrow neighborhood of γ in $\Sigma = \Sigma \times 0$ as a v-colored framed graph in $\Sigma \times [-1, 1]$. Ribboning this framed graph we obtain a v-colored ribbon graph in $\Sigma \times [-1, 1]$. The cylinder $\Sigma \times [-1, 1]$ with this v-colored ribbon graph inside is an extended 3-cobordism between two copies of (Σ, λ). Its operator invariant τ^e defined in Section IV.6 yields an endomorphism of $\mathscr{T}^e_{\mathscr{V}}(\Sigma, \lambda)$.

Proof of Lemma. Recall the non-degenerate anomaly-free TQFT $(\mathscr{T}^e, \tau^w) = (\mathscr{T}^e_{\mathscr{V}}, \tau^w_{\mathscr{V}})$ based on the cobordism theory $(\mathfrak{B}^w, \mathfrak{A}^e)$ formed by e-surfaces and weighted extended 3-manifolds (see Section IV.9). We define another TQFT based on the same cobordism theory. For an e-surface Υ, denote by $\neg\Upsilon$ the same e-surface with the opposite orientation (we keep the distinguished arcs with their orientations and marks). For a weighted extended 3-cobordism $M = (M, \partial_-(M), \partial_+(M))$, denote by $\neg M = (\neg M, \neg\partial_-(M), \neg\partial_+(M))$ the same cobordism with the opposite orientations in M and in ∂M (we keep the embedded v-colored ribbon graph and the weight). It is easy to verify that the rule $(\Upsilon \mapsto \mathscr{T}^e(\neg\Upsilon), M \mapsto \tau^w(\neg M))$ defines a TQFT based on $(\mathfrak{B}^w, \mathfrak{A}^e)$. Denote this TQFT by $(\mathscr{T}', \tau')$. Since the TQFT $(\mathscr{T}^e, \tau^w)$ is non-degenerate and anomaly-free, the TQFT $(\mathscr{T}', \tau')$ is also non-degenerate and anomaly-free. For any weighted extended 3-manifold M with empty boundary, we have

$$\tau'(M) = \tau^w(\neg M) = \tau^w_{\overline{\mathscr{V}}}(M) \tag{4.3.d}$$

where the last equality follows from formula (II.2.5.a). Note that the definition of the space-structures $\mathfrak{A}^e$ and $\mathfrak{B}^w$ involves only the underlying category of $\mathscr{V}$ (with its tensor product and duality) and does not involve the braiding and twist. Therefore the TQFT $(\mathscr{T}^e_{\overline{\mathscr{V}}}, \tau^w_{\overline{\mathscr{V}}})$ is based on the same cobordism theory $(\mathfrak{B}^w, \mathfrak{A}^e)$. Theorem III.3.3 and the identity (4.3.d) imply that the TQFT's $(\mathscr{T}^e_{\overline{\mathscr{V}}}, \tau^w_{\overline{\mathscr{V}}})$ and $(\mathscr{T}', \tau')$ are isomorphic. In particular, for any e-surface Υ, there is a K-isomorphism $\mathscr{T}^e_{\overline{\mathscr{V}}}(\Upsilon) = \mathscr{T}'(\Upsilon)$. By Theorem III.3.3, this isomorphism is natural with respect to e-homeomorphisms. Moreover, it is easy to show (using Lemma IV.6.8.2) that this isomorphism is also natural with respect to weak

e-homeomorphisms. When the family of distinguished arcs in Υ is void we have $\neg\Upsilon = -\Upsilon$ so that

$$\mathcal{T}^e_{\overline{\mathcal{V}}}(\Upsilon) = \mathcal{T}'(\Upsilon) = \mathcal{T}^e_{\mathcal{V}}(\neg\Upsilon) = \mathcal{T}^e_{\mathcal{V}}(-\Upsilon).$$

The last module is isomorphic to $(\mathcal{T}^e_{\mathcal{V}}(\Upsilon))^\star = \mathrm{Hom}_K(\mathcal{T}^e_{\mathcal{V}}(\Upsilon), K)$ via the isomorphism induced by the self-duality pairing defined in Section III.2. (This isomorphism is natural with respect to arbitrary orientation-preserving homeomorphisms of surfaces, cf. Exercise IV.7.3.) Combining these isomorphisms we get $\mathcal{T}^e_{\overline{\mathcal{V}}}(\Upsilon) = (\mathcal{T}^e_{\mathcal{V}}(\Upsilon))^\star$. Applying this equality to $\Upsilon = (\Sigma, \lambda)$ we get (4.3.b).

4.4. The case of non-oriented 3-manifolds. The results of the previous subsections may be extended to the non-orientable setting. For simplicity, we restrict ourselves to the case of closed 3-manifolds.

Let M be a closed connected non-orientable 3-manifold. Note first that the surgery invariant τ is not defined for non-orientable 3-manifolds. Therefore to compute $|M|$ via τ we have to appeal to other 3-manifolds. Consider the oriented 2-sheeted covering $\overline{M}$ of M determined by the first Stiefel-Whitney class $w^1(M) \in H^1(M; \mathbb{Z}/2\mathbb{Z})$. The manifold $\overline{M}$ is orientable. Moreover, the covering involution $\overline{M} \to \overline{M}$ inverts the orientation in $\overline{M}$ so that the orientation in $\overline{M}$ is unique up to homeomorphism. The following theorem is the analogue of Theorem 4.1.1 in the non-orientable setting.

4.4.1. Theorem. *For any closed connected non-orientable* 3*-manifold M, we have* $|M| = \tau(\overline{M})$.

Theorem 4.4.1 will be proven in Section X.3.

4.5. Exercise. Show (without the assumption that K is a field) that for any closed oriented surface Σ equipped with a Lagrangian subspace in 1-homologies, we have $\mathrm{Dim}(\mathcal{E}(\Sigma)) = (\mathrm{Dim}(\mathcal{T}^e(\Sigma)))^2$.

Notes

Sections 1 and 2. The state sum model for $|M|$ is similar to the lattice model of Ponzano and Regge [PR] known in physics literature in the context of Regge calculus. Ponzano and Regge used numerical $6j$-symbols associated to the Lie algebra $sl_2(\mathbb{C})$. These symbols are infinite in number and do not provide a finite model. (Indeed, the category of representations of $sl_2(\mathbb{C})$ is semisimple but not modular.) The attention of physicists concerned with the Ponzano-Regge model was focused on the asymptotic form, continuum limit, and physical interpretations of the corresponding state sum, see e.g. [HP], [Le], [Oo1]. A finite state sum model was introduced by Turaev and Viro [TV]. It is based on the $6j$-symbols associated with $U_q(sl_2(\mathbb{C}))$ where q is a complex root of unity. The state sum model introduced in Section 1 generalizes the model of [TV]. For other generalizations and reformulations of this model, see [DJN], [FG], [Ka6], [KL].

Section 3. The construction of the simplicial TQFT generalizes the construction introduced in [TV].

Section 4. Theorem 4.1.1 was independently obtained by the author (see [Tu7] - [Tu9]) and Walker [Wa] in the case of $6j$-symbols associated to $U_q(sl_2(\mathbb{C}))$ where q is a complex root of unity. Theorem 4.4.1 in this case is contained in [Tu8]. Theorem 4.2 is new.

Chapter VIII
Generalities on shadows

Fix, up to the end of the chapter, an abelian group A and an element $\omega \in A$.

1. Definition of shadows

1.0. Outline. We begin a study of two-dimensional geometric objects called shadows. The idea of a shadow is inspired by the properties of 2-skeletons of 4-manifolds. As we shall see in Chapter IX, every compact oriented piecewise-linear 4-manifold gives rise to a shadow.

1.1. Simple 2-polyhedra. The role of underlying topological spaces in the theory of shadows is played by so-called simple 2-polyhedra. The class of simple 2-polyhedra includes compact surfaces as well as 2-dimensional polyhedra with singular points. The singularities in question are rather mild and locally standard; they can not be destroyed by small deformations of polyhedra. By polyhedra we mean finite (compact) simplicial spaces.

We first define simple 2-polyhedra without boundary. A 2-dimensional polyhedron X is called a simple 2-polyhedron without boundary if every point of X has a neighborhood homeomorphic either to

(1.1.1) the plane $\mathbb{R}^2$, or to

(1.1.2) the union of three half-planes in $\mathbb{R}^3$ meeting along their common boundary line, or to

(1.1.3) the cone over the 1-skeleton of a 3-dimensional simplex.

More generally, a 2-dimensional polyhedron X is called a simple 2-polyhedron (with boundary) if each point of X has a neighborhood homeomorphic to one of the spaces (1.1.1), (1.1.2), (1.1.3), or to

(1.1.4) the halfplane $\mathbb{R}^2_+$, or to

(1.1.5) the union of three copies of the positive quadrant $\{(x, y) \in \mathbb{R}^2 : x \geq 0, y \geq 0\}$ meeting each other along the half-line $x = 0, y \geq 0$ (cf. Figure 1.1).

For example, any compact surface is a simple 2-polyhedron. A more elaborate example of a simple 2-polyhedron may be obtained by gluing a compact surface Σ to a compact surface Σ' along a generic immersion $\partial\Sigma \to \mathrm{Int}(\Sigma') = \Sigma' \setminus \partial\Sigma'$.

The points of a simple 2-polyhedron X which have neighborhoods homeomorphic to $\mathbb{R}^2$ form a surface, denoted by $\mathrm{Int}(X)$. Connected components of $\mathrm{Int}(X)$ are called 2-strata or regions of X. Compactness of X implies that the number of regions of X is finite. We say that X is orientable (oriented) if $\mathrm{Int}(X)$ is orientable (oriented). It is clear that any orientation of $\mathrm{Int}(X)$ induces orientations of all regions of X.

For a simple 2-polyhedron X, the set $X\setminus\mathrm{Int}(X)$ is a graph in the sense of Section VI.4.1. Its vertices are those points of X which either have a neighborhood of type (1.1.3) and correspond to the cone point or have a neighborhood of type (1.1.5) and correspond to the point $x = y = 0$. Each vertex of $X \setminus \mathrm{Int}(X)$ has valency four, i.e., it is incident to four edges of $X\setminus\mathrm{Int}(X)$ counted with multiplicities. The vertices and 1-strata of the graph $X \setminus \mathrm{Int}(X)$ will be called vertices and 1-strata of X. Some components of $X\setminus\mathrm{Int}(X)$ may be homeomorphic to the circle. These components do not have vertices.

The graph $X \setminus \mathrm{Int}(X)$ splits as a union of two subgraphs:

$$X \setminus \mathrm{Int}(X) = \mathrm{sing}(X) \cup \partial X.$$

The graph $\mathrm{sing}(X)$ consists of those points of X which have no neighborhoods homeomorphic to $\mathbb{R}^2$ or $\mathbb{R}^2_+$. The graph ∂X is called the boundary of X and consists of those points of X which have no neighborhoods homeomorphic to the spaces (1.1.1), (1.1.2), or (1.1.3). This graph is trivalent. The intersection $\mathrm{sing}(X) \cap \partial X$ coincides with the set of vertices of ∂X and equals the set of 1-valent vertices of $\mathrm{sing}(X)$. The graph ∂X has a collar in X, i.e., a neighborhood homeomorphic to $\partial X \times [0, 1]$. It is clear that X is a simple polyhedron without boundary in the sense specified above if and only if $\partial X = \emptyset$.

It is clear that the closure $\overline{Y}$ of a region $Y \subset X$ consists of Y and several vertices and 1-strata of X. Two regions Y_1, Y_2 are said to be adjacent if there is a 1-stratum of X contained both in $\overline{Y_1}$ and $\overline{Y_2}$.

It is obvious that any homeomorphism of simple polyhedra $X \to X'$ carries $\mathrm{Int}(X)$, ∂X, and $\mathrm{sing}(X)$ homeomorphically onto $\mathrm{Int}(X')$, $\partial X'$, and $\mathrm{sing}(X')$.

1.2. Shadowed 2-polyhedra. A shadowed polyhedron over (the fixed abelian group) A is an orientable simple 2-polyhedron such that each of its regions is endowed with an element of A. The element of A assigned to a region Y is called the gleam of Y and denoted by $\mathrm{gl}(Y)$.

By homeomorphisms of shadowed polyhedra we shall mean homeomorphisms of the underlying polyhedra preserving the gleams of regions.

The gleam of a region should be viewed as a sort of global measure of this region. Sometimes it is convenient to think that the gleam is concentrated in certain parts of the region or even in certain points of the region. This leads to the following convention always used in our pictures: if certain elements of A are attached to disjoint parts of a region then the gleam of this region equals the sum of these elements.

1.3. Basic shadow moves. We define three local moves P_1, P_2, P_3 on shadowed polyhedra over A. These moves are shown in Figure 1.1 where the symbols $a_1, a_2, a_3, \ldots, a, b, c, e, f$ stand for the gleams (or parts of gleams) of the corresponding regions. The symbol 0 stands for the zero of A. The move P_1 changes the way in which the bottom half-plane is glued to the rest of the polyhedron. After the application of P_1 this half-plane is glued along the curve drawn in bold. Note that the resulting 2-polyhedron can not be embedded into $\mathbb{R}^3$. The move P_2 (resp. P_3) pushes the top half-plane across a 1-stratum (resp. a vertex) of the polyhedron. For visual convenience, one may think that the moves P_2, P_3 proceed inside a 3-ball.

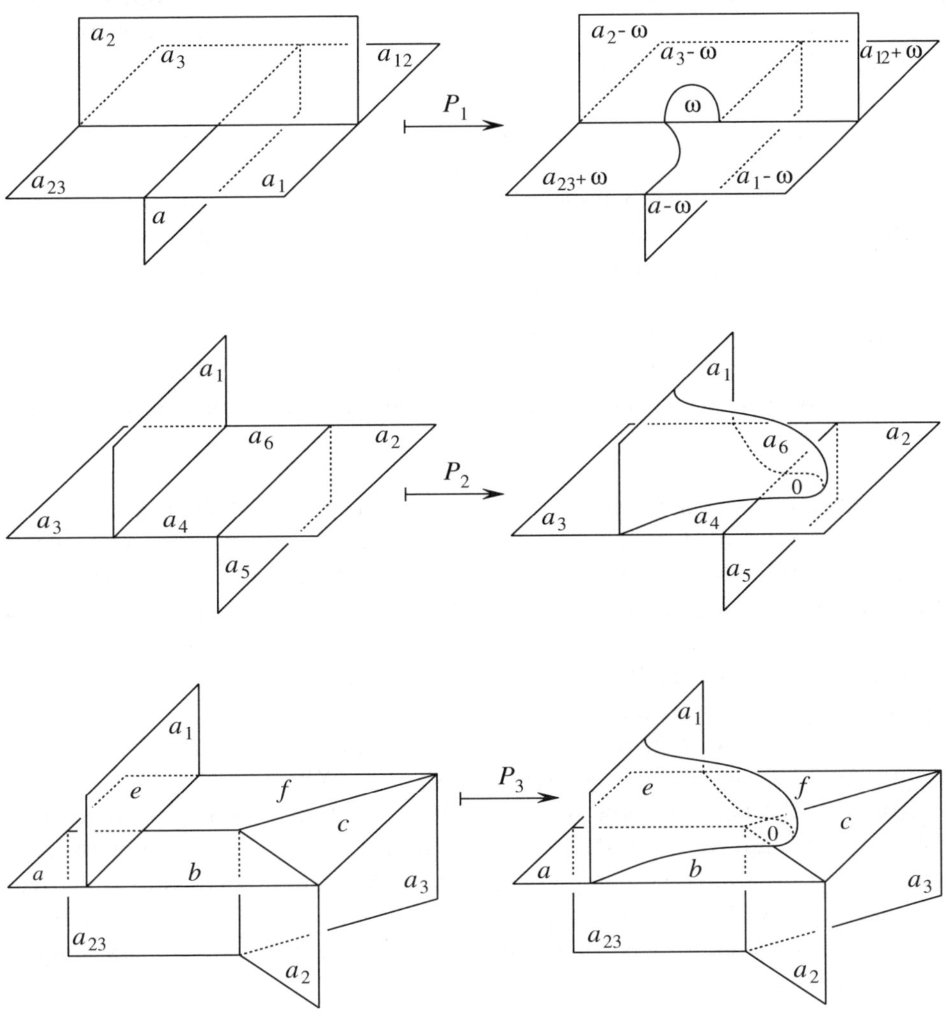

Figure 1.1

Figure 1.1 depicts local patterns for the moves P_1, P_2, P_3. When applying these moves one should keep in mind that certain distinct regions in the local models of Figure 1.1 may actually lie in the same region of a shadowed polyhedron. The convention of Section 1.2 concerned with the gleams should be always followed. For example, if we apply P_1 so that the domains marked by a_1, a_2 belong to one "big" region, then the total contribution of these two domains to the gleam of this region equals $a_1 + a_2$ before the move and $a_1 + a_2 - 2\omega$ after it. An example concerning P_2: the domains marked by a_4, a_6 belong to one region before the move and contribute $a_4 + a_6$ to the gleam of this region. After the move these two domains belong either to one region or two different regions. In the first case they contribute $a_4 + a_6$ to the gleam of the ambient region, in the second case they contribute a_4 and a_6 to the gleams of the ambient regions. This shows that the move P_2 is not determined solely by the underlying geometric transformation and may depend on the specific splitting of $a_4 + a_6 \in A$ as the sum of two elements $a_4, a_6 \in A$.

Each of the moves P_1, P_2, P_3 creates a new region which is an open disk with the gleam ω in the case of P_1 and zero gleam in the case of P_2, P_3. The moves P_2, P_3 do not change the total gleam whilst P_1 decreases it by ω.

The moves P_1, P_2, P_3 have obvious unique inverses denoted by $P_1^{-1}, P_2^{-1}, P_3^{-1}$. The geometric transformation underlying P_2^{-1} may break the orientability of simple 2-polyhedra. We shall speak of the move P_2^{-1} and apply this move to shadowed polyhedra only when it preserves orientability.

The moves $P_1^{\pm 1}, P_2^{\pm 1}, P_3^{\pm 1}$ are called basic shadow moves. All basic moves are performed away from the boundary of shadowed polyhedra and do not change the boundary.

1.4. Shadows. Two shadowed 2-polyhedra over A are called shadow equivalent if they can be obtained from each other by a homeomorphism and a sequence of basic moves $P_1^{\pm 1}, P_2^{\pm 1}, P_3^{\pm 1}$ performed in the class of shadowed 2-polyhedra over A. Equivalence classes of shadowed 2-polyhedra are called shadows over A. In order to emphasize the role of the distinguished element $\omega \in A$, which enters the definition of shadows via P_1, we shall sometimes speak of shadows over the pair (A, ω).

Notation: the shadow equivalence class of a shadowed polyhedron X is denoted by $[X]$. We shall say that X represents the shadow $[X]$.

Note that the basic shadow moves preserve the simple homotopy type of 2-polyhedra. Therefore, each shadow has a well-defined simple homotopy type which is just the simple homotopy type of the underlying 2-polyhedra. In particular, shadows have homotopy groups, homologies, and other standard topological invariants of polyhedra. This allows one to talk about connected shadows, simply-connected shadows, etc.

Since the basic moves preserve the boundary of shadowed polyhedra, each shadow α has a boundary $\partial\alpha$. Clearly, $\partial\alpha$ is a trivalent graph.

An important class of shadows is formed by compact orientable surfaces with gleams. Every compact connected orientable surface Σ endowed with a gleam $a \in A$ is a shadowed polyhedron. It is denoted by Σ_a. This polyhedron has no singular points so that the basic moves can not be applied to Σ_a. Hence, the shadow equivalence class $[\Sigma_a]$ consists solely of shadowed surfaces homeomorphic to Σ_a.

1.5. Remarks. 1. The geometric transformations of simple 2-polyhedra underlying the basic moves P_1, P_2, P_3 were studied by Matveev [Ma1] and Piergallini [Pi]. They showed that two simple 2-polyhedra may be related by such transformations if and only if they may be obtained from each other by a 3-deformation, i.e., a sequence of n-cell extensions and n-cell contractions with $n \leq 3$. Thus, the theory of shadows over the trivial group $A = 0$ is equivalent to the theory of simple 2-polyhedra considered up to 3-deformations.

2. If $\overline{A}$ is an abelian group containing A then each shadowed polyhedron over A may be considered as a shadowed polyhedron over $\overline{A}$. In this way, each shadow over (A, ω) gives rise to a shadow over $(\overline{A}, \omega)$. More generally, if $f : A \to \overline{A}$ is a group homomorphism of abelian groups, then any shadow α over (A, ω) in the obvious way produces a shadow $f_*(\alpha)$ over $(\overline{A}, f(\omega))$. For instance, if $\overline{A} = 0$, then $f_*(\alpha)$ is the underlying 2-polyhedron of α considered up to the transformations of simple 2-polyhedra underlying $P_1 - P_3$.

2. Miscellaneous definitions and constructions

2.0. Outline. We discuss a few technical points of the theory of shadows. The definitions and results of the first three subsections will be used systematically in what follows. The material of Sections 2.4 - 2.6 is less important and may be skipped.

2.1. Operations on shadows. We define three simple operations on shadows: negation, capping a circle component of the boundary, and summation.

Negation associates to any shadow α the "opposite" shadow $-\alpha$ obtained by negation of all gleams. More exactly, applying the involution $a \mapsto -a : A \to A$ to the gleams of regions of a shadowed polyhedron X we get the opposite shadowed polyhedron $-X$. The correspondence $X \mapsto -X$ commutes with P_2, P_3 and transforms P_1 into a composition of P_2 and P_1^{-1}, see Figure 2.1. (To simplify Figure 2.1 we assume that the pieces of X shown in the initial picture contribute zero to the gleams of the regions. In other words, these gleams are concentrated outside of the picture. In view of our conventions this does not spoil the generality of the argument. A similar simplification will be applied in the figures to follow.)

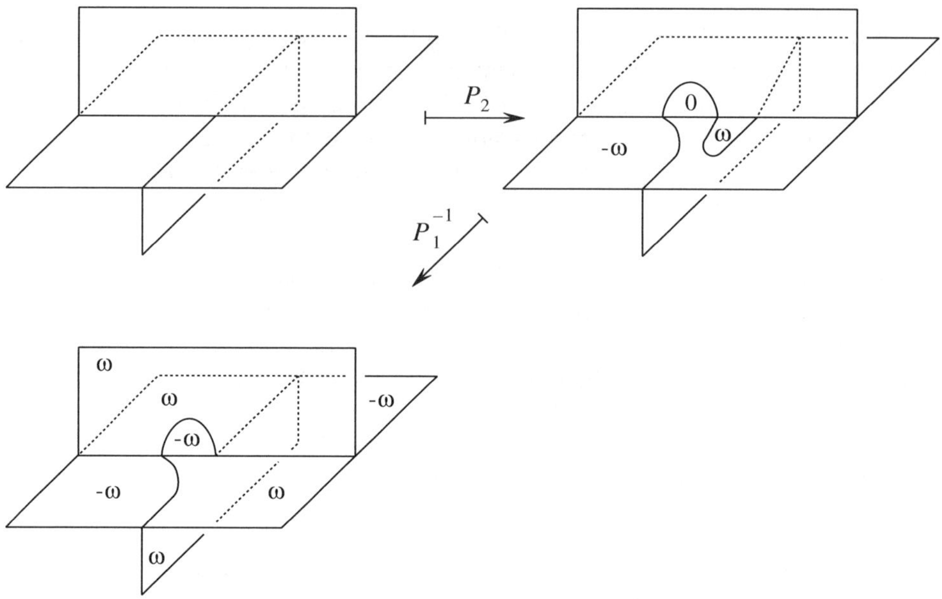

Figure 2.1

Therefore, the correspondence $X \mapsto -X$ preserves the shadow equivalence. For the shadow $\alpha = [X]$, set $-\alpha = [-X]$.

Let X be a shadowed 2-polyhedron and S be a circle component of ∂X. We form a shadowed 2-polyhedron $X' = X \cup_S D^2$ by gluing of the 2-disk D^2 to X along a homeomorphism $\partial D^2 \to S$. It is understood that all regions of X keep their gleams under the passage to X'. The region of X' containing D^2 inherits the gleam of the region of X adjacent to S. We say that X' is obtained from X by capping the circle S. Capping commutes with the basic moves so that we may apply it to shadows.

For any connected shadows α, β, we define their sum $\alpha + \beta$ as follows. Let X, Y be shadowed polyhedra representing α and β respectively. Consider the shadowed 2-polyhedron $X + Y$ obtained from the disjoint union $X \amalg Y$ by identifying two closed disks $D \subset \text{Int}(X)$ and $D' \subset \text{Int}(Y)$ via a homeomorphism $f : D \to D'$. It is clear that the open 2-disk $\text{Int}(D) = \text{Int}(D')$ is a region of $X+Y$. Its gleam is defined to be 0. The gleams of other regions of $X + Y$ are equal to the gleams of the corresponding regions of $X \amalg Y$ so that the gleam of any region of $X \amalg Y$ is equal to the gleam of its image in $X + Y$.

2.1.1. Lemma. *The shadow equivalence class of $X + Y$ does not depend on the choice of D, D', and f.*

This lemma guarantees that the shadow class of $X+Y$ depends only on $\alpha = [X]$ and $\beta = [Y]$. Set $\alpha + \beta = [X + Y]$. The summation of shadows is commutative and associative. For a shadow α and an integer $m \geq 1$, we shall denote by $m\alpha$ the sum of m copies of α.

Proof of Lemma. First consider the case when X is an (orientable) surface. The independence of $[X+Y]$ on the choice of D is clear since any two disks on a connected surface are ambient isotopic. The independence of $[X+Y]$ on the choice of f follows from the fact that the pair (X, D) admits a self-homeomorphism inverting orientation of D.

Consider the case $\operatorname{sing}(X) \neq \emptyset$. To show independence of $[X+Y]$ on the choice of D (with a fixed D') it suffices to consider the sums $X+Y$ corresponding to choices of D in adjacent regions of X. These sums may be related by a composition of P_2 and P_2^{-1}. Namely, if Z_1, Z_2, Z_3 are three regions of X attached to a 1-stratum of $\operatorname{sing}(X)$ then the sum $X+Y$ corresponding to $D \subset Z_2$ is obtained from the sum $X+Y$ corresponding to $D \subset Z_1$ by pushing Z_3 across D. This "pushing" is a composition of P_2 and P_2^{-1}.

Figure 2.2 shows schematically that using P_2 and P_2^{-1} we may relate the sums $X+Y$ obtained via different choices of f. Each transformation shown in Figure 2.2 corresponds to performing the P_2-move followed by P_2^{-1}; the composition of three transformations of Figure 2.2 keeps D, D' and replaces f by its composition with an orientation reversing homeomorphism $D \to D$.

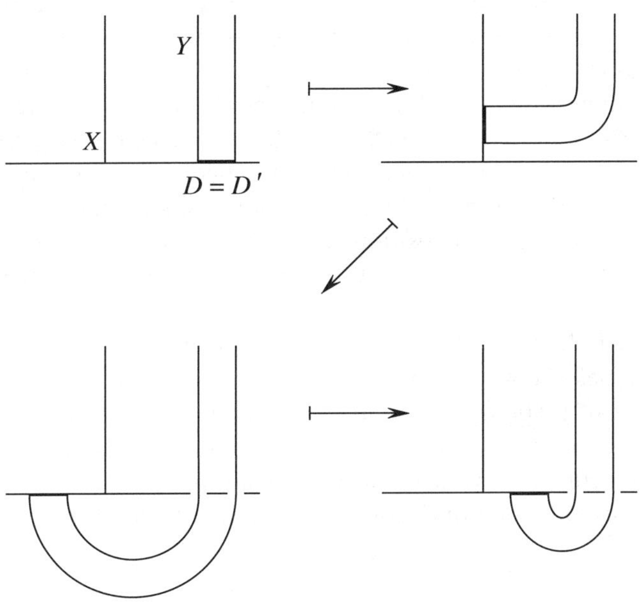

Figure 2.2

2.2. Suspension and stabilization. Recall that the symbol S_0^2 denotes the 2-sphere equipped with the zero gleam. By the suspension of a shadow α we mean the shadow $\alpha + [S^2_0]$. If α is represented by a shadowed polyhedron X then its suspension is represented by a shadowed polyhedron X' obtained by gluing a 2-disk D' to X along a homeomorphism of $\partial D'$ onto the boundary of a closed 2-disk $D \subset \text{Int}(X)$. The open 2-disks $\text{Int}(D)$, $\text{Int}(D')$ are regions of X' with zero gleams. The gleams of other regions of X' are equal to the gleams of the corresponding regions of X. In particular, if Y is the region of X containing D then the region $Y \backslash D$ of X' inherits the gleam of Y. There is a natural retraction $X' \to X$ projecting D' onto D.

Two connected shadowed polyhedra X, X' are called stably shadow equivalent if for certain integers $m, n \geq 0$,

$$[X] + m\,[S^2_0] \;=\; [X'] + n\,[S^2_0].$$

Non-connected shadowed polyhedra X, X' are called stably shadow equivalent if there exists a bijective correspondence between their connected components such that corresponding components are stably shadow equivalent to each other. By a stable shadow we mean a shadowed polyhedron considered up to stable shadow equivalence. We have the stabilization mapping, stab, which assigns to each shadow $[X]$ the stable shadow represented by X.

2.3. Simple deformations. If a shadowed polyhedron X' is obtained from a shadowed polyhedron X by a basic move P_i^ε ($i = 1, 2, 3$; $\varepsilon = \pm 1$), then there exist mutually inverse homotopy equivalences $X \to X'$, $X' \to X$ identical outside the neighborhood where the move proceeds. Similarly, if $X' = X + S^2_0$, then we have the inclusion $X \hookrightarrow X'$ and the natural retraction $X' \to X$ (see Section 2.2). These mappings $X \to X'$, $X' \to X$ are called elementary simple deformations. If $X_1, \ldots., X_n$ is a sequence of shadowed polyhedra such that each of them is obtained from the preceding one by a basic move, or suspension, or desuspension ($X + S^2_0 \mapsto X$), then the composition $X_1 \to X_n$ of the elementary simple deformations $X_1 \to X_2$, $X_2 \to X_3, \ldots.,$ $X_{n-1} \to X_n$ is called a simple deformation.

2.4. Augmented shadows. Let M be a topological space. By an M-augmented shadowed polyhedron we mean a shadowed 2-polyhedron X provided with a homotopy class of mappings $X \to M$. The mappings belonging to this class are called augmentation mappings. By moves P_1, P_2, P_3 in the class of augmented shadowed polyhedra we mean the same transformations $X \mapsto X'$ as in Section 1.3 such that the following diagrams are commutative up to homotopy:

$$\begin{array}{ccc} X & \longrightarrow & X' \\ \downarrow & & \downarrow \\ M & = & M, \end{array} \qquad \begin{array}{ccc} X' & \longrightarrow & X \\ \downarrow & & \downarrow \\ M & = & M. \end{array} \tag{2.4.a}$$

Here the vertical arrows are the given augmentation mappings and the horizontal arrows are the elementary simple deformations defined in Section 2.3. The elementary simple deformations at hand are homotopy equivalences and therefore the commutativity of the first diagram is equivalent to the commutativity of the second one. By a suspension in the class of augmented shadowed polyhedra we mean the transformation $X \mapsto X' = X + S_0^2$ such that diagrams (2.4.a) are commutative up to homotopy. The commutativity of these diagrams implies that the restriction of the augmentation mapping $X' \to M$ to $S_0^2 \subset X'$ is homotopically trivial.

We define M-augmented shadows and M-augmented stable shadows in the same way as in Sections 1.4, 2.2 with the only difference that instead of shadowed polyhedra we shall speak of M-augmented shadowed polyhedra. There is an obvious "forgetting of M" operation which assigns to each M-augmented shadow an ordinary shadow called its source.

2.5. Sliding of regions. In analogy with the addition of handles in the Morse-Smale theory, we define slidings of regions of a shadowed 2-polyhedron over certain disk regions. The sliding preserves the shadow equivalence class of the shadowed polyhedron. The construction of sliding is somewhat technical, it will be used a few times in Chapter IX.

Let X be a shadowed 2-polyhedron. We say that a region $Y \subset X$ is a disk region if its closure $\overline{Y} \subset X$ is a closed embedded 2-disk with interior Y and $\overline{Y} \cap \partial X = \emptyset$. We may regard X as the result of gluing the 2-disk $\overline{Y}$ to the simple 2-polyhedron $X \setminus Y$ along the inclusion $\partial\overline{Y} = \overline{Y} \setminus Y \hookrightarrow X \setminus Y$. It is easy to see that a regular neighborhood of the circle $\partial\overline{Y}$ in $X \setminus Y$ contains either an annulus or a Möbius band with core $\partial\overline{Y}$. In the former case the disk region Y is said to be untwisted. The annulus $S^1 \times [-1, 1] \subset X \setminus Y$ with core $S^1 \times 0 = \partial\overline{Y}$ is called a collar of Y. The loops $S^1 \times (-1), S^1 \times 1 \subset X \setminus \overline{Y}$ are called longitudes of Y.

Let Y, Z, T be three distinct regions of X such that $\overline{Y} \cap \overline{T} = \emptyset$ and Y is an untwisted disk region with zero gleam. We assume that Z is adjacent to both T and Y. Fix a narrow collar $S^1 \times [-1, 1] \subset X \setminus Y$ of Y such that the longitude $\ell = S^1 \times 1$ traverses Z. Consider an embedded band $\beta \subset \overline{Z}$ such that a base of β lies in a 1-stratum of X contained in $\overline{T} \cap \overline{Z}$ and the opposite base of β lies in ℓ. Otherwise, β should be disjoint from $\operatorname{sing}(X)$ and the collar of Y. With this data, we define a sliding of T over Y along β. (The reader is recommended to draw a corresponding picture.)

The sliding of T over Y produces a shadowed 2-polyhedron X' constructed as follows. Let a, c be the bases of β such that $a \subset \overline{T} \cap \overline{Z}$ and $c \subset \ell$. Let b, d be the sides of β so that $\partial\beta$ is formed by four consecutive intervals a, b, c, d. We cut T out of X and then glue it back along a mapping $\partial\overline{T} \to X \setminus T$ which is the identity on $\partial\overline{T} \setminus a$ and which maps a onto the arc a' going along $b, \ell \setminus c$, and d. This operation gives a simple 2-polyhedron X'. We equip the regions of X' with gleams as follows. Consider the 1-strata of $\operatorname{sing}(X)$ which cross $\partial\overline{Y}$ (transversally). They split the half-collar $S^1 \times [0, 1]$ into m rectangles where $m \geq 0$ is the number of vertices of X lying in $\partial\overline{Y}$. The interval c lies on the boundary of one of these rectangles, say R_1. (If $m = 0$ then the role of R_1 is played by the half-collar $S^1 \times [0, 1]$.) The remaining rectangles $R_2, ..., R_m$ give rise to regions $\text{Int}(R_2), ..., \text{Int}(R_m)$ of X'. The gleams of these $m - 1$ regions are taken to be zero. The regions Y, Z, T preserve their gleams. (If β splits Z into two regions, then they are provided with arbitrary gleams with their sum equal to $\text{gl}(Z)$.) If T_1 is the region of X attached to a and distinct from T, Z, then $T_1' = T_1 \cup \text{Int}(R_1) \cup \beta \setminus (b \cup d)$ is a region of X'. Set $\text{gl}(T_1') = \text{gl}(T_1)$. All other regions of X survive in X' (up to subtracting $R_2, ..., R_m$) and preserve the gleams. This makes X' a shadowed 2-polyhedron.

2.5.1. Theorem. *The shadowed polyhedra X, X' are shadow equivalent.*

Proof. There is an obvious homotopy of the interval a modulo its ends carrying a into a'. The homotopy proceeds in the union of β, Y, and the half-collar $S^1 \times [0, 1]$. This homotopy may be lifted to a compositon of the move P_2 pushing a into Y along β, the move P_3 applied m times to the arc under homotopy (where m is the same number as above), and the move P_2^{-1} pushing the arc out of Y in its final position. The intermediate 2-polyhedra may be equipped with gleams in a natural way so that the composition of the moves transforms X into X'. Thus, $[X] = [X']$.

2.6. Exercise. Show that the shadow equivalences over (A, ω) and $(A, -\omega)$ coincide. (Hint: use the arguments given in Section 2.1.)

3. Shadow links

3.0. Outline. We introduce shadow links in shadowed polyhedra. This notion is inspired by the classical theory of plane link diagrams. Instead of link diagrams in the plane we involve shadowed loops in 2-polyhedra and suitable analogues of the Reidemeister moves. The relationships with ordinary links in 3-manifolds and their diagrams will be clarified in Chapter IX.

Shadow links will be instrumental in the next section where we define surgery on shadowed polyhedra.

At first reading, the reader may confine himself to the simpler case where the underlying 2-polyhedron X is a surface.

3.1. Shadow links. Let X be a shadowed 2-polyhedron over A. A generic system of loops $L = L_1 \cup ... \cup L_m$ in X is a finite collection of loops $L_1, ..., L_m$ in $X \setminus \partial X$ such that:

(i) $L_1, ..., L_m$ have only double transversal crossings and self-crossings lying in $\text{Int}(X)$,

(ii) $L_1, ..., L_m$ do not meet vertices of X and meet 1-strata of X transversally.

Connected components of the surface $\mathrm{Int}(X) \setminus L$ are called regions of L (in X).

A shadowed system of loops in X (over A) is a generic system of loops $L = L_1 \cup ... \cup L_m$ in X such that :

(3.1.1) each loop L_j is provided with an element of A, called the pre-twist of L_j;

(3.1.2) each region of L is provided with an element of A, called the gleam of this region;

(3.1.3) for any region Y of X, the sum of the gleams of regions of L contained in Y equals $\mathrm{gl}(Y)$.

The remarks on the gleams of regions made in Section 1.2 will be applied in the setting of shadowed loops as well.

We introduce eight moves $S_1 - S_8$ on shadowed systems of loops in X, see Figures 3.1 - 3.3. In Figures 3.2, 3.3 the loops are drawn in bold in order to distinguish them from 1-strata of X. The moves S_1, S_2, S_3 proceed in $\mathrm{Int}(X)$ and $S_4 - S_8$ proceed in a neighborhood of $\mathrm{sing}(X)$ far away from ∂X. (In the case where X is a surface only three moves S_1, S_2, S_3 are relevant; these are shadow versions of the Reidemeister moves on link diagrams.) The symbols $a, b, \omega, -\omega, -2\omega$ in the figures stand for parts of the gleams of the corresponding regions contributed by the depicted domains. As above, we specify only changes in the gleams. One may think that before application of the moves the gleams of the "big" regions have been concentrated outside of the neighborhoods where the moves proceed, so that the displayed pieces of regions contribute 0 to the gleams (except in the case of S_2 and S_5 where one of "big" regions has the gleam $a + b$). Only the moves S_1, S_4 change the pre-twists. The moves S_1, S_4 decrease the pre-twist of the loop by 2ω and ω respectively. This is schematically shown by small boxes with the symbols -2ω and $-\omega$ inside attached to the loop. Note that the moves S_2, S_5 are not determined by the underlying geometric transformation and may depend on the specific splitting of $a + b \in A$ as the sum of two elements $a, b \in A$. The moves $S_1 - S_8$ have obvious unique inverses. By S_j^{-1} we mean an arbitrary inverse to S_j.

Two shadowed systems of loops in X are called isotopic if they can be obtained from each other by a sequence of moves $S_1 - S_8$, the inverse moves, and an ambient isotopy in X. Isotopy classes of shadowed systems of loops in X are called shadow links in X.

As we shall see in Chapter IX, shadow links are closely related to framed links in 3-manifolds. Forgetting of the pre-twists corresponds to forgetting of the framings. Thus, shadow links may be thought of as framed.

3.2. Cylinders and cones over shadow links. To each shadowed system of loops $L = L_1 \cup ... \cup L_m$ in a shadowed 2-polyhedron X we associate two shadowed

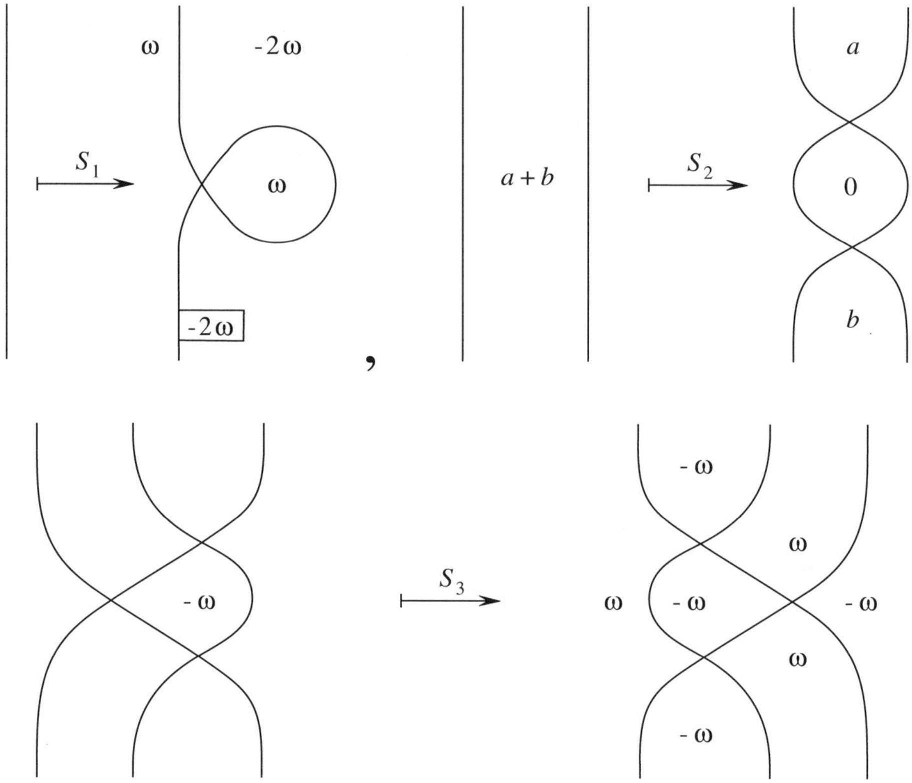

Figure 3.1

2-polyhedra CO_L and CY_L called the cone and the cylinder over L. Glue m copies of the 2-disk to X along arbitrary parametrizations $S^1 \to L_1, \ldots, S^1 \to L_m$. This yields a simple 2-polyhedron, say Z, with $\partial Z = \partial X$. The regions of Z are regions of L in X and m disks attached to the loops of L. The regions of the first type are already equipped with gleams. We provide each region of the second type with the gleam equal to the pre-twist of the corresponding loop. This makes Z a shadowed 2-polyhedron. It is denoted by CO_L and called the (shadow) cone over L.

Similarly, glue m copies of the cylinder $S^1 \times [0, 1]$ to X along parametrizations $S^1 \times 1 \to L_1, \ldots, S^1 \times 1 \to L_m$. This produces a simple 2-polyhedron, say C, with $\partial C = \partial X \amalg mS^1$. The regions of C are regions of L in X and m cylinders $S^1 \times (0, 1)$ attached to the loops of L. As above, the regions of the first type are already equipped with gleams. We provide each region of the second type with the gleam equal to the pre-twist of the corresponding loop. This makes C a shadowed 2-polyhedron. It is denoted by CY_L and called the (shadow) cylinder over L. It is obvious that capping the components of $mS^1 \subset \partial(CY_L)$ we get CO_L.

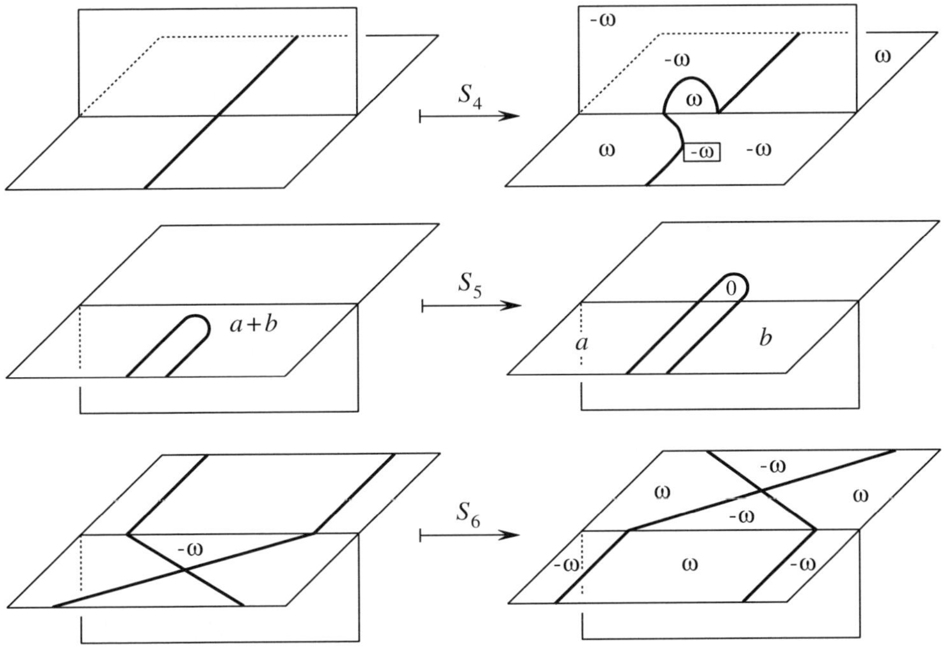

Figure 3.2

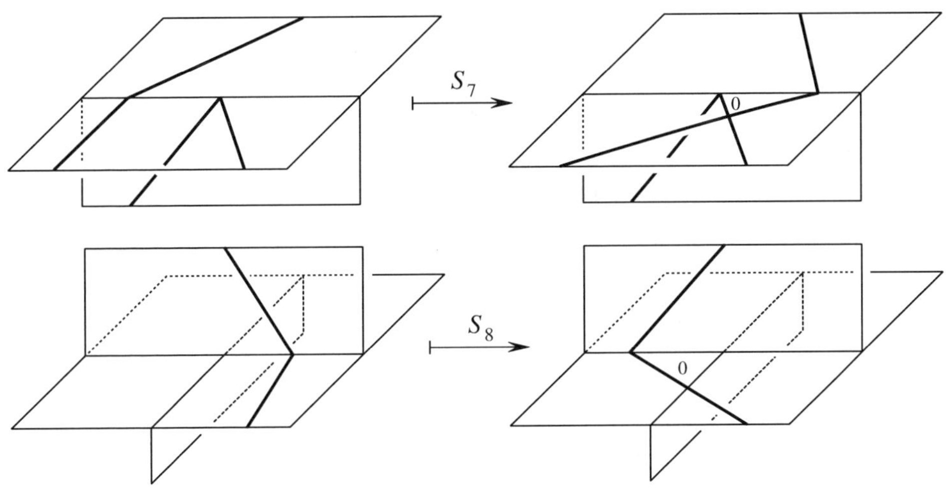

Figure 3.3

3.2.1. Theorem. *If two shadowed systems of loops K, L in X are isotopic, then the shadowed polyhedra CY_K, CY_L are shadow equivalent.*

Theorem 3.2.1 implies a similar assertion for cones. Theorem 3.2.1 shows that to each shadow link α in X we may associate two shadows $CO_\alpha = [CO_L], CY_\alpha =$

$[CY_L]$ where L is an arbitrary shadowed system of loops in X representing α. The shadows CO_α and CY_α are called the cone and the cylinder over α respectively. Clearly, $\partial(CY_\alpha) = \partial X \amalg mS^1$ where m is the number of components of α. The shadow CO_α is obtained from CY_α by capping all components of mS^1.

Proof of Theorem. It suffices to consider the case when K is obtained from L by the move S_i, $i = 1, ..., 8$. If $i = 2, 4, 5, 7, 8$, then CY_K is obtained from CY_L by P_2, P_1, P_2, P_3, P_3 respectively. Consider the move S_3. The shadowed polyhedron CY_L may be obtained from CY_K by a composition of P_1, P_3^{-1}, P_3, and P_1^{-1}: see Figure 3.4 where the horizontal plane presents a small disk neighborhood in X, and the two vertical half-planes present pieces of $S^1 \times [0,1]$ glued along two branches of L. The cylinder glued to the loop under isotopy is not drawn but its presence is understood. (The gleam of this cylinder decreases by ω under the first deformation and increases by ω under the fourth deformation; these changes are not indicated in Figure 3.4.) The move S_6 is treated similarly to S_3.

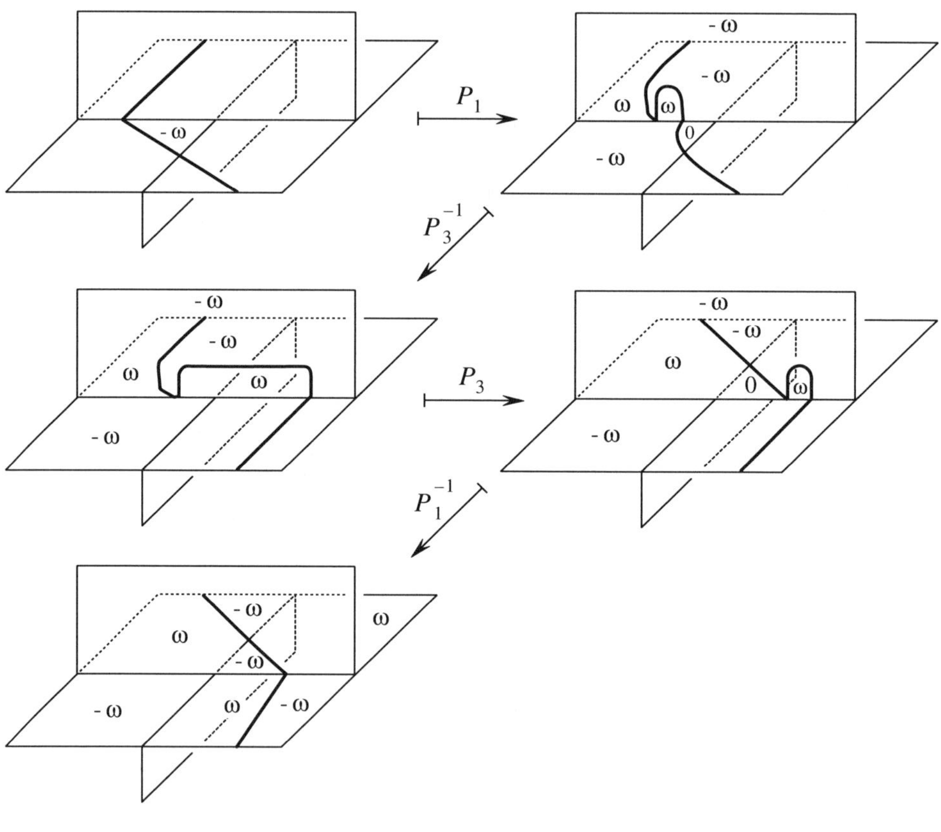

Figure 3.4

It remains to consider the move S_1. Let K be obtained from L by an application of S_1 to a small arc ℓ. Applying, if necessary, S_2 we may assume that ℓ hits another branch of L (possibly lying on the same loop as ℓ). See Figure 3.5 where the horizontal plane represents X, the vertical half-plane represents a piece of the cylinder glued along the branch of L crossed by ℓ, and ℓ is drawn in bold. Figure 3.5 shows that CY_K may be obtained from CY_L by a composition of P_1, P_1, P_3, and P_2^{-1}. Thus, CY_K is shadow equivalent to CY_L.

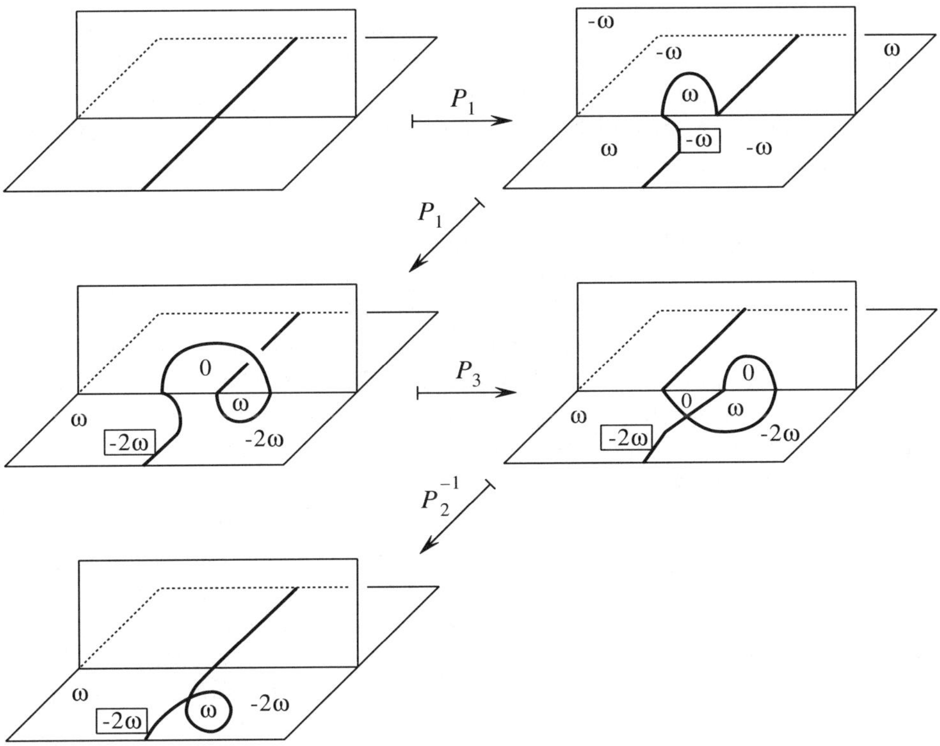

Figure 3.5

3.3. Remarks. 1. Figure 3.5 shows that the move S_1 applied inside a component of a shadowed polyhedron with non-empty singular set may be presented as a composition of S_4, S_4, S_7, S_2.

2. Eliminating certain loops from a shadowed system of loops (and adding appropriate gleams), we get shadowed subsystems of loops. This gives us the notion of shadow sublinks.

3. Each generic loop L in a simple 2-polyhedron X has a natural normal bundle which is a linear vector bundle over the circle parametrizing L. At points of $L \cap \text{Int}(X)$ this is the standard normal bundle. Along $L \cap \text{sing}(X)$ the usual

construction also applies since in a neighborhood of any such point the loop L traverses two branches of $\mathrm{Int}(X)$ locally forming a 2-disk.

4. Surgeries on shadows

4.0. Outline. We describe a cobordism theory for shadows. We do not attempt to introduce "shadowed 3-polyhedra" and to use them as cobordisms between 2-dimensional shadows. We instead introduce shadow analogues of Morse surgeries and define cobordisms as compositions of such surgeries. These cobordisms imitate the cobordisms of 4-manifolds; as we shall see in Chapter IX the shadows of cobordant 4-manifolds are cobordant.

4.1. Surgeries on shadowed polyhedra. A shadowed 2-polyhedron X over A may be surgered along an arbitrary shadow link α in X. The surgery, defined below, yields a shadow with the same boundary ∂X.

We first define the longitude-meridianal shadow link associated with α. Present α by a shadowed system of loops $L = L_1 \cup ... \cup L_m$ in X where m is the number of components of α. Let $B_1, ..., B_m$ be small disjoint closed 2-disks in $\mathrm{Int}(X)$ such that $B_j \cap L = B_j \cap L_j$ is a diameter of B_j for $j = 1, ..., m$. This diameter splits B_j into two half-disks, say B_j^1 and B_j^2. We define a shadowed system of loops $\tilde{L}$ in X with the underlying loops $L_1, ..., L_m, \partial B_1, ..., \partial B_m$. The loops $L_1, ..., L_m$ inherit their pre-twists from α; the loops $\partial B_1, ..., \partial B_m$ are equipped with zero pre-twists. The open disks $\mathrm{Int}(B_j^1)$ and $\mathrm{Int}(B_j^2)$ are regions of $\tilde{L}$, we endow them with the gleams 2ω and -2ω respectively. The gleams of other regions of $\tilde{L}$ are uniquely determined by the condition that L is a sublink of $\tilde{L}$, i.e., that the gleam of an arbitrary region Y of L is equal to the sum of the gleams of regions of $\tilde{L}$ contained in Y.

The isotopy class of $\tilde{L}$ does not depend on the order B_j^1, B_j^2. An isotopy inverting this order is shown in Figure 4.1 where the pre-twist of the circle under isotopy equals zero at the beginning and at the end of the isotopy. The isotopy class of $\tilde{L}$ does not depend on the choice of the disks $B_1, ..., B_m$. A typical isotopy pushing B_j along L_j across a self-intersection of L or across an intersection of L_j with $\mathrm{sing}(X)$ is shown in Figure 4.2. (As usual we indicate only changes in the gleams of "big" regions, the pre-twist of the circle under isotopy equals zero.) The isotopy class of $\tilde{L}$ also does not depend on the choice of the shadowed system of loops L representing α. Indeed, pushing $B_1, ..., B_m$ along L we may assume that these disks do not interfere with an isotopy of L.

The shadow link in X presented by $\tilde{L}$ is called the longitude-meridianal link associated with α. The shadow cone over this link is called the result of surgery on X along α. This cone is a shadow over A with boundary ∂X.

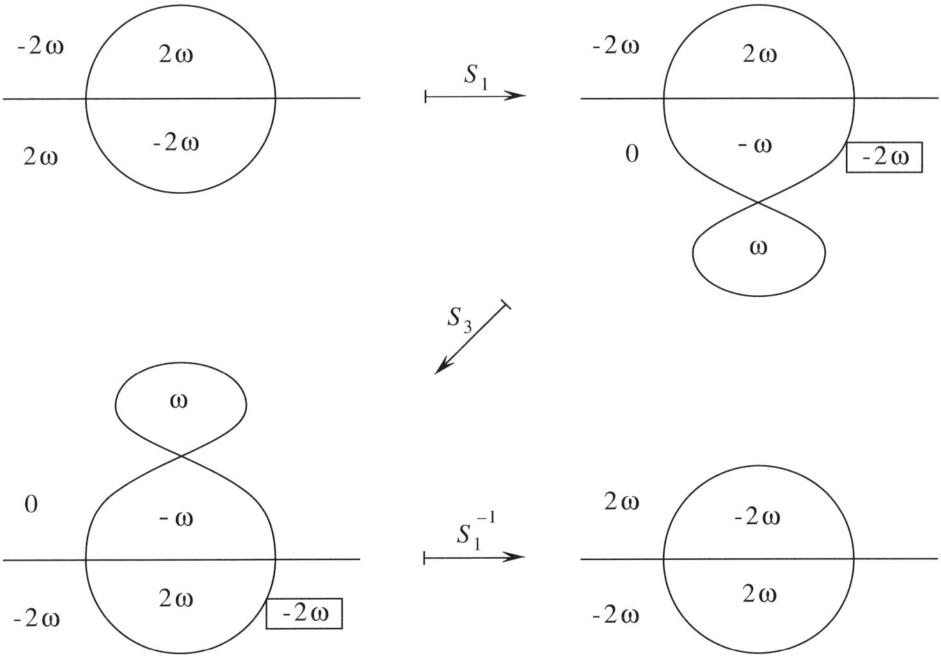

Figure 4.1

It is obvious that the surgery along a shadow link may be presented as a composition of surgeries along shadow knots (i.e., 1-component shadow links). We shall see in Chapter IX that surgeries along shadow knots may be regarded as shadow versions of the index 2 surgeries on 4-manifolds. The loops $L_1, \ldots, L_m$ play the role of the left-hand spheres of a surgery, the 2-spheres composed from $B_1, \ldots, B_m$ and the 2-disks attached to $\partial B_1, \ldots, \partial B_m$ in the cone play the role of the right-hand spheres.

4.2. Lemma. *Let X be a shadowed polyhedron. Let L be a shadowed simple closed curve in* $\mathrm{Int}(X)$ *bounding a small disk in* $\mathrm{Int}(X)$ *such that the pre-twist of L equals zero and the gleam of the disk region bounded by L equals 2ω. Then the surgery on X along L produces $X + S^2_{2\omega} + S^2_{-2\omega}$.*

Proof. The longitude-meridianal link associated to L is isotopic (via S_2) to the shadow link presented by two disjoint simple loops $\ell_1, \ell_2 \subset \mathrm{Int}(X)$ such that ℓ_1 bounds a disk in $\mathrm{Int}(X)$ containing ℓ_2, the pre-twists of both ℓ_1 and ℓ_2 are equal to zero, the gleam of the annulus region between ℓ_1 and ℓ_2 equals -2ω, the gleam of the disk region bounded by ℓ_2 equals 2ω. It is easy to observe that the cone over this shadowed system of loops represents $(X + S^2_{-2\omega}) + S^2_{2\omega}$. (The disks involved in these two addition operations are the disks glued to X along ℓ_1 and ℓ_2 to form the cone.)

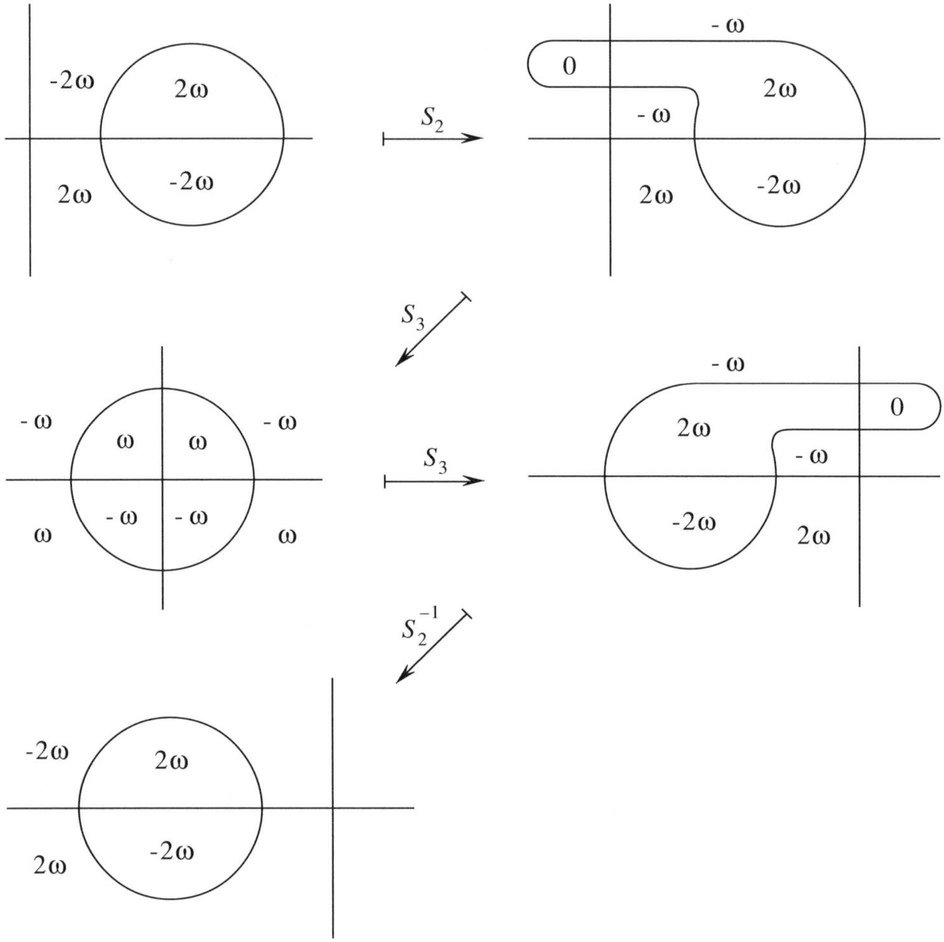

Figure 4.2

4.3. Cobordism of shadows. Two shadowed 2-polyhedra over A are called cobordant if they can be obtained from each other by a homeomorphism, the moves P_1, P_2, P_3, surgeries, and their inverse transformations. Two shadows are cobordant if certain shadowed polyhedra representing these shadows are cobordant.

By definition, cobordism is an equivalence relation in the class of shadows. This relation is rather coarse. For instance, the fundamental group does not survive surgeries: each shadow is cobordant to a simply-connected one. In fact, surgery on a shadowed polyhedron over a link whose components generate the fundamental group produces a simply-connected shadow.

Non-triviality of shadow cobordisms will be established in Section 5 and Chapter X where we introduce cobordism invariants of shadows.

The cobordism classes of connected shadows form an abelian semigroup under the addition of shadows. Lemma 4.2 shows that the class of $S^2_{2\omega} + S^2_{-2\omega} = S^2_{2\omega} + (-S^2_{2\omega})$ is the zero element of this semigroup. Warning: in general, the shadow $X + (-X)$ is not cobordant to zero.

We define a stable cobordism of shadows to be the equivalence relation generated by suspension and cobordism. Two shadows α and β are stably cobordant if and only if for some integers $m, n \geq 0$, the shadow $\alpha + m[S^2_0]$ is cobordant to $\beta + n[S^2_0]$. The relation of stable cobordism also applies to stable shadows in the obvious way.

4.4. Reformulation of surgeries. In Section X.2 we shall need the following equivalent description of surgeries. Let X be a shadowed polyhedron over A and $L = L_1 \cup ... \cup L_m$ be a shadowed system of loops in X. Consider the shadow cone $C = CO_L$ over L. Let $D_1, ..., D_m$ be the closed 2-disks glued to X along the components of L to form C. Choose inside $D_1, ..., D_m$ small closed disks $D'_1, ..., D'_m$ respectively. Glue to C another family of m disks $D''_1, ..., D''_m$ along homeomorphisms of their boundaries onto $\partial D'_1, ..., \partial D'_m$. This yields a simple 2-polyhedron containing C. Denote this polyhedron by Z. Homotopically, Z is the wedge of C and m copies of S^2. We equip the regions of Z with gleams. The regions of Z lying in X keep the gleams which they have as regions of L. For each $i = 1, ..., m$, set $\text{gl}(\text{Int}\, D'_i) = -2\omega$, $\text{gl}(\text{Int}\, D''_i) = 2\omega$, and set $\text{gl}(\text{Int}(D_i) \setminus D'_i)$ to be the pre-twist of L_i. This makes Z a shadowed polyhedron over A. (Note that C is a subpolyhedron but not a shadow subpolyhedron of Z.)

4.4.1. Lemma. *The shadowed polyhedron Z is shadow equivalent to the result of surgery on X along $L = L_1 \cup ... \cup L_m$.*

This lemma suggests an equivalent definition of surgery on L which does not involve the longitude-meridianal systems of loops and therefore does not involve the geometric argument given in Figures 4.1 and 4.2. However, our original definition fits better the surgery theory of manifolds (cf. Chapter IX).

Proof of Lemma. Let Z' be the shadow cone over a longitude-meridianal shadowed system of loops in X associated to L. The lemma follows from the fact that the shadowed polyhedron Z' may be obtained from Z by m consecutive applications of the basic move P_2. To see this, let us apply to Z the following transformations. Choose for each $i = 1, ..., m$, a region Y_i of Z lying in X and adjacent to L_i. Let us push Y_i across $\partial D'_i$ as follows. Choose a simple arc in $D_i \setminus D'_i$ connecting ∂Y_i with $\partial D'_i$. Deform a small segment of ∂Y_i along this arc to make it close to $\partial D'_i$, and apply P_2 to Z pushing this segment inside D'_i. After this transformation the segment in question will split D'_i into two half-disks with gleams 0 and -2ω. Applying to Z such transformations for all $i = 1, ..., m$, we get a shadowed polyhedron which is easily seen to be homeomorphic to Z'.

5. Bilinear forms of shadows

Everywhere in this section the coefficient group of homologies is $\mathbb{Z}$.

5.0. Outline. We define a symmetric A-valued bilinear form in the integer 2-homologies of a shadow over A. When A is a subgroup of the additive group of real numbers the signature of this form is a cobordism invariant of shadows. The bilinear form of shadows is closely related to the intersection form of 4-manifolds, see Chapter IX.

5.1. Bilinear forms of a shadowed polyhedron. Let X be a shadowed 2-polyhedron. Each region Y of X, provided with an orientation, gives rise to an additive homomorphism

$$h \mapsto \langle h|Y\rangle : H_2(X, \partial X) \to \mathbb{Z}.$$

This homomorphism is induced by the contraction $X/\partial X \to X/(X \setminus Y)$ and the identification $H_2(X/(X \setminus Y)) = \mathbb{Z}$ determined by the orientation of Y. For any $h_1, h_2 \in H_2(X, \partial X)$, the product $\langle h_1|Y\rangle\langle h_2|Y\rangle \in \mathbb{Z}$ does not depend on the choice of orientation of Y. Set

$$\tilde{Q}_X(h_1, h_2) = \sum_Y \langle h_1|Y\rangle\langle h_2|Y\rangle \mathrm{gl}(Y) \in A, \tag{5.1.a}$$

where Y runs over all regions of X. It is clear that $\tilde{Q}_X$ is an A-valued symmetric bilinear form in $H_2(X, \partial X)$. Restricting this form to $H_2(X) \subset H_2(X, \partial X)$ we get an A-valued symmetric bilinear form in $H_2(X)$ denoted by Q_X.

The annihilators $\mathrm{Ann}(Q_X) \subset H_2(X)$ and $\mathrm{Ann}(\tilde{Q}_X) \subset H_2(X, \partial X)$ of these forms may be non-trivial. Since $H_2(X)$ and $H_2(X, \partial X)$ are free abelian groups, these annihilators are also free abelian groups. The rank of $\mathrm{Ann}(Q_X)$ is called the nullity of X and denoted by $\mathrm{null}(X)$.

For example, if X is a compact connected orientable surface equipped with a gleam $a \in A$, then $\tilde{Q}_X$ is presented by the 1×1-matrix $[a]$. We have $Q_X = \tilde{Q}_X$ in the case $\partial X = \emptyset$ and $Q_X = 0$ in the case $\partial X \neq \emptyset$.

5.2. Theorem. *Let X, X' be shadow equivalent shadowed 2-polyhedra. Then any simple deformation $X \to X'$ induces isomorphisms of the forms $\tilde{Q}_X$ and Q_X onto $\tilde{Q}_{X'}$ and $Q_{X'}$ respectively.*

Theorem 5.2 shows that the bilinear forms $\tilde{Q}, Q$ are invariant under shadow equivalence. Therefore, we may speak of the bilinear forms $\tilde{Q}_\alpha, Q_\alpha$ of any shadow α. These are A-valued symmetric bilinear forms on $H_2(\alpha, \partial\alpha)$ and $H_2(\alpha)$ respectively. (These abelian groups and the forms $\tilde{Q}_\alpha, Q_\alpha$ on them are defined only up to isomorphism.) It is obvious that the forms $\tilde{Q}_\alpha, Q_\alpha$ are additive with respect to summation of shadows.

Theorem 5.2 implies that the nullity of shadowed polyhedra is preserved under shadow equivalence and yields an invariant of shadows. It is denoted by null.

The quotients of the bilinear forms $\tilde{Q}, Q$ by their annihilators are invariant under suspension of shadows and yield bilinear forms of stable shadows.

Proof of Theorem. We shall show that $\tilde{Q}_X$ is transformed into $\tilde{Q}_{X'}$ by the isomorphism

$$H_2(X, \partial X) \to H_2(X', \partial X') \tag{5.2.a}$$

induced by a simple deformation $X \to X'$. This would also imply the claim concerning Q.

It suffices to consider the case of elementary simple deformation where X' is obtained from X by the move P_i, $i = 1, 2, 3$. The cases $i = 2$ and $i = 3$ are obvious since the gleams of the small disk regions created by P_2, P_3 are equal to zero and therefore the expressions defining $\tilde{Q}_X, \tilde{Q}_{X'}$ are identical.

Let $i = 1$. Denote by Y'_{13} the small disk region of X' created by P_1. It is clear that regions of X' distinct from Y'_{13} correspond naturally and bijectively to regions of X. For a region Y of X, the corresponding region of X' will be denoted by Y'. For any $h \in H_2(X, \partial X)$, denote by h' the image of h under the isomorphism (5.2.a). Orient all regions of X' and equip the regions of X with the induced orientation. It is obvious that for any region Y of X, we have

$$\langle h' | Y' \rangle \;=\; \langle h | Y \rangle.$$

Let $Y_0, Y_1, Y_2, Y_3, Y_{12}, Y_{23}$ be the regions of X marked by $a, a_1, a_2, a_3, a_{12}, a_{23}$ respectively (see Figure 1.1). For $h \in H_2(X, \partial X)$, set $h_i = \langle h | Y_i \rangle$ and $h_{13} = \langle h' | Y'_{13} \rangle$, where $i = 0, 1, 2, 3, 12, 23$. For any $g, h \in H_2(X, \partial X)$, we have

$$\tilde{Q}_{X'}(g', h') - \tilde{Q}_X(g, h) = \tag{5.2.b}$$

$$= (g_{12}h_{12} + g_{23}h_{23} + g_{13}h_{13} - g_0h_0 - g_1h_1 - g_2h_2 - g_3h_3)\,\omega.$$

It remains to show that the number in parentheses is equal to 0.

Note that by inverting the orientation of Y_i we replace g_i and h_i with $-g_i$ and $-h_i$ while keeping g_j, h_j with $j \neq i$. This does not change the expression in parentheses. Therefore, we may assume that the orientations of regions correspond to the counterclockwise orientation in the plane of the picture under the projection of Figure 1.1. We have the following obvious compatibility conditions on the numbers $\{h_i\}$:

$$h_2 + h_3 - h_{23} = h_0 + h_{12} - h_3 = h_0 + h_{13} - h_2 = h_3 + h_{13} - h_1 = 0.$$

Thus,

$$h_0 = h_2 + h_3 - h_1,\;\; h_{12} = h_1 - h_2,\;\; h_{13} = h_1 - h_3,\;\; h_{23} = h_2 + h_3.$$

Similarly,

$$g_0 = g_2 + g_3 - g_1, \ g_{12} = g_1 - g_2, \ g_{13} = g_1 - g_3, \ g_{23} = g_2 + g_3.$$

Substituting these expressions in the right-hand side of (5.2.b) we get 0.

5.3. Signature of shadows. Let A be a subgroup of the additive group of real numbers and let $\omega \neq 0$. We define the signature $\sigma(\alpha)$ of a shadow α over A to be the signature of the form $\mathbb{R} \otimes_A Q_\alpha$. The signature is an additive invariant of shadows. Since $\sigma([S_0^2]) = 0$ this is an invariant of stable shadows.

It is a simple exercise to check that for cobordant shadows α, β over A, the bilinear forms $\mathbb{R} \otimes_A Q_\alpha$ and $\mathbb{R} \otimes_A Q_\beta$ are isomorphic up to factorizing out the annihilators and stabilization by the form

$$\begin{bmatrix} 0 & 2\omega \\ 2\omega & 0 \end{bmatrix}.$$

(It is here that we need the assumption $\omega \neq 0$.) This implies that the signature is a cobordism invariant of shadows.

5.4. Remark. For any 2-component shadow link $L = L_1 \cup L_2$ in a shadowed polyhedron X such that $Q_X = 0$ and L_1, L_2 are homologically trivial, we can define a linking number $\operatorname{lk}(L_1, L_2) \in A$. Namely, set $\operatorname{lk}(L_1, L_2) = \tilde{Q}_C(u_1, u_2)$ where C is the shadow cylinder of L and u_1, u_2 are arbitrary elements of $H_2(C, \partial C)$ carried by the boundary homomorphism $H_2(C, \partial C) \to H_1(\partial C)$ into the homology classes of the boundary circles of C corresponding to L_1, L_2. This is a shadow analogue of the standard linking number of knots in a homological 3-sphere (cf. Section IX.5).

5.5. Exercise. Show that $b_1(\alpha) + \operatorname{null}(\alpha)$ is an additive cobordism invariant of shadows. (Here b_1 is the first Betti number over $\mathbb{R}$.)

6. Integer shadows

6.0. Outline. For each region Y of a simple 2-polyhedron without boundary X, we define a residue $\eta(Y) \in \mathbb{Z}/2$ determined by topological properties of X in a neighborhood of Y. This suggests that we should consider shadowed 2-polyhedra whose gleams are integers representing these residues (mod 2). For technical reasons, it is more convenient to consider shadowed 2-polyhedra whose gleams are integers or half-integers determined (mod 1) by these residues. This leads to the notions of integer shadowed polyhedron and integer shadow without boundary. Similar ideas work for shadowed polyhedra with boundary, under the assumption that the boundary is framed, i.e., thickened into a framed graph.

We shall see in Chapter IX that the shadows of 4-manifolds are integer shadows.

6.1. Simple polyhedra with framed boundary. By a simple 2-polyhedron with framed boundary we mean a pair (X, Γ) where X is a simple 2-polyhedron and Γ is a framed graph with core ∂X. (For the definition of framed graphs see Section VI.4.2.) In other words, a simple 2-polyhedron with framed boundary is a simple 2-polyhedron X whose boundary is thickened into a framed graph Γ. It is understood that $X \cap \Gamma = \partial X$. If $\partial X = \emptyset$ then $\Gamma = \emptyset$.

For any simple 2-polyhedron with framed boundary (X, Γ), the union $X' = X \cup \Gamma$ is a simple 2-polyhedron. It is obvious that $\text{sing}(X') = \text{sing}(X) \cup \partial X$ and $\partial X' = \partial \Gamma$.

For instance, a shadowed polyhedron X whose boundary consists of circles admits a canonical framing $\Gamma = \partial X \times [-1, 1]$ where $\partial X \times 0 = \partial X$.

6.2. The residue η. The objective of this subsection is to associate to any region Y of an orientable simple 2-polyhedron with framed boundary (X, Γ) a residue $\eta(Y) \in \mathbb{Z}/2\mathbb{Z}$. This residue will be instrumental in the definition of integer shadows.

It is clear that Y is a region of $X' = X \cup \Gamma$ and its closure $\overline{Y} \subset X$ does not meet $\partial X' = \partial \Gamma$. Just to give the idea consider the case when $\overline{Y}$ is an embedded surface with the interior Y. A regular neighborhood of $\partial \overline{Y}$ in $X' \setminus Y$ contains several annuli and Möbius bands whose cores are components of $\partial \overline{Y}$. Set $\eta(Y) = n (\text{mod } 2)$ where n is the number of Möbius bands in this family. In particular, if Y is a closed surface then $\eta(Y) = 0$.

Here is a definition of $\eta(Y)$ in the general case. Let Y_0 be a compact subsurface of Y which is a deformation retract of Y so that $Y \setminus Y_0 = \partial Y_0 \times [0, 1)$. We may view $\overline{Y} \setminus Y_0$ as the cylinder of a local embedding $f : \partial Y_0 \to \text{sing}(X')$. The compact 1-manifold ∂Y_0 is determined by Y uniquely up to ambient isotopy in Y. This 1-manifold is the "infinity of Y".

We construct a line bundle $\nu = \nu(Y)$ over ∂Y_0 as follows. Denote by V the set of vertices of $\text{sing}(X')$. Since $f : \partial Y_0 \to \text{sing}(X')$ is a local embedding, the finite set $f^{-1}(V)$ splits ∂Y_0 into a finite number of open intervals which are mapped by f homeomorphically onto 1-strata of $\text{sing}(X')$. Each such interval, say e, determines a branch of $\text{Int}(X')$ attached to $f(e) \subset \text{sing}(X')$ (this branch lies in Y and hits $f(e)$ from the side of e). The other two branches of $\text{Int}(X')$ attached to $f(e)$ form together with $f(e)$ a 2-manifold. Take the normal line bundle of $f(e)$ in this 2-manifold and transfer it back to e via f. This yields a line bundle over $\partial Y_0 \setminus f^{-1}(V)$. It is easy to see that this line bundle uniquely extends to a line bundle ν over ∂Y_0 such that f extends to a local embedding of the total space of ν into X'. Set $\eta(Y) = n (\text{mod } 2)$ where n is the number of components of ∂Y_0 over which ν is non-orientable.

It is obvious that $\eta(Y)$ is the (only) obstruction to the extension of the linear bundle ν to Y_0.

Note the role of the framing Γ of ∂X: we use it to form X' and to define the residue η for the regions of X adjacent to ∂X.

6.3. Integer shadowed polyhedra and integer shadows. Denote the additive group of integers and half-integers by $(1/2)\mathbb{Z}$. An integer shadowed polyhedron X is a shadowed polyhedron over $(1/2)\mathbb{Z}$ provided with a framing of ∂X so that the gleam of any region Y of X is determined (mod $\mathbb{Z}$) by the residue $\eta(Y)$ as follows:

$$\mathrm{gl}(Y) + \eta(Y)/2 = 0 \pmod{\mathbb{Z}}. \tag{6.3.a}$$

In other words, an integer shadowed polyhedron is an orientable simple 2-polyhedron X, with framed boundary, such that all regions of X are provided with integer or half-integer gleams and the gleam of a region Y is an integer if and only if $\eta(Y) = 0$.

It is a simple exercise to check that condition (6.3.a) is preserved under the shadow moves $P_1^{\pm 1}$ where $\omega = 1/2$ and the moves $P_2^{-1}, P_3^{\pm 1}$. In the theory of integer shadowed polyhedra, by P_2 we mean any move of type P_2 preserving (6.3.a). Everywhere in this theory $\omega = 1/2$.

An integer shadow is an equivalence class of integer shadowed polyhedra under the equivalence relation generated by homeomorphism and the moves $P_1^{\pm 1}$, $P_2^{\pm 1}$, $P_3^{\pm 1}$. Note that these moves preserve the boundary and its framing. Thus, any integer shadow has a framed boundary.

Each integer shadowed polyhedron determines an underlying shadowed polyhedron over $(1/2)\mathbb{Z}$, it is obtained by forgetting the framing of the boundary. In this way each integer shadow determines an underlying shadow over the pair $((1/2)\mathbb{Z}, 1/2)$. Therefore all invariants of shadows over $((1/2)\mathbb{Z}, 1/2)$ can be viewed as invariants of integer shadows.

The definitions and results of Sections 1, 2, 5 extend to integer shadows with the obvious changes. In particular, by a stable integer shadow we shall mean an integer shadow considered up to stabilization.

6.4. Integer shadow links. Here we define integer shadow links and explain how to view the results of Sections 3, 4 in this setting.

Let X be an integer shadowed polyhedron. An integer shadowed system of loops L in X is a shadowed system of loops in X over $A = (1/2)\mathbb{Z}$ such that all regions of the shadow cone over L satisfy (6.3.a). For example, if $X = S^2$ then this condition means that the pre-twists of the loops are integers and the gleam of a region Y of L in X belongs to $\mathbb{Z}$ if the number of vertices of Y is even and to $(1/2) + \mathbb{Z}$ otherwise.

It is straightforward to check that the moves $S_i^{\pm 1}, i = 1, 3, 4, 6, 7, 8$ and S_2^{-1}, S_5^{-1} on integer shadowed system of loops preserve condition (6.3.a) on the shadow cone. By S_2, S_5 we shall mean in this setting the moves of type S_2, S_5 preserving (6.3.a).

Two integer shadowed systems of loops in X are called isotopic if they can be obtained from each other by a sequence of moves $S_i^{\pm 1}; i = 1, ..., 8$ and an ambient isotopy in X. Isotopy classes of integer shadowed systems of loops in X are called integer shadow links in X.

In the same way as in Sections 3, 4 we define shadow cones over integer shadow links and surgery and cobordism for integer shadows. The results of Section 4 extend to this setting with obvious changes.

6.5. Remarks. 1. Every orientable simple 2-polyhedron X without boundary may be transformed into a shadowed polyhedron over $\mathbb{Z}/2\mathbb{Z}$ by assigning to each region Y of X the residue $\eta(Y)$. The reader may verify that under the transformations of simple 2-polyhedra underlying $P_1 - P_3$, these residues behave in the way indicated in Figure 1.1 where $\omega = 1 (\mathrm{mod}\, 2)$. Thus, simple 2-polyhedra considered modulo these transformations determine shadows over $(\mathbb{Z}/2\mathbb{Z}, 1 (\mathrm{mod}\, 2))$.

Similarly, every generic system of loops in an orientable simple 2-polyhedron without boundary canonically upgrades to a shadowed system of loops over $\mathbb{Z}/2\mathbb{Z}$. (The pre-twists and gleams should be defined as the residues associated to the corresponding regions of the cone.) It is easy to check that homotopic systems of loops determine isotopic shadowed system of loops.

2. The relationships between shadows and 4-manifolds established in the next chapter may be exploited to study integer shadows. For example, the results of Section IX.5 imply that the bilinear form of any integer shadow takes values in $\mathbb{Z}$ (a priori its values lie in $(1/2)\mathbb{Z}$). It would be interesting to have a direct proof of this fact.

7. Shadow graphs

7.0. Outline. The notion of a shadow graph generalizes the notion of a shadow link. The only essential difference is that in the setting of graphs there are two additional moves, involving vertices of graphs.

The definitions and results of this section will be used in Section IX.8 and in the last sections of Chapter X.

7.1. Definition of shadow graphs. Consider a simple 2-polyhedron X and a graph γ (in the sense of Section VI.4.1). An immersion $f : \gamma \to X$ is said to be generic if all vertices of γ are mapped to $\mathrm{Int}(X)$, the image of f does not meet ∂X and the vertices of X, the image of f intersects 1-strata of X transversally, all self-crossing points of f are double transversal crossings of 1-strata of γ and lie in $\mathrm{Int}(X)$. The image $g = f(\gamma)$ of a generic immersion $f : \gamma \to X$ is called an immersed graph in X. We say that γ is the underlying abstract graph of g and f is a parametization of g. The connected components of the surface $\mathrm{Int}(X) \setminus g$ are called regions of g (in X).

Assume that X is a shadowed 2-polyhedron over A. A shadowed graph in X (over A) is an immersed graph g in X such that:

(7.1.1) each 1-stratum of the underlying abstract graph of g is provided with an element of A, called the pre-twist of this 1-stratum;

(7.1.2) each region of g is provided with an element of A, called the gleam of this region;

(7.1.3) for any region Y of X, the sum of the gleams of regions of g contained in Y equals $\mathrm{gl}(Y)$.

Besides the moves $S_1 - S_8$ introduced in Section 3 we may apply the moves S_9 and S_{10} shown in Figure 7.1 where the graphs are drawn in bold. The move S_9 proceeds in $\mathrm{Int}(X)$ and does not change the gleams of "big" regions, neither does S_{10}. The gleams of the small disk regions created by S_9 and S_{10} are equal to zero.

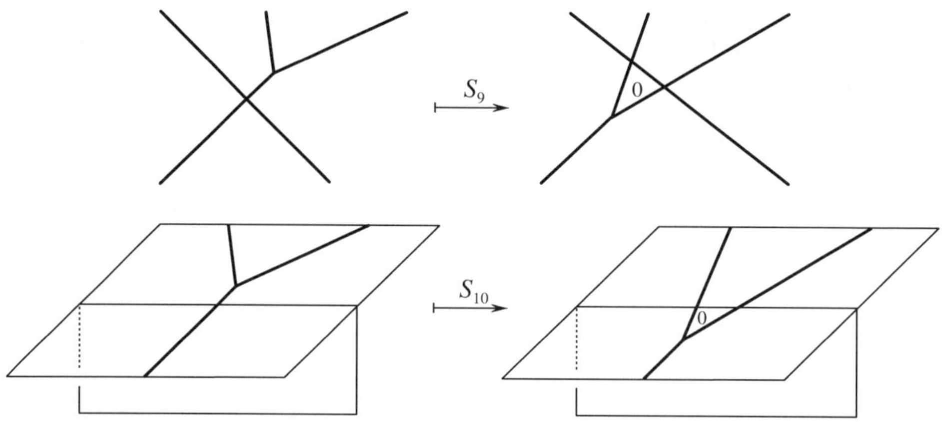

Figure 7.1

Two shadowed graphs in X are called isotopic if they can be obtained from each other by (a sequence of) moves $S_1 - S_{10}$, their inverses, and an ambient isotopy in X. Isotopy classes of shadowed graphs in X are called shadow graphs in X.

To each shadowed graph g in X we assign a shadowed 2-polyhedron CY_g called the shadow cylinder over g. Geometrically, this is the cylinder of the generic immersion $\gamma \to g$ parametrizing g. The regions of CY_g are the regions of g in X and the cylinders $e \times (0, 1)$ where e runs over 1-strata of γ. The regions of the first type are already equipped with gleams. Each region of the second type is provided with the gleam equal to the pre-twist of e. This makes CY_g a shadowed 2-polyhedron. It is obvious that $\partial(CY_g) = \partial X \amalg \gamma$. Theorem 3.2.1

directly generalizes to the setting of graphs. Namely, if two shadowed graphs in X are isotopic, then the shadow cylinders over these graphs are shadow equivalent. This allows us to speak of shadow cylinders of shadow graphs.

7.2. Integer shadow graphs. Consider a simple 2-polyhedron X and a framed graph Γ with core γ. Note that the cylinder C of a generic immersion $\gamma \to X$ is a simple 2-polyhedron with boundary $\partial C = \gamma$. This boundary is framed via the inclusion $\gamma \subset \Gamma$. Thus, the pair (C, Γ) is a simple 2-polyhedron with framed boundary.

Assume that X is an integer shadowed polyhedron. An integer shadowed graph in X (modeled on Γ) is the image of a generic immersion $\gamma \to X$ satisfying (7.1.1) - (7.1.3) with $A = (1/2)\mathbb{Z}$ such that the pair (the shadow cylinder C of this immersion, Γ) is an integer shadowed polyhedron. Thus, all regions of C should satisfy (6.3.a).

The moves $S_i^{\pm 1}, i = 1, 3, 4, 6, ..., 10$ and S_2^{-1}, S_5^{-1} on shadowed graphs preserve condition (6.3.a). By S_2, S_5, we shall mean the moves of type S_2, S_5 preserving (6.3.a).

Two integer shadowed graphs in X are called isotopic if they can be obtained from each other by a sequence of moves $S_i^{\pm 1}; i = 1, ..., 10$ and an ambient isotopy in X. Isotopy classes of integer shadowed graphs in X are called integer shadow graphs in X. To each shadow graph in X we associate the cylinder over this graph. In the notation above this is the integer shadow represented by (C, Γ).

Notes

Shadows were introduced in [Tu9], [Tu10]. The exposition in this chapter follows the lines of [Tu9] with minor modifications.

It seems appropriate to mention the original motivation for the introduction of shadow links, which were precursors of shadows. It is customary to use the orthogonal projection $\mathbb{R}^3 \to \mathbb{R}^2$ in order to present links in $\mathbb{R}^3$ by plane diagrams. There is another important mapping relating dimensions 3 and 2, namely, the Hopf fibration $S^3 \to S^2$ with fiber S^1. One may ask if it can be used to describe knots in S^3 in terms of some pictures in S^2. (The author's attention to this problem was attracted by the work of Thomas Fiedler [Fi1] on algebraic links in S^3 and their Hopf images in S^2.) The image of a link under the Hopf mapping is a system of loops in S^2 without overcrossings and undercrossings: the preimage of a crossing point consists of two points in a circle and there is no way to distinguish their roles. The solution lies in ascribing additional information (the gleams) not to crossing points but to components of the complement of the loops in S^2. This leads to a theory of shadowed loops in surfaces which describes links in 3-manifolds fibered over surfaces with the fiber S^1 and defines quantum invariants of such 3-manifolds and links, see [Tu10]. To generalize these constructions to arbitrary 3-manifolds we involve simple 2-polyhedra as is done in [Tu9] and here in Chapters VIII - X.

Chapter IX
Shadows of manifolds

1. Shadows of 4-manifolds

1.0. Outline. The objective of this section is to define the shadow of a compact oriented piecewise-linear 4-manifold. This definition underlies the role of shadows in the topology of low-dimensional manifolds. Using the shadows of 4-manifolds and the state sum invariants of shadows introduced in Chapter X we shall eventually develop a shadow approach to quantum invariants of 3-manifolds.

We begin with background material on 4-dimensional topology: categories of 4-manifolds, 2-polyhedra in 4-manifolds, skeletons of 4-manifolds, etc. Then we introduce shadows of 4-manifolds and state their main properties. They will be proven in Section 7.

1.1. Categories of 4-manifolds. In contrast to the theory of 3-manifolds, in dimension 4 there is an essential difference between smooth and topological settings (see [FrQ]). On the other hand, the categories of smooth and PL (piecewise-linear) 4-manifolds are equivalent. It is well known that every smooth manifold admits a natural C^1-smooth PL-structure. For 4-manifolds, this gives a bijective correspondence between isotopy classes of smooth and piecewise-linear structures (see [Ca1], [Ca2]).

We shall not consider topological 4-manifolds. Since the categories of piecewise-linear and smooth 4-manifolds are equivalent, the choice of a category is a technical matter. We shall work mainly in the piecewise-linear category occasionally involving smooth structures. In particular, by a 4-manifold we mean a piecewise-linear 4-manifold and by homeomorphisms of 4-manifolds we mean piecewise-linear homeomorphisms, unless the smooth setting is mentioned explicitly.

1.2. Locally flat and smooth 2-polyhedra in 4-manifolds. The definition of shadows of 4-manifolds is based on a study of 2-polyhedra in 4-manifolds. Here we make a few general remarks on polyhedra in 4-manifolds.

By a polyhedron in a PL-manifold W, we mean a subpolyhedron of a PL-triangulation of W. A 2-dimensional polyhedron $X \subset W$ is said to be flat in a point $x \in X$ if, in a neighborhood of x, the polyhedron X lies in a 3-dimensional (piecewise-linear) submanifold of W. This means that x has a neighborhood $U_x \subset W$ which can be piecewise-linearly embedded in $\mathbb{R}^n$ with $n = \dim W$ such that $U_x \cap X \subset \mathbb{R}^3 \subset \mathbb{R}^n$. The 2-polyhedron $X \subset W$ is said to be locally flat if it is flat at all points. This condition excludes local knottedness of X (cf. Remark 1.13 below).

In the smooth setting the notion parallel to that of a locally flat polyhedron is the notion of a smooth polyhedron. For simplicity, we restrict ourselves to simple 2-polyhedra with empty boundary (see Section VIII.1.1). A simple 2-polyhedron with empty boundary X lying in a smooth 4-manifold W is said to be smooth if every point $x \in X$ has a neighborhood $U_x \subset W$ which can be smoothly embedded in $\mathbb{R}^4$ such that the image of $U_x \cap X$ equals one of the standard subsets (VIII.1.1.1) - (VIII.1.1.3) of $\mathbb{R}^3 \times 0 \subset \mathbb{R}^4$. This generalizes the standard notion of a smooth surface in a 4-manifold. Note that if X is a smooth simple 2-polyhedron in a smooth 4-manifold W then $\operatorname{sing}(X)$ is a smooth graph in W and $\operatorname{Int}(X)$ is a smooth surface in W attached to $\operatorname{sing}(X)$ in a locally standard fashion.

The languages of locally flat and smooth 2-polyhedra are equivalent. If X is a smooth simple 2-polyhedron in a smooth 4-manifold W then W has a PL-triangulation containing X as a locally flat subpolyhedron. Conversely, if X is a locally flat simple 2-polyhedron in a piecewise-linear 4-manifold W then both X and the PL-structure in W are smooth with respect to a certain smooth structure in W. This follows from the smoothing results of Cairns [Ca1], [Ca2] and the Schoenflies theorem in dimension 3 due to Alexander [Al1].

The language of smooth polyhedra has the advantage of allowing us to use tangent and normal vector bundles on manifolds and other notions of differential topology. On the other hand, the language of locally flat 2-polyhedra is better suited to the combinatorial nature of 2-polyhedra. We shall freely switch between these two languages.

1.3. Handles and handlebodies. A 4-dimensional i-handle (or handle of index i) is the pair $(B^i \times B^{4-i}, \partial B^i \times B^{4-i})$ where B^n denotes a closed n-dimensional Euclidean ball. The space $\partial B^i \times B^{4-i}$ is called the base of this handle. In particular, a 4-handle is just a 4-ball whose base is its boundary sphere. We shall often use 3-handles $(B^3 \times [-1, 1], S^2 \times [-1, 1])$ where $S^2 = \partial B^3$ and 1-handles $([-1, 1] \times B^3, \{-1, 1\} \times B^3)$.

We say that a 4-manifold W is obtained from a 4-manifold W' by attaching an i-handle if W is a union of W' and an i-handle meeting along the base of this handle lying in $\partial W'$.

A (closed) 4-handlebody is a compact connected 4-manifold obtained by attaching several 1-handles to a closed 4-ball. The simplest handlebodies are the 4-ball itself and the product $B^3 \times S^1$. The interior of a closed 4-handlebody is called an open 4-handlebody.

1.4. Skeletons of 4-manifolds. A skeleton of a compact 4-manifold W is a locally flat orientable simple 2-polyhedron without boundary $X \subset \operatorname{Int}(W) = W \setminus \partial W$ such that

(∗) the manifold W may be obtained from a closed regular neighborhood of X in W by attaching 3-handles and 4-handles.

For instance, if W is the total space of a disk bundle over a closed orientable surface Σ then the zero section $\Sigma \subset W$ is a skeleton of W. In this case W itself is a closed regular neighborhood of Σ.

Condition (∗) implies that the part of X lying in a connected component of W is connected. Since $\dim(X) = 2$, the part of $W \setminus X$ lying in a connected component of W is also connected.

Using the duality between 1-handles and 3-handles in 4-manifolds we may reformulate condition (∗) in terms of $W \setminus X$. In the case of closed connected W, condition (∗) holds if and only if $W \setminus X$ is an open 4-handlebody, in other words, (∗) holds if and only if the complement in W of an open regular neighborhood of X is a closed 4-handlebody. For example, the 2-sphere $S^2 = \mathbb{C}P^1 \subset \mathbb{C}P^2$ is a skeleton of $\mathbb{C}P^2$ because its complement is an open 4-ball. Similarly, the 2-sphere lying in S^4 as the boundary of a 3-ball is a skeleton of S^4 because its complement is homeomorphic to $\mathrm{Int}(B^3) \times S^1$.

Let W be a compact connected 4-manifold with non-empty boundary. A collar of ∂W in W is a closed regular neighborhood of ∂W in W identified with the cylinder $W \times [0, 1]$ so that $\partial W = \partial W \times 0$. It is easy to see that a locally flat orientable simple 2-polyhedron without boundary $X \subset W \setminus \partial W$ satisfies condition (∗) if and only if the complement in W of an open regular neighborhood of X is obtained from a collar $\partial W \times [0, 1] \subset W$ by attaching 1-handles to $\partial W \times 1$.

The following theorem ensures the existence of skeletons.

1.5. Theorem. *Every compact oriented (piecewise-linear) 4-manifold has a skeleton.*

Theorem 1.5 follows from elementary properties of handle decompositions of 4-manifolds. We prove this theorem in Section 7.

1.6. Shadowing of skeletons. Let X be a skeleton of a compact oriented 4-manifold W. We shall equip X with the structure of an integer shadowed polyhedron. Since $\partial X = \emptyset$ we need only specify the gleams of regions of X. Note that the definition of gleams given below is local in the sense that it proceeds in a small neighborhood of X in W.

If Y is a connected component of X homeomorphic to a closed surface then Y is a region of X. For such a region, define $\mathrm{gl}(Y) \in \mathbb{Z} \subset (1/2)\mathbb{Z}$ to be the self-intersection number of Y in W. Thus, $\mathrm{gl}(Y) = u \cdot u$ where $u \in H_2(W; \mathbb{Z})$ is the homological class represented by the surface Y with a certain orientation and the dot $\cdot$ denotes the intersection product. Note that $(-u) \cdot (-u) = u \cdot u$ and therefore $\mathrm{gl}(Y) \in \mathbb{Z}$ does not depend on the choice of orientation of Y. The number $u \cdot u$ may be computed as follows: deform Y in W into a transversal position Y' and sum up the local intersection indices over all points of $Y \cap Y'$ where the orientation of Y' is induced by that of Y.

Let Y be a region of X not homeomorphic to a closed surface. Then Y is non-compact and contains a compact subsurface $Y_0 \subset Y$ as a deformation retract (cf. Section VIII.6.2). Denote by $\nu_W = \nu_W(Y_0)$ the normal bundle of Y_0 in W. This is a 2-dimensional real vector bundle over Y_0. Recall the line bundle $\nu = \nu(Y)$ over ∂Y_0 constructed in Section VIII.6.2. The bundle ν may be viewed as a subbundle of the restriction of ν_W to ∂Y_0. (This is clear from the description of $\nu(Y)$ as the normal bundle of ∂Y_0 in a disjoint union of annuli and Möbius bands lying in W and containing the components of ∂Y_0 as their cores. These annuli and Möbius bands are determined by a regular neighborhood of $\partial \overline{Y}$ in X.) Roughly speaking, the gleam of Y is the obstruction to the extension of ν to a line subbundle of ν_W. Here are the details. Let $\xi : E \mapsto Y_0$ be the projective bundle associated to ν_W, the fiber of ξ over a point $y \in Y_0$ being the set of all lines in the fiber of ν_W over y passing through the origin. The circle bundle ξ is trivial since Y_0 is homotopically 1-dimensional. Orient the surface Y_0 in an arbitrary way. This orientation together with the one of W induce orientations in ν_W and ξ. The line bundle ν induces a section of ξ over ∂Y_0. The obstruction to the extension of this section to Y_0 is an element of $H^2(Y_0, \partial Y_0; \pi_1(S^1)) = H^2(Y_0, \partial Y_0; \mathbb{Z}) = \mathbb{Z}$ where the former identification is induced by the orientation of ξ and the latter identification is induced by the orientation of Y_0. We define $\mathrm{gl}(Y)$ to be one half of the resulting integer. Note that $\mathrm{gl}(Y)$ does not depend on the choice of orientation in Y_0 (we have used this orientation twice). Clearly, $\mathrm{gl}(Y) \in (1/2)\mathbb{Z}$.

1.6.1. Lemma. *The skeleton X of W with the gleams of regions defined above is an integer shadowed polyhedron.*

Proof. We ought to show that for any region Y of X, we have $\mathrm{gl}(Y) + \eta(Y)/2 = 0 \pmod 1$ where $\eta(Y)$ is the residue mod 2 defined in Section VIII.6.2. If Y is a closed surface then this equality is obvious because $\mathrm{gl}(Y) \in \mathbb{Z}$ and $\eta(Y) = 0$. Assume that Y is non-compact. Let $\xi : E \mapsto Y_0$ be the circle bundle used above. To compute $\mathrm{gl}(Y)$ consider a trivialization $E = Y_0 \times S^1$ of ξ where the orientation in the fibers of E corresponds to a certain fixed orientation in S^1. The line bundle ν induces a section of ξ over ∂Y_0, i.e., a mapping $f : \partial Y_0 \to S^1$. The degree of this mapping with respect to the fixed orientation of S^1 and the orientation of ∂Y_0 induced by that of Y_0 is the obstruction used in the definition of $\mathrm{gl}(Y)$. Thus $\mathrm{gl}(Y) = \deg(f)/2$.

Let n be the number of components of ∂Y_0. Then $\deg(f) = \deg(f_1) + \dots + \deg(f_n)$ where f_i is the restriction of f to the i-th component of ∂Y_0. It is obvious that the integer $\deg(f_i)$ is even if and only if the mapping f_i lifts to the non-trivial 2-fold covering of S^1. This occurs when the restriction of ν to the i-th component of ∂Y_0 is a trivial line bundle. Therefore, by the definition of $\eta(Y)$, we have $\eta(Y) = \deg(f) \pmod 2$ and $\mathrm{gl}(Y) + \eta(Y)/2 = 0 \pmod 1$.

1.7. Theorem. *Let W be a compact oriented 4-manifold. Any two skeletons of W shadowed as above are stably shadow equivalent in the class of integer shadowed 2-polyhedra.*

Theorem 1.7 is the main result of this chapter. It allows us to define the shadow $\mathrm{sh}(W)$ of W to be the stable shadow equivalence class of any skeleton of W shadowed as in Section 1.6. Thus, $\mathrm{sh}(W)$ is an integer stable shadow. It follows from definitions that this shadow has empty boundary and commutes with disjoint union. It is obvious that $\mathrm{sh}(-W) = -\mathrm{sh}(W)$.

The shadow sh may be viewed as a geometric invariant of 4-manifolds which encodes 4-dimensional topology in terms of 2-dimensional polyhedra. This geometric invariant is by no means complete: in general it is impossible to reconstruct 4-manifolds from their shadows. For instance, the exterior of a 4-handlebody lying in a compact oriented 4-manifold W has the same shadow as W. (Indeed, a shadowed skeleton of this exterior is at the same time a shadowed skeleton of W.) As it will be clear below, this is the only kind of indeterminacy which may occur in the reconstruction problem. This shows that the shadow is a rather strong invariant of 4-manifolds. In particular, the shadow captures the intersection form and the signature of 4-manifolds. The problems concerning the reconstruction of 4-manifolds from their shadows and realization of shadows by 4-manifolds will be addressed in Section 6.

The simplest examples of 4-manifolds whose shadows may be explicitly computed are provided by disk bundles over surfaces. Let W be the total space of a 2-disk bundle over a closed connected orientable surface Σ. Then $\mathrm{sh}(W) = \mathrm{stab}\,([\Sigma_\chi])$ where χ is the self-intersection number of Σ in W. Since puncturing a 4-manifold (i.e., removing a small open 4-ball) preserves its shadow, we have

$$\mathrm{sh}(\mathbb{C}P^2) = \mathrm{stab}([S^2_1]) \quad \text{and} \quad \mathrm{sh}(-\mathbb{C}P^2) = \mathrm{stab}([S^2_{-1}]). \tag{1.7.a}$$

Here we use the fact that punctured $\mathbb{C}P^2$ (resp. $-\mathbb{C}P^2$) is the total space of a disk bundle over S^2 with the self-intersection number of S^2 being 1 (resp. -1). Another simple example: $\mathrm{sh}(S^4) = \mathrm{sh}(B^4) = \mathrm{stab}([S^2_0])$.

The definition of shadows of 4-manifolds may be extended to 4-manifolds with framed graphs in the boundary. The resulting shadows have non-empty boundary. We shall discuss this extension in Section 8.

Theorem 1.7 as well as Theorems 1.8 - 1.11 formulated below are proven in Section 7. It will be clear from the proof of Theorem 1.7 that the shadow moves relating skeletons of W may be performed inside W. This leads to a more precise version of Theorem 1.7 concerned with augmented skeletons. By an augmented shadowed skeleton of W we mean a pair (a shadowed skeleton X of W, the inclusion $X \hookrightarrow W$).

1.8. Theorem. *Let W be a compact oriented 4-manifold. Any two augmented shadowed skeletons of W are stably shadow equivalent in the class of W-augmented integer shadowed 2-polyhedra.*

This theorem allows us to upgrade $\mathrm{sh}(W)$ to a W-augmented shadow.
Now we state main properties of the shadows of 4-manifolds.

1.9. Theorem (additivity). *Let W_1 and W_2 be compact connected oriented 4-manifolds. Then $\mathrm{sh}(W_1 \# W_2) = \mathrm{sh}(W_1) + \mathrm{sh}(W_2)$. If $\partial W_1 \neq \emptyset$ and $\partial W_2 \neq \emptyset$ then $\mathrm{sh}(W_1 \#_\partial W_2) = \mathrm{sh}(W_1) + \mathrm{sh}(W_2)$ where $W_1 \#_\partial W_2$ is the result of gluing W_1 and W_2 along 3-balls lying in their boundaries.*

Theorem 1.9 is quite straightforward and the reader may prove it right away as an exercise. (We shall give a proof in Section 7.)

1.10. Theorem (comparison of bilinear forms). *Let W be a compact oriented 4-manifold. Let $X \subset W$ be a skeleton of W shadowed as in Section 1.6. Then the inclusion homomorphism $j : H_2(X; \mathbb{Z}) \to H_2(W; \mathbb{Z})$ is surjective and for any $u, v \in H_2(X; \mathbb{Z})$, we have $j(u) \cdot j(v) = Q_X(u, v)$.*

As above, the dot denotes the intersection product. For the definition of the bilinear form Q_X, see Section VIII.5. Theorem 1.10 shows that, considered modulo their annihilators, the intersection form of W and the bilinear form of $\mathrm{sh}(W)$ are isomorphic. In particular, they have equal signatures.
The next theorem justifies the notion of cobordism of shadows and relates it to cobordisms of 4-manifolds.

1.11. Theorem. *Let W, W' be compact connected oriented 4-manifolds with homeomorphic boundaries. If W, W' are cobordant modulo the boundary then their shadows $\mathrm{sh}(W), \mathrm{sh}(W')$ are cobordant.*

We say that W, W' are cobordant modulo the boundary if the 4-manifold obtained by gluing W and W' along $\partial W = \partial W'$ bounds a compact oriented 5-manifold. It is well known that compact oriented 4-manifolds with homeomorphic boundaries are cobordant modulo the boundary if and only if their signatures are equal.

1.12. Overview of further sections. To prove Theorems 1.7 - 1.11 we introduce another (independent) approach to shadows of 4-manifolds. We construct a shadow $\mathrm{sh}'(W)$ of a 4-manifold W using handle decompositions of W rather than skeletons of W. This construction given in Section 4 uses shadows of links in 3-manifolds defined in Sections 2 and 3. In Section 5 we relate the bilinear form of $\mathrm{sh}'(W)$ to the intersection form in $H_2(W)$. In Section 6 we show how to thicken integer shadows into 4-manifolds. This thickening is essentially inverse to the construction of sh$'$. In Section 7 we show that $\mathrm{sh}(W) = \mathrm{sh}'(W)$. This allows us to deduce Theorems 1.7 - 1.11 from the corresponding properties of $\mathrm{sh}'(W)$. In Section 8 we extend the shadow sh to 4-manifolds with framed graphs in the boundary.

1.13. Remark. The local flatness of skeletons of 4-manifolds is essential in the constructions of Section 1.6. It allows us to switch to the smooth category and use normal vector bundles. We may reformulate the local flatness in terms of polyhedral links of points. Let X be a 2-polyhedron in a 4-dimensional PL-manifold W. Every point $x \in X$ lies inside a closed 4-ball $U \subset W$ such that the 3-sphere ∂U intersects X transversally and $X \cap U$ is the cone over the 1-polyhedron $\partial U \cap X$ with the cone point x. It is well known that the pair $(\partial U, \partial U \cap X)$ does not depend on the choice of U up to PL-homeomorphisms (see [RS]). This pair is called the polyhedral link of x. It follows from definitions that the polyhedron X is flat in the point x if and only if the polyhedral link of x is unknotted in the sense that the 1-polyhedron $\partial U \cap X$ lies in a certain 2-sphere $S^2 \subset \partial U = S^3$. In the case where X is a simple 2-polyhedron, the homeomorphism type of $\partial U \cap X$ may be described explicitly. If $x \in \mathrm{Int}(X)$ then $\partial U \cap X = S^1$. If $x \in \mathrm{sing}(X)$ then $\partial U \cap X$ is a θ-curve, i.e., a union of three intervals with common vertices. Finally, if x is a vertex of X then $\partial U \cap X$ is homeomorphic to the 1-skeleton of a 3-simplex.

2. Shadows of 3-manifolds

2.0. Outline. We define shadows of 3-manifolds. This definition is independent of the material of Section 1 and proceeds entirely in the framework of 3-dimensional topology.

This section paves the way for a more elaborate setting of Section 3 where we define shadows of links in 3-manifolds. This will eventually lead to another definition of shadows of 4-manifolds and a proof of the theorems formulated in Section 1.

2.1. Skeletons of 3-manifolds. The skeleton of a compact 3-manifold M is an orientable simple 2-polyhedron (with empty boundary) $X \subset M \setminus \partial M$ such that $M \setminus X$ is a disjoint union of open 3-balls and an open collar $\partial M \times [0, 1) \subset M$ of ∂M. For instance, the standard 2-sphere $S^2 \subset S^3$ is a skeleton of S^3. The closed orientable surface Σ is a skeleton of $\Sigma \times [-1, 1]$.

2.1.1. Theorem. *Every compact 3-manifold has skeletons.*

The proof of Theorem 2.1.1 given in Section 2.3 uses the technique of dual cell subdivisions of triangulations due to H. Poincaré. This technique is outlined in Section 2.2.

Different skeletons of the same 3-manifold may be related by so-called Matveev-Piergallini moves. By Matveev-Piergallini moves we mean the geometric transformations of orientable simple 2-polyhedra underlying the basic shadow moves P_2, P_3, the suspension $X \mapsto X + S_0^2$, and the inverse transformations. As usual, we use inverse transformations only when they preserve the orientablility. Note that we do not include the transformation underlying P_1 and its inverse in the set of Matveev-Piergallini moves. There are no gleams in the Matveev-Piergallini theory.

2.1.2. Theorem. *Let M be a compact 3-manifold. Any two skeletons of M may be related by a finite sequence of Matveev-Piergallini moves in the class of skeletons of M.*

It is understood that the moves are performed inside small 3-balls in M which serve as the ambient 3-space in Figure VIII.1.1. It is obvious that each such move applied to a skeleton of M produces a skeleton of M.

Theorem 2.1.2 follows from the results of Casler, Matveev, and Piergallini concerned with spines of 3-manifolds. We recall these results in Section 2.4 and prove Theorem 2.1.2 in Section 2.5.

We upgrade any skeleton $X \subset M$ to a shadow by equipping all regions of X with zero gleams. This yields an integer shadowed polyhedron. (Condition (VIII.6.3.a) follows from the fact that the normal line bundle of Y_0 in M yields an extension of $\nu(Y)$ to Y_0 and therefore $\eta(Y) = 0$.) It is obvious that the Matveev-Piergallini moves on skeletons lift to the basic shadow moves P_2, P_3, the suspension, and the inverse moves. Note that neither non-zero gleams nor the move P_1 appear in this context. Ignoring this fact and considering the skeletons of M up to all shadow moves and suspension we get the following corollary.

2.1.3. Corollary. *Let M be a compact 3-manifold. Any two skeletons of M equipped with zero gleams are stably shadow equivalent in the class of integer shadowed 2-polyhedra.*

From now on we shall consider skeletons of 3-manifolds as integer shadowed polyhedra with the gleams of regions equal to zero. The stable integer shadow represented by a skeleton of M is called the (stable) internal shadow of M and denoted by $\mathrm{ish}(M)$. For example, $\mathrm{ish}(S^3) = \mathrm{stab}([S_0^2])$. The internal shadow is defined for non-oriented and even non-orientable 3-manifolds.

The language of shadowed polyhedra may look artificial in this setting since we encounter only zero gleams. This language will be justified in Section 3.

2.2. Dual cell subdivisions of manifolds. We outline the classical construction of the dual cell subdivision of a piecewise-linear triangulation. For more details, see [RS].

Let μ be a piecewise-linear triangulation of an m-dimensional manifold M, possibly with boundary. For a strictly increasing sequence $E_0 \subset E_1 \subset ... \subset E_n$ of (closed) simplices of μ, we denote by $[E_0, E_1, ..., E_n]$ the n-dimensional linear simplex in E_n whose vertices are the barycenters of $E_0, E_1, ..., E_n$. Such simplices corresponding to all sequences $E_0 \subset E_1 \subset ... \subset E_n$ with fixed E_n form a simplicial subdivision of E_n. Therefore the simplices $[E_0, E_1, ..., E_n]$ corresponding to all strictly increasing sequences of simplices of μ form a simplicial subdivision of μ. It is called the first barycentric subdivision of μ and denoted by μ^1.

Let us construct the cell subdivision μ^* of M dual to μ. (The reader unfamiliar with this construction is recommended to experiment with triangulations of the 2-disk and to draw a few figures.) For any simplex E of μ denote by E^* the union of all simplices $[E_0, E_1, ..., E_n]$ of μ^1 with $E_0 = E$. Each such simplex intersects E in one point which is the barycenter of E. Therefore E^* intersects E solely in the barycenter of E. Since μ is piecewise-linear, E^* is a closed PL-cell of dimension $m - \dim(E)$. (This means that E^* is piecewise-linearly homeomorphic to the standard closed PL-ball of dimension $m - \dim(E)$.) If E does not lie in ∂M then the sphere $\partial E^* \subset M \setminus \partial M$ is the union of the cells F^* where F runs over all simplices of μ containing E and distinct from E. Therefore, in the case $\partial M = \emptyset$ the cells $\{E^* \mid E \text{ is a simplex of } \mu\}$ form a cell subdivision of M. This is the cell subdivision μ^* dual to μ.

If $\partial M \neq \emptyset$ we use the triangulation $\partial\mu$ of ∂M induced by μ and the dual cell subdivision $(\partial\mu)^*$ of ∂M. To each simplex E of μ lying in ∂M we associate two cells: the cell $E^* \subset M$ defined above and the cell $E^*_\partial \subset \partial M$ dual to E with respect to $\partial\mu$. It is obvious that $E^*_\partial = E^* \cap \partial M \subset \partial E^*$. Moreover, the sphere ∂E^* is the union of E^*_∂ and the cells F^* where F runs over all simplices of μ containing E and distinct from E. This shows that the cells $\{E^* \mid E \text{ is a simplex of } \mu\}$ and $\{E^*_\partial \mid E \text{ is a simplex of } \partial\mu\}$ form a cell subdivision of M. This is the cell subdivision μ^* dual to μ. Note that the cell space $(\partial M, (\partial\mu)^*)$ is a cell subspace of (M, μ^*).

2.3. Proof of Theorem 2.1.1. Choose a triangulation μ of M and consider the dual cell subdivision μ^* of M. Denote by X the 2-skeleton of μ^*, i.e., the union of 0-cells, 1-cells, and 2-cells of μ^*. It is clear that X is a 2-dimensional polyhedron embedded in M. Its complement in M consists of open 3-cells of μ^*. An inspection of the local behaviour of X shows that X is a simple polyhedron whose regions are open 2-cells of μ^*. (The vertices and 1-strata of X are the 0-cells and open 1-cells of μ^*.) Thus, X is an orientable simple 2-polyhedron in M. If $\partial M = \emptyset$ then X is a skeleton of M. In the case $\partial M \neq \emptyset$ we have $\partial M \subset X$. Pushing X inside M slightly we get an orientable simple 2-polyhedron in M whose complement consists of open 3-balls and an open collar of ∂M. This 2-polyhedron is a skeleton of M.

2.4. Spines of 3-manifolds. A spine of a closed 3-manifold is a 2-polyhedron $X \subset M$ such that $M \setminus X$ is an open 3-ball. A spine of a compact 3-manifold M with $\partial M \neq \emptyset$ is a 2-polyhedron $X \subset M \setminus \partial M$ such that $M \setminus X$ is homeomorphic to $\partial M \times [0, 1)$.

A simple 2-polyhedron X is said to be special if $\partial X = \emptyset$, all regions of X are open 2-disks, and the graph $\mathrm{sing}(X)$ has no components homeomorphic to S^1. A spine X of a compact 3-manifold M is said to be special if it is a special simple 2-polyhedron.

According to Casler [Cas], every compact 3-manifold M has a special spine. This result is much deeper than the existence of skeletons. Explicit constructions of spines are rather involved, fortunately we shall not need them.

According to [Mat2] and [Pi], any two special spines of M may be related by a (finite) sequence of transformations underlying the shadow moves $P_2^{\pm 1}, P_3^{\pm 1}$. It is understood that these transformations are performed inside 3-balls in M and that all intermediate 2-polyhedra are special spines of M.

2.5. Proof of Theorem 2.1.2. Consider in more detail the suspension on a skeleton $X \subset M$. Take a region Y of X and a small closed 2-disk $D \subset Y$. The suspension attaches a closed 2-disk D' to X along the circle $\partial D' = \partial D$. The disk D' is attached inside a small 3-ball in M. The resulting 2-polyhedron $X' = X \cup D' \subset M$ is a skeleton of M. The region Y of X gives rise to a region $Y \setminus D$ of X'. Pushing D' across a 1-stratum of X adjacent to Y (via the move P_2) we get a skeleton $X'' \subset M$ whose regions are either open 2-disks or homeomorphic to regions of X. Thus, if X is special then so is X''. It is obvious that a closed regular neighborhood of X'' in M is homeomorphic to a closed regular neighborhood of X' in M. It is obtained from a closed regular neighborhood of X in M by puncturing (removing a small open 3-ball).

We may also follow the suspension $X \mapsto X'$ with a transformation pushing D across several 1-strata of X adjacent to Y. This gives a skeleton $\tilde{X} \subset M$ which has the same regions as X except that Y is replaced by one or several regions with simpler topology and a few open 2-disks. Iterating the construction $X \mapsto \tilde{X}$ we can transform any skeleton $X \subset M$ into a special skeleton of M.

Now we can prove the theorem. Let X_1, X_2 be two skeletons of M. Applying to X_1 and X_2 the construction $X \mapsto \tilde{X}$ described above, we transform both X_1 and X_2 into special skeletons of M. Applying the construction $X \mapsto X''$, we can transform the latter skeletons into special skeletons $Z_1, Z_2 \subset M$ such that the complements of their regular neighborhoods in M consist of an equal number of small open 3-balls and a collar of ∂M (if $\partial M \neq \emptyset$).

It is well known that any two collars of ∂M are ambiently isotopic in M. Similarly, any two systems of small disjoint open 3-balls consisting of an equal number of balls are ambiently isotopic in M. Therefore applying to Z_1 an ambient isotopy in M we may assume that certain regular neighborhoods of Z_1 and Z_2 in M coincide as sets. Denote this common regular neighborhood of Z_1 and Z_2 by U.

It is obvious that both Z_1 and Z_2 are special spines of U. The Matveev-Piergallini theorem quoted in Section 2.4 guarantees that Z_1 and Z_2 may be related by a sequence of Matveev-Piergallini moves inside U. This implies the claim of the theorem.

2.6. Remarks. 1. There is a stronger, non-stable version of the internal shadow of a 3-manifold represented by an arbitrary spine of this 3-manifold (with zero gleams of regions). However, the stable shadow is sufficient for our aims.

2. The proof of Theorem 2.1.1 derives a skeleton of a 3-manifold M from a triangulation of M. Theorem 2.1.2 implies that for any skeleton X obtained from a triangulation of M, the stable shadow $\text{stab}([X])$ does not depend on the choice of triangulation. This fact may be established by purely combinatorial methods using Pachner's theory of bistellar subdivisions (see Section VII.2.1) or Alexander's theory of elementary subdivisions of triangulations, see [Al2], [TV]. This yields a combinatorial construction of $\text{ish}(M)$ avoiding the theory of spines.

3. There is another approach to skeletons of 3-manifolds based on Heegaard surfaces. A Heegaard surface in a closed connected 3-manifold M is a closed surface $\Sigma \subset M$ splitting M in the union of two 3-dimensional handlebodies. Let $\{D_i\}_i$ and $\{D'_j\}_j$ be systems of meridianal disks in these two handlebodies. (A system of meridianal disks in a handlebody U is a collection of disjoint, properly embedded 2-disks in U such that splitting U along these disks gives a closed 3-ball.) Assume that the loops $\{\partial D_i\}_i$, $\{\partial D'_j\}_j$ lie in general position in Σ. Then $X = \Sigma \cup (\cup_i D_i) \cup (\cup_j D'_j) \subset M$ is a skeleton of M whose complement consists of two disjoint open 3-balls. Hence $\text{ish}(M) = \text{stab}([X])$. This formula may be used to compute the internal shadow of M. It may be used also as a definition of $\text{ish}(M)$ based on the theory of Heegaard decompositions. The independence of $\text{stab}([X])$ of the choice of Heegaard surface and meridianal disks can be deduced from the Singer theorem on stable equivalence of Heegaard decompositions of M and the fact that all systems of meridianal disks in a handlebody are related by disk slidings.

Similar constructions apply to 3-manifolds with boundary. A Heegaard surface in a compact connected 3-manifold M with $\partial M \neq \emptyset$ is a closed surface $\Sigma \subset M$ splitting M into a union of a 3-dimensional handlebody and a 3-manifold obtained by attaching several 3-dimensional handles of index 2 to the top base of the cylinder $\Sigma \times [0, 1]$. The cores of these handles play the role of meridianal disks in the constructions above. This approach is equivalent to the one based on self-indexing Morse functions $f : M \to \mathbb{R}$ with $f(\partial M) = 3$. The equivalence goes by assigning to such a function f, the Heegaard surface $f^{-1}(3/2)$.

The techniques of dual cell subdivisions and Heegaard splittings are more classical but slightly less satisfactory than the technique of spines, since they do not produce all skeletons of 3-manifolds. Any of these three approaches suits our aims. For the sake of practical computations, the technique of Heegaard splittings seems to be the most convenient.

2.7. Exercise. Show that for any compact 3-manifold M, we have $-\mathrm{ish}(M) = \mathrm{ish}(M)$. If M^0 is obtained from M by puncturing then $\mathrm{ish}(M^0) = \mathrm{ish}(M)$. For any compact connected 3-manifolds M, N, we have $\mathrm{ish}(M\#N) = \mathrm{ish}(M) + \mathrm{ish}(N)$.

3. Shadows of links in 3-manifolds

3.0. Outline. The objective of this section is to construct shadow cones of framed links in 3-manifolds. This is a decisive step towards a second definition of shadows of 4-manifolds given in Section 4.

First of all we extend the classical theory of planar link diagrams to framed links in compact oriented 3-manifolds. In the role of the plane we use skeletons of 3-manifolds. The link diagrams on skeletons are defined as in the classical theory with one essential difference: for each link component, we specify an integer or half-integer indicating the number of "additional" twists in the framing. Instead of three Reidemeister moves we have eight moves. The next step is to define the shadow projection of a link diagram. This projection replaces the information on over/undercrossings with numbers (gleams) located in the regions of the diagram. Combining these two constructions we pass from framed links in 3-manifolds to shadow links in 2-polyhedra. Taking the cone (as defined in Section VIII.3.2) we obtain the shadow cone for framed links in 3-manifolds.

3.1. Diagrams of links. Let M be a compact oriented 3-manifold with a skeleton $X \subset M$. Assume that the surface $\mathrm{Int}(X)$ is equipped with a normal direction in M. (A normal direction on $\mathrm{Int}(X)$ in M is an orientation of the normal line bundle of $\mathrm{Int}(X)$ in M.) A link diagram in X is a generic system of loops in X (cf. Section VIII.3.1) such that at each crossing point of the loops one of the two intersecting branches is distinguished and said to be the lower one, the second branch being the upper one. Recall that the crossings of generic loops may occur only inside $\mathrm{Int}(X)$. Pushing slightly all upper branches in $M \setminus X$ along the given normal direction we associate to every link diagram d a link $\ell = \ell(d) \subset M$. Its normal bundle $\nu_M(\ell)$ in M has two transversal 1-dimensional subbundles ν_1 and ν_2 defined as follows. The line bundle ν_1 on ℓ is induced by the normal bundle of $\mathrm{Int}(X) \subset M$ restricted to d. (The bundle ν_1 is well-defined in a point of $d \cap \mathrm{sing}(X)$ since, in a neighborhood of such a point, d traverses two regions of X locally forming a 2-disk.) The line bundle ν_2 on ℓ is induced by the normal bundle of d in X (cf. Remark VIII.3.3.3). The line bundles ν_1, ν_2 on ℓ are isotopic, an isotopy is obtained by rotation of ν_1 around ℓ in the positive direction determined by the orientation of M.

An enriched link diagram in X is a link diagram in X whose underlying loops are equipped with integers or half-integers called pre-twists. The pre-twist of a loop L belongs to $\mathbb{Z}$ if the normal line bundle of L in X is trivial and to $(1/2)+\mathbb{Z}$ otherwise. Thus, the pre-twist of L is an integer if and only if the normal line bundles $\nu_1 \approx \nu_2$ on the corresponding knot $\ell(L) \subset M$ are trivial.

An enriched link diagram in X gives rise to a framed link in M as follows. Consider first the link $\ell \subset M$ with normal line bundle ν_1 determined by the underlying link diagram with pre-twists forgotten. Then twist ν_1 around each component of ℓ as many times as the pre-twist of the corresponding loop. (The positive direction of the twist is determined by the orientation of M. For instance, a pre-twist of 1/2 gives rise to a positive half-twist of ν_1.) This produces a trivial line bundle on ℓ. Its non-zero sections yield the desired framing of ℓ.

It is easy to see that every framed link in M may be presented by an enriched link diagram in X. According to [Tu13] two such diagrams present isotopic framed links if and only if these diagrams may be obtained from each other by a sequence of moves $\Omega_1, \ldots, \Omega_8$ (see Figures 3.1 - 3.3), their inverses, and an ambient isotopy in X. The moves $\Omega_1, \Omega_2, \Omega_3$ proceed in $\mathrm{Int}(X)$, and $\Omega_4, \ldots, \Omega_8$ proceed in a

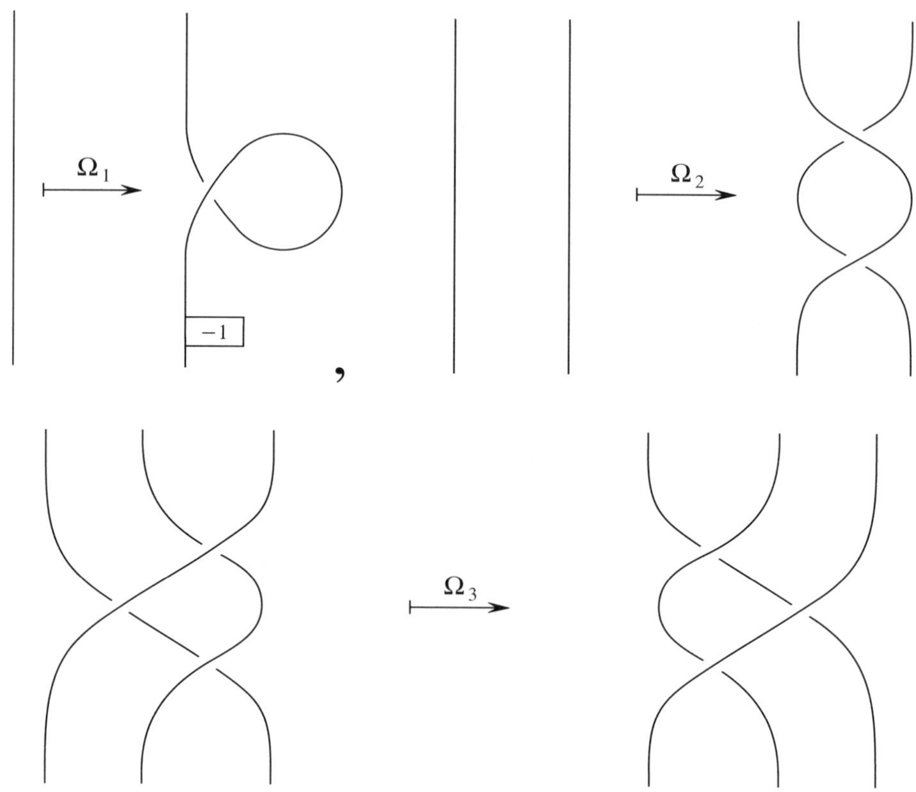

Figure 3.1

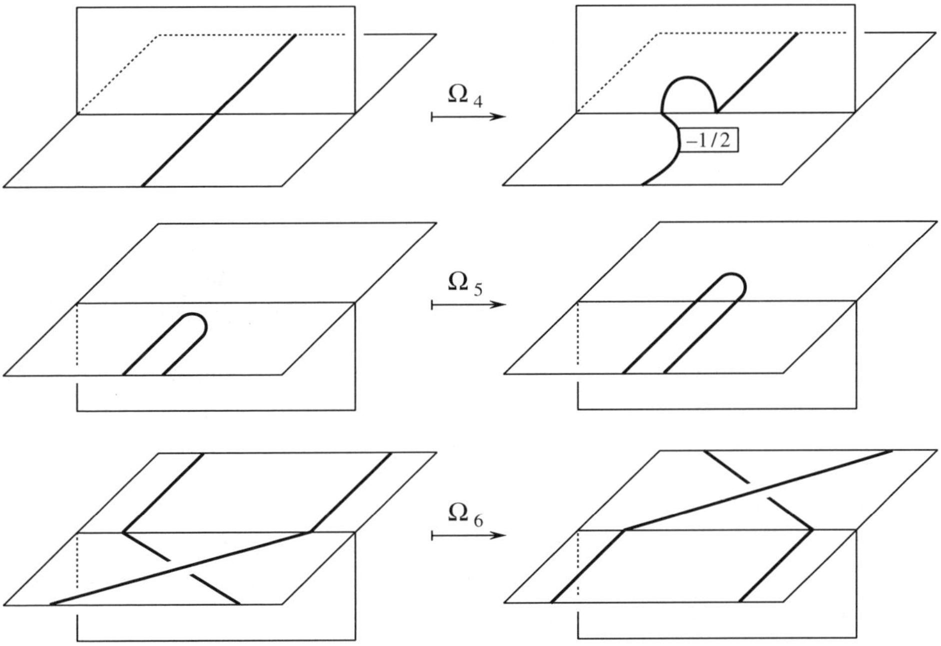

Figure 3.2

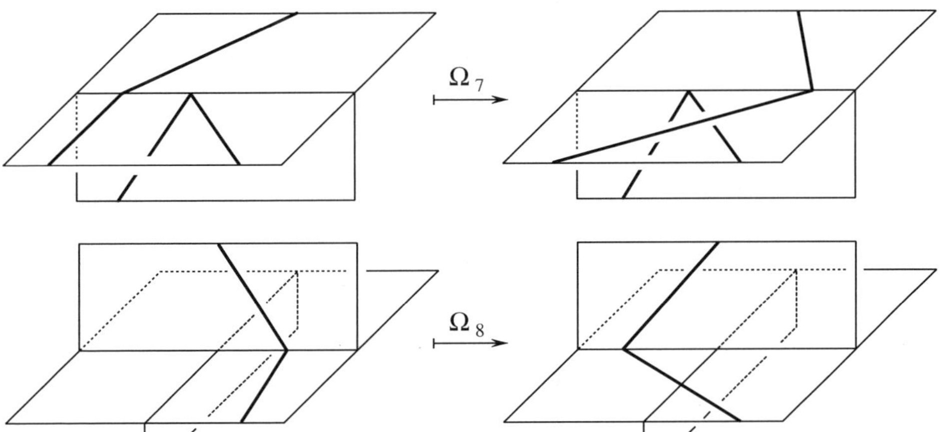

Figure 3.3

neighborhood of sing(X). In Figures 3.2, 3.3 the link diagrams on X are drawn in bold in order to distinguish them from 1-strata of X. The moves Ω_2, Ω_3, Ω_5, ..., Ω_8 do not change the pre-twists, while Ω_1 and Ω_4 decrease the pre-twist by 1 and 1/2 respectively. (We assume that in the pictures of Ω_1, Ω_4 the orientation in M

corresponds to right-handed orientation in $\mathbb{R}^3$.) This completes the description of framed links in M in terms of enriched link diagrams in X.

3.2. Shadow projections of links. Let M be a compact oriented 3-manifold and let $X \subset M \setminus \partial M$ be a skeleton of M. To each framed link $\ell \subset M$ we associate an integer shadow link in X. It is denoted by $\mathrm{sh}(\ell; X)$ and called the shadow projection of ℓ into X.

Fix an orientation in $\mathrm{Int}(X)$. This orientation, together with that of M, determines a normal direction on $\mathrm{Int}(X)$ in M. This allows us to present framed links in M by link diagrams in X.

Let D be an enriched link diagram in X presenting a framed link $\ell \subset M$. We define an integer shadowed system of loops $\mathrm{sh}(D)$ in X as follows. The loops of $\mathrm{sh}(D)$ and their pre-twists are those of D. Each self-crossing point of D contributes to the gleams of four adjacent regions as in Figure 3.4 where the orientation of $\mathrm{Int}(X)$ is counterclockwise. Each point of $D \cap \mathrm{sing}(X)$ contributes to the gleams of five adjacent regions as in Figure 3.5 where the orientation of M is the right-handed one. The gleam of a region of $\mathrm{sh}(D)$ is defined to be the sum of the contributions of adjacent crossing points.

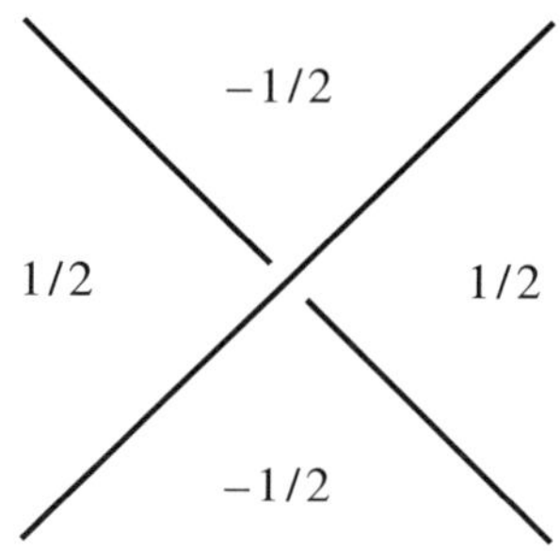

Figure 3.4

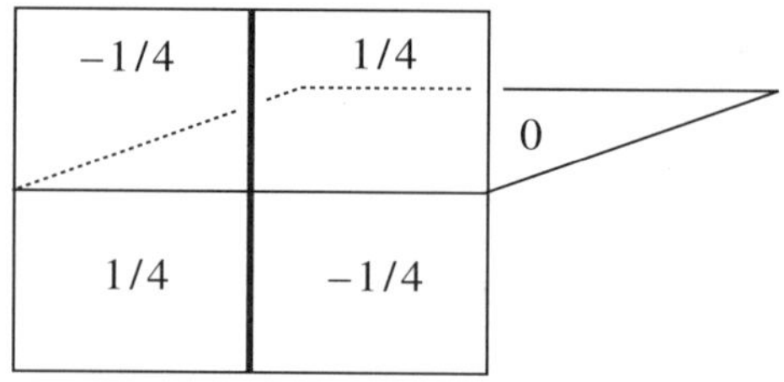

Figure 3.5

3.2.1. Lemma. $\mathrm{sh}(D)$ *is an integer shadowed system of loops in* X.

It is a simple exercise to check that each move Ω_i, $i = 1, ..., 8$ applied to a link diagram D induces the corresponding move S_i on $\mathrm{sh}(D)$. Therefore, the results of Section 3.1 imply that the shadow link in X represented by $\mathrm{sh}(D)$ depends only on the isotopy class of ℓ in M. This shadow link is the shadow projection $\mathrm{sh}(\ell; X)$ of ℓ into X.

The shadow link $\mathrm{sh}(\ell; X)$ does not depend on the choice of orientation in $\mathrm{Int}(X)$. The constructions of a diagram D of ℓ and the shadow $\mathrm{sh}(D)$ use this orientation but their composition is independent of it. Note that $\mathrm{sh}(\ell; X)$ depends on the orientation of M.

Proof of Lemma. It is obvious that the sum of the gleams of regions of $\mathrm{sh}(D)$ contained in any fixed region of X is equal to zero. Therefore $\mathrm{sh}(D)$ is a shadowed system of loops in X. It remains to show that the gleam of any region Y of $\mathrm{sh}(D)$ belongs to $(1/2)\mathbb{Z}$ and that the shadow cone over $\mathrm{sh}(D)$ satisfies (VIII.6.3.a). Recall the compact subsurface Y_0 of Y and the line bundle $\nu(Y)$ over ∂Y_0 defined in Section VIII.6.2. We may view the 1-manifold ∂Y_0 as a polygon whose vertices correspond in a natural way to (i) vertices of $\mathrm{sing}(X)$ adjacent to Y, (ii) self-crossings of D adjacent to Y, and (iii) crossing points of $\mathrm{sing}(X)$ with D adjacent to Y. The contributions of the vertices of types (i), (ii), (iii) to $\mathrm{gl}(Y)$ are equal to $0, \pm 1/2, \pm 1/4$ respectively. The edges of ∂Y_0 correspond to 1-strata of $\mathrm{sing}(X)$ and arcs of D adjacent to Y. The vertices of ∂Y_0 of type (iii) split ∂Y_0 into two parts corresponding to $\mathrm{sing}(X)$ and D respectively. Therefore such vertices are even in number and $\mathrm{gl}(Y) \in (1/2)\mathbb{Z}$.

Consider the shadow cone over $\mathrm{sh}(D)$. For the disk regions of this cone attached to the loops of $\mathrm{sh}(D)$, condition (VIII.6.3.a) follows from the condition on the pre-twists imposed in Section 3.1. Therefore it suffices to check (VIII.6.3.a) for the regions of $\mathrm{sh}(D)$. Let Y be a region of $\mathrm{sh}(D)$ in X. For any vertex of ∂Y_0 corresponding to a self-crossing point of D, we may smooth D at this point so that Y remains a region of the resulting diagram. The smoothing proceeds in a small disk neighborhood of the crossing point and replaces two intersecting branches of D with two disjoint branches with the same ends (cf. Figure XII.1.1). This transformation adds to $\mathrm{gl}(Y)$ either $+1/2$ or $-1/2$ and removes one half-twist from the line bundle $\nu(Y)$. Therefore this transformation keeps the residue $(\mathrm{gl}(Y) + \eta(Y)/2) (\mathrm{mod}\, 1)$ intact. Hence, it is enough to consider the case when there are no self-crossing points of D adjacent to Y, i.e., the polygon ∂Y_0 has no vertices of type (ii). Denote by a the part of this polygon formed by (closed) edges corresponding to 1-strata of $\mathrm{sing}(X)$. The normal orientation of $Y \subset \mathrm{Int}(X)$ in M induces a trivialization of $\nu(Y)$ over a. Every connected component b of $\partial Y_0 \setminus a$ corresponds to a simple arc in D adjacent to Y, connecting two points of $\mathrm{sing}(X) \cap D$. Each end of b contributes $\pm 1/4$ to $\mathrm{gl}(Y)$. It is a nice geometric exercise to verify that if the joint contribution of the ends of b is equal to zero then the trivialization of $\nu(Y)$ over a extends across b. If this joint contribution is equal to $\pm 1/2$ then the trivialization of $\nu(Y)$ over a does not extend across

b. This implies that the line bundle $\nu(Y)$ is trivial over a connected component $S^1 \subset \partial Y_0$ if and only if the vertices of this component of type (iii) contribute an integer amount to $\mathrm{gl}(Y)$. Summing up over all components of ∂Y_0 we get $\mathrm{gl}(Y) + \eta(Y)/2 \in \mathbb{Z}$.

3.3. Shadow cones of framed links. Let ℓ be a framed link in a compact oriented 3-manifold M. The shadow cone $CO(M, \ell)$ of ℓ is the cone over the shadow projection $\mathrm{sh}(\ell; X)$ of ℓ into a skeleton $X \subset M$. This is a stable integer shadow with empty boundary (cf. Sections VIII.3.2 and VIII.6.4). It does not depend on the choice of X. Indeed, any two skeletons of M may be related by a sequence of Matveev-Piergallini moves proceeding inside small 3-balls in M. At each step we may deform ℓ out of the 3-ball where the move proceeds so that the move does not interfere with a diagram of ℓ. Therefore this move may be viewed as a corresponding shadow move (or a suspension, or an inverse shadow move) on the shadow cone. The resulting shadow moves (together with the shadow moves induced by the isotopy of ℓ) establish stable shadow equivalence of the cones obtained from different skeletons.

Example: $CO(M, \emptyset) = \mathrm{ish}(M)$.

Similarly, we may define shadow cylinders of framed links. They will be discussed in a more general context in Section 8.

3.4. Exercise. Show additivity of the shadow cone: if k, ℓ are framed links in compact connected oriented 3-manifolds M, N then

$$CO(M \# N, k \amalg \ell) = CO(M, k) + CO(N, \ell). \tag{3.4.a}$$

Show that

$$CO(-M, \ell) = -CO(M, \ell). \tag{3.4.b}$$

4. Shadows of 4-manifolds via handle decompositions

4.0. Outline. We construct a shadow sh' of 4-manifolds using the technique of handle decompositions, link diagrams, and shadow projections. The advantage of this approach over the one of Section 1 stems from the fact that we can prove consistency of the definition of sh' directly. Equivalence of these two approaches will be established in Section 7.

4.1. Shadow of a 4-manifold. It is well known that any compact piecewise-linear 4-manifold may be obtained from a finite number of disjoint closed 4-balls by attaching i-handles with $i = 1, 2, 3, 4$. Such a decomposition of the manifold into handles is called a handle decomposition, the initial 4-balls are viewed as 0-handles.

Let W be a compact oriented (piecewise-linear) 4-manifold. Fix a handle decomposition of W and denote by H the union of 0-handles and 1-handles. It is clear that connected components of H are oriented 4-dimensional handlebodies. (Each component of W contains exactly one component of H.) The 2-handles of W are attached to H along certain disjoint solid tori $\{p_j : S^1 \times B^2 \hookrightarrow \partial H\}_j$. These solid tori give rise to a framed link $\ell \subset \partial H$ with components $\{p_j(S^1 \times x)\}_j$ and framing determined by the longitudes $\{p_j(S^1 \times y)\}_j$ where x and y are two distinct points of the 2-disk B^2. We provide ∂H with an orientation induced by that of H so that the pair (the tangent vector directed into $W \setminus H$, the orientation of ∂H) determines the given orientation of W. Consider the shadow cone $CO(\partial H, \ell)$ of ℓ (see Section 3). This is a stable integer shadow with empty boundary. Set $\mathrm{sh}'(W) = CO(\partial H, \ell)$.

4.2. Theorem. *For any compact oriented (piecewise-linear) 4-manifold W, the shadow $\mathrm{sh}'(W)$ does not depend on the choice of handle decomposition of W.*

This theorem shows that $\mathrm{sh}'(W)$ is well-defined. It is obvious that sh' commutes with disjoint union. We shall show in Section 7 that $\mathrm{sh}'(W) = \mathrm{sh}(W)$.

Proof of Theorem. It is well known that any two handle decompositions of W are related by a sequence of handle slidings and births and deaths of complementary handle pairs. Births and deaths of complementary (0,1)- or (3,4)-handle pairs do not change the topological type of the pair (H, ℓ). A birth of a complementary (2,3)-handle pair does not change H and adds to ℓ one unknotted unlinked component with the trivial framing. This results in a suspension of the shadow cone which does not change its stable type. A birth of a complementary (1,2)-handle pair replaces $(\partial H, \ell)$ by its connected sum with $(S^1 \times S^2, k)$ where $k = S^1 \times x$ with the framing corresponding to the longitude $S^1 \times y$; where x and y are two distinct points of S^2. Let α be the shadow cone of the knot $k \subset S^1 \times S^2$. In view of (3.4.a) it suffices to show that $\alpha = \mathrm{stab}\,([S^2_0])$. Let r, s be distinct points of S^1 and let B^2_+, B^2_- be complementary hemispheres of S^2 with $B^2_+ \cap B^2_- = S^1$. Set $X = (S^1 \times S^1) \cup (r \times B^2_+) \cup (s \times B^2_-) \subset S^1 \times S^2$. It is obvious that X is a skeleton of $S^1 \times S^2$. The shadow projection of k in X is presented by the loop $S^1 \times r \subset S^1 \times S^1 \subset X$ with the zero pre-twist and zero gleams of all regions. Therefore α is represented by the torus $S^1 \times S^1$ with three 2-disks glued to it along the circles $r \times S^1$, $s \times S^1$, and $S^1 \times r$, the gleams of all regions being equal to zero. This shadowed polyhedron, say X', may be embedded in S^3 so that its complement in S^3 consists of three disjoint open 3-balls. Hence X' is a skeleton of S^3. Since all skeletons of S^3 are stably shadow equivalent,

$$\alpha = \mathrm{stab}\,([X']) = \mathrm{stab}\,([S^2_0]).$$

Let us prove the invariance of $CO(\partial H, \ell)$ under the handle slidings. The slidings of handles of indices 1 and 3 do not change the topological type of the pair

$(\partial H, \ell)$. The slidings of handles of index 2 have the effect of band summation on the components of ℓ. A band summation replaces one of the components of ℓ by its band sum with another component, taking proper care of the framings. Replacing, if necessary, ℓ with an isotopic link in ∂H we may assume that the given band is narrow and short. Present ℓ by an enriched link diagram D on a skeleton X of ∂H. We may choose X small enough that it approximates ℓ, its framing, and the band very closely. Hence, we can assume that this band lies in $\mathrm{Int}(X)$ and only intersects D along its bases lying on two loops d and e of D. We may also assume that d and e have no intersections and self-intersections and that the pre-twist of e is equal to 0. This ensures that a regular neighborhood of e in X contains an annulus (and not a Möbius band). Therefore the 2-disk B_e^2 attached to e in the shadow cone of $\mathrm{sh}(D)$ is untwisted in the sense of Section VIII.2.5 and has zero gleam. The link obtained from ℓ by the band move may be presented by the enriched link diagram D' obtained from D by replacing the loop d with a loop d' which goes around d and a longitude of e in X. The pre-twist of d' equals that of d, other loops preserve their pre-twists. It is easy to see that the cone over $\mathrm{sh}(D')$ can be obtained from the cone over $\mathrm{sh}(D)$ by sliding the 2-disk region attached to d over the disk region $\mathrm{Int}(B_e^2)$. Theorem VIII.2.5.1 implies that these two cones are shadow equivalent. Therefore the shadow $CO(\partial H, \ell)$ is invariant under band summation. This completes the proof of the theorem.

4.3. The W-augmented shadow. Let W be a compact oriented 4-manifold. The construction of the shadow $\mathrm{sh}'(W)$ is carried out more precisely, it produces a W-augmented stable shadow with source $\mathrm{sh}'(W)$. Let H, ℓ be the same objects as in Section 4.1 and let X be a skeleton of ∂H. Let $\{\ell_i\}_i$ be the components of ℓ and $\{L_i\}_i$ be the corresponding components of the shadow projection of ℓ into X. For each L_i, consider the 2-disk $B_i \subset CO(\partial H, \ell)$ glued to X along L_i to form $CO(\partial H, \ell)$. We define a mapping $B_i \to W$ as follows: a small concentric subdisk B_i' is mapped homeomorphically onto the core of the 2-handle of W attached along a regular neighborhood of ℓ_i in ∂H; the annulus $B_i \setminus B_i'$ is mapped onto a narrow annulus in ∂H bounded by ℓ_i and L_i. (It is understood that these two mappings extend the same homeomorphism $\partial B_i' \to \ell_i$.) These mappings $\{B_i \to W\}_i$ extend the inclusion $X \hookrightarrow \partial H$ to a mapping $CO(\partial H, \ell) \to W$. An inspection of our construction shows that this last mapping considered up to homotopy determines a W-augmented stable shadow depending solely on W. Its source is the stable shadow $\mathrm{sh}'(W)$. Note that all mappings $C \to W$ representing this W-augmented shadow induce isomorphisms of fundamental groups and epimorphisms in the 2-dimensional homologies.

4.4. Examples. In many cases the shadow $\mathrm{sh}'(W)$ may be computed explicitly. For instance, if W is obtained by attaching 2-handles to the 4-ball B^4 along a framed link $\ell \subset S^3 = \partial B^4$ (and possibly some 3- and 4-handles), then by the very definition $\mathrm{sh}'(W) = \mathrm{stab}(CO(S^3, \ell))$. For instance, taking $\ell = \emptyset$ we get

$\mathrm{sh}'(B^4) = \mathrm{stab}([S_0^2])$. Another example is provided by $\mathbb{C}P^2$. Since punctured $\mathbb{C}P^2$ may be obtained by attaching a 2-handle to the 4-ball along the trivial knot with framing 1 we have $\mathrm{sh}'(\mathbb{C}P^2) = \mathrm{stab}([S_1^2])$. Similarly, $\mathrm{sh}'(-\mathbb{C}P^2) = \mathrm{stab}([S_{-1}^2])$.

4.5. Elementary properties of sh′. We list a few elementary properties of sh′.

4.5.1. Lemma. *Let W be a compact oriented 4-manifold. Then $\mathrm{sh}'(-W) = -\mathrm{sh}'(W)$. If W_0 is obtained by puncturing W then $\mathrm{sh}'(W_0) = \mathrm{sh}'(W)$. More generally, if W_0 is the closed exterior of a finite graph lying in $W \setminus \partial W$ then $\mathrm{sh}'(W_0) = \mathrm{sh}'(W)$.*

Proof. The equality $\mathrm{sh}'(-W) = -\mathrm{sh}'(W)$ follows from (3.4.b). Other equalities follow from the fact that a handle decomposition of W may be obtained from that of W_0 by attaching handles of indices 3 and 4.

4.5.2. Lemma. *Let W_1, W_2 be compact connected oriented 4-manifolds. Then $\mathrm{sh}'(W_1 \# W_2) = \mathrm{sh}'(W_1) + \mathrm{sh}'(W_2)$. If $\partial W_1 \neq \emptyset$ and $\partial W_2 \neq \emptyset$ then*

$$\mathrm{sh}'(W_1 \#_\partial W_2) = \mathrm{sh}'(W_1) + \mathrm{sh}'(W_2). \tag{4.5.a}$$

For the definition of $\#_\partial$, see Section 1.9.

Proof of Lemma. A handle decomposition of $W_1 \#_\partial W_2$ can be obtained from handle decompositions of W_1, W_2 by adding a 1-handle connecting W_1 to W_2. The corresponding handlebodies H, H_1, H_2 formed by 0- and 1-handles are related by the formula $H = H_1 \#_\partial H_2$. Hence, $\partial H = \partial H_1 \# \partial H_2$ and (4.5.a) follows from (3.4.a). The first claim of the Lemma follows from (4.5.a) and the previous lemma. Indeed, the puncturing transforms # into $\#_\partial$.

4.6. Exercises. 1. Show that for any compact oriented 4-manifold W with non-empty boundary, $\mathrm{sh}'(W \#_\partial (S^2 \times B^2)) = \mathrm{sh}'(W)$.

2. Show that the shadowed polyhedron X' constructed in the proof of Theorem 4.2 is shadow equivalent to $S_0^2 + S_0^2$.

5. Comparison of bilinear forms

5.0. Outline. The main result of this section (Theorem 5.1) relates the intersection form of a 4-manifold W to the bilinear form Q of $\mathrm{sh}'(W)$. This is a version of Theorem 1.10 for sh′.

5.1. Theorem. *Let W be a compact connected oriented 4-manifold. Let $f : Z \to W$ be a mapping representing the augmented shadow of W, where Z is an integer shadowed 2-polyhedron representing* $\mathrm{sh}'(W)$. *Then the induced homomorphism $f_* : H_2(Z) \to H_2(W)$ is surjective and for any $u, v \in H_2(Z)$,*

$$f_*(u) \cdot f_*(v) = Q_Z(u, v). \tag{5.1.a}$$

Here and below in this section, the coefficient group of homologies is $\mathbb{Z}$.

Theorem 5.1 will be proven in Section 5.3 using the shadow computation of linking numbers given in Section 5.2.

5.2. Linking numbers. Let $\ell = \ell_1 \cup ... \cup \ell_r$ be a framed oriented r-component link in a compact oriented 3-manifold M. A formal linear combination $m_1\ell_1 + ... + m_r\ell_r$ with $m_1, ..., m_r \in \mathbb{Z}$ is said to be homologically trivial if this combination, considered as a cycle, represents $0 \in H_1(M)$. For two homologically trivial combinations

$$\varepsilon = m_1\ell_1 + ... + m_r\ell_r \quad \text{and} \quad \eta = n_1\ell_1 + ... + n_r\ell_r, \tag{5.2.a}$$

we define their linking number $\mathrm{lk}(\varepsilon, \eta)$ as follows. Consider the longitudes $\ell'_1, ..., \ell'_r$ of $\ell_1, ..., \ell_r$ determined by the framing. For any compact oriented (possibly singular) surface $\Sigma \subset M$ bounded by the cycle $n_1\ell'_1 + ... + n_r\ell'_r$, set:

$$\mathrm{lk}(\varepsilon, \eta) = \sum_{i=1}^{r} m_i(\ell_i \cdot \Sigma).$$

Here $\ell_i \cdot \Sigma$ is the intersection product of ℓ_i with Σ. Homological triviality of ε ensures the independence of $\mathrm{lk}(\varepsilon, \eta)$ of the choice of Σ. The linking number is symmetric: $\mathrm{lk}(\varepsilon, \eta) = \mathrm{lk}(\eta, \varepsilon)$. This linking number generalizes the usual one defined in the case where $\varepsilon = \ell_1$ and $\eta = \ell_2$. (In this case the linking number is independent of the framing of ℓ.)

We shall give a "shadow" computation of $\mathrm{lk}(\varepsilon, \eta)$. Let C be the cone over a shadow projection of ℓ into a skeleton $X \subset M$. Provide the 2-disks glued to X to form C with the orientation induced by the given orientation of ℓ. It is obvious that the quotient C/X obtained from C by contracting $X \subset C$ into a point is a wedge of r oriented 2-spheres. Denote by $s_i \in H_2(C/X)$ the homological class of the i-th oriented 2-sphere where $i = 1, ..., r$. It is obvious that $s_1, ..., s_r$ is a basis in $H_2(C/X) = \mathbb{Z}^r$.

Let j denote the projection $C \to C/X$ and let j_* denote the induced homomorphism $H_2(C) \to H_2(C/X)$. The homological triviality of ε and η implies the existence of $u, v \in H_2(C)$ such that

$$j_*(u) = m_1s_1 + ... + m_rs_r \quad \text{and} \quad j_*(v) = n_1s_1 + ... + n_rs_r. \tag{5.2.b}$$

5.2.1. Lemma. *For any $u, v \in H_2(C)$ satisfying* (5.2.b),

$$\text{lk}(\varepsilon, \eta) = Q_C(u, v) \tag{5.2.c}$$

Proof. Theorem VIII.5.2 and the argument of Section 3.3 imply that if (5.2.c) holds for one choice of a skeleton $X \subset M$ then it holds for any other choice. We choose X as follows. Let $U_1, ..., U_r$ be disjoint closed regular neighborhoods of $\ell_1, ..., \ell_r$ in $M \setminus \partial M$. Set $M_0 = \overline{M \setminus (U_1 \cup ... \cup U_r)}$. Clearly, M_0 is a compact 3-manifold with $\partial M_0 = \partial M \amalg \partial U_1 \amalg ... \amalg \partial U_r$. Fix a triangulation of M_0 and consider the 2-skeleton Z of the dual cell-subdivision. It was already noted in Section 2.3 that Z is a simple 2-polyhedron in M_0 with $\partial M_0 \subset Z$. The complement of Z in M_0 consists of disjoint open 3-balls.

Shifting ℓ_i along its framing, we get a simple closed curve $\ell'_i \subset \partial U_i$. Consider a meridianal disk D_i of the solid torus U_i, $i = 1, ..., r$. We assume that ℓ'_i intersects ∂D_i transversally in one point lying inside a region of Z. We also assume that both ℓ'_i and ∂D_i traverse this region exactly once and split it into four subregions F_i, G_i, H_i, I_i as shown in Figure 5.1. (In Figure 5.1 we are looking at Z from a point inside U_i.)

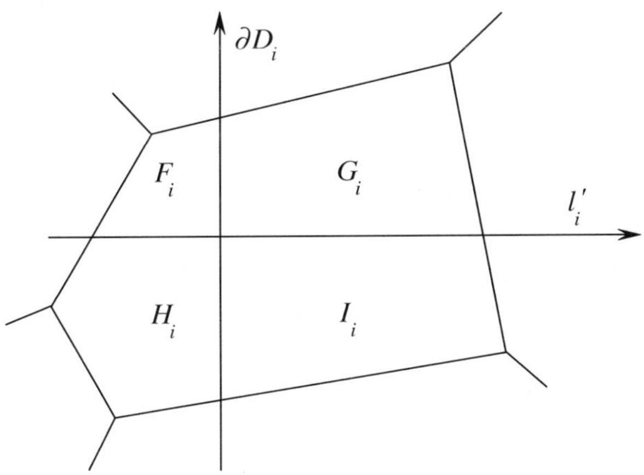

Figure 5.1

It is obvious that the simple 2-polyhedron $X = Z \cup D_1 \cup ... \cup D_r \subset M$ is a skeleton of M. The loops $\ell'_1, ..., \ell'_r$ (equipped with zero pre-twists) represent the shadow projection of ℓ into X. We take C to be the cone over this shadowed system of loops in X. The gleams of the regions of C are computed directly from definitions:

$$\text{gl}(F_i) = \text{gl}(I_i) = -1/2, \quad \text{gl}(G_i) = \text{gl}(H_i) = 1/2 \quad (i = 1, ..., r);$$

all other regions of C have zero gleam.

Equip the regions F_i, G_i, H_i, I_i with orientations corresponding to the counterclockwise orientation in the plane of Figure 5.1. Equip D_i with an orientation such that $\ell_i \cdot D_i = 1$.

Take arbitrary $u, v \in H_2(C)$ satisfying (5.2.b). Formula (VIII.5.1.a) implies that $Q_C(u, v) = (\mu_1 + ... + \mu_r)/2$ where

$$\mu_i = \langle u|G_i\rangle\langle v|G_i\rangle + \langle u|H_i\rangle\langle v|H_i\rangle - \langle u|F_i\rangle\langle v|F_i\rangle - \langle u|I_i\rangle\langle v|I_i\rangle.$$

For any $u \in H_2(C)$, we have $\langle u|G_i\rangle = \langle u|F_i\rangle + \langle u|D_i\rangle$. Since $j_*(u) = m_1 s_1 + ... + m_r s_r$, we have

$$\langle u|H_i\rangle = \langle u|F_i\rangle + m_i, \quad \langle u|I_i\rangle = \langle u|G_i\rangle + m_i = \langle u|F_i\rangle + \langle u|D_i\rangle + m_i.$$

The formula $j_*(v) = n_1 s_1 + ... + n_r s_r$ implies similar equalities with u replaced by v and m_i replaced by n_i. Substituting these expressions into the formula $Q_C(u, v) = (\mu_1 + ... + \mu_r)/2$ we get

$$Q_C(u, v) = -\frac{1}{2}\sum_{i=1}^{r}(m_i\langle v|D_i\rangle + n_i\langle u|D_i\rangle).$$

To compute the right-hand side we keep the orientation of the regions F_i, G_i, H_i, I_i, D_i and orient all other regions of C in an arbitrary way. Consider the following integer 2-chains in C

$$\Sigma_u = -\sum_Y \langle u|Y\rangle[Y], \quad \Sigma_v = -\sum_Y \langle v|Y\rangle[Y]$$

where Y runs over all regions of C except the 2-disks attached to $\ell'_1, ..., \ell'_r$. It follows from (5.2.b) that these 2-chains are bounded by the 1-cycles $m_1\ell'_1 + ... + m_r\ell'_r$ and $n_1\ell'_1 + ... + n_r\ell'_r$ respectively. It is obvious that $\ell_i \cdot \Sigma_u = -\langle u|D_i\rangle$ and $\ell_i \cdot \Sigma_v = -\langle v|D_i\rangle$ for any i. Therefore

$$\mathrm{lk}(\varepsilon, \eta) = \sum_{i=1}^{r} m_i(\ell_i \cdot \Sigma_v) = -\sum_{i=1}^{r} m_i\langle v|D_i\rangle.$$

Similarly,

$$\mathrm{lk}(\eta, \varepsilon) = \sum_{i=1}^{r} n_i(\ell_i \cdot \Sigma_u) = -\sum_{i=1}^{r} n_i\langle u|D_i\rangle.$$

Hence

$$Q_C(u, v) = (\mathrm{lk}(\varepsilon, \eta) + \mathrm{lk}(\eta, \varepsilon))/2 = \mathrm{lk}(\varepsilon, \eta).$$

5.3. Proof of Theorem 5.1. Fix a handle decomposition of W and denote by H and ℓ the same objects as in Section 4.1. Let C be the cone over the shadow projection of ℓ into a skeleton of ∂H. Let $g : C \to W$ be the augmentation mapping constructed in Section 4.3. It follows from the definition of augmented shadows that there are simple deformations $h : C \to Z$ and $h' : Z \to C$ such that up to homotopy $f \circ h = g$ and $g \circ h' = f$. The first equality and surjectivity of $g_* : H_2(C) \to H_2(W)$ imply the surjectivity of f_*. The equality $g \circ h' = f$ and Theorem VIII.5.2 show that it suffices to consider the case $Z = C$, $f = g$.

Let $\ell_1, ..., \ell_r$ be the components of ℓ, and let $B_1, ..., B_r$ be the 2-disks glued to the skeleton of ∂H to form C. Orient these disks in an arbitrary way. Fix $u, v \in H_2(C)$ and set $m_i = \langle u | B_i \rangle$, $n_i = \langle v | B_i \rangle$ where $i = 1, ..., r$. It is obvious that the homological class $g_*(v) \in H_2(W)$ may be represented by a closed oriented surface $\Sigma(v) \subset W$ lying in the boundary of the union of H and 2-handles. We may assume that the part of $\Sigma(v)$ lying in the i-th 2-handle consists of $|n_i|$ disks parallel to $g(B_i)$ with the orientation of B_i if $n_i > 0$ and the opposite orientation if $n_i < 0$. Denote the part of $\Sigma(v)$ lying in ∂H by Σ_v. This is a compact oriented surface obtained from $\Sigma(v)$ by puncturing it $|n_1| + ... + |n_r|$ times. Provide $\ell_1, ..., \ell_r$ with orientations opposite to those induced by the orientations in $B_1, ..., B_r$. It is clear that the surface Σ_v is bounded by the 1-cycle $n_1\ell_1 + ... + n_r\ell_r$. Similar constructions apply to u replacing n_i with m_i. Therefore the 1-cycles $\varepsilon = m_1\ell_1 + ... + m_r\ell_r$ and $\eta = n_1\ell_1 + ... + n_r\ell_r$ are homologically trivial in ∂H. Deforming $\Sigma(u)$ in a general position with respect to $\Sigma(v)$ in W we easily compute that

$$g_*(u) \cdot g_*(v) = \Sigma(u) \cdot \Sigma(v) = \sum_{i=1}^{r} m_i(\ell_i \cdot \Sigma_v) = \mathrm{lk}(\varepsilon, \eta).$$

It is clear that $u, v \in H_2(C)$ satisfy the conditions of Lemma 5.2.1. Therefore $g_*(u) \cdot g_*(v) = \mathrm{lk}(\varepsilon, \eta) = Q_C(u, v)$.

6. Thickening of shadows

6.0. Outline. We show how to thicken integer shadows into 4-manifolds. This thickening will be used in the next section to prove equivalence of the shadows sh and sh′ of 4-manifolds.

The idea of thickening is very simple and well known in dimension 1. Any graph lying in $\mathbb{R}^2$ may be thickened into a framed graph in $\mathbb{R}^2$ just by extending its vertices and edges to 2-disks and bands. We may additionally twist these bands in $\mathbb{R}^3$ around their cores. To this end we should specify for each edge the number of twists. The thickening of shadowed polyhedra is similar in spirit but technically more complicated because there is no ambient space playing the role of $\mathbb{R}^2$ or $\mathbb{R}^3$. Instead of numbers assigned to the edges of a graph we use the gleams of regions. The reader may skip the technical part of this section without much harm.

6.1. Thickening of shadowed polyhedra. Let X be a integer shadowed polyhedron with empty boundary. The objective of this subsection is to construct in a canonical way a compact oriented 4-manifold W_X containing X as a deformation retract.

We define W_X for connected X and then extend it to non-connected X via disjoint union. Assume from now on that X is connected. If $\text{sing}(X) = \emptyset$, i.e., if X is a closed connected orientable surface then W_X is the total space of the 2-disk bundle over X with Euler class equal to the gleam $\text{gl}(X) \in \mathbb{Z}$ of the only region of X. The 4-manifold W_X contains X as the zero section. We provide W_X with an orientation such that the self-intersection number of $X \subset W_X$ is equal to $\text{gl}(X)$.

Consider the case $\text{sing}(X) \neq \emptyset$. The idea is to compose W_X from pieces obtained from the vertices, 1-strata, and regions of X. We first prepare a piece M_e associated to a 1-stratum e of X. This is a 3-manifold containing a regular neighborhood of e in X and constructed as follows. By a topological tripod we shall mean a union of three closed intervals meeting only in a common vertex. To each point $x \in e$ we associate a tripod formed by three intervals meeting each other in their common vertex x and lying in the three regions of X attached to e. We assume that these segments touch e transversally in x. When x moves in e the associated tripod moves along, forming the normal bundle of e in X with the tripod as a fiber. Using the embedding of the tripod into a plane triangle shown in Figure 6.1 we extend this normal bundle of e to a fibration ξ over e with the triangle as a fiber. We distinguish two cases: e is a circle and e is an open interval whose endpoints are vertices of X. In the first case M_e is the total space of ξ. It is obvious that M_e is homeomorphic either to the product $S^1 \times B^2$ or to the twisted product $S^1 \tilde{\times} B^2$. Consider the case where e is an open interval with endpoints a and b (possibly $a = b$). Let us restrict ξ to a long closed subinterval of e which almost exhausts e. Denote by M_e the total space of the resulting (trivial) bundle over this subinterval. This 3-manifold is homeomorphic to $[0, 1] \times B^2$ and may be regarded as a 3-dimensional polytope with two triangular faces and three square faces. The triangular faces are the fibers of ξ over the endpoints of the subinterval of e. We shall denote these triangles by $V'_e(a)$ and $V'_e(b)$ where it is understood that $V'_e(a)$ (resp. $V'_e(b)$) is the fiber of ξ lying over the endpoint in question close to a (resp. to b). The vertices of $V'_e(a)$ and $V'_e(b)$ correspond in the obvious way to the (germs of) three regions of X attached to e.

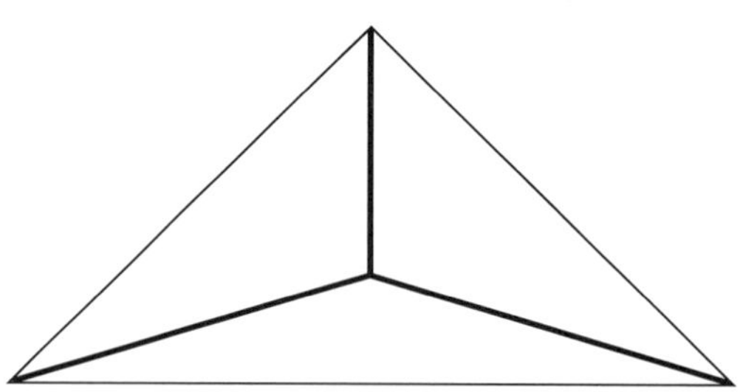

Figure 6.1

Now we introduce a "detail" K_a associated to a vertex a of X. Denote by T_a the 3-dimensional tetrahedron whose vertices are numerated by the four (germs of) edges of X incident to a. The 1-skeleton of T_a is just the polyhedral link of a in X. Denote by K_a the convex 3-dimensional polytope obtained from T_a by cutting out small disjoint tetrahedral neighborhoods of its four vertices. The boundary of K_a consists of 4 hexagons and 4 triangles. These triangles are called the bases of K_a. They correspond in the obvious way to the vertices of T_a, i.e., to the (germs of) edges of X incident to a. The base of K_a corresponding to the edge e is denoted by $V_e(a)$. The three vertices of $V_e(a)$ correspond in the obvious way to the (germs of) three regions of X attached to e. The detail K_a has six "long" edges lying on the edges of T_a and twelve "short" edges bounding the bases. The long edges of K_a correspond to the (germs of) six regions of X adjacent to a. A regular neighborhood of a in X lies in K_a as a cone whose cone point a is the barycenter of K_a and whose base is formed by 6 long edges of K_a together with four tripods inscribed in the triangular bases of K_a. There is a retraction of K_a on this cone such that the preimage of a consists of 4 intervals connecting a to the barycenters of the hexagonal faces of K_a.

At the next step we construct a compact 3-manifold M_X depending on the germ of the graph $\text{sing}(X)$ in X. Consider the disjoint union of the 3-manifolds M_e and K_a corresponding to all 1-strata e and vertices a of $\text{sing}(X)$. For each edge e of $\text{sing}(X)$ and for each endpoint a of e, identify the triangles $V_e(a) \subset \partial K_a$ and $V'_e(a) \subset \partial M_e$ along a homeomorphism which identifies vertices corresponding to the same regions of X attached to e. (Such a homeomorphism is unique up to an isotopy constant on the vertices of triangles.) Denote by M_X the compact 3-manifold obtained from the disjoint union of $\{M_e\}_e$ and $\{K_a\}_a$ by these identifications. This manifold contains a closed regular neighborhood, say u, of the graph $\text{sing}(X)$ in X so that $\partial u \subset \partial(M_X)$. Note that ∂u consists of a finite number of disjoint embedded circles in $\partial(M_X)$. It is obvious that M_X is a disjoint union of 3-dimensional handlebodies and that u is a deformation retract of M_X. We note that M_X may be non-orientable.

It is well known that all 3-manifolds may be smoothed in an essentially unique manner. The construction of M_X may be easily adopted to the smooth category. This allows us to view M_X as a smooth manifold and to use its tangent bundle.

Now we derive from the pair $(M_X, \partial u)$ a 4-manifold $H = H_X$. Consider the determinant line bundle of the tangent vector bundle of M_X. Choose a Euclidean metric in this line bundle. Let $H \to M_X$ be the subbundle formed by the vectors of length ≤ 1. This is a locally trivial bundle over M_X with the fiber $[-1, 1]$ called the determinant segment bundle of M_X. It is clear that H is a compact 4-manifold containing M_X as the zero section. The manifold H possesses a canonical orientation defined in a point $x \in M_X \subset H$ by the 4-tuple $(\alpha, \beta, \gamma, \alpha \wedge \beta \wedge \gamma)$ where α, β, γ are linearly independent vectors tangent to M_X in x. The 1-manifold $\partial u \subset \partial(M_X) \subset \partial H$ is a link in ∂H. We endow ∂u with its normal line bundle in $\partial(M_X)$.

The piece of W_X associated to a region Y of X is the 4-manifold $Y_0 \times B^2$ where $Y_0 = \overline{Y \setminus u}$ and B^2 is the 2-disk. Here Y_0 is a compact surface lying in Y as a deformation retract. The surface Y_0 lies in $Y_0 \times B^2$ as the zero section of the trivial disk bundle $Y_0 \times B^2 \to Y_0$. We provide the 4-manifold $Y_0 \times B^2$ with a framed link in the boundary. The link is presented by $\partial Y_0 \subset \partial(Y_0 \times B^2)$, its framing is determined by any non-zero section of the bundle $Y_0 \times B^2 \to Y_0$ restricted to ∂Y_0.

The next and most important step is to glue $Y_0 \times B^2$ to $H = H_X$. We shall glue these 4-manifolds along solid tori in their boundaries obtained as regular neighborhoods of the links specified above. More exactly, let $\ell_1, \ldots, \ell_m$ be the components of ∂Y_0. Each circle ℓ_i lies as a knot in $\partial(Y_0 \times B^2)$ where it is equipped with a framing. Since $\partial Y_0 = \partial u \cap Y \subset \partial u$, a copy of ℓ_i lies as a knot in ∂H where it is endowed with a normal line bundle. We twist this line bundle as follows. Choose for each i a number r_i which belongs to $\mathbb{Z}$ if the normal line bundle on $\ell_i \subset \partial H$ is orientable, and to $(1/2) + \mathbb{Z}$ if not. It follows from the definition of integer shadowed polyhedra that

$$\sum_{i=1}^{m} r_i = \eta(Y)/2 = \mathrm{gl}(Y) \pmod{\mathbb{Z}}.$$

Therefore we may choose $\{r_i\}_i$ so that

$$\sum_{i=1}^{m} r_i = \mathrm{gl}(Y).$$

For every $i = 1, \ldots, m$, twist the normal line bundle on ℓ_i in ∂H exactly r_i times around ℓ_i. (Note that a (1/2)-twist inserts a rotation by the angle π.) The positive direction of the twist corresponding to $r_i > 0$ is determined by the orientation of ∂H induced by that of H. This twist produces an orientable normal line bundle over $\ell_1 \cup \ldots \cup \ell_m \subset \partial H$. A non-singular section of this bundle yields a framing of this link. It is obvious that the identity mapping $\ell_i \to \ell_i$ extends to a homeomorphism of regular neighborhoods of ℓ_i in ∂H and $\partial(Y_0 \times B^2)$ inverting the orientation and identifying the specified framings. We glue $Y_0 \times B^2$ to H along these homeomorphisms corresponding to $i = 1, \ldots, m$. It turns out that the resulting 4-manifold does not depend on the choice of $\{r_i\}_i$ up to a homeomorphism which is the identity on H. To prove this, it suffices to consider the case where r_i, r_j with certain $i \neq j$ are traded for $r_i \pm 1, r_j \mp 1$ respectively. We construct a self-homeomorphism of $Y_0 \times B^2$ which conjugates the corresponding two gluings of $Y_0 \times B^2$ to H. Choose a simple arc in Y_0 connecting ℓ_i with ℓ_j. The self-homeomorphism of $Y_0 \times B^2$ in question is the identity in those points of $Y_0 \times B^2$ whose projection to Y_0 lies outside a narrow neighborhood of this arc. When a moving point y in Y_0 crosses this arc transversally, the 2-disk lying over y in $Y_0 \times B^2$ is rotated by 2π. Two possible directions of this rotation correspond to replacing r_i, r_j with $r_i + 1, r_j - 1$ or with $r_i - 1, r_j + 1$. This yields a self-homeomorphism of $Y_0 \times B^2$ conjugating the two gluings.

Let us glue to H the manifolds $\{Y_0 \times B^2\}_Y$ corresponding to all regions Y of X. This gives a compact connected orientable 4-manifold. The orientation of H extends to a certain orientation of this manifold. Denote the resulting oriented 4-manifold by W_X. It is clear that the polyhedron X lies in W_X as a deformation retract. Moreover, there is a retraction $f = f_X : W_X \to X$ such that: (i) the restriction of f to $f^{-1}(\text{Int}(X))$ is a locally trivial 2-disk bundle over $\text{Int}(X)$; (ii) for any point a of a 1-stratum of X, the set $f^{-1}(a)$ is a product of a tripod and a closed interval; (iii) for any vertex a of X, the set $f^{-1}(a)$ is the union of four squares along one common edge.

The construction of W_X may be carried out in both piecewise-linear and smooth categories. The manifold W_X is determined by X up to orientation-preserving piecewise-linear (resp. smooth) homeomorphisms.

The next theorem relates the thickening of 2-polyhedra to the theory of shadows.

6.2. Theorem. *Let X and Y be integer shadowed 2-polyhedra with empty boundary. If X and Y are shadow equivalent then the manifolds W_X, W_Y are piecewise-linearly homeomorphic. If $Y = X + S^2_0$ then $W_Y = W_X \#_\partial (S^2 \times B^2)$.*

Proof. To prove the first claim it suffices to consider the case when Y is obtained from X by the basic move P_i where $i = 1, 2, 3$. For $i = 2, 3$, the claim is essentially obvious. Indeed, let U, V be the subsets of X, Y shown in Figure VIII.1.1 so that the move P_i replaces U with V. We may regard U and V as subsets of a 3-ball B^3 with $\partial U = \partial V \subset \partial B^3$. The part of W_X lying "over" U (i.e., the set $f_X^{-1}(U) \subset W_X$) may be identified with a closed regular neighborhood of U in B^3 multiplied by $[-1, 1]$. Similarly, the part of W_Y lying over V may be identified with a closed regular neighborhood of V in B^3 multiplied by $[-1, 1]$. It is obvious that there exists an isotopy of B^3 in itself constant on ∂B^3 and carrying the regular neighborhood of U onto that of V. Therefore there exists an isotopy of $B^3 \times [-1, 1]$ in itself constant on the boundary and carrying $f_X^{-1}(U)$ homeomorphically onto $f_Y^{-1}(V)$. The resulting self-homeomorphism of $B^3 \times [-1, 1]$ can be extended by the identity on the complement of $B^3 \times [-1, 1]$ to a homeomorphism $W_X \to W_Y$.

The case of P_1 is more complicated since then Y can not be embedded in Euclidean 3-space. We need the following auxiliary definition. Let K be a detail described in Section 6.1. Denote by $E(K)$ the subsurface of ∂K formed by 4 triangular bases of K and narrow closed regular neighborhoods of 6 long edges of K. By its very definition, the surface $E(K)$ is subdivided into 4 disjoint triangles and 6 bands connecting these triangles. The complement of $E(K)$ in the 2-sphere ∂K consists of four open 2-disks. We call $E(K)$ the frame of K. Topologically the surface $E(K)$ is a 2-sphere with four holes.

Let us analyze W_X more closely. Let K be the detail corresponding to the vertex of U shown in the first picture of Figure VIII.1.1. The detail K is a closed regular neighborhood of this vertex in the 3-manifold M_X constructed in Section 6.1. We may assume that $K \cap X = U$. The manifold W_X is obtained by gluing the 4-ball $f_X^{-1}(U) = K \times [-1, 1]$ to $f_X^{-1}(\overline{X \setminus U})$ along $E(K) \times [-1, 1]$. The gluing is determined by the gleams of six regions of X adjacent to a and the position of the frame $E(K)$ in $S^3 = \partial(K \times [-1, 1])$. (The gluing is performed along a closed regular neighborhood of $E(K)$ in this 3-sphere, to determine the gluing we need $E(K)$ itself.) Note that the frame $E(K)$ is unknotted in S^3, i.e., lies on the boundary of a 3-ball in S^3.

Let us analyze M_Y and W_Y. To form M_Y we ought to glue two details, K_1, K_2, corresponding to two vertices of Y created by P_1 and two polytopes $\{M_e\}_e$ associated to the short edges connecting these vertices (see Figure VIII.1.1). Each M_e is the cylinder over a triangle, it is glued to K_1, K_2 along its triangular bases. We may ignore these two polytopes M_e and simply glue the corresponding triangular bases of K_1, K_2 to each other. Thus, K_1, K_2 are glued along two pairs of bases corresponding to two short edges of Y. Denote the result of this gluing by Φ. Clearly, Φ is the twisted (non-orientable) product $S^1 \tilde{\times} B^2$. Figure 6.2 shows the details K_1, K_2 with the gluing performed along one pair of bases. Here the long edges of K_1, K_2 correspond to the regions of Y marked by the same labels in Figure VIII.1.1. The long edges of K_1, K_2 marked by 0 and 13 correspond to the regions marked by $a - \omega$ and ω respectively. The four triangular bases of K_1, K_2 which are not glued in Φ form together with band neighborhoods in $\partial\Phi$ of the six long edges marked by 0, 1, 2, 3, 12, 23 a subsurface of $\partial\Phi$. Denote this subsurface equipped with this splitting into 4 triangles and 6 bands by E. Clearly, E is homeomorphic to the frame of a detail.

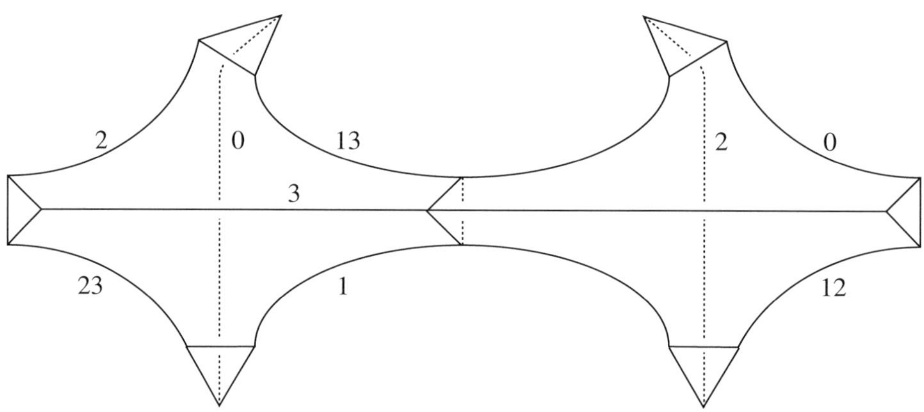

Figure 6.2

Denote by Z the small disk region created by P_1 (marked in Figure VIII.1.1 by ω). Denote by Z_0 a small closed 2-disk in Z. To construct W_Y we consider the determinant segment bundle over Φ and glue the 4-ball $Z_0 \times B^2$ to its total space following the instructions of Section 6.1. Denote the resulting compact oriented 4-manifold by B. It is easy to see (and will be shown below) that B is a 4-ball. Note that $E \subset \partial B$. The manifold W_Y may be obtained by gluing B and $f_X^{-1}(\overline{X \setminus U})$ along a regular neighborhood of E in ∂B. The gluing mapping is determined by the gleams of six regions of Y adjacent to Z. These gleams differ from the gleams of the corresponding regions of X by $\pm\omega = \pm 1/2$. To handle this difference we apply a positive (resp. negative) half-twist to the bands of $E \subset \partial B = S^3$ marked by 12, 23 (resp. 0, 1, 2, 3). Here the positive direction of twisting is determined by the orientation of ∂B induced by that of B. Denote by E' the embedded frame in ∂B obtained from E by these twists. Now, the manifold W_Y may be obtained by gluing B and $f_X^{-1}(\overline{X \setminus U})$ along a regular neighborhood of E' in ∂B where the gluing mapping is determined by the gleams of regions of X in the same way as in the construction of W_X. Therefore to conclude the proof it suffices to show that the frame E' is unknotted in $\partial B = S^3$.

We construct an embedding $\Phi \hookrightarrow \mathbb{R}^4$ as follows. Embed the details K_1, K_2 in $\mathbb{R}^3 = \mathbb{R}^3 \times 0 \subset \mathbb{R}^4$ as in Figure 6.2. Let b_1, b_2 be the top triangular bases of K_1, K_2. To obtain Φ we should glue b_1 to b_2. We assume that: b_1, b_2 lie in distinct planes parallel to the plane $0 \times \mathbb{R}^2$; b_2 may be obtained from b_1 by a parallel translation along the horizontal line $\mathbb{R} \times 0 \times 0$; the top edges of b_1, b_2 are parallel to the line $0 \times \mathbb{R} \times 0$; the triangle b_1 is symmetric with respect to the vertical line (parallel to $0 \times 0 \times \mathbb{R}$) passing through the bottom vertex of b_1. Call the last line the axis of b_1. Orient the top edge of b_1 towards the reader, i.e., in the direction leading from the end lying on the edge marked by 0 to the end lying on the edge marked by 2. Rotate b_1 around its axis in $x \times \mathbb{R}^3 \subset \mathbb{R}^4$ and at the same time increase the first (horizontal) coordinate x so that when the triangle b_1 performs a π-rotation it will coincide with b_2. The trace of this deformation of b_1 together with $K_1 \cup K_2 \subset \mathbb{R}^3$ form the image of an embedding $\Phi \hookrightarrow \mathbb{R}^4$. Note that there are two possible senses for the rotation of b_1: the positive one where the oriented edge of b_1 rotates towards $0 \times 0 \times 0 \times \mathbb{R}_+$ and the opposite (negative) one. It is understood that we use the negative rotation.

To specify the position of Φ in $\mathbb{R}^4$ consider the solid torus $t = S^1 \times B^2 \subset \mathbb{R}^3$ formed by $K_1 \cup K_2$ and a short prismatic cylinder with the bases b_1 and b_2. Let us thicken t to a 4-dimensional solid torus $\Psi = S^1 \times B^3 \subset \mathbb{R}^4$ so that $\Psi \cap (\mathbb{R}^3 \times 0) = t$. The boundary of Ψ consists of the positive and negative parts:

$$\partial_+\Psi = \partial\Psi \cap (\mathbb{R}^3 \times \mathbb{R}_+) \ , \ \partial_-\Psi = \partial\Psi \cap (\mathbb{R}^3 \times \mathbb{R}_-).$$

We assume that the projection $\mathbb{R}^4 \to \mathbb{R}^3 \times 0$ along the fourth coordinate maps both $\partial_+\Psi$ and $\partial_-\Psi$ homeomorphically onto t. Let us provide t and Ψ with the orientations induced by the canonical orientations in $\mathbb{R}^3$ and $\mathbb{R}^4$. Provide $\partial\Psi$ with the orientation induced by that of Ψ. It follows from the definitions that the projections $\partial_+\Psi \to t$ and $\partial_-\Psi \to t$ invert and preserve the orientation, respectively.

We may assume that Φ lies in Ψ so that $\partial\Phi \subset \partial\Psi$. This enables us to identify Ψ with the total space of the determinant segment bundle over Φ.

It is obvious that the long edges of K_1, K_2 marked by 13 form a circle $S^1 \subset \partial\Phi$. Its regular neighborhood in $\partial\Phi$ is a Möbius band. The sense of the rotation of b_1 chosen above ensures that this is a $(-1/2)$-twisted Möbius band in $\partial\Psi$ so that a positive half-twist of this band around its core produces a "vertical" annulus whose core lies in $\mathbb{R}^3 \times 0$ and whose bases lie in $\mathbb{R}^3 \times \mathbb{R}_+$ and $\mathbb{R}^3 \times \mathbb{R}_-$. According to the definition of W_Y, we have to apply this half-twist before gluing Ψ to the 4-ball $Z_0 \times B^2$ corresponding to the disk region Z created by P_1. Therefore this gluing may be performed inside $\mathbb{R}^4$. It amounts to attaching to Ψ the 2-handle obtained as a closed regular neighborhood in $\mathbb{R}^4$ of the plane disk in $\mathbb{R}^3 \setminus \mathrm{Int}(t)$ bounded by the circle $S^1 \subset \partial\Phi$ mentioned above. This shows that the 4-manifold B produced by the gluing in question is a 4-ball.

The position of the frame E in $\partial\Psi$ is shown in Figure 6.3. Here the band drawn in bold presents the projection in $\mathbb{R}^3$ of a band in $\partial_+\Psi$ and the "dotted" band presents the projection in $\mathbb{R}^3$ of a band in $\partial_-\Psi$. The remaining part of E lies in the 2-torus ∂t. (This torus is not drawn but its presence is understood.)

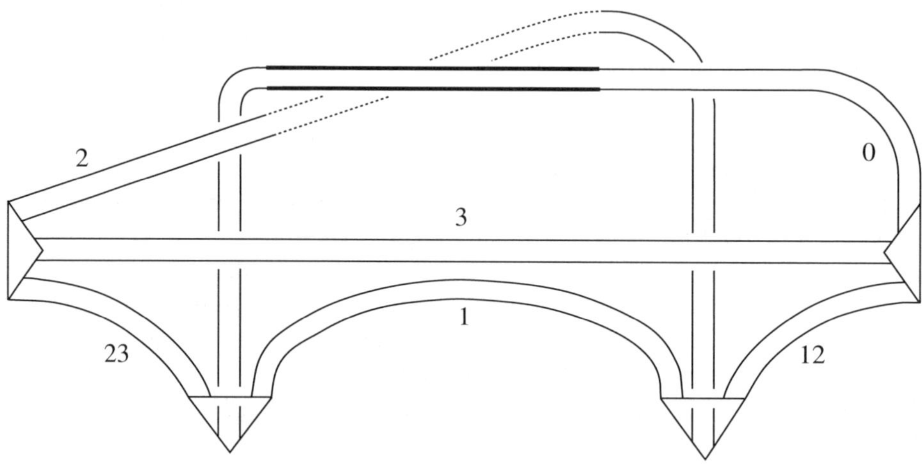

Figure 6.3

Our next aim is to deform $E' \subset \partial\Psi$ into $\partial_+\Psi$ and to visualize the image of the resulting frame under the projection $\partial_+\Psi \to t$. Note that with the exception of the broken band the frame E lies in $\partial_+\Psi$. Therefore we may ignore the distinction between the boldface and ordinary lines in Figure 6.3 and treat this figure as the image of E (minus the broken band) under the projection $\partial_+\Psi \to t$. The broken band connects in $\partial_-\Psi$ two segments lying in $\partial t = \partial(\partial_-\Psi)$. This band can not be deformed into ∂t. Recall that to construct E' we apply a negative half-twist to this band. The resulting band is isotopic (modulo its bases lying in ∂t) to a band in ∂t. This last band is dashed in Figure 6.4. It remains to insert positive (resp. negative) half-twists in the bands of E marked by 12, 23 (resp. 0, 1, 3). We should be cautious here: since these bands lie in $\partial_+\Psi$ and the projection $\partial_+\Psi \to t$ inverts orientation we have to invert the sense of these additional half-twists in our picture. This yields the picture of E' in Figure 6.4. This figure depicts the image under the projection $\partial_+\Psi \to t$ of a frame in $\partial_+\Psi$ isotopic to E' in $\partial\Psi$. Since this frame lies away from the 2-handle attached to Ψ to produce B, the same picture presents the frame E' in $\partial B = S^3$. It remains to observe that Figure 6.4 presents an unknotted frame. One way to see this is to note that the core of this frame is an unknotted embedding of the 1-skeleton of a tetrahedron in $\mathbb{R}^3$. The only possible knottedness may arise from the bands which may be twisted around their cores. The corresponding twisting numbers are completely determined by the linking numbers of the four boundary components of the frame. It is easy to draw the link formed by the boundary components and to see that it is trivial. Hence the linking numbers in question are equal to zero and the frame in Figure 6.4 is unknotted. This completes the proof of the first claim of the theorem. The second claim of the theorem follows from definitions.

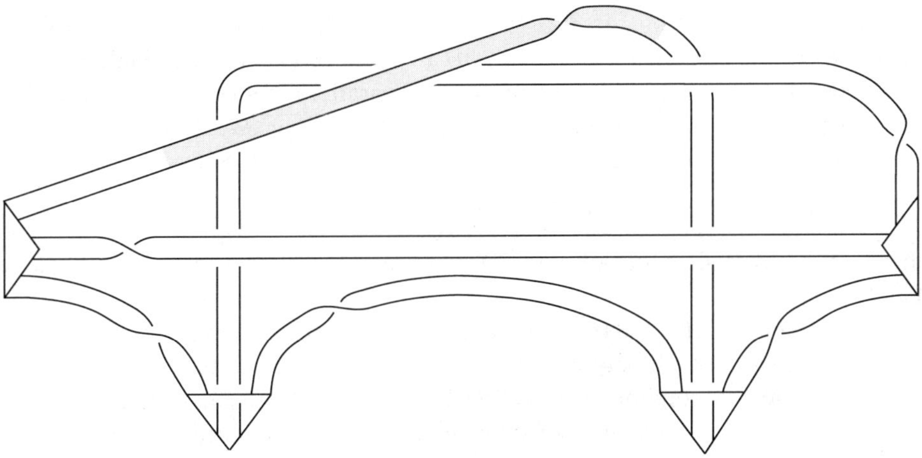

Figure 6.4

6.3. Theorem. *Let X be an integer shadowed polyhedron with empty boundary. Then*

$$\mathrm{sh}'(W_X) = \mathrm{stab}([X]).$$

Proof. It suffices to consider the case where X is connected. It follows from the previous theorem and Exercise 4.6.1 that the shadow $\mathrm{sh}'(W_X)$ is preserved under suspension of X. Suspending if necessary X and applying the basic move P_2 we may guarantee that all regions of X are 2-disks and the graph $\mathrm{sing}(X)$ is non-empty, connected, distinct from S^1, and has no loops (i.e., edges with coinciding vertices). It is clear that the vertices and edges of $\mathrm{sing}(X)$ together with the regions of X form a cell decomposition of X. Thickening this decomposition of $X \subset W_X$ in the obvious way we get a handle decomposition of W_X. The union of 0-handles and 1-handles of this handle decomposition coincides with the manifold $H = H_X$ obtained as the total space of a segment bundle over a 3-dimensional handlebody $M = M_X$ (see Section 6.1). The 2-handles of W_X are attached to H along a framed link $\ell \subset \partial M \subset \partial H$ whose components are the circles $\{\partial Y_0\}_Y$, where Y runs over regions of X. The framing of the circle $\partial Y_0 \subset \partial H$ is obtained by twisting its normal line bundle in ∂M exactly $\mathrm{gl}(Y)$ times and choosing a non-singular section of the resulting trivial line bundle. (In the notation of Section 6.1 we have $\ell = \partial u$.)

To compute $\mathrm{sh}'(W_X)$ we construct a skeleton of the closed 3-manifold ∂H as follows. For each edge e of $\mathrm{sing}(X)$, consider the common base δ_e of two details corresponding to the endpoints of e. This base is a proper 2-disk in M that intersects $e \subset \mathrm{sing}(X) \subset M$ transversally in one point. The complement of these disks in M consists of disjoint 3-balls centered at the vertices of $\mathrm{sing}(X)$.

The manifold ∂H supports an involution j induced by multiplication by -1 in the fibers of the determinant line bundle over M. It is obvious that $M = (\partial H)/j$ and the fixed point set of j is ∂M. Each disk $\delta_e \subset M$ lifts to ∂H in two different ways. We choose one such lifting $\tilde{\delta}_e \subset \partial H$ for each e. This is an embedded 2-disk in ∂H bounded by $\partial \delta_e \subset \partial M$. Denote by δ'_e a parallel copy of $\tilde{\delta}_e$, i.e., a 2-disk in ∂H obtained by slightly pushing $\tilde{\delta}_e$ into $\partial H \setminus \tilde{\delta}_e$ so that $\partial \delta'_e \subset \partial M$. We assume that the disks corresponding to different edges are disjoint. Set

$$Z = \partial M \cup \cup_e (\tilde{\delta}_e \cup j(\delta'_e)) \subset \partial H$$

where e runs over all edges of $\mathrm{sing}(X)$. It is easy to observe that Z is a skeleton of ∂H.

The framed link $\ell \subset \partial M \subset Z$ represents its own shadow projection in Z where the gleams of all regions of ℓ in Z are equal to zero. The pre-twist of the loop ∂Y_0 corresponding to a region Y of X equals $\mathrm{gl}(Y)$. Denote by C the shadow cone of this shadowed system of loops in Z. According to definitions, $\mathrm{sh}'(W_X) = \mathrm{stab}([C])$.

Let us show that $\mathrm{stab}([C]) = \mathrm{stab}([X])$. For any edge e of $\mathrm{sing}(X)$, the circles $\partial \delta_e \subset \partial M$ and $\partial \delta'_e \subset \partial M$ bound a thin annulus Δ_e in ∂M. The link ℓ intersects Δ_e in three short segments connecting $\partial \delta_e$ with $\partial \delta'_e$ and splitting Δ_e into three rectangular regions. Set $\Sigma_e = \tilde{\delta}_e \cup j(\delta'_e) \cup \Delta_e$. It is obvious that Σ_e is a 2-sphere embedded in Z.

The complement in Z of the open annuli $\{\mathrm{Int}(\Delta_e)\}_e$ (where e runs over all edges of $\mathrm{sing}(X)$) consists of disjoint 2-spheres corresponding to the vertices of $\mathrm{sing}(X)$. The 2-sphere corresponding to a vertex $a \in \mathrm{sing}(X)$ bounds a 3-ball in M containing a and contained in the detail K_a. Denote this 2-sphere by ∇_a.

Note that the gleams of the regions of C lying in the 2-spheres Σ_e, ∇_a are equal to zero. Therefore we may remove from C an arbitrary disk region lying in any of these spheres without changing the stable shadow type of C. We remove from every sphere Σ_e a rectangular region lying in Δ_e. This transformation keeps the spheres $\{\nabla_a\}_a$ intact. We remove from each of the latter spheres one arbitrary region. It remains to observe that the resulting shadowed polyhedron is homeomorphic to X. Thus $\mathrm{sh}'(W_X) = \mathrm{stab}([C]) = \mathrm{stab}([X])$.

6.4. Exercise. Show that for any shadowed 2-polyhedron X, the closed 3-manifold $\partial(W_X)$ is connected and the inclusion homomorphism of fundamental groups $\pi_1(\partial(W_X)) \to \pi_1(W_X)$ is surjective.

7. Proof of Theorems 1.5 and 1.7 - 1.11

7.1. Proof of Theorem 1.5. Let W be a compact connected oriented 4-manifold. Fix a handle decomposition of W. Let H be the 4-dimensional submanifold of W formed by the 0-handles and 1-handles. Let $\ell \subset \partial H$ be the link formed by the cores of the bases of 2-handles. Take a skeleton X of ∂H so small that ℓ lies in X as a generic system of disjoint simple loops. The components of ℓ bound disjoint 2-disks in the corresponding 2-handles of W. These disks only meet X along their boundaries. Denote by X' the union of these disks with X. It is obvious that X' is a locally flat orientable simple 2-polyhedron embedded in W. We shall show that X' is a skeleton of W.

We should verify that the manifold W may be obtained from a closed regular neighborhood of X' in W by attaching 3-handles and 4-handles. Let U be a closed regular neighborhood of X' in W formed by the 2-handles of W and a closed regular neighborhood of X in W. Since X is a skeleton of ∂H its complement in ∂H consists of several open 3-balls $B_1, ..., B_m$. We attach to U handles of index 3 with the cores $B_1, ..., B_m$. This operation performed inside W yields a submanifold U' of W formed by the 2-handles of W and a closed regular neighborhood $\partial H \times [-1, 1]$ of $\partial H = \partial H \times 0$ in W. Note that the handlebody H may be obtained from $\partial H \times [0, 1]$ by attaching 3-handles and 4-handles. Therefore adding to U' these 3-handles and 4-handles (inside H) as well as 3-handles and 4-handles of the given handle decomposition of W we fill in W. Hence, X' is a skeleton of W.

7.2. Lemma. *Let W be a compact oriented 4-manifold. For any shadowed skeleton X of W, we have* $\mathrm{stab}([X]) = \mathrm{sh}'(W)$.

Proof. Let W_X be the 4-manifold obtained by thickening X. We shall extend the embedding $X \hookrightarrow W$ to a homeomorphism of W_X onto a closed regular neighborhood U of X in W. This would imply our claim. Indeed, Theorem 6.3 would imply that $\mathrm{sh}'(U) = \mathrm{sh}'(W_X) = \mathrm{stab}([X])$. Since X is a skeleton of W, the manifold W may be obtained by attaching 3-handles and 4-handles to U. Therefore $\mathrm{sh}'(W) = \mathrm{sh}'(U) = \mathrm{stab}([X])$.

Let M_X be the 3-manifold constructed in Section 6.1 and let u be a closed regular neighborhood of $\mathrm{sing}(X)$ in X properly contained in M_X. First, we extend the embedding $u \hookrightarrow W$ to an embedding $M_X \hookrightarrow W$.

Consider the component M_e of M_X derived from a circle 1-stratum e of $\mathrm{sing}(X)$. The manifold M_e is the product (possibly twisted) of $e = S^1$ and a closed 2-disk. Let u_e be the component of u containing e. Thus, u_e is a closed regular neighborhood of $e \subset X$ such that $(u_e, \partial u_e) \subset (M_e, \partial M_e)$. We treat u_e as a union of topological tripods associated to the points of e. Consider a neighborhood $e \times B^3$ of e in W with $e = e \times 0$ where 0 is the center of B^3. We assume that the tripod associated to any point $x \in e$ lies in $x \times B^3$ with its three endpoints in the 2-sphere $x \times \partial B^3$. Cutting out the solid torus $e \times B^3$ along a fiber $\{pt\} \times B^3$ we get the cylinder $[0, 1] \times B^3$ with a continuous family of triples of points in the 2-spheres $t \times \partial B^3$, $t \in [0, 1]$. All such families are homeomorphic to a standard (product) one. We conclude that the topological type of the pair $(e \times B^3, u_e)$ is determined by the monodromy $S^2 \to S^2$ which preserves three distinguished points of S^2 as a set.

7.2.1. Claim. *Any degree* 1 *self-homeomorphism of S^2 preserving three distinct points $a, b, c \in S^2$ is isotopic to the identity in the class of self-homeomorphisms of S^2 preserving a, b, and c.*

This claim follows from the fact (due to M. Dehn) that any degree 1 self-homeomorphism of a surface of finite topological type may be presented as a composition of Dehn twists along simple closed curves in this surface. There are three isotopy classes of simple closed curves in $S^2 \setminus \{a, b, c\}$. They are represented by small loops encircling a, b, c. It is easy to see that the Dehn twists along these loops are isotopic to the identity in the class of self-homeomorphisms of S^2 preserving a, b, c.

Claim 7.2.1 implies that the position of X in a neighborhood of $e = S^1$ is determined up to homeomorphism by the permutation in the set of 3 distinguished points induced by the monodromy. If this permutation is trivial then the pair $(e \times B^3, u_e)$ is homeomorphic to the product of e and the pair (B^3, a tripod lying in an equatorial 2-disk $B^2 \subset B^3$). In this case we extend the embedding $u_e \hookrightarrow W$ to an embedding $M_e \hookrightarrow W$ whose image is $e \times B^2$. If the permutation mentioned above is a transposition of two points then the embedding $M_e \hookrightarrow W$ is obtained via a π-rotation of the equatorial 2-disk $B^2 \subset B^3$ around its diameter. Finally, if the permutation in question acts cyclically then the embedding $M_e \hookrightarrow W$ is obtained via the rotation of B^2 in itself by the angle $\pm 2\pi/3$. In all cases we get an embedding $M_e \hookrightarrow W$ extending the embedding $u_e \hookrightarrow W$.

Recall that M_X is glued from the polytopes $\{K_a\}_a$ and $\{M_e\}_e$ where a runs over vertices of $\mathrm{sing}(X)$ and e runs over 1-strata of $\mathrm{sing}(X)$. Since X is locally flat in

W we may extend the embedding $K_a \cap X \hookrightarrow W$ to an embedding $K_a \hookrightarrow W$ whose image lies in a small neighborhood of $a \in W$. Let e be an (open) edge of sing(X). Let a and b be the endpoints of e. Consider a neighborhood $e \times B^3$ of e in W where $e = e \times 0$. We assume that the tripod in u associated to point $x \in e$ lies in $x \times B^3$ with its three endpoints in the 2-sphere $x \times \partial B^3$. The triangular base $V_e(a)$ of K_a gives rise to a triangle in the fiber $x \times \partial B^3$ where x is a point of e lying close to a. (Here by triangle we mean three embedded intervals AB, BC, CA intersecting only in their vertices A, B, C.) The vertices of the triangle in question are the endpoints of the tripod in u centered in x. Similarly, the triangular base $V_e(b)$ of K_b gives rise to a triangle in $y \times \partial B^3$ where y is a point of e lying close to b. The vertices of this triangle are the endpoints of the tripod centered in y. Moving along e from x towards y and keeping track of the endpoints of tripods we get a homeomorphism $x \times \partial B^3 \to y \times \partial B^3$ carrying the vertices of the first triangle into the vertices of the second one. It follows from Claim 7.2.1 that freezing the images of these vertices we may deform this homeomorphism $x \times \partial B^3 \to y \times \partial B^3$ so that it transforms the first triangle onto the second one. Now it is easy to embed M_e into W in the way compatible with the gluings $V_e(a) = V'_e(a)$ and $V_e(b) = V'_e(b)$ and the chosen embeddings $V_e(a) \subset K_a \hookrightarrow W$ and $V_e(b) \subset K_b \hookrightarrow W$. This completes the construction of an embedding $M_X \hookrightarrow W$ extending the embedding $u \hookrightarrow W$.

Let us identify the 3-manifold M_X with its image in W. Using the orientation of W we may identify the normal line bundle of M_X in W with the determinant line bundle of the tangent vector bundle of M_X. This allows us to identify a closed regular neighborhood of M_X in W with the oriented 4-manifold $H = H_X$ constructed in Section 6.1. As in Section 6.1, the 1-manifold $\partial u \subset \partial(M_X) \subset \partial H$ presents a link in ∂H.

Let Y be a region of X and let $Y_0 = \overline{Y \setminus u}$ be a compact subsurface of Y. Let ν be the normal line bundle of $\partial Y_0 \subset \partial u$ in $\partial(M_X)$. We may identify ν with the line bundle $\nu(Y)$ constructed in Section VIII.6.2 and used in Section 1.6 to define the gleam of Y. When we twist ν around the components of ∂Y_0 as in Section 6.1, we get a normal line bundle over ∂Y_0 which extends to a normal line bundle over Y_0 in W. Therefore the gluing of $Y_0 \times B^2$ to H used in Section 6.1 may be carried out inside W so that $Y_0 = Y_0 \times 0$. This shows that the embedding $X \hookrightarrow W$ extends to an embedding of W_X onto a closed regular neighborhood of X in W.

7.3. Proof of Theorem 1.7. Theorem 1.7 follows from Lemma 7.2 and the fact that the shadow sh$'(W)$ is well-defined.

7.4. Proof of Theorem 1.8. Repeating the proof of Theorem 1.7 while keeping track of the inclusion $X \hookrightarrow W$ we obtain Theorem 1.8.

7.5. Remark. Lemma 7.2 implies that $\mathrm{sh}(W) = \mathrm{sh}'(W)$ for any compact oriented 4-manifold W. Similar arguments show that the augmented stable shadows of W introduced in Sections 1.8 and 4.3 are equal.

7.6. Proof of Theorem 1.9. Theorem 1.9 follows from the equality $\mathrm{sh} = \mathrm{sh}'$ and (4.5.2).

7.7. Proof of Theorem 1.10. Theorem 1.10 follows from the equality $\mathrm{sh} = \mathrm{sh}'$ and Theorem 5.1.

7.8. Lemma. *Let W, W' be compact connected oriented 4-manifolds with homeomorphic boundaries. If W, W' are cobordant modulo the boundary then the shadows $\mathrm{sh}'(W), \mathrm{sh}'(W')$ are cobordant.*

Proof. If W is cobordant to W' then there exists a sequence of surgeries of indices 2 and 3 which transforms W into W'. (Surgeries of indices 1 and 4 may be replaced by surgeries of indices 3 and 2 along small unknotted spheres.) By duality, it is enough to consider the case when W' is obtained from W by a surgery of index 2 along a 1-dimensional knot $k \subset W$.

Let H and ℓ be the objects used in the definition of $\mathrm{sh}'(W)$. Since the inclusion homomorphism $\pi_1(\partial H \setminus \ell) \to \pi_1(W)$ is surjective and homotopic 1-knots in W are isotopic, we may assume that $k \subset \partial H \setminus \ell$. Equip k with a framing in ∂H such that this framing together with the vector field on k directed inside H determine the surgery at hand. Denote by m a small meridianal loop of k in $\partial H \setminus \ell$ equipped with the trivial framing.

7.8.1. Claim. *The shadow $\mathrm{sh}'(W')$ is represented by the shadow cone of the framed link $k \cup \ell \cup m \subset \partial H$.*

This claim, proven below, implies the lemma. Indeed, consider a shadow projection of $k \cup \ell \cup m$ into a skeleton of ∂H. Capping the loops corresponding to ℓ yields a 2-component shadow link, say, $k' \cup m'$ in a shadowed polyhedron C. The shadowed polyhedron C represents the shadow cone $CO(\partial H, \ell) = \mathrm{sh}'(W)$. The shadow link $k' \cup m'$ in C is the longitude-meridianal pair of the shadow knot k' in C (see Figure 7.1). Therefore the shadow cone of $k \cup \ell \cup m \subset \partial H$ is obtained from C by a surgery along k'. Claim 7.8.1 would imply that $\mathrm{sh}'(W')$ is obtained from $\mathrm{sh}'(W)$ by a surgery.

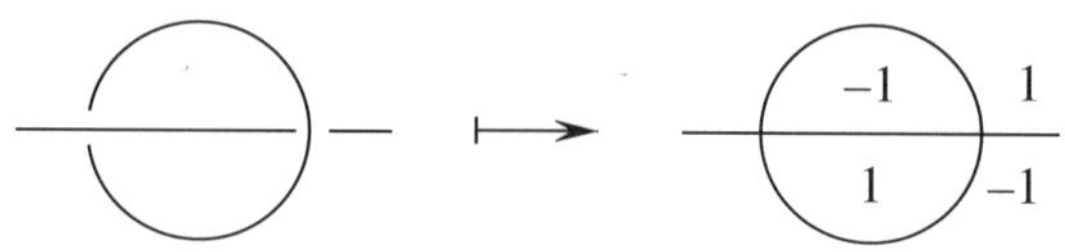

Figure 7.1

To prove (7.8.1) we construct a handle decomposition of W'. We first construct a handle decomposition of the (closed) exterior of k in W. Let t be a solid torus $k \times B^2 \subset \partial H \setminus \ell$ with core k. Let $k \times \partial B^2 \times [0, 1]$ be a collar of ∂t in t. Denote by E the 4-manifold obtained from H by attaching $k \times B^2 \times [0, 1]$ along $k \times \partial B^2 \times [0, 1] \subset \partial H$. It is easy to construct a relative handle decomposition of $E(\operatorname{mod} H)$. Namely, E may be obtained from H by attaching a 2-handle along m and attaching a 3-handle. On the other hand, it is obvious that E is diffeomorphic to the exterior of the 1-knot in H obtained by pushing k into $H \setminus \partial H$. Therefore, the exterior of k in W may be obtained from H by attaching 2-handles along m and components of ℓ, followed by attaching handles of indices 3 and 4. To obtain W' we also add a 2-handle along k and a 4-handle. This implies that W' may be obtained from H by attaching 2-handles along k, m, and components of ℓ followed by attaching handles of indices 3 and 4. This implies Claim 7.8.1 and the lemma.

7.9. Proof of Theorem 1.11. Theorem 1.11 follows from the equality $\mathrm{sh} = \mathrm{sh}'$ and Lemma 7.8.

8. Shadows of framed graphs

8.0. Outline. As a generalization of the shadow of a 4-manifold, we define the shadow of a 4-manifold with a framed graph sitting in its boundary. This shadow has a boundary which is the core of the given framed graph. On the whole the theory runs parallel to the one developed in previous sections. We indicate only the main changes leaving the details to the reader.

The results of this section will be used in Chapter X.

8.1. Four-dimensional graph pairs and their shadows. By a 4-dimensional graph pair we mean a pair (a compact oriented piecewise-linear 4-manifold W, a framed graph Γ in ∂W). The boundary of such a pair is the pair $(\partial W, \Gamma \subset \partial W)$. By homeomorphisms of 4-dimensional graph pairs we mean their orientation-preserving piecewise-linear homeomorphisms.

A skeleton of a 4-dimensional graph pair (W, Γ) is a locally flat orientable simple 2-polyhedron $X \subset W$ with $\partial X = X \cap \partial W = X \cap \Gamma$ being a core of Γ such that the manifold W may be obtained from a closed regular neighborhood of X in W by attaching handles of indices 3 and 4. It follows from the last condition that the part of X lying in a connected component of W is connected.

8.1.1. Theorem. *Every compact oriented 4-dimensional graph pair (W, Γ) has a skeleton.*

It is obvious that for any skeleton X of (W, Γ), the pair (X, Γ) is an orientable simple 2-polyhedron with framed boundary. This polyhedron may be upgraded to an integer shadowed polyhedron in the same way as in Section 1.6, although to compute the gleam of a region of X adjacent to ∂X we should apply the constructions of Section 1.6 to the simple 2-polyhedron $X \cup \Gamma = X \cup_{\partial X} \Gamma \subset W$.

8.1.2. Theorem. *Any two skeletons of* (W, Γ) *shadowed as in Section* 1.6 *are stably shadow equivalent in the class of integer shadowed 2-polyhedra.*

This theorem allows us to define $\mathrm{sh}(W, \Gamma)$ to be the stable shadow equivalence class of a skeleton of (W, Γ) shadowed as in Section 1.6. Thus, $\mathrm{sh}(W, \Gamma)$ is an integer stable shadow whose boundary is the core of Γ. It is easy to check that $\mathrm{sh}(-W, \Gamma) = -\mathrm{sh}(W, \Gamma)$. Clearly, $\mathrm{sh}(W, \emptyset) = \mathrm{sh}(W)$.

As in the case of 4-manifolds without graphs the shadow moves relating different skeletons of (W, Γ) may be performed inside W. As an exercise the reader may formulate analogues of Theorem 1.8 and 1.9 in this setting. Here are analogues of Theorems 1.10 and 1.11.

8.1.3. Theorem. *Let* (W, Γ) *be a* 4*-dimensional graph pair. Let* $X \subset W$ *be a skeleton of* (W, Γ) *shadowed as in Section* 1.6. *Then the inclusion homomorphism* $j : H_2(X; \mathbb{Z}) \to H_2(W; \mathbb{Z})$ *is surjective and for any* $u, v \in H_2(X; \mathbb{Z})$, *we have* $j(u) \cdot j(v) = Q_X(u, v)$.

8.1.4. Theorem. *Let* (W_1, Γ_1) *and* (W_2, Γ_2) *be* 4*-dimensional graph pairs with homeomorphic boundaries. If* W_1, W_2 *are cobordant modulo the boundary then the stable shadows* $\mathrm{sh}(W_1, \Gamma_1)$ *and* $\mathrm{sh}(W_2, \Gamma_2)$ *are cobordant.*

The definitions and arguments of Sections 2 - 7 directly generalize to the setting of 4-dimensional graph pairs. We outline below the technique of graph diagrams on skeletons of 3-manifolds, the construction of shadow cylinder for framed graphs in 3-manifolds, the definition of shadow sh' for 4-dimensional graph pairs, and the construction of thickening for 2-polyhedra with boundary.

8.2. Graph diagrams. Let M be an oriented compact 3-manifold with a skeleton $X \subset M$. Assume that the surface $\mathrm{Int}(X)$ is equipped with a normal direction in M. A graph diagram in X is a generic immersion of a graph γ in X (see Section VIII.7.1) such that at each crossing point one of the two intersecting branches is distinguished and said to be the lower one, the second branch is said to be the upper one. Here all crossings are transversal intersections of 1-strata in $\mathrm{Int}(X)$. Each graph diagram d in X determines a framed graph in M as follows. Thicken d to an immersion of a framed graph in X. Each crossing point of d gives rise to a crossing of two bands (or annuli) of this framed graph. Pushing slightly all "upper" bands in $M \setminus X$ along the specified normal direction we get a framed graph $\Gamma(d) \subset M$ with the core γ.

An enriched graph diagram in X is a graph diagram in X such that all 1-strata of the underlying abstract graph γ are equipped with integer or half-integer pre-twists. (We do not impose any conditions on these pre-twists, in particular for loops we get a slightly more general theory than the one in Section 3.1.) Every enriched graph diagram D in X gives rise to a framed graph $\Gamma(D) \subset M$ obtained as follows. Consider the framed graph determined by the underlying graph diagram (with the pre-twists forgotten) and twist its bands and annuli along their cores as dictated by the corresponding pre-twists.

The graph diagrams are subject to the moves $\Omega_1 - \Omega_8$ shown in Figures 3.1 - 3.3. (The comments on these moves made in Section 3.1 apply here as well.) There are two additional moves Ω_9 and Ω_{10} specific to graph diagrams. For Ω_9, see Figure 8.1 where the orientation of M may correspond to left-handed or right-handed orientation in $\mathbb{R}^3$. For Ω_{10}, see the picture of S_{10} (Figure VIII.7.1) where the symbol 0 should be omitted.

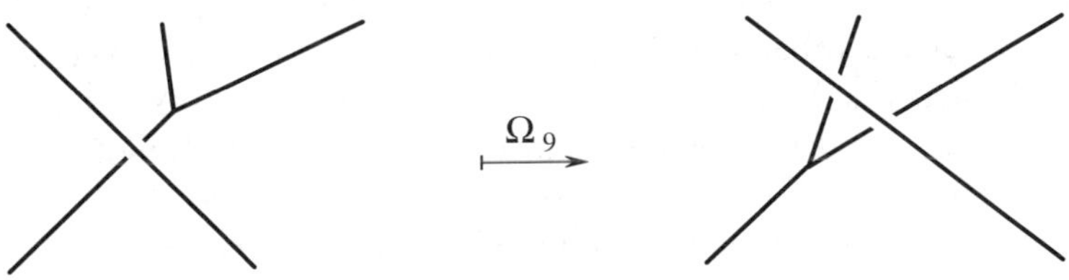

Figure 8.1

It is easy to see that every framed graph in M may be presented by an enriched graph diagram in X. Two graph diagrams present isotopic framed graphs if and only if these diagrams may be obtained from each other by a sequence of moves $\Omega_1 - \Omega_{10}$, their inverses, and an ambient isotopy in X. This yields a description of framed graphs in M in terms of enriched graph diagrams in X.

8.3. Shadow projections and cylinders of framed graphs. Let M be a compact oriented 3-manifold with skeleton $X \subset M$. Assume that the surface $\mathrm{Int}(X)$ is equipped with a normal direction in M. Let D be an enriched graph diagram in X with underlying abstract graph γ. The diagram D determines an integer shadowed graph in X modelled on the framed graph $\Gamma(D)$. This shadowed graph is geometrically represented by the immersion $\gamma \to X$ underlying D. The pre-twists of the 1-strata are provided by D. The gleams of regions are defined in exactly the same way as in the theory of link diagrams. (The vertices of γ do not contribute to gleams.) It is left to the reader to verify condition (VIII.6.3.a). The resulting integer shadowed graph in X modelled on $\Gamma(D)$ is called the shadow projection of D. It is easy to check that graph diagrams related by the moves $\Omega_1 - \Omega_{10}$ have shadow projections related by the moves $S_1 - S_{10}$.

In contrast with the link theory, framed graphs in 3-manifolds do not generally give rise to shadow cones. Instead they give rise to shadow cylinders defined as follows. Consider a framed graph $\Gamma \subset M$. Present Γ by an enriched diagram D in X. The shadow projection of D in X is an integer shadowed graph in X modelled on Γ. The cylinder over this shadowed graph is an integer shadowed polyhedron (C, Γ) whose boundary is the core of Γ, see Section VIII.7. We define the shadow cylinder $CY(M, \Gamma)$ of Γ to be the stable integer shadow represented by (C, Γ). This stable shadow depends only on the isotopy class of Γ in M and does not depend on the choice of D, X, and normal direction on $\operatorname{Int}(X)$ (cf. Section 3.3).

If Γ arises from a framed link $\ell \subset M$ via thickening of the components of ℓ into annuli, then the boundary of $CY(M, \Gamma)$ consists of circles. Capping these circles yields the shadow cone $CO(M, \ell)$ defined in Section 3.3.

8.4. Shadow sh′ for 4-dimensional graph pairs. The definition of the shadow sh′ given in Section 4 extends to 4-dimensional graph pairs. Let (W, Γ) be a 4-dimensional graph pair. Fix a handle decomposition of W and consider the same objects H, ℓ as in Section 4.1. The manifold $W \setminus \operatorname{Int}(H)$ is obtained from a collar $\partial W \times [0, 1]$ of ∂W by attaching handles of indices 0, 1, 2 to $\partial W \times 1$. We may assume that these handles do not meet $\Gamma \times 1 \subset \partial W \times 1$. Therefore $\Gamma \times 1$ presents a framed graph, say Γ', in ∂H. This framed graph is determined by Γ up to isotopy and band summation with components of ℓ. Denote by $\mathrm{sh}'(W, \Gamma)$ the stable integer shadow obtained from the shadow cylinder $CY(\partial H, \Gamma' \cup \ell)$ by capping those components of its boundary which correspond to components of ℓ. Arguments similar to those in Section 4.2 show that $\mathrm{sh}'(W, \Gamma)$ is independent of the choices made in the course of the construction. The boundary of $\mathrm{sh}'(W, \Gamma)$ may be identified with the core of Γ.

8.5. Thickening of integer shadowed polyhedra with boundary. To extend the results of Sections 6 and 7 to 4-dimensional graph pairs we define thickening for integer shadowed polyhedra with boundary. Let (X, Γ) be an integer shadowed polyhedron. Recall that ∂X is a core of Γ so that $X' = X \cup \Gamma$ is a simple 2-polyhedron. Since the graph $\operatorname{sing}(X') = \operatorname{sing}(X) \cup \partial X$ does not meet $\partial X' = \partial \Gamma$ we may apply the same constructions as in Section 6.1 to $\operatorname{sing}(X')$ and define a 3-manifold $M = M_{X'}$ properly containing a closed regular neighborhood of $\operatorname{sing}(X')$ in X. An inspection shows that there is a copy of Γ sitting in ∂M. Now we repeat the constructions of Section 6.1 word for word and obtain a 4-manifold W_X with a copy of Γ in $\partial(W_X)$.

Theorems 6.2 and 6.3 extend directly to this setting. This allows us to show that $\mathrm{sh}'(W, \Gamma) = \mathrm{sh}(W, \Gamma)$ and to prove Theorems 8.1.1 - 8.1.4 following the lines of Section 7.

Notes

The exposition follows the lines of [Tu9] with some modifications.

Chapter X
State sums on shadows

Fix a strict unimodular category $(\mathcal{V}, \{V_i\}_{i\in I})$ with ground ring K and rank $\mathcal{D} \in K$. We adhere to Conventions VI.3.1. Throughout the chapter the ground group of shadows is the additive group of integers and half-integers $(1/2)\mathbb{Z}$ with the distinguished element $1/2$.

1. State sum models on shadowed polyhedra

1.0. Outline. We introduce a state sum model on shadowed 2-polyhedra. This model will be instrumental throughout the chapter. It belongs to the class of face models in the sense that the states (colors) are assigned to the faces (2-dimensional regions) of 2-polyhedra. The main algebraic ingredients of the model are the normalized $6j$-symbols derived from $\mathcal{V}$.

The state sum on shadowed 2-polyhedra is somewhat similar to the state sum on triangulated 3-manifolds introduced in Chapter VII. The roles of tetrahedra, 2-faces, and edges of triangulations are played here by the vertices, edges, and regions of shadowed 2-polyhedra respectively. A sketchy description of the model may be given as follows. Consider a shadowed polyhedron X without boundary. Fix a coloring of the regions of X by elements of the set I. We associate to a vertex of X the normalized $6j$-symbol determined by the colors of 6 adjacent regions. This $6j$-symbol lies in the tensor product of 4 symmetrized multiplicity modules determined by the 4 edges of the graph $\text{sing}(X)$ attached to the vertex at hand. Every edge connects two (possibly coinciding) vertices and gives rise to dual factors of the corresponding tensor products. We form the tensor product of the $6j$-symbols associated to vertices of X and contract it along the dual pairs of factors determined by the edges. This gives an element of K because every factor of the tensor product in question is contracted with some other factor. Multiplying this element of K by an expression computed from the Euler characteristics, gleams, and colors of regions of X we obtain the term of the state sum corresponding to the given coloring. For shadowed polyhedra with boundary, we construct in a similar way an element of a certain module determined by the boundary.

Although the model is quite involved, it becomes simpler for many concrete polyhedra and does allow explicit computations.

1.1. Preliminaries on trivalent graphs. Let γ be a trivalent graph in the sense of Section VI.4.1. To each coloring λ of γ we assign the K-module

$$H(\gamma, \lambda) = \bigotimes_x H_x(\gamma, \lambda) = \bigotimes_x H(i_x, j_x, k_x)$$

where x runs over the vertices of γ and i_x, j_x, k_x are the λ-colors of three 1-strata of γ attached to x and oriented towards x (cf. Section VI.4.1). Here $H(i, j, k)$ is the symmetrized multiplicity module defined in Section VI.3.

Set

$$\mathscr{E}(\gamma) = \bigoplus_{\lambda \in \mathrm{col}(\gamma)} H(\gamma, \lambda)$$

where $\mathrm{col}(\gamma)$ denotes the set of all colorings of γ. It follows from Lemma II.4.2.1 that $\mathscr{E}(\gamma)$ is a projective K-module. For instance, $\mathscr{E}(S^1) = K^{\mathrm{card}(I)}$ and $\mathscr{E}(\emptyset) = H(\emptyset) = K$. It is obvious that for any 3-valent graphs γ_1, γ_2 with colorings λ_1, λ_2 respectively, we have $H(\gamma_1 \amalg \gamma_2, \lambda_1 \amalg \lambda_2) = H(\gamma_1, \lambda_1) \otimes_K H(\gamma_2, \lambda_2)$. Therefore $\mathscr{E}(\gamma_1 \amalg \gamma_2) = \mathscr{E}(\gamma_1) \otimes_K \mathscr{E}(\gamma_2)$.

For each coloring $\lambda \in \mathrm{col}(\gamma)$, we define the dual coloring $\lambda^* \in \mathrm{col}(\gamma)$ by the formula $\lambda^*(\vec{e}) = (\lambda(\vec{e}))^*$ where $\vec{e}$ runs over oriented 1-strata of γ and $*$ is the canonical involution in the set I, see Section II.1.4. Obviously, $\lambda^{**} = \lambda$. The module $H(\gamma, \lambda^*)$ is dual to $H(\gamma, \lambda)$, the duality pairing

$$\langle .,. \rangle_\lambda : H(\gamma, \lambda^*) \otimes_K H(\gamma, \lambda) \to K$$

is induced by the duality $H(i^*, j^*, k^*) \otimes_K H(i, j, k) \to K$ with $i, j, k \in I$. The pairings $\langle .,. \rangle_\lambda$ corresponding to colorings of γ determine a pairing $\langle .,. \rangle : \mathscr{E}(\gamma) \otimes_K \mathscr{E}(\gamma) \to K$ by the formula

$$\langle \bigoplus_{\lambda \in \mathrm{col}(\gamma)} x_\lambda, \bigoplus_{\lambda \in \mathrm{col}(\gamma)} y_\lambda \rangle = \sum_{\lambda \in \mathrm{col}(\gamma)} \langle x_\lambda, y_{\lambda^*} \rangle_\lambda$$

where $x_\lambda, y_\lambda \in H(\gamma, \lambda)$ for $\lambda \in \mathrm{col}(\gamma)$. It is obvious that the form $\langle .,. \rangle$ in $\mathscr{E}(\gamma)$ is symmetric and non-degenerate; the summands $H(\gamma, \lambda)$, $H(\gamma, \lambda')$ of $\mathscr{E}(\gamma)$ are orthogonal unless $\lambda' = \lambda^*$ in which case they are matched via $\langle .,. \rangle_\lambda$. For instance, the form $\langle .,. \rangle$ in $\mathscr{E}(S^1) = K^{\mathrm{card}(I)}$ is given by the formula

$$\langle \bigoplus_{i \in I} x_i, \bigoplus_{i \in I} y_i \rangle = \sum_{i \in I} x_i y_{i^*} \in K$$

where $x_i, y_i \in K$ for $i \in I$.

1.2. Face models on shadowed 2-polyhedra. Let X be a shadowed 2-polyhedron over $(1/2)\mathbb{Z}$. Recall that ∂X is a trivalent graph so that the definitions of Section 1.1 apply to ∂X. We construct a state sum model on X producing an element $|X| = |X|_{(\mathcal{V}, \mathcal{D})}$ of the module $\mathscr{E}(\partial X)$. (In particular, if $\partial X = \emptyset$ then $|X| \in K$.) We shall see in Section 2 that $|X|$ is a cobordism invariant of stable shadows.

The model introduced here is rather heavy due to possibly complicated topological structure of X. At first reading, the reader may assume that all regions of X are open 2-disks and all 1-stata of the graph $\text{sing}(X)$ are edges with at least one end not lying in ∂X. This simplifies the model and *a posteriori* does not lead to a loss of generality because any shadowed polyhedron is stably shadow equivalent to a polyhedron satisfying this assumption.

Denote by $\text{Reg}(X)$ the set of oriented regions of X. Inversion of orientation defines an involution $Y \mapsto Y^*$ in $\text{Reg}(X)$. A coloring of X is a mapping $\varphi : \text{Reg}(X) \to I$ such that $\varphi(Y^*) = (\varphi(Y))^*$ for any oriented region Y of X. The set of colorings of X is denoted by $\text{col}(X)$. This set has $(\text{card}(I))^r$ elements where r is the number of (non-oriented) regions of X.

Any oriented 1-stratum $\vec{e}$ of ∂X gives rise to an orientation in the (unique) region Y of X attached to this stratum. This orientation in Y is the one inducing the given orientation in $\vec{e}$; it is determined by a pair (a positively oriented tangent vector of $\vec{e}$, a vector pointing into Y). Denote the region Y with this orientation by $Y_{\vec{e}}$. It is obvious that inversion of orientation of $\vec{e}$ corresponds to inversion of orientation of $Y_{\vec{e}}$. Therefore any coloring φ of X induces a coloring $\partial\varphi$ of ∂X by the formula $\partial\varphi\,(\vec{e}) = \varphi(Y_{\vec{e}})$. We say that φ extends $\partial\varphi$ and that $\partial\varphi$ is the restriction of φ to ∂X.

Fix a coloring $\varphi \in \text{col}(X)$. We shall define an element $|X|_\varphi \in H(\partial X, \partial\varphi)$. The construction of $|X|_\varphi$ is as follows. First of all, we assign to each vertex x of $X \setminus \partial X$ a module $G(X, \varphi, x)$ and an element $|x|_\varphi$ of this module. Secondly, we assign to each 1-stratum e of the graph $\text{sing}(X)$ a module $G(X, \varphi, e)$ and an element $|e|_\varphi$ of this module. Then we show that the tensor product of these modules over all x and certain 1-strata e may be canonically mapped into $H(\partial X, \partial\varphi)$. Multiplying the image of $(\otimes_x |x|_\varphi) \otimes (\otimes_e |e|_\varphi)$ under this mapping by a numerical factor we obtain $|X|_\varphi \in H(\partial X, \partial\varphi)$.

To every oriented 1-stratum $\vec{e}$ of $\text{sing}(X)$ we assign the module $G_{\vec{e}} = H(i, j, k)$ where i, j, k are φ-colors of three regions of X attached to $\vec{e}$ and equipped with orientations inducing the one in $\vec{e}$. For each non-oriented 1-stratum e of $\text{sing}(X)$, set

$$G(X, \varphi, e) = G_{\vec{e}_1} \otimes_K G_{\vec{e}_2}$$

where $\vec{e}_1$ and $\vec{e}_2$ denote e with two opposite orientations. The modules $G_{\vec{e}_1}$ and $G_{\vec{e}_2}$ are dual to each other. We define $|e|_\varphi \in G(X, \varphi, e)$ to be the canonical element determined by duality.

Let x be a vertex of X not lying in ∂X. We define the module $G(X, \varphi, x)$ to be the tensor product of four modules $G_{\vec{e}}$ corresponding to four oriented edges of $\text{sing}(X)$ incident to x and oriented towards x. (Recall that an edge of a graph is a 1-stratum of this graph non-homeomorphic to a circle.) Our immediate aim is to define $|x|_\varphi \in G(X, \varphi, x)$. Note that x has a closed neighborhood U of type (VIII.1.1.3) where x corresponds to the cone point. Set $L = U \setminus \text{Int}(U)$ so that L is the polyhedral link of x in X. It is clear that L is homeomorphic to the

1-skeleton of a 3-simplex and U is the cone over L with the cone point x. Four vertices of L lie on four arcs issuing from x in $\mathrm{sing}(X)$. Six edges of L lie in six (germs of) regions of X adjacent to x. Embed L into the unit 2-sphere $S^2 \subset \mathbb{R}^3$ in an arbitrary way. Denote by Γ_x the framed graph in S^2 obtained by thickening the vertices and edges of L into 2-disks and bands inside S^2. (This is the same framed graph as in Figure VI.5.1.) The coloring φ of X induces a coloring of Γ_x as follows. Every edge h of L is contained in a region of X, every orientation α of h gives rise to an orientation of this region determined by the pair (α, a tangent vector based at a point of h and pointing out of U). We assign to the oriented edge (h, α) the value of φ on this oriented region of X. This yields a coloring of Γ_x. Denote by Γ_x^φ the resulting colored framed graph in $S^2 \subset \mathbb{R}^3$. Considered up to isotopy in $\mathbb{R}^3$ this colored framed graph does not depend on the choice of embedding $L \to S^2$ used in the construction. It is easy to deduce from definitions that $G(X, \varphi, x) = H^\star(\Gamma_x^\varphi)$. Set

$$|x|_\varphi = \left|\Gamma_x^\varphi\right| \in G(X, \varphi, x).$$

For the definition of $\left|\Gamma_x^\varphi\right|$, see Section VI.4; in the terminology of Chapter VI, $\left|\Gamma_x^\varphi\right|$ is a normalized $6j$-symbol.

Consider the module

$$G = \bigotimes_x G(X, \varphi, x) \otimes_K \bigotimes_f G(X, \varphi, f)$$

where x runs over all vertices of X not lying in ∂X and f runs over edges of $\mathrm{sing}(X)$ with both endpoints in ∂X. Since each factor of G is a tensor product of modules of type $G_{\vec{e}}$, the module G is decomposed into a tensor product of such modules. Every edge e of $\mathrm{sing}(X)$ with both ends in $X \setminus \partial X$ gives rise to two dual factors in this decomposition corresponding to the germs of e in its endpoints (or, equivalently, to two orientations of e). Denote by cntr_e the tensor contraction of G along this pair of dual factors. The contractions cntr_e associated to different edges involve disjoint pairs of factors and commute with each other. Denote by cntr the composition of these contractions corresponding to all edges of $\mathrm{sing}(X)$ with both ends in $X \setminus \partial X$. The module obtained from G via cntr may be canonically identified with $H(\partial X, \partial\varphi)$. Indeed, the factors of G surviving cntr correspond to the edges of $\mathrm{sing}(X)$ with either one or both endpoints in ∂X. Each vertex of ∂X is incident to one such edge. If e is an edge of $\mathrm{sing}(X)$ with endpoints $x \in X \setminus \partial X$ and $y \in \partial X$ then the factor of $G(X, \varphi, x)$ corresponding to e may be identified with $H_x(\partial X, \partial\varphi)$. If f is an edge of $\mathrm{sing}(X)$ with endpoints $x, y \in \partial X$ then $G(X, \varphi, f) = H_x(\partial X, \partial\varphi) \otimes_K H_y(\partial X, \partial\varphi)$. Therefore the factors of G surviving cntr may be identified with the modules $\{H_x(\partial X, \partial\varphi)\}_x$ where x runs over all vertices of ∂X. The tensor product of these factors is nothing but $H(\partial X, \partial\varphi)$. In this way the tensor contraction cntr may be regarded as a homomorphism $G \to H(\partial X, \partial\varphi)$.

We now define numerical coefficients associated to regions of X, edges of ∂X, and circle strata of $\mathrm{sing}(X)$. For a region Y of X, set $\dim(\varphi(Y)) = \dim(\varphi(\vec{Y}))$ and $v'_\varphi(Y) = v'_{\varphi(\vec{Y})}$ where $\vec{Y}$ denotes Y with a certain orientation. The equalities $\dim(i) = \dim(i^*)$, $v'_i = v'_{i^*}$ for $i \in I$ ensure the independence of $\dim(\varphi(Y))$ and $v'_\varphi(Y)$ on the choice of orientation in Y. Similarly, for a (non-oriented) edge e of ∂X, set $\dim'(\partial\varphi(e)) = \dim'(\partial\varphi(\vec{e}))$ where $\vec{e}$ is e with a certain orientation. The equality $\dim'(i) = \dim'(i^*)$ for $i \in I$ ensures independence of $\dim'(\partial\varphi(e))$ on the choice of orientation in e.

For a circle 1-stratum g of $\mathrm{sing}(X)$, set

$$h_\varphi(g) = h^{ijk} = \mathrm{Dim}\,(\mathrm{Hom}_{\mathcal{V}}(\mathbb{1}, V_i \otimes V_j \otimes V_k)) \in K$$

where i, j, k are φ-colors of three regions of X attached to g and equipped with the orientations inducing the same orientation in g. It follows from the results of Section II.4.5 that $h_{\varphi(g)}$ does not depend on the choice of orientation of g.

Set

$$|X|_\varphi = |X|_1^\varphi\, |X|_2^\varphi\, |X|_3^\varphi\, |X|_4^\varphi\, |X|_5^\varphi \in H(\partial X, \partial\varphi)$$

where

$$|X|_1^\varphi = \Big(\prod_e \dim'(\partial\varphi(e))\Big)^{-1} \in K, \quad |X|_2^\varphi = \prod_Y \dim(\varphi(Y))^{\chi(Y)} \in K,$$

$$|X|_3^\varphi = \prod_Y (v'_\varphi(Y))^{2\mathrm{gl}(Y)} \in K, \quad |X|_4^\varphi = \prod_g h_\varphi(g) \in K,$$

$$|X|_5^\varphi = \mathrm{cntr}((\otimes_x |x|_\varphi) \otimes (\otimes_f |f|_\varphi)) \in H(\partial X, \partial\varphi).$$

Here e runs over edges (but not circle 1-strata) of ∂X, Y runs over regions of X, g runs over circle 1-strata of $\mathrm{sing}(X)$, x runs over vertices of X not lying in ∂X, and f runs over edges of $\mathrm{sing}(X)$ with both endpoints in ∂X. As usual, gl denotes the gleam and χ denotes the Euler characteristic. Note that the Euler characteristics of non-compact regions of X are well defined because every region is homeomorphic to the interior of a compact surface and has the same Euler characteristic as this surface.

Recall that $\mathcal{E}(\partial X) = \oplus_\lambda H(\partial X, \lambda)$ where λ runs over all colorings of ∂X. We define a state sum $|X|$ by the formula

$$|X| = \mathcal{D}^{-b_2(X)-\mathrm{null}(X)} \sum_{\varphi\in\mathrm{col}(X)} |X|_\varphi \in \mathcal{E}(\partial X).$$

Here $b_2(X) = \dim H_2(X;\mathbb{R})$ is the second Betti number of X and $\mathrm{null}(X)$ is the nullity (i.e., the rank of the annihilator) of the bilinear form Q_X defined in Section VIII.5.1. The choice of the normalizing factor ensures convenient properties of $|X|$, as will become clear below.

For $\lambda \in \mathrm{col}(\partial X)$, we denote by $|X, \lambda|$ the projection of $|X|$ into the summand $H(\partial X, \lambda)$ of $\mathscr{E}(\partial X)$. In other words,

$$|X, \lambda| = \mathscr{D}^{-b_2(X)-\mathrm{null}(X)} \sum_{\varphi \in \mathrm{col}(X), \partial\varphi=\lambda} |X|_\varphi \in H(\partial X, \lambda).$$

We shall regard $|X, \lambda|$ as an invariant of the shadowed polyhedron X with boundary colored via λ. It is clear that $|X| = \oplus_\lambda |X, \lambda|$ where λ runs over all colorings of ∂X. We may also consider analogous state sum invariants of shadowed polyhedra with partially colored boundary. (This means that certain components of the boundary are colored and others are not.)

1.3. Remarks and examples. 1. The definition of $|X|$ is ready for explicit computations. The amount of computations may be somewhat reduced due to the following observation. Assume that for certain $i, j, k \in I$, we have $\mathrm{Hom}(\mathbb{1}, V_i \otimes V_j \otimes V_k) = 0$. Assume that a coloring φ of X attains the values i, j, k on three oriented regions of X adjacent to a certain 1-stratum of $\mathrm{sing}(X)$ so that their orientations induce the same orientation in this 1-stratum. Then $|X|_\varphi = 0$. Indeed, if the 1-stratum in question is a circle then $h^{ijk} = 0$ explicitly appears as a factor of $|X|_\varphi$. If this 1-stratum is an edge of $\mathrm{sing}(X)$ with at least one endpoint x not lying in ∂X then the module $\mathrm{Hom}(\mathbb{1}, V_i \otimes V_j \otimes V_k) = 0$ appears as a factor of the ambient module of $|x|_\varphi$ so that $|x|_\varphi = 0$. If both endpoints of this 1-stratum lie in ∂X then $\mathrm{Hom}(\mathbb{1}, V_i \otimes V_j \otimes V_k) = 0$ appears as a factor of $H(\partial X, \partial\varphi)$ so that this module equals 0. Therefore such a coloring φ does not contribute to $|X|$.

2. For any closed connected oriented surface Σ of genus g, we have

$$\left|\Sigma_0\right| = \mathscr{D}^{-2} \sum_{i \in I} (\dim(i))^{2-2g}$$

and

$$\left|\Sigma_n\right| = \mathscr{D}^{-1} \sum_{i \in I} (v_i')^{2n} (\dim(i))^{2-2g}$$

for any non-zero gleam $n \in (1/2)\mathbb{Z}$. It is easy to compute (cf. formula (II.2.4.a)) that

$$\left|S_0^2\right| = 1, \ \left|S_1^2\right| = \mathscr{D}\Delta_{\mathcal{V}}^{-1}, \ \left|S_{-1}^2\right| = \mathscr{D}^{-1}\Delta_{\mathcal{V}}.$$

1.4. Multiplicativity with respect to gluing. We show that the state sum $|X|$ is multiplicative with respect to the gluing of shadowed 2-polyhedra along the boundary. This property of $|X|$ will play important role in the proof of the invariance of $|X|$ under shadow moves. The multiplicativity of $|X|$ is similar to the multiplicativity of the 3-manifold invariant $|M|$.

Let X_1 and X_2 be two shadowed 2-polyhedra meeting along a trivalent graph $\gamma = X_1 \cap X_2$. Assume that for both $r = 1$ and $r = 2$ the graph γ is formed by several connected components of ∂X_r. It is clear that $X = X_1 \cup X_2$ is a simple

2-polyhedron with $\partial X = (\partial X_1 \setminus \gamma) \amalg (\partial X_2 \setminus \gamma)$. We define the gleam of any region Y of X to be the sum of the gleams of regions of X_1 and X_2 contained in Y. This makes X a shadowed 2-polyhedron.

For $r = 1, 2$, consider the state sum

$$|X_r| \in \mathscr{E}(\partial X_r) = \mathscr{E}(\partial X_r \setminus \gamma) \otimes_K \mathscr{E}(\gamma).$$

Denote by ξ the homomorphism

$$\mathscr{E}(\partial X_1 \setminus \gamma) \otimes_K \mathscr{E}(\gamma) \otimes_K \mathscr{E}(\partial X_2 \setminus \gamma) \otimes_K \mathscr{E}(\gamma) \to \mathscr{E}(\partial X) = \mathscr{E}(\partial X_1 \setminus \gamma) \otimes_K \mathscr{E}(\partial X_2 \setminus \gamma)$$

induced by the bilinear form $\mathscr{E}(\gamma) \otimes_K \mathscr{E}(\gamma) \to K$ constructed in Section 1.1.

1.4.1. Lemma (the multiplicativity formula). *We have*

$$|X| = \mathscr{D}^c \, \xi(|X_1| \otimes |X_2|) \tag{1.4.a}$$

where $c = b_2(X_1) + b_2(X_2) - b_2(X) + \text{null}(X_1) + \text{null}(X_2) - \text{null}(X)$.

To prove Lemma 1.4.1 we need the following assertion which follows directly from Lemmas VII.2.2.1 and VII.2.2.2.

1.4.2. Lemma. *Let Z be a projective K-module and $\delta \in Z^* \otimes Z$ be the canonical element determined by duality. Let $Z_1, \dots, Z_n$ be n copies of Z and $\delta^{\otimes n} \in \otimes_{i=1}^n (Z_i \otimes_K Z_i^*)$ be the tensor product of n copies of δ. Then the homomorphism $\otimes_{i=1}^n (Z_i \otimes_K Z_i^*) \to K$ obtained by tensor contraction of the pairs $(Z_i^* = Z^*, Z_{i+1} = Z)$ for $i = 1, \dots, n-1$ followed by tensor contraction of the pair (Z_1, Z_n^*) carries $\delta^{\otimes n}$ into* $\text{Dim}(Z) \in K$.

1.4.3. Proof of Lemma 1.4.1. It is obvious that the powers of $\mathscr{D}$ appearing on the right-hand and left-hand sides of (1.4.a) are the same. Each coloring φ of X restricts to X_1, X_2 and yields colorings φ_1, φ_2 of these polyhedra. This establishes a bijective correspondence between the colorings of X and pairs of colorings ($\varphi_1 \in \text{col}(X_1)$, $\varphi_2 \in \text{col}(X_2)$) such that φ_1, φ_2 induce dual colorings of γ. Therefore it suffices to prove that for any $\varphi \in \text{col}(X)$,

$$|X|_\varphi = \xi(|X_1|_{\varphi_1} \otimes |X_2|_{\varphi_2}). \tag{1.4.b}$$

To prove (1.4.b) we compare the factors contributed to the left-hand side and the right-hand side by regions, vertices, and 1-strata. More explicitly, we shall prove that

$$|X|_1^\varphi \, |X|_2^\varphi = |X_1|_1^{\varphi_1} \, |X_1|_2^{\varphi_1} \, |X_2|_1^{\varphi_2} \, |X_2|_2^{\varphi_2}, \tag{1.4.c}$$

$$|X|_3^\varphi = |X_1|_3^{\varphi_1} \, |X_2|_3^{\varphi_2}, \tag{1.4.d}$$

$$|X|_4^\varphi \, |X|_5^\varphi = |X_1|_4^{\varphi_1} \, |X_2|_4^{\varphi_2} \, \xi(|X_1|_5^{\varphi_1} \otimes |X_2|_5^{\varphi_2}). \tag{1.4.e}$$

Every region Y of X may be decomposed as a disjoint union

$$Y_1 \amalg \dots \amalg Y_m \amalg e_1 \amalg \dots \amalg e_n$$

where $Y_1, \dots, Y_m$ are regions of X_1, X_2, and $e_1, \dots, e_n$ are 1-strata of γ. Since the Euler characteristic of a circle 1-stratum equals 0 and the Euler characteristic of an (open) edge equals -1, we have

$$\chi(Y) = \chi(Y_1) + \dots + \chi(Y_m) - r$$

where r is the number of 1-strata, say $e_{j_1}, \dots, e_{j_r}$, among $e_1, \dots, e_n$ which are edges and not circles. Therefore

$$\dim(\varphi(Y))^{\chi(Y)} = \prod_{k=1}^{m} \dim(\varphi(Y))^{\chi(Y_k)} \left(\prod_{l=1}^{r} \dim(\lambda(e_{j_l}))\right)^{-1} \tag{1.4.f}$$

where $\lambda = \partial\varphi_1 = (\partial\varphi_2)^*$. For any edge e of γ, we have $\lambda(e) = \partial\varphi_1(e) = (\partial\varphi_2(e))^*$ and therefore

$$\dim(\lambda(e)) = \dim'(\partial\varphi_1(e)) \dim'(\partial\varphi_2(e)).$$

Substituting this expression into (1.4.f) and multiplying the resulting equalities over all regions Y of X we get a formula equivalent to (1.4.c).

Equality (1.4.d) follows from the fact that the gleam of any region Y of X is equal to the sum of the gleams of regions of X_1, X_2 contained in Y.

Let us prove (1.4.e). First consider the case $\partial X = \emptyset$ so that $\partial X_1 = \partial X_2 = \gamma$. The L.H.S. and the R.H.S. of (1.4.e) have many equal factors. Each circle 1-stratum of X_1 or X_2 is a circle 1-stratum of X and contributes the same factor to the L.H.S. and the R.H.S. Each vertex of X contributes the same factors to the L.H.S. and the R.H.S. (Note that there are no vertices of X lying in γ.) Each edge of X_1, X_2 whose endpoints do not lie in γ gives rise to the same tensor contractions on the L.H.S. and on the R.H.S. There are factors which appear on the L.H.S. but not on the R.H.S. Namely, the polyhedron X may have circle 1-strata formed by edges of both $\operatorname{sing}(X_1)$ and $\operatorname{sing}(X_2)$, the corresponding factors appear only on the L.H.S. The edges of X which pierce γ are formed from several edges of X_1, X_2 and yield contractions on the L.H.S. which do not appear on the R.H.S. Similarly, there are factors which appear on the R.H.S. but not on the L.H.S. Every edge f of X_1 or X_2 with both endpoints in γ contributes a δ-type factor to the R.H.S. which does not appear on the L.H.S. Every vertex of γ gives rise to a tensor contraction on the R.H.S. (via ξ) which has no counterpart on the L.H.S. We have to match these different contributions.

Consider the following equivalence relation in the set of edges of $\operatorname{sing}(X_1) \amalg \operatorname{sing}(X_2)$: two edges are equivalent if they lie on the same 1-stratum of X. The edges belonging to one equivalence class J lie on a 1-stratum $g = g_J$ of X which may be either a circle stratum or an edge. We shall show that in the first case the joint contribution of the edges of the class J and the endpoints of these edges to the R.H.S. of (1.4.e) is equal to $h_\varphi(g)$. In the second case the joint contribution of the edges of the class J and the endpoints of these edges to the R.H.S of (1.4.e) is equal to $\mathrm{cntr}_g(|x|_\varphi \otimes |y|_\varphi)$ where x, y are the endpoints of g. This would imply (1.4.e).

Consider the case where g is a circle 1-stratum of X meeting both X_1 and X_2. This 1-stratum contributes $h_\varphi(g)$ to the L.H.S. of (1.4.e). The circle g may be decomposed as a union of vertices $x_1, \ldots, x_{2n}$ of γ, edges $e_1, \ldots, e_n$ of $\mathrm{sing}(X_1)$, and edges $f_1, \ldots, f_n$ of $\mathrm{sing}(X_2)$ such that x_{2j-1}, x_{2j} are the endpoints of e_j, and x_{2j}, x_{2j+1} are the endpoints of f_j for $j = 1, \ldots, n$ (where $x_{2n+1} = x_1$). Each edge $e_1, \ldots, e_n$, $f_1, \ldots, f_n$ contributes a δ-type factor to the R.H.S. of (1.4.e). The vertices $x_1, \ldots, x_{2n}$ give rise to tensor contractions of matched modules. It is easy to see that these contractions follow the pattern of Lemma 1.4.2 where $Z = H(i, j, k)$ with i, j, k being the $\partial\varphi_1$-colors of the edges of γ incident to x_1 and oriented towards x_1. Lemma 1.4.2 implies that the joint contribution of the edges $e_1, \ldots, e_n$, $f_1, \ldots, f_n$ and the vertices $x_1, \ldots, x_{2n}$ to the R.H.S. is equal to $\mathrm{Dim}(H(i, j, k)) = h_\varphi(g)$, i.e., to the contribution of g to the L.H.S.

Let g be an edge of X which pierces γ. Let x, y be the endpoints of g (possibly, $x = y$). Since x, y are vertices of X they do not lie on γ. We can decompose g as a union of vertices $x_1, \ldots, x_m$ of γ and edges $e_0, e_1, \ldots, e_m$ of $\mathrm{sing}(X_1)$, $\mathrm{sing}(X_2)$ such that x_j, x_{j+1} are the endpoints of e_j for $j = 1, \ldots, m-1$, e_0 connects x with x_1, and e_m connects x_m with y. All edges e_j with even j lie in one of the two sets $\mathrm{sing}(X_1)$, $\mathrm{sing}(X_2)$ and all edges e_j with odd j lie in the second set. The joint contribution of x, y, g to the L.H.S. is equal to $\mathrm{cntr}_g(|x|_\varphi \otimes |y|_\varphi)$. The joint contribution of $x, x_1, \ldots, x_m, y, e_0, e_1, \ldots, e_m$ to the R.H.S. is obtained from

$$|x|_\varphi \otimes |y|_\varphi \otimes (\otimes_{j=1}^{m-1} |e_j|_\varphi)$$

by tensor contractions in $x_1, \ldots, x_m$ used in the definition of ξ. The edge e_0 contributes to the R.H.S. an identification of a factor of $G(X_r, \varphi_r, x)$ with $H_{x_1}(\gamma, \partial\varphi_r)$ where $r \in \{1, 2\}$ such that $x \in X_r$. This identification allows us to apply the tensor contraction in x_1 to $|x|_\varphi \otimes |e_1|_\varphi$. Similarly, e_m contributes an identification of a factor of $G(X_s, \varphi_s, y)$ with $H_{x_m}(\gamma, \partial\varphi_s)$ where $s \in \{1, 2\}$ such that $y \in X_s$. This identification allows us to apply the tensor contraction in x_m to $|y|_\varphi \otimes |e_{m-1}|_\varphi$. It follows from Lemma VII.2.2.2 that the contribution of $x, x_1, \ldots, x_m, y, e_0, e_1, \ldots, e_m$ to the R.H.S. of (1.4.e) is equal to the same expression $\mathrm{cntr}_g(|x|_\varphi \otimes |y|_\varphi)$ as in the L.H.S. This completes the proof of (1.4.e) and (1.4.b) in the case $\partial X = \emptyset$.

In the case $\partial X \neq \emptyset$ the argument is similar, although we have to consider two more possibilities for the 1-stratum g: it may connect two vertices of ∂X or a vertex of ∂X with a vertex of $X \setminus \partial X$.

1.5. Exercise. Show that $|X|_{(\overline{\mathcal{V}}, \mathcal{D})} = |-X|_{(\mathcal{V}, \mathcal{D})}$.

2. State sum invariants of shadows

2.0. Outline. We show that the state sum introduced in Section 1 is invariant under shadow moves, suspension, and cobordism. The Racah identity, the orthogonality relation, and the Biedenharn-Elliott identity ensure invariance of the state sum under the basic shadow moves. The cobordism invariance is deduced from formula (II.4.5.g). The reader should pay attention to Lemma 2.4 which often serves as a convenient computational tool.

At the end of the section we use these results to define a cobordism invariant of stable shadows. This accomplishes our study of abstract shadows and their state sum invariants. In further sections we shall apply this technique to shadows of manifolds.

2.1. Theorem. *For any shadowed polyhedron X, the state sum $|X|$ is invariant under basic shadow moves on X.*

Note that the shadow moves do not change the boundary of X so that the state sum $|X|$ considered before and after the moves belongs to the same module.

Proof of Theorem. Consider first the move P_2 which is slightly more subtle than P_1 and P_3 since under this move one of the regions may split into two regions. The move P_2 applies inside a standard subpolyhedron of X shown in Figure VIII.1.1. The remaining part of X is not changed under the move. Lemma 1.4.1 implies that it suffices to prove P_2-invariance of $|X|$ in the case when X *is* the shadowed 2-polyhedron drawn in Figure VIII.1.1 (to the left of the P_2-arrow) with arbitrary integer or half-integer gleams $a_1, ..., a_6$. Let X' be the shadowed 2-polyhedron drawn to the right of the P_2-arrow in Figure VIII.1.1. We have $\partial X' = \partial X$. Denote by $D, Y_1, Y_2, ..., Y_6$ the regions of X' marked in Figure VIII.1.1 by $0, a_1, a_2, ..., a_6$ respectively.

Two colorings ψ_1 and ψ_2 of X' are said to be compatible if they coincide on all oriented regions of X' except, possibly, D. Compatibility is an equivalence relation in the set of colorings of X'. The colorings in one equivalence class are distinguished by their values on D. Each equivalence class consists of card(I) elements.

Let G be an equivalence class of compatible colorings of X'. Pick $\psi \in G$. Denote by i, j, k, k', m the ψ-colors of the regions Y_3, Y_1, Y_4, Y_6, Y_5 respectively, provided with orientations corresponding to the counterclockwise orientation in the plane of Figure VIII.1.1. Denote by l, n the ψ-colors of the regions Y_2 and D provided with orientations corresponding to the clockwise orientation in the plane of Figure VIII.1.1. By our assumptions, only the color n depends on the choice of $\psi \in G$. It follows from definitions that

$$\left|X'\right|_1^{\psi} = (\dim'(i)\dim'(j)\dim'(k)\dim'(k')\dim'(l)\dim'(m))^{-1},$$

$$\left|X'\right|_2^{\psi}=\dim(i)\dim(j)\dim(k)\dim(k')\dim(l)\dim(m)\dim(n),$$

$$\left|X'\right|_3^{\psi}=(v_i')^{2a_3}(v_j')^{2a_1}(v_k')^{2a_4}(v_{k'}')^{2a_6}(v_l')^{2a_2}(v_m')^{2a_5}.$$

Denote the right-hand side of the last formula by v_G. Note that v_G does not depend on the choice of ψ in G. We also have $\left|X'\right|_4^{\psi}=1$ and

$$\left|X'\right|_5^{\psi}=*_{im^*n}*_{jln^*}\left(\begin{vmatrix} i^* & j^* & k^* \\ l^* & m^* & n^* \end{vmatrix}\otimes\begin{vmatrix} i & j & k' \\ l & m & n \end{vmatrix}\right).$$

Therefore

$$\left|X'\right|_{\psi}=v_G\dim'(i)\dim'(j)\dim'(k)\dim'(k')\dim'(l)\dim'(m)\dim(n)\times$$

$$\times *_{im^*n}*_{jln^*}\left(\begin{vmatrix} i^* & j^* & k^* \\ l^* & m^* & n^* \end{vmatrix}\otimes\begin{vmatrix} i & j & k' \\ l & m & n \end{vmatrix}\right).$$

If $k\neq k'$ then the orthonormality relation for $6j$-symbols implies that

$$\sum_{\psi\in G}\left|X'\right|_{\psi}=0. \tag{2.1.a}$$

Thus the total contribution of the colorings of class G to $\left|X'\right|$ vanishes. If $k=k'$ then the colorings $\psi\in G$ induce, in an obvious way, a coloring $\varphi=\varphi_G$ of X. A similar computation shows that

$$|X|_{\varphi}=v_G\dim'(i)\dim'(j)\dim'(l)\dim'(m)\,\mathrm{Id}(i,j,k^*)\otimes\mathrm{Id}(k,l,m^*)$$

(for the definition of $\mathrm{Id}(i,j,k^*)$, see Section VI.5.3). The orthonormality relation for $6j$-symbols implies that

$$\sum_{\psi\in G}\left|X'\right|_{\psi}=|X|_{\varphi}. \tag{2.1.b}$$

Summing up the equalities (2.1.a), (2.1.b) over all equivalence classes G and taking into account the formulas $b_2(X')=b_2(X)$ and $\mathrm{null}(X')=\mathrm{null}(X)$ we get $\left|X'\right|=|X|$.

Invariance of $|X|$ under the moves P_1, P_3 is proven in a similar way using the Racah identity and the Biedenharn-Elliott identity.

2.2. Theorem. *For any connected shadowed polyhedra X_1, X_2, we have*

$$|X_1+X_2|=|X_1|\otimes|X_2|.$$

If the boundary of X_1 (or X_2, or both) is empty then the tensor product in the last formula amounts to multiplication by an element of the ground ring K.

Proof of Theorem. Let Z be the shadowed polyhedron representing $X_1 + X_2$ and obtained by gluing of X_1 to X_2 along a homeomorphism of closed 2-disks $D_1 \subset \text{Int}(X_1)$ and $D_2 \subset \text{Int}(X_2)$. For $r = 1, 2$ denote by Y_r the region of X_r containing D_r. We provide Y_1, Y_2, D_1, D_2 with orientations which induce the same orientation in the circle $\partial D_1 \approx \partial D_2$.

We may use the inclusion $X_r \subset Z$ with $r = 1, 2$ to restrict any coloring $\psi \in \text{col}(Z)$ to a coloring $\psi_r \in \text{col}(X_r)$. It is understood that $\psi_r(Y_r) = \psi(Y_r \setminus D_r)$. The formula $\psi \mapsto (\psi_1, \psi_2)$ defines a surjective mapping $\text{col}(Z) \to \text{col}(X_1) \times \text{col}(X_2)$. Let G be a fiber of this mapping consisting of all $\psi \in \text{col}(Z)$ with certain given ψ_1, ψ_2. The colorings belonging to G are distinguished by their values on $\text{Int}(D_1) = \text{Int}(D_2)$. We claim that

$$\sum_{\psi \in G} |Z|_\psi = |X_1|_{\psi_1} \otimes |X_2|_{\psi_2} . \tag{2.2.a}$$

This follows directly from the definitions and equality (II.4.5.b) where $i = \psi_1(Y_1)$, $j = \psi_2(Y_2)$, and k is the ψ-color of $\text{Int}(D_1) = \text{Int}(D_2)$. Summing up the equalities (2.2.a) over all fibers G and taking into account the formulas $b_2(Z) = b_2(X_1) + b_2(X_2)$ and $\text{null}(Z) = \text{null}(X_1) + \text{null}(X_2)$ we get $|Z| = |X_1| \otimes |X_2|$.

2.3. Corollary. *The state sum $|X|$ is invariant under suspension.*

This follows from Theorem 2.2 and the equality $\left|S_0^2\right| = 1$.

2.4. Lemma. *Let X be a shadowed polyhedron with colored boundary. Let L be a shadowed loop in X and C be the shadow cylinder over L. Endow ∂C with the coloring extending the one of ∂X and assigning $0 \in I$ to the circle $\partial C \setminus \partial X \approx L$. Then $|C| = |X|$.*

Note that if λ^0 is the coloring of ∂C that is the extension of a coloring λ of ∂X and assigns 0 to $\partial C \setminus \partial X$ then $H(C, \lambda^0) = H(X, \lambda) \otimes_K K = H(X, \lambda)$. Therefore $|C|$ and $|X|$ belong to the same module so that the equality $|C| = |X|$ makes sense.

Proof of Lemma. Suspend X if necessary to ensure that $\text{sing}(X) \neq \emptyset$. Then apply several P_2-moves to X so that the resulting shadowed polyhedron, again denoted by X, has no circle components in $\text{sing}(X)$. (The loop L does not interfere with these moves since we may always deform it out of the neighborhoods where the moves proceed.) Theorem 2.1 and Corollary 2.3 guarantee that both $|C|$ and $|X|$ are invariant under these transformations. Thus, it suffices to consider the case when $\text{sing}(X)$ has no circle components. Similarly, we may assume that L crosses $\text{sing}(X)$ at least once (so that $\text{sing}(C)$ also has no circle components) and that no edge of $\text{sing}(X)$ has both ends in ∂X. The polyhedron C automatically inherits this property. These assumptions simplify the expressions for $|C|, |X|$.

It is obvious that the regions of C are exactly the regions of L in X and one additional region Y_0 adjacent to the boundary circle $\partial C \setminus \partial X$. To compute $|C|$ we ought to consider the colorings of C extending the given coloring of ∂C and in particular assigning $0 \in I$ to Y_0. As we know, the module $H(0, i, j)$ is non-trivial if and only if $j = i^*$. Remark 1.3.1 shows that it is enough to consider only those colorings of C which satisfy the following condition: any two regions of L in X attached from different sides to a small subarc of L (containing no crossing points) and equipped with orientations inducing opposite orientations in this arc have equal colors. In other words, it is enough to consider only those colorings of C which are induced by colorings of X.

Let φ be an arbitrary coloring of X and let ψ be the induced coloring of C. (The coloring ψ assigns 0 to Y_0 and assigns to any oriented region Y of L the value of φ on the region of X containing Y and provided with the induced orientation.) We claim that

$$|C|_\psi = |X|_\varphi \,. \tag{2.4.a}$$

Since $b_2(C) = b_2(X)$ and $\operatorname{null}(C) = \operatorname{null}(X)$, equality (2.4.a) would imply the claim of the lemma.

Let us prove (2.4.a). It is obvious that $|C|_1^\psi = |X|_1^\varphi$. By definition of shadowed loops, the gleam of any region Y of X is equal to the sum of the gleams of regions of L in X contained in Y. This implies that $|C|_3^\psi = |X|_3^\varphi$. Note also that under our assumptions, we have $|C|_4^\psi = |X|_4^\varphi = 1$.

It follows from Lemma VI.5.3.1 that for any crossing point x of L with an arc of $\operatorname{sing}(X)$, the associated $6j$-symbol $|x|_\psi$ is the product of a δ-type tensor and a numerical factor $(\dim'(i)\dim'(j))^{-1}$ where i, j are the φ-colors of two regions of X adjacent to the arc of $\operatorname{sing}(X)$ in question and containing the branch of L crossing this arc. Similarly, for any self-crossing point x of L, the associated symbol $|x|_\psi$ is the product of a δ-type tensor and a numerical factor $(\dim(i))^{-1} = (\dim'(i))^{-2}$ where i is the φ-color of the region of X containing x. Denote by q the product of these numerical factors over all self-crossings of L and crossings of L with $\operatorname{sing}(X)$. We claim that

$$q\,|C|_2^\psi = |X|_2^\varphi \quad \text{and} \quad q^{-1}\,|C|_5^\psi = |X|_5^\varphi \,.$$

These equalities would imply (2.4.a).

For any region Y of X, we may compute the total number of entries $\dim'(\varphi(Y))$ contributed to q by the self-crossing points of L lying inside Y and by the crossings of L with the boundary of $\overline{Y}$. An elementary counting of Euler characteristics shows that this number is equal to $2(\chi(Y) - \sum_Z \chi(Z))$ where Z runs over all regions of L contained in Y. This implies that $q\,|C|_2^\psi = |X|_2^\varphi$.

Let us show that $q^{-1}\,|C|_5^\psi = |X|_5^\varphi$. Since there are no edges of $\operatorname{sing}(X)$ with both ends in ∂X we have $|X|_5^\varphi = \operatorname{cntr}(\otimes_x |x|_\varphi)$ where x runs over all 4-valent vertices of $\operatorname{sing}(X)$ (i.e., over vertices of $\operatorname{sing}(X)$ not lying in ∂X) and cntr denotes the composition of tensor contractions corresponding to all edges of sing(X) with both ends of valency 4. Similarly, $|C|_5^\psi = \mathrm{cntr}(\otimes_x |x|_\psi)$ where x runs over all 4-valent vertices of sing(C) and cntr denotes the composition of tensor contractions corresponding to all edges of sing(C) with both ends of valency 4. The vertices of sing(C) are just the vertices of sing(X), the self-crossings of L, and the crossings of L with sing(X). The presence of the factor q^{-1} in $q^{-1}\,|C|_5^\psi$ signifies that instead of $6j$-symbols $|x|_\psi$ associated to self-crossings of L and the crossings of L with sing(X) we simply involve the corresponding δ-type tensors. In the case of self-crossings of L these δ-type tensors are nothing but units in different copies of the ground ring K. Indeed, each self-crossing of L is traversed by the boundary of Y_0 twice so that the associated 6-tuple of colors contains 0 twice. This leads to the degeneration of the associated modules into the rank 1 module K (cf. Section VI.3.5). The tensor contractions along the edges of sing(C) lying in L identify these copies of K. Therefore we may safely forget about these factors and their tensor contractions. The only remaining distinction between $q^{-1}\,|C|_5^\psi$ and $|X|_5^\varphi$ amounts to the δ-type factors in $q^{-1}\,|C|_5^\psi$ arising from the crossing points of L with the edges of sing(X). Each such crossing point x splits the ambient edge of sing(X) into two subedges. The tensor contractions along these subedges completely eliminate the δ-type factor derived from x so that the contribution of x together with these two subedges is the same as if there were no crossing at x. In this way we may inductively eliminate all factors corresponding to the crossings of L with sing(X). This shows that $q^{-1}\,|C|_5^\psi = |X|_5^\varphi$ and completes the proof of the lemma.

2.5. Theorem. *The state sum $|X|$ is invariant under surgeries on X.*

Proof. Consider first the simple 2-polyhedron R obtained from two closed 2-disks B and D by gluing along an embedding $\partial B \to \mathrm{Int}(D)$. Denote by D' the closed disk in $\mathrm{Int}(D)$ bounded by the image of $\partial B \approx S^1$ under this embedding. Provide the regions $\mathrm{Int}(B)$, $\mathrm{Int}(D')$, and $\mathrm{Int}(D) \setminus D'$ with the gleams 1, -1, and 0 respectively. It is clear that R is a shadowed polyhedron with the boundary $\partial R = \partial D = S^1$. For $k \in I$, denote by R^k the shadowed polyhedron R whose boundary circle is oriented in an arbitrary way and labelled with k. It is clear that $b_2(R) = \mathrm{null}(R) = 1$. It follows from definitions that

$$\left|R^k\right| = \mathcal{D}^{-2}(\dim'(k))^{-1} \sum_{i,j \in I} h^{ijk} v_i v_j^{-1} \dim(i)\dim(j) \in K.$$

Formula (II.4.5.g) (with k replaced with k^*) shows that $\left|R^k\right| = \delta_k^0$.

Let us prove the theorem for the case $\partial X = \emptyset$. It is enough to prove the invariance of $|X|$ under surgeries along shadow knots in X. Let L be a shadowed loop in X. Consider the shadow cylinder $C = CY_L$ over L and orient the circle ∂C in an arbitrary way. For $k \in I$, label ∂C with k and denote the resulting shadowed polyhedron with colored boundary by C^k. The description of surgery given in Section VIII.4.4 shows that the shadow produced by surgery on X along L may be obtained by gluing C and R along a homeomorphism $\partial C \to \partial R$. Denote the shadowed polyhedron obtained via this gluing by Z. Lemma 1.4.1 implies that

$$|Z| = \mathcal{D}^c \langle |C|, |R| \rangle = \mathcal{D}^c \sum_{k \in I} |C^k| \, |R^{k^*}| = \mathcal{D}^c \, |C^0|$$

where $\langle ., . \rangle$ is the bilinear form in $\mathscr{E}(S^1) = K^{\mathrm{card}(I)}$ discussed in Section 1.1 and

$$c = b_2(C) + b_2(R) - b_2(Z) + \mathrm{null}(C) + \mathrm{null}(R) - \mathrm{null}(Z) =$$

$$= 2 + b_2(X) - b_2(Z) + \mathrm{null}(X) - \mathrm{null}(Z).$$

It is easy to compute that if L is homologically trivial in X over $\mathbb{R}$ then $b_2(Z) = b_2(X) + 2$ and $\mathrm{null}(Z) = \mathrm{null}(X)$ so that $c = 0$. If L presents a non-trivial element of $H_1(X; \mathbb{R})$ then $b_2(Z) = b_2(X) + 1$ and $\mathrm{null}(Z) = \mathrm{null}(X) + 1$ so that again $c = 0$. Therefore in both cases $|Z| = |C^0|$. According to Lemma 2.4 we have $|C^0| = |X|$. Thus $|Z| = |X|$ which completes the proof of the theorem in the case $\partial X = \emptyset$.

The case $\partial X \neq \emptyset$ is similar. The only difference is that instead of the pairing $\langle ., . \rangle$ we should use the homomorphism ξ introduced in Section 1.4.

2.6. Invariants of shadows. Theorem 2.1 shows that the state sum $|X|$ is an invariant of shadows (over $((1/2)\mathbb{Z}, 1/2)$). For any shadow α, set $|\alpha| = |X| \in \mathscr{E}(\partial\alpha)$ where X is an arbitrary shadowed polyhedron representing α. More generally, for any coloring λ of $\partial\alpha$, the state sum $|X, \lambda|$ is an invariant of the pair (α, λ). This invariant will be denoted by $|\alpha, \lambda|$. It is clear that $|\alpha| = \sum_{\lambda \in \mathrm{col}(\partial\alpha)} |\alpha, \lambda|$.

Theorem 2.2 implies that the invariant $|...|$ is multiplicative under summation of connected shadows. Corollary 2.3 and Theorem 2.5 show that $|\alpha|$ is a cobordism invariant of stable shadows.

There is a convenient renormalization of $|\alpha|$ involving the signature $\sigma(\alpha) \in \mathbb{Z}$ (see Section VIII.5). For any shadow α, set

$$||\alpha|| = (\mathcal{D}\Delta_{\mathcal{V}}^{-1})^{-\sigma(\alpha)} \, |\alpha| \, .$$

Since $\sigma(\alpha)$ is a cobordism invariant of stable shadows, $||\alpha||$ is also a cobordism invariant of stable shadows. This invariant is multiplicative in the class of connected shadows. The renormalized invariant $||...||$ has one technical advantage over $|...|$: it takes the value 1 not only on $[S^2_0]$, as $|...|$ does, but also on $[S^2_1]$ and $[S^2_{-1}]$:

$$\left|\left| [S^2_0] \right|\right| = \left|\left| [S^2_1] \right|\right| = \left|\left| [S^2_{-1}] \right|\right| = 1.$$

For any coloring λ of $\partial\alpha$, set

$$||\alpha, \lambda|| = (\mathcal{D}\Delta_{\mathcal{V}}^{-1})^{-\sigma(\alpha)} \, |\alpha, \lambda| \, .$$

It is clear that $||\alpha|| = \sum_{\lambda\in \mathrm{col}(\partial\alpha)} ||\alpha, \lambda||$.

2.7. Remarks. 1. Applying the invariant $|...|$ to cylinders over shadow links and shadow graphs, we get isotopy invariants of links and graphs on shadowed polyhedra. For instance, consider a shadow link β on a shadowed polyhedron without boundary. Assume that the components of β are oriented and colored with elements of the set I. This data induces a coloring, say λ, of $\partial(CY_\beta)$. Thus, we have a numerical invariant $\left|CY_\beta, \lambda\right| \in H(\partial(CY_\beta), \lambda) = K$ of β. This invariant is preserved under simultaneous inversion of orientation in all components of β and trading their colors for the dual ones.

2. An essential feature of our constructions is the finiteness of the state sum introduced in Section 1. It is ensured by the fundamental assumption that the given family $\{V_i\}_{i\in I}$ of simple objects of $\mathcal{V}$ is finite. However, this finiteness is not necessary to define colorings of shadowed polyhedra and individual terms of the state sums. We may repeat word for word the definitions of Section 1 using an arbitary unimodal semisimple category instead of $\mathcal{V}$. Generally speaking, this gives an infinite number of colorings and an infinite number of terms of the state sum. There is a remarkable geometric situation where the number of colorings contributing non-zero terms is finite; this is the case for (cylinders over) colored shadow links in the 2-disk with colored boundary. This fact follows from the rudimentary finiteness property of semisimple categories mentioned at the beginning of Section II.4.5. Therefore, for such links, we may derive state sum invariants from unimodal semisimple categories generalizing the invariants derived above from unimodular categories. More generally, the same construction applies to colored shadow links (and shadow graphs) in connected shadowed surfaces with non-empty colored boundary.

3. Invariants of 3-manifolds from the shadow viewpoint

3.0. Outline. We apply the technique of state sum invariants to the shadows of 3-manifolds. The invariant of 3-manifolds obtained via the internal shadow is equivalent to the state sum invariant defined in Chapter VII. This is due to a duality between the models introduced in Chapter VII and in Section 1.

A deep result formulated in this section (Theorem 3.3) affirms that the state sum invariants of shadows of 4-manifolds bounded by a 3-manifold M are equivalent to the invariant $\tau(M)$ introduced in Chapter II. Theorem 3.3 allows us to relate two different viewpoints: one based on surgery, Kirby moves, and splitting of link diagrams into elementary pieces and one based on shadows and $6j$-symbols. This relationship yields a key to the proof of Theorems VII.4.1.1

and VII.4.4.1 given at the end of the section. The fundamental Theorem 3.3 will be proven in Sections 4 - 6.

3.1. Theorem. *For any closed* 3*-manifold* M*, we have*

$$|| \, ish(M) || = | \, ish(M) | = \mathcal{D}^{2(b_3(M)-b_2(M))} \, |M| \tag{3.1.a}$$

where $|M|$ *is the simplicial state sum introduced in Section* VII.1.

The expression $b_3(M) - b_2(M)$ may be rewritten as $b_0(M) - b_1(M)$. Indeed, it follows from Poincaré duality that the Euler characteristic of any closed 3-manifold M is equal to zero and therefore $b_3(M) - b_2(M) = b_0(M) - b_1(M)$. For orientable M, $b_3(M) = b_0(M)$ is the number of connected components of M and $b_2(M) = b_1(M)$.

Proof of Theorem. It is obvious that the bilinear form of the shadow $\mathrm{ish}(M)$ is equal to zero. Therefore the signature of $\mathrm{ish}(M)$ equals 0 and $|| \, \mathrm{ish}(M) || = | \, \mathrm{ish}(M) |$.

Let X be the skeleton of M derived from a triangulation μ of M as in Section IX.2.3. We consider X as a shadowed polyhedron over $(1/2)\mathbb{Z}$ with zero gleam in every region. Each coloring φ of M induces a dual coloring $\varphi^{\perp}$ of X by the formula $\varphi^{\perp}(e) = \varphi(e^{\perp})$ where e is an oriented 2-cell of X and $e^{\perp}$ is the dual normally oriented edge of μ. This establishes a bijective correspondence between the colorings of (M, μ) and the colorings of X. We have

$$|X|_{\varphi^{\perp}} = \mathcal{D}^{2(b_3(M)-b_2(M))} \, |M|_{\varphi} \, . \tag{3.1.b}$$

This follows directly from definitions. We should take into account that X has no boundary and $\mathrm{sing}(X)$ has no circle strata, all 2-regions of X are 2-disks, and $b_2(X) = a + b_2(M) - b_3(M)$ where a is the number of vertices of M. Summing up equalities (3.1.b) over all colorings of (M, μ) we get $|X| = \mathcal{D}^{2(b_3(M)-b_2(M))} \, |M|$. This implies (3.1.a).

3.2. External approach. The external approach involves 4-manifolds bounded by a 3-manifold. Let M be a closed oriented 3-manifold and let W be a compact oriented 4-manifold bounded by M. (Existence of W is ensured by the Rokhlin theorem mentioned in Section II.2.1.) Set

$$\tau^0(M) = \tau^0_{(\mathcal{V},\mathcal{D})}(M) = || \, \mathrm{sh}(W) ||_{(\mathcal{V},\mathcal{D})} \in K.$$

It is well known that the manifold W is unique up to cobordism and a connected sum with several copies of $\mathbb{C}P^2$. (This follows from the fact that compact oriented 4-manifolds with the same boundary and equal signatures are cobordant.) Therefore the shadow $\mathrm{sh}(W)$ considered up to cobordism and summation with $[S^2_1]$ depends solely on M. The results of Section 2.6 imply that $\tau^0(M)$ is a well-defined invariant of M. It is clear that τ^0 is multiplicative with respect to disjoint union.

3.3. Theorem. *For any closed oriented* 3-*manifold* M, *we have*

$$\tau(M) = \mathcal{D}^{b_2(M)-b_3(M)}\,\tau^0(M)$$

where $\tau(M)$ *is the invariant of* M *defined in Section* II.2.2.

The proof of Theorem 3.3 is given in Sections 4 - 6. It is based on the passage to the shadow world discussed in Sections 5 and 6.

Theorem 3.3 may be used as an alternative definition of $\tau(M)$ based on the technique of $6j$-symbols and shadows. This alternative definition is slightly less general because we have to assume unimodality of $\mathcal{V}$ in order to use $6j$-symbols. On the other hand, such a definition avoids the Kirby calculus.

3.4. Proof of Theorem VII.4.1.1. Let X be a skeleton of M. It is easy to see (cf. Section IX.7.1) that $X \times (1/2)$ is a skeleton of $M \times [0, 1]$. Since this skeleton admits a global section of its "normal bundle" in $M \times [0, 1]$ the shadowing of Section IX.1.6 yields identically zero gleams. Therefore

$$\mathrm{sh}(M \times [0, 1]) = \mathrm{stab}([X]) = \mathrm{ish}(M).$$

Hence

$$||\,\mathrm{ish}(M)|| = ||\,\mathrm{sh}(M \times [0, 1])|| = \tau^0((M \times 1) \amalg (-M \times 0)) = \tau^0(M)\,\tau^0(-M).$$

Theorem 3.3 implies that the last expression is equal to

$$\mathcal{D}^{2(b_3(M)-b_2(M))}\tau(M)\,\tau(-M).$$

On the other hand, Theorem 3.1 implies that $||\,\mathrm{ish}(M)|| = \mathcal{D}^{2(b_3(M)-b_2(M))}\,|M|$. Hence $|M| = \tau(M)\,\tau(-M)$.

3.5. Proof of Theorem VII.4.4.1. The proof is analogous to the proof of Theorem VII.4.1.1 with the only difference that the role of $M \times [0, 1]$ is played by the cylinder of the projection $\overline{M} \to M$. (This cylinder is an orientable 4-manifold bounded by $\overline{M}$ and containing M as a deformation retract.)

3.6. Remark. We may consider the product $\mathcal{D}^{2(b_2(M)-b_3(M))}\,|\,\mathrm{ish}(M)| \in K$ for any compact 3-manifold M. Theorem 3.1 shows that this product generalizes the invariant $|M|$ of closed 3-manifolds.

4. Reduction of Theorem 3.3 to a lemma

4.0. Outline. We reduce Theorem 3.3 to Lemma 4.1 concerned with invariants of framed links in S^3.

4.1. Lemma. *Let Γ be a colored framed graph in S^3 consisting of annuli. Let Γ^* be the same framed graph with the dual coloring. Then*

$$\left|CY(S^3, \Gamma)\right| = \mathbb{F}(\Gamma^*).$$

This statement needs a few comments. For the definitions of colored framed graphs and their invariant $\mathbb{F} = \mathbb{F}_{\mathcal{V}}$, see Section VI.4. For the definition of the dual coloring, see Section 1.1. Since Γ^* has no 2-disks, $\mathbb{F}(\Gamma^*) \in K$. The shadow cylinder $CY(S^3, \Gamma)$ of Γ is a stable integer shadow whose boundary is identified with the disjoint union of the cores of annuli of Γ (see Section IX.8.3). The given coloring of these cores allows us to treat $CY(S^3, \Gamma)$ as a shadow with colored boundary. Since $\gamma = \partial(CY(S^3, \Gamma))$ consists of colored circles, we have $\left|CY(S^3, \Gamma)\right| \in \mathscr{E}(\gamma) = K$. Therefore the equality in the statement of the lemma makes sense.

Lemma 4.1 is the key to the proof of Theorem 3.3. This lemma allows us to compare invariants of framed links derived from the functor F of Chapter I with those derived from $6j$-symbols. We generalize Lemma 4.1 to framed tangles and graphs in Section 5 and prove this generalized result in Section 6. The remaining part of Section 4 is devoted to a deduction of Theorem 3.3 from Lemma 4.1.

4.2. Lemma. *Let X be a shadowed 2-polyhedron over $(1/2)\mathbb{Z}$ whose boundary consists of circles $L_1, \ldots, L_m$. Let Z be the shadowed 2-polyhedron over $(1/2)\mathbb{Z}$ obtained from X by capping all the components of ∂X. Then*

$$|Z| = \mathscr{D}^{b_2(X)+\mathrm{null}(X)-b_2(Z)-\mathrm{null}(Z)} \sum_{\lambda \in \mathrm{col}(\partial X)} \left(\prod_{n=1}^{m} \dim(\lambda(L_n)) \right) |X, \lambda| .$$

Here we view each term $|X, \lambda|$ as an element of the ground ring K so that both the right-hand and the left-hand sides belong to K.

Lemma 4.2 follows directly from definitions.

4.3. Proof of Theorem 3.3. It is enough to consider the case where M is connected so that $b_3(M) = 1$. Present M as the result of surgery on S^3 along a framed link $L \subset S^3$ with components $L_1, \ldots, L_m$. Fix an orientation of L. We shall use this orientation to identify colorings of L with mappings from the set of components of L into the set I. Pick a coloring $\lambda \in \mathrm{col}(L)$ of L. Thickening (L, λ) along the framing of L we get a colored framed graph Γ_λ consisting of m annuli. Since the components of L are oriented we may also view Γ_λ as a colored ribbon graph without coupons. By the very definition of $\mathbb{F}$ we have $F(\Gamma_\lambda) = \mathbb{F}(\Gamma_\lambda)$ where on the left-hand side Γ_λ is regarded as a ribbon graph and on the right-hand side Γ_λ is regarded as a framed graph. Lemma 4.1 implies that $F(\Gamma_\lambda) = \left|CY(S^3, \Gamma_{\lambda^*})\right|$.

It follows from this formula and the definition of $\tau(M)$ that

$$\tau(M) = \Delta^{\sigma(L)} \mathcal{D}^{-\sigma(L)-m-1} \sum_{\lambda \in \operatorname{col}(L)} \dim(\lambda) \left| CY(S^3, \Gamma_{\lambda^*}) \right| .$$

Exchanging λ and λ^* in this formula and using the equality $\dim(i^*) = \dim(i)$ we obtain

$$\tau(M) = \Delta^{\sigma(L)} \mathcal{D}^{-\sigma(L)-m-1} \sum_{\lambda \in \operatorname{col}(L)} \dim(\lambda) \left| CY(S^3, \Gamma_{\lambda}) \right| . \tag{4.3.a}$$

Let us compute $\tau^0(M)$. Recall that $\tau^0(M) = ||\operatorname{sh}(W)|| = \left|\left| \operatorname{sh}'(W) \right|\right|$ where W is an arbitrary compact oriented 4-manifold bounded by M. We take $W = W_L$ to be the 4-manifold obtained by attaching 2-handles to the 4-ball B^4 along regular neighborhoods of $L_1, \ldots, L_m$ in $S^3 = \partial B^4$ (cf. Section II.2.1). It follows from the definition of the shadow sh' (see Section IX.4.1) that $\operatorname{sh}'(W_L) = CO(S^3, L)$. Hence $\tau^0(M) = \left|\left| CO(S^3, L) \right|\right|$. To compute $\left|\left| CO(S^3, L) \right|\right|$, we consider the shadow projection of L into $S^2 \subset S^3$. More exactly we take an enriched diagram D of L in S^2 and consider its shadow projection $\operatorname{sh}(D)$, cf. Section IX.3.2. Denote by X the shadow cylinder of $\operatorname{sh}(D)$. As usual we may identify ∂X with L. Denote by Z the shadowed polyhedron obtained from X by capping all components of ∂X. By definition, Z represents the cone $CO(S^3, L)$ of L. Lemma 4.2 implies that

$$\tau^0(M) = \left|\left| CO(S^3, L) \right|\right| = (\mathcal{D}\Delta^{-1})^{-\sigma(Z)} \, |Z| =$$

$$= \Delta^{\sigma(Z)} \mathcal{D}^{-\sigma(Z)+b_2(X)+\operatorname{null}(X)-b_2(Z)-\operatorname{null}(Z)} \sum_{\lambda \in \operatorname{col}(L)} \dim(\lambda) \, |X, \lambda| .$$

It is easy to compute that $b_2(X) = \operatorname{null}(X) = 1$ and $b_2(Z) = m+1$. It follows from Theorem IX.5.1 and the equality $\operatorname{sh}'(W_L) = \operatorname{stab}([Z])$ that the manifold W_L and the shadowed polyhedron Z have equal signatures. Thus $\sigma(Z) = \sigma(L)$. Note that the inclusion homomorphism $H_2(Z; \mathbb{R}) \to H_2(W_L; \mathbb{R})$ is surjective with kernel $\mathbb{R}$. Therefore Theorem IX.5.1 implies that $\operatorname{null}(Z) - 1$ equals the nullity of the intersection form in $H_2(W_L; \mathbb{R})$. Poincaré duality and the formula $H_1(W_L; \mathbb{R}) = 0$ imply that the nullity of this intersection form equals $b_1(M) = b_2(M)$. Therefore $\operatorname{null}(Z) = b_2(M) + 1$.

By definition, the pair (X, λ) represents the shadow cylinder (with colored boundary) $CY(S^3, \Gamma_\lambda)$. Hence $|X, \lambda| = \left| CY(S^3, \Gamma_\lambda) \right|$. Substituting these expressions in the formula for $\tau^0(M)$, we obtain

$$\tau^0(M) = \Delta^{\sigma(L)} \mathcal{D}^{-\sigma(L)-m-b_1(M)} \sum_{\lambda \in \operatorname{col}(L)} \dim(\lambda) \left| CY(S^3, \Gamma_\lambda) \right| =$$

$$= \mathcal{D}^{1-b_2(M)} \tau(M) = \mathcal{D}^{b_3(M)-b_2(M)} \tau(M).$$

4.4. Remark. The proof of Theorem 3.3 sheds more light (or, so to say, more gleam) on the role of the factor $\dim(\lambda)$ in the definition of $\tau(M)$ given in Chapter II. From the shadow viewpoint this factor is contributed by the 2-disks glued to the shadow cylinder of the link to form the shadow cone. These disks constitute a shadow version of 2-handles attached to the 4-ball along the link components.

5. Passage to the shadow world

5.0. Outline. The goal of this section is to formulate a local version of Lemma 4.1 or, in other words, to generalize this lemma to tangles and graphs. To this end we introduce framed graphs with free ends in $\mathbb{R}^3$. In analogy with the category of v-colored ribbon graphs $\mathrm{Rib}_{\mathcal{V}}$ (see Chapter I) we define a category of v-colored framed graphs $\mathrm{Fr}_{\mathcal{V}}$. We introduce two linear representations of $\mathrm{Fr}_{\mathcal{V}}$ using the functor F of Chapter I and the state sums based on $6j$-symbols. The main result formulated in this section (Theorem 5.6) establishes equivalence of these two representations. Finally, we deduce Lemma 4.1 from Theorem 5.6. Note that to prove Lemma 4.1 we do not need the full generality of Theorem 5.6, it is sufficient to consider framed tangles, i.e., framed graphs without 2-disks.

5.1. Framed graphs with free ends. Let k, l be non-negative integers. A framed (k, l)-graph in $\mathbb{R}^3$ is a compact surface Γ embedded in $\mathbb{R}^2 \times [0, 1]$ and decomposed into a union of finite number of annuli (twisted and untwisted), bands, and 2-disks satisfying the following conditions: Γ meets the planes $\mathbb{R}^2 \times 0$ and $\mathbb{R}^2 \times 1$ orthogonally along the intervals

$$\{ [i - (1/10), i + (1/10)] \times 0 \times 0 \mid i = 1, ..., k \}, \tag{5.1.a}$$

$$\{ [j - (1/10), j + (1/10)] \times 0 \times 1 \mid j = 1, ..., l \} \tag{5.1.b}$$

which are bases of certain bands of Γ; other bases of bands lie in the boundary of the 2-disks, otherwise the bands, 2-disks, and annuli are disjoint; each of the 2-disks is incident to exactly three bands (counted with multiplicities). Any framed (k, l)-graph in $\mathbb{R}^3$ has $k + l$ "free ends" attached to the intervals (5.1.a) and (5.1.b). For example, a framed $(0, 0)$-graph in $\mathbb{R}^3$ is nothing but a framed graph in $\mathbb{R}^2 \times [0, 1]$ in the sense of Section VI.4.2.

Colorings and v-colorings of framed (k, l)-graphs are defined in exactly the same way as colorings and v-colorings of framed graphs without free ends (see Section VI.4.2). The notion of an isotopy readily extends to framed (k, l)-graphs (the isotopies proceed in $\mathbb{R}^2 \times [0, 1]$, do not move the bases (5.1.a) and (5.1.b), and keep colors fixed). A framed (k, l)-graph in $\mathbb{R}^3$ is said to be orientable if its surface may be oriented so that in the points of the intervals (5.1.a) and (5.1.b) the orientation is determined by the pair of tangent vectors $((1, 0, 0), (0, 0, 1))$.

The technique of diagrams easily extends to framed (k, l)-graphs. Let γ be an (abstract) graph with $k+l$ vertices of valency 1 and remaining vertices of valency 3. A graph (k, l)-diagram with underlying graph γ is an immersion $\gamma \to \mathbb{R}\times[0, 1]$ such that

(i) the image of γ meets the lines $\mathbb{R}\times 0$ and $\mathbb{R}\times 1$ transversally in the points

$$(1, 0), (2, 0), ..., (k, 0), (1, 1), (2, 1), ..., (l, 1) \tag{5.1.c}$$

which are images of the 1-valent vertices of γ,

(ii) all self-crossing points of the image of γ are transversal intersections of two 1-strata of γ,

(iii) at each crossing point one of the two intersecting 1-strata is distinguished and said to be the lower one, the second 1-stratum being the upper one.

Every graph (k, l)-diagram determines a framed (k, l)-graph in $\mathbb{R}^3$ via thickening of the image of γ in the strip $\mathbb{R}\times[0, 1]$ and pushing all upper branches up. (In the figures to follow the plane $\mathbb{R}\times 0\times\mathbb{R}$ is identified with the plane of the picture and "up" means towards the reader.) The resulting framed (k, l)-graph contains γ as the core. The edges of γ incident to its 1-valent vertices are called free ends of the diagram.

An enriched graph (k, l)-diagram is a graph (k, l)-diagram such that all 1-strata of the underlying abstract graph are equipped with integer or half-integer pre-twists. Every enriched diagram gives rise to a framed graph in $\mathbb{R}^2\times[0, 1]$: consider the framed graph determined by the underlying diagram with the pre-twists forgotten and twist its bands and annuli along their cores as many times as the corresponding pre-twists. It is easy to see that any framed (k, l)-graph Γ in $\mathbb{R}^3$ may be presented by an enriched diagram in $\mathbb{R}\times[0, 1]$.

5.2. Category of framed graphs. Following the lines of Section I.2 we define a category $\mathrm{Fr}_{\mathcal{V}}$ of v-colored orientable framed graphs in $\mathbb{R}^3$. Its objects are finite sequences of elements of the set I including the empty sequence. A morphism $i \to j$ in $\mathrm{Fr}_{\mathcal{V}}$ is an isotopy class of v-colored orientable framed (k, l)-graphs in $\mathbb{R}^3$, such that i (resp. j) is the sequence of colors of their bottom (resp. top) free ends directed downwards. Composition of morphisms is defined in the same way as in the setting of ribbon graphs. The identity morphisms are presented by framed graphs consisting of vertical untwisted unlinked bands.

We can provide $\mathrm{Fr}_{\mathcal{V}}$ with a tensor product in the same way as in the theory of ribbon graphs but we do not need this.

5.3. Functor $\mathbb{F}$: $\mathrm{Fr}_{\mathcal{V}} \to \mathcal{V}$. We define a covariant functor $\mathbb{F} : \mathrm{Fr}_{\mathcal{V}} \to \mathcal{V}$. For any object $i = (i_1, ..., i_k)$ of $\mathrm{Fr}_{\mathcal{V}}$, set $\mathbb{F}(i) = V_{i_1}\otimes ... \otimes V_{i_k}$. To define the action of $\mathbb{F}$ on the morphisms note that the ribboning of framed graphs defined in Section VI.4.3 may be extended to orientable framed graphs with free ends word for word. If Γ is a v-colored orientable framed (k, l)-graph in $\mathbb{R}^3$ representing a morphism $(i_1, ..., i_k) \to (j_1, ..., j_l)$ then the ribboning of Γ yields a ribbon graph, say Ω_Γ, whose bottom free ends are directed downwards and colored with $V_{i_1}, \ldots, V_{i_k}$ respectively and whose top free ends are directed upwards and colored with $V_{j_1^*}, \ldots, V_{j_l^*}$ respectively. As is clear from the definition of ribboning, the ribbon graph Ω_Γ is determined by Γ non-uniquely. However the same arguments as in Section VI.4.3 show that the morphism

$$F(\Omega_\Gamma) : V_{i_1} \otimes \ldots \otimes V_{i_k} \to (V_{j_1^*})^* \otimes \ldots \otimes (V_{j_l^*})^*$$

does not depend on the indeterminacy in the definition of Ω_Γ. Set

$$\mathbb{F}(\Gamma) = (w_{j_1}^{-1} \otimes \ldots \otimes w_{j_l}^{-1})\, F(\Omega_\Gamma) : V_{i_1} \otimes \ldots \otimes V_{i_k} \to V_{j_1} \otimes \ldots \otimes V_{j_l}.$$

(For the definition of homomorphisms $\{w_j\}_{j\in I}$, see Section VI.3.) It follows from the definition of ribboning and Theorem I.2.5 that $\mathbb{F}$ is a covariant functor $\mathrm{Fr}_{\mathcal{V}} \to \mathcal{V}$.

5.4. Digression: h_0 and the tensor product. Recall the category $\mathrm{Proj}(K)$ of projective K-modules and K-linear homomorphisms. Denote by h_0 the covariant functor $\mathcal{V} \to \mathrm{Proj}(K)$ transforming an object V into $h_0(V) = \mathrm{Hom}_{\mathcal{V}}(\mathbb{1}, V)$ and transforming a morphism $f : V \to W$ into the homomorphism $g \mapsto fg : \mathrm{Hom}_{\mathcal{V}}(\mathbb{1}, V) \to \mathrm{Hom}_{\mathcal{V}}(\mathbb{1}, W)$ (cf. Section IV.2.2). For any sequence $i = (i_1, \ldots, i_k) \in I^k$, the module $h_0(V_{i_1} \otimes \ldots \otimes V_{i_k})$ may be described in terms of multiplicity modules introduced in Sections VI.1.2 and VI.3.2. Let us call a sequence $m = (m_0, m_1, \ldots, m_k) \in I^{k+1}$ admissible if $m_0 = m_k = 0$. For any admissible sequence $m = (m_0, \ldots, m_k)$, set

$$G(i, m) = \bigotimes_{p=1}^{k} H(m_{p-1}, i_p, m_p^*).$$

This is a projective K-module determined by i and m. If $k = 0$ then $G(i, m) = K$. If $k = 1$ then $G(i, m) = 0$ unless $i_1 = 0$ in which case $G(i, m) = K$. If $k = 2$ then $G(i, m) = 0$ unless $m_1 = i_1 = i_2^*$ in which case $G(i, m) = K$.

Consider the K-linear homomorphism

$$\bigotimes_{p=1}^{k} H_{m_p}^{m_{p-1} i_p} \to h_0(V_{i_1} \otimes \ldots \otimes V_{i_k}) \tag{5.4.a}$$

defined by the colored ribbon graph presented by the diagram in Figure 5.1. Here the module

$$\bigotimes_{p=1}^{k} H_{m_p}^{m_{p-1} i_p} \tag{5.4.b}$$

appears as the module of all possible colorings of coupons in Figure 5.1. Lemma VI.1.1.2 implies that the direct sum of the homomorphisms (5.4.a) over all admissible sequences $m \in I^{k+1}$ yields an isomorphism onto $h_0(V_{i_1} \otimes \ldots \otimes V_{i_k})$.

Thus,

(5.4.c) $$h_0(V_{i_1} \otimes \ldots \otimes V_{i_k}) = \bigoplus_{m=(m_0,\ldots,m_k)\in I^{k+1}, m_0=m_k=0} \left(\bigotimes_{p=1}^{k} H_{m_p}^{m_{p-1} i_p} \right).$$

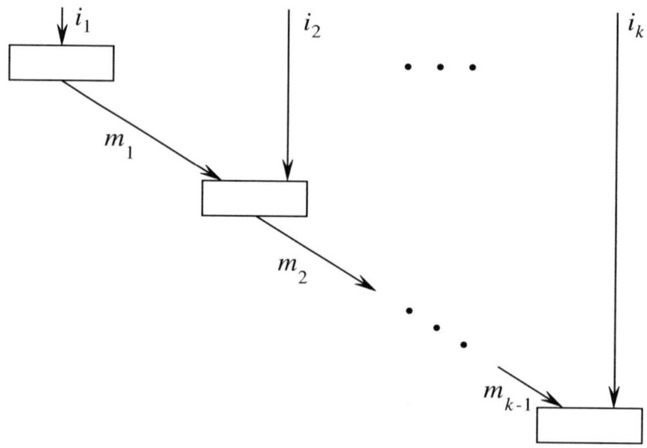

Figure 5.1

Identifying $H_{m_p}^{m_{p-1} i_p}$ with $H(m_{p-1}, i_p, m_p^*)$ along the canonical isomorphism we get a splitting

(5.4.d) $$h_0(V_{i_1} \otimes \ldots \otimes V_{i_k}) = \bigoplus_{m=(m_0,\ldots,m_k)\in I^{k+1}, m_0=m_k=0} G(i, m).$$

Although we shall not need it, note that the projection

$$h_0(V_{i_1} \otimes \ldots \otimes V_{i_k}) \to \bigotimes_{p=1}^{k} H_{m_p}^{m_{p-1} i_p}$$

carries any $f \in h_0(V_{i_1} \otimes \ldots \otimes V_{i_k})$ into $\left(\prod_{p=1}^k \dim(m_p)\right) F(\Omega_f)$ where Ω_f is the colored ribbon (0,0)-graph with one colored and k uncolored coupons drawn in Figure 5.2 (cf. Lemmas IV.10.6.1 and IV.10.6.2). Here $F(\Omega_f)$ is a K-linear functional

$$\bigotimes_{p=1}^{k} H^{m_p}_{m_{p-1} i_p} \to K,$$

i.e., an element of the module (5.4.b).

5.5. Functor $\mathbb{G} : \mathbf{Fr}_{\mathcal{V}} \to \mathbf{Proj}(K)$. Our aim is to give a "shadow" computation of the composition $h_0\mathbb{F} : \mathrm{Fr}_{\mathcal{V}} \to \mathrm{Proj}(K)$. With this in mind, we introduce another

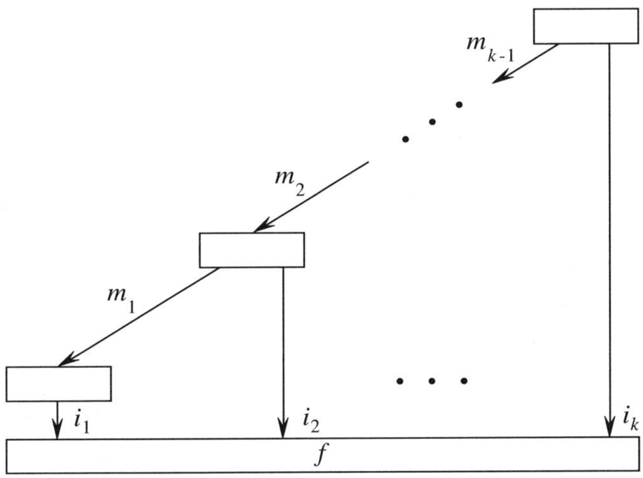

Figure 5.2

covariant functor $\mathbb{G} : \mathrm{Fr}_{\mathcal{V}} \to \mathrm{Proj}(K)$. Its definition is in terms of $6j$-symbols and state sum models on shadowed polyhedra. *A posteriori* we shall see that $\mathbb{G} = h_0\mathbb{F}$.

The functor $\mathbb{G}$ acts on the objects of $\mathrm{Fr}_{\mathcal{V}}$ in the same way as $h_0\mathbb{F}$. Namely, $\mathbb{G}(i_1, ..., i_k) = h_0(V_{i_1} \otimes ... \otimes V_{i_k})$. Let us define the action of $\mathbb{G}$ on morphisms in $\mathrm{Fr}_{\mathcal{V}}$. Let Γ be a v-colored orientable framed (k, l)-graph in $\mathbb{R}^3$. Let $i = (i_1, ..., i_k)$ and $j = (j_1, ..., j_l)$ be the sequences of colors of the bottom and top ends of Γ directed downward. Consider the decomposition (5.4.d) and a similar decomposition

$$h_0(V_{j_1} \otimes ... \otimes V_{j_l}) = \bigoplus_{n=(n_0,...,n_l)\in I^{l+1}, n_0=n_l=0} G(j, n). \tag{5.5.a}$$

For any sequences m, n entering into these decompositions, we shall construct a K-homomorphism $\Gamma^n_m : G(i, m) \to G(j, n)$. These homomorphisms form a block-matrix determining $\mathbb{G}(\Gamma) : \oplus_m G(i, m) \to \oplus_n G(j, n)$.

We define $\Gamma^n_m : G(i, m) \to G(j, n)$ as follows. Consider an enriched diagram D of Γ in $\mathbb{R} \times [0, 1]$. Choose a big positive number R so that D is contained in $(-R, R) \times [0, 1]$. (In particular, $R > \max(k, l)$.) Denote the rectangle $[-R, R] \times [0, 1]$ by X. We provide the only region of X with the zero gleam and regard X as a shadowed polyhedron. We regard D as a diagram of Γ in X. Consider the shadow projection $\mathrm{sh}(D)$ of D in X. (It is defined as in Section IX.8, the free ends of D do not contribute to the gleams of regions.) The shadow projection $\mathrm{sh}(D)$ is a shadowed graph in X with $k+l$ ends lying in ∂X. Consider the shadow cylinder C of $\mathrm{sh}(D)$ whose underlying 2-polyhedron is obtained by gluing $\gamma \times [0, 1]$ to X along an immersion $\gamma \times 1 \to X$ parametrizing D; here γ denotes the core of Γ. The boundary of C is formed by the topological circle ∂X and a graph obtained from $\gamma \times 0$ by attaching the intervals $a \times [0, 1]$ where $a = a \times 0$ runs over

1-valent vertices of γ. We identify this graph with γ in the obvious way. Thus, $\partial C = \gamma \cup \partial X$ where the set $\gamma \cap \partial X$ coincides with the set of 1-valent vertices of γ and with the set (5.1.c). The gleams of regions of C are defined in exactly the same way as in Section VIII.7.1.

We endow ∂C with a coloring $\lambda = \lambda_{m,n}$ as follows. The 1-strata of $\gamma \subset \partial C$ inherit a coloring from Γ. For each $p = 1, ..., k-1$, color the edge $[p, p+1] \times 0$ of ∂C directed to the right by m_p. For each $q = 1, ..., l-1$, color the edge $[q, q+1] \times 1$ of ∂C directed to the left by n_q. The edges of ∂C containing the vertical intervals $R \times [0, 1]$ and $(-R) \times [0, 1]$ are colored with 0.

Recall the module $H(\partial C, \lambda)$ which is the tensor product of the symmetrized multiplicity modules associated to vertices of ∂C. The given coloring of the 2-disks of Γ induces a coloring of the vertices of $\gamma \subset \partial C$. This means that each of these vertices is provided with a preferred element of the corresponding multiplicity module. The remaining vertices of ∂C are the points (5.1.c) of ∂X. These vertices are not colored. It is clear that the multiplicity modules associated to the vertices $(p, 0)$ and $(q, 1)$ are $H(m_{p-1}, i_p, m_p^*)$ and $H(n_{q-1}^*, j_q^*, n_q)$ respectively. (Here $p = 1, 2, ..., k$ and $q = 1, 2, ..., l$.) Therefore

$$H(\partial C, \lambda) = H(\gamma) \otimes_K (\otimes_{p=1}^k H(m_{p-1}, i_p, m_p^*)) \otimes_K (\otimes_{q=1}^l H(n_{q-1}^*, j_q^*, n_q)) =$$

$$= H(\gamma) \otimes_K \otimes_K G(i, m) \otimes_K (G(j, n))^*.$$

Set

$$\dim'(\Gamma) = \prod_e \dim'(\lambda(e)) \tag{5.5.b}$$

where e runs over all (non-oriented) edges of γ (including the edges incident to the 1-valent vertices). Here $\dim'(\lambda(e)) = \dim'(\lambda(\vec{e}))$ where $\vec{e}$ denotes the edge e with some orientation. Note that the circle 1-strata of γ do not contribute to $\dim'(\Gamma)$. Set

$$\dim_m^n(\Gamma) = \prod_{q=1}^{l} \dim'(n_q) \prod_{p=1}^{k} (\dim'(m_p))^{-1}.$$

The state sum invariant $|C, \lambda^*| \in H(\partial C, \lambda^*)$ gives rise to the following K-linear functional $\Gamma(m, n) : G(i, m) \otimes_K (G(j, n))^* \to K$. For any $y \in G(i, m) \otimes_K (G(j, n))^*$,

$$\Gamma(m, n)(y) = (\dim'(\Gamma))^{-1} \dim_m^n(\Gamma) \langle \left|C, \lambda^*\right|, x \otimes y\rangle \in K$$

where $x \in H(\gamma)$ is the tensor product of the given colors of trivalent vertices of γ so that $x \otimes y \in H(\partial C, \lambda)$. For the definition of the bilinear symmetric form $\langle ., . \rangle : H(\partial C, \lambda^*) \otimes_K H(\partial C, \lambda) \to K$, see Section 1.1. In particular, if Γ does not have 2-disks then

$$\Gamma(m, n) = (\dim'(\Gamma))^{-1} \dim_m^n(\Gamma) \left|C, \lambda^*\right| \in H(\partial C, \lambda^*) = (G(i, m))^* \otimes_K G(j, n).$$

Denote by Γ^n_m the K-homomorphism $G(i,m) \to G(j,n)$ corresponding to $\Gamma(m,n)$. It follows from the results of Sections VIII.7 and IX.8 that Γ^n_m does not depend on the choice of D and only depends on the isotopy class of Γ.

5.6. Theorem. $\mathbb{G} : \mathrm{Fr}_{\mathcal{V}} \to \mathrm{Proj}(K)$ *is a covariant functor and* $h_0\mathbb{F} = \mathbb{G}$.

Theorem 5.6 is proven in Section 6.

By Theorem 5.6, for any v-colored orientable framed (k,l)-graph $\Gamma \subset \mathbb{R}^3$, we have $h_0\mathbb{F}(\Gamma) = \mathbb{G}(\Gamma)$. In particular, for graphs without free ends (i.e., with $k = l = 0$), we have $\mathbb{F}(\Gamma) = \mathbb{G}(\Gamma) \in K$.

5.7. Corollary. *Let* Γ *be a colored framed graph in* S^3 *with oriented surface (and without free ends). Let* Γ^* *be the same framed graph with the dual coloring. Then*

$$\big|CY(S^3,\Gamma)\big| = \dim'(\Gamma)\,\mathbb{F}(\Gamma^*) \in H(\Gamma). \tag{5.7.a}$$

Here $\dim'(\Gamma) = \dim'(\Gamma^*) \in K$ is defined by formula (5.5.b). For the definition of $\mathbb{F}(\Gamma^*) \in H^\star(\Gamma^*)$, see Section VI.4.3. We identify $H^\star(\Gamma^*) = \mathrm{Hom}_K(H(\Gamma^*), K)$ with $H(\Gamma)$ using the duality pairing introduced in Section 1.1. Under this identification both sides of (5.7.a) belong to $H(\Gamma)$.

Proof of Corollary. The definition of the shadow cylinder $CY(S^3,\Gamma)$ involves diagrams of Γ in a skeleton of S^3. On the other hand, Theorem 5.6 deals with diagrams of Γ in the plane or in a plane rectangle. We have to relate these two settings.

Let D be an enriched diagram of Γ in the skeleton $S^2 \subset S^3$. Let B be a 2-disk in S^2 containing the diagram D. Denote by C the cylinder of the shadow graph $\mathrm{sh}(D)$ in B. It is clear that $\partial C = \partial B \amalg \gamma$ where γ is the colored core of Γ. Assigning $r \in I$ to the circle ∂B oriented counterclockwise, we transform C into a shadowed polyhedron with colored boundary. Denote this polyhedron by C_r.

The shadow cylinder $CY(S^3,\Gamma)$ may be represented by the shadowed polyhedron CY obtained from C by capping of ∂B. In analogy with Lemma 4.2 we have

$$\big|CY(S^3,\Gamma)\big| = |CY| = \mathcal{D}^{b_2(C_r)+\mathrm{null}(C_r)-b_2(CY)-\mathrm{null}(CY)} \sum_{r\in I} \dim(r)\,|C_r|\,.$$

It is easy to compute that $\mathrm{null}(C_r) = b_2(C_r) = 0$ and $\mathrm{null}(CY) = b_2(CY) = 1$. Hence

$$\big|CY(S^3,\Gamma)\big| = \mathcal{D}^{-2} \sum_{r\in I} \dim(r)\,|C_r| \in H(\gamma) = H(\Gamma). \tag{5.7.b}$$

For each 2-disk d of Γ^*, fix an element x_d of the module $H(i,j,k)$ where i,j,k are the colors of the cores of three bands of Γ^* attached to d and directed towards d. Denote the family $\{x_d\}_d$ by x. The pair (Γ^*, x) is a v-colored framed graph. Consider the disjoint union $(\Gamma', x) = (\Gamma^*, x) \amalg \mathbb{O}_r \subset S^3$ where $\mathbb{O}_r$ is an unknotted untwisted annulus in S^3 whose core is oriented and colored with $r \in I$. We claim that

$$\mathbb{G}(\Gamma', x) = (\dim'(\Gamma))^{-1} \langle |C_r| , \otimes_d x_d \rangle \tag{5.7.c}$$

where $\langle ., . \rangle$ is the duality pairing $H(\gamma) \otimes_K H(\gamma^*) \to K$. Equality (5.7.c) would imply the claim of the Corollary. Indeed, Theorem 5.6 guarantees that $\mathbb{G}(\Gamma', x) = \mathbb{F}(\Gamma', x)$. It is obvious that $\mathbb{F}(\Gamma', x) = \dim(r)\, \mathbb{F}(\Gamma^*, x)$. Therefore

$$\langle |C_r| , \otimes_d x_d \rangle = \dim(r) \dim'(\Gamma)\, \mathbb{F}(\Gamma^*, x).$$

Substituting this formula in (5.7.b) we get

$$\langle \left| CY(S^3, \Gamma) \right| , \otimes_d x_d \rangle = \dim'(\Gamma)\, \mathbb{F}(\Gamma^*, x).$$

This implies (5.7.a).

To prove (5.7.c) we may assume that B lies inside the rectangle

$$X = [-R, R] \times [0, 1] \subset \mathbb{R}^2 = S^2 \setminus \{\infty\}$$

for large positive R. Provide the circle ∂X with the color $0 \in I$. We present Γ' by the diagram $D' = (D, \lambda^*) \cup \partial B$ in $\mathrm{Int}(X)$ where D is the colored enriched diagram of Γ used above, λ denotes the given coloring of Γ, and the circle ∂B is provided with zero pre-twist and color r. We regard D' as a colored enriched graph diagram in X. Let C' be the cylinder of the shadowed graph $\mathrm{sh}(D')$ in X. Since Γ' has no free ends $\dim_0^0(\Gamma) = 1$. By definition,

$$\mathbb{G}(\Gamma', x) = (\dim'(\Gamma))^{-1} \langle \left| C', (\lambda^*)^* \right| , \otimes_d x_d \rangle = \tag{5.7.d}$$

$$= (\dim'(\Gamma))^{-1} \langle \left| C', \lambda \right| , \otimes_d x_d \rangle.$$

The cylinder C' differs from C only in that it has one additional region $\mathrm{Int}(X) \setminus B$ and one additional boundary component ∂X. Since this component is colored with $0 \in I$ the computation of $\left| C', \lambda \right|$ involves only those colorings of C' which assign 0 to $\mathrm{Int}(X) \setminus B$. Such colorings restricted to C yield exactly the colorings of C_r assigning r to the outer region of D in B (with counterclockwise orientation). This implies that $\left| C', \lambda \right| = |C_r|$ (cf. Lemma 2.4). Substituting this expression in (5.7.d) we get (5.7.c).

5.8. Proof of Lemma 4.1. Lemma 4.1 is a special case of Corollary 5.7 where Γ has no 2-disks so that $H(\Gamma) = K$ and $\dim(\Gamma) = 1$.

5.9. Remark. The constructions and results of this section generalize directly to framed graphs with a non-orientable surface.

6. Proof of Theorem 5.6

6.0. Outline. We first show that $\mathbb{G}$ is a covariant functor and then compare the values of $h_0\mathbb{F}$ and $\mathbb{G}$ on generators of the category of framed graphs.

6.1. Lemma. *$\mathbb{G} : \mathrm{Fr}_{\mathcal{V}} \to \mathrm{Proj}(K)$ is a covariant functor.*

Proof. Let $i = (i_1, ..., i_k)$ be an object of $\mathrm{Fr}_{\mathcal{V}}$. Let us show that $\mathbb{G}(\mathrm{id}_i) = \mathrm{id}_{\mathbb{G}(i)}$. The identity morphism id_i is the isotopy class of the colored framed (k, k)-graph Γ formed by k untwisted unlinked vertical bands whose cores are oriented downward and have colors $i_1, ..., i_k$ respectively. We present Γ by a diagram in $\mathbb{R}\times[0, 1]$ consisting of k disjoint vertical intervals equipped with zero pre-twists. The polyhedron C used in Section 5.5 to define Γ_m^n may be described explicitly. It has no vertices and has $2k + 1$ rectangular regions whose gleams are equal to zero. Let $m = (m_0, ..., m_k)$ and $n = (n_0, ..., n_k)$ be two admissible sequences. If $m \neq n$ then $\Gamma_m^n = 0$ because, in this case, C has no colorings that extend the coloring $\lambda = \lambda_{m,n}$ of ∂C. In the case $m = n$ there is exactly one such coloring which attains the values $i_1, ..., i_k, m_0, m_1, ..., m_k$ on the corresponding regions of C. We may compute $|C, \lambda^*|$ directly from definitions (see Section 1.2). This gives

$$\left|C, \lambda^*\right| = \left(\prod_{p=1}^{k} \dim'(i_p)\right) \bigotimes_{p=1}^{k} \mathrm{Id}(m_{p-1}, i_p, m_p^*) = \dim'(\Gamma) \bigotimes_{p=1}^{k} \mathrm{Id}(m_{p-1}, i_p, m_p^*).$$

(Recall that $\mathrm{Id}(i, j, k)$ is the canonical element of $H(i, j, k) \otimes_K H(i^*, j^*, k^*)$ determined by the duality defined in Section VI.3.3.) Therefore

$$\Gamma(m, m) = \otimes_{p=1}^{k} \mathrm{Id}(m_{p-1}, i_p, m_p^*)$$

and Γ_m^m is the identity endomorphism of $G(i, m)$.

Let us show that $\mathbb{G}$ transforms composition into composition. Let Γ be a v-colored orientable framed (k, l)-graph in $\mathbb{R}^3$ representing a morphism $(i_1, ..., i_k) \to (j_1, ..., j_l)$. Let Γ' be a v-colored orientable framed (l, l')-graph in $\mathbb{R}^3$ representing a morphism $(j_1, ..., j_l) \to (j'_1, ..., j'_{l'})$. Denote by Γ'' the composition $\Gamma'\Gamma$ obtained by placing Γ' on the top of Γ and compressing the result into $\mathbb{R}^2 \times [0, 1]$.

Consider first the case when Γ and Γ' are framed tangles, i.e., have no 2-disks. Let $m \in I^{k+1}$ and $n' \in I^{l'+1}$ be admissible sequences of colors. Denote by cntr the tensor contraction

$$\bigoplus_n \big((G(i, m))^* \otimes_K G(j, n) \otimes_K (G(j, n))^* \otimes_K G(j', n')\big) \to (G(i, m))^* \otimes_K G(j', n')$$

induced by the evaluation pairing $G(j, n) \otimes_K (G(j, n))^* \to K$. Here n runs over admissible sequences of colors of length $l+1$. To show that $\mathbb{G}(\Gamma'') = \mathbb{G}(\Gamma')\,\mathbb{G}(\Gamma)$

it is enough to show that

$$\text{(6.1.a)} \qquad \Gamma''(m,n') = \operatorname{cntr}\Big(\sum_n \Gamma(m,n)\otimes\Gamma'(n,n')\Big).$$

Let us present Γ and Γ' by enriched graph diagrams D and D' respectively. The composition Γ'' is presented by the diagram $D'' = D'D$ obtained by putting D' on the top of D and compressing into $\mathbb{R}\times[0,1]$. We assume that all three diagrams D, D', and $D'D$ are contained in the rectangle $[-R,R]\times[0,1]$ for large positive R. We form the shadow cylinders C, C', C'' of $\mathrm{sh}(D)$, $\mathrm{sh}(D')$, $\mathrm{sh}(D'')$ as in Section 5.5. We may regard C'' as the result of gluing C and C' along the obvious homeomorphism of the "top" part of C onto the "bottom" part of C'. Here the top part of C consists of the interval $[-R,R]\times 1$ and l intervals $a\times[0,1]$ with $a=(1,1),\dots,(l,1)$ glued to $[-R,R]\times 1$ along $a = a\times 1$. Similarly, the bottom part of C' consists of a copy of $[-R,R]\times 0$ and l intervals $a\times[0,1]$ with $a=(1,0),\dots,(l,0)$. Note that the gleam of any region Y of C'' is equal to the sum of the gleams of regions of C and C' contained in Y.

The given coloring of Γ,Γ' and the admissible sequences m, n' determine a coloring of $\partial C''$ in the same way as in Section 5.5. Denote this coloring by $\lambda''=\lambda_{m,n'}$. Similarly, for any admissible sequence $n\in I^{l+1}$, we have colorings $\lambda=\lambda_{m,n}$ of ∂C and $\lambda'=\lambda_{n,n'}$ of $\partial C'$.

We have

$$\Gamma''(m,n') = (\dim'(\Gamma''))^{-1}\dim_m^{n'}(\Gamma'')\left|C'',(\lambda'')^*\right|.$$

For any admissible sequence $n\in I^{l+1}$,

$$\Gamma(m,n) = (\dim'(\Gamma))^{-1}\dim_m^{n}(\Gamma)\left|C,\lambda^*\right|,$$

$$\Gamma'(n,n') = (\dim'(\Gamma'))^{-1}\dim_n^{n'}(\Gamma')\left|C',(\lambda')^*\right|.$$

It is easy to compute that

$$\dim'(\Gamma'') = \Big(\prod_{q=1}^{l}\dim'(j_q)\Big)^{-1}\dim'(\Gamma)\dim'(\Gamma'),$$

$$\dim_m^{n'}(\Gamma'') = \dim_m^{n}(\Gamma)\dim_n^{n'}(\Gamma').$$

Therefore to prove (6.1.a) it is enough to show that

$$\text{(6.1.b)} \quad \left|C'',(\lambda''_{m,n'})^*\right| = \Big(\prod_{q=1}^{l}\dim'(j_q)\Big)^{-1}\operatorname{cntr}\Big(\sum_n\big(\left|C,(\lambda_{m,n})^*\right|\otimes\left|C',(\lambda'_{n,n'})^*\right|\big)\Big).$$

To compute $\left|C'',(\lambda'')^*\right|\in H(\partial C'',(\lambda'')^*)$ we should consider all colorings φ'' of C'' such that $\partial\varphi''=\lambda''$. Each such coloring φ'', restricted to C (resp. C'), yields a coloring of C (resp. of C') denoted by φ (resp. by φ'). Denote by $n=n(\varphi'')=(n_0,n_1,\dots,n_l)$ the sequence of values of $\partial\varphi'$ on the bottom intervals

$[0, 1]\times 0, [1, 2]\times 0, \ldots, [l, l+1]\times 0 \subset \partial C'$ oriented to the right. The same sequence may be obtained as the values of $\partial\varphi$ on the top intervals $[0, 1] \times 1, [1, 2] \times 1, \ldots, [l, l+1] \times 1 \subset \partial C$ oriented to the left. It follows from definitions that $\partial\varphi = \lambda_{m,n}$ and $\partial\varphi' = \lambda'_{n,n'}$. Therefore to prove (6.1.b) it is enough to show that for any $\varphi'' \in \mathrm{col}(C'')$, we have

$$|C''|_{\varphi''} = \prod_{q=1}^{l} (\dim'(j_q))^{-1}\,\mathrm{cntr}\,(|C|_{\varphi} \otimes |C'|_{\varphi'}). \tag{6.1.c}$$

It is easy to compute that

$$|C''|_1^{\varphi''} = \prod_{q=1}^{l} (\dim'(j_q)\dim(n_q))\,|C|_1^{\varphi}\,|C'|_1^{\varphi'},$$

$$|C''|_2^{\varphi''} = (\prod_{q=1}^{l} \dim(j_q)\dim(n_q))^{-1}\,|C|_2^{\varphi}\,|C'|_2^{\varphi'},$$

$$|C''|_3^{\varphi''} = |C|_3^{\varphi}\,|C'|_3^{\varphi'}, \quad |C''|_4^{\varphi''} = |C|_4^{\varphi}\,|C'|_4^{\varphi'}.$$

Using Lemma VII.2.2.2 we may compute that

$$|C''|_5^{\varphi''} = \mathrm{cntr}\,(|C|_5^{\varphi} \otimes |C'|_5^{\varphi'}).$$

Multiplying these equalities we get (6.1.c). This completes the proof of the lemma in the case when Γ and Γ' are framed tangles. The general case is considered similarly. Instead of cntr, we should use its tensor product with the obvious isomorphism $H(\Gamma) \otimes_K H(\Gamma') = H(\Gamma'')$.

6.2. Lemma. *Let $r, m \in I$ and $V = V_m \otimes V_r$. Let*

$$u : \bigoplus_{i\in I} H^i_{mr} \otimes_K H_i^{mr} \to \mathrm{Hom}(V, V)$$

be the isomorphism (II.4.2.b) *applied to $V = W = V_m \otimes V_r$. Let δ^i_{mr} be the element of $H^i_{mr} \otimes_K H_i^{mr}$ determined by duality. Then*

$$u(\sum_{i\in I} \dim(i)\,\delta^i_{mr}) = \mathrm{id}_V.$$

Proof. Set

$$u^{-1}(\mathrm{id}_V) = \sum_{i\in I}\sum_{k\in R_i} a_{i,k} \otimes b_{i,k}$$

where $a_{i,k} \in H^i_{mr}$, $b_{i,k} \in H_i^{mr}$ and k runs over a finite set of indices R_i that depends on i. Fix $i \in I$. We shall prove that

$$(\dim(i))^{-1} \sum_{k \in R_i} a_{i,k} \otimes b_{i,k} = \delta^i_{mr}. \tag{6.2.a}$$

This formula would imply the claim of the lemma.

Recall that the duality pairing $H^i_{mr} \otimes_K H_i^{mr} \to K$ introduced in Section VI.1.2 carries a pair $(x \in H^i_{mr}, y \in H_i^{mr})$ into $\operatorname{tr}(xy) \in K$ where xy is the composition of $x : V_m \otimes V_r \to V_i$ and $y : V_i \to V_m \otimes V_r$. To prove (6.2.a) it is enough to show that for arbitrary $x \in H^i_{mr}$ and $y \in H_i^{mr}$,

$$\operatorname{tr}(xy) = (\dim(i))^{-1} \sum_{k \in R_i} \operatorname{tr}(xb_{i,k}) \operatorname{tr}(a_{i,k} y). \tag{6.2.b}$$

We have

$$\operatorname{tr}(xy) = \operatorname{tr}(x \operatorname{id}_V y) = \sum_{s \in I, k \in R_s} \operatorname{tr}(x\, b_{s,k}\, a_{s,k}\, y).$$

The homomorphism $xb_{s,k} : V_s \to V_i$ is non-zero only if $s = i$ in which case this homomorphism is multiplication by $(\dim(i))^{-1}\operatorname{tr}(xb_{s,k})$. Similar remarks apply to $a_{s,k} y : V_i \to V_s$. Therefore $xy : V_i \to V_i$ is multiplication by

$$(\dim(i))^{-2} \sum_{k \in R_i} \operatorname{tr}(xb_{i,k}) \operatorname{tr}(a_{i,k} y).$$

This implies (6.2.b) and (6.2.a).

6.3. Lemma. *Let $r, m \in I$. There is a decomposition into a finite sum*

$$\operatorname{id}_{V_m} \otimes \left((\operatorname{id}_{V_r} \otimes w_{r^*}^{-1})\, b_{V_r} \right) = \sum_{g \in I, t \in T_g} (\beta_{g,t} \otimes \operatorname{id}_{V_{r^*}}) \alpha_{g,t} \tag{6.3.a}$$

where $\alpha_{g,t} \in H_m^{gr^*}$, $\beta_{g,t} \in H_g^{mr}$ and t runs over a finite set of indices T_g that depends on g (see Figure 6.1). For any such decomposition and any $g \in I$, the elements of the symmetrized multiplicity modules $\alpha'_{g,t} \in H(g, r^*, m^*)$, $\beta'_{g,t} \in H(m, r, g^*)$ corresponding to the elements $\alpha_{g,t}$ and $\beta_{g,t}$ under the isomorphisms $H_m^{gr^*} = H(g, r^*, m^*)$, $H_g^{mr} = H(m, r, g^*)$ satisfy the equalities

$$\sum_{t \in T_g} \alpha'_{g,t} \otimes \beta'_{g,t} = \dim(g) \operatorname{Id}(m^*, r^*, g).$$

Proof. Existence of decomposition (6.3.a) follows from Lemma VI.1.1.2. For any decomposition (6.3.a), we have

$$\operatorname{id}_{V_m \otimes V_r} = \sum_{g \in I, t \in T_g} \beta_{g,t}\, \eta_{g,t} \tag{6.3.b}$$

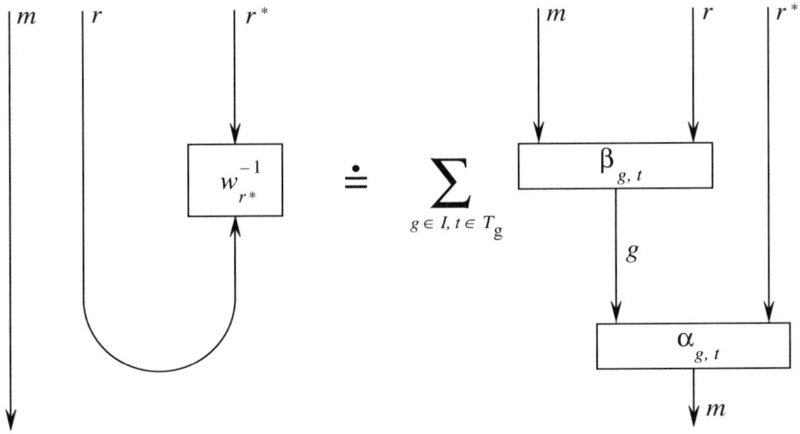

Figure 6.1

with

$$\eta_{g,t} = (\mathrm{id}_{V_g} \otimes d_{V_r})(\mathrm{id}_{V_g} \otimes w_{r^*} \otimes \mathrm{id}_{V_r})(\alpha_{g,t} \otimes \mathrm{id}_{V_r}) \in H^g_{mr}.$$

(To see this draw the corresponding pictures.) Formula (6.3.b) and the previous lemma imply that for any $g \in I$,

$$\sum_{t \in T_g} \eta_{g,t} \otimes \beta_{g,t} = \dim(g)\, \delta^g_{mr}. \tag{6.3.c}$$

It follows from definitions that the same $\alpha'_{g,t} \in H(g, r^*, m^*)$ corresponds to both $\eta_{g,t}$ and $\alpha_{g,t}$ under the identification isomorphisms $H^g_{mr} = H(g, r^*, m^*) = H^{gr^*}_m$. Therefore the claim of the lemma follows from (6.3.c) and the fact that the identification isomorphisms $H^g_{mr} = H(g, r^*, m^*)$, $H^{mr}_g = H(m, r, g^*)$ preserve the duality pairing.

6.4. Proof of Theorem. By Lemma 6.1, $\mathbb{G} : \mathrm{Fr}_{\mathcal{V}} \to \mathrm{Proj}(K)$ is a covariant functor that coincides with $h_0\mathbb{F}$ on the objects of $\mathrm{Fr}_{\mathcal{V}}$. To show that these two functors are equal we shall use the technique of generators of categories discussed in Section I.3.1. Here we need a slightly different notion of generators. We say that a family of morphisms in $\mathrm{Fr}_{\mathcal{V}}$ generates $\mathrm{Fr}_{\mathcal{V}}$ if any morphism in this category may be obtained from these generators and identity endomorphisms of objects using composition. Here we do not involve tensor products of morphisms.

A morphism in the category $\mathrm{Fr}_{\mathcal{V}}$ is the isotopy class of a v-colored orientable framed graph in $\mathbb{R}^3$. An orientable framed graph may be presented by an enriched diagram in $\mathbb{R} \times [0, 1]$ such that the pre-twists of all 1-strata are integers. Introducing into this diagram an appropriate number of positive or negative curls we may always replace it by another diagram, say D, of the same framed graph such that the pre-twists of all 1-strata are equal to zero. Slightly deforming D

in a neighborhood of crossing points, we may assume that near each such point the upper branch goes from south-west to north-east as in Figure 6.2.a. Similarly, deforming D in a neighborhood of vertices, we may assume that near any vertex the diagram looks like Figure 6.2.b, i.e., one edge lies below the vertex and two edges lie above the vertex. Using the same argument as in the proof of Lemma I.3.1.1 we may split any such diagram as a composition of diagrams shown in Figure 6.2 where the pre-twists of all 1-strata are equal to 0. The colors of intervals and vertices are not indicated in Figure 6.2, it is understood that they vary in the set I and the corresponding symmetrized multiplicity modules. Therefore the v-colored framed graphs presented in Figure 6.2 generate $\mathrm{Fr}_{\mathcal{V}}$. To prove the equality $h_0\mathbb{F} = \mathbb{G}$ it remains to show that both $h_0\mathbb{F}$ and $\mathbb{G}$ take the same values on these generators.

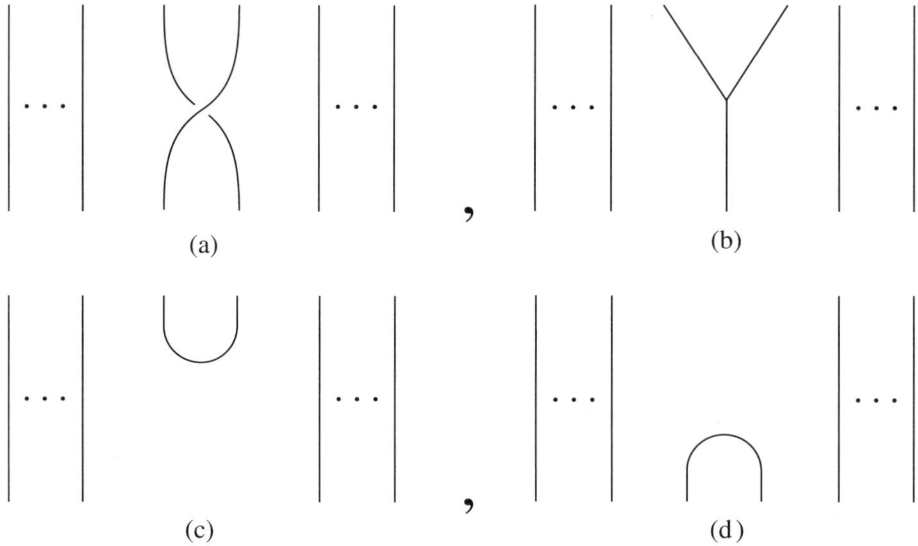

Figure 6.2

We shall explicitly compute the values of $h_0\mathbb{F}$ and $\mathbb{G}$ on the colored framed graph Γ presented in Figure 6.2.c. Let s (resp. $k - s$) be the number of vertical strings lying to the left (resp. to the right) of the cup. Let $i_1, \ldots, i_k$ be the colors of vertical strings oriented downward. Let r be the color of the cup-like string oriented to the right. The ribboning of Γ yields a colored ribbon tangle Ω_Γ with one coupon on each string. By definition (see Section 5.3),

$$\text{(6.4.a)} \qquad \mathbb{F}(\Gamma) = (w_{i_1}^{-1} \otimes \ldots \otimes w_{i_s}^{-1} \otimes w_r^{-1} \otimes w_{r^*}^{-1} \otimes w_{i_{s+1}}^{-1} \ldots \otimes w_{i_k}^{-1})\, F(\Omega_\Gamma).$$

It is obvious that the color w_{i_n} of the coupon on the n-th vertical string cancels with $w_{i_n}^{-1}$. Therefore we may cancel these coupons and morphisms. Similarly,

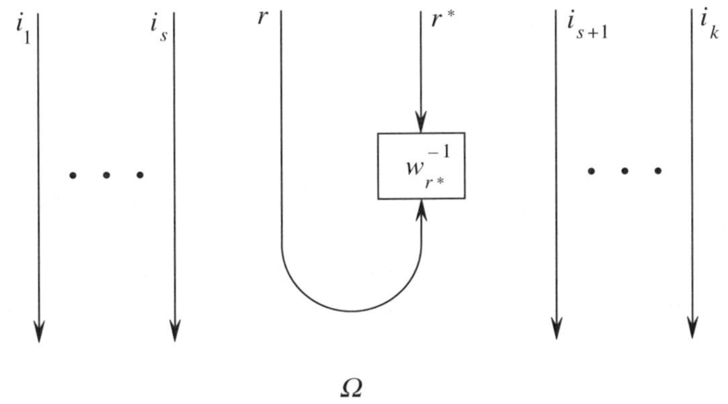

Figure 6.3

we cancel w_r^{-1} with the color of the coupon of the cup-like string. Therefore $\mathbb{F}(\Gamma) = F(\Omega)$ where Ω is the v-colored ribbon graph shown in Figure 6.3.

Set $i = (i_1, ..., i_k)$ and $j = (i_1, ..., i_s, r, r^*, i_{s+1}, ..., i_k)$. We present the homomorphism $h_0\mathbb{F}(\Gamma) = h_0F(\Omega)$ by a block-matrix $\{\gamma_m^n\}_m^n : \oplus_m G(i, m) \to \oplus_n G(j, n)$ where $m = (m_0, ..., m_k)$ and $n = (n_0, ..., n_{k+2})$ run over admissible sequences. The homomorphism $\gamma_m^n : G(i, m) \to G(j, n)$ may be computed as follows. Fix an admissible sequence $m = (m_0, ..., m_k) \in I^{k+1}$.

Denote by Ω' the colored ribbon graph obtained by attaching the diagram in Figure 5.1 to the diagram of Ω from below. The operator invariant $F(\Omega') : G(i, m) \to \mathbb{G}(j) = \oplus_n G(j, n)$ coincides with the homomorphism $\{\gamma_m^n\}^n : G(i, m) \to \oplus_n G(j, n)$. Here the module $G(i, m)$ appears as the module of colorings of uncolored coupons of Ω'. In order to compute $F(\Omega')$, present Ω' as a composition of three colored ribbon graphs

$$\Omega' = \Omega_1 \circ (\Omega_2 \otimes \downarrow_{i_{s+1}} \otimes ... \otimes \downarrow_{i_k}) \circ \Omega_3$$

where Ω_2 is the ribbon graph on the left-hand side of Figure 6.1 with $m = m_s$. Here Ω_3 has $k-s$ coupons and looks as the graph in Figure 5.1; Ω_1 has s coupons and looks as the graph in Figure 5.1 with a vertical band attached to the bottom coupon from below and $k-s+2$ vertical bands added on the right. Using Lemma 6.3 we may replace Ω_2 with the ribbon graph on the right-hand side of Figure 6.1. This yields a decomposition of $F(\Omega')$ into a linear combination of operator invariants of ribbon graphs similar to the one in Figure 5.1. More exactly, set

$$n_g = (m_0, ..., m_s, g, m_s, m_{s+1}, ..., m_k) \in I^{k+3}$$

and denote by $\Omega(j, n_g)$ the colored ribbon graph shown in Figure 5.1 where k is replaced with $k + 2$ and the sequences $(i_1, ..., i_k)$ and $(m_0 = 0, m_1, ..., m_{k-1},$

$m_k = 0$) are replaced by the sequences j and n_g respectively. Then

$$F(\Omega') = \sum_{g\in I, t\in T_g} F(\Omega(j, n_g; t))$$

where $\Omega(j, n_g; t)$ is the colored ribbon graph $\Omega(j, n_g)$ whose two coupons obtained from Ω_2 are colored with $\alpha_{g,t}, \beta_{g,t}$. This implies that the image of $F(\Omega') : G(i, m) \to \mathbb{G}(j) = \oplus_n G(j, n)$ is contained in $\oplus_{g\in I} G(j, n_g)$. Thus, $\gamma_m^n = 0$ unless $n = n_g$ for some $g \in I$. Lemma 6.3 implies that for any $g \in I$, the homomorphism

$$\gamma_m^{n_g} = \sum_{t\in T_g} F(\Omega(j, n_g; t))$$

carries any $x \in G(i, m)$ into $\dim(g)x \otimes \mathrm{Id}(m_s, r, g^*)$. (If $s = 0$, we use the case $m = 0$ of Lemma 6.3.)

Let us compute the homomorphism $\Gamma_m^n : G(i, m) \to G(j, n)$ defined in Section 5.5. Let C be the shadow cylinder over the diagram in Figure 6.2.c. As a polyhedron, C consists of a big plane rectangle $[-R, R] \times [0, 1]$ and $k + 1$ rectangles which are glued to the big one along homeomorphisms of their bottom bases onto $k + 1$ arcs shown in Figure 6.2.c. The gleams of all regions of C are equal to zero. The coloring $\lambda = \lambda_{m,n}$ of ∂C is determined by the sequences i, m, n as above. If $n \neq n_g$ for all $g \in I$ then C has no colorings extending $\lambda_{m,n}$. Therefore $\Gamma_m^n = 0 = \gamma_m^n$. Assume that $n = n_g$ for a certain $g \in I$. Then there is exacty one coloring φ of C extending λ^*. This coloring assigns to the regions of C (with appropriate orientations) the colors $m_0, m_1, ..., m_k, i_1, ..., i_k, r, g$. It is easy to compute that

$$|C|_1^\varphi = (\dim'(m_s)\dim'(g)\dim'(r))^{-1}\prod_{p=1}^{k}(\dim(m_p)\dim'(i_p))^{-1},$$

$$|C|_2^\varphi = \dim(g)\dim(r)\prod_{p=1}^{k}(\dim(m_p)\dim(i_p)),\quad |C|_3^\varphi = |C|_4^\varphi = 1,$$

and

$$|C|_5^\varphi = \mathrm{Id}(m_s, r, g^*) \otimes \bigotimes_{p=0}^{k-1} \mathrm{Id}(m_p, m_{p+1}^*, i_p).$$

We also have

$$\dim'(\Gamma) = \dim'(r)\prod_{p=1}^{k}(\dim'(i_p)),\qquad \dim_m^n(\Gamma) = \dim'(m_s)\dim'(g).$$

Multiplying these equalities we get

$$\Gamma(m,n) = \dim(g)\,\mathrm{Id}(m_s, r, g^*) \otimes \bigotimes_{p=0}^{k-1} \mathrm{Id}(m_p, m_{p+1}^*, i_p). \tag{6.4.b}$$

This implies that for any $x \in G(i, m)$,

$$\Gamma_m^n(x) = \dim(g)\, x \otimes \mathrm{Id}(m_s, r, g^*) = \gamma_m^n(x).$$

Therefore $h_0\mathbb{F}(\Gamma) = \mathbb{G}(\Gamma)$.

Let Γ_d be the colored framed graph in Figure 6.2.d. The values of $h_0\mathbb{F}$ and $\mathbb{G}$ on Γ_d may be computed explicitly in a similar way. To compute $h_0\mathbb{F}(\Gamma_d)$ we use the following assertion which is analogous to Lemma 6.3. Let

$$H_g^{mr^*} \otimes_K H_m^{gr} \to \mathrm{Hom}(V_m, V_m) \tag{6.4.c}$$

be the pairing represented by the ribbon graph in Figure 6.4. Then the composition of this pairing with the isomorphisms $\mathrm{Hom}(V_m, V_m) = K \cdot \mathrm{id}_{V_m} = K$ and

$$H(m, r^*, g^*) \otimes_K H(g, r, m^*) \to H_g^{mr^*} \otimes_K H_m^{gr} \tag{6.4.d}$$

is the standard duality pairing

$$H(m, r^*, g^*) \otimes_K H(g, r, m^*) \to K \tag{6.4.e}$$

multiplied by $(\dim(m))^{-1}$. To see this, consider the closure of the ribbon graph in Figure 6.4. This closure represents the same pairing (6.4.c) composed with the trace $x \mapsto \mathrm{tr}(x) : \mathrm{Hom}(V_m, V_m) \to K$. Now our assertion follows from the definitions of the isomorphism (6.4.d) and the pairing (6.4.e). The factor $(\dim(m))^{-1}$ appears here because the isomorphism $\mathrm{Hom}(V_m, V_m) = K \cdot \mathrm{id}_{V_m} = K$ transforms any $x \in \mathrm{Hom}(V_m, V_m)$ into $(\dim(m))^{-1}\mathrm{tr}(x)$. The computation of $\mathbb{G}(\Gamma_d)$ is quite similar to the computation of $\mathbb{G}(\Gamma)$ given above. (The shadow cylinders used to compute $\mathbb{G}(\Gamma_d)$ and $\mathbb{G}(\Gamma)$ are homeomorphic.) The only essential difference is that

$$\dim_m^n(\Gamma) = (\dim'(m_{s+1})\dim'(g))^{-1}.$$

This yields a formula analogous to (6.4.b) with $\dim(g)$ replaced by $(\dim(m_{s+1}))^{-1}$. This implies the equality $h_0\mathbb{F}(\Gamma_d) = \mathbb{G}(\Gamma_d)$.

Let Γ be the colored framed graph shown in Figure 6.2.a. The value of $\mathbb{G}$ on Γ may be computed as follows. Let k be the number of strings of the diagram. Let s be the number of strictly vertical strings of the diagram lying to the left of the crossing. Let $i = (i_1, ..., i_k)$ be the sequence of colors of the strings in Figure 6.2.a oriented downward and ordered in accordance with the order of bottom ends. Set $j = (i_1, ..., i_s, i_{s+2}, i_{s+1}, i_{s+3}, ..., i_k)$. Take arbitrary admissible $m = (m_0, ..., m_k) \in I^{k+1}$ and $n = (n_0, ..., n_k) \in I^{k+1}$. It is obvious that the homomorphism $\Gamma_m^n : G(i, m) \to G(j, n)$ is zero unless $n = n(g) = (m_0, ..., m_s, g, m_{s+2}, ..., m_k)$ with $g \in I$. Assume that $n = n(g)$ for some $g \in I$. Let C be the shadow cylinder over

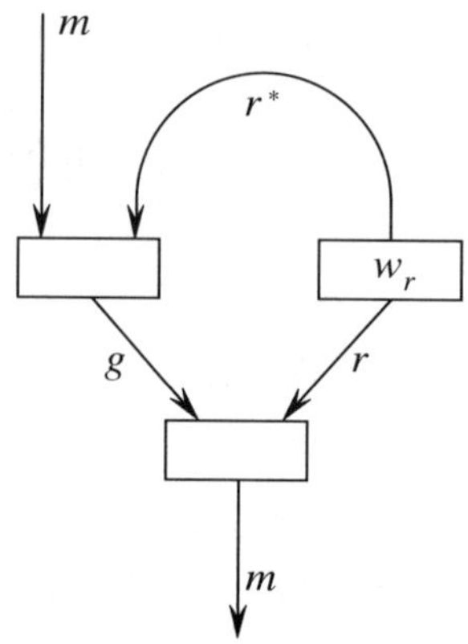

Figure 6.4

the shadow projection of the diagram in Figure 6.2.a. This polyhedron consists of a big plane rectangle $[-R, R] \times [0, 1]$ and k rectangles which are glued to the big one along homeomorphisms of (one of) their bases onto the k arcs shown in Figure 6.2.a. The polyhedron C has one vertex located in the crossing point of the diagram. The gleams of two triangular regions of C adjacent to the vertex are equal to $-1/2$. The gleams of two pentagonal regions of C adjacent to the vertex are equal to 1/2. Other regions of C have zero gleam. The coloring $\lambda = \lambda_{m,n}$ of ∂C is determined by the sequences i, m, n as above. There is only one coloring φ of C that extends λ^*. This coloring assigns the colors $m_0, m_1, ..., m_k, i_1, ..., i_k, g$ to the regions of C (with appropriate orientations). A direct computation shows that

$$|C|_1^\varphi = \dim'(m_{s+1})(\dim'(g))^{-1}\prod_{p=1}^{k}(\dim(m_p)\dim'(i_p))^{-1},$$

$$|C|_2^\varphi = \dim(g)\prod_{p=1}^{k}(\dim(m_p)\dim(i_p)),\ \ |C|_3^\varphi = v'_{m_s}v'_{m_{s+2}}(v'_{m_{s+1}}v'_g)^{-1},\ \ |C|_4^\varphi = 1.$$

Since C has only one vertex it is easy to compute that

$$|C|_5^\varphi = \begin{vmatrix} i_{s+1} & m_s & m_{s+1} \\ i_{s+2} & m_{s+2} & g \end{vmatrix} \otimes \bigotimes_{0\leq p\leq k-1,\, p\neq s,s+1} \mathrm{Id}(m_p, m^*_{p+1}, i_p).$$

We also have

$$\dim'(\Gamma) = \prod_{p=1}^{k}\dim'(i_p),\quad \dim^n_m(\Gamma) = (\dim'(m_{s+1}))^{-1}\dim'(g).$$

Multiplying these equalities we get

$$\Gamma(m,n) = v'_{m_s} v'_{m_{s+2}} (v'_{m_{s+1}} v'_g)^{-1} \dim(g) \begin{vmatrix} i_{s+1} & m_s & m_{s+1} \\ i_{s+2} & m_{s+2} & g \end{vmatrix} \otimes$$

$$\otimes \bigotimes_{0 \leq p \leq k-1, p \neq s, s+1} \mathrm{Id}(m_p, m^*_{p+1}, i_p).$$

The computation of $h_0\mathbb{F}(\Gamma)$ is analogous to computations made above with the following changes. The ribboning of Γ yields a colored ribbon tangle with one coupon on each string. The colors of these coupons cancel with the morphisms w_i^{-1} entering into the definition of $\mathbb{F}(\Gamma)$. Now we use the equality in Figure VI.5.4 where the indices i, j, k, l, m, n should be replaced with $m_s, i_{s+1}, m_{s+1}, i_{s+2}, m_{s+2}, g$ respectively. The $6j$-symbol appearing on the right-hand side in Figure VI.5.5 is computed in Section VI.5.6. Namely, under the passage to symmetrized multiplicity modules this symbol is transformed into

$$v'_{m_s} v'_{m_{s+2}} (v'_{m_{s+1}} v'_g)^{-1} \dim(g) \begin{vmatrix} i_{s+1} & m_s & m_{s+1} \\ i_{s+2} & m_{s+2} & g \end{vmatrix}.$$

This implies the equality $h_0\mathbb{F}(\Gamma) = \mathbb{G}(\Gamma)$.

The generator shown in Figure 6.2.b is considered in a similar way. If $i = (i_1, ..., i_k)$ and $j = (i_1, ..., i_s, x, y, i_{s+3}, ..., i_k)$ are the source and the target of the corresponding morphism in $\mathrm{Fr}_{\mathcal{V}}$ then for any admissible

$$m = (m_0, ..., m_k) \in I^{k+1}, \quad n = (m_0, ..., m_s, g, m_{s+1}, ..., m_k) \in I^{k+2},$$

the corresponding block $G(i, m) \to G(j, n)$ of the block-matrix for both $h_0\mathbb{F}(\Gamma)$ and $\mathbb{G}(\Gamma)$ is equal to

$$\dim(g) \begin{vmatrix} m_s^* & x^* & g^* \\ y^* & m^*_{s+1} & i^*_{s+1} \end{vmatrix}.$$

(To see this for $h_0\mathbb{F}(\Gamma)$, analyze the proof of the orthonormality relation for $6j$-symbols.)

7. Invariants of framed graphs from the shadow viewpoint

7.0. Outline. The technique of state sum invariants is applied here to shadows of framed graphs in 3-manifolds. We first develop the approach involving 4-manifolds bounded by a given 3-manifold. This yields a shadow computation of the invariant τ of framed graphs introduced in Chapter VI. We also give a shadow computation of the invariant of 3-manifolds with colored triangulated boundary introduced in Chapter VII. The relevant results (Theorems 7.1.1 and 7.2.1) will be instrumental in Section 8. At the end of the section we study internal invariants of framed graphs in 3-manifolds. They are shown to be essentially equivalent to the invariant τ.

7.1. External invariants of framed graphs. Let M be a closed oriented 3-manifold and let Γ be a colored framed graph in M. Let W be a compact oriented 4-manifold bounded by M. Recall the shadow $\mathrm{sh}(W,\Gamma)$ of the pair (W,Γ) (see Section IX.8). Set

$$\tau^0(M,\Gamma) = \tau^0_{(\mathcal{V},\mathcal{D})}(M,\Gamma) = ||\,\mathrm{sh}(W,\Gamma)||_{(\mathcal{V},\mathcal{D})} \in H(\Gamma).$$

The same arguments as in Section 3.2 show that $\tau^0(M,\Gamma)$ is a well-defined invariant of the pair (M,Γ) independent of the choice of W.

By Theorem 3.3, $\tau^0(M,\emptyset) = \tau^0(M) = \mathcal{D}^{b_3(M)-b_2(M)}\tau(M)$. The following theorem generalizes this result.

7.1.1. Theorem. *For any closed oriented 3-manifold M and any colored framed graph $\Gamma \subset M$, we have*

$$\tau^0(M,\Gamma) = \mathcal{D}^{b_3(M)-b_2(M)}\,\dim'(\Gamma)\,\tau(M,\Gamma^*)$$

where $\tau(M,\Gamma^*)$ is the invariant of (M,Γ^*) defined in Section VI.4.7.

Here $\dim'(\Gamma)$ is defined by formula (5.5.b) where λ denotes the given coloring and e runs over edges (but not circle 1-strata) of the core of Γ. Note that $\tau(M,\Gamma^*) \in H^\star(\Gamma^*) = H(\Gamma)$ so that the equality in the statement of the theorem makes sense. The proof of Theorem 7.1.1 is analogous to the proof of Theorem 3.3 with the obvious changes. Instead of Lemma 4.1 we use Corollary 5.7.

7.2. Comparison with the simplicial model. Let M be a compact oriented 3-manifold with non-empty triangulated boundary and let $\psi \in \mathrm{col}(\partial M)$. We reformulate here the invariant $|M,\psi|$ introduced in Section VII.1.6 in terms of τ^0.

Denote by γ the trivalent graph in ∂M dual to the 1-skeleton of the triangulation of ∂M. The graph γ is formed by vertices and edges of the cell subdivision of ∂M dual to the triangulation, cf. Remark VII.3.5. The coloring ψ of ∂M induces a coloring $\psi^\perp$ of γ by the formula $\psi^\perp(e) = \psi(e^\perp)$ where e is an oriented edge of γ and $e^\perp$ is the dual edge of ∂M with the corresponding normal orientation.

The double $\overline{M}$ of M is obtained by gluing of a copy of $-M$ to M along the identity homeomorphism of their boundaries. (If $\partial M = \emptyset$ then $\overline{M} = M \amalg (-M)$.) A regular neighborhood of γ in the surface $\partial M \subset \overline{M}$ may be regarded as a framed graph in $\overline{M}$ with the core γ. Denote this framed graph endowed with the coloring $\psi^\perp$ by Γ^ψ. Note that both $|M,\psi|$ and $\tau^0(\overline{M},\Gamma^\psi)$ belong to the module $H_\psi(\partial M) = H(\Gamma^\psi)$.

7.2.1. Theorem. *Let c be the number of vertices of the given triangulation of ∂M. Then*

$$|M, \psi| = \mathcal{D}^{2b_2(M)-2b_3(M)-c}\, \tau^0(\overline{M}, \Gamma^\psi).$$

Proof. Let μ be a triangulation of M extending the given triangulation of ∂M. Consider the cell subdivision μ^* of M dual to μ. Let $X \subset M$ be the union of closed 2-cells of this subdivision dual to the edges of μ. It is easy to see that X is a simple 2-polyhedron with boundary $\gamma = X \cap \partial M$. This polyhedron has nice topological properties simplifying the calculation of the state sums. Namely, it has neither circle 1-strata nor edges with both ends in ∂X. All regions of X are open 2-disks. The complement of X in M consists of disjoint open 3-balls centered in the vertices of μ not lying in ∂M and of 3-dimensional semi-balls centered in the vertices of ∂M.

Consider the 4-manifold $W = M \times [-1, 1]$ with an orientation induced by that of M. It is clear that $\partial W = \overline{M}$. The polyhedron $X = X \times 0 \subset W$ is a locally flat orientable simple 2-polyhedron with boundary $\gamma \times 0$. It is easy to see that W may be obtained from a closed regular neighborhood of X by attaching 3-handles whose cores are 3-balls in $M \setminus X$ centered in the vertices of μ not lying in ∂M. In the terminology of Section IX.8 the polyhedron X is a skeleton of the 4-dimensional graph pair (W, Γ^ψ).

The shadowing of X defined in Sections IX.1.6 and IX.8.1 yields identically zero gleams because the normal bundle of $X \cup_\gamma \Gamma^\psi$ in W admits a non-singular section. This, in particular, implies that the signature of X is equal to zero. We have

$$\tau^0(\overline{M}, \Gamma^\psi) = \big|\big| \operatorname{sh}(W, \Gamma^\psi)\big|\big| = \big|X, \psi^\perp\big|$$

where $\psi^\perp$ is the coloring of $\partial X = \gamma$ determined by ψ. It remains to show that

$$|M, \psi| = \mathcal{D}^{2b_2(M)-2b_3(M)-c} \left|X, \psi^\perp\right|. \tag{7.2.a}$$

As in the proof of Theorem 3.1, the formula $\varphi^\perp(e) = \varphi(e^\perp)$ establishes a bijective correspondence $\varphi \mapsto \varphi^\perp$ between colorings of μ extending ψ and colorings of X extending $\psi^\perp$. (Here e is an oriented disk region of X and $e^\perp$ is the dual edge of μ with the induced normal orientation.) It follows from definitions that for any $\varphi \in \operatorname{col}(M, \mu)$, we have

$$|M|_\varphi = \mathcal{D}^{-2a-c}\, |X|_{\varphi^\perp}$$

where a is the number of vertices of μ not lying in ∂M. Summing up over all $\varphi \in \operatorname{col}(M, \mu)$ with $\partial\varphi = \psi$ we get

$$|M, \psi| = \mathcal{D}^{b_2(X)+\operatorname{null}(X)-2a-c} \left|X, \psi^\perp\right|.$$

It is straightforward to compute that $\operatorname{null}(X) = b_2(X) = b_2(M) - b_3(M) + a$. This yields (7.2.a).

7.3. Internal invariants. Let M be a compact oriented 3-manifold and let Γ be a colored framed graph in $\mathrm{Int}(M)$ (without free ends). The internal invariant of (M, Γ) is defined by the formula

$$|M, \Gamma| = \mathcal{D}^{2(b_2(M)-b_3(M))} (\dim'(\Gamma))^{-1} |CY(M, \Gamma)| \in H(\Gamma).$$

Note that the definition of $|M, \Gamma|$ appeals to graph diagrams in skeletons of M and does not involve 4-manifolds. The invariant $|M, \Gamma|$ generalizes the simplicial invariant $|M|$ corresponding to closed M and empty Γ. Indeed, $CY(M, \emptyset) = \mathrm{ish}(M)$ and by Theorem 3.1

$$|M, \emptyset| = \mathcal{D}^{2(b_2(M)-b_3(M))} |\mathrm{ish}(M)| = |M| .$$

The following two theorems relate $|M, \Gamma|$ to invariants defined in Chapters VI and VII. The first of these theorems shows that the invariant $|M, \Gamma|$ coincides with the invariant $\tau(M, \Gamma^*)$ up to a factor independent of the choice of Γ.

7.3.1. Theorem. *Let Γ be a colored framed graph in an oriented closed 3-manifold M. Then $|M, \Gamma| = \tau(M, \Gamma^*)\, \tau(-M)$.*

Proof. Generalizing the argument given in Section 3.4 we may verify that $\mathrm{sh}(M \times [0, 1], \Gamma \times 1) = CY(M, \Gamma)$. Therefore

$$\dim'(\Gamma)\, |M, \Gamma| = \mathcal{D}^{2(b_2(M)-b_3(M))} \,|\, \mathrm{sh}(M \times [0, 1], \Gamma \times 1)| =$$

$$= \mathcal{D}^{2(b_2(M)-b_3(M))} \,||\, \mathrm{sh}(M \times [0, 1], \Gamma \times 1)|| =$$

$$= \mathcal{D}^{2(b_2(M)-b_3(M))}\, \tau^0((M, \Gamma) \amalg (-M, \emptyset)) = \dim'(\Gamma)\, \tau(M, \Gamma^*)\, \tau(-M).$$

Here the second equality follows from the fact that the signature of $\mathrm{sh}(M \times [0, 1], \Gamma \times 1)$ is equal to the signature of the 4-manifold $M \times [0, 1]$ and therefore equals 0.

7.3.2. Theorem. *Let M be a compact oriented 3-manifold with non-empty triangulated boundary and let $\psi \in \mathrm{col}(\partial M)$. Let c be the number of vertices of the given triangulation of ∂M. Let γ be the trivalent graph in ∂M dual to the 1-skeleton of the given triangulation of ∂M. Let Γ_ψ be a regular neighborhood of γ in ∂M slightly pushed into* Int(M) *and regarded as a framed graph in M with core γ and coloring $\psi^\perp$. Then $|M, \psi| = \mathcal{D}^{-c}\dim'(\Gamma_\psi)\, \left|M, \Gamma_\psi\right|$.*

Proof. Let μ be a triangulation of M extending the given triangulation of ∂M. Subdividing μ, if necessary, we may assume that ∂M is a full subcomplex of μ, i.e., that any simplex of μ with vertices in ∂M lies in ∂M. Let X be the shadowed polyhedron with zero gleams and $\partial X = \gamma$ constructed in the proof of Theorem 7.2.1. We claim that X represents the shadow cylinder of Γ_ψ in M, i.e., that $CY(M, \Gamma_\psi) = \mathrm{stab}([X])$. This claim would imply the theorem: by (7.2.a) we have

$$|M, \psi| = \mathcal{D}^{2(b_2(M)-2b_3(M)-c} \left|X, \psi^{\perp}\right| = \mathcal{D}^{2b_2(M)-2b_3(M)-c} \left|CY(M, \Gamma_\psi)\right|.$$

By definition of $|M, \Gamma_\psi|$, the last expression is equal to $\mathcal{D}^{-c}\dim'(\Gamma)\ |M, \Gamma_\psi|$.

Let $C \subset X$ be the union of closed 2-cells of μ^* dual to the edges of μ not lying in ∂M. It is easy to see that C is a skeleton of M. Its complement consists of an open collar of ∂M and disjoint open 3-balls centered in the vertices of μ not lying in ∂M (cf. [RS, Corollary 3.9]).

Let $\Sigma \subset C$ be the union of closed 2-cells of μ^* dual to those edges of μ which have one endpoint in ∂M and the other endpoint in $M \setminus \partial M$. Analyzing the position of Σ in 3-simplices of μ, it is easy to see that Σ is a surface embedded in $\text{Int}(M)$ and parallel to ∂M. More exactly, the surfaces Σ and ∂M cobound a collar $\partial M \times [0, 1] \subset M$ of ∂M so that $\partial M = \partial M \times 0$ and $\Sigma = \partial M \times 1$. This collar intersects C along Σ. The intersection of this collar with X consists of Σ and the union of closed 2-cells of μ^* dual to edges of ∂M. Each such 2-cell may be considered as a topological rectangle with bottom base in ∂M and top base in Σ. Indeed, if e is an edge of ∂M then the dual 2-cell $e^* \subset M$ may be regarded as a rectangle whose bottom base is the 1-cell of γ dual to e, whose sides are the 1-cells of μ^* dual to two 2-simplices of ∂M adjacent to e, and whose top base is the union of closed 1-cells of μ^* dual to 2-simplices of μ intersecting ∂M along e. This shows that the intersection $X \cap (\partial M \times [0, 1])$ is a union of Σ and a cylinder $\gamma \times [0, 1]$ over $\gamma = \gamma \times 0$. Consider the base $\gamma' = \gamma \times 1 \subset \Sigma$ of this cylinder. This is an embedded 3-valent graph in Σ formed by closed 1-cells of μ^* dual to 2-simplices of μ intersecting ∂M along one edge.

We equip all edges of γ' with zero pre-twists. The coloring $\psi^{\perp}$ of γ induces a coloring of γ' in the obvious way. Thus, γ' becomes an enriched graph diagram of Γ'_ψ in the skeleton C of M. The cylinder over this graph diagram in C may be canonically identified with X. Note that the gleams of all regions of $\text{sh}(\gamma')$ in C are equal to 0. This follows from the fact that γ' has no self-crossings and C lies on one side of the surface $\Sigma \subset M$ so that the contributions $\pm 1/4$ of the crossing points of γ' with $\text{sing}(C)$ cancel. Therefore the polyhedron X with zero gleams of regions represents the shadow cylinder of Γ_ψ in M.

8. Proof of Theorem VII.4.2

8.1. Plan of the proof. The proof of Theorem VII.4.2 proceeds in three steps. First of all, we construct a non-degenerate anomaly-free TQFT $(\mathcal{T}_1, \tau_1)$ based on 3-cobordisms with pairs of v-colored framed graphs sitting inside. This TQFT is a version of the TQFT $(\widehat{\mathcal{T}}, \widehat{\tau})$ involved in the theorem. Secondly, we use shadows to construct another anomaly-free TQFT $(\mathcal{T}_2, \tau_2)$ based on the same cobordism theory as $(\mathcal{T}_1, \tau_1)$ and extending the simplicial TQFT involved in the Theorem.

Finally, we show that the values of τ_1 and τ_2 on closed 3-manifolds coincide. This allows us to apply Theorem III.3.7 and to conclude that the TQFT's $(\mathcal{T}_1, \tau_1)$ and $(\mathcal{T}_2, \tau_2)$ are isomorphic. Restricting their isomorphism to 3-cobordisms with empty framed graphs accomplishes the proof.

8.2. The TQFT $(\mathcal{T}_1, \tau_1)$. Consider the space-structure $\mathfrak{A}$ of an oriented closed surface and the space-structure $\mathfrak{B}$ of an oriented compact 3-manifold endowed with a pair of v-colored framed graphs (without free ends) lying in this manifold. These graphs are completely independent of each other and in particular may intersect each other. The gluing, boundary, and cylinders are defined in the obvious way. Note that under the gluing of 3-cobordisms we form the disjoint union of first (resp. second) framed graphs lying in these cobordisms. The cylinder over a surface is provided with empty framed graphs. It is clear that $(\mathfrak{B}, \mathfrak{A})$ is a cobordism theory in the sense of Section III.1.3.

We shall define a TQFT $(\mathcal{T}_1, \tau_1)$ based on the cobordism theory $(\mathfrak{B}, \mathfrak{A})$. We employ the same method as in Section IV.8. Namely, we first construct a TQFT $(\mathcal{T}_1^s, \tau_1^s)$ based on the space-structures $(\mathfrak{B}, \mathfrak{A})$ where additionally all surfaces are provided with Lagrangian spaces in real 1-homologies. Then we dispense with Lagrangian spaces.

The modular functor $\mathcal{T}_1^s$ is the modular functor $\mathcal{T}^s$ (defined in Section IV.8.1) restricted to surfaces without distinguished arcs. Thus, for any closed oriented surface Σ with a Lagrangian space $\lambda \subset H_1(\Sigma; \mathbb{R})$, we have

$$\mathcal{T}_1^s(\Sigma, \lambda) = \mathcal{T}_{\mathcal{V}}^e(\Sigma, \lambda) \otimes_K \mathcal{T}_{\overline{\mathcal{V}}}^e(\Sigma, \lambda). \tag{8.2.a}$$

Let M be a compact oriented 3-cobordism between oriented closed surfaces with distinguished Lagrangian spaces $(\partial_- M, \lambda_-)$ and $(\partial_+ M, \lambda_+)$. Let $\Gamma_1, \Gamma_2 \subset \mathrm{Int}(M)$ be v-colored framed graphs. Ribboning Γ_1, Γ_2 as in Section VI.4.3 we get v-colored ribbon graphs $\Omega_1, \Omega_2 \subset \mathrm{Int}(M)$. Set

$$\tau_1^s(M, \Gamma_1, \Gamma_2) = \tau_{\mathcal{V}}^e(M, \Omega_1) \otimes \tau_{\overline{\mathcal{V}}}^e(M, \Omega_2) : \mathcal{T}_1^s(\partial_- M, \lambda_-) \to \mathcal{T}_1^s(\partial_+ M, \lambda_+).$$

As in Section VI.4, neither of the factors on the right-hand side depend on the indeterminacy in the construction of Ω_1, Ω_2. The arguments of Section IV.8.1 apply in this setting and show that $(\mathcal{T}_1^s, \tau_1^s)$ is an anomaly-free TQFT admitting an action of weak e-homeomorphisms. Therefore, as in Section IV.8.2, we may eliminate the Lagrangian spaces. In other words, the TQFT $(\mathcal{T}_1^s, \tau_1^s)$ induces an anomaly-free TQFT based on $(\mathfrak{B}, \mathfrak{A})$. Denote this TQFT by $(\mathcal{T}_1, \tau_1)$.

The construction of $(\mathcal{T}_1, \tau_1)$ differs from the constructions of Section IV.8 in several points. First, we have restricted ourselves to surfaces without distinguished arcs. (An extension to the general case is straightforward but we do not consider it.) Secondly, we use framed graphs rather than ribbon graphs. This is a matter of language, framed graphs are better suited for our aims because we shall eventually involve shadows. A really important point is that we apply the operators $\tau_{\mathcal{V}}^e$ and

$\tau^e_{\mathcal{V}}$ to two mutually independent ribbon graphs. This leads to the following crucial result.

8.2.1. Lemma. *The TQFT* $(\mathcal{T}_1, \tau_1)$ *is non-degenerate.*

Proof. Let $\Sigma = \Sigma_{(g;)}$ be the standard closed oriented surface of genus g bounding the standard handlebody $U = U_{(g;)}$ (see Section IV.1). Provide Σ with the Lagrangian space

$$\lambda = \operatorname{Ker}(\operatorname{incl}_* : H_1(\Sigma; \mathbb{R}) \to H_1(U; \mathbb{R})).$$

By definition,

$$\mathcal{T}^e_{\mathcal{V}}(\Sigma, \lambda) = \mathcal{T}_{\mathcal{V}}(\Sigma, \operatorname{id}_\Sigma) = \bigoplus_{j \in I^g} \operatorname{Hom}(\mathbb{1}, \Phi((g;); j))$$

where $\operatorname{id}_\Sigma : \Sigma \to \Sigma$ is the identity mapping regarded as a parametrization of Σ by itself and $\Phi((g;); j) = \otimes_{r=1}^g (V_{j_r} \otimes V_{j_r}^*)$.

Recall the standard ribbon graph $R = R_{(g;)} \subset \operatorname{Int}(U)$ and the decorated handlebody $H(U, R, j, x)$ where $j = (j_1, ..., j_g) \in I^g$ and $x \in \operatorname{Hom}(\mathbb{1}, \Phi((g;); j))$ (see Section IV.1). Lemma IV.2.1.3 implies that the elements $\tau^e(H(U, R, j, x)) \in \mathcal{T}^e_{\mathcal{V}}(\Sigma, \lambda)$ corresponding to arbitrary j and x generate the module $\mathcal{T}^e_{\mathcal{V}}(\Sigma, \lambda)$.

We replace the coupon of R with the colored ribbon graph shown in Figure IV.2.2 where $n = 2g$ and $W_{2r-1} = V_{j_r}, W_{2r} = V_{j_r}^* \approx V_{j_r^*}$ for $r = 1, ..., g$. Lemma IV.2.2.3 implies that varying $i_1, ..., i_n$ and varying the colors of coupons of this ribbon graph we obtain v-colored ribbon graphs whose operator invariants generate $\operatorname{Hom}(\mathbb{1}, \Phi((g;); j))$. At the price of changing the colors of coupons via the canonical isomorphisms

$$\operatorname{Hom}(\mathbb{1}, V_i \otimes V_j \otimes V_k) = \operatorname{Hom}(V_k^*, V_i \otimes V_j) = \operatorname{Hom}(V_{k^*}, V_i \otimes V_j)$$

with $i, j, k \in I$ (cf. Exercise I.1.8.1), we may modify each trivalent coupon in Figure IV.2.2 such that its top base is incident to two incoming bands and its bottom base is incident to one outgoing band. This does not change the operator invariant. If $j_1 = i_1$ then the leftmost (2-valent) coupon in Figure IV.2.2 may be eliminated without changing the operator invariant (up to a non-zero scalar factor). If $j_1 \neq i_1$ then the operator invariant in question is equal to zero. In a similar way we may eliminate the rightmost (2-valent) coupon in Figure IV.2.2. We conclude that the module $\mathcal{T}^e_{\mathcal{V}}(\Sigma, \lambda)$ is generated by elements $\tau^e(U, \Omega)$ where Ω runs over v-colored ribbon graphs in $\operatorname{Int}(U)$ whose bands and annuli are colored with objects $\{V_j\}_{j \in I}$ and such that every coupon of Ω has two incoming bands attached to the top base and one outgoing band attached to the bottom base. Rounding the corners of such a graph Ω we get a v-colored framed graph $\Gamma \subset \operatorname{Int}(U)$ which, in turn, gives rise via ribboning to a v-colored ribbon graph $\Omega_\Gamma \subset \operatorname{Int}(U)$, see Section VI.4. It follows from the results of Section II.4.5 that $\tau^e(U, \Omega_\Gamma) = \tau^e(U, \Omega) \in \mathcal{T}^e_{\mathcal{V}}(\Sigma, \lambda)$. Therefore the module $\mathcal{T}^e_{\mathcal{V}}(\Sigma, \lambda)$ is generated by elements $\tau^e_{\mathcal{V}}(U, \Omega)$ corresponding to v-colored ribbon graphs $\Omega \subset \mathrm{Int}(U)$ obtained from v-colored framed graphs in $\mathrm{Int}(U)$ by ribboning. A similar assertion holds for $\overline{\mathcal{V}}$. This implies that the module $\mathcal{T}_1^s(\Sigma, \lambda)$ given by (8.2.a) is generated by elements $\tau_1^s(U, \Gamma_1, \Gamma_2)$ where Γ_1, Γ_2 run independently over v-colored framed graphs in $\mathrm{Int}(U)$. Therefore the TQFT $(\mathcal{T}_1^s, \tau_1^s)$ is non-degenerate. Non-degeneracy of $(\mathcal{T}_1^s, \tau_1^s)$ implies non-degeneracy of $(\mathcal{T}_1, \tau_1)$.

8.3. The TQFT $(\mathcal{T}_2, \tau_2)$. The underlying cobordism theory of $(\mathcal{T}_2, \tau_2)$ is the cobordism theory $(\mathfrak{B}, \mathfrak{A})$ introduced at the beginning of Section 8.2. To construct $(\mathcal{T}_2, \tau_2)$ we employ the method of Section VII.3. Namely, we first construct a TQFT $(\mathcal{T}_2^s, \tau_2^s)$ based on the space-structures $(\mathfrak{B}, \mathfrak{A})$ where, in addition, all surfaces are triangulated. Then we dispense with triangulations.

Let M be a compact oriented 3-cobordism with triangulated bases $\partial_- M, \partial_+ M$. Let $\Gamma_1, \Gamma_2 \subset \mathrm{Int}(M)$ be v-colored framed graphs in $\mathrm{Int}(M)$. Recall that to any triangulated compact surface Σ, we associate a projective K-module

$$\tilde{\mathcal{E}}(\Sigma) = \bigoplus_{\psi \in \mathrm{col}(\Sigma)} H_\psi(\Sigma), \tag{8.3.a}$$

see Section VII.1.5. We shall define a K-homomorphism

$$\tilde{e}(M, \Gamma_1, \Gamma_2) : \tilde{\mathcal{E}}(\partial_- M) \to \tilde{\mathcal{E}}(\partial_+ M). \tag{8.3.b}$$

With respect to the splitting (8.3.a) this homomorphism is presented by a block-matrix

$$\{e_\psi^\varphi(M, \Gamma_1, \Gamma_2) : H_\psi(\partial_- M) \to H_\varphi(\partial_+ M)\}_{\varphi,\psi}$$

where ψ runs over colorings of $\partial_- M$ and φ runs over colorings of $\partial_+ M$.

Consider the trivalent graphs $\gamma_- \subset \partial_- M$ and $\gamma_+ \subset \partial_+ M$ dual to the given triangulations. As in Section 7.2, regular neighborhoods of these graphs in ∂M give rise to framed graphs in the double $\overline{M} = M \cup_{\partial M} (-M)$ of M. The colorings ψ and φ induce colorings $\psi^\perp$ and $\varphi^\perp$ of these framed graphs. Denote the resulting colored framed graphs in $\overline{M}$ by Γ_-^ψ and Γ_+^φ respectively. We consider four disjoint colored framed graphs in $\overline{M}$:

$$(\Gamma_-^\psi)^* = \Gamma_-^{\psi^*} \subset \partial M, \quad \Gamma_+^\varphi \subset \partial M, \quad \Gamma_1^* \subset M, \quad \Gamma_2^* \subset -M.$$

Note that Γ_1^*, Γ_2^* lie in two different copies of M inside $\overline{M}$ and that $\Gamma_-^{\psi^*}, \Gamma_+^\varphi$ lie in the common part ∂M of these copies of M. Denote the union of these four colored framed graphs in $\overline{M}$ by Γ. Consider the tensor

$$\tau^0(\overline{M}, \Gamma) \in H(\Gamma) = H(\Gamma_-^{\psi^*}) \otimes_K H(\Gamma_+^\varphi) \otimes_K H(\Gamma_1^*) \otimes_K H(\Gamma_2^*).$$

(For the definition of τ^0, see Section 7.1.)

Denote by c the number of 2-simplices in the given triangulation of ∂M (or, what is the same, the number of vertices of $\gamma_- \amalg \gamma_+$). The given coloring of the 2-disks of Γ_1 and Γ_2 gives rise via the tensor product to elements of $H(\Gamma_1)$ and

$H(\Gamma_2)$. Coupling

$$\mathcal{D}^{2b_2(M)-2b_3(M)-c}\,(\dim'(\Gamma_1)\dim'(\Gamma_2))^{-1}\,\tau^0(\overline{M},\Gamma)\in H(\Gamma)$$

with these two elements we get an element of the module

$$H(\Gamma_-^{\psi^*})\otimes_K H(\Gamma_+^{\varphi})=(H_\psi(\partial_- M))^\star\otimes_K H_\varphi(\partial_+ M).$$

The corresponding K-linear homomorphism $H_\psi(\partial_- M)\to H_\varphi(\partial_+ M)$ is denoted by $e_\psi^\varphi(M,\Gamma_1,\Gamma_2)$. Varying φ and ψ we get a block-matrix of the homomorphism (8.3.b).

It is obvious that the homomorphism (8.3.b) is multiplicative with respect to disjoint union of 3-cobordisms and natural with respect to homeomorphisms of cobordisms preserving the orientation, the triangulation of the boundary, and the given v-colored framed graphs. The following lemma establishes functoriality of the homomorphism (8.3.b).

8.3.1. Lemma. *Let Σ be a triangulated closed surface splitting an oriented compact 3-cobordism M with triangulated bases $\partial_- M,\partial_+ M$ into two cobordisms $(M_1,\partial_- M,\Sigma)$ and $(M_2,\Sigma,\partial_+ M)$. Let $\Gamma_{1,r},\Gamma_{2,r}\subset \mathrm{Int}(M_r)$ be v-colored framed graphs (without free ends) in M_r where $r=1,2$. Set $\Gamma_1=\Gamma_{1,1}\cup\Gamma_{1,2}\subset M$ and $\Gamma_2=\Gamma_{2,1}\cup\Gamma_{2,2}\subset M$. Then*

$$\tilde e(M,\Gamma_1,\Gamma_2)=\tilde e(M_2,\Gamma_{1,2},\Gamma_{2,2})\,\tilde e(M_1,\Gamma_{1,1},\Gamma_{2,1}).$$

This lemma allows us to reproduce the constructions of Sections VII.3.2 - VII.3.4 in this setting. (Here, however, we consider oriented surfaces and oriented 3-cobordisms with a pair of v-colored framed graphs inside.) This gives a TQFT $(\mathcal{T}_2,\tau_2)$ based on oriented closed (non-triangulated) surfaces and oriented compact 3-cobordisms endowed with a pair of embedded v-colored framed graphs.

Proof of Lemma. Let us first consider the case where $\Gamma_{1,r}=\Gamma_{2,r}=\emptyset$ for both $r=1$ and $r=2$. (We shall drop these empty graphs from the notation for the operator invariants of 3-cobordisms.)

Consider the trivalent graphs $\gamma_-\subset\partial_- M,\gamma_+\subset\partial_+ M,\gamma_0\subset\Sigma$ dual to the given triangulations of the surfaces $\partial_- M,\partial_+ M,\Sigma$. Their regular neighborhoods $\Gamma_-,\Gamma_+,\Gamma_0$ in these surfaces can be viewed as framed graphs. Any colorings $\psi\in\mathrm{col}(\partial_- M),\varphi\in\mathrm{col}(\partial_+ M),\nu\in\mathrm{col}(\Sigma)$ induce colorings $\psi^\perp,\varphi^\perp$, and $\nu^\perp$ of Γ_-,Γ_+, and Γ_0 respectively. Denote the resulting colored framed graphs by $\Gamma_-^\psi,\Gamma_+^\varphi$, and Γ_0^ν. Denote by c_-,c_+, and c_0 the number of vertices of $\partial_- M,\partial_+ M$, and Σ respectively.

Fix a coloring ψ of $\partial_- M$ and a coloring φ of $\partial_+ M$. We must show that

$$e_\psi^\varphi(M) = \sum_{\nu \in \mathrm{col}(\Sigma)} e_\nu^\varphi(M_2)\, e_\psi^\nu(M_1). \tag{8.3.c}$$

By definition,

$$e_\psi^\varphi(M) = \mathscr{D}^{2b_2(M)-2b_3(M)-c_--c_+}\, \tau^0(\overline{M}, \Gamma_-^{\psi^*} \cup \Gamma_+^\varphi)$$

is an element of the module

$$H(\Gamma_-^{\psi^*}) \otimes_K H(\Gamma_+^\varphi) = \mathrm{Hom}_K(H_\psi(\partial_- M), H_\varphi(\partial_+ M)).$$

Similar formulas hold for $e_\psi^\nu(M_1)$ and $e_\nu^\varphi(M_2)$. Therefore formula (8.3.c) may be rewritten as follows:

$$\tau^0(\overline{M}, \Gamma_-^{\psi^*} \cup \Gamma_+^\varphi) = \tag{8.3.d}$$

$$= \mathscr{D}^C \sum_{\nu \in \mathrm{col}(\Sigma)} \mathrm{cntr}\left(\tau^0(\overline{M_1}, \Gamma_-^{\psi^*} \cup \Gamma_0^\nu) \otimes \tau^0(\overline{M_2}, \Gamma_0^{\nu^*} \cup \Gamma_+^\varphi)\right)$$

where

$$C = 2\,(b_2(M_1) - b_3(M_1) + b_2(M_2) - b_3(M_2) - b_2(M) + b_3(M) - c_0)$$

and the symbol cntr denotes the tensor contraction induced by the duality pairing $H(\Gamma_0^\nu) \otimes_K H(\Gamma_0^{\nu^*}) \to K$.

To compute the invariants τ^0 which appear in this formula we should use 4-manifolds bounded by $\overline{M}$, $\overline{M_1}$, and $\overline{M_2}$. For these manifolds we take

$$W = M \times [-1, 1], \quad W_1 = M_1 \times [-1, 1], \quad W_2 = M_2 \times [-1, 1].$$

It is clear that W may be obtained from W_1 and W_2 by gluing the copies of $\Sigma \times [0, 1]$ contained in $\partial(W_1)$ and $\partial(W_2)$.

Let X_1 be a skeleton of the 4-dimensional graph pair $(W_1, \Gamma_- \cup \Gamma_0 \subset \overline{M_1} = \partial(W_1))$ (see Section IX.8.1) so that $\partial(X_1) = \gamma_- \cup \gamma_0$. We assume that the closure of every region of X_1 contains at most one edge of γ_0. This may be ensured by applying suspension and the shadow move P_2 to X_1 inside W_1. Let X_2 be an arbitrary skeleton of the 4-dimensional graph pair $(W_2, \Gamma_0 \cup \Gamma_+ \subset \overline{M_2} = \partial(W_2))$. We have $\partial(X_2) = \gamma_0 \cup \gamma_+$ and $X_1 \cap X_2 = \gamma_0$. We claim that the 2-polyhedron $X = X_1 \cup X_2 = X_1 \cup_{\gamma_0} X_2 \subset W$ is a skeleton of the 4-dimensional graph pair $(W, \Gamma_- \cup \Gamma_+ \subset \overline{M} = \partial W)$. It is obvious that X is a locally flat simple 2-polyhedron with $\partial X = \gamma_- \cup \gamma_+$. The assumption on X_1 excludes Möbius bands in $\mathrm{Int}(X)$ and guarantees that the polyhedron X is orientable, together with X_1 and X_2. Note finally that W may be obtained from any closed regular neighborhood of X in W by attaching several handles of indices 3 and 4. We first attach 3-handles whose cores are disks of the cell subdivision of Σ dual to the given triangulation. (This is possible because these disks are properly embedded in (W, X).) These 3-handles together with the closed regular neighborhood of X split W into two pieces, the r-th piece being the complement of a regular neighborhood of X_r in W_r for $r = 1, 2$. Since X_r is a skeleton of W_r we may fill in this complement with 3-handles and 4-handles attached to the regular neighborhood of X_r.

By definition,

$$\tau^0(\overline{M}, \Gamma_-^{\psi^*} \cup \Gamma_+^{\varphi}) = ||X|| = |X| \in H(\gamma_-, (\psi^\perp)^*) \otimes_K H(\gamma_+, \varphi^\perp).$$

Similarly,

$$\tau^0(\overline{M}_1, \Gamma_-^{\psi^*} \cup \Gamma_0^{\nu}) = |X_1| \in H(\gamma_-, (\psi^\perp)^*) \otimes_K H(\gamma_0, \nu^\perp)$$

and

$$\tau^0(\overline{M}_2, \Gamma_0^{\nu^*} \cup \Gamma_+^{\varphi}) = |X_2| \in H(\gamma_0, (\nu^\perp)^*) \otimes_K H(\gamma_+, \varphi^\perp).$$

To prove (8.3.d) we may directly apply Lemma 1.4.1 to the shadowed polyhedra X_1, X_2, X. We need only to show that the number C in (8.3.d) coincides with the number c in (1.4.a). Since $b_3(X) = b_2(W, X) = 0$ and $b_i(W) = b_i(M)$ for all i, the exact homology sequence of the pair (W, X) yields

$$b_2(X) = b_2(M) - b_3(M) + b_3(W, X)$$

where $b_3(W, X) = \dim H_3(W, X; \mathbb{R})$. Theorem 8.1.3 and the fact that the intersection form in $H_2(W)$ is zero imply that the bilinear form Q_X in $H_2(X)$ is zero. Hence $\mathrm{null}(X) = b_2(X)$ and

$$\mathrm{null}(X) + b_2(X) = 2\, b_2(X) = 2\,(b_2(M) - b_3(M) + b_3(W, X)).$$

Similarly, for $r = 1, 2$, we have

$$\mathrm{null}(X_r) + b_2(X_r) = 2\,(b_2(M_r) - b_3(M_r) + b_3(W_r, X_r)).$$

Substituting these expressions in the formula for c we get

$$c - C = 2\,(b_3(W_1, X_1) + b_3(W_2, X_2) - b_3(W, X) + c_0).$$

The right-hand side of this expression equals 0 due to the Mayer-Vietoris homology sequence of the triple of pairs $((W, X), (W_1, X_1), (W_2, X_2))$ and the formulas

$$(W_1, X_1) \cap (W_2, X_2) = (\Sigma \times [-1, 1], \gamma_0 \times 0),$$

$$b_2(\Sigma \times [-1, 1], \gamma_0 \times 0) = b_2(\Sigma, \gamma_0) = c_0,$$

$$b_3(\Sigma \times [-1, 1], \gamma_0 \times 0) = b_2(W_1, X_1) = b_2(W_2, X_2) = 0.$$

In the case where the framed graphs Γ_1, Γ_2 are non-empty, the proof is similar, the only difference being that instead of the 4-dimensional graph pairs

$$(M_1 \times [-1, 1], \Gamma_- \cup \Gamma_0), \quad (M_2 \times [-1, 1], \Gamma_0 \cup \Gamma_+), \quad (M \times [-1, 1], \Gamma_- \cup \Gamma_+)$$

(where $\Gamma_- = \Gamma_- \times 0$, etc.) we consider the graph pairs

$$(M_1 \times [-1, 1],\ \Gamma_- \cup \Gamma_0 \cup (\Gamma_{1,1}^* \times (-1)) \cup (\Gamma_{2,1}^* \times 1)), \tag{8.3.e}$$

(8.3.f) $$(M_2 \times [-1,1],\ \Gamma_0 \cup \Gamma_+ \cup (\Gamma^*_{1,2} \times (-1)) \cup (\Gamma^*_{2,2} \times 1)),$$

(8.3.g) $$(M \times [-1,1],\ \Gamma_- \cup \Gamma_+ \cup (\Gamma^*_1 \times (-1)) \cup (\Gamma^*_2 \times 1)).$$

8.4. Remark. Homological computations in the proof of Lemma 8.3 become simpler if we choose X_1, X_2 to be the skeletons obtained from triangulations of M_1, M_2 as in Section 7.2.1. In the case of non-empty Γ_1, Γ_2 a derivation of skeletons of graph pairs (8.3.e) - (8.3.g) from triangulations of M_1, M_2 is more involved, this is why we prefer to deal in this proof with arbitrary skeletons.

8.5. Lemma. *If the ground ring K of the category $\mathcal{V}$ is a field of zero characteristic then the TQFT's $(\mathcal{T}_1, \tau_1)$ and $(\mathcal{T}_2, \tau_2)$ are isomorphic.*

Proof. These two TQFT's are based on the same cobordism theory. Let us show that they coincide on closed spaces. Let M be a closed oriented 3-manifold with v-colored framed graphs $\Gamma_1, \Gamma_2 \subset M$. By definition,

$$\tau_1(M, \Gamma_1, \Gamma_2) = \tau_{\mathcal{V}}(M, \Gamma_1)\, \tau_{\overline{\mathcal{V}}}(M, \Gamma_2).$$

It follows from the equality $\overline{M} = M \amalg (-M)$ and the multiplicativity of τ^0 with respect to disjoint union that

$$\tau_2(M, \Gamma_1, \Gamma_2) = \mathcal{D}^{2(b_2(M)-b_3(M))}\, (\dim'(\Gamma_1)\dim'(\Gamma_2))^{-1}\, \tau^0_{\mathcal{V}}(M, \Gamma^*_1)\, \tau^0_{\mathcal{V}}(-M, \Gamma^*_2).$$

Theorem 7.1.1 and the result of Exercise VI.4.8 imply that

$$\tau_2(M, \Gamma_1, \Gamma_2) = \tau_{\mathcal{V}}(M, \Gamma_1)\, \tau_{\mathcal{V}}(-M, \Gamma^*_2) = \tau_{\mathcal{V}}(M, \Gamma_1)\, \tau_{\overline{\mathcal{V}}}(M, \Gamma_2).$$

Therefore

$$\tau_1(M, \Gamma_1, \Gamma_2) = \tau_2(M, \Gamma_1, \Gamma_2).$$

By Lemma 8.2.1, the TQFT $(\mathcal{T}_1, \tau_1)$ is non-degenerate. Theorem III.3.7 implies that the TQFT's $(\mathcal{T}_1, \tau_1)$ and $(\mathcal{T}_2, \tau_2)$ are isomorphic.

8.6. Proof of Theorem VII.4.2 completed. It follows directly from definitions that restricting the TQFT $(\mathcal{T}_1, \tau_1)$ to 3-cobordisms with empty framed graphs we get the TQFT $(\widehat{\mathcal{T}}, \widehat{\tau})$. Theorem 7.2.1 implies that restricting the TQFT $(\mathcal{T}_2, \tau_2)$ to 3-cobordisms with empty framed graphs we get the TQFT $(\mathcal{E}, |...|)$. Therefore Theorem VII.4.2 follows from Lemma 8.5.

9. Computations for graph manifolds

9.0. Outline. The state sum on a shadowed 2-polyhedron becomes considerably simpler when the polyhedron has no vertices. Such 2-polyhedra arise as shadows of 4-manifolds bounded by so-called graph 3-manifolds. This enables us to compute explicitly the invariant $\tau = \tau_{(\mathcal{V},\mathcal{D})}$ for these 3-manifolds.

9.1. Invariants of fiber bundles over surfaces. The simplest 3-dimensional graph manifolds are circle bundles over oriented surfaces. In this case the technique of shadows is especially efficient.

Let M be the total space of an oriented circle bundle over a closed connected oriented surface Σ. Denote the genus of Σ by g and the Euler number of the fibration $M \to \Sigma$ by n. The manifold M bounds the total space W_n of the 2-disk fibration over Σ where the self-intersection index of the zero section $\Sigma \subset W_n$ is equal to n. We have $\mathrm{sh}(W_n) = \mathrm{stab}([\Sigma_n])$. By Poincaré duality $b_3(M) = b_0(M) = 1$ and $b_2(M) = b_1(M)$. By Theorem 3.3

$$\tau(M) = \mathcal{D}^{b_1(M)-1}\tau^0(M) = \mathcal{D}^{b_1(M)-1} \left|\left| [\Sigma_n] \right|\right| = \Delta^{\sigma(W_n)}\mathcal{D}^{b_1(M)-1-\sigma(W_n)} \left|\Sigma_n\right|$$

where $\sigma(W_n)$ is the signature of W_n. The computation of $|\Sigma_n|$ is straightforward, see Section 1.3. We distinguish three cases. If $n = 0$, i.e., if $M = \Sigma \times S^1$ then $\sigma(W_n) = 0$, $b_1(M) = 2g + 1$, and

$$\tau(M) = \mathcal{D}^{2g-2} \sum_{i \in I} (\dim(i))^{2-2g}.$$

This yields another computation of $\tau(\Sigma \times S^1)$ (cf. Section IV.7.2). Together with Theorem III.2.1.3 this gives another proof of the Verlinde formula for the dimension of the module of states of Σ.

If $n > 0$ then $\sigma(W_n) = 1$, $b_1(M) = 2g$, and

$$\tau(M) = \Delta\mathcal{D}^{2g-3} \sum_{i \in I} v_i^n (\dim(i))^{2-2g}.$$

If $n < 0$ then $\sigma(W_n) = -1$, $b_1(M) = 2g$, and

$$\tau(M) = \Delta^{-1}\mathcal{D}^{2g-1} \sum_{i \in I} v_i^n (\dim(i))^{2-2g}.$$

In the case $\Sigma = S^2$ this gives another computation of the invariant τ for $S^1 \times S^2$ and lens spaces of type $(n, 1)$ with $n \geq 2$ (cf. Section II.2.2).

9.2. Graph manifolds. A 3-dimensional graph manifold is determined by a symmetric square matrix $A = (a_{p,q})_{p,q=1}^m$ over $\mathbb{Z}$ and a sequence of non-negative integers $g_1, \ldots, g_m$. It is constructed as follows. For $p = 1, \ldots, m$, denote by Σ_p a closed connected oriented surface of genus g_p. Let W_p be the total space of a 2-disk fibration over Σ_p with the self-intersection index of $\Sigma_p \subset W_p$ equal to $\pm a_{p,p}$. Orient W_p so that this intersection index equals $a_{p,p}$. For each pair $p, q \in \{1, \ldots, m\}$ with $p \neq q$, glue W_p to W_q along $|a_{p,q}|$ balls. More exactly, choose $|a_{p,q}|$ disjoint small 2-disks in Σ_p and identify the part of W_p lying over each of these disks with $B^2 \times B^2$, where B^2 is a standard 2-disk and the fibers of W_p

have the form $B^2 \times \{pt\}$. Assume that under these identifications the orientation of W_p corresponds to the product orientation in $B^2 \times B^2$, and the orientation of Σ_p corresponds to some given orientation in $B^2 = \{pt\} \times B^2$. Similarly, choose $|a_{p,q}|$ disjoint small 2-disks in Σ_q and identify the part of W_q lying over each of these disks with $B^2 \times B^2$ where the fibers of W_q have the form $B^2 \times \{pt\}$ and the orientation of W_p corresponds to the product orientation in $B^2 \times B^2$. Assume that the orientation of Σ_p corresponds to the given orientation in $B^2 = \{pt\} \times B^2$ if $a_{p,q} \geq 0$ and to the opposite orientation if $a_{p,q} < 0$. We glue W_p to W_q along the homeomorphism of the specified $|a_{p,q}|$ balls $B^2 \times B^2 \subset W_p$ onto the specified $|a_{p,q}|$ balls $B^2 \times B^2 \subset W_q$ induced by the permutation $(x, y) \mapsto (y, x) : B^2 \times B^2 \to B^2 \times B^2$. We apply such gluings corresponding to all non-ordered pairs (p, q) to the disjoint union $\amalg_{p=1}^m W_p$. It is understood that the gluings corresponding to different pairs (p, q) are performed along disjoint copies of $B^2 \times B^2$. This results in a compact oriented (possibly, non-connected) 4-manifold W. It is obvious that the intersection form in $H_2(W; \mathbb{Z}) = \mathbb{Z}^m$ is given by the matrix A.

The 3-manifold $M = \partial W$ is called the 3-dimensional graph manifold determined by $A = (a_{p,q})_{p,q=1}^m$ and $g_1, ..., g_m$. It is clear that if A is a direct sum of two square matrices then M is the disjoint union of two corresponding 3-dimensional graph manifolds. If A is a 1×1-matrix then M is a circle bundle over a closed connected orientable surface of genus g_1.

9.3. Computation of $\tau(M)$. Let M be the 3-dimensional graph manifold determined by a symmetric integer square matrix $A = (a_{p,q})_{p,q=1}^m$ and non-negative integers $g_1, ..., g_m$. Denote by $\sigma(A)$ and $\mathrm{null}(A)$ the signature and the nullity of the symmetric bilinear form in $\mathbb{R}^m$ determined by A. For each $p \in I$, set

$$a_p = \sum_{q \neq p} |a_{p,q}| \in \mathbb{Z}.$$

Recall the matrix $S = \{S_{i,j}\}_{i,j \in I}$ involved in the definition of the modular category $\mathcal{V}$. For each mapping $\varphi : \{1, ..., m\} \to I$ and each pair $p, q = 1, ..., m$, set $s^{\varphi}_{p,q} = S_{\varphi(p),\varphi(q)}$ if $a_{p,q} \geq 0$ and $s^{\varphi}_{p,q} = S_{\varphi(p)^*,\varphi(q)}$ if $a_{p,q} < 0$. It follows from the properties of S that the matrix $\{s^{\varphi}_{p,q}\}_{p,q=1}^m$ is symmetric. Denote the set of mappings $\{1, ..., m\} \to I$ by I^m.

9.3.1. Theorem.

$$\tau(M) = \Delta^{\sigma(A)} \mathcal{D}^b \sum_{\varphi \in I^m} \left(\prod_{p=1}^m v_{\varphi(p)}^{a_{p,p}} (\dim(\varphi(p)))^{2-2g_p-a_p} \prod_{p<q} (s^{\varphi}_{p,q})^{|a_{p,q}|} \right)$$

where $b = b_1(M) - b_0(M) - m - \mathrm{null}(A) - \sigma(A)$.

Proof. We first describe the shadow of the 4-manifold W constructed in Section 9.2. For each $p \in \{1, ..., m\}$, the surface Σ_p embeds in W as the zero section of the corresponding 2-disk bundle. For any pair $p, q \in \{1, ..., m\}$ with $p \neq q$, the surfaces Σ_p and Σ_q intersect each other transversally in $|a_{p,q}|$ points with the sign of every intersection being 1 if $a_{p,q} \geq 0$ and -1 if $a_{p,q} < 0$. The union $\cup_p \Sigma_p$ of these surfaces is a 2-polyhedron lying in W as a deformation retract. This 2-polyhedron is not simple in the sense of Section VIII.1.1 because the crossing points of surfaces do not have neighborhoods required by the definition of simple 2-polyhedra. To construct a skeleton of W, we modify $\cup_p \Sigma_p$ as follows.

Let U be a closed regular neighborhood of the set of double points of $\cup_p \Sigma_p$ in W. This neighborhood consists of a disjoint closed 4-balls where

$$a = \sum_{q=1}^{m} a_p/2 = \sum_{p<q} |a_{p,q}|$$

is the number of double points of $\cup_p \Sigma_p$. It is clear that $Z = (\cup_p \Sigma_p) \setminus \mathrm{Int}(U)$ is a compact oriented surface consisting of m connected components where the p-th component is the surface Σ_p punctured a_p times. We provide the p-th component of Z with the gleam $a_{p,p}$ and regard Z as a shadowed polyhedron with boundary.

A connected component of U, containing a double point of $\cup_p \Sigma_p$, is a 4-ball $B^4 \subset W$ whose boundary 3-sphere intersects Z along two components of ∂Z. These two circles form the Hopf link in $S^3 = \partial B^4$. We provide both components of this link with zero framing as in Figure I.2.12. Consider the shadow projection of this link into the skeleton $S^2 \subset S^3$ and denote by Y the shadow cylinder over this projection. Glue Y to Z along a homeomorphism of $\partial Y = S^1 \amalg S^1$ onto the two components of ∂Z in question. Performing such gluings for all double points of $\cup_p \Sigma_p$, we get a shadowed polyhedron X over $(1/2)\mathbb{Z}$. Clearly, $X = Z \cup \overline{(X \setminus Z)}$ where $\overline{(X \setminus Z)}$ consists of a disjoint copies of Y. The embedding $Z \subset W \setminus U$ extends to an embedding $X \subset W$. It is easy to verify that the shadowed polyhedron X represents $\mathrm{sh}(W)$, the key observation is that the shadow cylinder Y represents the shadow of the 4-dimensional graph pair (B^4, the Hopf link in ∂B^4). Note that $\mathrm{null}(Y) = b_2(Y) = 1$.

By Theorem 3.3

$$\tau(M) = \mathcal{D}^{b_1(M)-b_0(M)}\, \tau^0(M) =$$

$$= \mathcal{D}^{b_1(M)-b_0(M)}\, ||X|| = \Delta^{\sigma(A)} \mathcal{D}^{b_1(M)-b_0(M)-\sigma(A)}\, |X|\,.$$

It remains to compute $|X|$. We first compute the state sum invariant of Y corresponding to a coloring of $\partial Y = S^1 \amalg S^1 \subset S^3$. Orient these two circles so that their linking number in S^3 is equal to 1. Assigning to these oriented circles certain $i, j \in I$ we obtain a coloring of ∂Y denoted by $\lambda(i, j)$. By Lemma 4.1

$$|Y, \lambda(i, j)| = S_{i^*, j^*} = S_{i,j} \in H(Y, \lambda(i, j)) = K.$$

We shall use the following equivalent formula

$$\text{(9.3.a)} \qquad \sum_{\psi\in \mathrm{col}(Y),\, \partial\psi=\lambda(i,j)} |Y|_\psi = \mathcal{D}^{b_2(Y)+\mathrm{null}(Y)}\, |Y, \lambda(i,j)| = \mathcal{D}^2 S_{i,j}.$$

Using the given orientations of the surfaces $\Sigma_1, \ldots, \Sigma_m$ we identify the set of colorings of Z with I^m. Recall that the polyhedron X consists of Z and a copies of Y glued to Z along their boundaries. A coloring $\varphi \in I^m$ of Z restricts to a coloring of ∂Z and induces a coloring of the boundaries of these copies of Y. More exactly, consider a copy of Y corresponding to an intersection point of Σ_p and Σ_q in W where $p < q$. If $a_{p,q} > 0$ then the orientations of Σ_p, Σ_q induce an orientation in ∂Y such that the linking number of the components of ∂Y in S^3 is equal to 1. Therefore φ induces the coloring $\lambda(\varphi(p), \varphi(q))$ of ∂Y. If $a_{p,q} < 0$ then the orientations of Σ_p, Σ_q induce an orientation in ∂Y such that the linking number of the components of ∂Y in S^3 is equal to -1. Therefore φ induces the coloring $\lambda((\varphi(p))^*, \varphi(q))$ of ∂Y. These facts and formula (9.3.a) allow us to compute the sum $\sum_\psi |X|_\psi$ where ψ runs over colorings of X extending φ. Namely,

$$\sum_{\psi\in \mathrm{col}(X),\, \psi|_Z=\varphi} |X|_\psi = |Z|_\varphi \prod_{p<q} (\mathcal{D}^2 s^\varphi_{p,q})^{|a_{p,q}|} =$$

$$= \mathcal{D}^{2a} \prod_{p=1}^{m} v_{\varphi(p)}^{a_{p,p}} (\dim(\varphi(p)))^{2-2g_p-a_p} \prod_{p<q} (s^\varphi_{p,q})^{|a_{p,q}|}.$$

It is clear that

$$|X| = \mathcal{D}^{-b_2(X)-\mathrm{null}(X)} \sum_{\varphi\in I^m} \sum_{\psi\in \mathrm{col}(X),\, \psi|_Z=\varphi} |X|_\psi .$$

We have $b_2(X) = m + a$ and $\mathrm{null}(X) = \mathrm{null}(A) + a$. Combining these formulas we get the claim of the theorem.

9.4. Estimates in the unitary case. Assume that the modular category $\mathcal{V}$ is unitary and $\mathcal{D} > 0$. We may use Theorem 9.3.1 to estimate from above the absolute value of $\tau(M)$ for the graph manifold M, determined by a symmetric integer square matrix $A = (a_{p,q})_{p,q=1}^m$ and non-negative integers $g_1, \ldots, g_m$. Using the equalities $|\mathcal{D}\Delta^{-1}| = |v_i| = 1$ and the inequality $|S_{i,j}| \leq \dim(i)\dim(j)$ for any $i, j \in I$, we get

$$|\tau(M)| \leq \mathcal{D}^{b_1(M)-b_0(M)-m-\mathrm{null}(A)} \prod_{p=1}^{m} \left(\sum_{i\in I} (\dim(i))^{2-2g_p} \right).$$

Notes

Section 1. The $6j$-symbols were first used in connection with link invariants by Kirillov and Reshetikhin [KR1]. They introduced a state sum model on link diagrams in $\mathbb{R}^2$ computing the Jones polynomial of links in $\mathbb{R}^3$ via the numerical $6j$-symbols associated to $U_q(sl_2(\mathbb{C}))$. In [Tu10] it was shown that the height function may be eliminated and the information on diagram crossings may be replaced with gleams. This yields a state sum model on shadow links which was later on generalized to arbitrary shadows (see [Tu9]). Here we give a more general formulation based on an arbitrary unimodular category.

Sections 2 and 3. The results of these sections generalize the results of [Tu9].

Sections 4 - 6. Lemma 4.1 and Theorem 5.6 formalize and generalize the results of Kirillov and Reshetikhin [KR1] who computed the $U_q(sl_2(\mathbb{C}))$-invariants of framed graphs in $\mathbb{R}^3$ in terms of $6j$-symbols.

Sections 7 - 9. The results of these sections are new.

Part III

Towards Modular Categories

Chapter XI
An algebraic construction of modular categories

1. Hopf algebras and categories of representations

1.0. Outline. A rich source of monoidal categories is provided by the theory of Hopf algebras. Indeed, the category of finite dimensional modules over a Hopf algebra has the natural structure of a monoidal category. In this sense monoidal categories are dual to Hopf algebras. Here we recall basic definitions of the theory of Hopf algebras and discuss the associated monoidal categories.

1.1. Definition of Hopf algebras. Let A be an algebra with unit 1_A over a commutative ring with unit, K. Assume that A is provided with multiplicative K-linear homomorphisms $\Delta : A \to A^{\otimes 2} = A \otimes_K A$ and $\varepsilon : A \to K$, called the comultiplication and the counit respectively, and a K-linear homomorphism $s : A \to A$, called the antipode. It is understood that $\Delta(1_A) = 1_A \otimes 1_A$ and $\varepsilon(1_A) = 1$. We say that $(A, \Delta, \varepsilon, s)$ or, briefly, A is a Hopf algebra if these homomorphisms satisfy together with the multiplication $m : A \times A \to A$ the following identities:

(1.1.a) $$(\mathrm{id}_A \otimes \Delta)\Delta = (\Delta \otimes \mathrm{id}_A)\Delta,$$

(1.1.b) $$m(s \otimes \mathrm{id}_A)\Delta = m(\mathrm{id}_A \otimes s)\Delta = \varepsilon \cdot 1_A,$$

(1.1.c) $$(\varepsilon \otimes \mathrm{id}_A)\Delta = (\mathrm{id}_A \otimes \varepsilon)\Delta = \mathrm{id}_A.$$

Note that to write down (1.1.a) we identify $(A \otimes A) \otimes A = A \otimes (A \otimes A)$ via $(a \otimes b) \otimes c = a \otimes (b \otimes c)$ where $a, b, c \in A$. Similarly, to write down (1.1.c) we identify $K \otimes A = A \otimes K = A$ via $1 \otimes a = a \otimes 1 = a$. These identifications are incorporated in the theory to follow.

The axioms imply that the antipode s is an antiautomorphism of both the algebra and the coalgebra structures in A (see [Ab], [Sw]). This means that

$$m(s \otimes s) = s \circ m \circ P_A : A^{\otimes 2} \to A, \quad P_A(s \otimes s)\Delta = \Delta \circ s : A \to A^{\otimes 2}$$

where P_A denotes the flip $a \otimes b \mapsto b \otimes a : A^{\otimes 2} \to A^{\otimes 2}$. It also follows from the axioms that $s(1_A) = 1_A$ and $\varepsilon \circ s = \varepsilon : A \to K$.

If the Hopf algebra A viewed as a K-module is projective (of finite type) then we can define a Hopf algebra structure in the linear dual $A^* = \mathrm{Hom}_K(A, K)$. The multiplication, comultiplication, and antipode in A^* are dual to the comultiplication, multiplication, and antipode in A. The roles of unit and counit in A^* are played by the counit of A and the homomorphism $y \mapsto y(1_A) : A^* \to K$ respectively.

1.2. Examples. 1. A simple example of a Hopf algebra is provided by the group ring $K[G]$ of a group G. The homomorphisms Δ, ε, s are defined on the additive generators $g \in G$ by the formulas $\Delta(g) = g \otimes g$, $\varepsilon(g) = 1$, and $s(g) = g^{-1}$.

2. Let G be a finite group. We define a Hopf algebra A as follows. As a K-module this algebra is freely generated by the set $\{\delta_g\}_{g\in G}$. Multiplication in A is defined by the formulas $\delta_g\delta_h = 0$ if $g \neq h$ and $\delta_g\delta_h = \delta_g$ if $g = h$. The comultiplication $\Delta : A \to A^{\otimes 2}$ and the antipode $s : A \to A$ are defined by

$$\Delta(\delta_g) = \sum_{h\in G} \delta_h \otimes \delta_{h^{-1}g} \quad \text{and} \quad s(\delta_g) = \delta_{g^{-1}}.$$

The counit $\varepsilon : A \to K$ is defined by $\varepsilon(\delta_g) = 1$ if g is the unit element of G and $\varepsilon(\delta_g) = 0$ otherwise. Note that $\sum_{g\in G} \delta_g$ is the unit of A. Equalities (1.1.a) - (1.1.c) follow from definitions. The Hopf algebra A is dual to the Hopf algebra $K[G]$ of Example 1.

3. An especially important example is provided by the universal enveloping algebra $U(\mathfrak{g})$ of a Lie algebra $\mathfrak{g}$. The homomorphisms Δ, ε, s are defined on the multiplicative generators $g \in \mathfrak{g}$ by the formulas $\Delta(g) = g \otimes 1 + 1 \otimes g$, $\varepsilon(g) = 0$, and $s(g) = -g \in \mathfrak{g}$.

1.3. Category of representations. Let A be a Hopf algebra over K. By an A-module of finite rank we mean a left A-module whose underlying K-module is projective (of finite type). We consider the "category of representations" Rep(A) whose objects are A-modules of finite rank and whose morphisms are A-linear homomorphisms. We show that Rep(A) has a natural structure of a monoidal Ab-category with duality.

For objects V, W of Rep(A) set $V \otimes W = V \otimes_K W$ where the action of A is obtained from the obvious product action of $A^{\otimes 2}$ in $V \otimes_K W$ via the comultiplication $\Delta : A \to A^{\otimes 2}$. The tensor product of morphisms is the standard tensor product of homomorphisms. Associativity of the tensor product follows from (1.1.a). We note that this tensor product is not strictly associative. Indeed, the modules $(U \otimes V) \otimes W$ and $U \otimes (V \otimes W)$ are isomorphic but not identical. To be precise we should involve the associativity isomorphisms

$$(u \otimes v) \otimes w \mapsto u \otimes (v \otimes w) : (U \otimes V) \otimes W \to U \otimes (V \otimes W).$$

It is customary in linear algebra to supress these isomorphisms and to write simply $U \otimes V \otimes W$. We shall follow this tradition tacitly incorporating the associativity isomorphisms.

The ring K considered as an A-module, where the action of A is via the counit ε, is denoted by the same symbol K. It follows from (1.1.c) that K is the unit object of $\operatorname{Rep}(A)$. (As above we supress the identification isomorphisms $K \otimes V = V \otimes K = V$.) The axioms of monoidal category listed in Section I.1.1 follow directly from definitions. The usual addition of homomorphisms makes $\operatorname{Rep}(A)$ a monoidal Ab-category in the sense of Section II.1.1.

Using the antipode, s in A, we can provide $\operatorname{Rep}(A)$ with a duality. For an object V of $\operatorname{Rep}(A)$, set $V^* = \operatorname{Hom}_K(V, K)$ where the action of A is defined by $(ay)(x) = y(s(a)x)$ where $a \in A$, $x \in V$, $y \in V^*$. In other words, $ay = (\rho(s(a)))^*(y)$ where $\rho(a)$ is the homomorphism $x \mapsto ax : V \to V$. Here the star denotes the standard dualization transforming a K-homomorphism $f : U \to V$ into the homomorphism $x \mapsto fx : V^* \to U^*$.

The duality homomorphism $d_V : V^* \otimes V \to K$ is just the evaluation pairing $(y, x) \mapsto y(x)$. Set

$$b_V = (d_V)^* : K = K^* \to (V^* \otimes V)^* = V^{**} \otimes V^* = V \otimes V^*$$

where $V^{**} = V$ is the standard identification. It is clear that for any $k \in K$, we have $b_V(k) = k\delta_V$ where $\delta_V = b_V(1)$ is the canonical element of $V \otimes V^*$. This element is uniquely characterized by the following property. Expand δ_V as a finite sum

$$\delta_V = \sum_i g_i \otimes g^i \tag{1.3.a}$$

where $\{g_i\}_i$ and $\{g^i\}_i$ belong to V and V^* respectively. Then for any $x \in V$, $y \in V^*$, we have

$$y(x) = \sum_i y(g_i)\, g^i(x).$$

It is easy to verify that for any K-homomorphism $f : V \to V$, we have

$$(\mathrm{id}_V \otimes f^*)(\delta_V) = (f \otimes \mathrm{id}_{V^*})(\delta_V).$$

In the case where V is a free K-module, $\delta_V = \sum_i e_i \otimes e^i$ where $\{e_i\}_i$ is an arbitrary basis of V and $\{e^i\}_i$ is the dual basis of V^*.

1.3.1. Lemma. *The category* $\operatorname{Rep}(A)$ *is a monoidal* Ab-*category with duality.*

Proof. We have to show that the homomorphisms d_V and b_V are A-linear and satisfy the identities $(\mathrm{id}_V \otimes d_V)(b_V \otimes \mathrm{id}_V) = \mathrm{id}_V$ and $(d_V \otimes \mathrm{id}_{V^*})(\mathrm{id}_{V^*} \otimes b_V) = \mathrm{id}_{V^*}$. Both d_V and b_V are defined in terms of general linear algebra and do not depend on the A-linear structure in V. Therefore the identities in question can be verified by a direct computation.

The formula $m(s \otimes \mathrm{id}_A)\Delta = \varepsilon \cdot 1_A$ directly implies that d_V is A-linear. Let us show that b_V is A-linear. Set $\delta = b_V(1) \in V \otimes V^*$. For $a \in A$, denote the homomorphism $x \mapsto ax : V \to V$ by $\rho(a)$. Note that $\rho(1_A) = \mathrm{id}_V$.

We should show that for any $a \in A$, we have $a\delta = \varepsilon(a)\delta$. Let $\Delta(a) = \sum_j a_j \otimes b_j$ where j runs over a certain finite set of indices and $a_j, b_j \in A$. Set $b'_j = s(b_j)$. We have

$$a\delta = \Delta(a)\delta = \sum_j (a_j \otimes 1_A)(1_A \otimes b_j)\delta = \sum_j (\rho(a_j) \otimes \mathrm{id}_{V^*})(\mathrm{id}_V \otimes (\rho(b'_j))^*)(\delta) =$$

$$= \sum_j (\rho(a_j) \otimes \mathrm{id}_{V^*})(\rho(b'_j) \otimes \mathrm{id}_{V^*})(\delta) = (\rho(\sum_j a_j b'_j) \otimes \mathrm{id}_{V^*})(\delta).$$

The formula $m(\mathrm{id}_A \otimes s)\Delta = \varepsilon \cdot 1_A$ ensures that $\sum_j a_j b'_j = \varepsilon(a)1_A$. This implies that $a\delta = \varepsilon(a)\delta$.

1.4. Remark. The category $\mathrm{Rep}(A)$ is monoidal but not strict monoidal. We can easily pass from $\mathrm{Rep}(A)$ to a strict monoidal category. In fact, there is a general procedure transforming any monoidal category $\mathcal{R}$ into a strict monoidal category $\mathcal{V} = \mathcal{V}(\mathcal{R})$. The objects of $\mathcal{V}$ are finite sequences $(V_1, ..., V_k)$ of objects of $\mathcal{R}$ (including the empty sequence). The morphisms from $(V_1, ..., V_k)$ to $(W_1, ..., W_l)$ are $\mathcal{R}$-morphisms $V_1 \otimes ... \otimes V_k \to W_1 \otimes ... \otimes W_l$. The tensor product of objects of $\mathcal{V}$ is the juxtaposition of sequences, the tensor product of morphisms is obtained by the obvious application of the tensor product in $\mathcal{R}$. It is easy to check that $\mathcal{V}$ is a strict monoidal category with the unit object being the empty sequence. There is a covariant inclusion functor $\mathcal{R} \to \mathcal{V}$ assigning to any object V of $\mathcal{R}$ the 1-term sequence V and assigning to any morphism $f : V \to W$ the same morphism in $\mathcal{V}$. This inclusion is an equivalence of categories.

It may be verified that any braiding (resp. twist, duality) in $\mathcal{R}$ induces a braiding (resp. twist, duality) in $\mathcal{V}$. In particular, $(V_1, ..., V_k)^* = (V_k^*, ..., V_1^*)$. In this way any ribbon category gives rise to a strict ribbon category. The same is true for modular and semisimple categories.

2. Quasitriangular Hopf algebras

2.0. Outline. The results of Section 1 strongly suggest that we should look for ribbon categories in the class of representation categories of Hopf algebras. Here we focus our attention on braidings postponing the study of twists to Section 3. We formulate natural conditions on a Hopf algebra which ensure the existence of a braiding in its representation category. This leads to the fundamental notions of a quasitriangular Hopf algebra and a universal R-matrix.

2.1. Definition of quasitriangular Hopf algebras. The notion of a quasitriangular Hopf algebra is dual to the notion of a braided category. The role of the braiding for a Hopf algebra A is played by an element $R \in A^{\otimes 2}$.

Let $(A, \Delta, \varepsilon, s)$ be a Hopf algebra over a commutative ring with unit, K. For $a \in A$, set $\Delta'(a) = P_A(\Delta(a)) \in A^{\otimes 2}$ where $P_A = \mathrm{Perm}_A$ is the flip $a \otimes b \mapsto b \otimes a : A^{\otimes 2} \to A^{\otimes 2}$. (The homomorphism $\Delta' : A \to A^{\otimes 2}$ is called the opposite comultiplication in A.) For any $R \in A^{\otimes 2}$, set $R_{12} = R \otimes 1_A \in A^{\otimes 3}$, $R_{23} = 1_A \otimes R \in A^{\otimes 3}$, and

$$R_{13} = (\mathrm{id}_A \otimes P_A)(R_{12}) = (P_A \otimes \mathrm{id}_A)(R_{23}) \in A^{\otimes 3}.$$

Let R be an invertible element of the algebra $A^{\otimes 2}$. The pair (A, R) is called a quasitriangular Hopf algebra if for any $a \in A$,

$$\Delta'(a) = R\,\Delta(a)\,R^{-1}, \tag{2.1.a}$$

$$(\mathrm{id}_A \otimes \Delta)(R) = R_{13}R_{12}, \tag{2.1.b}$$

$$(\Delta \otimes \mathrm{id}_A)(R) = R_{13}R_{23}. \tag{2.1.c}$$

Note that on the right-hand sides of these formulas we use multiplications in $A^{\otimes 2}$, $A^{\otimes 3}$ induced by the one in A. The invertible element $R \in A^{\otimes 2}$ satisfying (2.1.a), (2.1.b), (2.1.c) is called a universal R-matrix of A.

The next lemma yields a few simple corollaries to the axioms.

2.1.1. Lemma. *Let (A, R) be a quasitriangular Hopf algebra. Then*

$$(\varepsilon \otimes \mathrm{id}_A)(R) = (\mathrm{id}_A \otimes \varepsilon)(R) = 1_A, \tag{2.1.d}$$

$$(s \otimes \mathrm{id}_A)(R) = (\mathrm{id}_A \otimes s^{-1})(R) = R^{-1}, \tag{2.1.e}$$

$$(s \otimes s)(R) = R. \tag{2.1.f}$$

Proof. Set $i = \mathrm{id}_A : A \to A$. It follows from (1.1.c) and (2.1.c) that

$$R = (\varepsilon \otimes i \otimes i)(\Delta \otimes i)(R) = (\varepsilon \otimes i \otimes i)(R_{13}R_{23}) = (1_A \otimes (\varepsilon \otimes i)(R))\,R.$$

Since R is invertible, these equalities imply that $1_A \otimes (\varepsilon \otimes i)(R) = 1_A \otimes 1_A$. Applying the algebra multiplication $m : A \otimes A \to A$ to both sides we obtain $(\varepsilon \otimes i)(R) = 1_A$. Similarly, it follows from (1.1.c) and (2.1.b) that

$$R = (i \otimes \varepsilon \otimes i)(i \otimes \Delta)(R) = (i \otimes \varepsilon \otimes i)(R_{13}R_{12}) = R\,((i \otimes \varepsilon)(R) \otimes 1_A).$$

The same argument as above shows that this implies $(i \otimes \varepsilon)(R) = 1_A$.

Set $m_{12} = m \otimes i : A^{\otimes 3} \to A^{\otimes 2}$ and $m_{23} = i \otimes m : A^{\otimes 3} \to A^{\otimes 2}$. It follows from (1.1.b) and (2.1.d) that

$$m_{12}((s \otimes i \otimes i)(\Delta \otimes i)(R)) = (\varepsilon \otimes i)(R) = 1_A.$$

On the other hand, (2.1.c) implies that

$$m_{12}((s \otimes i \otimes i)(\Delta \otimes i)(R)) = m_{12}((s \otimes i \otimes i)(R_{13}R_{23})) = (s \otimes i)(R)\,R.$$

Therefore $(s \otimes i)(R) = R^{-1}$.

It follows from (1.1.b) and the definition of Δ' that

$$m(i \otimes s^{-1})\Delta' = s^{-1} \circ (\varepsilon \cdot 1_A) = \varepsilon \cdot 1_A : A \to A.$$

Therefore

$$m_{23}((i \otimes i \otimes s^{-1})(i \otimes \Delta')(R)) = (i \otimes \varepsilon)(R) = 1_A.$$

It follows from (2.1.b) that $(i \otimes \Delta')(R) = R_{12}R_{13}$. Hence

$$m_{23}((i \otimes i \otimes s^{-1})(i \otimes \Delta')(R)) = m_{23}((i \otimes i \otimes s^{-1})(R_{12}R_{13})) = R\,(i \otimes s^{-1})(R).$$

Therefore $(i \otimes s^{-1})(R) = R^{-1}$. Finally, we have

$$(s \otimes s)(R) = (i \otimes s)(s \otimes i)(R) = (i \otimes s)(R^{-1}) = (i \otimes s)(i \otimes s^{-1})(R) = R.$$

2.2. The element *u*. The universal R-matrix of any quasitriangular Hopf algebra (A, R) gives rise to a certain element of A denoted by u_R or simply by u (see [Dr3, Proposition 2.1]). By definition, $u = m(s \otimes \mathrm{id}_A)(P_A(R))$. If R is expanded as a finite sum

(2.2.a) $$R = \sum_j \alpha_j \otimes \beta_j$$

with $\alpha_j, \beta_j \in A$, then

(2.2.b) $$u = \sum_j s(\beta_j)\,\alpha_j \in A.$$

It is known that u is invertible in A, $\Delta(u) = (u \otimes u)(P_A(R)R)^{-1}$, and for any $a \in A$,

(2.2.c) $$s^2(a) = uau^{-1}.$$

There is an explicit formula computing u^{-1} from R^{-1}: if

$$R^{-1} = \sum_k \mu_k \otimes \nu_k$$

with $\mu_k, \nu_k \in A$, then

(2.2.d) $$u^{-1} = \sum_k s^{-1}(\nu_k)\,\mu_k \in A.$$

2.3. From quasitriangular Hopf algebras to braided categories. Let (A, R) be a quasitriangular Hopf algebra. We provide the category $\mathrm{Rep}(A)$ with a braiding as follows. For any objects V, W of $\mathrm{Rep}(A)$, define $c_{V,W} : V \otimes W \to W \otimes V$ to be the composition of the multiplication by $R \in A^{\otimes 2}$ and the flip $P_{V,W} : V \otimes W \to W \otimes V$.

If R is expanded as a sum (2.2.a) then for any $x \in V$, $y \in W$,

$$c_{V,W}(x \otimes y) = \sum_j \beta_j y \otimes \alpha_j x. \tag{2.3.a}$$

2.3.1. Lemma. *The family of homomorphisms* $\{c_{V,W} : V \otimes W \to W \otimes V\}_{V,W}$ *is a braiding in* $\operatorname{Rep}(A)$.

Proof. Let us show that $c_{V,W}$ is a morphism in $\operatorname{Rep}(A)$, i.e., that it is A-linear. Take arbitrary $z \in V \otimes W$ and $a \in A$. We have

$$c_{V,W}(az) = P_{V,W}(R\,\Delta(a)z) = P_{V,W}(\Delta'(a)Rz) = \Delta(a)P_{V,W}(Rz) = a\,c_{V,W}(z).$$

Invertibility of R in $A^{\otimes 2}$ implies that $c_{V,W}$ is an isomorphism.

Let us verify the first braiding identity (I.1.2.b) which we reproduce here:

$$c_{U,V\otimes W} = (\mathrm{id}_V \otimes c_{U,W})(c_{U,V} \otimes \mathrm{id}_W). \tag{2.3.b}$$

For any $\alpha \in U \otimes V \otimes W$, we have

$$(\mathrm{id}_V \otimes c_{U,W})(c_{U,V} \otimes \mathrm{id}_W)(\alpha) = (\mathrm{id}_V \otimes c_{U,W})(P_{12}(R_{12}\alpha)) = P_{23}(R_{23}(P_{12}(R_{12}\alpha)))$$

where P_{12} is the flip $x \otimes y \otimes z \mapsto y \otimes x \otimes z$ and P_{23} is the flip $x \otimes y \otimes z \mapsto x \otimes z \otimes y$. Denote by $P_{1,23}$ the permutation homomorphism

$$x \otimes y \otimes z \mapsto y \otimes z \otimes x : U \otimes V \otimes W \to V \otimes W \otimes U.$$

A simple computation using the expansion (2.2.a) shows that

$$P_{23}(R_{23}(P_{12}(R_{12}(\alpha)))) = P_{1,23}(R_{13}R_{12}\alpha).$$

On the other hand, it follows from the definitions of $c_{U,V\otimes W}$ and the action of A in $V \otimes W$ that

$$c_{U,V\otimes W}(\alpha) = P_{1,23}((\mathrm{id}_A \otimes \Delta)(R)\alpha).$$

Therefore (2.3.b) follows from the equality $(\mathrm{id}_A \otimes \Delta)(R) = R_{13}R_{12}$. Similarly, the second braiding identity $c_{U\otimes V,W} = (c_{U,W} \otimes \mathrm{id}_V)(\mathrm{id}_U \otimes c_{V,W})$ follows from the equality $(\Delta \otimes \mathrm{id}_A)(R) = R_{13}R_{23}$.

2.4. The Drinfel'd double construction. There is a general method, due to V. Drinfel'd, producing quasitriangular Hopf algebras. This method is called the double construction. It starts with a Hopf algebra A over a field and produces the structure of a quasitriangular Hopf algebra in the vector space $A \otimes A^*$. The double construction plays a fundamental role in the theory of quantum groups. We confine ourselves to a brief description of this construction in the setting of finite dimensional Hopf algebras.

Let A be a finite dimensional Hopf algebra over a field K. Denote by A^0 the Hopf algebra obtained from the dual Hopf algebra A^* by replacing the comultiplication with the opposite one and replacing the antipode with its inverse. Thus, the comultiplication Δ in A^0 is defined by the formula $\Delta(y)(a \otimes b) = y(ba)$ for any $y \in A^0$ and $a, b \in A$. Note that the tensor product $A \otimes_K A^0$ (viewed as a vector space) contains a canonical element δ_A determined by duality.

2.4.1. Theorem. *Let A be a finite dimensional Hopf algebra over a field K. There exists a unique quasitriangular Hopf algebra $(D(A), R)$ which coincides as a coalgebra with $A \otimes_K A^0$ such that* (i) *the inclusions $A = A \otimes 1_{A^0} \hookrightarrow D(A)$ and $A^0 = 1_A \otimes A^0 \hookrightarrow D(A)$ are Hopf algebra homomorphisms and* (ii) *R is equal to the image of δ_A under the tensor product $A \otimes_K A^0 \hookrightarrow D(A) \otimes_K D(A)$ of these two inclusions.*

For a proof, see [Dr1], [Dr2]. Note that $R = \sum_i (e_i \otimes 1_{A^0}) \otimes (1_A \otimes e^i) \in D(A) \otimes_K D(A)$ for any choice of dual bases $\{e_i\}_i$ and $\{e^i\}_i$ in A and A^0.

3. Ribbon Hopf algebras

3.0. Outline. We formulate natural conditions on a quasitriangular Hopf algebra which ensure the existence of a twist in the representation category.

3.1. Definition of ribbon Hopf algebras. A ribbon Hopf algebra is a triple (A, R, v) consisting of a quasitriangular Hopf algebra (A, R) and an invertible element v of the center of A such that

(3.1.a) $$\Delta(v) = P_A(R)R\,(v \otimes v) \quad \text{and} \quad s(v) = v.$$

The element v is called a universal twist of A.

For an object V of $\mathrm{Rep}(A)$, we define the twist $\theta_V : V \to V$ to be multiplication by $v \in A$.

3.1.1. Lemma. *Let (A, R, v) be a ribbon Hopf algebra. The family of homomorphisms $\{\theta_V : V \to V\}_V$ is a twist in the braided monoidal category* $\mathrm{Rep}(A)$. *This twist is compatible with duality.*

Proof. Since v is an invertible central element of A, the homomorphism $\theta_V : V \to V$ is an A-linear isomorphism. It is easy to check that for any A-modules V, W, the composition $c_{W,V}c_{V,W} : V \otimes W \to V \otimes W$ is multiplication by $P_A(R)R$. Therefore the equality $\Delta(v) = P_A(R)R(v \otimes v)$ implies the twist identity $\theta_{V \otimes W} = c_{W,V}c_{V,W}(\theta_V \otimes \theta_W)$.

It remains to prove that $(\theta_V \otimes \mathrm{id}_{V^*})b_V = (\mathrm{id}_V \otimes \theta_{V^*})b_V$. This is equivalent to $(v \otimes v^{-1})\delta = \delta$ where $\delta = \delta_V = b_V(1)$. We shall use the expansion (1.3.a) of δ. The equality $s(v) = v$ implies that for any $x \in V, y \in V^*$, we have $(vy)(x) =$

$y(vx)$. Replacing x, y with $v^{-1}x, v^{-1}y$ we get $y(v^{-1}x) = (v^{-1}y)(x)$. Therefore

$$\sum_i y(vg_i) \otimes (v^{-1}g^i)(x) = \sum_i (vy)(g_i) \otimes g^i(v^{-1}x) = (vy)(v^{-1}x) = y(vv^{-1}x) = y(x).$$

Here the second equality follows from the characteristic property of δ mentioned in Section 1.3. Hence, the sum $\sum_i vg_i \otimes v^{-1}g^i$ satisfies the same property as δ and therefore coincides with δ.

3.2. Theorem. *For any ribbon Hopf algebra A, the category* Rep(A) *is a ribbon* Ab-*category.*

Theorem 3.2 follows directly from Lemmas 2.3.1 and 3.1.1.

Theorem 3.2 allows us to apply the definitions and results of Chapter I to Rep(A). We may consider the K-valued isotopy invariant F of framed oriented links in S^3 whose components are colored with A-modules of finite rank. We may also consider the trace tr of endomorphisms of objects of Rep(A) introduced in Section I.1.5. In this context, the trace tr is called quantum trace and denoted by tr_q. This trace takes values in $\text{End}_A(K)$. Note that the A-linear homomorphisms $K \to K$ are nothing but multiplications by elements of K. Thus, we identify $\text{End}_A(K) = K$ so that tr_q takes values in K. For an object V of Rep(A), its quantum dimension is defined to be $\dim_q(V) = \text{tr}_q(\text{id}_V) \in K$. For completeness, we give an explicit computation of tr_q via the usual trace of endomorphisms of modules.

3.3. Lemma. *Let (A, R, v) be a ribbon Hopf algebra over a commutative ring with unit, K. For any morphism $f : V \to V$ in* Rep(A), *we have* $\text{tr}_q(f) = \text{Tr}(uvf)$ *where* Tr *is the usual trace of endomorphisms, $u = u_R \in A$ is given by* (2.2.b) *and uvf is the composition of f and multiplication by uv.*

For the definition of Tr in the case where V is not free, see Section I.1.5.1, cf. also Lemma II.4.3.1.

Observe that the function $f \mapsto \text{Tr}(uvf)$ is not a trace on the full algebra of K-linear homomorphisms $V \to V$, but it is a trace if restricted to the subalgebra of A-linear homomorphisms.

Proof of Lemma. We first compute the standard trace Tr of any K-homomorphism $h : V \to V$ in terms of the decomposition (1.3.a) of $\delta_V = b_V(1) \in V \otimes V^*$. By definition (cf. Sections I.1.5 and I.1.7)

$$\text{Tr}(h) = d_V P_{V,V^*}(h \otimes \text{id}_{V^*})(\delta_V) = d_V P_{V,V^*}(h \otimes \text{id}_{V^*})(\sum_i g_i \otimes g^i) =$$

$$= d_V P_{V,V^*}(\sum_i h(g_i) \otimes g^i) = d_V(\sum_i g^i \otimes h(g_i)) = \sum_i g^i(h(g_i)).$$

Now we compute $\mathrm{tr}_q(f)$ using (2.2.a):

$$\mathrm{tr}_q(f) = d_V c_{V,V^*}((\theta_V f) \otimes \mathrm{id}_{V^*})(\delta_V) = d_V(\sum_{i,j} \beta_j g^i \otimes \alpha_j v f(g_i)) =$$

$$= \sum_{i,j} \beta_j g^i(\alpha_j v f(g_i)) = \sum_i g^i(\sum_j s(\beta_j)\alpha_j v f(g_i)) = \sum_i g^i(uvf(g_i)).$$

Therefore $\mathrm{tr}_q(f) = \mathrm{Tr}(uvf)$.

3.4. Examples. Fundamental examples of quasitriangular Hopf algebras are provided by the theory of quantum groups, see Sections 6, 7. Here we consider two elementary examples intended to illustrate the concepts of quasitriangular and ribbon Hopf algebras. In both examples the symbol K denotes a commutative ring with unit.

1. Provide K with the identity comultiplication $\Delta = \mathrm{id}_K : K \to K \otimes_K K = K$, the identity counit $\varepsilon = \mathrm{id}_K : K \to K$, and the identity antipode $s = \mathrm{id}_K : K \to K$. It is obvious that in this way K acquires the structure of Hopf algebra. Set $R = 1 \in K \otimes_K K = K$ and $v = 1 \in K$. The triple (K, R, v) is a ribbon Hopf algebra. Its category of representations coincides with the ribbon category $\mathrm{Proj}(K)$ described in Section I.1.7.1.

2. Let G be a finite abelian group endowed with a bilinear pairing $c : G \times G \to K^*$ and a group homomorphism $\varphi : G \to K^*$ such that $\varphi(g^2) = 1$ for any $g \in G$. (We use the multiplicative notation for the group operation in G.) Consider the Hopf algebra A of Example 1.2.2. Set

$$R = \sum_{g,h \in G} c(g, h)\, \delta_g \otimes \delta_h \in A^{\otimes 2} \quad \text{and} \quad v = \sum_{g \in G} \varphi(g)\, c(g, g)\, \delta_g \in A.$$

It turns out that the triple (A, R, v) is a ribbon Hopf algebra. Indeed, both R and v are invertible, the inverse elements being

$$R^{-1} = \sum_{g,h \in G} (c(g, h))^{-1} \delta_g \otimes \delta_h, \quad v^{-1} = \sum_{g \in G} \varphi(g)(c(g, g))^{-1} \delta_g.$$

Formula (2.1.a) follows from the commutativity of A and the equality $\Delta' = \Delta$. Formulas (2.1.b), (2.1.c), (3.1.a) follow from the bilinearity of c and definitions.

By Theorem 3.2, the category $\mathrm{Rep}(A)$ is a ribbon Ab-category. This category includes the category $\mathcal{V} = \mathcal{V}(G, K, c, \varphi)$ defined in Section I.1.7.2 as a ribbon subcategory. To see this, consider for each $g \in G$, the A-module V_g defined to be K where δ_g acts as id_K and δ_h with $h \in G \setminus \{g\}$ acts as 0. It is easy to check that $V_g \otimes V_h = V_{gh}$ and $(V_g)^* = V_{g^{-1}}$. Therefore the formula $g \mapsto V_g$ defines a covariant embedding $\mathcal{V} \to \mathrm{Rep}(A)$ commuting with tensor product, braiding, twist, and duality.

This example shows that in the case of finite G the ribbon category $\mathcal{V}(G, K, c, \varphi)$ may be obtained as (a subcategory of) the representation category of a ribbon Hopf algebra. The construction of $A(G, K, c, \varphi)$ does not apply to the case of infinite G because both the comultiplication and the R-matrix would be given by infinite sums. This phenomenon is typical: many difficulties with infinite sums in Hopf algebras disappear on the level of categories.

3.5. Remarks. 1. A construction, due to Reshetikhin and Turaev [RT1], extends any quasitriangular Hopf algebra to a ribbon Hopf algebra. Combining this extension with the Drinfel'd double construction we may produce vast families of examples of ribbon Hopf algebras and ribbon categories. However, it is only under very special circumstances that these categories are semisimple or modular.

Parallel constructions may be applied to categories. Namely, any monoidal category with duality can be "doubled" into a ribbon category (see [JS4], [Maj2], [KT]).

2. The constructions of Sections 1 - 3 directly extend to quasi-Hopf algebras in the sense of Drinfel'd [Dr4]. The only essential difference is that in the definition of Rep(A) the associativity isomorphisms are determined by the Drinfel'd associator.

3.6. Exercise. Let (A, R) be a quasitriangular Hopf algebra. Let v be an invertible element of the center of A such that $\Delta(v) = P_A(R)R\,(v \otimes v)$. Show that $(s(v) - v)A$ is a coideal of A. Deduce that the quotient $A/(s(v) - v)A$ is a ribbon Hopf algebra. (Hint: verify that $\Delta(s(v)) = P_A(R)R\,(s(v) \otimes s(v))$.)

4. Digression on quasimodular categories

4.0. Outline. By weakening the axiom of domination in the definition of a modular category we obtain the more general notion of a quasimodular category. We describe a procedure of "purification" which transforms any quasimodular category into a modular category.

Quasimodular categories naturally arise in the theory of quantum groups and in the skein theory (Chapter XII).

4.1. Negligible morphisms. Let $\mathcal{V}$ be a ribbon Ab-category. A morphism $f : V \to W$ in $\mathcal{V}$ is said to be negligible if for any morphism $g : W \to V$, we have $\operatorname{tr}(fg) = 0$. Until now we have not met negligible morphisms because in semisimple categories (in particular in modular categories) all negligible morphisms are equal to zero. This is not the case in arbitrary ribbon Ab-categories.

It is obvious that the sum of two negligible morphisms $V \to W$ is negligible and that the composition of a negligible morphism with any other morphism is negligible.

4.1.1. Lemma. *The tensor product of a negligible morphism in $\mathcal{V}$ with an arbitrary morphism in $\mathcal{V}$ is negligible.*

This lemma shows that negligible morphisms in $\mathcal{V}$ form a two-sided ideal with respect to composition and tensor product.

Proof of Lemma. Let $f : V \to W$ be a negligible morphism in $\mathcal{V}$. Let $h : X \to Y$ be an arbitrary morphism in $\mathcal{V}$. To prove that $f \otimes h$ is negligible we should show that for any morphism $g : W \otimes Y \to V \otimes X$, the trace of $(f \otimes h)g \in \operatorname{End}(W \otimes Y)$ is equal to 0. This follows from the negligibility of f and the identity

$$\operatorname{tr}((f \otimes h)g) = \operatorname{tr}(f(\mathrm{id}_V \otimes F(\cap_Y^-))(\mathrm{id}_V \otimes h \otimes \mathrm{id}_{Y^*})(g \otimes \mathrm{id}_{Y^*})(\mathrm{id}_W \otimes F(\cup_Y))).$$

Here F is the operator invariant of v-colored ribbon graphs given by Theorem I.2.5, $\cap_Y^-$ and $\cup_Y$ are the ribbon graphs defined in Section I.2.3. The identity in question follows directly from the geometric interpretation of the trace given in Corollary I.2.7.2, it suffices to draw the corresponding picture. A similar argument shows that the morphism $h \otimes f$ is negligible.

4.2. Purification. A ribbon Ab-category is said to be pure if all negligible morphisms in this category are equal to zero. We describe a purification procedure transforming any ribbon Ab-category $\mathcal{V}$ into a pure ribbon Ab-category $\mathcal{V}_p$.

For objects V, W of $\mathcal{V}$, denote the set of negligible morphisms $V \to W$ by $\operatorname{Negl}(V, W)$. This is a subgroup of $\operatorname{Hom}(V, W)$ so that we may consider the quotient

$$\operatorname{hom}(V, W) = \operatorname{Hom}(V, W)/\operatorname{Negl}(V, W).$$

For instance, by Lemma I.1.5.1 we have $\operatorname{Negl}(\mathbb{1}, \mathbb{1}) = 0$ so that $\operatorname{hom}(\mathbb{1}, \mathbb{1}) = \operatorname{Hom}(\mathbb{1}, \mathbb{1})$.

We define a new category $\mathcal{V}_p$ whose objects are the same as in $\mathcal{V}$. The morphisms $V \to W$ in $\mathcal{V}_p$ are elements of $\operatorname{hom}(V, W)$. The composition, tensor product, and adddition of morphisms are induced by the corresponding operations in $\mathcal{V}$. Similarly, the braiding, twist, and duality in $\mathcal{V}_p$ are induced by the braiding, twist, and duality in $\mathcal{V}$. It is easy to verify that $\mathcal{V}_p$ is a pure ribbon Ab-category with the same ground ring as $\mathcal{V}$. We have an obvious covariant functor $\mathcal{V} \to \mathcal{V}_p$ identical on the objects and transforming each morphism in $\mathcal{V}$ into its class modulo negligible morphisms.

The following lemma will be useful in Chapter XII.

4.2.1. Lemma. *Any Hermitian structure on a ribbon* Ab-*category* $\mathcal{V}$ *induces a Hermitian structure on* $\mathcal{V}_p$.

Proof. It suffices to show that for any negligible morphism $f : V \to W$, the conjugated morphism $\overline{f} : W \to V$ is also negligible. For any morphism $g : V \to W$, we have $\operatorname{tr}(\overline{f}g) = \operatorname{tr}(\overline{\overline{g}f}) = \overline{\operatorname{tr}(\overline{g}f)} = 0$. Here the second equality follows from Corollary II.5.1.4 and the third equality follows from the negligibility of f.

4.3. Quasimodular categories. Roughly speaking, quasimodular categories are generalizations of modular categories admitting negligible morphisms. Such morphisms appear in the following weakened version of domination (cf. Section II.1.3). We say that a family $\{V_i\}_{i\in I}$ of objects of a ribbon Ab-category $\mathcal{V}$ quasidominates an object V of $\mathcal{V}$ if there exist a finite set $\{V_{i(r)}\}_r$ of objects of this family (possibly with repetitions) and a family of morphisms $\{f_r : V_{i(r)} \to V, g_r : V \to V_{i(r)}\}_r$ such that

$$\mathrm{id}_V - \sum_r f_r g_r \in \operatorname{Negl}(V, V).$$

If $\mathcal{V}$ is pure then quasidomination coincides with domination.

A quasimodular category is a pair consisting of a ribbon Ab-category $\mathcal{V}$ and a finite family $\{V_i\}_{i\in I}$ of simple objects of $\mathcal{V}$ satisfying axioms (II.1.4.1), (II.1.4.2), (II.1.4.4), and the following weakened version of the axiom of domination:

(4.3.1). All objects of $\mathcal{V}$ are quasidominated by $\{V_i\}_{i\in I}$.

The following lemma shows that each quasimodular category $(\mathcal{V}, \{V_i\}_{i\in I})$ gives rise in a canonical way to a modular category.

4.3.2. Lemma. *Let* $(\mathcal{V}, \{V_i\}_{i\in I})$ *be a quasimodular category. Then* $(\mathcal{V}_p, \{V_i\}_{i\in I})$ *is a modular category.*

Proof. Let us show first that $\operatorname{Negl}(V_i, V_i) = 0$ for any $i \in I$. Since the object V_i is simple, $\operatorname{Hom}(V_i, V_i) = K \cdot \mathrm{id}_{V_i}$ where K is the ground ring of $\mathcal{V}$. For $k \in K$, the trace of $k \cdot \mathrm{id}_{V_i}$ is equal to $k \dim(V_i)$. As it was remarked in Section II.1.4, the non-degeneracy axiom (II.1.4.4) implies that $\dim(V_i)$ is invertible in K. Hence the trace of $k \cdot \mathrm{id}_{V_i}$ is equal to 0 if and only if $k = 0$. This implies that $\operatorname{Negl}(V_i, V_i) = 0$ so that $\hom(V_i, V_i) = \operatorname{Hom}(V_i, V_i) = K$. Thus, V_i is a simple object of $\mathcal{V}_p$. The axioms of modular category for $\mathcal{V}_p$ follow directly from the corresponding axioms for $\mathcal{V}$ and the fact that all negligible morphisms in $\mathcal{V}$ are annihilated in $\mathcal{V}_p$.

5. Modular Hopf algebras

5.0. Outline. Following the lines of Sections 1 - 3 we dualize the axioms of quasimodular category. This leads to the notion of a modular Hopf algebra. We show that the representation category of a modular Hopf algebra is a quasimodular category. Its purification yields a modular category.

5.1. Preliminaries on modules. Let A be an algebra over a commutative ring with unit, K. We need a few preliminary definitions and remarks concerning A-modules. We say that an A-module V is simple if every A-linear endomorphism of V is multiplication by an element of K. Simple modules should be contrasted with irreducible modules which are defined as A-modules without non-trivial A-submodules. If K is an algebraically closed field then any irreducible A-module is simple. For, if $f : V \to V$ is an A-linear endomorphism of an irreducible A-module V then any non-zero eigenspace of f is an A-submodule of V and therefore coincides with V.

Assume that A is a ribbon Hopf algebra over K. We say that an A-module of finite rank, V, is negligible if $\mathrm{tr}_q(f) = 0$ for any $f \in \mathrm{End}_A(V)$. Equivalently, V is negligible if its identity endomorphism id_V is negligible. It is clear that if V is negligible then any morphism $f : W \to V$ or $g : V \to W$ is negligible. (For, $f = \mathrm{id}_V f$, $g = g\,\mathrm{id}_V$ and the composition of a negligible morphism with any other morphism is negligible.) For any negligible module V, we have $\dim_q(V) = \mathrm{tr}_q(\mathrm{id}_V) = 0$.

5.2. Modular Hopf algebras. Let K be a commutative ring with unit. A modular Hopf algebra over K is a pair (a ribbon Hopf K-algebra A, a finite family of simple A-modules of finite rank $\{V_i\}_{i\in I}$) satisfying the following conditions.

(5.2.1) For some $0 \in I$, we have $V_0 = K$ (where A acts in K via the counit).

(5.2.2) For each $i \in I$, there exists $i^* \in I$ such that V_{i*} is isomorphic to $(V_i)^*$.

(5.2.3) For any $k, l \in I$, the tensor product $V_k \otimes V_l$ splits as a finite direct sum of certain V_i, $i \in I$ (possibly with multiplicities) and a negligible A-module.

To formulate the last condition consider the homomorphism $x \mapsto P_A(R)R\,x : V_i \otimes V_j \to V_i \otimes V_j$ (it is A-linear due to (2.1.a)) and denote its quantum trace by $S_{i,j}$.

(5.2.4) The square matrix $[S_{i,j}]_{i,j\in I}$ is invertible over K.

A slight change in comparison with the definition of modular categories is the appearance of direct sums and negligible modules. Otherwise this is a literal translation of the definition of modular categories to the setting of ribbon Hopf algebras. We have to involve negligible modules because they do appear in the theory of quantum groups. The direct sums may be avoided but this does not seem to lead to any greater generality.

The ribbon Hopf algebras constructed in Section 3.5 extend naturally to modular Hopf algebras. In Example 3.5.1 the distinguished family of modules consists of one module K. In Example 3.5.2 we distingush the modules $\{V_g\}_{g\in G}$. In both cases negligible modules do not arise, due to the elementary nature of these examples.

5.3. Construction of modular categories. For any modular Hopf algebra $(A, \{V_i\}_{i\in I})$, we define a subcategory $\mathcal{V}$ of $\mathrm{Rep}(A)$. The objects of $\mathcal{V}$ are A-modules of finite rank quasidominated by $\{V_i\}_{i\in I}$. The morphisms in $\mathcal{V}$ are all A-homomorphisms of such modules. The composition, tensor product, and addition of morphisms in $\mathcal{V}$ are induced by the corresponding operations in $\mathrm{Rep}(A)$. We say that $\mathcal{V}$ is the subcategory of $\mathrm{Rep}(A)$ quasidominated by $\{V_i\}_{i\in I}$.

5.3.1. Lemma. *The category $\mathcal{V}$ is a monoidal subcategory of* $\mathrm{Rep}(A)$.

We prove Lemma 5.3.1 at the end of this subsection. It is clear that the braiding, twist, and duality in $\mathrm{Rep}(A)$ defined in Sections 1 - 3 induce a braiding, twist, and duality in $\mathcal{V}$. The axioms of ribbon category for $\mathcal{V}$ follow from the corresponding properties of $\mathrm{Rep}(A)$. By definition, all objects of $\mathcal{V}$ are quasidominated by $\{V_i\}_{i\in I}$. The other axioms of a quasimodular category for $(\mathcal{V}, \{V_i\}_{i\in I})$ follow from the corresponding axioms of a modular Hopf algebra. We have the following theorem.

5.3.2. Theorem. *Let $(A, \{V_i\}_{i\in I})$ be a modular Hopf algebra. Then the pair $(\mathcal{V}, \{V_i\}_{i\in I})$ is a quasimodular category.*

Purifying $(\mathcal{V}, \{V_i\}_{i\in I})$ as in Section 4 we get a modular category $(\mathcal{V}_p, \{V_i\}_{i\in I})$. In this way each modular Hopf algebra gives rise to a modular category. For example, the modular Hopf algebras discussed at the end of Section 5.2 give rise to the modular categories introduced in Section II.1.7.

5.3.3. Proof of Lemma 5.3.1. It is clear that K is an object of $\mathcal{V}$ because $V_0 = K$. We should verify the following two claims: (i) if V is an object of $\mathcal{V}$ then so is V^*, (ii) if V, W are objects of $\mathcal{V}$ then so is $V \otimes W$. We need a few preliminary remarks. For morphisms $f, g : U \to V$ with negligible difference $f - g : U \to V$, we shall write $f \simeq g$. In particular, $f \simeq 0$ if and only if f is negligible. Recall the dual morphism $f^* : V^* \to U^*$ defined in Section I.1.8 by the formula $f^* = (d_V \otimes \mathrm{id}_{U^*})(\mathrm{id}_{V^*} \otimes f \otimes \mathrm{id}_{U^*})(\mathrm{id}_{V^*} \otimes b_U)$. The results of Section 4.1 imply that if f is negligible then so is f^*. Note that $f^* - g^* = (f - g)^*$. Therefore if $f \simeq g$ then $f^* \simeq g^*$. Similarly, for any morphisms $f, g : U \to V$ and $f', g' : V \to W$, the equality $f'f - g'g = f'(f - g) + (f' - g')g$ shows that $f \simeq g$, $f' \simeq g'$ imply that $f'f \simeq g'g$.

Now we are ready to prove claim (i). Since V is an object of $\mathcal{V}$ there is a family of morphisms $\{f_r : V_{i(r)} \to V, g_r : V \to V_{i(r)}\}_r$ such that $\mathrm{id}_V \simeq \sum_r f_r g_r$. Then

$$\mathrm{id}_{V^*} = (\mathrm{id}_V)^* \simeq (\sum_r f_r g_r)^* = \sum_r (g_r)^*(f_r)^*.$$

The composition $(g_r)^*(f_r)^* : V^* \to V^*$ passes through $(V_{i(r)})^* \approx V_{i(r)^*}$. Therefore $(g_r)^*(f_r)^*$ may be presented as a certain composition $V^* \to V_{i(r)^*} \to V^*$. This shows that V^* is quasidominated by $\{V_i\}_{i\in I}$.

To prove claim (ii) consider a decomposition $\mathrm{id}_V \simeq \sum_r f_r g_r$ as above and a similar decomposition $\mathrm{id}_W \simeq \sum_s f'_s g'_s$. Then

$$\mathrm{id}_{V\otimes W} = \mathrm{id}_V \otimes \mathrm{id}_W \simeq \sum_{r,s} f_r g_r \otimes f'_s g'_s = \sum_{r,s} (f_r \otimes f'_s)(g_r \otimes g'_s). \tag{5.3.a}$$

The target of each morphism $g_r \otimes g'_s$ is an object $V_k \otimes V_l$ with certain $k, l \in I$ depending on r and s. By (5.2.3) the identity endomorphism of $V_k \otimes V_l$ splits as a sum of several compositions $V_k \otimes V_l \to V_i \to V_k \otimes V_l$ and a projection onto a negligible summand. The last projection is a negligible morphism. Therefore the morphism $g_r \otimes g'_s = \mathrm{id}_{V_k\otimes V_l}(g_r \otimes g'_s)$ expands as a sum of compositions $V \otimes W \to V_i \to V_k \otimes V_l$ with $i \in I$ (modulo negligible morphisms). Substituting this decomposition of $g_r \otimes g'_s$ into (5.3.a) we get a splitting of $\mathrm{id}_{V\otimes W}$ as a sum of morphisms which pass through $\{V_i\}_{i\in I}$ (modulo negligible morphisms). Therefore $V \otimes W$ is quasidominated by $\{V_i\}_{i\in I}$.

6. Quantum groups at roots of unity

6.0. Outline. The theory of quantum groups was originally conceived as a machinery that produces solutions to the Yang-Baxter equation. Later on, quantum groups found applications in several areas including topology of knots and 3-manifolds. The importance of quantum groups for our study is that they give rise to non-trivial modular categories.

A comprehensive exposition of the relevant aspects of the theory of quantum groups would take a full-size monograph. We give a very sketchy introduction to quantum groups aimed at a construction of modular categories.

6.1. Definition of quantum groups. Quantum groups are Hopf algebras obtained by a 1-parameter deformation of the universal enveloping algebras of simple complex Lie algebras. We shall restrict ourselves to quantum groups corresponding to simple Lie algebras of types A, D, E. (Other simple Lie algebras may be treated similarly, cf. Remark 6.4.2.)

Let $\mathfrak{g}$ be a simple Lie algebra of type A, D, or E with the Cartan matrix $(a_{ij})_{1\leq i,j\leq m}$ with integer $m \geq 1$. Recall that $a_{ii} = 2$ for all $i = 1, \ldots, m$ and

$a_{ij} = a_{ji} \in \{0, -1\}$ for $i \neq j$. Fix a non-zero complex number $q \neq \pm 1$. The quantum group $U_q\mathfrak{g}$ is the algebra over $\mathbb{C}$ defined by $4m$ generators E_i, F_i, K_i, K_i^{-1} where $i = 1, ..., m$ and relations

(6.1.a) $$K_iK_j = K_jK_i, \quad K_iK_i^{-1} = K_i^{-1}K_i = 1,$$

(6.1.b) $$K_iE_j = q^{a_{ij}}E_jK_i, \quad K_iF_j = q^{-a_{ij}}F_jK_i,$$

(6.1.c) $$E_iF_j - F_jE_i = \delta_{ij}\frac{K_i - K_i^{-1}}{q - q^{-1}}$$

(6.1.d) $$E_iE_j = E_jE_i, \quad F_iF_j = F_jF_i \quad \text{if} \quad a_{ij} = 0,$$

(6.1.e) $$E_i^2E_j - (q + q^{-1})E_iE_jE_i + E_jE_i^2 = 0 \quad \text{if} \quad a_{ij} = -1,$$

$$F_i^2F_j - (q + q^{-1})F_iF_jF_i + F_jF_i^2 = 0 \quad \text{if} \quad a_{ij} = -1$$

where $i, j = 1, ..., m$. We provide $U_q\mathfrak{g}$ with the structure of a Hopf algebra. The comultiplication Δ in $U_q\mathfrak{g}$ is defined on the generators by

$$\Delta(E_i) = E_i \otimes 1 + K_i \otimes E_i, \ \Delta(F_i) = F_i \otimes K_i^{-1} + 1 \otimes F_i, \ \Delta(K_i) = K_i \otimes K_i.$$

The antipode s is defined by

$$s(K_i) = K_i^{-1}, \ s(E_i) = -K_i^{-1}E_i, \ s(F_i) = -F_iK_i.$$

The counit ϵ is defined by

$$\epsilon(K_i) = 1 \quad \text{and} \quad \epsilon(E_i) = \epsilon(F_i) = 0.$$

It may be verified by direct computations that $U_q\mathfrak{g}$ is a Hopf algebra.

Setting $K_i = \exp(-hH_i/2), q = \exp(-h/2)$ in the formulas above and passing to the limit $h \to 0$ we obtain the standard relations among the Chevalley generators of the universal enveloping algebra $U(\mathfrak{g})$ of $\mathfrak{g}$. The comultiplication, antipode, and counit in $U_q\mathfrak{g}$ degenerate to the standard comultiplication, antipode, and counit in $U(\mathfrak{g})$. This allows us to regard $U_q\mathfrak{g}$ as a q-deformation of $U(\mathfrak{g})$.

The study of quantum groups splits into two cases, one where the parameter q is a root of unity and one where the parameter is generic. If q is a root of unity of even order then $U_q\mathfrak{g}$ gives rise to a modular category (see Section 6.3). For a generic parameter, we derive from the quantum group a semisimple category with an infinite number of simple objects (see Section 7). Note that the treatments of these two cases are essentially independent.

6.2. Preliminary definitions and notation. We shall need the braid group $B(\mathfrak{g})$ and the Weyl group $W(\mathfrak{g})$ associated to $\mathfrak{g}$. The braid group $B(\mathfrak{g})$ is generated by m generators $T_1, ..., T_m$ subject to the following relations: $T_iT_j = T_jT_i$ if $a_{ij} = 0$; $T_iT_jT_i = T_jT_iT_j$ if $a_{ij} = -1$. The Weyl group $W(\mathfrak{g})$ is the quotient of $B(\mathfrak{g})$ by the relations $T_1^2 = 1, \dots, T_m^2 = 1$. The images of $T_1, \dots, T_m$ in $W(\mathfrak{g})$ are denoted respectively by $r_1, \dots, r_m$. Any $w \in W(\mathfrak{g})$ may be expressed as a product $r_{i_1} r_{i_2} \dots r_{i_k}$. The minimal possible number k in such an expression is called the length of w. The Weyl group $W(\mathfrak{g})$ is finite and contains a unique element $w_0 \in W(\mathfrak{g})$ of maximal length.

For an integer n, set

$$[n]_q = \frac{q^n - q^{-n}}{q - q^{-1}} = q^{-n+1} + q^{-n+3} + \dots + q^{n-3} + q^{n-1}.$$

Note that $[n]_1 = n$ so that we may regard $[n]_q$ as a q-deformation of n. For $n \geq 0$, set $[n]_q! = [1]_q [2]_q \dots [n]_q$. In particular, $[0]! = [1]! = 1$.

Denote by $(b_{ij})_{1 \leq i,j \leq m}$ the matrix over $\mathbb{Q}$ that is inverse to the Cartan matrix $(a_{ij})_{i,j}$. For $i = 1, \dots, m$, set $b_i = \sum_{j=1}^m b_{ij}$. It is known that the numbers $b_1, \dots, b_m$ are integers or half-integers.

6.3. Construction of a modular category. Assume that q is a primitive l-th root of unity with $l \geq 3$. Set $l' = l$ if l is odd and $l' = l/2$ if l is even.

The algebra $U_q\mathfrak{g}$ gives rise to a finite dimensional ribbon Hopf algebra $\tilde{U}_q\mathfrak{g}$. It is obtained as the quotient of $U_q\mathfrak{g}$ by the two-sided ideal generated by $E_i^{l'}, F_i^{l'}, K_i^l - 1$ with $i = 1, \dots, m$. This ideal is also a coideal annihilated by the counit and invariant under the antipode. Therefore $\tilde{U}_q\mathfrak{g}$ is a Hopf algebra over $\mathbb{C}$. It is known to be finite dimensional (see [Lu2], [Lu3]). In this section we discuss (i) the universal R-matrix and the universal twist of $\tilde{U}_q\mathfrak{g}$, (ii) the representation theory of $\tilde{U}_q\mathfrak{g}$, and (iii) the modular category of $\tilde{U}_q\mathfrak{g}$-modules.

(i). Assume first that l is coprime with the determinant of the Cartan matrix $(a_{ij})_{1 \leq i,j \leq m}$. Then the Hopf algebra $\tilde{U}_q\mathfrak{g}$ possesses a universal R-matrix $R \in \tilde{U}_q\mathfrak{g} \otimes_{\mathbb{C}} \tilde{U}_q\mathfrak{g}$ (see [Ro4]). The construction of R is based on the double construction outlined in Section 2.4. Namely, denote by U_+ (resp. U_-) the subalgebra of $U_q\mathfrak{g}$ generated by E_i, K_i, K_i^{-1} (resp. F_i, K_i, K_i^{-1}) with $i = 1, \dots, m$. It is easy to verify that U_+ and U_- are Hopf subalgebras of $\tilde{U}_q\mathfrak{g}$. It turns out that the Hopf algebra U_- is canonically isomorphic to $(U_+)^0$. Moreover, the inclusions $U_+ \hookrightarrow \tilde{U}_q\mathfrak{g}$ and $U_- \hookrightarrow \tilde{U}_q\mathfrak{g}$ induce a Hopf algebra epimorphism $D(U_+) \to \tilde{U}_q\mathfrak{g}$. It is clear that a Hopf algebra quotient of a quasitriangular Hopf algebra is itself quasitriangular. Therefore $\tilde{U}_q\mathfrak{g}$ is quasitriangular. Moreover, one can explicitly compute the universal R-matrix R of $\tilde{U}_q\mathfrak{g}$. We give here an expression for R referring for a proof to [Ro4].

The braid group $B(\mathfrak{g})$ acts in $\tilde{U}_q\mathfrak{g}$ in the following way. If $a_{ij} = 0$ then

$$T_i(E_j) = E_j, \ T_i(F_j) = F_j, \ T_i(K_j) = K_j.$$

If $a_{ij} = -1$ then

$$T_i(E_j) = q^{-1} E_j E_i - E_i E_j, \ T_i(F_j) = q F_i F_j - F_j F_i, \ T_i(K_j) = K_i K_j.$$

Finally, $T_i(E_i) = -F_i K_i$, $T_i(F_i) = -K_i^{-1} E_i$, $T_i(K_i) = K_i^{-1}$.

We fix a decomposition $w_0 = r_{i_1} r_{i_2} ... r_{i_N}$ where w_0 is the element of $W(\mathfrak{g})$ of maximal length, N. For $s = 1, ..., N$, set

(6.3.a) $\quad e_s = T_{i_1} T_{i_2} ... T_{i_{s-1}}(E_{i_s}) \in \tilde{U}_q\mathfrak{g}$ and $f_s = T_{i_1} T_{i_2} ... T_{i_{s-1}}(F_{i_s}) \in \tilde{U}_q\mathfrak{g}$.

Denote by $(.,.)$ the bilinear form in $\mathbb{Z}^m$ defined (in the standard basis) by the Cartan matrix $(a_{ij})_{i,j}$. For each $\beta = (\beta_1, ..., \beta_m) \in \mathbb{Z}^m$, set $K_\beta = K_1^{\beta_1} K_2^{\beta_2} ... K_m^{\beta_m} \in \tilde{U}_q\mathfrak{g}$. Denote by $\tilde{I}$ the set of sequences $(\beta_1, ..., \beta_m) \in \mathbb{Z}^m$ such that $l > \beta_i \geq 0$ for all $i = 1, ..., m$. The set $\tilde{I}$ is a cube in $\mathbb{Z}^m$ consisting of l^m elements.

The formula for the universal R-matrix $R \in \tilde{U}_q\mathfrak{g} \otimes_{\mathbb{C}} \tilde{U}_q\mathfrak{g}$ obtained via the double construction is:

$$R = l^{-m} \prod_{s=1}^{N} \left(\sum_{n=0}^{l'-1} q^{-\frac{n(n-1)}{2}} \frac{(1-q^2)^n}{[n]!} e_s^n \otimes f_s^n \right) \left(\sum_{\beta,\gamma \in \tilde{I}} q^{(\beta,\gamma)} K_\beta \otimes K_\gamma \right).$$

Note that the left-hand side and hence also the right-hand side do not depend on the choice of decomposition $w_0 = r_{i_1} r_{i_2} ... r_{i_N}$.

We construct a universal twist of $\tilde{U}_q\mathfrak{g}$ as follows. Set

$$\mu = K_1^{-2b_1} K_2^{-2b_2} ... K_m^{-2b_m} \in \tilde{U}_q\mathfrak{g}.$$

A direct computation on generators shows that $s^2(a) = \mu a \mu^{-1}$ for any $a \in \tilde{U}_q\mathfrak{g}$. By the general theory of quasitriangular Hopf algebras we have $s^2(a) = uau^{-1}$ where $u = u_R$. Therefore $v = u^{-1}\mu$ lies in the center of $\tilde{U}_q\mathfrak{g}$. It follows from the definition of comultiplication in $\tilde{U}_q\mathfrak{g}$ that $\Delta(\mu) = \mu \otimes \mu$. By the general theory $\Delta(u^{-1}) = \mathrm{Perm}(R)R\,(u^{-1} \otimes u^{-1})$. Therefore $\Delta(v) = \mathrm{Perm}(R)R\,(v \otimes v)$. It is known that $s(v) = v$ ([Ro5]). Both u and μ are invertible, therefore v is invertible in $\tilde{U}_q\mathfrak{g}$. This shows that v is a universal twist of $\tilde{U}_q\mathfrak{g}$. Thus, the triple $(\tilde{U}_q\mathfrak{g}, R, v)$ is a ribbon Hopf algebra.

In the case where l is not coprime with $\det(a_{ij})$ there is a natural extension of $\tilde{U}_q\mathfrak{g}$ to a finite dimensional ribbon Hopf algebra, see [RT2] for the case $\mathfrak{g} = sl_2(\mathbb{C})$ and [Ro5] for the general case.

(ii). The representation theory of $\tilde{U}_q\mathfrak{g}$ is not fully understood. We present here a few results relevant to our study, for details and proofs see [APW], [CK], [Lu3].

Let $\lambda_1, ..., \lambda_m$ be non-negative integers. Under certain conditions on $\lambda = (\lambda_1, ..., \lambda_m)$ there exists an irreducible finite dimensional (non-trivial) $\tilde{U}_q\mathfrak{g}$-module, V_λ, generated by a certain $x \in V_\lambda$ such that

$$E_i x = 0, \;\; F_i^{\lambda_i+1} x = 0, \;\; K_i x = q^{\lambda_i} x$$

for all $i = 1, ..., m$. The module V_λ is unique up to isomorphism. The condition on λ ensuring the existence and irreducibility of V_λ is usually expressed by saying that λ belongs to the Weyl alcove $WA(\mathfrak{g}, l)$. This alcove is determined by the inequality $\sum_{i=1}^m (\lambda_i + 1) p_i < l'$ where $p_1, ..., p_m$ are coefficients in the decomposition of the highest root of $\mathfrak{g}$ as a linear combination of simple roots.

For $\mathfrak{g} = sl_{m+1}(\mathbb{C})$, this inequality has the form $\sum_{i=1}^{m}(\lambda_i + 1) < l'$. For $\lambda = 0$, we have $V_\lambda = \mathbb{C}$.

(iii). Assume that the order l of q is even and that the Weyl alcove $WA(\mathfrak{g}, l)$ is non-empty. (The last condition is equivalent to the inequality $l' \geq r$ where r is the Coxeter number of $\mathfrak{g}$.) Then we can derive a modular Hopf algebra from $\tilde{U}_q\mathfrak{g}$ at least in the case where $\mathfrak{g}$ is a Lie algebra of types A, D. This is the tuple $(\tilde{U}_q\mathfrak{g}, R, v, \{V_\lambda\}_\lambda)$ where R is the universal R-matrix of $\tilde{U}_q\mathfrak{g}$, v is the universal twist of $\tilde{U}_q\mathfrak{g}$, and $\{V_\lambda\}_\lambda$ is the family of irreducible $\tilde{U}_q\mathfrak{g}$-modules constructed above. (In the case where l is not coprime with $\det(a_{ij})$, we use the finite dimensional extension of $\tilde{U}_q\mathfrak{g}$ mentioned above.) The tuple $(\tilde{U}_q\mathfrak{g}, R, v, \{V_\lambda\}_\lambda)$ satisfies the axioms of a modular Hopf algebra. Indeed, the modules $\{V_\lambda\}_\lambda$ are irreducible and therefore simple. Axiom (5.2.1) is obvious. Validity of (5.2.2) is well known, see, for instance, [An] or [TW]. The fundamental axiom (5.2.3) was verified by Andersen [An] and Andersen and Paradowski [AP] using ideas of S. Donkin and C.M. Ringel on modules with good filtrations. Validity of (5.2.4) was verified in [TW] for Lie algebras of types A, B, C, D using results of Kac-Peterson [KP] and Kac-Wakimoto [KW]. (In the case $\mathfrak{g} = sl_2(\mathbb{C})$ the axioms of modular Hopf algebra for $(\tilde{U}_q\mathfrak{g}, R, v, \{V_\lambda\}_\lambda)$ were first verified in [RT2].)

By the general procedure described in Section 5 we can derive a modular category over $\mathbb{C}$ from the modular Hopf algebra $(\tilde{U}_q\mathfrak{g}, R, v, \{V_\lambda\}_\lambda)$. Its objects are finite dimensional $\tilde{U}_q\mathfrak{g}$-modules quasidominated by the family $\{V_\lambda\}_\lambda$. The morphisms are $\tilde{U}_q\mathfrak{g}$-homomorphisms considered up to the addition of negligible homomorphisms. The braiding and twist are induced by the action of R and v in the usual way. The distinguished simple objects are the modules $\{V_\lambda\}_\lambda$. This modular category is denoted by $\mathcal{V}_q\mathfrak{g}$.

6.4. Remarks. 1. In order to construct the modular category $\mathcal{V}_q\mathfrak{g}$ we imposed several conditions on the order l of the root of unity q. Namely we assumed that l is even and $l \geq \max(3, r)$ where r is the dual Coxeter number of $\mathfrak{g}$. The condition $l \geq \max(3, r)$ seems to be natural, there are no indications that it may be dropped. The evenness of l is used to verify (5.2.4). It would be interesting to construct a similar modular category in the case of odd l.

2. All the results formulated above extend to simple complex Lie algebras of type B, C, although the formulas in the definition of quantum groups and the formula for the universal R-matrix become slightly more complicated. It seems that the only missing step in the derivation of modular categories from exceptional simple Lie algebras is a verification of (5.2.4).

3. For simple Lie algebras $\mathfrak{g}$ of type A, B, C, D, the category $\mathcal{V}_q\mathfrak{g}$ is known to be unimodal, see [RT1, p. 582] for the case $\mathfrak{g} = sl_2(\mathbb{C})$ and [TW, Corollary 5.3.3] for the general case. This allows us to apply the techniques of Chapters VI - X to this category.

4. It is implicit in the papers of Wenzl [We2], [We4] (though the details have not been worked out) that the category $\mathcal{V}_q\mathfrak{g}$ is Hermitian for any root of unity q and unitary for $q = \exp(\pm 2\pi\sqrt{-1}/l)$.

5. Some authors use the symbol $U_q\mathfrak{g}$ for a slightly different Hopf algebra depending on the choice of a square root t of q. This algebra is defined by $4m$ generators $X_i^+, X_i^-, k_i, k_i^{-1}$ where $i = 1, ..., m$ and relations obtained from (6.1.a) - (6.1.e) as follows. E_r, F_r for $r = 1, ..., m$ should be replaced with X_r^+, X_r^- respectively; K_r, K_r^{-1} should be replaced with k_r, k_r^{-1} except in (6.1.c) where K_i, K_i^{-1} should be replaced with k_i^2, k_i^{-2}; q, q^{-1} should be replaced with t^2, t^{-2} except in (6.1.b) where q should be replaced with t. The comultiplication, antipode, and counit are defined by the formulas $\Delta(X_i^\pm) = k_i \otimes X_i^\pm + X_i^\pm \otimes k_i^{-1}, \Delta(k_i) = k_i \otimes k_i, s(k_i) = k_i^{-1}, s(X_i^\pm) = -q^{\mp 1}X_i^\pm, \epsilon(k_i) = 1, \epsilon(X_i^\pm) = 0$. The passage from $U_q\mathfrak{g}$ to this Hopf algebra is obtained via the substitution $K_r = k_r^2, E_r = k_r X_r^+, F_r = X_r^- k_r^{-1}$.

7. Quantum groups with generic parameter

7.0. Outline. We define a version $\mathcal{U}_h\mathfrak{g}$ of $U_q\mathfrak{g}$ which is an h-adic Hopf algebra over the ring of formal power series $\mathbb{C}[[h]]$. We derive from $\mathcal{U}_h\mathfrak{g}$ a semisimple ribbon category with an infinite number of simple objects. Alhough this category can not be used to construct a 3-dimensional TQFT or invariants of shadows, it is interesting from the algebraic point of view. We can apply the techniques of Chapters I and VI to this category and obtain invariants of links in $\mathbb{R}^3$ and $6j$-symbols.

7.1. Completions and h-adic Hopf algebras. Set $K = \mathbb{C}[[h]]$. The h-adic completion of a K-module V is the inverse limit

$$\widehat{V} = \lim_{\leftarrow} V/h^n V$$

where $n = 0, 1, 2, ...$ There is a natural K-linear completion homomorphism $V \to \widehat{V}$ which is, in general, neither injective nor surjective. The completed tensor product $V \widehat{\otimes}_K W$ of K-modules V, W is the h-adic completion of $V \otimes_K W$, i.e.,

$$V \widehat{\otimes}_K W = \lim_{\leftarrow} (V \otimes_K W)/h^n (V \otimes_K W).$$

For example, if $V = K$ or V is a direct sum of finite number of copies of K then $\widehat{V} = V$. If V is a direct sum of infinite number of copies of K then $\widehat{V}$ is larger than V.

For algebras over the ring of formal power series $K = \mathbb{C}[[h]]$, we may generalize the definitions of comultiplication, universal R-matrix, and universal twist.

Instead of tensor squares and cubes of algebras we consider their h-adic completions. This yields the notions of h-adic Hopf algebras, h-adic quasitriangular Hopf algebras, and h-adic ribbon Hopf algebras over K.

Let A be an h-adic Hopf algebra over K. By an A-module of finite rank we mean a left A-module whose underlying K-module is a direct sum of a finite number of copies of K. Consider the category $\mathrm{Rep}(A)$ whose objects and morphisms are A-modules of finite rank and A-linear homomorphisms. For objects V, W of $\mathrm{Rep}(A)$ set $V \otimes W = V \otimes_K W = V \widehat{\otimes}_K W$ where the action of A is obtained from the obvious product action of $A \widehat{\otimes}_K A$ in $V \widehat{\otimes}_K W$ via the comultiplication $A \to A \widehat{\otimes}_K A$. Using the equalities $\widehat{V} = V$, $V \widehat{\otimes}_K W = V \otimes_K W$ for objects of $\mathrm{Rep}(A)$, we define the braiding and twist in $\mathrm{Rep}(A)$ as in Sections 2 and 3.

Here is an analogue of Theorem 3.2 in this setting.

7.1.1. Theorem. *For any h-adic ribbon Hopf algebra A over $\mathbb{C}[[h]]$, the category* $\mathrm{Rep}(A)$ *is a ribbon* Ab-*category with ground ring* $\mathbb{C}[[h]]$.

7.2. The h-adic quantum group. We introduce a version $\mathcal{U}_h\mathfrak{g}$ of $U_q\mathfrak{g}$ which is an h-adic Hopf algebra. The main idea is to replace $q \in \mathbb{C}$ (or rather $-2\log(q)$) in the definition of $U_q\mathfrak{g}$ with an indeterminate h and to apply the h-adic completion.

The algebra $\mathcal{U}_h\mathfrak{g}$ is the h-adic completion of the $\mathbb{C}[[h]]$-algebra generated by E_i, F_i, H_i with $i = 1, \ldots, m$ subject to relations

$$H_iH_j - H_jH_i = 0, \quad H_iE_j - E_jH_i = a_{ij}E_j, \quad H_iF_j - F_jH_i = -a_{ij}F_j$$

for $i, j = 1, \ldots, m$, the relation obtained from (6.1.c) by the substitutions

$$K_i = \exp(-hH_i/2), \quad q = \exp(-h/2),$$

the relation (6.1.d), and the relation obtained from (6.1.e) by the substitution $q = \exp(-h/2)$. The comultiplication, antipode, and counit in $\mathcal{U}_h\mathfrak{g}$ are defined on E_i, F_i by the same formulas as above and on H_i by the formulas

$$\Delta(H_i) = H_i \otimes 1 + 1 \otimes H_i, \quad s(H_i) = -H_i, \quad \varepsilon(H_i) = 0.$$

A generic element of $\mathcal{U}_h\mathfrak{g}$ may be expanded as a formal sum $\sum_{n=0}^{\infty} P_n h^n$ where $P_0, P_1, \ldots$ are polynomials in E_i, F_i, H_i with $i = 1, \ldots, m$.

The main result about $\mathcal{U}_h\mathfrak{g}$ is that it has the natural structure of an h-adic ribbon Hopf algebra over $\mathbb{C}[[h]]$. Here we give a brief decription of the universal R-matrix and twist.

Note first that the braid group $B(\mathfrak{g})$ acts on $\mathcal{U}_h\mathfrak{g}$. The action of the generators $T_1, \ldots, T_m$ is defined as follows. If $a_{ij} = 0$ then $T_i(E_j) = E_j, T_i(F_j) = F_j, T_i(H_j) = H_j$. If $a_{ij} = -1$ then

$$T_i(E_j) = \exp(\frac{h}{2})E_jE_i - E_iE_j, \ T_i(F_j) = \exp(-\frac{h}{2})F_iF_j - F_jF_i, \ T_i(H_j) = H_i + H_j.$$

Finally,

$$T_i(E_i) = -F_i \exp(-hH_i/2), \quad T_i(F_i) = -\exp(hH_i/2)E_i, \quad T_i(H_i) = -H_i.$$

We fix a decomposition $w_0 = r_{i_1} r_{i_2} ... r_{i_N}$ where N is the length of $w_0 \in W(\mathfrak{g})$. For $s = 1, ..., N$, set

$$e_s = T_{i_1} T_{i_2} ... T_{i_{s-1}}(E_{i_s}) \in \mathcal{U}_h\mathfrak{g} \quad \text{and} \quad f_s = T_{i_1} T_{i_2} ... T_{i_{s-1}}(F_{i_s}) \in \mathcal{U}_h\mathfrak{g}.$$

The R-matrix $R \in \mathcal{U}_h\mathfrak{g} \widehat{\otimes}_{\mathbb{C}[[h]]} \mathcal{U}_h\mathfrak{g}$ is given by the formula

$$R = \prod_{s=1}^{N} \left(\sum_{n=0}^{\infty} \exp\left(\frac{n(n-1)h}{4}\right) \frac{(1-\exp(-h))^n}{[n]!} e_s^n \otimes f_s^n \right) \exp\left(\frac{h}{2} \sum_{i,j=1}^{m} b_{ij} H_i \otimes H_j\right).$$

Here we use the notation of Section 6.2. The symbol $[n]!$ denotes the formal power series obtained from the expression for $[n]_q!$ by the substitution $q = \exp(-h/2)$. Note that the formal power series $(1 - \exp(-h))^n$ is divisible by h^n so that for every given n, the coefficient of h^n in the expression for R is given by a finite sum. Therefore R is an element of the completed tensor square of $\mathcal{U}_h\mathfrak{g}$.

The universal twist in $\mathcal{U}_h\mathfrak{g}$ is defined by the formula

(7.2.a) $$v = u_R^{-1} \exp(hb_1H_1 + ... + hb_mH_m) \in \mathcal{U}_h\mathfrak{g}$$

where $u_R \in \mathcal{U}_h\mathfrak{g}$ is determined by the R-matrix and the numbers $b_1, ..., b_m$ are the same as in Section 6.2.

7.2.1. Theorem. *The triple $(\mathcal{U}_h\mathfrak{g}, R, v)$ is an h-adic ribbon Hopf algebra.*

For more details and a proof of identities (2.1.a) - (2.1.c), (3.1.a), see [KR2], [LS], [Dr3].

7.3. The category $\mathcal{V}_q\mathfrak{g}$. An $\mathcal{U}_h\mathfrak{g}$-module of finite rank V is said to be regular if the operators $x \mapsto H_1x, ..., x \mapsto H_mx : V \to V$ may be simultaneously diagonalised with respect to a basis of V over $\mathbb{C}[[h]]$. (One can show that in this case the eigenvalues of these operators are non-negative integers.) For instance, the ring $\mathbb{C}[[h]]$ where $\mathcal{U}_h\mathfrak{g}$ acts via the counit is regular.

It is obvious that the regularity of $\mathcal{U}_h\mathfrak{g}$-modules is preserved under tensor product over $\mathbb{C}[[h]]$ and under passage to the dual module. Therefore regular $\mathcal{U}_h\mathfrak{g}$-modules and $\mathcal{U}_h\mathfrak{g}$-homomorphisms form a monoidal category with duality. Denote this category by $\mathcal{V}_q\mathfrak{g}$. It is a subcategory of the category $\text{Rep}(\mathcal{U}_q\mathfrak{g})$ defined in Section 7.1. The braiding and twist in $\text{Rep}(\mathcal{U}_q\mathfrak{g})$ determined by the universal R-matrix and twist in $\mathcal{U}_h\mathfrak{g}$ may be restricted to $\mathcal{V}_q\mathfrak{g}$ and yield a braiding and a twist in this category.

7.3.1. Theorem. *The category $\mathcal{V}_q\mathfrak{g}$ is a ribbon* Ab*-category with ground ring* $\mathbb{C}[[h]]$.

This theorem summarizes the results stated above.

7.4. Classification of regular $\mathcal{U}_h\mathfrak{g}$-modules. We formulate here the classification of regular $\mathcal{U}_h\mathfrak{g}$-modules of finite rank due to Lusztig [Lu1] and Rosso [Ro1].

Note first that a regular $\mathcal{U}_h\mathfrak{g}$-module of finite rank splits as a direct sum of a finite number of indecomposable regular $\mathcal{U}_h\mathfrak{g}$-modules. (A module is indecomposable if it can not be presented as a direct sum of non-trivial modules.)

The classification of indecomposable regular $\mathcal{U}_h\mathfrak{g}$-modules of finite rank proceeds in terms of highest weights. Each such module V contains an element x such that x generates V over $\mathcal{U}_h\mathfrak{g}$ and $E_i x = 0$, $H_i x = n_i x$ for all $i = 1, ..., m$ and certain $n_1, ..., n_m \in \mathbb{Z}$. The element x is called a highest weight vector of V. It is unique up to multiplication by an invertible element of $\mathbb{C}[[h]]$. Therefore the sequence $(n_1, ..., n_m)$, called the highest weight of V, does not depend on the choice of x.

Assigning to indecomposable regular $\mathcal{U}_h\mathfrak{g}$-modules of finite rank their highest weights, we obtain a bijection between the set of isomorphism classes of such modules and the set of sequences of m non-negative integers.

Each indecomposable regular $\mathcal{U}_h\mathfrak{g}$-modules of finite rank is simple in the sense that all its $\mathcal{U}_h\mathfrak{g}$-endomorphisms are multiplications by elements of $\mathbb{C}[[h]]$. Indeed, each such endomorphism has to preserve the highest weight vector at least up to multiplication by an element of $\mathbb{C}[[h]]$.

Summing up the results formulated above we obtain the following theorem.

7.4.1. Theorem. *The category $\mathcal{V}_q\mathfrak{g}$ is a semisimple ribbon* Ab-*category.*

As distinguished simple objects we may choose arbitrary representatives of the isomorphism classes of indecomposable regular $\mathcal{U}_h\mathfrak{g}$-modules.

7.5. Remarks. 1. Remarks 6.4.2 and 6.4.3 apply to $\mathcal{V}_q\mathfrak{g}$ word for word.

2. We may apply the technique of Chapter I to the category $\mathcal{V}_q\mathfrak{g}$. Assigning sequences of m non-negative integers to the components of a framed oriented link $L \subset \mathbb{R}^3$ we get a link colored over $\mathcal{V}_q\mathfrak{g}$. Applying the operator invariant F we obtain a $\mathbb{C}[[h]]$-valued isotopy invariant of L. It is not difficult to show that this invariant is a Laurent polynomial in the variable $q = \exp(-h/2)$.

The colorings of links as above are infinite in number so that we obtain an infinite family of polynomial link invariants. At a certain stage this created a psychological problem for topologists who are more accustomed to work under lack of invariants. Later on it was understood that these invariants may be packed together to define 3-dimensional TQFT's corresponding to roots of unity.

3. The classification of indecomposable regular $\mathcal{U}_h\mathfrak{g}$-modules of finite rank by sequences of m non-negative integers coincides with the standard classification of irreducible $\mathfrak{g}$-modules of finite complex dimension. This gives a canonical

bijective correspondence between these two families of isomorphism classes of modules.

These observations show that assigning irreducible finite dimensional $\mathfrak{g}$-modules to the components of a framed oriented link in $\mathbb{R}^3$ we get a colored link over $\mathcal{V}_q\mathfrak{g}$. The invariant F which arises when all link components are colored with the fundamental representation of $\mathfrak{g}$ is especially important. As above, this invariant is a Laurent polynomial in the variable $q = \exp(-h/2)$. For instance, if $\mathfrak{g} = sl_2(\mathbb{C})$ then this is a version of the Jones polynomial of links in $\mathbb{R}^3$. (We discuss the Jones polynomial in Chapter XII from a more elementary viewpoint.) For $\mathfrak{g} = sl_{m+1}(\mathbb{C})$, this gives a 1-variable reduction of the 2-variable Homfly polynomial. Varying $m = 0, 1, 2, \ldots$ we get a sequence of 1-variable polynomial invariants of links equivalent to the Homfly polynomial. When $\mathfrak{g}$ belongs to the series B, C, D this gives a 1-variable reduction of the Kauffman 2-variable polynomial. Varying $\mathfrak{g}$ in any of these series we obtain a sequence of 1-variable polynomial invariants of links equivalent to the Kauffman polynomial. For more on this, see [AW], [ADO], [ADW], [Tu4] and the excellent reviews [Li2], [Mo1].

Notes

Sections 1 - 3. Quasitriangular Hopf algebras were introduced by Drinfel'd, see [Dr1], [Dr2]. Ribbon Hopf algebras were introduced by Reshetikhin and Turaev, see [RT1].

Example 3.5.2 was communicated to the author by C. Cibils.

Section 4. The material of this section is inspired by the properties of quantum groups at roots of unity.

Section 5. Modular Hopf algebras were introduced in [RT2] with the view to deriving 3-manifold invariants from quantum groups. The definitions of ribbon and modular Hopf algebra given here simplify the ones in [RT1], [RT2]. (Note also that our twist v corresponds to v^{-1} in [RT2].)

Section 6 and 7. The study of quantum groups was inspired by the works of physicists on the Yang-Baxter equation, see [Fa], [FRT], [FST]. The definition of the quantum deformation of $sl_2(\mathbb{C})$ was given by Kulish and Reshetikhin [KuR], the comultiplication in this deformation was defined by E. Sklyanin. Quantum deformations of arbitrary simple Lie algebras were introduced by Drinfel'd [Dr1], [Dr2] and independently Jimbo [Ji1], [Ji2] (see also [Wo]). The representation theory of quantum groups with generic parameter has been studied by Lusztig [Lu1] and Rosso [Ro1]. For the case of roots, see [APW], [CK], [We2], [We4]. For a comprehensive introduction to the theory of quantum groups, see [Kas] and [Lu4].

Chapter XII
A geometric construction of modular categories

Throughout this chapter the symbol K denotes a commutative associative ring with unit and the symbol a denotes an invertible element of K.

1. Skein modules and the Jones polynomial

1.0. Outline. We introduce a family of K-modules called skein modules. They are generated by tangle diagrams in the plane quotiented by Kauffman's skein identity. In further sections the skein modules are used to construct skein categories which will eventually give rise to a modular category.

Kauffman's skein identity gives immediate access to the Jones polynomial of links in $\mathbb{R}^3$. We briefly discuss this polynomial.

1.1. The skein module $E_{k,l}$. Let k, l be non-negative integers. We shall consider tangle diagrams in the strip $\mathbb{R} \times [0, 1]$ with k inputs and l outputs. A tangle (k, l)-diagram consists of a finite number of arcs and loops in $\mathbb{R} \times [0, 1]$, the end points of the arcs being the k fixed points in the line $\mathbb{R} \times 0$ and l fixed points in the line $\mathbb{R} \times 1$. At each self-crossing point of the diagram one of the two crossing branches should be distinguished and said to be the lower one, the second branch being the upper one. We always assume that the arcs and loops of tangle diagrams lie in general position. On the other hand, we do not assume them to be oriented or colored. As usual, any tangle (k, l)-diagram in $\mathbb{R} \times [0, 1]$ represents a framed (k, l)-tangle in $\mathbb{R}^2 \times [0, 1]$. To visualize this tangle, thicken all arcs and loops into bands and annuli taking into account the overcrossing/undercrossing information. (It is understood that the bands and annuli are close to and almost parallel with the plane of the diagram.)

Let $E_{k,l} = E_{k,l}(a)$ be the K-module generated by all tangle (k, l)-diagrams quotiented by

(i) ambient isotopy in $\mathbb{R} \times [0, 1]$ fixed on the boundary of $\mathbb{R} \times [0, 1]$;

(ii) the relation $D \amalg \mathbb{O} = -(a^2 + a^{-2})D$, where D is an arbitrary diagram and $\mathbb{O}$ is a simple closed curve in $\mathbb{R} \times [0, 1]$ bounding a disk in the complement of D;

(iii) the identity in Figure 1.1.

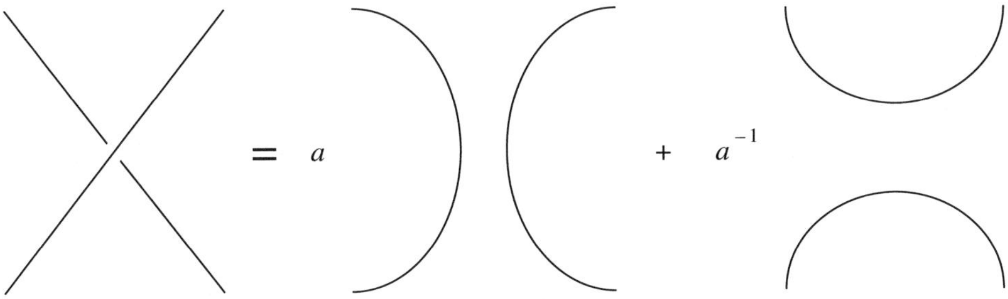

Figure 1.1

The formula in Figure 1.1 is Kauffman's celebrated skein relation. This relation involves three tangle diagrams identical except in a small 2-disk where they appear as depicted. Two diagrams on the right-hand side are obtained by smoothing a crossing of the diagram on the left-hand side. These two smoothed diagrams acquire coefficients a and a^{-1}. In order to determine which coefficient corresponds to which diagram we use the following rule. Let us move towards the crossing point along the upper branch. Just before reaching the crossing point turn to the left (in the clockwise direction) and move until attaining the lower branch, then move along this lower branch away from the crossing. In this way we get one of the two smoothings. It acquires the coefficient a and the second possible smoothing acquires the coefficient a^{-1}.

The module $E_{k,l}$ is called the skein (k, l)-module corresponding to $a \in K$. Each tangle (k, l)-diagram D represents an element of $E_{k,l}$ denoted by $\langle D \rangle$ and called the skein class of D. Note that if $k + l$ is odd then there are no tangle (k, l)-diagrams and $E_{k,l} = 0$.

Applying the skein relation to all crossing points of a tangle (k, l)-diagram D we may expand D as a formal linear combination of the classes of diagrams without crossing points (with coefficients in K). It is easy to check that this linear combination does not depend on the order in which we apply the skein relation to the crossing points of D. Further, using the relation (ii) we may get rid of loops inductively. This gives a canonical expansion of D as a linear combination of diagrams consisting of $(k+l)/2$ disjoint simple arcs. We shall call such (k, l)-diagrams simple. This argument shows that $E_{k,l}$ is a free K-module with a basis represented by simple (k, l)-diagrams. In particular, $E_{0,0} = K$ is generated by the skein class $\langle \emptyset \rangle = 1$ of the empty diagram. Although it is not needed, observe that the number of simple (k, l)-diagrams is equal to the Catalan number $\binom{2m}{m}/(m+1)$ where $m = (k + l)/2$.

1.2. Theorem. *The skein class of any tangle (k, l)-diagram is invariant under three moves shown in Figure* 1.2.

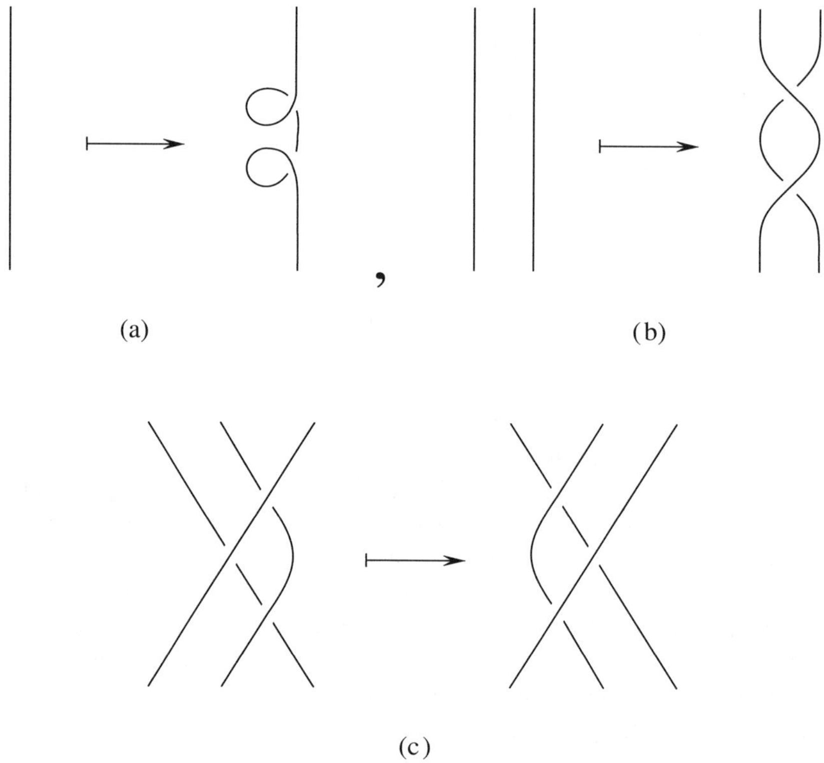

Figure 1.2

This fundamental (though elementary) theorem underlies the theory of skein modules. It is well known that two tangle (k, l)-diagrams represent isotopic framed tangles in $\mathbb{R}^2 \times [0, 1]$ if and only if these diagrams may be related by the moves shown in Figure 1.2 and the inverse moves. (We used a similar fact in the setting of colored oriented diagrams in Chapter I.) Therefore the skein class $\langle D \rangle \in E_{k,l}$ of a tangle diagram D yields an isotopy invariant of framed (k, l)-tangles. It is called the skein class of the tangle.

Proof of Theorem. Let D' be a tangle diagram obtained from a tangle diagram D by inserting one small positive curl. (The upper curl in Figure 1.2.a is positive, the lower curl is negative.) The computations in Figure 1.3 show that $\langle D' \rangle = -a^3 \langle D \rangle$. Similarly, if D' is a tangle diagram obtained from a tangle diagram D by inserting one small negative curl then $\langle D' \rangle = -a^{-3} \langle D \rangle$. This implies the invariance of $\langle D \rangle$ under the move in Figure 1.2.a. It should be stressed that the equalities in this and further figures mean the equalities of skein classes in the corresponding skein modules.

The invariance of $\langle D \rangle$ under the move shown in Figure 1.2.b is verified in Figure 1.4. Here we apply the Kauffman relation twice: first in the lower crossing

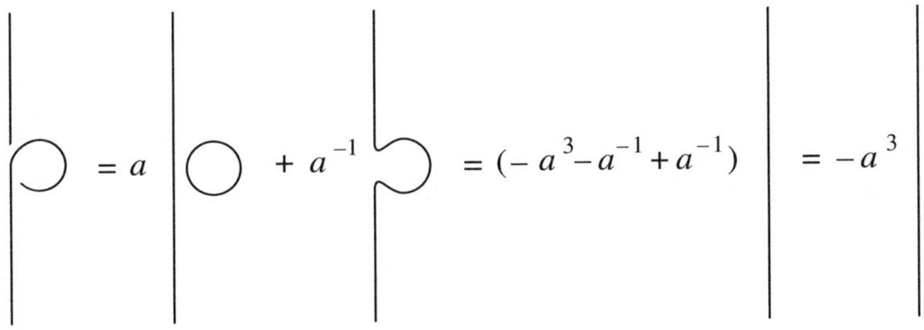

Figure 1.3

point and then in the second crossing point. We also use the result of the previous paragraph concerning positive curls. In Figure 1.5 we check the invariance of $\langle D\rangle$ under the move in Figure 1.2.c. The second and third equalities in Figure 1.5 follow from the invariance of $\langle D\rangle$ under the move in Figure 1.2.b.

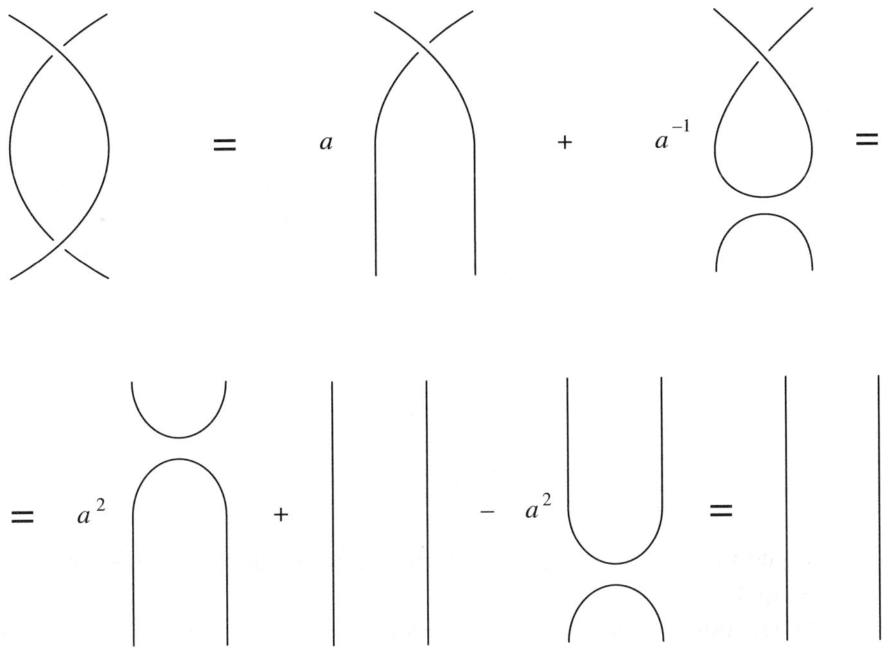

Figure 1.4

1.3. The bracket polynomial. Take $K = \mathbb{R}$. For any framed link $L \subset \mathbb{R}^3$, its skein class $\langle L\rangle \in E_{0,0} = \mathbb{R}$ provides a numerical isotopy invariant of L. It is easy to deduce from Kauffman's recursive formula that $\langle L\rangle$ may be expressed as a Laurent polynomial in $a \in \mathbb{R}$. The resulting one-variable Laurent polynomial $\langle L\rangle(a)$ is an isotopy invariant of L. It is called the bracket polynomial of L. The

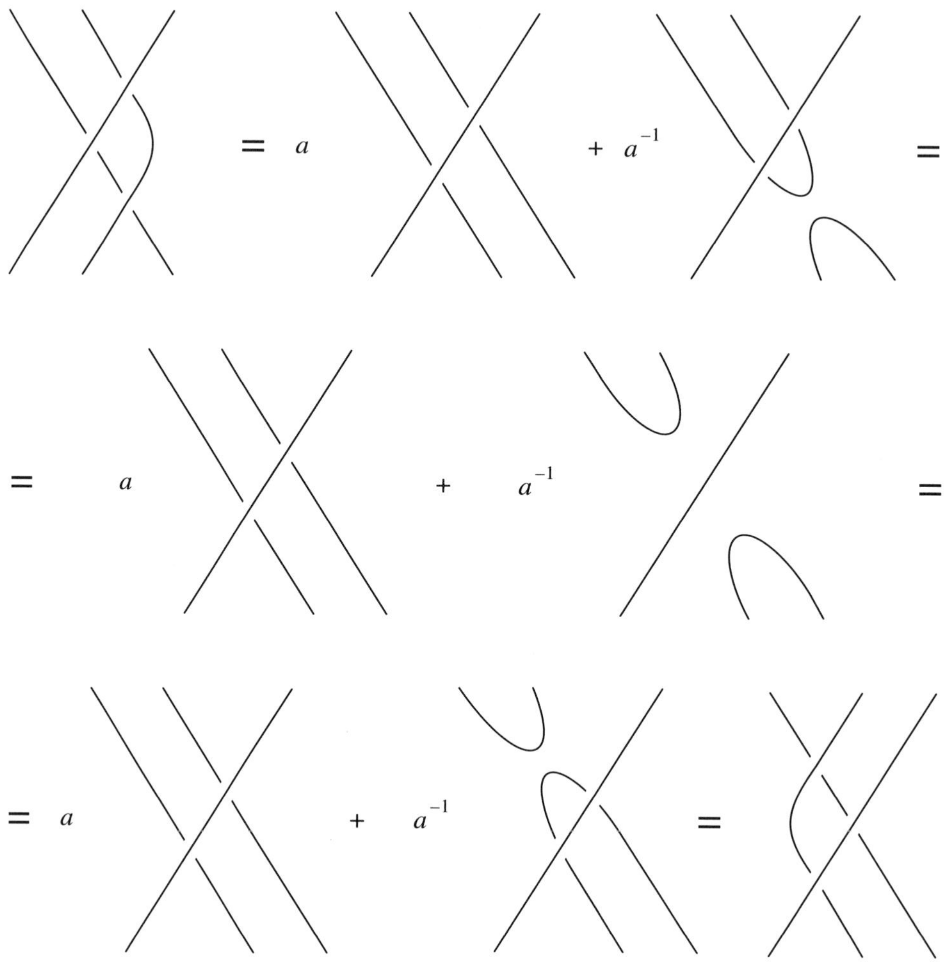

Figure 1.5

bracket polynomial is normalized so that its value for the trivial knot with zero framing is equal to $-(a^2 + a^{-2})$.

The bracket polynomial of a framed link does depend on the choice of the framing. Twisting the framing m times around a component leads to multiplication of the bracket polynomial by $(-a^3)^m$. The bracket polynomial regarded up to multiplication by integer powers of $-a^3$ is an invariant of non-framed links.

1.4. The Jones polynomial. Using the bracket polynomial, we define a polynomial invariant of oriented (non-framed) links in $\mathbb{R}^3$. This is a version of the Jones polynomial of oriented links.

Every oriented link $L \subset \mathbb{R}^3$ admits a framing such that the linking number of L with the link obtained by pushing L along the corresponding normal vector field is equal to 0. Such a framing is not unique (unless L is a knot) but the bracket polynomials corresponding to such framings are equal. This yields a polynomial invariant of L denoted by $V_L(a)$. Note that if L is a knot then the framing in question is given by the framing number 0. For a link L with $m \geq 2$ components $L_1, \ldots, L_m$, we can use the following framing on L satisfying the condition above. Provide each L_i with the framing number $-\sum_{j \neq i} \operatorname{lk}(L_i, L_j)$ where lk is the linking number. Denote the resulting framed link by L'. Clearly, $\operatorname{lk}(L, L') = 0$. Thus, $V_L(a) = \langle L' \rangle (a)$.

The polynomial $V_L(a)$ may be computed from any diagram D of L as follows. The orientation of L determines signs ± 1 of all crossing points of D (by definition, the crossings in Figure I.1.6 have signs $+1$ and -1, respectively). Let $w(D) \in \mathbb{Z}$ denote the sum of these signs over all crossing points of D. (The number $w(D)$ is called the writhe of D.) Then

$$V_L(a) = (-a^3)^{-w(D)} \langle D \rangle (a).$$

The polynomial $V_L(a)$ is the Jones polynomial of L, at least up to normalization. To obtain the original Jones polynomial, we should substitute $a = t^{-1/4}$ in $V_L(a)$ and divide by $-(t^{1/2} + t^{-1/2})$. For more details and references, see [Ka5], [Li2].

1.5. Exercise. Show that if the framed link $L \subset \mathbb{R}^3$ is the disjoint union of two framed links L_1 and L_2 then $\langle L \rangle = \langle L_1 \rangle \langle L_2 \rangle$. If L' is the mirror image of L then $\langle L' \rangle (a) = \langle L \rangle (a^{-1})$.

2. Skein category

2.0. Outline. We use the skein modules $\{E_{k,l}\}_{k,l}$ introduced in Section 1 to construct an Ab-category $\mathcal{S} = \mathcal{S}(a)$. We provide $\mathcal{S}$ with a braiding, twist, and duality so that it becomes a ribbon Ab-category. This is a preliminary version of a more elaborate skein category introduced in Section 6.

2.1. The category $\mathcal{S}$. The objects of $\mathcal{S}$ are non-negative integers $0, 1, 2, \ldots$. A morphism $k \to l$ of $\mathcal{S}$ is just an element of $E_{k,l}$. The composition fg of morphisms $g : k \to l, f : l \to m$ represented by tangle diagrams is defined by attaching a diagram of f on the top of a diagram of g and compressing the result into $\mathbb{R} \times [0, 1]$. This composition extends to arbitrary morphisms by linearity. The diagram consisting of k disjoint vertical arcs represents the identity morphism $\mathrm{id}_k : k \to k$. (In particular, the identity morphism $\mathrm{id}_0 : 0 \to 0$ is represented by the empty diagram.)

We provide the category $\mathcal{S}$ with a tensor product. The tensor product of the objects $k, l \in \{0, 1, 2, \ldots\}$ is the object $k+l$. The tensor product of two morphisms represented by tangle diagrams D, D' is the juxtaposition obtained by placing D' to the right of D. This tensor product extends to arbitrary morphisms by linearity. It is clear that $\mathcal{S}$ is a strict monoidal category with the unit object $\mathbb{1}_{\mathcal{S}} = 0$. The addition in the modules $\{\mathrm{Hom}_{\mathcal{S}}(k, l) = E_{k,l}\}_{k,l}$ makes $\mathcal{S}$ a monoidal Ab-category with ground ring $\mathrm{End}(\mathbb{1}_{\mathcal{S}}) = E_{0,0} = K$.

2.2. Braiding, twist, and duality in $\mathcal{S}$. Let $k, l \in \{0, 1, 2, \dots\}$. To define the braiding morphism $c_{k,l} : k \otimes l \to l \otimes k$ we take a bunch of k vertical arcs representing id_k and place it from above across a bunch of l vertical arcs representing id_l. This results in a tangle $(k + l, k + l)$-diagram with kl crossing points, see Figure 2.1.a where $k = 3, l = 2$. We define $c_{k,l} : k \otimes l \to l \otimes k$ to be the skein class of this diagram, i.e., the element of $E_{k+l,l+k} = E_{k+l,k+l}$ represented by this diagram.

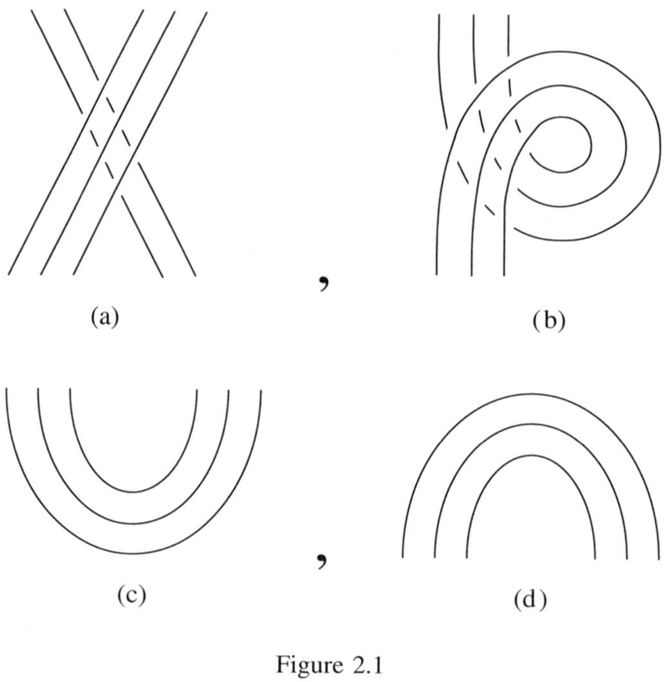

Figure 2.1

To define the twist $\theta_k : k \to k$ it is convenient to use framed tangles rather than diagrams. Take a bunch of k vertical unlinked untwisted bands and apply to it one full right-hand twist. The skein class of the resulting framed (k, k)-tangle is the twist $\theta_k : k \to k$. Here we use Theorem 1.2 which allows us to consider skein classes of framed tangles. A diagram representing θ_3 is drawn in Figure 2.1.b.

We agree that all objects of $\mathcal{S}$ are self-dual: for $k = 0, 1, 2, \dots$, set $k^* = k$. The duality morphism $b_k : 0 \to k \otimes k^* = 2k$ is represented by the diagram formed by k concentric cup-like arcs, see Figure 2.1.c where $k = 3$. The morphism $d_k : k^* \otimes k \to 0$ is represented by the diagram formed by k concentric cap-like arcs, see Figure 2.1.d where $k = 3$.

2.3. Theorem. *The monoidal category $\mathcal{S}$ with braiding, twist, and duality as defined above is a strict ribbon* Ab-*category.*

Proof. We should verify the identities specified in Sections I.1.2 and I.1.3 and establish invertibility of the twist and braiding. Consider the first braiding identity $c_{U,V\otimes W} = (\mathrm{id}_V \otimes c_{U,W})(c_{U,V} \otimes \mathrm{id}_W)$. Note that the braiding morphisms in $\mathcal{S}$ are represented by tangle diagrams without appealing to formal linear combinations. Therefore both sides of this identity are represented by tangle diagrams. It is enough to draw these diagrams in order to see that they are ambient isotopic in $\mathbb{R} \times [0, 1]$. Therefore they belong to the same skein class. This yields the first braiding identity. The second braiding identity (I.1.2.c) and the axioms of duality (I.1.3.b), (I.1.3.c) are verified in a similar way.

Let us prove the naturality of the braiding. We should verify that for any morphisms $f : V \to V'$, $g : W \to W'$ in $\mathcal{S}$,

$$(g \otimes f)\, c_{V,W} = c_{V',W'}(f \otimes g). \tag{2.3.a}$$

Since the tensor product and composition in $\mathcal{S}$ are bilinear, it is enough to consider the case where f, g are represented by tangle diagrams. The diagrams appearing on the left-hand and right-hand sides of (2.3.a) are not, in general, ambient isotopic in $\mathbb{R} \times [0, 1]$. However, the framed tangles represented by these diagrams are isotopic in $\mathbb{R}^2 \times [0, 1]$. Therefore (2.3.a) follows from Theorem 1.2. Formulas (I.1.2.h) and (I.1.3.d) are proven similarly.

Invertibility of the twist is obvious because the right-hand twist cancels the left-hand twist up to isotopy. Invertibility of the braiding morphism $c_{k,l}$ is also obvious: the inverse morphism is presented by the tangle diagram obtained from that of $c_{l,k}$ by taking the mirror image with respect to the plane of the diagram.

2.4. Overview of further sections. The construction of the category $\mathcal{S}$ is too elementary to produce a modular or even semisimple category. The category $\mathcal{S}$ has only two simple objects 0 and 1 which do not dominate the remaining objects. We shall modify $\mathcal{S}$. To this end we study the algebras $\{\mathrm{End}_{\mathcal{S}}(k) \mid k = 0, 1, ...\}$ in more detail (Sections 3 - 5). In Section 6 we derive another ribbon category $\mathcal{V}$ from $\mathcal{S}$. Its properties are studied in Section 7 where we show that if a is a primitive $4r$-th root of unity then $\mathcal{V}$ is quasimodular and if a is generic then $\mathcal{V}$ is semisimple with an infinite number of simple objects. We also study the multiplicity modules of $\mathcal{V}$ (Section 8) and Hermitian and unitary structures in $\mathcal{V}$ (Section 9).

2.5. Remarks. 1. Since $\mathcal{S}$ is a ribbon category, we may use it along the lines of Chapter I. Thus, we may color ribbon graphs in $\mathbb{R}^3$ with objects and morphisms of $\mathcal{S}$ and consider the corresponding operator invariant F. The resulting invariants may be obtained from the Kauffman bracket by simple geometric manipulations. For example, if a knot $L \subset \mathbb{R}^3$ is colored with an object $k = 0, 1, 2, \ldots$ of $\mathcal{S}$ then to compute $F(L) \in \mathrm{End}_{\mathcal{S}}(0) = K$ we can take k longitudes of L determined by the framing and apply the bracket polynomial to this link. In the case $k = 1$ we recover the Jones polynomial (up to normalization).

2. The ideas of Section 2.2 may be applied to other categories of tangles and graphs, for instance, to the category of v-colored ribbon graphs $\mathrm{Rib}_{\mathcal{V}}$ studied in Chapter I. In this way we may transform $\mathrm{Rib}_{\mathcal{V}}$ to a ribbon category. This category has too many morphisms to be interesting from the algebraic point of view. In contrast to $\mathrm{Rib}_{\mathcal{V}}$, the definition of $\mathcal{S}$ involves a subtle factorization (passage to the skein classes) which makes it suitable for an algebraic study.

3. The Temperley-Lieb algebra

3.0. Outline. We study the algebras of endomorphisms $\{\mathrm{End}_{\mathcal{S}}(k) = E_{k,k} \,|\, k = 0, 1, 2, \ldots\}$ of the objects of $\mathcal{S}$. The algebra $E_{k,k}$ is a geometric version of an algebra introduced (in terms of generators and relations) by H. Temperley and E. Lieb. At the end of the section we discuss a trace $E_{k,k} \to K$.

3.1. Definition of the Temperley-Lieb algebra. For any $k \geq 0$, the module $\mathrm{End}_{\mathcal{S}}(k) = E_{k,k}$ acquires the structure of an algebra over K. The product xy of $x, y \in E_{k,k}$ is just the composition of x and y viewed as morphisms $k \to k$. (If x, y are represented by tangle diagrams then to compute xy we attach a diagram of x on top of a diagram of y and compress the result into $\mathbb{R} \times [0, 1]$. This product extends to arbitrary $x, y \in E_{k,k}$ by linearity.) It is obvious that the algebra $E_{k,k}$ is associative and has a unit 1_k represented by the diagram consisting of k disjoint vertical arcs. We call this algebra the k-th Temperley-Lieb algebra and denote it by E_k. For instance, $E_0 = K$ and $E_1 = K$. For $k \geq 3$, the algebra E_k is non-commutative.

The algebra E_k regarded as a K-module admits a basis whose elements are represented by simple tangle (k, k)-diagrams (with k fixed endpoints in the bottom boundary line and k fixed endpoints in the top boundary line). The algebra E_k admits a rather simple family of multiplicative generators. Namely, for $i = 1, \ldots, k - 1$, denote by e_i the diagram shown in Figure 3.1. By abuse of notation, we shall denote the skein class $\langle e_i \rangle \in E_k$ by the same symbol e_i.

3.2. Theorem. *The elements $1_k, e_1, e_2, \ldots, e_{k-1} \in E_k$ generate E_k as a K-algebra.*

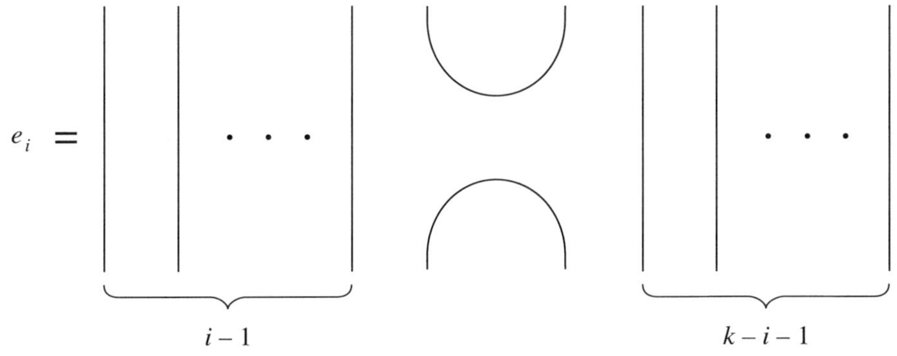

Figure 3.1

Proof. It is enough to prove the following claim.

Claim. *Let $D \subset \mathbb{R} \times [0, 1]$ be a simple (k, k)-diagram that is not a diagram of 1_k. Then D is ambient isotopic in $\mathbb{R} \times [0, 1]$ to a diagram decomposable as a product of $e_1, e_2, \ldots, e_{k-1}$.*

For the definition of simple diagrams, see Section 1.1. To get some geometric feeling the reader may prove this claim for $k = 2, 3, 4$. We shall prove the claim by induction on k. If $k = 2$ then D is ambient isotopic to e_1 and the claim is true. Assume that the claim is true for simple diagrams with $k - 1$ strands and prove it for simple diagrams with k strands. Let $x_1, \ldots, x_k$ be the inputs (bottom endpoints) of D enumerated from left to right. Since D is not a diagram of 1_k, there is a strand of D connecting two inputs of D. Since D is simple, there is a strand of D connecting two inputs which are immediate neighbors of each other. Denote by $i = i(D)$ the minimal $i = 1, 2, \ldots, k$ such that there is an arc of D connecting x_i and x_{i+1}. Now we use induction on $i(D)$. If $i(D) > 1$ then $D = D'e_i$ where D' is a simple (k, k)-diagram with $i(D') = i(D) - 1$. The diagram D' is obtained from D by the following transformation in a neighborhood of x_{i-1}, x_i, x_{i+1}. We slightly deform the strand of D issuing from x_{i-1} to produce a mutually cancelling local maximum and local minimum of the height function. We may assume that the local minimum lies strictly above the arc of D connecting x_i and x_{i+1}. Now we may strip off e_i and present D as $D = D'e_i$ with $i(D') = i(D) - 1$. (Example of such a decomposition: $(\cup \otimes |)(| \otimes \cap) = e_1e_2$.) It remains to consider the case $i(D) = 1$. We decompose D as $D = D''e_1$ where D'' is a simple (k, k)-diagram constructed as follows. Consider the strand of D descending from the leftmost output (top endpoint) in $\mathbb{R} \times 1$. We take a small arc on this strand lying close to this output and push it down close to the strand of D connecting x_1 and x_2. This allows us to strip off e_1 and to present D in the form $D = D''e_1$. (Example of such a decomposition: $(| \otimes \cup)(\cap \otimes |) = e_2e_1$.) It is clear that D'' contains a strand joining the leftmost input to the leftmost output. In other words, D'' is obtained by adding a vertical interval from the left to a simple $(k - 1, k - 1)$-diagram.

The inductive assumption implies that D'' is a product of diagrams $\{e_i\}_{i=2}^k$. This implies our claim and completes the proof of the theorem.

3.3. Lemma. *The elements $1_k, e_1, e_2, \ldots, e_{k-1} \in E_k$ satisfy the following relations:*

$$e_i^2 = -(a^2 + a^{-2})\, e_i, \tag{3.3.a}$$

$$e_i e_j = e_j e_i \quad \text{if } |i - j| \geq 2, \tag{3.3.b}$$

$$e_i e_{i\pm 1} e_i = e_i \tag{3.3.c}$$

where $i, j = 1, \ldots, k-1$.

It is understood that (3.3.c) holds whenever $i \pm 1 \in \{1, \ldots, k-1\}$. The proof of Lemma 3.3 is a simple geometric exercise. The relations (3.3.b) and (3.3.c) hold on the level of ambient isotopy classes of diagrams; to prove (3.3.a) one should involve the second relation in the definition of skein modules.

Although it is not needed, note that the relations (3.3.a), (3.3.b), (3.3.c) generate all relations between $e_1, e_2, \ldots, e_{k-1} \in E_k$.

3.4. Trace tr : $E_k \to K$. Theorem 2.3 allows us to apply to $\mathcal{S}$ the theory of ribbon categories. In particular, we may consider the trace of morphisms $E_k = \mathrm{End}_{\mathcal{S}}(k) \to \mathrm{End}_{\mathcal{S}}(0)$ introduced in Section I.1.5. Using the identification $\mathrm{End}_{\mathcal{S}}(0) = K$ determined by the Kauffman bracket we get a trace $E_k \to K$. For convenience of the reader, we give here a direct (equivalent) definition of this trace.

For any $f \in E_k$, define $\mathrm{tr}(f) \in K$ as follows. Assume first that f may be presented by a tangle diagram D. Let D' be the closure of D obtained by joining the bottom endpoints of D with the top endpoints of D by k arcs unlinked with each other and with D (cf. Figure I.2.10). It is clear that D' is a link diagram. Set $\mathrm{tr}(f) = \langle D' \rangle \in K$. This trace extends to arbitrary $f \in E_k$ by linearity.

For instance, the closure of the diagram e_i consists of $k-1$ disjoint simple loops. Therefore $\mathrm{tr}(e_i) = (-a^2 - a^{-2})^{k-1}$ for any $i = 1, \ldots, k-1$. Similarly, $\mathrm{tr}(\mathrm{id}_k) = \mathrm{tr}(1_k) = (-a^2 - a^{-2})^k$.

3.4.1. Lemma. (i) *For any $f, g \in E_k$, we have* $\mathrm{tr}(fg) = \mathrm{tr}(gf)$.

(ii) *For any $k, l = 0, 1, \ldots$ and any $f \in E_k, g \in E_l$, we have* $\mathrm{tr}(f \otimes g) = \mathrm{tr}(f)\,\mathrm{tr}(g)$.

(iii) *For any element f of the subalgebra $K(1_k, e_1, e_2, \ldots, e_{k-2}) \subset E_k$ generated by $1_k, e_1, e_2, \ldots, e_{k-2}$, we have* $\mathrm{tr}(fe_{k-1}) = (-a^2 - a^{-2})^{-1}\mathrm{tr}(f)$.

Proof. It is enough to consider the case where f, g are skein classes of tangle diagrams. The link diagrams representing $\mathrm{tr}(fg)$ and $\mathrm{tr}(gf)$ are ambient isotopic, therefore $\mathrm{tr}(fg) = \mathrm{tr}(gf)$. The framed link representing $\mathrm{tr}(f \otimes g)$ is the disjoint union of framed links representing $\mathrm{tr}(f)$ and $\mathrm{tr}(g)$. Therefore, (ii) follows from the result of Exercise 1.5.

For any $f \in K(1_k, e_1, e_2, ..., e_{k-2})$, the diagram used to compute $\mathrm{tr}(f)$ splits as a disjoint union of a small circle and the diagram used to compute $\mathrm{tr}(fe_{k-1})$. Therefore, (iii) follows from the second relation in the definition of skein classes.

3.4.2. Lemma. *Modulo the identification $End_{\mathcal{G}}(0) = K$ the trace $tr : E_k \to K$ defined above coincides with the trace $End_{\mathcal{G}}(k) \to End_{\mathcal{G}}(0)$ introduced in Section* I.1.5.

In further sections we shall only use the definition of the trace given above in this section. This makes the exposition independent of the results of Chapter I and in particular of Theorem I.2.5.

Proof of Lemma. Consider the skein class $f = \langle D \rangle \in E_k$ of a tangle (k, k)-diagram D. The trace of f in the sense of Section I.1.5 is equal to the composition $d_k c_{k,k}((\theta_k f) \otimes \mathrm{id}_k) b_k : 0 \to 0$. Using the definitions of $d_k, b_k, c_{k,k}, \theta_k, \mathrm{id}_k$ given in Section 1 we may visualize the framed link in $\mathbb{R}^3$ representing this composition. It is easy to check that this framed link is isotopic to the one represented by the closure of D. Therefore they belong to the same skein class. This proves the lemma.

3.5. Canonical inclusions. For each pair $k, l = 0, 1, 2, ...$ with $k < l$, there is a canonical inclusion $E_k \hookrightarrow E_l$. Algebraically, it is defined by the formula $f \mapsto f \otimes \mathrm{id}_{l-k}$. Geometrically, it is defined by adding $l - k$ vertical intervals from the right. It is obvious that for any $f \in E_k$, we have $\mathrm{tr}(f \otimes \mathrm{id}_{l-k}) = (-a^2 - a^{-2})^{l-k}\mathrm{tr}(f)$.

3.6. Exercise. Show that $E_k = K \cdot 1_k \oplus K(e_1, ..., e_{k-1})$ where $K(e_1, ..., e_{k-1})$ is the subalgebra of E_k generated by $e_1, ..., e_{k-1}$.

4. The Jones-Wenzl idempotents

4.0. Outline. We consider certain idempotents in the Temperley-Lieb algebras introduced by V. Jones and studied by H. Wenzl. These idempotents will play a crucial role in further sections.

4.1. Theorem. *Assume that $a^{4n} - 1 \in K^*$ for $n = 1, 2, ..., k$ where K^* is the group of invertible elements of K. Then there exists a unique element $f_k \in E_k$ such that $f_k - 1_k \in K(e_1, ..., e_{k-1})$ and $e_i f_k = f_k e_i = 0$ for all $i = 1, ..., k - 1$.*

Here, as above, $K(e_1, ..., e_{k-1})$ is the subalgebra of E_k generated by $e_1, ..., e_{k-1}$.

It follows from the properties of $f = f_k$ that $f^2 - f = (f-1)f = 0$. Thus, f_k is an idempotent. It is called the Jones-Wenzl idempotent of E_k. Since f_k commutes with all multiplicative generators of E_k it belongs to the center of E_k. The idempotent f_k annihilates all additive generators of E_k mentioned in Section 3.1 except 1_k. (Indeed, all these generators are monomials in $e_1, ..., e_{k-1}$.)

Proof of Theorem. If f and f' are two elements of E_k satisfying the conditions of the theorem then $(f-1)f' = 0 = (f'-1)f$ and therefore $f' = ff' = f'f = f$. This shows the uniqueness of f_k.

Let us construct f_k. For any integer n, set

$$[n] = [n]_{a^2} = (a^{2n} - a^{-2n})(a^2 - a^{-2})^{-1} \in K.$$

In particular, $[1] = 1$ and $[2] = a^2 + a^{-2}$. Our assumptions ensure that $[1], [2], ..., [k]$ are invertible in K.

Set $f^{(1)} = 1_k \in E_k$ and inductively

$$f^{(n+1)} = f^{(n)} + \frac{[n]}{[n+1]} f^{(n)} e_n f^{(n)} \in E_k$$

where $n = 1, ..., k-1$. An easy induction shows that $f^{(n)} - 1_k \in K(e_1, ..., e_{n-1})$. This implies that $f^{(n)}$ commutes with $e_{n+1}, ..., e_{k-1}$.

Claim. *For each $n = 1, 2, ..., k$, we have*

(4.1.a$_n$) $$f^{(n)} f^{(n)} = f^{(n)},$$

(4.1.b$_n$) $$e_i f^{(n)} = 0 \quad \textit{for all} \ i < n.$$

For $n = 1, 2, ..., k-1$, we have

(4.1.c$_n$) $$(e_n f^{(n)})^2 = -([n+1]/[n]) e_n f^{(n)}.$$

We first prove this claim and then proceed to the proof of the lemma. In the case $n = 1$ the claim is obvious. Inductively suppose that this claim is true for a certain n and prove it for $n+1$. Set $\alpha = [n]/[n+1]$ so that $f^{(n+1)} = f^{(n)} + \alpha f^{(n)} e_n f^{(n)}$.

Proof of (4.1.a$_{n+1}$):

$$f^{(n+1)} f^{(n+1)} = f^{(n)} + 2\alpha f^{(n)} e_n f^{(n)} + \alpha^2 f^{(n)} (e_n f^{(n)})^2 =$$

$$= f^{(n)} + \alpha f^{(n)} e_n f^{(n)} = f^{(n+1)}$$

where the first equality follows from the inductive assumption (4.1.a$_n$) and the second equality follows from (4.1.c$_n$).

Proof of (4.1.b_{n+1}). It follows directly from (4.1.b_n) and the definition of $f^{(n+1)}$ that $e_i f^{(n+1)} = 0$ for $i < n$. By (4.1.c_n), we have

$$e_n f^{(n+1)} = e_n f^{(n)} + \alpha(e_n f^{(n)})^2 = e_n f^{(n)} - e_n f^{(n)} = 0.$$

Proof of (4.1.c_{n+1}). It follows from (4.1.a_n) that $f^{(n)} f^{(n+1)} = f^{(n+1)}$. Recall that $f^{(n)}$ commutes with e_{n+1}. Therefore

$$(e_{n+1} f^{(n+1)})^2 = e_{n+1}(f^{(n)} + \alpha f^{(n)} e_n f^{(n)}) e_{n+1} f^{(n+1)} =$$

$$= e_{n+1}^2 f^{(n+1)} + \alpha f^{(n)} e_{n+1} e_n e_{n+1} f^{(n+1)} =$$

$$= -(a^2 + a^{-2}) e_{n+1} f^{(n+1)} + \alpha f^{(n)} e_{n+1} f^{(n+1)} =$$

$$= -(a^2 + a^{-2} - \alpha) e_{n+1} f^{(n+1)} = -([n+2]/[n+1]) e_{n+1} f^{(n+1)}$$

where the last equality follows from the recursive formula

$$[n+2] = (a^2 + a^{-2})[n+1] - [n]. \tag{4.1.d}$$

Set $f_k = f^{(k)} \in E_k$. We show that f_k satisfies the conditions of the lemma. By (4.1.b_k), we have $e_i f_k = 0$ for all $i = 1, 2, ..., k-1$. It was already mentioned above that $f_k - 1_k \in K(e_1, ..., e_{k-1})$. It remains to show that $f_k e_i = 0$ for $i = 1, 2, ..., k-1$. Consider a K-linear antiautomorphism rot : $E_k \to E_k$ defined as follows. Let us rotate the strip $\mathbb{R} \times [0, 1] \times 0$ around the horizontal line $\mathbb{R} \times (1/2) \times 0$ in $\mathbb{R}^3$. Rotation by the angle π carries this strip into itself. Any tangle (k, k)-diagram is carried by this rotation into another tangle (k, k)-diagram. It is easy to check that this transformation preserves the Kauffman relation and induces a K-linear antiautomorphism, rot, of E_k. It is obvious that $\mathrm{rot}(1_k) = 1_k$ and $\mathrm{rot}(e_i) = e_i$ for $i = 1, ..., k-1$. An easy induction shows that for any $n = 1, 2, ..., k$, we have $\mathrm{rot}(f^{(n)}) = f^{(n)}$. In particular, $\mathrm{rot}(f_k) = f_k$. Hence, for any generator e_i, we have $f_k e_i = \mathrm{rot}(f_k)\mathrm{rot}(e_i) = \mathrm{rot}(e_i f_k) = 0$.

4.2. Remarks. 1. The tensor product in the category $\mathcal{S}$ gives rise to a bilinear form $(f, g) \mapsto f \otimes g : E_k \otimes E_l \to E_{k+l}$. It follows directly from the inductive construction of f_k given in the proof of Theorem 4.1 that for any $k \geq 0$,

$$f_{k+1} = (f_k \otimes 1_1) + \frac{[k]}{[k+1]} (f_k \otimes 1_1)\, e_k\, (f_k \otimes 1_1) \in E_{k+1} \tag{4.2.a}$$

where $1_1 = \mathrm{id}_1$ is the unit of E_1 represented by a vertical arc. Of course, $f_k \otimes 1_1$ is just the image of f_k under the canonical inclusion $E_k \hookrightarrow E_{k+1}$.

2. The proof of Theorem 4.1 shows that the Jones-Wenzl idempotent $f_k \in E_k$ is invariant under rotation around a horizontal line. As a matter of fact, it is also invariant under rotation around a vertical line. Namely, rotate the strip $\mathbb{R} \times [0, 1] \times 0$ around the line $0 \times (1/2) \times \mathbb{R}$ orthogonal to this strip in $\mathbb{R}^3$. Rotation by the angle π carries the strip into itself. Any tangle (k, k)-diagram is carried by this rotation into another tangle (k, k)-diagram. It is easy to check that this transformation

preserves the Kauffman relation and induces a K-linear antiautomorphism $\mathrm{rot}' : E_k \to E_k$. Clearly, $\mathrm{rot}'(e_i) = e_{k-i}$ for any i. This implies that $\mathrm{rot}'(f_k)$ satisfies the conditions of Theorem 4.1. The uniqueness in this theorem ensures that $\mathrm{rot}'(f_k) = f_k$.

4.3. Roots of unity versus generic elements. The sequence of idempotents $f_0 \in E_0(a), f_1 \in E_1(a), ...$ may be infinite or finite depending on the properties of $a \in K$. We say that the element $a \in K$ is generic if $a^{4n} - 1 \in K^*$ for all $n \geq 1$. For any generic $a \in K$, we have an infinite sequence of the corresponding idempotents $f_0, f_1, ...$

We say that $a \in K$ is a primitive k-th root of unity if $a^n - 1 \in K^*$ for $n = 1, 2, ..., k-1$ and $a^k = 1$. For any primitive $4r$-th root of unity $a \in K$, we have exactly r idempotents $f_0, f_1, ..., f_{r-1}$. Note that if K is a field then our notion of a primitive root of unity coincides with the standard one.

4.4. The Jones-Wenzl idempotents as generators. We may view each idempotent f_k as a morphism $k \to k$ in the category $\mathcal{S}$. It turns out that in the case of generic a the sequence $f_0, f_1, f_2, ...$ generates the set of all morphisms in $\mathcal{S}$ as a two-sided ideal. (For a precise formulation, see Lemma 4.4.1 below.) In the case when a is a primitive $4r$-th root of unity there is a similar result *modulo negligible morphisms* (see Lemma 4.4.3). These fundamental facts will be crucial in Section 6. To prove Lemma 4.4.3 we need to compute the trace of f_k. This computation is given in Lemma 4.4.2.

4.4.1. Lemma. *If $a \in K$ is generic then any morphism $k \to l$ in $\mathcal{S} = \mathcal{S}(a)$ may be expressed as a sum $\sum_s x_s f_{i_s} y_s$ where s runs over a finite set of indices, $i_s \in \{0, 1, 2, ...\}$, and $x_s : i_s \to l$, $y_s : k \to i_s$ are morphisms in $\mathcal{S}$.*

Proof. Denote by G the set of morphisms in $\mathcal{S}$ which may be expressed as in the statement of the lemma. Every Jones-Wenzl idempotent $f_k = \mathrm{id}_k f_k \mathrm{id}_k : k \to k$ belongs to G. It is clear that the composition of any morphism belonging to G with any other morphism in $\mathcal{S}$ belongs to G. Since any morphism $f : k \to l$ is equal to $f\,\mathrm{id}_k$, it is enough to prove that for each $k = 0, 1, 2, ...$ the identity morphism $\mathrm{id}_k = 1_k \in E_k$ belongs to G. For $k = 0$ and $k = 1$, this is true because $1_k = f_k$ in these cases. Inductively, assume that $\mathrm{id}_k \in G$ for all $k < N$ and show that $\mathrm{id}_N \in G$, where $N \geq 2$. By the very definition of f_N we have $f_N - 1_N \in \sum_{i=1}^{N-1} e_i E_N$. The morphism $e_i : N \to N$ of $\mathcal{S}$ may be expressed in the obvious way as a composition of a morphism $e_i' : N \to N-2$ and a morphism $e_i'' : N-2 \to N$. Thus $e_i = e_i''\,\mathrm{id}_{N-2}\, e_i'$. By the inductive assumption $\mathrm{id}_{N-2} \in G$ so that $e_i \in G$. Hence $f_N - 1_N \in G$ and $1_N \in G$.

4.4.2. Lemma. *If $a^{4n} - 1 \in K^*$ for $n = 1, 2, ..., k$ then $tr(f_k) = (-1)^k[k+1]$.*

Proof. For $k = 1$, we have

$$\mathrm{tr}(f_k) = \mathrm{tr}(1_1) = \langle \mathbb{O} \rangle = -a^2 - a^{-2} = -[2]$$

where $\mathbb{O}$ is a simple closed curve in the plane. We shall inductively assume that $\mathrm{tr}(f_k) = (-1)^k[k+1]$ and show that $\mathrm{tr}(f_{k+1}) = (-1)^{k+1}[k+2]$. Set $f = f_k \otimes 1_1 \in E_{k+1}$. We have

$$\mathrm{tr}(f_{k+1}) = \mathrm{tr}(f) + ([k]/[k+1])\,\mathrm{tr}(f e_k f) = \mathrm{tr}(f) + ([k]/[k+1])\,\mathrm{tr}(f f e_k) =$$

$$= \mathrm{tr}(f) + ([k]/[k+1])\,\mathrm{tr}(f e_k) = \mathrm{tr}(f) - ([k]/[k+1])(a^2 + a^{-2})^{-1}\mathrm{tr}(f).$$

Here we use Lemma 3.4.1 (i) and (iii). Substituting in the last formula the expression

$$\mathrm{tr}(f) = \mathrm{tr}(f_k)\,\mathrm{tr}(1_1) = (-1)^{k+1}(a^2 + a^{-2})[k+1]$$

and applying (4.1.d) we obtain $\mathrm{tr}(f_{k+1}) = (-1)^{k+1}[k+2]$.

4.4.3. Lemma. *If $a \in K$ is a primitive $4r$-th root of unity then any morphism $k \to l$ in $\mathcal{S} = \mathcal{S}(a)$ may be expressed as a sum $z + \sum_s x_s f_{i_s} y_s$ where s runs over a finite set of indices, $i_s \in \{0, 1, 2, ..., r-2\}$, $x_s : i_s \to l$, $y_s : k \to i_s$ are morphisms in $\mathcal{S}$ and $z : k \to l$ is a negligible morphism in $\mathcal{S}$.*

Proof. The reader may notice that we do not include the idempotent $f_{r-1} \in E_{r-1}$ in the set of generators. The reason is that this idempotent, viewed as a morphism $(r-1) \to (r-1)$, is negligible. (This means that for any $g \in E_{r-1}$, we have $\mathrm{tr}(f_{r-1}g) = 0$.) This fact underlies the whole proof and we establish it first. Since f_{r-1} annihilates all generators of E_{r-1} except 1_{r-1}, it is enough to show that $\mathrm{tr}(f_{r-1}1_{r-1}) = \mathrm{tr}(f_{r-1}) = 0$. This follows directly from Lemma 4.4.2 since $[r] = (a^{2r} - a^{-2r})(a^2 - a^{-2})^{-1} = 0$.

Now we proceed to the proof of the lemma. Denote by G the set of morphisms in $\mathcal{S}$ which may be expressed as in the statement of the lemma. In particular every idempotent $f_k = \mathrm{id}_k f_k \mathrm{id}_k$ with $k = 0, 1, 2, ..., r-2$ viewed as a morphism $k \to k$ belongs to G. It is clear that the composition of any morphism belonging to G with any other morphism in $\mathcal{S}$ belongs to G. Since any morphism $f : k \to l$ is equal to $f\,\mathrm{id}_k$, it is enough to prove that for each $k = 0, 1, 2, ...$ the identity endomorphism $\mathrm{id}_k = 1_k \in E_k$ belongs to G. For $k = 0$ and $k = 1$, this is true because $1_k = f_k$ in these cases. Inductively, assume that $\mathrm{id}_k \in G$ for all $k < N$ and show that $\mathrm{id}_N \in G$ where $N \geq 2$. If $N \leq r-2$ then the idempotent $f_N \in E_N$ is well-defined and belongs to the set $\{f_i\}_{i=0}^{r-2}$ of generators of G. The same argument as in the proof of Lemma 4.4.1 shows that $\mathrm{id}_N \in G$. If $N = r-1$ then the idempotent $f_N \in E_N$ is also well-defined but does not belong to this set of generators. Still, the argument used in the proof of Lemma 4.4.1 shows that $f_N - 1_N \in G$. Since the morphism $f_N = f_{r-1}$ is negligible, we get $1_N \in G$. Consider the case $N \geq r$. By the inductive assumption there is a decomposition $\mathrm{id}_{N-1} = z + \sum_s x_s f_{i_s} y_s$ with negligible z and $i_s \in \{0, 1, 2, ..., r-2\}$. It is clear that

$$\mathrm{id}_N = \mathrm{id}_{N-1} \otimes \mathrm{id}_1 = z \otimes \mathrm{id}_1 + \sum_s x_s f_{i_s} y_s \otimes \mathrm{id}_1 .$$

It follows from Lemma XI.4.1.1 that the morphism $z \otimes \mathrm{id}_1$ is negligible. We also have

$$x_s f_{i_s} y_s \otimes \mathrm{id}_1 = (x_s \otimes \mathrm{id}_1)(f_{i_s} \otimes \mathrm{id}_1)(y_s \otimes \mathrm{id}_1).$$

Note that $f_{i_s} \otimes \mathrm{id}_1$ is a morphism $i_s + 1 \to i_s + 1$ where $i_s + 1 \leq r - 1 < N$. Therefore by the inductive assumption the morphism $f_{i_s} \otimes \mathrm{id}_1 = (f_{i_s} \otimes \mathrm{id}_1)\mathrm{id}_{i_s+1}$ belongs to G. This implies that $\mathrm{id}_N \in G$ and completes the proof.

4.5. Graphical calculus for the Jones-Wenzl idempotents. We shall graphically present the Jones-Wenzl idempotent $f_k \in E_{k,k}$ by a box with k inputs and k outputs. This box, depicted as an empty square, represents a formal linear combination of tangle (k, k)-diagrams whose skein class equals f_k. Note that in our pictures the number of inputs of a box is always equal to the number of outputs. It is always understood that the corresponding Jones-Wenzl idempotent is well-defined.

In order to simplify the pictures, we shall represent bunches of parallel curves in the plane by single curves labelled by a non-negative integer specifying the number of curves in the bunch. Thus, an integer n beside a curve indicates the presence of n copies of that curve, all parallel in the plane of the picture.

Inductive formula (4.2.a) is shown in the graphical notation in Figure 4.1. As usual, the equality here means equality of skein classes.

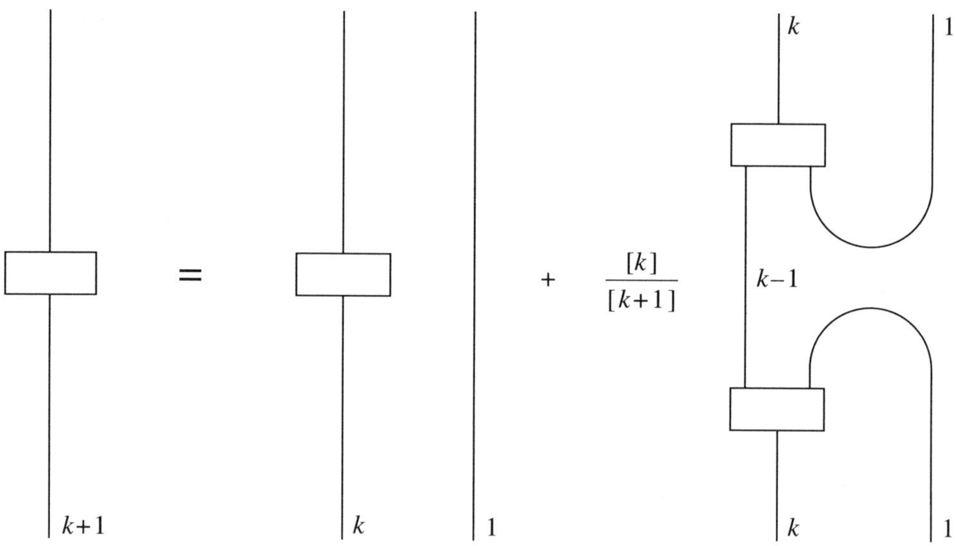

Figure 4.1

The pictorial calculus for the Jones-Wenzl idempotents outlined above will be systematically used throughout the chapter. It should not be confused with the graphical calculus used in Parts I and II to study ribbon and framed graphs. These are two independent (though related) systems of graphical notation used for different purposes.

4.6. Exercise. Verify the identities in Figure 4.2. (In the second formula i, j, k are arbitrary non-negative integers such that the idempotent f_{i+j+k} is well-defined. If $i=k=0$ then this formula means just that $f_j^2 = f_j$.)

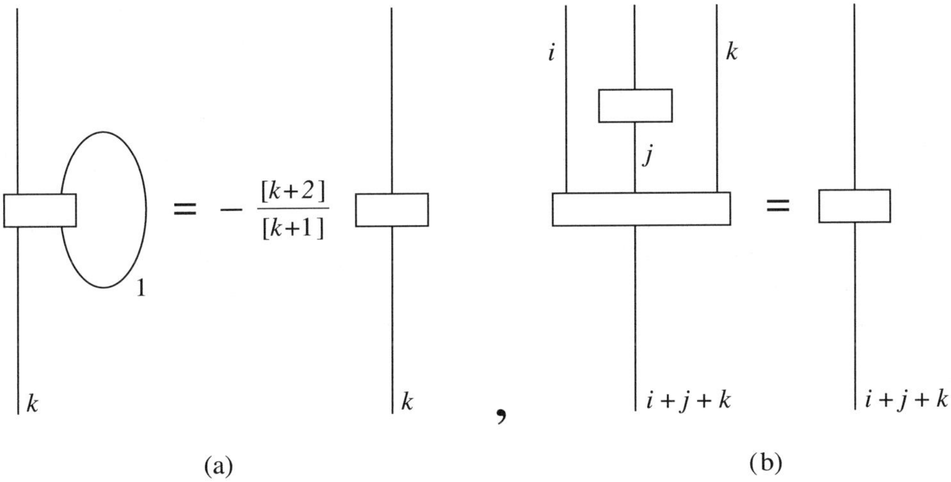

Figure 4.2

5. The matrix S

5.0. Outline. We continue preparations for the construction of a modular category. We compute the traces of certain morphisms in $\mathcal{S}$ and show that the matrix formed by these traces is invertible over K.

The results of this section will play a technical role in Section 7. We shall see that the matrix computed here coincides with the matrix S in the non-degeneracy axiom for modular and quasimodular categories. The reader may safely skip this section and come back to it when necessary.

5.1. Lemma. *If $a^{4n}-1 \in K^*$ for $n = 1, 2, \ldots, k$ then for any $i, j = 0, 1, \ldots, k$,*

$$\operatorname{tr}(c_{j,i}c_{i,j}(f_i \otimes f_j)) = (-1)^{i+j}[(i+1)(j+1)] = (-1)^{i+j}\frac{a^{2(i+1)(j+1)} - a^{-2(i+1)(j+1)}}{a^2 - a^{-2}}.$$

Here $c_{j,i}c_{i,j}(f_i \otimes f_j)$ is an endomorphism of the object $i+j$ of $\mathcal{S}$ so that we may consider its trace.

Proof of Lemma. The proof goes in 7 steps. The first three steps deal with an augmentation $\varepsilon : E_i \to K$. Steps 4 - 6 are concerned with a certain homomorphism $R_i : \oplus_j E_j \to E_i$. The seventh step completes the proof.

Step 1. Any $x \in E_i$ may be written as a linear combination of the basis elements of E_i mentioned in Section 3.1. Denote by $\varepsilon_i(x)$ or $\varepsilon(x)$ the coefficient of the unity 1_i in this expansion. It is clear that ε is a K-linear homomorphism $E_i \to K$. It is important that ε is an algebra homomorphism, i.e., that $\varepsilon(xy) = \varepsilon(x)\,\varepsilon(y)$ for any $x, y \in E_i$. Indeed, if $x = y = 1_i$ then $\varepsilon(xy) = \varepsilon(x)\,\varepsilon(y) = 1$. For all other choices of basis elements, we have $\varepsilon(xy) = \varepsilon(x)\,\varepsilon(y) = 0$. By Theorem 4.1 we have $xf_i = f_i x = \varepsilon(x) f_i$ for any $x \in E_i$.

Step 2. Denote by y_i the element of E_i shown at the beginning of Figure 5.1. This element is represented by the tangle (i, i)-diagram consisting of $i-1$ vertical strands and one strand that links them. We claim that $\varepsilon(y_i) = a^{2(i-1)}$. Indeed, applying the Kauffman skein relation in two crossing points of the strings labelled by 1 we get the equality in Figure 5.1. Any further expansion of the second term on the right-hand side will never achieve i arcs going from the bottom to the top. Therefore $\varepsilon(y_i) = a^2 \varepsilon_{i-1}(y_{i-1})$. For $i = 1$, we have $y_1 = 1_1$ and $\varepsilon_1(y_1) = 1$. Hence $\varepsilon(y_i) = a^{2(i-1)}$.

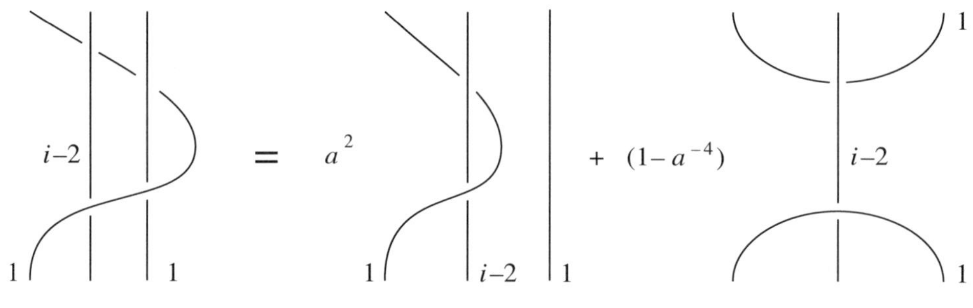

Figure 5.1

Step 3. Denote by x_i the element of E_i shown at the beginning of Figure 5.2. This element is represented by a tangle (i, i)-diagram consisting of i vertical strands and one circle that links them. We claim that $\varepsilon(x_i) = -a^{2(i+1)} - a^{-2(i+1)}$. Indeed, applying the skein relation in two crossing points of the strings labelled by 1 we get the equality in Figure 5.2. Therefore

$$\text{(5.1.a)}\quad \varepsilon(x_i) = (1-a^4)\,\varepsilon(y_i) + a^{-2}\varepsilon_{i-1}(x_{i-1}) = a^{2(i-1)}(1-a^4) + a^{-2}\varepsilon_{i-1}(x_{i-1}).$$

For $i = 0$, we have $\varepsilon(x_i) = -a^2 - a^{-2}$. The solution to the recurrence relation (5.1.a) is $\varepsilon(x_i) = -a^{2(i+1)} - a^{-2(i+1)}$.

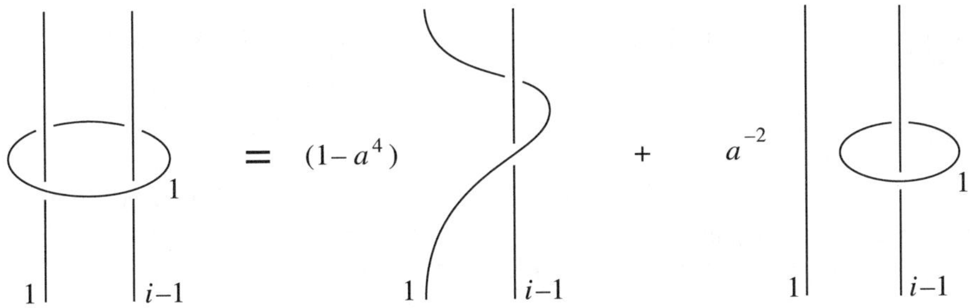

Figure 5.2

Step 4. For any $i = 0, 1, \ldots, k$, we define a K-linear homomorphism $R = R_i : \oplus_{j\geq 0} E_j \to E_i$. It carries $x \in E_j$ in the skein class of the (i, i)-diagram in Figure 5.3. For instance, $R(1_1) = x_i$. It is easy to check that the homomorphism R transforms the tensor product into the product in E_i: for any $x \in E_j$ and $y \in E_k$, we have $R(x \otimes y) = R(x)R(y)$. (Although it is not needed here, note that the image of R lies in the center of E_i.)

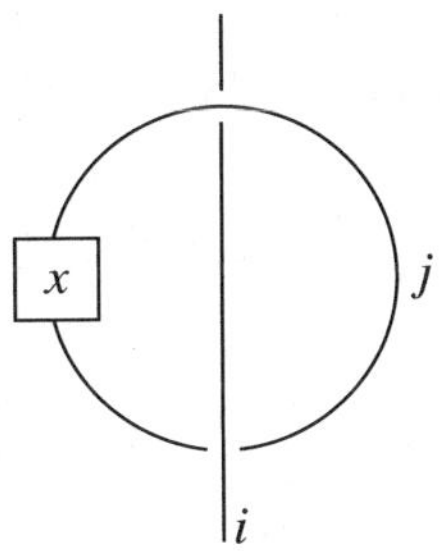

Figure 5.3

The multiplication in the algebra E_j also behaves nicely under R: for any $x, y \in E_j$, we have $R(xy) = R(yx)$. Indeed, to compute $R(xy)$ we have to take a kind of closure of xy. This implies that for x, y presented by tangle (j, j)-diagrams, the skein classes $R(xy)$ and $R(yx)$ are represented by isotopic framed tangles and therefore coincide with each other.

Our next aim is to compute $R_i(f_j) \in E_i$ for all $i, j = 0, 1, \ldots, k$.

Step 5. We need the notion of the j-th Chebyshev polynomial $S_j(x)$ where j is a non-negative integer. This is a monic polynomial of degree j with integer coefficients. It is defined inductively by $S_0(x) = 1, S_1(x) = x$, and $S_{j+1}(x) = xS_j(x) - S_{j-1}(x)$. It is easy to verify that

$$S_j(t + t^{-1}) = \frac{t^{j+1} - t^{-j-1}}{t - t^{-1}}. \tag{5.1.b}$$

Step 6. For $i, j = 0, 1, ..., k$, we compute $R_i(f_j) \in E_i$ as a polynomial in $x_i \in E_i$. We claim that

$$R_i(f_j) = S_j(x_i). \tag{5.1.c}$$

Fix i and set $R = R_i : \oplus_{j \geq 0} E_j \to E_i$. For $j = 0, 1$, equality (5.1.c) holds because $R(f_0) = R(1_0) = 1_i$ and $R(f_1) = R(1_1) = x_i$. Thus, it is enough to show that $R(f_{j+1}) = x_i R(f_j) - R(f_{j-1})$ for any j. It follows from (4.2.a) that

$$R(f_{j+1}) = x_i R(f_j) + ([j]/[j+1])\, R((f_j \otimes 1_1) e_j (f_j \otimes 1_1)). \tag{5.1.d}$$

Here

$$R((f_j \otimes 1_1) e_j (f_j \otimes 1_1)) = R(e_j (f_j \otimes 1_1)^2) = R(e_j (f_j \otimes 1_1)) =$$

$$= R(e_j (f_{j-1} \otimes 1_2)) + ([j-1]/[j])\, R(e_j (f_{j-1} \otimes 1_2)(e_{j-1} \otimes 1_1)(f_{j-1} \otimes 1_2)).$$

Both terms in the last expression may be computed explicitly. It follows from definitions that

$$R(e_j (f_{j-1} \otimes 1_2)) = (-a^2 - a^{-2})\, R(f_{j-1}).$$

The reader may draw a picture for $e_j (f_{j-1} \otimes 1_2)(e_{j-1} \otimes 1_1)(f_{j-1} \otimes 1_2) \in E_j$ and check that its closure is the same as the closure of $f_{j-1} f_{j-1} = f_{j-1} \in E_{j-1}$. Hence

$$R(e_j (f_{j-1} \otimes 1_2)(e_{j-1} \otimes 1_1)(f_{j-1} \otimes 1_2)) = R(f_{j-1}).$$

This gives

$$R((f_j \otimes 1_1) e_j (f_j \otimes 1_1)) = (-a^2 - a^{-2}) R(f_{j-1}) + ([j-1]/[j]) R(f_{j-1}) =$$

$$= -([j+1]/[j]) R(f_{j-1}).$$

Together with (5.1.d) this gives $R(f_{j+1}) = x_i R(f_j) - R(f_{j-1})$. This implies (5.1.c) and completes Step 6.

Step 7. Now we can bring all the pieces together and finish the proof of the lemma. To compute $\operatorname{tr}(c_{j,i} c_{i,j} (f_i \otimes f_j))$ we should draw a picture for $c_{j,i} c_{i,j} (f_i \otimes f_j) \in E_{i+j}$ and take the skein class of its closure. It is obvious that $c_{j,i} c_{i,j} (f_i \otimes f_j)$ and the product $R_i(f_j)\, f_i \in E_i$ have isotopic closures. Therefore

$$\operatorname{tr}(c_{j,i} c_{i,j} (f_i \otimes f_j)) = \operatorname{tr}(R_i(f_j)\, f_i) = \operatorname{tr}(\varepsilon(R_i(f_j))\, f_i) = \varepsilon(R_i(f_j)) \operatorname{tr}(f_i).$$

Since $\varepsilon : E_i \to K$ is an algebra homomorphism, we have

$$\varepsilon(R_i(f_j)) = \varepsilon(S_j(x_i)) = S_j(\varepsilon(x_i)) =$$

$$= S_j(-a^{2(i+1)} - a^{-2(i+1)}) = (-1)^j [(i+1)(j+1)]/[i+1]$$

where the last equality follows from (5.1.b). Multiplying this formula by $\operatorname{tr}(f_i) = (-1)^i\, [i+1]$ we get the claim of the lemma.

5.2. Lemma. *Let $a \in K$ be a primitive $4r$-th root of unity with $r \geq 2$. If $2r \in K^*$ then the matrix*

$$S = \{S_{i,j}\}_{i,j=0}^{r-2} = \{(-1)^{i+j}[(i+1)(j+1)]\}_{i,j=0}^{r-2}$$

is invertible over K.

Proof. We shall prove that $(a^2 - a^{-2})^2 S^2 = (-2r)\delta_i^j$. Since the scalar matrix on the right-hand side is invertible over K, the matrix S is also invertible over K.

To simplify notation set $b = a^2$, $i' = i+1$, $j' = j+1$, $k' = k+1$. By definition, $(a^2 - a^{-2})S_{i,j} = (-1)^{i+j}(b^{i'j'} - b^{-i'j'})$. Therefore

$$(a^2 - a^{-2})^2 \sum_{j=0}^{r-2} S_{i,j}S_{j,k} = (-1)^{i+k} \sum_{j'=1}^{r-1} (b^{(i'+k')j'} - b^{(i'-k')j'} - b^{(k'-i')j'} + b^{-(i'+k')j'}).$$

If $i \neq k$ then we have here four geometric progressions. A direct computation shows that

$$(5.2.a) \qquad (a^2 - a^{-2})^2 \sum_{j=0}^{r-2} S_{i,j}S_{j,k} =$$

$$= (-1)^{i+k} \left[\frac{b^{(i'+k')r} - b^{i'+k'}b^{-(i'+k')r}}{b^{i'+k'} - 1} - \frac{b^{(i'-k')r} - b^{i'-k'}b^{-(i'-k')r}}{b^{i'-k'} - 1} \right].$$

Note that $(a^{2r} - 1)(a^{2r} + 1) = a^{4r} - 1 = 0$. Since a is a primitive $4r$-th root of unity, $a^{2r} - 1$ is invertible in K. Therefore $a^{2r} = -1$ and $b^r = a^{2r} = -1$. If $i' \equiv k' \pmod 2$ then $b^{(i'+k')r} = b^{(i'-k')r} = 1$ and both fractions on the right-hand side of (5.2.a) are equal to -1. If $i' \equiv k' + 1 \pmod 2$ then both fractions are equal to 1. In any case the right-hand side is equal to 0. In the case $i = k$ the computation is even easier and gives $-2r$.

6. Refined skein category

6.0. Outline. We derive a ribbon Ab-category $\mathcal{V}$ from $\mathcal{S}$. Roughly speaking, we shall increase the number of objects and decrease the number of morphisms. The Jones-Wenzl idempotents, discussed in Section 4, will play a main role in the construction of $\mathcal{V}$. We shall study the category $\mathcal{V}$ further in Sections 7 - 9.

6.1. The category $\mathcal{V}(a)$. In this subsection we define $\mathcal{V} = \mathcal{V}(a)$ as an Ab-category. Throughout the construction we assume that $a \in K$ is either a generic element or a primitive $4r$-th root of unity with $r \geq 2$ (see Section 4.3). We first define a set $J = J(a)$ as follows. If a is generic then J is the set of all positive integers, $J = \{1, 2, ...\}$. If a is a primitive $4r$-th root of unity then $J = \{1, 2, ..., r-2\}$. (If $r = 2$ then $J = \emptyset$.)

The objects of $\mathcal{V}$ are arbitrary finite sequences of elements of J. Thus, an object V of $\mathcal{V}$ is a sequence $(j_1, j_2, ..., j_l) \in J^l$ with $l \geq 0$. In particular, the empty sequence is an object of $\mathcal{V}$ corresponding to $l = 0$.

To define the morphisms in $\mathcal{V}$ we need more notation. For any sequence $V = (j_1, ..., j_l) \in J^l$, set $|V| = j_1 + j_2 + ... + j_l$ and

$$\mathcal{F}_V = f_{j_1} \otimes ... \otimes f_{j_l} \in E_{|V|} = \mathrm{End}_{\mathcal{S}}(|V|).$$

Here we regard each idempotent $f_{j_n} \in E_{|V_n|}$ as a morphism $|j_n| \to |j_n|$ in $\mathcal{S}$. The symbol $\otimes$ denotes the tensor product of morphisms in $\mathcal{S}$. It follows from the identity $(f_k)^2 = f_k$ that

$$\mathcal{F}_V \mathcal{F}_V = \mathcal{F}_V. \tag{6.1.a}$$

Note that $|V| = 0$ if and only if V is the empty sequence $\emptyset$. It is clear that $\mathcal{F}_\emptyset = \mathrm{id}_0 = 1 \in E_0 = K$.

Let U and V be objects of $\mathcal{V}$. Any $\mathcal{S}$-morphism $x : |U| \to |V|$ may be composed from the left with $\mathcal{F}_V$ and from the right with $\mathcal{F}_U$. Set

$$x^f = \mathcal{F}_V \, x \, \mathcal{F}_U \in \mathrm{Hom}_{\mathcal{S}}(|U|, |V|) = E_{|U|,|V|}.$$

Since $\mathcal{F}_V$ and $\mathcal{F}_U$ are idempotents, we have $(x^f)^f = x^f$. Therefore the formula $x \mapsto x^f$ defines a projection $E_{|U|,|V|} \to E_{|U|,|V|}$. Denote the image of this projection by $E^f_{|U|,|V|}$. This is a submodule of $E_{|U|,|V|}$ that may be alternatively described by the formula

$$E^f_{|U|,|V|} = \{x \in E_{|U|,|V|} \mid x^f = x\}.$$

Formula (6.1.a) implies that for any $x \in E^f_{|U|,|V|}$, we have

$$\mathcal{F}_V \, x = x \, \mathcal{F}_U = x. \tag{6.1.b}$$

6.1.1. Lemma. *For any objects U, V, W of $\mathcal{V}$ and any $x \in E^f_{|U|,|V|}$, $y \in E^f_{|V|,|W|}$, we have $yx \in E^f_{|U|,|W|}$.*

Proof. It follows from the formulas $x^f = x$, $y^f = y$ and equality (6.1.a) that $(yx)^f = (y^f x^f)^f = y^f x^f = yx$.

Now we may define morphisms in $\mathcal{V}$. For any objects U, V of $\mathcal{V}$, set

$$\mathrm{Hom}_{\mathcal{V}}(U, V) = E^f_{|U|,|V|} \subset E_{|U|,|V|}.$$

Thus, a $\mathcal{V}$-morphism $U \to V$ is just a skein class $x \in E_{|U|,|V|}$ such that $x^f = x$. The composition of morphisms is the same as in $\mathcal{S}$. Lemma 6.1.1 ensures that the composition of morphisms in $\mathcal{V}$ is a morphism in $\mathcal{V}$.

It remains to define the identity morphism $\mathrm{id}_V : V \to V$. Note that the identity morphism $\mathrm{id}_{|V|} : |V| \to |V|$ in the category $\mathcal{S}$ can not be used as id_V because in general it does not belong to $E^f_{|V|,|V|}$. On the other hand, the $\mathcal{S}$-morphism $\mathcal{F}_V : |V| \to |V|$ belongs to $E^f_{|V|,|V|}$ because $(\mathcal{F}_V)^f = \mathcal{F}_V\mathcal{F}_V\mathcal{F}_V = \mathcal{F}_V$. This allows us to regard $\mathcal{F}_V$ as a morphism $V \to V$ in $\mathcal{V}$. Formula (6.1.b) shows that this morphism is the identity endomorphism of V in $\mathcal{V}$. Thus, $\mathrm{id}_V = \mathcal{F}_V$. For instance, if V is a one-term sequence j then $\mathrm{id}_V = \mathcal{F}_V = f_j$.

Note that the set $\mathrm{Hom}(U, V) = E^f_{|U|,|V|}$ is a K-module and in particular an abelian group. In this way the category $\mathcal{V}$ becomes an Ab-category.

6.2. Monoidal structure on $\mathcal{V}$. We equip $\mathcal{V}$ with a monoidal structure. The tensor product of objects is the juxtaposition of finite sequences. The tensor product of morphisms is induced by the one in $\mathcal{S}$ in the obvious way. The following lemma shows that the tensor product of morphisms in $\mathcal{V}$ is a morphism in $\mathcal{V}$.

6.2.1. Lemma. *For any objects U, V, U', V' of $\mathcal{V}$ and any $x \in E^f_{|U|,|V|}$, $x' \in E^f_{|U'|,|V'|}$, we have $x \otimes x' \in E^f_{|U\otimes U'|,|V\otimes V'|}$.*

Proof. $(x \otimes x')^f = x^f \otimes (x')^f = x \otimes x'$.

It is clear that $\mathcal{V}$ is a strict monoidal Ab-category. As usual, the role of the unit object $\mathbb{1} = \mathbb{1}_{\mathcal{V}}$ is played by the empty sequence. The ground ring of $\mathcal{V}$ is $\mathrm{Hom}_{\mathcal{V}}(\mathbb{1}, \mathbb{1}) = E^f_{0,0} = E_{0,0} = K$.

We shall often use the identity

$$\mathcal{F}_{U\otimes V} = \mathcal{F}_U \otimes \mathcal{F}_V$$

which follows directly from definitions.

6.3. Pictorial calculus for $\mathcal{V}$. We may use pictures to present morphisms in $\mathcal{V}$. For objects $U = (i_1, ..., i_k), V = (j_1, ..., j_l)$ of $\mathcal{V}$ and any $x \in E_{|U|,|V|}$, we may present $x^f \in E^f_{|U|,|V|}$ graphically as follows. Assume first that x is the skein class of a tangle diagram D with $|U|$ inputs and $|V|$ outputs. We attach to D from above the juxtaposition of boxes representing $f_{j_1}, ..., f_{j_l}$ and attach from below the juxtaposition of boxes representing $f_{i_1}, ..., f_{i_k}$. This gives a diagram of x^f. Thus, we may present a morphism in $\mathcal{V}$ by a picture consisting of a tangle diagram lying in the middle, a row of boxes attached to its top boundary line, and a row of boxes attached to the bottom boundary line. To present a generic morphism in $\mathcal{V}$ we should involve formal linear combinations of such pictures with the same rows of boxes. The composition is defined by putting one picture on the top of the other while keeping the bottom and top boxes and replacing the boxes in the middle with the corresponding linear combinations of tangle diagrams. Recall that tangle diagrams are considered up to relations (i) - (iii) of Section 1.1.

In this graphical notation the tensor product of two morphisms in $\mathcal{V}$ is obtained by placing the picture of the second morphism to the right of the picture of the first morphism without any intersection or linking.

6.4. Braiding and twist in $\mathcal{V}$. We provide $\mathcal{V}$ with a braiding c and a twist θ. For any objects U and V of $\mathcal{V}$, set

$$c_{U,V} = (c_{|U|,|V|})^f \in E^f_{|U\otimes V|,|V\otimes U|} = \mathrm{Hom}_{\mathcal{V}}(U \otimes V, V \otimes U).$$

Here $c_{|U|,|V|} \in E_{|U\otimes V|,|V\otimes U|}$ is the braiding in $\mathcal{S}$ defined in Section 2.2. For any object V of $\mathcal{V}$, set

$$\theta_V = (\theta_{|V|})^f \in E^f_{|V|,|V|} = \mathrm{Hom}_{\mathcal{V}}(V, V)$$

where $\theta_{|V|} \in E_{|V|,|V|}$ is the twist in $\mathcal{S}$.

6.4.1. Lemma. *The morphisms $\{c_{U,V}\}_{U,V}$ and $\{\theta_V\}_V$ form a braiding and a twist in $\mathcal{V}$.*

Proof. Consider first the twist θ_V of an object $V = (j_1, ..., j_l)$ of $\mathcal{V}$. By definition, $\theta_V = \mathcal{F}_V \theta_{|V|} \mathcal{F}_V$. (The reader is recommended to draw the pictures corresponding to this and further equalities. It should be stressed that $\theta_{|V|}$ is a morphism in the category $\mathcal{S}$ but not a morphism in $\mathcal{V}$.)

We have the following equalities of morphisms in $\mathcal{S}$:

$$\theta_V = \mathcal{F}_V \, \theta_{|V|} = \theta_{|V|} \, \mathcal{F}_V. \tag{6.4.a}$$

Indeed, to compute θ_V we take the framed tangle formed by $|V|$ vertical (unlinked untwisted) strands and apply to it a full right-hand twist. Then we attach from above and below a row of boxes representing $f_{j_1}, ..., f_{j_l}$. The boxes attached from below may be pulled up towards the upper boxes along the strands of the tangle. At the last moment each pair of boxes corresponding to the same family of strands may be replaced by one box because $(f_{j_k})^2 = f_{j_k}$ for every k. This proves the first equality stated above. Similarly, pushing down the upper boxes we get the second equality. In other words, we may safely remove from the picture of θ_V either the upper or the lower row of boxes without changing the morphism represented by the picture. Similarly,

$$\mathcal{F}_V \theta^{-1}_{|V|} \mathcal{F}_V = \mathcal{F}_V \theta^{-1}_{|V|} = \theta^{-1}_{|V|} \mathcal{F}_V.$$

This implies that the morphisms θ_V and $\mathcal{F}_V \theta^{-1}_{|V|} \mathcal{F}_V$ are mutually inverse in $\mathcal{V}$. Indeed,

$$\theta_V \mathcal{F}_V \theta^{-1}_{|V|} \mathcal{F}_V = \mathcal{F}_V \theta_{|V|} \theta^{-1}_{|V|} \mathcal{F}_V = \mathcal{F}_V = \mathrm{id}_V$$

and

$$\mathcal{F}_V \theta_{|V|}^{-1} \mathcal{F}_V \theta_V = \mathcal{F}_V \theta_{|V|}^{-1} \theta_{|V|} \mathcal{F}_V = \mathcal{F}_V = \mathrm{id}_V .$$

This implies that the twist $\theta_V : V \to V$ is invertible. We can establish naturality of the twist analogously: for objects U, V of $\mathcal{V}$ and any $x \in E^f_{|U|,|V|}$, we have

$$\theta_V x = \theta_{|V|} \mathcal{F}_V x = \theta_{|V|} x = x \theta_{|U|} = x \mathcal{F}_U \theta_{|U|} = x \theta_U .$$

Here we use (6.1.b), (6.4.a), and naturality of the twist in $\mathcal{S}$.

Let us prove the first braiding identity (I.1.2.b). The same argument as above shows that for any objects U, V of $\mathcal{V}$, we have

$$c_{U,V} = \mathcal{F}_{V\otimes U}\, c_{|U|,|V|}\, \mathcal{F}_{U\otimes V} = \mathcal{F}_{V\otimes U}\, c_{|U|,|V|} = c_{|U|,|V|}\, \mathcal{F}_{U\otimes V} . \tag{6.4.b}$$

For any objects U, V, W of $\mathcal{V}$, we have

$$c_{U,V\otimes W} = \mathcal{F}_{V\otimes W\otimes U}\, c_{|U|,|V\otimes W|}\, \mathcal{F}_{U\otimes V\otimes W} =$$

$$= \mathcal{F}_{V\otimes W\otimes U}\, (\mathrm{id}_{|V|} \otimes c_{|U|,|W|})\, (c_{|U|,|V|} \otimes \mathrm{id}_{|W|})\, \mathcal{F}_{U\otimes V\otimes W}$$

where the last equality holds because the morphisms $\{c_{k,l}\}_{k,l}$ form a braiding in $\mathcal{S}$. We have

$$\mathcal{F}_{V\otimes W\otimes U}\, (\mathrm{id}_{|V|} \otimes c_{|U|,|W|}) = (\mathcal{F}_V\, \mathrm{id}_{|V|}) \otimes (\mathcal{F}_{W\otimes U}\, c_{|U|,|W|}) = \mathrm{id}_V \otimes c_{U,W} .$$

Similarly,

$$(c_{|U|,|V|} \otimes \mathrm{id}_{|W|})\, \mathcal{F}_{U\otimes V\otimes W} = c_{U,V} \otimes \mathrm{id}_W .$$

This completes the proof of (I.1.2.b). Formulas (I.1.2.c) and (I.1.2.h) are proven similarly. Naturality and invertibility of $c_{U,V}$ are established in the same way as naturality and invertibility of the twist. (Invertibility of $c_{U,V}$ also follows from invertibility of the twist and formula (I.1.2.h).) This completes the proof of the lemma.

6.5. Duality in $\mathcal{V}$. For any object $V = (j_1, \dots, j_l) \in J^l$ of $\mathcal{V}$, we define the dual object to be $V^* = (j_l, j_{l-1}, \dots, j_1)$. Set

$$b_V = (b_{|V|})^f = \mathcal{F}_{V\otimes V^*} b_{|V|} \in E^f_{0,|V\otimes V^*|} = \mathrm{Hom}_{\mathcal{V}}(\mathbb{1}_{\mathcal{V}}, V \otimes V^*)$$

and

$$d_V = (d_{|V|})^f = d_{|V|} \mathcal{F}_{V^*\otimes V} \in E^f_{|V^*\otimes V|,0} = \mathrm{Hom}_{\mathcal{V}}(V^* \otimes V, \mathbb{1}_{\mathcal{V}}).$$

Here $b_{|V|} \in E_{0,|V\otimes V^*|}$ and $d_{|V|} \in E_{|V^*\otimes V|,0}$ are the duality morphisms in $\mathcal{S}$ defined in Section 2.2. For a picture of b_V with $l = 2$, see Figure 6.1.

6.5.1. Lemma. *The family of morphisms $\{b_V, d_V\}_V$ is a duality in $\mathcal{V}$ compatible with braiding and twist.*

Proof. We have to verify the axioms of duality (I.1.3.b), (I.1.3.c), and the compatibility axiom (I.1.3.d). They are deduced from the corresponding formulas in

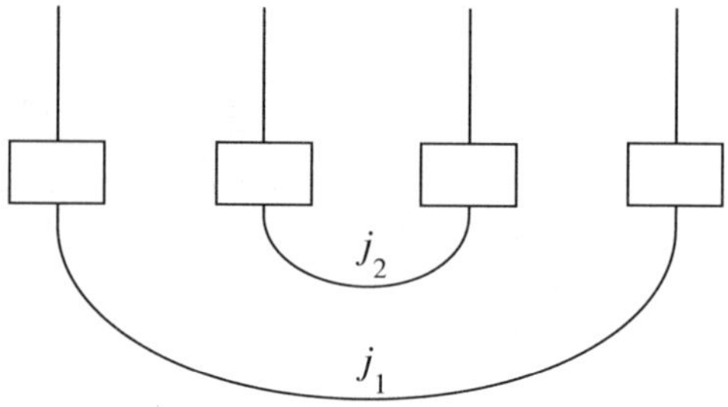

Figure 6.1

$\mathcal{S}$. The method is the same as in the proof of Lemma 6.4.1 since all boxes representing $\{f_k\}_k$ may be pushed along the corresponding strands. There is a minor complication which appears in the proof of (I.1.3.b) and (I.1.3.c). When we push the boxes representing $\mathcal{F}_{V^*}$ along the strands of the diagram of $b_{|V|}$ to the left they arrive at the end turned upside down. More exactly, under this transformation each box is rotated in the plane of the picture by the angle of π. By Remark 4.2.2, the resulting box presents the same idempotent. Therefore, we may safely remove the boxes representing $\mathcal{F}_{V^*}$ from the picture of b_V without changing b_V. In other words,

$$b_V = (\mathcal{F}_V \otimes \mathcal{F}_{V^*})\, b_{|V|} = (\mathrm{id}_{|V|} \otimes \mathcal{F}_{V^*})\, b_{|V|} = (\mathcal{F}_V \otimes \mathrm{id}_{|V^*|})\, b_{|V|}.$$

Similar arguments apply to d_V. The rest of the proof is analogous to the proof of Lemma 6.4.1.

6.6. Theorem. *Let $a \in K$ be a generic element or a primitive $4r$-th root of unity with $r \geq 2$. The category $\mathcal{V}(a)$ with the braiding, twist, and duality defined above is a ribbon* Ab*-category with ground ring K.*

This theorem summarizes the results of previous subsections. In Section 7 we shall show that the category $\mathcal{V}$ is either semisimple or quasimodular depending on the choice of a.

6.7. Comparison of traces. Theorem 6.6 allows us to apply to $\mathcal{V}$ the theory of ribbon categories. In particular, for any object V of $\mathcal{V}$, we have the trace $\mathrm{tr}_{\mathcal{V}} : \mathrm{End}_{\mathcal{V}}(V) \to \mathrm{End}_{\mathcal{V}}(\mathbb{1}) = K$ introduced in Section I.1.5. Note that $\mathrm{End}_{\mathcal{V}}(V) = E^f_{|V|,|V|} \subset E_{|V|}$. The next lemma shows that the trace $\mathrm{tr}_{\mathcal{V}}$ is the restriction of the trace $\mathrm{tr} : E_{|V|} \to K$ defined in Section 3.4.

6.7.1. Lemma. *For any $x \in E^f_{|V|,|V|}$, we have* $\mathrm{tr}_{\mathcal{V}}(x) = \mathrm{tr}(x)$.

Proof. The proof is based on the same observations as the proofs of Lemmas 6.4.1 and 6.5.1. By definition

$$\mathrm{tr}_{\mathcal{V}}(x) = d_V c_{V,V^*}((\theta_V x) \otimes \mathrm{id}_{V^*}) b_V.$$

Set $k = |V|$. In the graphical notation, each of the morphisms $d_V, c_{V,V^*}, \theta_V, \mathrm{id}_{V^*}$, b_V differs from the morphisms $d_k, c_{k,k}, \theta_k, \mathrm{id}_k, b_k$ respectively by a row of boxes representing the Jones-Wenzl idempotents. All these boxes may be pushed along the strands towards x where they finally disappear because $\mathcal{F}_V x = x\mathcal{F}_V = x$. This gives the equality

$$\mathrm{tr}_{\mathcal{V}}(x) = d_k c_{k,k}((\theta_k x) \otimes \mathrm{id}_k) b_k.$$

On the right-hand side we have the expression defining the trace of x in $\mathcal{S}$. According to Lemma 3.4.2 this expression equals $\mathrm{tr}(x)$.

6.8. Simple objects of $\mathcal{V}$. Every $i \in J = J(a)$ regarded as a one-term sequence of elements of J determines an object of $\mathcal{V}$. It is denoted by V_i. This object is simple (for the definition of simple objects, see Section II.1.2). Indeed,

$$\mathrm{Hom}(V_i, V_i) = E^f_{i,i} = f_i E_i f_i = f_i f_i E_i = f_i E_i = K \cdot f_i = K \cdot \mathrm{id}_{V_i}.$$

Here the third equality follows from the fact that f_i lies in the center of the Temperley-Lieb algebra E_i, the fifth equality follows from the fact that f_i annihilates all generators of E_i except the unit, and the last equality follows from the formula $\mathrm{id}_{V_i} = f_i$. Thus we have a family of simple objects $\{V_i\}_{i\in J}$ numerated by elements of J. We add to this family the unit object $V_0 = \mathbb{1}_{\mathcal{V}}$ represented by the empty sequence. In this way we obtain a family of simple objects $\{V_i\}_{i\in I}$ where $I = J \cup \{0\}$. In the case of generic a we have $I = \{0, 1, 2, \ldots\}$. In the case where a is a primitive $4r$-th root of unity, $I = \{0, 1, 2, \ldots, r-2\}$.

It is easy to compute the dimension $\dim(V_i) = \dim_{\mathcal{V}}(V_i) \in K$ where $i \in I$. According to Lemmas 6.7.1 and 4.4.2

$$\dim(V_i) = \mathrm{tr}_{\mathcal{V}}(\mathrm{id}_{V_i}) = \mathrm{tr}(\mathrm{id}_{V_i}) = \mathrm{tr}(f_i) = (-1)^i\,[i+1]$$

for $i \neq 0$. The formula $\dim(V_i) = (-1)^i\,[i+1]$ holds also for $i = 0$ since $\dim(V_0) = \dim(\mathbb{1}) = 1$.

6.9. Remarks. 1. The construction of skein categories involves geometric objects (framed tangles, tangle diagrams, colored boxes) very similar to those used in Chapter I. To avoid possible confusion we emphasize once more the difference between our present setting and the one of Parts I and II. In Parts I and II we started with a ribbon (or modular) category and studied topological invariants derived from it. Here we begin with tangles and construct ribbon and modular categories from them. The resulting categories are interesting for topologists because they may be plugged in the machinery of Parts I and II to produce invariants of

3-manifolds. These categories are also important in their own right from an algebraic point of view.

2. All the arguments given in this section can be reformulated in purely algebraic terms without appealing to strands and boxes. This is left to the reader as an exercise.

6.10. Exercises. 1. Show that the twist $V_i \to V_i$ is equal to $(-1)^i a^{i(i+2)}\, \mathrm{id}_{V_i}$. (Hint: use the results of Steps 1 and 2 of the proof of Lemma 5.1, do not forget the contribution $-a^3$ of every strand.)

2. Show that the mirror ribbon categories $\overline{\mathcal{S}(a)}$ and $\overline{\mathcal{V}(a)}$ are isomorphic to $\mathcal{S}(a^{-1})$ and $\mathcal{V}(a^{-1})$ respectively.

3. Show that $V_1 = (1)$ is a fundamental object of $\mathcal{V}(a)$.

7. Modular and semisimple skein categories

7.0. Outline. We study the category $\mathcal{V}(a)$ introduced in the previous section. If $a \in K$ is a primitive $4r$-th root of unity then this category is shown to be quasimodular in the sense of Section XI.4.3. If a is generic then $\mathcal{V}(a)$ is semisimple in the sense of Section II.4.1. We also discuss unimodality of $\mathcal{V}(a)$.

7.1. Theorem. *Let $r \geq 2$ be an integer such that $2r$ is invertible in K. Let $a \in K$ be a primitive $4r$-th root of unity. Then the pair $(\mathcal{V}(a), \{V_i\}_{i\in I(a)})$ is a quasimodular category with ground ring K.*

Theorem 7.1 is the central result of this chapter. Combining Theorem 7.1 with the purification technique of Section XI.4 we get a modular category $\mathcal{V}_p = \mathcal{V}_p(a)$. This is a deep but quite a tangible example of a modular category.

In order to apply the technique of Parts I and II to $\mathcal{V}_p$ we need to have at our disposal a rank of $\mathcal{V}_p$, i.e., an element $\mathcal{D} \in K$ such that $\mathcal{D}^2 = \sum_{i=0}^{r-2} (\dim(V_i))^2$. A direct computation shows that

$$\sum_{i=0}^{r-2} (\dim(V_i))^2 = \sum_{i=0}^{r-2} [i+1]^2 = (-2r)(a^2 - a^{-2})^{-2}.$$

(This follows also from formula (II.3.8.a) and computations in the proof of Lemma 5.2.) Thus, for the rank to exist, the element $-2r \in K$ must admit a square root in K.

The category $\mathcal{V}_p$ is closely related to the modular categories derived from quantum groups. The fundamental observation is that for $K = \mathbb{C}$ (and $a \in \mathbb{C}$ a primitive $4r$-th root of unity), the category $\mathcal{V}_p(a)$ is equivalent to the purified representation category of $U_q(sl_2(\mathbb{C}))$ with $q = -a^2$. This follows from the fact that these categories have isomorphic multiplicity modules and the same $6j$-symbols (cf. Section 8.5). Thus, the theory of skein modules and Jones-Wenzl idempotents yields an alternative, geometric construction of the representation category of $U_q(sl_2(\mathbb{C}))$.

The skein construction of modular categories has a number of strong points. First of all it is quite elementary and straightforward. Secondly, it immediately gives a *strict* modular category. Although the passage from non-strict monoidal categories to the strict ones presents no difficulty, it makes the theory more cumbersome. It is a pleasure to have an approach that avoids this passage.

In the skein theory there is a vast choice of the ground rings K. For instance, we may take K to be a finite field, the field of p-adic numbers, etc. (We need only that $2r$ is invertible in K and $-2r$ admits a square root in K.) This may lead to a study of arithmetic properties of the corresponding invariants of 3-manifolds.

The skein construction of modular categories has not yet reached its complete generality. Geometric counterparts of quantum groups other than $U_q(sl_2(\mathbb{C}))$ have yet to be developed.

The 3-dimensional TQFT's derived from the modular category $\mathcal{V}_p(a)$ have interesting properties. The corresponding invariants $\tau(M)$ and $|M|$ of closed 3-manifolds may be refined to invariants of 3-manifolds with a spin structure or with a fixed element of $H^1(M;\mathbb{Z}/2\mathbb{Z})$. The original invariants $\tau(M)$ and $|M|$ are sums of those refined invariants over all spin structures (elements of $H^1(M;\mathbb{Z}/2\mathbb{Z})$). Similar splittings are available for the modules of states of surfaces and for the TQFT's themselves. For more on this, see [Bl], [BM], [KM], [Tu7], [Tu8], [TV].

Proof of Theorem. We have to verify the axioms of a quasimodular category. Axiom (II.1.4.1) follows directly from definitions. Axiom (II.1.4.2) also follows from definitions since $(V_i)^* = V_i$. (Thus, $i^* = i$ for all $i \in I$.)

Let us verify the axiom of weak domination (XI.2.4.1) which says that every object V of $\mathcal{V}$ is quasidominated by $\{V_i\}_{i\in I}$. By Lemma 4.4.3, the identity endomorphism of the object $|V|$ of $\mathcal{S}$ may be decomposed as follows:

$$\mathrm{id}_{|V|} = z + \sum_s x_s f_{i_s} y_s$$

where $z : |V| \to |V|$ is a negligible morphism in $\mathcal{S}$, the index s runs over a finite set, $i_s \in \{0, 1, 2, ..., r-2\}$, and $x_s : i_s \to |V|$, $y_s : |V| \to i_s$ are certain morphisms in $\mathcal{S}$. Consider the following morphisms in the category $\mathcal{V}$:

$$z^f = \mathcal{F}_V z \mathcal{F}_V : V \to V,\ (x_s)^f = \mathcal{F}_V x_s f_{i_s} : V_{i_s} \to V,\ (y_s)^f = f_{i_s} y_s \mathcal{F}_V : V \to V_{i_s}.$$

(We should be careful with the case $i_s = 0$. By definition, in this case $(x_s)^f = \mathcal{F}_V x_s \in E^f_{0,|V|} = \mathrm{Hom}_{\mathcal{V}}(V_0, V)$. However, the formula $(x_s)^f = \mathcal{F}_V x_s f_{i_s}$ is true because f_0 is represented by the empty diagram. Similar remarks apply to $(y_s)^f$.)

We have

$$\mathrm{id}_V = \mathcal{F}_V = \mathcal{F}_V \mathrm{id}_{|V|} \mathcal{F}_V = \mathcal{F}_V z \mathcal{F}_V + \sum_s (\mathcal{F}_V x_s f_{i_s})(f_{i_s} y_s \mathcal{F}_V) = z^f + \sum_s (x_s)^f (y_s)^f.$$

The morphism z^f is negligible in $\mathcal{V}$: for any morphism $g : V \to V$ in $\mathcal{V}$, we have by Lemma 6.7.1

$$\mathrm{tr}_{\mathcal{V}}(g z^f) = \mathrm{tr}(g \mathcal{F}_V z \mathcal{F}_V) = \mathrm{tr}(\mathcal{F}_V g \mathcal{F}_V z) = 0.$$

The last equality holds because z is negligible in $\mathcal{S}$. We conclude that V is quasidominated by $\{V_i\}_{i \in I}$.

It remains to verify the non-degeneracy axiom (II.1.4.4). Fix $i, j = 0, 1, ..., r-2$. Denote the braiding morphism (in $\mathcal{V}$) $V_i \otimes V_j \to V_j \otimes V_i$ by $c^{\mathcal{V}}_{i,j}$. Denote the braiding morphism (in $\mathcal{S}$) $i \otimes j \to j \otimes i$ by $c_{i,j}$. By (6.4.b) $c^{\mathcal{V}}_{i,j} = c_{i,j}(f_i \otimes f_j)$. Naturality of the braiding in $\mathcal{S}$ implies that

$$c^{\mathcal{V}}_{j,i} c^{\mathcal{V}}_{i,j} = c_{j,i}(f_j \otimes f_i)\, c_{i,j}(f_i \otimes f_j) = c_{j,i} c_{i,j}(f_i \otimes f_j)(f_i \otimes f_j) = c_{j,i} c_{i,j}(f_i \otimes f_j).$$

By definition and Lemma 6.7.1,

$$S_{i,j} = \mathrm{tr}_{\mathcal{V}}(c^{\mathcal{V}}_{j,i} c^{\mathcal{V}}_{i,j}) = \mathrm{tr}(c^{\mathcal{V}}_{j,i} c^{\mathcal{V}}_{i,j}) = \mathrm{tr}(c_{j,i} c_{i,j}(f_i \otimes f_j)).$$

Lemma 5.1 shows that $S_{i,j} = (-1)^{i+j}[(i+1)(j+1)]$. Lemma 5.2 implies that the matrix S is invertible.

7.2. Theorem. *If $a \in K$ is a primitive $4r$-th root of unity then the pair $(\mathcal{V}(a), \{V_i\}_{i \in I})$ is unimodal.*

For the definition of unimodality, see Section VI.2. Theorem 7.2 allows us to derive from $\mathcal{V}$ the normalized $6j$-symbols and to use them along the lines of Part II.

Proof of Theorem. All elements of the set I are self-dual. Fix $i \in I$ and set $V = V_i$. The condition of unimodality says that for any $x \in \mathrm{Hom}_{\mathcal{V}}(V \otimes V, \mathbb{1})$, we have $x(\mathrm{id}_V \otimes \theta_V) c_{V,V} = x$. If $i = 0$ then $V = \mathbb{1}, V \otimes V = \mathbb{1}$ and $\theta_V = c_{V,V} = \mathrm{id}_{\mathbb{1}}$ so that the claim is obvious. Assume that $i \neq 0$. In this case $\mathrm{Hom}_{\mathcal{V}}(V \otimes V, \mathbb{1}) = E^f_{2i,0} \subset E_{2i,0}$. Let us show that the module $E^f_{2i,0}$ is generated by $(d_i)^f$. Here $(d_i)^f = d_i(f_i \otimes f_i)$ where $d_i : 2i \to 0$ is the duality morphism in $\mathcal{S}$ represented by i concentric cap-like arcs. We compute the projection $y \mapsto y^f : E_{2i,0} \to E^f_{2i,0}$ as follows. Let $y = \langle D \rangle \in E_{2i,0}$ be the basis element represented by a simple $(2i, 0)$-diagram D (so that D has neither crossing points nor loops). Enumerate the $2i$ inputs of D from left to right by the numbers $1, 2, ..., 2i$. If there is an arc of D relating two inputs with numbers $\leq i$ then there is an arc ℓ of D that relates two consecutive inputs of D with numbers $j, j+1 \leq i$. Consider the arc of D which hits the $(j-1)$-th (or the $(j+2)$-th) input of D. Generating on this arc a mutually cancelling pair of a local maximum and a local minimum near this input we may arrange that there is a local minimum of D lying strictly above ℓ. It is clear now that $D = D'e_j$ where D' is a simple $(2i, 0)$-diagram and e_j is the j-th canonical generator of $E_{2i,2i}$. Since $j < i$, we have $e_j(f_i \otimes \mathrm{id}_i) = 0$. Therefore

$$y^f = y\,(f_i \otimes f_i) = \langle D' \rangle\, e_j\,(f_i \otimes \mathrm{id}_i)\,(\mathrm{id}_i \otimes f_i) = 0.$$

If there is no arc of D relating two inputs with numbers $\leq i$ then D is a diagram of d_i. Therefore $E^f_{2i,0} = K(d_i)^f$.

It remains to show that $(d_i)^f(\mathrm{id}_V \otimes \theta_V)c_{V,V} = (d_i)^f$. Each morphism appearing in this formula is obtained from the corresponding morphism in the category $\mathcal{S}$ by attaching boxes with a picture of f_i. As in the proof of Lemma 6.4.1 we may push all these boxes down to the bottom of the diagram. Therefore this formula is equivalent to

$$d_i(\mathrm{id}_i \otimes \theta_i)c_{i,i}(f_i \otimes f_i) = d_i(f_i \otimes f_i).$$

We claim that even the stronger formula

$$d_i(\mathrm{id}_i \otimes \theta_i)c_{i,i} = d_i \tag{7.2.a}$$

is true. Here d_i, θ_i, and $c_{i,i}$ are the duality, twist, and braiding in $\mathcal{S}$ defined in Section 2. We may represent both parts of (7.2.a) by tangle diagrams. It is easy to observe that these diagrams present isotopic framed tangles. Therefore the skein classes of these diagrams are equal which proves (7.2.a) and the theorem.

7.3. Remark. If the ground ring K is a field then any simple object of $\mathcal{V}_p$ is isomorphic to one of the objects $\{V_i\}_{i\in I}$ (cf. Remark II.1.8.2). This gives an invariant description of the family $\{V_i\}_{i\in I}$ as a family of representatives of the isomorphism classes of simple objects of $\mathcal{V}_p$.

7.4. Theorem. *Let $a \in K$ be a generic element. Then the pair $(\mathcal{V}(a), \{V_i\}_{i\in I(a)})$ is a unimodal semisimple category with ground ring K.*

Although the category $\mathcal{V}(a)$ with generic a cannot be used to construct a 3-dimensional TQFT or invariants of shadows, it is interesting from the algebraic point of view. The discussions that follow the statements of Theorems 7.1 and 7.2 apply to the case of generic a with the obvious changes.

Proof of Theorem. Axioms (II.1.4.1) and (II.1.4.2) follow from definitions. The axiom of domination (II.1.4.3) is verified in the same way as the axiom of weak domination in the proof of Theorem 7.1. The only difference is that we should omit z and use Lemma 4.4.1 instead of Lemma 4.4.3.

It remains to verify the Schur axiom (II.4.1.1), i.e., to show that for distinct $k, l \in I$, we have $\mathrm{Hom}_{\mathcal{V}}(V_k, V_l) = 0$. We should prove that $E^f_{k,l} = 0$, i.e., that $x^f = 0$ for any $x \in E_{k,l}$. Recall that the module $E_{k,l}$ is generated by the skein classes of simple (k, l)-diagrams. Let us prove that for the skein class $x = \langle D \rangle$

of any simple diagram D, we have $x^f = 0$. By definition, $x^f = f_l x f_k \in E_{k,l}$. We establish below that if $k > l$ then $x f_k = 0$ and if $k < l$ then $f_l x = 0$. In both cases $x^f = f_l x f_k = 0$. Assume for concreteness that $k > l$. Then the diagram D contains an arc ℓ which connects two consecutive inputs of D. Let $j, j+1$ be the numbers of these inputs in the row of inputs of D counted from the left. Consider the arc of D which hits the $(j-1)$-th (or the $(j+2)$-th) input of D. Generating on this arc a mutually cancelling pair of a local maximum and a local minimum near this input we may arrange that there is a local minimum of D lying strictly above ℓ. It is clear now that $D = D'e_j$ where D' is a simple (k, l)-diagram and e_j is the j-th canonical generator of $E_{k,k}$. Since $e_j f_k = 0$, we get $x f_k = \langle D' \rangle e_j f_k = 0$. This completes the verification of the Schur axiom.

The proof of unimodality of $\mathcal{V}(a)$ is the same as the proof of unimodality in Theorem 7.2.

7.5. Even subcategories. In the case where a is a primitive $4r$-th root of unity with odd $r \geq 3$ the same constructions yield another example of a modular category. This is the subcategory of $\mathcal{V}_p(a)$ consisting of objects V with even $|V|$.

The categories $\mathcal{S}(a), \mathcal{V}(a), \mathcal{V}_p(a)$ have a canonical $\mathbb{Z}/2\mathbb{Z}$-grading. An object k of $\mathcal{S}$ is even (odd) if k is an even (odd) number. An object V of $\mathcal{V}$ or $\mathcal{V}_p$ is even (odd) if $|V|$ is an even (odd) number. It is obvious that the morphisms between objects of different parity are equal to zero. Consider the subcategories $\mathcal{S}^e(a), \mathcal{V}^e(a), \mathcal{V}^e_p(a)$ of $\mathcal{S}(a), \mathcal{V}(a), \mathcal{V}_p(a)$ respectively consisting of even objects and arbitrary morphisms (in these categories). The same braiding, twist, and duality as above make these subcategories ribbon Ab-categories.

Here are analogues of Lemmas 4.4.1 and 4.4.3 in this setting.

7.5.1. Lemma. *If $a \in K$ is generic then any morphism $k \to l$ in $\mathcal{S}^e(a)$ may be expressed as a sum $\sum_s x_s f_{i_s} y_s$ where s runs over a finite set of indices, $i_s \in \{0, 2, 4, ...\}$, and $x_s : i_s \to l$, $y_s : k \to i_s$ are morphisms in $\mathcal{S}^e(a)$.*

7.5.2. Lemma. *If $a \in K$ is a primitive $4r$-th root of unity with odd $r \geq 3$ then any morphism $k \to l$ in $\mathcal{S}^e(a)$ may be expressed as a sum $z + \sum_s x_s f_{i_s} y_s$ where s runs over a finite set of indices, $i_s \in \{0, 2, 4, ..., r-3\}$, $x_s : i_s \to l$, $y_s : k \to i_s$ are morphisms in $\mathcal{S}^e(a)$ and $z : k \to l$ is a negligible morphism in $\mathcal{S}^e(a)$.*

The proofs of these lemmas are the same as the proofs of Lemmas 4.4.1 and 4.4.3. (At the end of the proof of Lemma 4.4.3 we should use id_2 and id_{N-2} instead of id_1 and id_{N-1}. The argument works for all even N except $N = r$. To exclude this value of N we assume that r is odd.)

Set $I^e(a) = \{0, 2, 4, ..., r-3\}$ if a is a primitive $4r$-th root of unity with $r \geq 3$ and $I^e(a) = \{0, 2, 4, ...\}$ if a is generic.

Here are the analogues of Theorems 7.1, 7.2, and 7.4.

7.5.3. Theorem. *Let $r \geq 3$ be an odd integer invertible in K. Let $a \in K$ be a primitive $4r$-th root of unity. Then the pair $(\mathcal{V}_p^e(a), \{V_i\}_{i \in I^e(a)})$ is a unimodular category with ground ring K.*

The invariant $\tau(M)$ of a closed oriented 3-manifold M derived from this modular category is called the quantum $SO(3)$-invariant of M.

7.5.4. Theorem. *For any generic $a \in K$, the pair $(\mathcal{V}(a), \{V_i\}_{i \in I^e(a)})$ is a unimodal semisimple category with ground ring K.*

The proofs of these theorems are the same as the proofs of Theorems 7.1, 7.2, 7.4. The matrix S appearing in the proof of Theorem 7.5.3 is the matrix $[(i+1)(j+1)]$ where i, j run over $\{0, 2, 4, \ldots, r-3\}$. A direct computation shows that $(a^2 - a^{-2})^2 S^2 = -r\delta_i^j$. This implies invertibility of S. Note also that for $\mathcal{V}_p^e(a)$ to have a rank, the element $-r \in K$ should have a square root. If this is the case then $\sqrt{-r}\,(a^2 - a^{-2})^{-1}$ is a rank of $\mathcal{V}_p^e(a)$.

8. Multiplicity modules

8.0. Outline. We compute the multiplicity modules of the modular category $\mathcal{V}_p = \mathcal{V}_p(a)$ where $a \in K$ is a root of unity.

8.1. Admissible triples and multiplicity modules. Let $a \in K$ be a primitive $4r$-th root of unity with $r \geq 2$. Set $I = I(a) = \{0, 1, \ldots, r-2\}$. The multiplicity module of $\mathcal{V}_p$ corresponding to $i, j, k \in I$ is defined by the formula

$$H^{ijk} = \operatorname{Hom}_{\mathcal{V}_p}(\mathbb{1}, V_i \otimes V_j \otimes V_k) = \operatorname{Hom}_{\mathcal{V}}(\mathbb{1}, V_i \otimes V_j \otimes V_k)/\operatorname{Negl}(\mathbb{1}, V_i \otimes V_j \otimes V_k).$$

Similarly, consider the module

$$H_{ijk} = \operatorname{Hom}_{\mathcal{V}_p}(V_i \otimes V_j \otimes V_k, \mathbb{1}) = \operatorname{Hom}_{\mathcal{V}}(V_i \otimes V_j \otimes V_k, \mathbb{1})/\operatorname{Negl}(V_i \otimes V_j \otimes V_k, \mathbb{1}).$$

By the general theory of modular categories (Lemma II.4.2.3), the formula $(x, y) \mapsto \operatorname{tr}(yx)$ defines a non-degenerate bilinear form $H^{ijk} \otimes_K H_{ijk} \to K$.

To compute the multiplicity modules we need the notion of an admissible triple of indices. We say that a triple $(i, j, k) \in I^3$ is admissible if $|i - j| \leq k \leq |i + j|$ and $i + j + k$ is even. We say that the triple $(i, j, k) \in I^3$ is strictly admissible if it is admissible and $i + j + k \leq 2r - 4$.

For an admissible triple (i, j, k), set

$$x = (i + j - k)/2, \quad y = (j + k - i)/2, \quad z = (i + k - j)/2. \tag{8.1.a}$$

Denote by $Q = Q(i, j, k)$ the element of the module

$$\mathrm{Hom}_{\mathcal{V}}(\mathbb{1}, V_i \otimes V_j \otimes V_k) = E^f_{0,i+j+k}$$

shown in Figure 8.1.a. Here we have three boxes containing the diagrams (which are formal linear combinations of tangle diagrams) of the Jones-Wenzl idempotents $f_i = f_{x+z}$, $f_j = f_{x+y}$, and $f_k = f_{y+z}$. Set

$$q(i, j, k) = Q(i, j, k) \ \big(\mathrm{mod}\,\mathrm{Negl}(\mathbb{1}, V_i \otimes V_j \otimes V_k)\big) \in H^{ijk}.$$

Denote by $\tilde{Q} = \tilde{Q}(i, j, k)$ the element of the module

$$\mathrm{Hom}_{\mathcal{V}}(V_i \otimes V_j \otimes V_k, \mathbb{1}) = E^f_{i+j+k,0}$$

shown in Figure 8.1.b. Set

$$\tilde{q}(i, j, k) = \tilde{Q}(i, j, k) \ \big(\mathrm{mod}\,\mathrm{Negl}(V_i \otimes V_j \otimes V_k, \mathbb{1})\big) \in H_{ijk}.$$

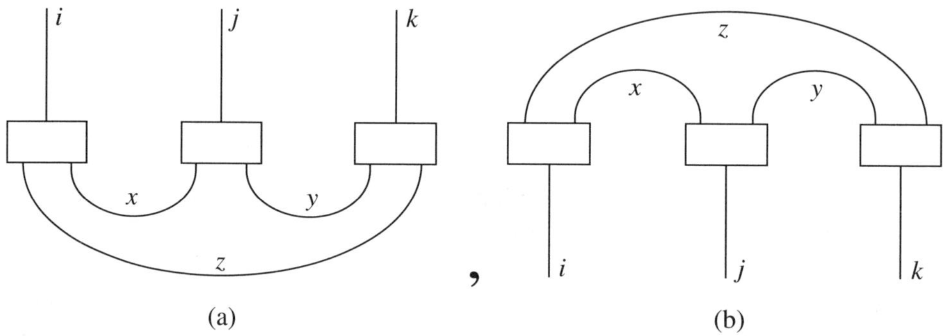

Figure 8.1

8.2. Theorem. *Let $a \in K$ be a primitive $4r$-th root of unity with $r \geq 2$. For any strictly admissible triple $(i, j, k) \in I^3$, we have $H^{ijk} = K \cdot q(i, j, k)$ and $H_{ijk} = K \cdot \tilde{q}(i, j, k)$. The multiplicity modules corresponding to other triples are equal to 0.*

The formula $H^{ijk} = K \cdot q(i, j, k)$ means that H^{ijk} is a free K-module generated by $q(i, j, k)$. Thus, the multiplicity modules of $\mathcal{V}(a)$ are either isomorphic to K or trivial. The multiplicity $h^{ijk} = \mathrm{Dim}(H^{ijk})$ is equal to 1 or 0.

The next lemma is needed to prove Theorem 8.2. We shall use the following notation: for an integer $n \geq 0$, set $[n]! = [1][2]...[n] \in K$. For instance, $[0]! = 1$, $[1]! = 1$, $[2]! = [2] = a^2 + a^{-2}$. Since $a \in K$ is a primitive $4r$-th root of unity, we have $[n]! \in K^*$ for $n = 0, 1, ..., r - 1$ and $[n]! = 0$ for $n \geq r$.

8.3. Lemma. *Let $a \in K$ be a primitive $4r$-th root of unity with $r \geq 2$. For any admissible triple $(i, j, k) \in I^3$, we have*

$$\text{(8.3.a)}\quad \operatorname{tr}(\tilde{Q}(i, j, k)Q(i, j, k)) = (-1)^{\frac{i+j+k}{2}} \frac{[\frac{i+j+k}{2} + 1]!\,[\frac{i+j-k}{2}]!\,[\frac{i+k-j}{2}]!\,[\frac{j+k-i}{2}]!}{[i]!\,[j]!\,[k]!}.$$

Proof. It is convenient to replace the indices i, j, k with the variables x, y, z given by (8.1.a). It is clear that $i = x + z$, $j = x + y$, $k = y + z$. The triple (i, j, k) is admissible if and only if x, y, z are non-negative integers such that $x + y \leq r - 2$, $x + z \leq r - 2$, $y + z \leq r - 2$. The right-hand side of (8.3.a) may be rewritten in terms of x, y, z as

$$\text{(8.3.b)}\qquad (-1)^{x+y+z} \frac{[x + y + z + 1]!\,[x]!\,[y]!\,[z]!}{[x + y]!\,[x + z]!\,[y + z]!}.$$

Set $\Gamma(x, y, z) = \operatorname{tr}(\tilde{Q}(i, j, k)Q(i, j, k))$ where $i = x + z$, $j = x + y$, $k = y + z$. To compute $\Gamma(x, y, z)$ we should attach the diagram in Figure 8.1.b on the top of the diagram in Figure 8.1.a and consider the skein class in $E_{0,0} = K$ of the resulting diagram. The equalities $f_i^2 = f_i$, $f_j^2 = f_j$, $f_k^2 = f_k$ imply that $\Gamma(x, y, z) \in E_{0,0} = K$ can be also represented by the diagram in Figure 8.2.

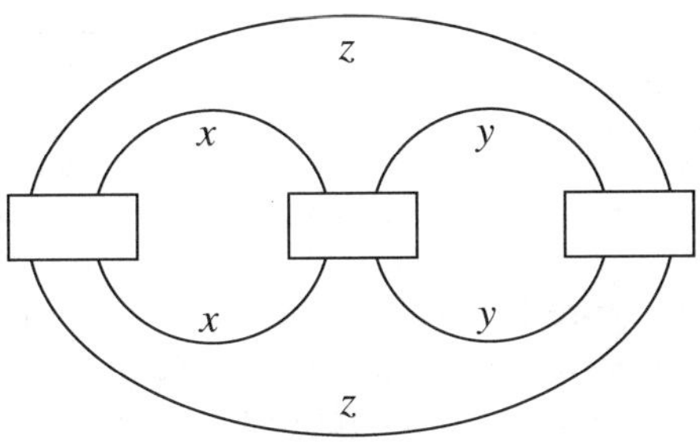

Figure 8.2

We shall establish the following recurrence relation:

$$\Gamma(x, y, z) = -\frac{[x + z + 1]}{[x + z]}\Gamma(x, y, z-1) + \frac{[y]^2}{[y + z - 1][y + z]}\Gamma(x+1, y-1, z-1).$$

It is straightforward to verify that the expression (8.3.b) satisfies the same relation. (This reduces to a demonstration of the equality

$$[x + y + z + 1][z] = [x + z + 1][y + z] - [x + 1][y]$$

which can be checked directly from the definition of $[n]$.) Thus, we may prove the lemma by induction on z. The base of the induction is easy. It follows from the equality in Figure 4.2.b that in the case $z = 0$ we may erase in Figure 8.2 the boxes containing f_{y+z} and f_{x+z} without changing the skein class. This implies that $\Gamma(x, y, 0) = \mathrm{tr}(f_{x+y}) = (-1)^{x+y}[x+y+1]$. This coincides with (8.3.b) for $z = 0$.

It remains to prove the recurrence relation formulated above. The main steps in the proof are the equalities in Figures 8.3 and 8.4. To save space, we have turned all diagrams in a counterclockwise direction by an angle of 90°. The reader should do the inverse operation.

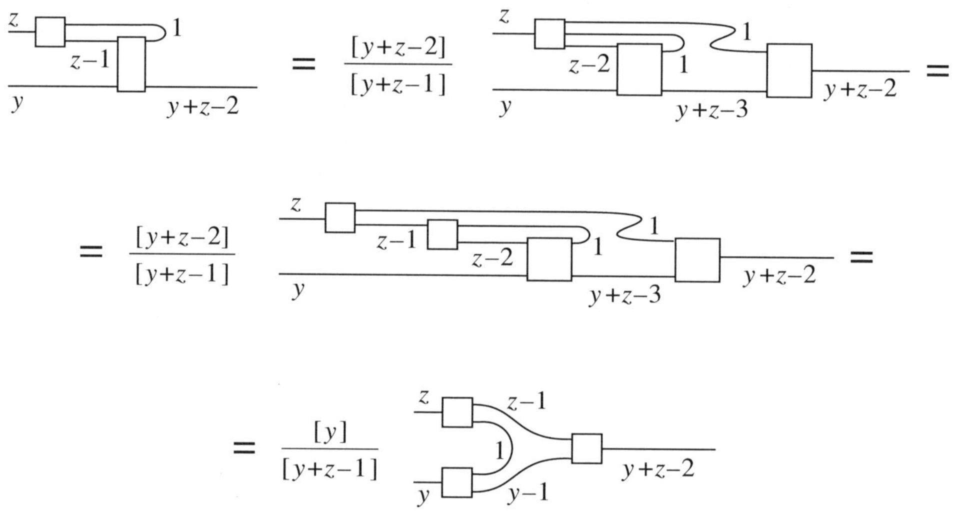

Figure 8.3

The first equality in Figure 8.3 follows from the relation in Figure 4.1 applied to the right box of the first diagram. (Here we use $f_z e_{z-1} = 0$ to annihilate one of the terms.) The second equality in Figure 8.3 follows from the one in Figure 4.2.b. Note that the third diagram in Figure 8.3 contains a fragment identical to the first diagram where z is replaced with $z-1$. Iterating this argument and using the formula in Figure 4.2.b we get the third equality.

The defining relation in Figure 4.1 followed by a double application of the equality in Figure 8.3 produces the formula in Figure 8.4. Now apply this formula to the box in Figure 8.2 containing f_{y+z}. Using the identities of Figure 4.2 we obtain the desired recurrence relation.

8.4. Proof of Theorem 8.2. Consider in more detail the projection $x \mapsto x^f = (f_i \otimes f_j \otimes f_k)x$ acting in the skein module $E_{0,i+j+k}$. Recall that a basis of this module is represented by simple diagrams, i.e., tangle diagrams without crossing points and without loops. The diagrams in question have 0 inputs and $i+j+k$ outputs. (For such diagrams to exist, the number $i+j+k$ should be even.) We split the outputs into three families consisting of the first (from the left) i outputs, the next j outputs, and the remaining k outputs. If a simple diagram D contains

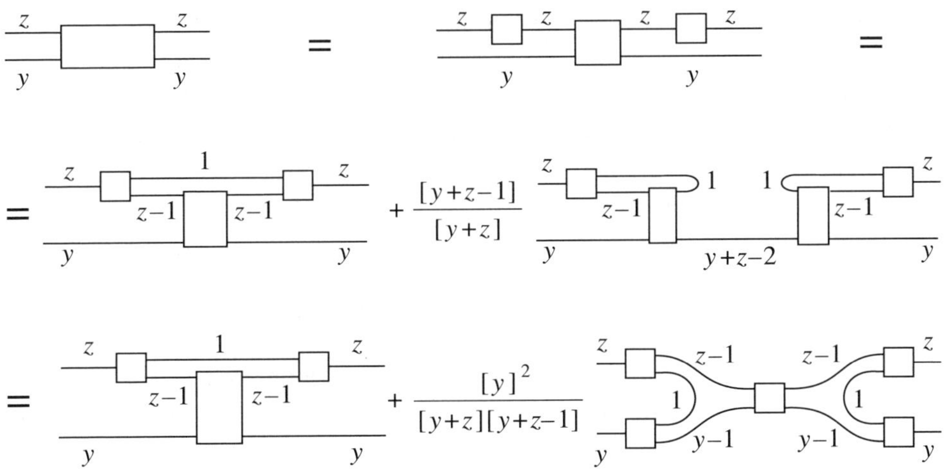

Figure 8.4

an arc connecting two outputs in the first family then it contains an arc connecting two outputs in this family which are immediate neighbors of each other. Since $f_i e_l = 0$ for any generator $e_l \in E_i$, we have $(f_i \otimes \mathrm{id}_j \otimes \mathrm{id}_k)\langle D\rangle = 0$ for any such D. Therefore $(f_i \otimes f_j \otimes f_k)\langle D\rangle = 0$. Similarly if D contains an arc connecting two outputs in the second (or third) family then $\langle D\rangle$ is annihilated by the projection $x \mapsto x^f$. In the case where $i+j<k$, or $i+k<j$, or $j+k<i$, any simple diagram as above contains an arc connecting two outputs of the same family. If $i+j \geq k, i+k \geq j, j+k \geq i$ then there exists exactly one simple diagram without such arcs. The projection $x \mapsto x^f$ carries the skein class of this diagram into $Q(i,j,k)$. This shows that $\mathrm{Hom}_{\mathcal{V}}(\mathbb{1}, V_i \otimes V_j \otimes V_k) = 0$ unless the triple (i,j,k) is admissible in which case the module $\mathrm{Hom}_{\mathcal{V}}(\mathbb{1}, V_i \otimes V_j \otimes V_k)$ is generated by $Q(i,j,k)$. A similar argument shows that the module $\mathrm{Hom}_{\mathcal{V}}(V_i \otimes V_j \otimes V_k, \mathbb{1})$ is trivial unless the triple (i,j,k) is admissible in which case this module is generated by $\tilde{Q}(i,j,k)$.

Now we may compute the module H^{ijk}. If the triple (i,j,k) is not admissible then $H^{ijk} = 0$. Assume that (i,j,k) is admissible but not strictly admissible, i.e., that $i+j+k \geq 2r-2$. To show triviality of H^{ijk} we should show that all morphisms $\mathbb{1} \to V_i \otimes V_j \otimes V_k$ in the category $\mathcal{V}$ are negligible. It is enough to show that the morphism $Q(i,j,k)$ is negligible. Since the module $\mathrm{Hom}_{\mathcal{V}}(V_i \otimes V_j \otimes V_k, \mathbb{1})$ is generated by $\tilde{Q}(i,j,k)$, it is enough to show that $\mathrm{tr}(\tilde{Q}(i,j,k)Q(i,j,k)) = 0$. This follows from Lemma 8.3 since $[\frac{i+j+k}{2}+1]! = 0$ for $i+j+k \geq 2r-2$.

Consider the case of a strictly admissible triple (i,j,k). Since $a \in K$ is a primitive $4r$-th root of unity, the expression on the right-hand side of (8.3.a) is invertible in K. This implies that

$$\mathrm{Hom}_{\mathcal{V}}(\mathbb{1}, V_i \otimes V_j \otimes V_k) = K \cdot Q(i,j,k), \quad \mathrm{Hom}_{\mathcal{V}}(V_i \otimes V_j \otimes V_k, \mathbb{1}) = K \cdot \tilde{Q}(i,j,k),$$

and there are no non-trivial negligible morphisms $V_i \otimes V_j \otimes V_k \to 1\!\!1$ or $1\!\!1 \to V_i \otimes V_j \otimes V_k$. Hence $H^{ijk} = K \cdot q(i,j,k)$ and $H_{ijk} = K \cdot \tilde{q}(i,j,k)$.

8.5. Symmetrized multiplicity modules and $6j$-symbols. Let $a \in K$ be a primitive $4r$-th root of unity with $r \geq 2$. Fix a square root $a' \in K$ of a. It follows from the result of Exercise 6.10.1 that $v'_i = (\sqrt{-1})^i (a')^{i(i+2)}$ is a square root of the twist $V_i \to V_i$ for any $i \in I$. For any triple $(i,j,k) \in I^3$, we have K-isomorphisms $\sigma_1(ijk) : H^{ijk} \to H^{jik}$ and $\sigma_2(ijk) : H^{ijk} \to H^{ikj}$ constructed in Section VI.3.2 using the braiding and the half-twists $\{v'_i\}_{i\in I}$. The symmetrized multiplicity module $H(i,j,k)$ is obtained by an identification of the modules $H^{ijk}, H^{jik}, H^{ikj}, H^{jki}, H^{kij}, H^{kji}$ along such K-isomorphisms, see Section VI.3.2. If the triple $(i,j,k) \in I^3$ is not strictly admissible then $H(i,j,k) = 0$. Assume that the triple $(i,j,k) \in I^3$ is strictly admissible. By Theorem 8.2, the module $H(i,j,k)$ is isomorphic to K. It can be computed that $\sigma_1(ijk)(q(i,j,k)) = q(j,i,k)$ and $\sigma_2(ijk)(q(i,j,k)) = q(i,k,j)$. (This follows, for instance, from [KL, Proposition 3 in Chapter 4] or [MV1, Theorem 3].) Thus, the generators $q(i,j,k) \in H^{ijk}$, $q(j,i,k) \in H^{jik}$, $q(i,k,j) \in H^{ikj}$, etc. represent the same free generator, say q, of the module $H(i,j,k) \approx K$.

We need another generator of $H(i,j,k)$ obtained by normalization of q. Recall the duality pairing $H(i,j,k) \otimes_K H(i,j,k) \to K$ defined in Section VI.3.3 (here $i^* = i, j^* = j, k^* = k$). It follows from Lemma 8.3 that this pairing carries $q \otimes q$ into $\Gamma(i,j,k)$ where $\Gamma(i,j,k)$ is the right-hand side of (8.3.a). Note that $\Gamma(i,j,k)$ does not depend on the order in the triple (i,j,k). Assume that for each (non-ordered) triple $\{i,j,k\}$, we have a square root $\Gamma'(i,j,k) \in K$ of $\Gamma(i,j,k)$. Both $\Gamma(i,j,k)$ and $\Gamma'(i,j,k)$ are invertible in K. We consider $(\Gamma'(i,j,k))^{-1}q$ as a preferred generator of the module $H(i,j,k)$ and identify $H(i,j,k)$ with K via $(\Gamma'(i,j,k))^{-1}q = 1 \in K$. This identification considerably simplifies the computations with the symmetrized multiplicity modules and $6j$-symbols. Under this identification, the duality pairing $H(i,j,k) \otimes_K H(i,j,k) \to K$ is given by the unit (1×1)-matrix.

Consider an ordered 6-tuple $(i,j,k,l,m,n) \in I^6$ and the corresponding normalized $6j$-symbol

$$\begin{vmatrix} i & j & k \\ l & m & n \end{vmatrix} \tag{8.5.a}$$

(see Section VI.5). If at least one of the triples $(i,j,k), (i,m,n), (j,l,n), (k,l,m)$ is not strictly admissible then the ambient module of this $6j$-symbol is equal to 0. Assume that the triples $(i,j,k), (i,m,n), (j,l,n), (k,l,m)$ are strictly admissible. Under the identifications of the corresponding symmetrized multiplicity modules with K described in the previous paragraph, the $6j$-symbol (8.5.a) becomes an element of K. The following explicit formula for this element can be deduced from [MV1, Theorem 2]. For a strictly admissible (non-ordered) triple $\{i,j,k\}$, set

$$\langle i, j, k\rangle = [(i+j-k)/2]!\,[(i+k-j)/2]!\,[(j+k-i)/2]!\,(\Gamma'(i, j, k))^{-1}.$$

Then

$$\begin{vmatrix} i & j & k \\ l & m & n \end{vmatrix} = \frac{\langle i, j, k\rangle\,\langle i, m, n\rangle\,\langle j, l, n\rangle\,\langle k, l, m\rangle}{[i]!\,[j]!\,[k]!\,[l]!\,[m]!\,[n]!} \times$$

$$\times \sum_z (-1)^z\,[z+1]!\,([z-i-j-k]!\,[z-i-m-n]!\,[z-j-l-n]!\,[z-k-l-m]! \times$$

$$\times\,[i+j+l+m-z]!\,[j+k+m+n-z]!\,[i+k+l+n-z]!)^{-1}$$

where z runs over integers such that the expressions in the square brackets are non-negative. A similar formula in the context of the representation theory of $U_q sl_2(\mathbb{C})$ was obtained in [KR1]. It generalizes the computation of the classical $6j$-symbols due to G. Racah. The formula in [KR1] differs from the one given above by the factor $\sqrt{-1}^{\,-(i+j+k+l+m+n)}$ introduced in [TV].

8.6. Exercise. Assume that $a \in K$ is generic. Show that for any $i, j, k \in I$ such that $|i-j| \leq k \leq |i+j|$ and the sum $i+j+k$ is even, the module $\mathrm{Hom}_{\mathcal{V}}(\mathbb{1}, V_i \otimes V_j \otimes V_k)$ is isomorphic to K. Show that for all other triples $i, j, k \in I$, this module is 0.

9. Hermitian and unitary skein categories

9.0. Outline. Under natural assumptions on the ground ring K and the element $a \in K$ the skein categories $\mathcal{V}(a)$ and $\mathcal{V}_p(a)$ may be endowed with a Hermitian structure. The main result says that when $K = \mathbb{C}$ and $a = \sqrt{-1}e^{\pm\pi\sqrt{-1}/2r}$ with $r \geq 2$ then the modular category $\mathcal{V}_p(a)$ is unitary. (For the definition of Hermitian and unitary categories, see Section II.5.)

9.1. Hermitian structure on the skein categories. Assume that the ground ring K is endowed with an involutive automorphism $x \mapsto \overline{x}$ such that $\overline{a} = a^{-1}$. Under these assumptions we construct a Hermitian structure on $\mathcal{V}(a)$ and $\mathcal{V}_p(a)$.

We first construct a Hermitian structure on the category $\mathcal{S}(a)$. Consider the mirror reflection of the strip $\mathbb{R}\times[0, 1]$ with respect to the horizontal line $\mathbb{R}\times(1/2)$. This reflection carries the strip into itself and transforms a tangle (k, l)-diagram D into a tangle (l, k)-diagram denoted by $-D$. Under this reflection the positive and negative smoothings at a crossing point of D are transformed respectively into the negative and positive smoothings of $-D$. Therefore there is no natural way to extend the correspondence $D \mapsto -D$ to a K-linear homomorphism $E_{k,l} \to E_{l,k}$. However, it is possible to extend this correspondence to an additive homomorphism $f \mapsto \overline{f} : E_{k,l} \to E_{l,k}$ linear over the involution $x \mapsto \overline{x} : K \to K$. This homomorphism is defined by the formula $\overline{\sum_i k_i \langle D_i \rangle} = \sum_i \overline{k_i} \langle -D_i \rangle$. It is compatible with the Kauffman skein relation because both the crossings and the coefficients a, a^{-1} in this relation exchange their places under the reflection so that, as a whole, the relation is preserved. It is clear that $\overline{kf} = \overline{k}\,\overline{f}$ for any $k \in K$, $f \in E_{k,l}$. It follows from the definitions that the mappings $\{f \mapsto \overline{f} : E_{k,l} \to E_{l,k}\}$ form a conjugation in $\mathcal{S}$ in the sense of Section II.5.1. Conditions (II.5.1.1) and (II.5.1.2) also follow from definitions.

The identity $-e_i = e_i$ and an easy induction show that any Jones-Wenzl idempotent f_i is invariant under the conjugation: $\overline{f_i} = f_i$. This implies that for any morphism $x : k \to l$ in $\mathcal{S}$, we have $\overline{x^f} = \overline{x}^f$. Therefore the set of morphisms in $\mathcal{V}$ is invariant under the conjugation in $\mathcal{S}$. This yields a conjugation and a Hermitian structure on $\mathcal{V}$.

By Lemma XI.4.2.1, the Hermitian structure on $\mathcal{V}$ survives the purification and induces a Hermitian structure on $\mathcal{V}_p$. Thus, for any primitive $4r$-th root of unity $a \in K$, the modular category $\mathcal{V}_p$ acquires a Hermitian structure.

9.2. Theorem. *Let $K = \mathbb{C}$ with the usual complex conjugation and $a = \sqrt{-1}\, e^{\pm\pi\sqrt{-1}/2r}$ with $r \geq 2$. Then the modular category $(\mathcal{V}_p(a), \{V_i\}_{i \in I(a)})$ is unitary.*

Proof. For any $n = 1, \ldots, r-1$, the number $[n] = (a^{2n} - a^{-2n})(a^2 - a^{-2})^{-1}$ is real and non-zero. It is easy to see that its sign is equal to $(-1)^{n+1}$. (It is here that we use the choice of a.) This implies that for any $i \in I$, the dimension $\dim(V_i) = (-1)^i [i+1]$ is strictly positive.

We should prove that for any object W of $\mathcal{V}_p$, the Hermitian form $(f, g) \mapsto \operatorname{tr}(f\overline{g})$ in $\operatorname{Hom}(\mathbb{1}, W)$ is positive definite (cf. Exercise II.5.6.2). By the very definition of $\mathcal{V}$ and $\mathcal{V}_p$, the object W is the tensor product of several simple objects V_i with $i \in I$. Lemmas IV.10.6.1, IV.10.6.2 imply that it is enough to show the positivity of the Hermitian forms in all multiplicity modules $H^{ijk} = \operatorname{Hom}(\mathbb{1}, V_i \otimes V_j \otimes V_k)$ with $i, j, k \in I$. By Theorem 8.2, it is enough to show that for any strictly admissible triple $(i, j, k) \in I^3$, we have $\operatorname{tr}(\overline{q}q) = \operatorname{tr}(q\overline{q}) > 0$ where $q = q(i, j, k)$ is the generator of H^{ijk} specified in Theorem 8.2. It is obvious that $\overline{q} = \tilde{q}(i, j, k) \in H_{ijk}$. Therefore $\operatorname{tr}(\overline{q}q) = \operatorname{tr}(\tilde{q}(i, j, k)q(i, j, k))$ is exactly the value computed in Lemma 8.3. Observe that for any $n = 0, 1, \ldots, r-1$, the number $[n]!$ is real and non-zero with the sign $(-1)^{n(n+1)/2}$. A computation of signs shows that the right-hand side of (8.3.a) is a positive real number. This proves the theorem.

9.3. Remark. It is easy to compute that $\mathcal{D} = \sqrt{r/2}/\sin(\pi/r)$ is a rank of the modular category $(\mathcal{V}_p(a), \{V_i\}_{i \in I(a)})$ where $a = \sqrt{-1}\, e^{\pi\sqrt{-1}/2r}$. Indeed, $\mathcal{D}^2 = (-2r)(a^2 - a^{-2})^{-2}$. We may apply the results of Section IV.11 to the corresponding invariant of 3-manifolds τ_r. For instance, for any closed oriented

3-manifold M of genus g, we have

$$|\tau_r(M)| \leq \mathcal{D}^{g-1} = \left(\frac{\sqrt{r/2}}{\sin(\pi/r)}\right)^{g-1}.$$

Notes

Sections 1 - 3. The Jones polynomial of links was discovered by Vaughan Jones in 1984. The original approach of Jones was inspired by his work on von Neumann algebras and subfactors (see [Jo1], [Jo2]). Soon afterwards L. Kauffman found an elementary definition of the Jones polynomial based on the skein relation (see [Ka2]). This led him to a study of skein modules and to a tangle-theoretic interpretation of the Temperley-Lieb algebra (see [Ka3] - [Ka6]). The skein category $\mathcal{S}$ is implicit in the very first works of Kauffman on this subject.

Section 4. Theorem 4.1 is due to Jones [Jo1, Lemma 4.2.1]. The proof given here and the inductive formula (4.2.a) are due to Wenzl [We1]. The fundamental Lemmas 4.4.1 and 4.4.3 (usually formulated in a different language that avoids categories) are well known to specialists.

The role of the Jones-Wenzl idempotents in the theory of 3-manifold invariants was first emphasized by W.B.R. Lickorish (see [Li3] - [Li7]). He defined invariants of closed 3-manifolds following the scheme of [RT2] (discussed in this book in Chapter II) but replacing Hopf algebras and irreducible modules with Temperley-Lieb algebras and Jones-Wenzl idempotents. The algebra used by Lickorish is more tangible for mathematicians without a special taste for representation theory. However, Lickorish's approach yields just the same invariants as in [RT2], specifically, the ones derived from the quantum group $U_q(sl_2(\mathbb{C}))$ at roots of unity. It is not known if it is possible to generalize this approach to obtain the invariants of 3-manifolds derived from other quantum groups (see [TW]).

The work of Lickorish suggested to the author that the theory of Temperley-Lieb algebras and Jones-Wenzl idempotents should give rise to modular categories. This relationship explains why this theory gives rise to 3-manifold invariants and places it in a more general framework.

Section 5. The material of Section 5 is due to Lickorish, see [Li3] - [Li5].

Sections 6, 7. The definitions and results of these sections are new.

Section 8. The definitions and results of this section are new except Lemma 8.3. This lemma is well known to specialists (see [KL], [Li7], [MV1]), our proof follows the one in [Li7].

Section 9. The material of this section is new.

Appendix I
Dimension and trace re-examined

1. We give a direct definition of the dimension Dim of projective modules and the trace Tr of endomorphisms of projective modules independent of the results of Chapter I.

Let K be a commutative ring with unit. For a K-linear endomorphism $f : X \to X$ of a free K-module of finite rank X, we have the usual trace $\operatorname{Tr}(f) \in K$ defined as the trace of the matrix of f with respect to a basis of X. Invariance of the trace of matrices under conjugation implies that $\operatorname{Tr}(f)$ does not depend on the choice of the basis.

Consider a K-linear endomorphism $f : X \to X$ of a projective K-module X. Let Y be a K-module such that the direct sum $X \oplus Y$ is a free K-module of finite rank. Let p and j be the projection $X \oplus Y \to X$ and embedding $X = X \oplus 0 \hookrightarrow X \oplus Y$ respectively. Consider the composition $jfp : X \oplus Y \to X \oplus Y$ and set $\operatorname{Tr}(f) = \operatorname{Tr}(jfp)$.

2. Lemma. $\operatorname{Tr}(f)$ *does not depend on the choice of* Y.

This lemma, proven below, shows that the trace $\operatorname{Tr}(f) \in K$ is well defined. It is clear that $\operatorname{Tr}(f)$ is additive with respect to direct summation of endomorphisms and multiplicative with respect to tensor multiplication of endomorphisms.

We define the dimension $\operatorname{Dim}(X) \in K$ of a projective K-module X by the formula $\operatorname{Dim}(X) = \operatorname{Tr}(\mathrm{id}_X)$. It is clear that $\operatorname{Dim}(X \oplus X') = \operatorname{Dim}(X) + \operatorname{Dim}(X')$, $\operatorname{Dim}(X \otimes_K X') = \operatorname{Dim}(X) \operatorname{Dim}(X')$, and $\operatorname{Dim}(K^n) = n$ for any integer $n \geq 0$. It is left to the reader to verify that these definitions of Tr and Dim are equivalent to those of Section I.1.

Proof of Lemma. Let Y and Y' be two K-modules such that $X \oplus Y$ and $X \oplus Y'$ are free modules of finite rank. Denote by p, p', j, j' the projections $X \oplus Y \to X$, $X \oplus Y' \to X$ and the embeddings $X \hookrightarrow X \oplus Y$, $X \hookrightarrow X \oplus Y'$ respectively. We shall show that $\operatorname{Tr}(jfp) = \operatorname{Tr}(j'fp')$.

Consider the free module of finite rank $V = X \oplus Y \oplus X \oplus Y'$. Composing the projection onto the first summand $V \to X$ with $f : X \to X$ and with the embedding onto the first summand $X \hookrightarrow V$ we get an endomorphism, say F, of V. If we choose a basis in V which is a juxtaposition of bases in $X \oplus Y$ and $X \oplus Y'$ then the matrix of F is obtained from the matrix of jfp by adding zero columns and zero rows. Therefore $\operatorname{Tr}(F) = \operatorname{Tr}(jfp)$. Similarly, composing the projection onto the third summand $V \to X$ with $f : X \to X$ and with the embedding onto the third summand $X \hookrightarrow V$ we get an endomorphism, F', of V. As above, $\mathrm{Tr}(F') = \mathrm{Tr}(j' f p')$. It remains to show that $\mathrm{Tr}(F) = \mathrm{Tr}(F')$.

It follows from definitions that for any $x \in X, y \in Y, z \in X, t \in Y'$,

$$F(x \oplus y \oplus z \oplus t) = f(x) \oplus 0 \oplus 0 \oplus 0, \quad F'(x \oplus y \oplus z \oplus t) = 0 \oplus 0 \oplus f(z) \oplus 0.$$

Define an involutive homomorphism $g : V \to V$ by the formula

$$g(x \oplus y \oplus z \oplus t) = z \oplus y \oplus x \oplus t.$$

It is straightforward to verify that $F' = gFg = gFg^{-1}$. Therefore $\mathrm{Tr}(F) = \mathrm{Tr}(F')$.

Appendix II
Vertex models on link diagrams

1. The operator invariant $F(\Omega)$ of a v-colored framed oriented link $\Omega \subset \mathbb{R}^3$ may often be computed as a vertex state sum on a diagram of Ω in $\mathbb{R}^2$. Vertex models on link diagrams were first used by L. Kauffman to compute the Jones polynomial of links in $\mathbb{R}^3$ (see [Ka2], cf. Section XII.1). The Kauffman model may be considered as a topological counterpart of the Potts model known in statistical mechanics (see [Ba1]). More general vertex models for link polynomials were introduced in [AW], [ADW], [Jo4], [Tu4], [Tu7]. Vertex models are less universal than the face models because they do not apply to link diagrams in surfaces of non-zero genus or in 2-polyhedra. (The only known exception is the Kauffman model which works for link diagrams in oriented surfaces). Vertex models played an important role at the early stages of the theory of quantum link invariants.

2. Consider a ribbon category $\mathcal{V}$ whose objects are free modules of finite rank over a commutative ring with unit, K, and whose morphisms are K-homomorphisms. In other words, $\mathcal{V}$ is a subcategory of the category of free modules of finite rank over K. We assume that $\mathbb{1}_{\mathcal{V}} = K$, the tensor product in $\mathcal{V}$ coincides with the ordinary tensor product over K, and that $V^* = \mathrm{Hom}_K(V, K)$ for any object V of $\mathcal{V}$.

Let $\Omega \subset \mathbb{R}^3$ be a colored framed oriented link. Present Ω by a generic diagram D in $\mathbb{R} \times [0, 1]$. The computation of $F(\Omega) \in K$ as a state sum on D is based on the following idea. Expand $F(\Omega)$ via the standard operators associated to the singular points of D, cf. Section I.3.1.1. Present these standard operators by matrices and compute $F(\Omega)$ from these matrices. We obtain an expression of $F(\Omega)$ as a polynomial in the matrix elements of the standard operators. This polynomial may be interpreted as a state sum on D. Since the matrix elements of standard operators are associated to the singular points ("vertices") of D, we obtain a "vertex" state sum model on D. Similar models are used in statistical mechanics (see [Ba1]; in contrast to the knot theory there are no under/overcrossings in statistical mechanics).

Here are the details. Fix bases in the K-modules used as the colors of the components of Ω. Denote by $\#D$ the set of singular points of D (cf. Section I.3.1.1). This set splits D into a finite number of disjoint arcs called the edges of D. Since the extremal points of the height function $\mathbb{R} \times [0, 1] \to [0, 1]$ are included in $\#D$, each edge projects injectively in $[0, 1]$. The orientation of the edge determines its vertical direction: either "up" or "down". For an edge e of D, denote by $V(e)$ the color of the component of Ω containing e. A state on D is a function assigning to every edge e of D an element of the fixed basis of $V(e)$, if e is directed downwards, and an element of the dual basis of $(V(e))^*$, if e is directed upwards. For each point $a \in \#D$, a part of D lying in a small neighborhood of a in $\mathbb{R}^2$ represents one of the elementary (2,2)-tangles drawn in Figure I.2.5, a cup-like tangle $\cup_V, \cup_{\bar{V}}$, or a cap-like tangle $\cap_V, \cap_{\bar{V}}$. Denote this tangle by D_a. The operator $F(D_a)$ acts in K-modules with distinguished bases obtained via tensoring. For any state λ on D and any $a \in \#D$, define the Boltzmann weight $\langle\lambda|a\rangle \in K$ to be the matrix element of $F(D_a)$ corresponding to the basis vectors provided by λ (or, more precisely, by the values of λ on the edges of D incident to a).

3. Theorem.

$$F(\Omega) = \sum_{\lambda} \left(\prod_{a \in \#D} \langle\lambda|a\rangle \right) \in K$$

where λ runs over all states on D.

Proof. Let $a_1, ..., a_n$ be the singular points of D enumerated in accordance with their decreasing height. Regarding Ω as a morphism $\emptyset \to \emptyset$ in the category of colored ribbon tangles, we obtain

$$\Omega = (I_1 \otimes D_{a_1} \otimes I'_1) \circ (I_2 \otimes D_{a_2} \otimes I'_2) \circ ... \circ (I_n \otimes D_{a_n} \otimes I'_n)$$

where $I_1, I'_1, I_2, I'_2, ..., I_n, I'_n$ are colored tangles consisting of vertical unlinked unwisted bands. Therefore

$$F(\Omega) = (\mathrm{id}_1 \otimes F(D_{a_1}) \otimes \mathrm{id}'_1)(\mathrm{id}_2 \otimes F(D_{a_2}) \otimes \mathrm{id}'_2) ... (\mathrm{id}_n \otimes F(D_{a_n}) \otimes \mathrm{id}'_n)$$

where $\mathrm{id}_1, \mathrm{id}'_1, ..., \mathrm{id}_n, \mathrm{id}'_n$ are the identity endomorphisms of certain K-modules. Here $F(\Omega)$ is a K-linear endomorphism of $F(\emptyset) = K$. It is easy to compute that this endomorphism is multiplication by

$$\sum_{\lambda} \left(\prod_{i=1}^{n} \langle\lambda|a_i\rangle \right) = \sum_{\lambda} \left(\prod_{a \in \#D} \langle\lambda|a\rangle \right).$$

4. Theorem 3 may be generalized to v-colored ribbon (0,0)-graphs. We just include the matrix elements of the homomorphisms which color the coupons in the set of Boltzmann weights. For a v-colored ribbon (k, l)-graph Ω, there is a similar formula for the matrix elements of $F(\Omega)$. The only difference is that in this case we sum up over states with fixed values on the boundary arcs of D.

The vertex model described above may be generalized to ribbon categories endowed with a covariant functor into a category of modules.

Appendix III
Gluing re-examined

In this Appendix we analyze in more detail the gluing of m-surfaces used in Chapter V.

1. Let Σ be a marked surface over a monoidal class $\mathscr{C}$. Let X, Y be two components of $\partial\Sigma$ marked with the same element of $\mathscr{C}$ and the signs $+1$ and -1. In Chapter V we consider the surface $\Sigma' = \Sigma/[X = Y]$ obtained from Σ by the identification $X = Y$. Strictly speaking, Σ' depends on the choice of an orientation-reversing base point preserving homeomorphism $X \to Y$. The goal of this Appendix is to show that the dependence of Σ' on this choice may be disregarded. We shall show that the surfaces Σ' corresponding to different homeomorphisms $X \to Y$ may be treated as copies of one m-surface. This allows us to speak of the module of states $\mathscr{H}(\Sigma')$ independent of the choice of a homeomorphism $X \to Y$.

2. Lemma. *Let $f : X \to X$ be a homeomorphism preserving the orientation and the base point $x \in X$. There exists a homeomorphism $\psi : \Sigma \to \Sigma$ such that* (i) *$\psi|_X = f$,* (ii) *ψ is the identity on $(\Sigma \setminus \mathrm{Int}(U)) \cup (x \times [0, 1])$ where $U = X \times [0, 1]$ is a cylindrical neighborhood of $X = X \times 0$ in Σ,* (iii) *ψ carries $X \times t \subset U$ into $X \times t$ for all $t \in [0, 1]$. Any two such homeomorphisms $\psi : \Sigma \to \Sigma$ are isotopic via an isotopy constant on $\partial\Sigma$.*

Proof. We shall show that (*) there exists an isotopy $f_t : (X, x) \to (X, x), t \in [0, 1]$ relating $f_0 = f$ and $f_1 = \mathrm{id}$, (**) such an isotopy is unique up to isotopy. Claim (*) allows us to define a homeomorphism $\psi : \Sigma \to \Sigma$ as follows. Choose a cylindrical neighborhood $U = X \times [0, 1]$ of $X = X \times 0$ in Σ. For a point $a \in \Sigma \setminus U$, set $\psi(a) = a$; for $a = z \times t \in U$ with $z \in X, t \in [0, 1]$, set $\psi(a) = f_t(z) \times t$. The properties (i) - (iii) of ψ claimed by Lemma follow from definitions. It follows from (**) that the isotopy type of ψ does not depend on the choice of $\{f_t\}_t$. It is clear that the homeomorphisms $\psi : \Sigma \to \Sigma$ corresponding to ambient isotopic cylindrical neighborhoods U are isotopic. Since any two cylindrical neighborhoods of X in Σ are isotopic, we obtain that the isotopy class of ψ modulo $\partial\Sigma$ is determined solely by f. It is obvious that any homeomorphism $\psi : \Sigma \to \Sigma$ satisfying (i) - (iii) can be obtained by this construction. Therefore the isotopy class of such ψ is determined solely by f.

It remains to verify (*) and (**). Fix a homeomorphism $h : X \setminus x \to (0, 1)$ so that the given orientation in X corresponds to the right-hand direction of the interval. It is clear that $hfh^{-1} : (0,1) \to (0,1)$ is an orientation-preserving homeomorphism. Conversely, for any orientation-preserving homeomorphism $g : (0,1) \to (0,1)$, the mapping $h^{-1}gh : X \setminus x \to X \setminus x$ uniquely extends to a homeomorphism $(X, x) \to (X, x)$. Therefore it is enough to show that any orientation-preserving self-homeomorphism of (0,1) is isotopic to the identity, and any two such isotopies are themselves isotopic. The orientation-preserving self-homeomorphisms of (0,1) are just strictly increasing functions $(0,1) \to (0,1)$ whose values converge to 0 (resp. 1) when the parameter converges to 0 (resp. 1). The space of such functions is convex. This implies our claim.

3. Let $\mathcal{G}$ denote the set of homeomorphisms $\{g : X \to Y\}$ inverting the orientation and preserving the base point. The identification $X = Y$ along $g \in \mathcal{G}$ yields an m-surface $\Sigma'(g)$ and a gluing projection $\Sigma \to \Sigma'(g)$. We construct a commuting system of m-homeomorphisms relating the surfaces $\{\Sigma'(g) \mid g \in \mathcal{G}\}$. For any pair $g_0, g_1 \in \mathcal{G}$, Lemma 2 yields an isotopy class of homeomorphisms $\psi(g_1^{-1}g_0) : \Sigma \to \Sigma$ that extend $g_1^{-1}g_0 : X \to X$. The homeomorphisms of this class induce isotopic m-homeomorphisms $\Sigma'(g_0) \to \Sigma'(g_1)$. Denote the resulting isotopy class of m-homeomorphisms $\Sigma'(g_0) \to \Sigma'(g_1)$ by $\psi(g_0, g_1)$. It is obvious that

$$\psi(g_0, g_0) = \mathrm{id}, \quad \psi(g_0, g_2) = \psi(g_1, g_2)\, \psi(g_0, g_1)$$

for any $g_0, g_1, g_2 \in \mathcal{G}$. Taking $g_2 = g_0$ we obtain $\psi(g_1, g_0) = (\psi(g_0, g_1))^{-1}$. These identities show that the surfaces $\{\Sigma'(g)\}_{g \in \mathcal{G}}$ may be viewed as different copies of a certain m-surface Σ' "obtained from Σ by identification $X = Y$".

4. For any 2-dimensional modular functor $\mathcal{H}$, we can treat the modules of states $\{\mathcal{H}(\Sigma'(g)) \mid g \in \mathcal{G}\}$ as different copies of one and the same module. Indeed, the m-homeomorphisms $\Sigma'(g_0) \to \Sigma'(g_1)$ belonging to the isotopy class $\psi(g_0, g_1)$ induce an isomorphism $\mathcal{H}(\Sigma'(g_0)) \to \mathcal{H}(\Sigma'(g_1))$ denoted by $\Psi(g_0, g_1)$. We have

$$\Psi(g_0, g_0) = \mathrm{id}, \ \Psi(g_1, g_0) = (\Psi(g_0, g_1))^{-1}, \ \Psi(g_0, g_2) = \Psi(g_1, g_2)\Psi(g_0, g_1)$$

for any $g_0, g_1, g_2 \in \mathcal{G}$. Thus we can use the isomorphisms $\{\Psi(g_0, g_1) \mid g_0, g_1 \in \mathcal{G}\}$ to identify the modules $\{\mathcal{H}(\Sigma'(g)) \mid g \in \mathcal{G}\}$. The resulting module is denoted by $\mathcal{H}(\Sigma')$. It is canonically isomorphic to each of the modules $\{\mathcal{H}(\Sigma'(g))\}_g$. Moreover, there is a well defined gluing homomorphism $\mathcal{H}(\Sigma) \to \mathcal{H}(\Sigma')$. Indeed, by axiom (V.1.5.4) we have for each $g \in \mathcal{G}$, a gluing homomorphism $\mathcal{H}(\Sigma) \to \mathcal{H}(\Sigma'(g))$. Denote this homomorphism by $q(g)$. For any $g_0, g_1 \in \mathcal{G}$, condition (i) of axiom (V.1.5.4) yields the following commutative diagram

$$\begin{array}{ccc}
\mathcal{H}(\Sigma) & \xrightarrow{(\psi(g_1^{-1}g_0))_\#} & \mathcal{H}(\Sigma) \\
{\scriptstyle q(g_0)}\downarrow & & \downarrow{\scriptstyle q(g_1)} \\
\mathcal{H}(\Sigma'(g_0)) & \xrightarrow{\Psi(g_0,g_1)} & \mathcal{H}(\Sigma'(g_1)).
\end{array}$$

It is easy to verify that the homeomorphism $\psi(g_1^{-1}g_0) : \Sigma \to \Sigma$ is isotopic to the identity in the class of homeomorphisms $\Sigma \to \Sigma$ preserving the base points of the boundary components. Therefore $(\psi(g_1^{-1}g_0))_{\#} = \mathrm{id}$ and

$$q(g_1) = \Psi(g_0, g_1)\, q(g_0).$$

Hence the isomorphisms $\{\Psi(g_0, g_1) \mid g_0, g_1 \in \mathcal{G}\}$ are compatible with the gluing homomorphisms. This yields a gluing homomorphism $\mathcal{H}(\Sigma) \to \mathcal{H}(\Sigma')$.

These results allow us to forget about the indeterminacy in the definition of gluing. We can simply speak of the m-surface $\Sigma' = \Sigma/[X = Y]$ obtained from Σ by identification $X = Y$, the associated module $\mathcal{H}(\Sigma')$, and the gluing homomorphism $\mathcal{H}(\Sigma) \to \mathcal{H}(\Sigma')$.

Appendix IV
The signature of closed 4-manifolds from a state sum

We show that the signature of a closed oriented piecewise-linear 4-manifold W may be computed via a state sum on a skeleton of W. The state sum in question is the one introduced in Section X.1 and based on the normalized $6j$-symbols derived from a strict unimodular category $(\mathcal{V}, \{V_i\}_{i\in I})$ with ground ring K and rank $\mathcal{D} \in K$. In this appendix we adhere to Conventions VI.3.1.

1. Theorem. *Let W be a closed oriented piecewise-linear 4-manifold with signature $\sigma = \sigma(W) \in \mathbb{Z}$. Let $X \subset W$ be a skeleton of W shadowed as in Section* IX.1.6. *Then*

$$(\mathcal{D}\Delta_{\mathcal{V}}^{-1})^{\sigma} = |X| = \mathcal{D}^{-b_2(X)-\mathrm{null}(X)} \sum_{\varphi\in \mathrm{col}(X)} |X|_{\varphi} \in K.$$

Proof. Let $W_0 = W \setminus \mathrm{Int}(B^4)$ where B^4 is a small closed 4-ball lying in $W\setminus X$. It is clear that $\partial W_0 = S^3$ and X is a skeleton of W_0. By Theorem X.3.3, $\mathcal{D}^{-1}\tau^0(S^3) = \tau(S^3) = \mathcal{D}^{-1}$, so that $\tau^0(S^3) = 1$. By definition (see Section X.3.2),

$$\tau^0(S^3) = ||\,\mathrm{sh}(W_0)|| = (\mathcal{D}\Delta_{\mathcal{V}}^{-1})^{-\sigma(X)}\,|\,X|$$

where $\sigma(X)$ is the signature of X. By Theorem IX.1.10, $\sigma(X) = \sigma(W) = \sigma$. This implies the claim of the theorem.

2. Theorem 1 expresses the signature of a closed 4-manifold W in terms of combinatorial structure of a 2-skeleton $X \subset W$ and the self-intersection numbers of the regions of X. Deriving a 2-skeleton of W from a triangulation of W, we can deduce from Theorem 1 the Crane-Yetter-Roberts computation of $\sigma(W)$ in terms of state sums on triangulations of W (see [CY], [Rob1], [Br], [CKY]).

3. Consider the modular category $\mathcal{V}_p(a)$ over $\mathbb{C}$ with rank $\mathcal{D} = \sqrt{-2r}(a^2-a^{-2})^{-1}$ constructed in Section XII.7.1. By definition,

$$\Delta_{\mathcal{V}} = \sum_{i=0}^{r-2} v_i^{-1}(\dim(V_i))^2 = \sum_{i=0}^{r-2}(-1)^i a^{-i(i+2)}(a^{2i+2} - a^{-2i-2})(a^2 - a^{-2})^{-2}.$$

In the case $a = \exp(\pi\sqrt{-1}/2r)$ the number $\mathcal{D}\Delta_{\mathcal{V}}^{-1}$ can be computed explicitly. A direct computation using the Gauss sums (see, for instance, [KL, Sections 12.7, 12.8]) shows that

$$\Delta_{\mathcal{V}} = \mathcal{D}\sqrt{-1}^{\,-r} e^{\sqrt{-1}\pi\frac{3(2-r)}{4r}}.$$

This implies that varying r we can compute the signature of a closed oriented 4-manifold from the corresponding state sum on its skeletons.

4. In the examples of modular categories known to the author, the number $\mathcal{D}\Delta^{-1}$ is a root of unity. Is it always the case?

Problems

In the main body of this book we have formulated several open questions directly related to our study. We reproduce them here and add a few more questions. They are formulated in a relatively vague way leaving space to the reader's imagination.

1. Give elementary examples of modular categories other than those in Section II.1.7.

2. Give a direct proof of the invariance of $\tau(M)$ under the Kirby moves without appealing to τ'.

3. Give an explicit formula for the invariant $\tau_{\mathcal{V}}$ of closed oriented 3-manifolds corresponding to the modular category $\mathcal{V}(G, K, c, \varphi)$ introduced in Section I.1.7. Give an explicit description of the corresponding 2-DMF. (Such a description may contribute to clarification of relationships between 3-dimensional TQFT's and the Segal-Shale-Weil representations of the metaplectic groups.)

4. Do there exist non-isomorphic modular categories $\mathcal{V}$, $\mathcal{V}'$ such that $\tau_{\mathcal{V}}(M) = \tau_{\mathcal{V}'}(M)$ for every closed oriented 3-manifold M?

5. Do there exist non-homeomorphic closed oriented 3-manifolds M, M' such that $\tau_{\mathcal{V}}(M) = \tau_{\mathcal{V}}(M')$ for every modular category $\mathcal{V}$? Does there exist a closed oriented 3-manifold $M \neq S^3$ such that $\tau_{\mathcal{V}}(M) = \tau_{\mathcal{V}}(S^3)$ for every modular category $\mathcal{V}$?

6. Give an intrinsic computation of the quantum invariant τ of closed oriented 3-manifolds.

7. Is the (projective) action of the mapping class group of a closed oriented surface Σ in the module of states of Σ irreducible? Consider the kernels of these actions corresponding to all modular categories. Is the intersection of these kernels non-trivial? (This would be hard to believe.)

8. Do there exist unitary (or at least Hermitian) modular categories that are not unimodal?

9. Is the TQFT $(\widehat{\mathcal{T}}, \widehat{\tau})$, constructed in Section IV.8, non-degenerate? Is the TQFT $(\mathcal{T}^e, \tau^e)$, restricted to e-surfaces without distinguished arcs and to 3-manifolds with empty ribbon graphs, non-degenerate?

10. Extend the 2-DMF's (as defined in Section V.1) to surfaces with ≥ 1 distinguished marked points in boundary components. Define gluing homomorphisms for gluings of surfaces along arcs in their boundaries. Express the modules of states of surfaces via the modules of states of triangles (= disks with three distinguished arcs in the boundary).

11. Give an intrinsic formulation of the non-degeneracy axiom (V.1.6.4) without appealing to underlying ribbon categories. Do there exist 2-DMF's satisfying (V.1.6.1) - (V.1.6.3) but not (V.1.6.4)? Same questions for weak 2-DMF's.

12. Describe the behavior of the duality pairings (see Section III.2) and the Hermitian forms (see Section IV.10) on the modules of states of surfaces under the gluing homomorphisms introduced in Chapter V.

13. Verify the duality axiom for the mirror modular functor $\overline{\mathcal{H}}$ (see Section V.3.3).

14. Is it true that for a rational weak 2-DMF $\mathcal{H}$ with charge ∇, the product $\nabla\Delta_{\mathcal{V}}$ is a rank of the underlying modular category $\mathcal{V}$? Is it true that rational weak 2-DMF's over the same commutative ring are isomorphic if and only if their underlying modular categories are isomorphic and charges are equal?

15. Do there exist ribbon Ab-categories which can not be realized as the underlying categories of 2-DMF's?

16. Give an axiomatic treatment of "weak" homeomorphisms and "weak" modular functors generalizing the definitions of Chapters IV and V to the abstract setting of Chapter III. (Idea: consider $\mathfrak{A}$-structures in the cylinder $X \times [0, 1]$ different from those induced by the $\mathfrak{A}$-structures in X.)

17. Formulate the properties of multiplicity modules and $6j$-symbols associated to a semisimple ribbon Ab-category $\mathcal{V}$ which would allow to reconstruct $\mathcal{V}$ from the $6j$-symbols (at least up to some natural equivalence).

18. What are the exact relationships between the theory of modular categories (in our sense) and the tensor setting of Moore and Seiberg?

19. Compute explicitly the isomorphism of TQFT's provided by Theorem VII.4.2 (cf. the conjecture formulated in Section VII.4.3). Extend Theorem VII.4.2 to fields of non-zero characteristic.

20. Give a direct proof of the fact (established in Chapters VIII, IX via thickenings) that the bilinear form of any integer shadow takes values in $\mathbb{Z}$.

21. Compute the classical invariants of knots and links in S^3 (the Alexander polynomial, the signatures, the Milnor invariants,...) in terms of shadow projections of knots and links in $S^2 \subset S^3$.

22. Do there exist non-trivial state sums on shadows whose partition functions are invariant under basic shadow moves and stabilization but not invariant under cobordism? (This could give invariants of 4-manifolds.)

23. Prove that the category $\mathcal{V}_q\mathfrak{g}$ (introduced in Section XI.6.3) is Hermitian for any root of unity q and unitary for $q = \exp(\pm 2\pi\sqrt{-1}/l)$ (for a possible approach see [We2], [We4]).

24. Prove that the skein category $\mathcal{V}_p(a)$ over $K = \mathbb{C}$ with $a \in \mathbb{C}$ a primitive $4r$-th root of unity is equivalent to the purified representation category of $U_q(sl_2(\mathbb{C}))$ with $q = -a^2$. (The equivalence should respect the braiding, twist, etc.) Give an interpretation of the even subcategory of $\mathcal{V}_p(a)$ in terms of quantum groups.

References

[Ab] Abe, E., Hopf algebras. Cambridge University Press, Cambridge 1980

[AW] Akutsu, Y., Wadati, M., Knot invariants and the critical statistical systems. J. Phys. Soc. Japan 56 (1987), 839-842

[ADO] Akutsu, Y., Deguchi, T., Ohtsuki, T., Invariants of colored links. J. Knot Theory and its Ramifications 1 (1992), 161-184

[ADW] Akutsu, Y., Deguchi, T., Wadati, M., Exactly solvable models and new link polynomials I-V. J. Phys. Soc. Japan 56 (1987), 3039-3051, 3464-3479; ibid. 57 (1988), 757-776, 1173-1185, 1905-1923

[Al1] Alexander, J.W., On the subdivision of a 3-space by a polyhedron. Proc. Nat. Acad. Sci. U.S.A. 10 (1924), 6-8

[Al2] —, The combinatorial theory of complexes. Ann. Math. 31 (1930), 294-322

[AC1] Altschuler, D., Coste, A., Quasi-Quantum Groups, Knots, Three-Manifolds and Topological Field Theory. Comm. Math. Phys. 150 (1992), 83-107

[AC2] —, Invariants of 3 manifolds from finite group cohomology. J. Geom. and Phys. 11 (93), 191

[An] Andersen, H. H., Tensor products of quantized tilting modules. Comm. Math. Phys. 149 (1991), 149-159

[AP] Andersen, H. H., Paradowski, J., Fusion categories arising from semisimple Lie algebras. Aarhus preprint 1993

[APK] Andersen, H. H., Polo, P., Wen, K., Representations of quantum algebras. Invent. Math. 104 (1991), 1-59

[AW] Archer, F., Williams, R.M., The Turaev-Viro state sum model and 3-dimensional quantum gravity. Phys. Lett. B 273:4 (1991), 438-444

[Ar] Arnold, V.I., Characteristic class entering in quantization conditions. Funct. Analysis and its Applications 1 (1967), 1-13

[AsW] Askey, R., Wilson, J., A Set of Orthogonal Polynomials that Generalize the Racah Coefficients or $6-j$ Symbols. SIAM J. Math. Anal. 10, n. 5 (1979), 1008-1016

[At1] Atiyah, M., Topological quantum field theories. Publ. Math. IHES 68 (1989), 175-186

[At2] —, On framings of 3-manifolds. Topology 29 (1990), 1-7

[At3] —, The Geometry and Physics of Knots. Cambridge Univ. Press, Cambridge 1990

[ADWi] Axelrod, S., Dellapietra, S., Witten, E., Geometric quantization of Chern-Simons gauge theory. J. Diff. Geom. 33 (1991), 787-902

[Bar] Bar-Natan, D., On the Vassiliev knot invariants. Topology (to appear)

[Ba1] Baxter, R.J., Exactly solved models in statistical mechanics. Academic Press, New York-London 1982

[Ba2] —, Partition function of the eight-vertex lattice model. Ann. Phys. 70 (1992), 193-228

[BD] Beliakova, A., Durhuus, B., Topological quantum field theory and invariants of graphs for quantum groups. Preprint 1993

[BL1] Biedenharn, L.C., Louck, J.D., Angular Momentum in Quantum Physics. Encyclopedia of mathematics and its applications 8. Addison-Wesley 1981

[BL2] —, The Racah-Wigner Algebra in Quantum Theory. Encyclopedia of mathematics and its applications 9. Addison-Wesley 1981

[BW] Birman, J., Wenzl, H., Braids, link polynomials and a new algebra. Trans. Amer. Math. Soc. 313 (1989), 249-273

[Bl] Blanchet, C., Invariants of three-manifolds with spin structure. Comm. Math. Helv. 67 (1992), 406-427

[BHMV1] Blanchet, C., Habegger, N., Masbaum, G., Vogel, P., Three-manifold invariants derived from the Kauffman bracket. Topology 31 (1992), 685-699

[BHMV2] —, Topological Quantum Field Theories Derived from the Kauffman Bracket. Preprint 1993

[Br] Broda, B., Surgical invariants of four-manifolds. Preprint 1993

[BZ] Burde, G., Zieschang, H., Knots, (de Gruyter Stud. Math. 5). Walter de Gruyter, Berlin-New York 1985

[Ca1] Cairns, S.S., Introduction of a Riemannian geometry on a triangulable 4-manifold. Ann. Math. 45 (1944), 218-219

[Ca2] —, The manifold smoothing problem. Bull. Amer. Math. Soc. 67 (1961), 237-238

[CLM] Cappell, S.E., Lee R., Miller, E.Y., Invariants of 3-Manifolds from Conformal Field Theory. In: Topology '90 (ed. by B. Apanasov et al.), 91-109. De Gruyter, Berlin-New York 1992

[Car] Cartier, P., Développements récents sur les groupes de tresses; applications à l'algèbre et à la topologie. Sém. Bourbaki, exp. 716, Astérisque 189-190 (1991), 17-67

[Cas] Casler, B.G., An embedding theorem for connected 3-manifolds with boundary. Proc. Amer. Math. Soc. 16 (1965), 559-566

[CK] Concini, C. de, Kac, V.G., Representations of quantum groups at roots of 1. Colloque Dixmier 1990, 471-506. Progress in Math. 92, Birkhäuser, Boston-Basel-Stuttgart 1990

[Cr1] Crane, L., 2-d Physics and 3-d Topology. Comm. Math. Phys. 135 (1991), 615-640

[Cr2] —, Conformal field theory, spin geometry, and quantum gravity. Phys. Lett. B 259 (1991), 243-248

[CKY] Crane, L., Kauffman, L.H., Yetter, D., Evaluating the Crane-Yetter invariant. Preprint 1993

[CY] Crane, L., Yetter, D., A categorical construction of 4d topological quantum field theories. Preprint 1992

[CLS] Couture, M., Lee, H.C., Schmeing, N.C., A new family of N-state representations of the braid group. In: Physics, Geometry, and Topology (ed. by H.C. Lee), 573-582. Plenum Press, New York 1990

[Deg] Degiovanni, P., Moore and Seiberg's equations and 3-d topological quantum field theory. Comm. Math. Phys. 145 (1992), 459-505

[De] Deligne, P., Private letters. Princeton, April 5 and April 29 (1991)

[DW] Dijkgraaf, R., Witten, E., Topological gauge theories and group cohomology. Comm. Math. Phys. 129 (1990), 393-429

[Dr1] Drinfel'd, V.G., Hopf algebras and the quantum Yang-Baxter equation. Dokl. Akad. Nauk SSSR 283 (1985), 1060-1064. English transl.: Soviet Math. Dokl. 32 (1985), 254-258

[Dr2] —, Quantum groups, Proc. of ICM Berkeley 1986. Providence, R.I. vol. 1 (1987), 798-820

[Dr3] —, On almost cocommutative Hopf algebras. Algebra i Analiz 1 (1989), 30-46. English transl.: Leningrad Math. J. 1 (1990), 321-342

[Dr4] —, Quasi-Hopf algebras. Algebra i Analiz 1 (1989), 114-148. English transl.: Leningrad Math. J. 1 (1990), 1419-1457

[Du] Durhuus, B., A discrete approach to topological quantum field theories. J. Geom. and Phys. 11 (1993), 155-180

[DJN] Durhuus, B., Jacobsen, H.P., Nest, R., Topological quantum field theories from generalized $6j$-symbols. Rev. Math. Phys. 5 (1991), 1

[EK] Evans, D., Kawahigashi, Y., From subfactors to 3-dimensional topological quantum field theories and back. Preprint 1992

[Fa] Faddeev, L.D., Integrable models in (1+1)-dimensional quantum field theory, Les Houches 1982. Elsevier Science Publishers 1984

[FRT] Faddeev, L.D., Reshetikhin, N.Y., Takhtajan, L.A., Quantization of Lie groups and Lie algebras. Leningrad Math. J. 1 (1990), 193-226

[FST] Faddeev, L.D., Sklyanin, E.K., Takhtajan, L.A., The quantum inverse problem. Theor. Math. Phys. 40 (1979), 194-220

[FG] Felder, G., Grandjean, O., On combinatorial three-manifold invariant. Preprint 1992

[FR] Fenn, R., Rourke, C., On Kirby's calculus of links. Topology 18 (1979), 1-15

[Fe] Ferguson, K., Link Invariants Associated to TQFT's with Finite Gauge Group. J. of Knot Theory and its Ramifications 2 (1993), 11-36

[Fi1] Fiedler, T., Algebraic links and the Hopf fibration. Topology 30 (1991), 259-265

[Fi2] —, A small state sum for knots. Topology 32 (1993), 281-294

[FrG] Freed, G., Gompf, R., Computer calculations of Witten's 3-manifold invariant. Comm. Math. Phys. 141 (1991), 79-117

[FQ] Freed, G., Quinn, F., Chern-Simons theory with Finite Gauge Group. Preprint 1991

[FrQ] Freedman, M.H., Quinn, F., Topology of 4-manifolds. Princeton University Press, Princeton, N. J. 1990

[FKe] Fröhlich, J., Kerler, T., Quantum groups, quantum categories and quantum field theory. Lect. Notes Math. 1542. Springer-Verlag, New York-Heidelberg-Berlin 1993

[FK1] Fröhlich, J., King, C., The Chern-Simons Theory and Knot Polynomials. Comm. Math. Phys. 126 (1989), 167-199

[FK2] —, Two-dimensional conformal field theory and three-dimensional topology. Int. J. Mod. Phys. A 4 (1989), 5321-5399

[FY1] Freyd, P., Yetter, D., Braided compact closed categories with applications to low-dimensional topology. Adv. Math. 77 (1989), 156-182

[FY2] —, Coherence Theorems via Knot Theory. J. Pure Appl. Algebra 78 (1992), 49-76

[Fu] Funar, L., Représentations du groupe symplectique et variétés de dimension 3. C.R. Ac. Sci. Paris 316 (1993), 1067-1072

[Ga] Garoufalidis, S., Relations among 3-manifold invariants. Thesis, The University of Chicago 1992

[GK] Gelfand, S., Kazhdan, D., Examples of Tensor Categories. Invent. Math. 109 (1992), 595-617

[Gi] Gilmer, P., Invariants for 1-dimensional cohomology classes arising from TQFT. Preprint 1994

[Go] Gocho, T., The topological invariant of three-manifolds based on the $U(1)$-gauge theory. J. Fac. Sci. Univ. Tokyo, Sect. 1A, Math. 39 (1992), 169-184

[Gu] Guadagnini, E., The Link Invariants of the Chern-Simons Field Theory, (de Gruyter Exp. in Math. 10). Walter de Gruyter, Berlin-New York 1993

[GS] Guillemin, V., Sternberg, S., Geometric asymptotics. Math. Surveys 14, A.M.S., Providence, R.I. 1977

[HP] Hasslacher, B., Perry, M., Spin Networks and Simplicial Quantum Gravity. Phys. Lett. 103B (1981), 21-24

[He] Hempel, J., 3-manifolds. Annals of Math. Studies, 86. Princeton University Press, Princeton, N. J. 1976

[Hen] Hennings, M.A., Hopf algebras and regular isotopy invariants for link diagrams. Math. Proc. Camb. Phil. Soc. 109 (1991), 59-77

[Hi] Hitchin, N., Flat connections and geometric quantization. Comm. Math. Phys. 131 (1990), 347-380

[Je] Jeffrey, L., Chern-Simons-Witten invariants of lens spaces and torus bundles, and the semi-classical approximation. Comm. Math. Phys. 147 (1992), 563-604

[Ji1] Jimbo, M., A q-difference analogue of $U(\mathfrak{g})$ and the Yang-Baxter equation. Letters Math. Phys. 10 (1985), 63-69

[Ji2] —, A q-analogue of $U(\mathfrak{g}(N+1))$, Hecke algebra, and the Yang-Baxter equation. Letters Math. Phys. 11 (1986), 247-252

[Ji3] —, Quantum R-matrix for the generalized Toda system. Comm. Math. Phys. 102 (1986), 537-547

[Jo1] Jones, V.F.R., Index for subfactors. Invent. Math. 72 (1983), 1-25

[Jo2] —–, A polynomial invariant of knots via von Neumann algebras. Bull. Amer. Math. Soc. 12 (1985), 103-112

[Jo3] —–, Hecke algebra representations of braid groups and link polynomials. Ann. Math. 126 (1987), 335-388

[Jo4] —–, On knot invariants related to some statistical mechanical models. Pacif. J. Math. 137 (1989), 311-334

[Jo5] —–, Subfactors and knots. CBMS Reg. Conf. Ser. in Math. 80 (1991)

[JL] Joseph, A., Letzter, G., Local finiteness of the adjoint action for quantized enveloping algebras. J. of Algebra 153 (1992), 289-318

[JS1] Joyal, A., Street, R., Braided monoidal categories. Mathematics Reports 850067, Macquarie University 1985

[JS2] —–, Braided monoidal categories. Mathematics Reports 860081, Macquarie University 1986

[JS3] —–, The Geometry of Tensor Calculus, I. Adv. Math. 88, n. 1 (1991), 55-112

[JS4] —–, Tortile Yang-Baxter operators in tensor categories. J. Pure Appl. Algebra 71 (1991), 43-51

[JS5] —–, An introduction to Tannaka duality and quantum groups. Lect. Notes Math. 1488, 413-492. Springer-Verlag, New York-Heidelberg-Berlin 1991

[KP] Kac, V.G., Peterson, D.H., Infinite dimensional Lie algebras, theta functions and modular forms. Adv. Math. 53 (1984), 125-264

[KW] Kac, V.G., Wakimoto, M., Modular and conformal invariance constraints in representation theory of affine algebras. Adv. Math. 70 (1988), 156-236

[K-B] Kania-Bartoszynska, J., Examples of different 3-manifolds with the same invariants of Witten and Reshetikhin-Turaev. Topology 32 (1993), 47-54

[Ka] Kaplan, S., Constructing framed 4-manifolds with given almost framed boundaries. Trans. Amer. Math. Soc. 254 (1979), 237-263

[KMS] Karowski, M., Müller, W., Schrader, R., State sum invariants of compact 3-manifolds with boundary and $6j$-symbols. J. Phys. A. 25 (1992), 4847

[KS] Karowski, M., Schrader, R., A Combinatorial Approach to Topological Quantum Field Theories and Invariants of Graphs. Comm. Math. Phys. 151 (1993), 355-402

[Kas] Kassel, C., Quantum groups. Graduate Texts in Math., Springer-Verlag, New York-Heidelberg-Berlin 1994

[KT] Kassel, C., Turaev, V. Double construction for monoidal categories. Acta Math. (to appear)

[Ka1] Kauffman, L., Formal Knot Theory. Mathematical Notes 30. Princeton University Press, Princeton, N. J. 1983

[Ka2] —–, State Models and the Jones Polynomial. Topology 26 (1987), 395-407

[Ka3] —–, Spin networks and knot polynomials. Int. J. of Modern Physics A5, n. 1 (1990), 93-115

[Ka4] —–, An invariant of regular isotopy. Trans. Amer. Math. Soc. 318 (1990), 417-471

[Ka5] —, Knots and Physics. World Scientific, Singapore 1991

[Ka6] —, Computing Turaev-Viro Invariants for 3-Manifolds. Manuscripta Math. 72 (1991), 81-94

[Ka7] —, Map Coloring, q-Deformed Spin Networks, and the Turaev-Viro Invariants for 3-Manifolds. Int. J. of Modern Physics B 6, n. 11/12 (1992), 1765-1794

[KL] Kauffman, L., Lins, S.L., Temperley-Lieb recoupling theory and invariants of 3-manifolds. Princeton University Press, Princeton, N.J. 1994

[Kaw] Kawauchi, A., Topological Imitation, Mutation, and the Quantum $SU(2)$ Invariant. J. of Knot Theory and its Ramifications 3 (1994), 25-39

[KWe] Kazhdan, D., Wenzl, H., Reconstructing monoidal categories. Advances in Soviet Math. 16, n. 2 (1993), 111-136

[KeL] Kelly, G., Laplaza, M., Coherence for compact closed categories. J. Pure Appl. Algebra 19 (1980), 193-213

[Ki] Kirby, R., A calculus of framed links in S^3. Invent. Math. 45 (1978), 35-56

[KM] Kirby, R., Melvin, P., On the 3-manifold invariants of Witten and Reshetikhin-Turaev for $sl_2(\mathbb{C})$. Invent. Math. 105 (1990), 473-545

[KMZ] Kirby, R., Melvin, P., Zhang, X., Quantum Invariants at the Sixth Root of Unity. Comm. Math. Phys. 151 (1993), 607-617

[KR1] Kirillov, A.N., Reshetikhin, N.Y., Representations of the algebra $U_q(sl_2)$, q-orthogonal polynomials and invariants of links. In: Infinite dimensional Lie algebras and groups, (ed. by V.G. Kac), 285-339. Adv. Ser. in Math. Phys. 7, World Scientific, Singapore 1988

[KR2] —, q-Weyl group and a multiplicative formula for universal R-matrices. Comm. Math. Phys. 134 (1990), 421-431

[Ko1] Kohno, T., Topological invariants for 3-manifolds using representations of mapping class groups I. Topology 31 (1992), 203-230

[Ko2] —, Three-manifold invariants derived from conformal field theory and projective representations of modular groups. Int. J. of Modern Phys. B 6 (1992), 1795-1805

[Ko3] —, Topological invariants for 3-manifolds using representations of mapping class groups II: Estimating tunnel numbers of knots. Preprint 1992

[KoT] Kohno, T., Takata, T., Symmetry of Witten's 3-Manifold Invariants for $sl(n, \mathbb{C})$. J. of Knot Theory and its Ramifications 2 (1993), 149-169

[Kon] Kontsevich, M., Rational conformal field theory and invariants of 3-dimensional manifolds. Preprint 1989

[KuR] Kulish, P.P., Reshetikhin, N. Y., The quantum linear problem for the sine-Gordon equation and higher representations. Zap. Nauchn. Sem. LOMI 101 (1980), 101-110. English transl.: J. Soviet Math. 23 (1983), 2435-2441

[KuS] Kulish, P.P., Sklyanin, E.K., Solutions of the Yang-Baxter equation. J. Soviet Math. 19 (1982), 1596-1620

[Kup1] Kuperberg, G., Involutory Hopf Algebras and 3-Manifold Invariants. Int. J. Math. 2 (1990), 41-66

[Kup2] —, The Quantum G_2 Link Invariant. Int. J. Math. 5 (1994), 61-85

[LP] Laudenbach, F., Poenaru, V., A note on 4-dimensional handlebodies. Bull. Soc. Math. France 100 (1972), 337-347

[La1] Lawrence, R.J., A functorial approach to the one-variable Jones polynomial. J. Diff. Geom. 37 (1993), 689-710

[La2] —, On algebras and triangle relations. In: Topological and Geometrical Methods in Field Theory (ed. by J. Mickelsson, O. Pekonen), 429-447. World Scientific, Singapore 1992

[La3] —, Triangulations, Categories and Extended Topological Field Theories. Harvard preprint 1992

[LS] Levendorskii, S.Z., Soibelman, Y.S., Some applications of quantum Weyl group. J. Geom. and Phys. 7, n. 2 (1991), 241-254

[Ler] Leray, J., Analyse lagrangienne et mécanique quantique. Publ. Math. IRMA 25 (1978), Strasbourg (English translation: MIT Press)

[Le] Lewis, S.M., Three dimensional Regge Quantum Gravity and $6j$ symbols. Phys. Lett. 122 B (1983), 265-267

[Li1] Lickorish, W.B.R., A representation of orientable combinatorial 3-manifolds. Ann. Math. 76 (1962), 531-540

[Li2] —, Polynomials for links. Bull. London Math. Soc. 20 (1988), 558-588

[Li3] —, Three-manifold invariants and the Temperley-Lieb algebra. Math. Ann. 290 (1991), 657-670

[Li4] —, Calculations with the Temperley-Lieb algebra. Comm. Math. Helv. 67 (1992), 571-591

[Li5] —, The Skein Method for Three-Manifold Invariants. J. of Knot Theory and its Ramifications 2 (1993), 171-194

[Li6] —, Distinct 3-manifolds with all $SU(2)_q$-invariants the same. Preprint 1992

[Li7] —, Skeins and Handlebodies. Preprint 1992

[Lins] Lins, S., Tables of Turaev-Viro Invariants for 3-manifolds. Preprint 1990

[LV] Lion, G., Vergne, M., The Weil representation, Maslov index and Theta series. Progress in Math. 6. Birkhäuser, Boston-Basel-Stuttgart 1980

[Lu] Lu, N., A simple proof of the fundamental theorem of Kirby calculus on links. Trans. Amer. Math. Soc. 331 (1992), 143-156

[Lu1] Lusztig, G., Quantum deformations of certain simple modules over enveloping algebras. Adv. Math. 70 (1988), 237-249

[Lu2] —, Finite dimensional Hopf algebras arising from quantized universal enveloping algebras. J. Amer. Math. Soc. 3 (1990), 257-296

[Lu3] —, Quantum groups at roots of 1. Geom. Dedicata 35 (1990), 89-113

[Lu4] —, Introduction to quantum groups. Progress in Math. 110. Birkhäuser, Boston-Basel-Stuttgart 1993

[Ly1] Lyubashenko, V.V., Tensor categories and RCFT I, II. Preprints of Inst. Theor. Phys. Ukrainian Ac. Sci. 1990

[Ly2] Lyubashenko, V.V., Ribbon categories as modular categories. Preprint 1993

[Ma1] MacLane, S., Natural associativity and commutativity. Rice University Studies 49 (1963), 28-46

[Ma2] —–, Categories for the working mathematician. Graduate Texts in Math. 5. Springer-Verlag, New York-Heidelberg-Berlin 1971

[Maj1] Majid, S., Quasitriangular Hopf algebras and Yang-Baxter equation. Int. J. Modern Phys. A 5 (1990), 1-91

[Maj2] —–, Representations, duals and quantum doubles of monoidal categories. Suppl. Rend. Circ. Math. Palermo, Ser. II, 26 (1991), 197-206

[MV1] Masbaum, G., Vogel, P., 3-valent graphs and the Kauffman bracket. Pac. J. Math. (to appear)

[MV2] —–, Verlinde Formulae for Surfaces with Spin Structure. Prepublications de l'Univ. Paris VII, 1993

[Mat1] Matveev, S.V., Zeeman's conjecture for unthickable special polyhedra is equivalent to the Andrews-Curtis conjecture. Siberian Math. Journal 28 (1988), 917-928

[Mat2] —–, Transformations of special spines and the Zeeman conjecture. Math. USSR Izvestia 31 (1988), 423-434

[MM] Melvin, P.M., Morton, H.R., The coloured Jones function. Preprint 1993

[MT] Mizoguchi, S., Tada, T., Three-dimensional gravity from the Turaev-Viro invariant. Phys. Rev. Lett. 68:12 (1992), 1795-1798

[Mo] Moise, E.E., Affine structures on 3-manifolds. Ann. Math., (2) 56 (1952), 96-114

[Mon] Montesinos, J.M., Heegaard diagrams for closed 4-manifolds. In: Geometric Topology (1979), 219-237

[MS1] Moore, G., Seiberg, N., Classical and quantum conformal field theory. Comm. Math. Phys. 123 (1989), 177-254

[MS2] —–, Lectures on RCFT. In: Physics, Geometry, and Topology (ed. by H.C. Lee), 263-361. Plenum Press, New York 1990

[Mo1] Morton, H.R., Invariants of links and 3-manifolds from skein theory and from quantum groups, to appear in: Topics in knot theory. Proc. of the NATO Summer Institute in Erzurum 1992, Kluwer

[Mo2] —–, Quantum Invariants Given by Evaluation of Knot Polynomials. J. of Knot Theory and its Ramifications 2 (1993), 195-209

[MSt] Morton, H.R., Strickland, P., Jones Polynomial Invariants for Knots and satellites. Math. Proc. Camb. Phil. Soc. 109 (1991), 83-103

[Mur1] Murakami, H., Quantum $SU(2)$-invariants dominate Casson's $SU(2)$-invariant. Math. Proc. Camb. Phil. Soc. 115 (1994), 253-281

[Mur2] —–, Quantum $SO(3)$-invariants dominate the $SU(2)$-invariant of Casson and Walker. Preprint 1993

[MOO] Murakami, H., Ohtsuki, T., Okada, M., Invariants of three-manifolds derived from linking matrices of framed links. Osaka J. Math. 29 (1992), 545-572

[Mu1] Murakami, J., On derivatives of the Jones polynomial. Kobe J. Math. 3 (1986), 61-64

[Mu2] —–, The representations of the q-analogue of Brauer's centralizer algebras and the Kauffman polynomial of links. Preprint 1988

[Mu3] —–, A state model for the multi-variable Alexander polynomial. Pacif. J. Math. 157 (1993), 109-135

[Mu4] —–, The Yamada Polynomial of Spacial Graphs and Knit Algebras. Comm. Math. Phys. 155 (1993), 511-522

[Oc] Ocneanu, A., Operator algebras, 3-manifolds, and quantum field theory (1991, unpublished)

[Oh1] Ohtsuki, T., A polynomial invariant of integral homology 3-spheres. Preprint 1993

[Oh2] —–, A polynomial invariant of rational homology 3-spheres. Preprint 1994

[Oo1] Ooguri, H., Partition Functions and Topology-Changing Amplitudes in the $3D$ Lattice Gravity of Ponzano and Regge. Nucl. Phys. B 382 (1992), 276-304

[Oo2] —–, Topological Lattice Models in Four Dimensions. Mod. Phys. Lett. A 7, n. 30 (1992), 2799-2810

[OS] Ooguri, H., Sasakura, N., Discrete and Continuum Approaches to Three-Dimensional Quantum Gravity. Mod. Phys. Lett. A 6, n. 39 (1991), 3591-3600

[Pa1] Pachner, U., Bistellare Aquivalenz kombinatorischer Mannigfaltigkeiten. Arch. Math. 30 (1978), 89-91

[Pa2] —–, PL-homeomorphic manifolds are equivalent by elementary shelling. J. Combinatorics 12 (1991), 129-145

[Pas] Pasquier, V., Etiology of IRF-models. Comm. Math. Phys. 118 (1988), 355-364

[Pe] Penrose, R., Applications of negative dimensional tensors. In: Combinatorial Mathematics and its Applications (ed. by D.J.A. Welsh), 221-244. Academic Press, New York 1971

[Pi] Piergallini, R., Standard moves for standard polyhedra and spines, Rend. Circ. Mat. Palermo 37, suppl. 18 (1988), 391-414

[Piu1] Piunikhin, S., Turaev-Viro and Kauffman invariants for 3-manifolds coincide. J. Knot Theory and its Ramifications 1 (1992), 105-136

[Piu2] —–, State sum models for trivalent knotted graph invariants using quantum group $sl_q(2)$. J. Knot Theory and its Ramifications 1 (1992), 253-278

[Piu3] —–, Reshetikhin-Turaev and Kontsevich-Kohno-Crane 3-manifold invariants coincide. J. Knot Theory and its Ramifications 2 (1993), 65-95

[Po1] Polyak, M., Graphical approach to the 3-manifold invariant of Turaev-Viro. J. Knot Theory and its Ramifications 1 (1992), 219-240

[Po2] —–, Invariants of 3-manifolds derived from conformal field theories. Preprint 1991

[PR] Ponzano, G., Regge, T., Semiclassical Limit of Racah Coefficients. In: Spectroscopic and Group Theoretical Methods In Physics (ed. by F. Block et al.), 1-58. North Holland, Amsterdam 1968

[Pr1] Przytycki, J.H., Skein modules of 3-manifolds. Bull. Pol. Acad. Sci. Math. 39 (1991), 91-100

[Pr2] —, Skein module of links in a handlebody. Topology '90 (ed. by B. Apanasov et al.), 381-390. De Gruyter, Berlin-New York 1992

[Pr3] —, Quantum group of links in a handlebody. Contemp. Math., Deformation Theory and Quantum groups with Applications to Math. Physics. (ed. by M. Gerstenhaber, J.D. Stasheff) 134 (1992), 235-245

[Qu1] Quinn, F., Topological Foundations of Topological Quantum Field Theory. Preprint 1991

[Qu2] —, Lectures on Axiomatic Topological Quantum Field Theory. Preprint 1992

[Ra] Racah, G., Theory of complex spectra, II, III. Phys. Rev. 62 (1942), 438-462; ibid. 63 (1943), 367-382

[Rei] Reidemeister, K., Knotentheorie. Chelsea Publishing Co., New York 1948. Copyright Julius Springer, Berlin 1932

[Reg] Regge, T., General Relativity Without Coordinates. Nuovo Cimento 19 (1961), 558-571

[Re1] Reshetikhin, N., Quantized universal enveloping algebras, the Yang-Baxter equation and invariants of links, I, II. LOMI-preprints E-4-87, E-17-87 Leningrad (1988)

[Re2] —, Quasitriangular Hopf algebras and invariants of tangles. Algebra i Analiz 1 (1988), 169-188. English transl.: Leningrad Math. J. 1 (1990), 491-513

[RT1] Reshetikhin, N., Turaev, V., Ribbon graphs and their invariants derived from quantum groups. Comm. Math. Phys. 127 (1990), 1-26

[RT2] —, Invariants of 3-manifolds via link polynomials and quantum groups. Invent. Math. 103 (1991), 547-598

[Rob1] Roberts, J., Skein theory and Turaev-Viro invariants. Preprint 1993

[Rob2] —, Skeins and mapping class groups. Math. Proc. Camb. Phil. Soc. 115 (1994), 53-77

[Rolf] Rolfsen, D., Knots and Links. Publish or Perish, Berkeley 1976

[Ro1] Rosso, M., Finite dimensional representations of the quantum analog of the enveloping algebra of a complex simple Lie algebra. Comm. Math. Phys. 117 (1988), 581-593

[Ro2] —, An analogue of P.B.W. theorem and the universal R-matrix for $U_h sl(N+1)$. Comm. Math. Phys. 124 (1989), 307-318

[Ro3] —, Représentations des groupes quantiques. Sém. Bourbaki exp. 744 (1990/91)

[Ro4] —, Quantum groups at a root of 1 and tangle invariants. In: Topological and Geometrical Methods in Field Theory, 347-358. World Scientific, Singapore 1992

[Ro5] —, private communication (1993)

[Rov] Rovelli, C., The basis of the Ponzano-Regge-Turaev-Viro-Ooguri quantum-gravity model in the loop representation basis. Preprint 1993

[RS] Rourke, C.P., Sanderson, B.J., Introduction to Piecewise-Linear Topology. Ergebnisse der Math. 69. Springer-Verlag, New York-Heidelberg-Berlin 1972

[Se] Segal, G., The definition of conformal field theory. Preprint 1989

[Sh] Shum, M.C., Tortile tensor categories. Ph. D. Thesis, Macquarie University (1989)

[Sw] Sweedler, M.E., Hopf algebras. W.A. Benjamin, Inc., New York 1969

[Ta] Takata, T., Invariants of 3-manifolds associated with quantum groups and Verlinde's formula. Publ. Res. Inst. Math. Kyoto 28 (1992), 139-167

[Tr] Trace, B., On the Reidemeister moves of a classical knots. Proc. Amer. Math. Soc. 89 (1983), 722-724

[Tu1] Turaev, V., A cocycle for the first symplectic Chern class and Maslov indices. Funkts. Analiz i ego Pril. 18, n. 1 (1984), 43-48. English transl.: Funct. Analysis and its Appl. 18 (1984), 35-39

[Tu2] —, First symplectic Chern class and Maslov indices. Zap. Nauchn. Sem. LOMI 143 (1985), 110-129. English transl.: J. Soviet Math. 37 (1987), 1115-1127

[Tu3] —, Reidemeister torsion in knot theory. Uspekhi Mat. Nauk 41 (1986), 97-147. English transl.: Russian Math. Surveys 41 (1986), 119-182

[Tu4] —, The Yang-Baxter equation and invariants of links. Invent. Math. 92 (1988), 527-553

[Tu5] —, Euler structures, non-singular vector fields, and Reidemeister type torsions. Izvest. AN SSSR 53 (1989), 607-643. English transl.: Math. USSR Izvestia

[Tu6] —, Operator invariants of tangles and R-matrices. Izvestia Ac. Sci. USSR 53 (1989). English transl.: Math. USSR Izvestia 35 (1990), 411-444

[Tu7] —, State sum models in low-dimensional topology. Proc. ICM, Kyoto, 1990 vol. 1 (1991), 689-698

[Tu8] —, Quantum invariants of 3-manifolds and a glimpse of shadow topology. C. R. Acad. Sci. Paris 313, Ser. I (1991), 395-398

[Tu9] —, Topology of shadows. Preprint, 1991

[Tu10] —, Shadow links and IRF-models of statistical mechanics, J. Diff. Geometry 36 (1992), 35-74

[Tu11] —, Modular categories and 3-manifold invariants. Int. J. of Modern Phys. B, 6 (1992), 1807-1824

[Tu12] —, Quantum Representations of Modular Groups and Maslov Indices. Topology '90 (ed. by B. Apanasov et al.), 381-390. De Gruyter, Berlin-New York 1992

[Tu13] —, Quantum invariants of links and 3-valent graphs in 3-manifolds. Publ. Math. IHES 77 (1993), 121-171

[Tu14] —, Quantum invariants of 3-manifolds. Publ. Math. IRMA, Strasbourg 1992

[Tu15] —, Axioms for topological quantum field theories. Ann. Fac. Sci. Toulouse 3, n. 1 (1994), 1-18

[TV] Turaev, V., Viro, O.Y., State sum invariants of 3-manifolds and quantum $6j$-symbols. Topology 31 (1992), 865-902

[TW] Turaev, V., Wenzl, H., Quantum Invariants of 3-manifolds Associated with Classical Simple Lie Algebras. Int. J. of Modern Math. 4 (1993), 323-358

[Ve] Verdier, J.-L., Groupes quantiques (d'après V.G. Drinfel'd). Sém. Bourbaki, exp. 685, Astérisque 152/153 (1987), 305-319

[Ver] Verlinde, E., Fusion rules and modular transformations in $2D$ conformal field theory. Nucl. Phys. B300 (1988), 360-376

[Wa] Walker, K., On Witten's 3-manifold invariants. Preprint 1991

[Wall] Wall, C.T.C., Non-additivity of the signature. Invent. Math. 7 (1969), 269-274

[Wal] Wallace, A.H., Modifications and cobounding manifolds. Can. J. Math. 12 (1960), 503-528

[We1] Wenzl, H., On sequences of projections. C. R. Math. Rep. Acad. Sci. Canada 9, n. 1 (1987), 5-9

[We2] —–, Hecke algebra of type A_n and subfactors. Invent. Math. 92 (1988), 349-383

[We3] —–, Representations of braid groups and the quantum Yang-Baxter equation. Pacif. J. Math. 145 (1990), 153-180

[We4] —–, Quantum groups and subfactors of type B, C, and D. Comm. Math. Phys. 133 (1990), 383-433

[We5] —–, Braids and invariants of 3-manifolds. Invent. Math. 114 (1993), 235-275

[Wh] Whitehead, J.H.C., On C^1-complexes. Ann. Math. 41 (1940), 804-824

[Wig] Wigner, E.P., On the matrices which reduce the Kronecker products of representations of S.R. groups (1940). Published in: Quantum Theory of Angular Momentum (ed. by L.C. Biedenharn and H. van Dam), 87-133. Academic Press, New York 1965

[Wi1] Witten, E., Topological quantum field theory. Comm. Math. Phys. 117 (1988), 353-386

[Wi2] —–, Quantum field theory and the Jones polynomial. Comm. Math. Phys. 121 (1989), 351-399

[Wo] Woronowicz, S.L., Compact matrix pseudogroups. Comm. Math. Phys. 111 (1987), 613-655

[Wr] Wright, C., The Reshetikhin-Turaev representation of the mapping class group. Ph. D. Thesis (1992), University of Michigan

[Ya] Yamada, S., An Invariant of Spatial Graphs. J. of Graph Theory 13 (1989), 537-551

[Ye1] Yetter, D.N., Markov algebras. In: Braids, AMS Contemp. Math. 78 (1988), 705-730

[Ye2] —–, Quantum groups and representations of monoidal categories. Math. Proc. Camb. Phil. Soc. 108 (1990), 261-290

[Ye3] —–, Topological quantum field theories associated to finite groups and crossed G-sets. J. Knot Theory and its Ramifications 1 (1992), 1-20

[Ye4] —–, State-Sum Invariants of 3-Manifolds Associated to Artinian Semisimple Tortile Categories. Preprint 1992

Subject index